AF657560

Handbook of Polymer Science and Technology

Volume 2

Polymer Rheology, Properties, Applications, Testing and Recycling of Polymers

Handbook of Polymer Science and Technology

Volume 2

Polymer Rheology, Properties, Applications, Testing and Recycling of Polymers

M.H. FERRY
C.Eng.

A.V. BECKER
M.Sc., Ph.D.

Edited by

S.C. BHATIA
B.E. (Chem), M.B.A.

CBS

CBS PUBLISHERS & DISTRIBUTORS PVT. LTD.

NEW DELHI • BENGALURU • PUNE • KOCHI • CHENNAI

ISBN : 81-239-1133-5

First Edition : 2004
Reprint : 2012

Published by Satish Kumar Jain and produced by V.K. Jain for
CBS Publishers & Distributors Pvt. Ltd.,
CBS Plaza, 4819/XI Prahlad Street, 24 Ansari Road, Daryaganj,
New Delhi - 110002, India. • Website: www.cbspd.com
e-mail: delhi@cbspd.com, cbspubs@airtelmail.in
Ph.: 23289259, 23266861, 23266867 • Fax: 011-23243014

Branches:

- ***Bengaluru:*** Seema House, 2975, 17th Cross, K.R. Road,
  Bansankari 2nd Stage, Bengaluru - 560070 Ph.: +91-80-26771678/79
  • E-mail: cbsbng@gmail.com, bangalore@cbspd.com
- ***Pune:*** Bhuruk Prestige, Sr. No. 52/12/2+1+3/2,
  Narhe, Haveli (Near Katraj-Dehu Road By-pass), Pune - 411051
  Ph.: +91-20-64704058/59, 32342277 • E-mail: pune@cbspd.com
- ***Kochi:*** 36/14, Kalluvilakam, Lissie Hospital Road,
  Kochi - 682018, Kerala • Ph.: +91-484-4059061-65
  Fax: +91-484-4059065 • E-mail: cochin@cbspd.com
- ***Chennai:*** 20, West Park Road, Shenoy Nagar, Chennai - 600030
  Ph.: +91-44-26260666, 26208620 E-mail: chennai@cbspd.com

Printed at :
India Binding House, Noida (UP)

Preface

This volume, dealing with basic concepts of polymer rheology, properties and testing of polymers, applications of polymers, environmental aspects of polymers and recycling of polymers, is the second volume of 'Handbook of Polymer Science and Technology'.

Every industry uses rubber for conveyor belts, hoses, tubings or as insulation in heavy cables. Every vehicle, such as motor car, bus and aeroplane moves on rubber tyres and consists of other parts made out of rubber such as anti-vibration mountings, window and door seals, radiation hoses or fan belts, and every home uses polymer-based household equipments.

Recognising the great importance of polymers, this reference text book is intended for the practising polymer technologists, rubber and plastic technologists, chemical and mechanical engineers, industrial chemists and research students. This book also provides an upto date information for the users of polymer products in other industries.

This volume is divided into six sections. Section I deals with basic concepts of polymer rheology. Rheology is the study of deformation of flow behaviour exhibited by materials. The rheological behaviour of polymers involves several widely different phenomena, which can be related to some extent to different molecular mechanisms. These phenomena and their associated major mechanism are — viscous flow, rubber like elasticity, visco-elasticity, Hookena elasticity. Chapter 1 deals with rheology and its importance. Chapter 2 is devoted to conventional viscometers and general concepts, while chapter 3 focuses on viscometric techniques and analysis of steady state flow. Chapter 4 explains normal stress measurements. In this chapter, discussed topics are analysis of the capillary jet, parallel plate configuration, elongation flows etc. In chapter 5, discussed topics are, various rheometric techniques of polymer melts. Chapter 6 concentrates on torque rheometers and processability testing, while chapter 7 is devoted to on-line measurement of rheological properties.

Section II is devoted to identification and testing of polymers. With the advent of science and technology, the concept of testing is an integral part of research and development, product design and manufacturing. Chapter 8 focuses on identification of polymers. This chapter gives general methods commonly employed in polymer industry for spot testing of end-products.

The major reasons for testing are to prove design concepts, to provide basic for reliability, safety, quality control and to meet standards and specifications. Keeping this in mind chapter 9 deals with basic concepts of testing polymers. Chapter 10 focuses on physical tests of polymers such as tensile strength and modulus, compression and shear, crepe, hardness, electrical tests, specific gravity, thermal conductivity etc. Chapter 11 focuses on chemical analysis of polymers, various quantitative analysis of individual polymers are also discussed. In order to appreciate fully the techniques which have been developed for the analysis of additives in polymers, it is necessary to be familiar with physical and chemical properties of additives used in plastics. Chapter 12 is devoted to chemical analysis of additives in polymer where as chapter 13 focuses on applications of gas chromatography. A full knowledge of all types of additives present in the polymer helps in selection of test methods, keeping this in mind chapter 14 explains the application of chromatography for analysis of unknown additives.

Section III explains the various uses of polymer products in automobiles, aerospace, railways, cables, bearing and bridges, piers and fenders, machinery and equipments, and packaging. The automobile combines many different materials that must be integrated with the ever changing design and manufacturing processes. Without polymer components, the automobile in its current state of technology cannot exist. Keeping this in mind chapter 15 is devoted to application of polymers in automobile. Chapter 16 describes in detail the uses of polymers and composite in aerospace and railway engineering, as typical uses in dry bearings and structural composites. Shock and vibrations in rail vehicles are much more severe due to rail-wheel interactions, track irregularities, crossings and rail joints, firing impulses of engine. This chapter also focuses on some novel applications of polymers in railway engineering. Chapter 17 describes in detail special applications of polymers and rubber.

Section IV is devoted to environmental aspects related to polymers. Green chemistry employs the fundamental principles of chemistry in reducing toxicity and waste from polymer at its source. It embraces chemistry as the solution to many environmental problems. Keeping this in mind chapter 18 is devoted to green chemistry and its role in designing safer polymer. Chapter 19 focuses on environmental considerations in designing with polymers. Weathering of polymers and role of polymers in society and the environment is also discussed. Chapter 20 explains various methods to control biological degradation of polymers by means of biocides. Chapter 21 deals with the effect of ionising radiation on polymers and packing materials. Various sterilising methods, sources, and radiation are discussed in detail. Chapter 22 focuses on recovery of carbon black from polymer waste by vacuum pyrolysis. Biodegradable polymers offer an attractive alternative to tradition petroleum based non biodegradable polymers, from an environmental perspective, keeping this in mind chapter 23 is devoted to environmental implications of polymer degradation.

Section V describes the various recycling methods of polymers. Chapter 24 gives an overview of various recycling methods, while chapter 25 describes the various equipments and machinery for the recycling of polymers. Chapter 26 explains various processes for recycling of poly(ethylene terephthalate) and polyurethane. Method of obtaining polyurethane from waste vegetable oil is also discussed. Chapter 27 focuses on feed stocks from plastic waste, and chapter 28 explains the recycling of various miscellaneous polymers, such as HDPE, polystyrene, nylon-6 and epoxy based materials.

Section VI is devoted to some special topics which are of great importance to polymer, rubber and plastics. Chapter 29 focuses on ferroelectric polymers its properties and applications. Blowing agents are substances that can produce pores or cell in polymeric composition, keeping this in mind chapter 30 is devoted to blowing agent for polymer. Chapter 31 explains the various annealing aspects of plastics.

The text in this book is throughout supplemented with diagrams and tables. The treatment of all topics is in a cogent, lucid style aimed at enabling the reader to grasp the information quickly and easily. The references and index are also given at the end of the book.

While painstaking care has gone into producing a useful and exhaustive reference textbook, the authors would welcome any constructive criticism and creative feedbacks from the students, teachers and professionals in the industry.

We wish to thank all those colleagues who provided figures and photographs and offered important advice during its preparation.

M.H. Ferry
A.V. Becker

Contents at a Glance

SECTION I

SECTION II

SECTION III

SECTION IV

SECTION V

SECTION VI

Contents

SECTION II

SECTION IV

SECTION V

SECTION VI

SECTION I

Polymer Rheology

CHAPTER 1

Rheology: An Overview

INTRODUCTION

It is anticipated that many readers will have little previous knowledge about rheology but will wish to find out how it can be useful to them in solving practical problems involving the flow of molten plastics. For this reason, it is our intention to supply sufficient basic information about rheology to enable the reader to understand and make use of the methods described. With this in mind, we begin at the beginning, with a definition of rheology. *Rheology* is the science that deals with the way materials deform when forces are applied to them. The term is most commonly applied to the study of liquids and liquid-like materials such as paint, catsup, oil well drilling mud, blood, polymer solutions and molten plastics, i.e., materials that flow, although rheology also includes the study of the deformation of solids such as occurs in metal forming and the stretching of rubber.

The two key words in the above definition of rheology are deformation and force. To learn anything about the rheological properties of a material, we must either measure the deformation resulting from a given force or measure the force required to produce a given deformation. For example, let us say you wish to evaluate the relative merits of several foam rubber pillows. Instinctively, you would squeeze (deform) the various products offered, noting the force required to deform the samples.

A pillow that required a high force to compress would be considered 'hard', and you probably wouldn't buy it, because it would be painful to sleep on. On the other hand, if it required too little force (too 'soft') it would not provide adequate support for your weary head. Foam rubber is a lightly cross-linked elastomer, and in squeezing it you would be evaluating it primarily on the basis of its resistance to deformation, i.e., its elastic modulus, which is a *rheological property*.

Now let us say that you are looking for a lubricating oil for some household applications, and that your local hardware store has a display in which samples of several oils are contained in glass bottles. You can pick up the bottles and invert them to make the oil flow from one end to the other. The 'light' oils flow very rapidly and splash as they hit the end of the bottle. The 'heavy' oils creep slowly in response to your test and take several seconds to accumulate at the end of the bottle. This is a rheological experiment. You have used the earth's gravitational field to supply the force, and the speed with which the oils flow is a measure of their rates of deformation. Lubricating oils are viscous liquids, and in tipping the bottles you would be evaluating them on the basis of their viscosity, i.e., their resistance to flow, and this is another example of a rheological property.

There are two principal aspects of rheology. One involves the development of quantitative relationships between deformation and force for a material of interest. The information for the development of such a relationship is obtained from experimental measurements. For example, for the foam pillow it might be observed that the force required to compress it a certain distance is proportional to the distance. In the case of the lubricating oil, it might be found that the speed with which it flows through a small hole in the bottom of a can is proportional to the height of the oil remaining in the can.

For a linear elastic material or a Newtonian fluid, such simple observations are sufficient to establish a general equation describing how such a material will respond to any type of deformation. Such an equation is called a 'constitutive equation' or a 'rheological equation of state'. However, for more complex materials such as molten plastics, the development of a constitutive equation is a much more difficult task, which may require the results of many types of experiment.

The second aspect of rheology is the development of relationships that show how rheological behaviour is influenced by the structure and composition of the material and the temperature and pressure. Ideally, one would like to know how these parameters affect the constitutive equation, but this has not been accomplished at the present time, except for very simple materials such as Newtonian fluids. In the case of more complex materials, one can at least develop relationships showing how specific rheological properties such as the viscosity and the relaxation modulus are influenced by molecular structure, composition, temperature and pressure.

Molten plastics are rheologically complex materials that can exhibit both viscous flow and elastic recoil. A truly general constitutive equation has not been developed for these materials, and our present knowledge of their rheological behaviour is largely empirical. This complicates the description and measurement of their rheological properties but makes polymer rheology an interesting and challenging field of study.

IMPORTANCE OF RHEOLOGICAL PROPERTIES

The forces that develop when a lubricant is subjected to a high-speed shearing deformation are obviously of central importance to mechanical engineers. The rheological property of interest in this application is the viscosity. The stiffness of a steel beam used to construct a building is of great importance to civil engineers, and the relevant property here is the modulus of elasticity.

The viscoelastic properties of molten polymers are of importance to plastics engineers, because it is these properties that govern flow behaviour whenever plastics are processed in the molten state. For example, in order to optimise the design of an extruder, the viscosity must be known as a function of temperature and shear rate. In injection moulding, the same information is necessary in order to design the mould in such a way that the melt will completely fill it in every shot. In blow moulding, the processes of parison sag and swell are governed entirely by the rheological properties of the melt.

STRESS AS A MEASURE OF FORCE

It has already been emphasised that force and deformation are the key words in the definition of rheology. In order to describe the rheological behaviour of a material in a quantitative way, i.e., to define rheological material constants (such as the viscosity of a Newtonian fluid) or material functions (such as the relaxation modulus of a rubber), it is necessary to establish clearly defined and quantitative measures of force and deformation. Furthermore, it is necessary to define these measures in such a way that they describe the state of the material of interest without detailed reference to the procedure used to make the rheological measurement.

For example, in the case of the evaluation of the pillows already discussed above, one way to quantify the test results would be to place the pillow on a table, place a flat board on top, and measure the distance between the board and the table both before and after a weight of a certain mass was placed on top of the board, as shown in Fig. 1.1. If our objective is simply to compare several pillows of the same size, it would be sufficient to simply list the amount of compression, in centimetres, caused by a weight having a mass of 1 kg.

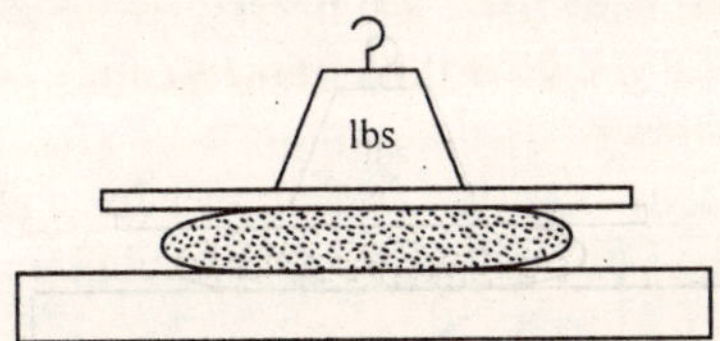

Fig. 1.1. Setup for testing pillows.

However, if our objective is to make a quantitative determination of the elastic properties of the foam rubber, the reporting of the test results is awkward. We must report the size and shape of the sample (the pillow), the mass of the weight applied, and the amount of compression. If one wishes to compare the behaviour of this foam with that of a second foam, the second material must be tested in exactly the same way as the first. It would be advantageous to be able to describe the elastic behaviour of the rubber using physical quantities which are defined so that they describe the state of the material under test, without reference to the details of the test procedure.

First let's look at force. Two types of force can act on a fluid element. A 'body force' acts directly on the mass of the element as the result of a force field. Usually only gravity need be considered, but a magnetic field can also generate a body force. A surface force is the result of the contact of a fluid element with a solid wall or with the surrounding fluid elements. It is the surface forces that are of interest in rheology.

The force exerted by a weight sitting on top of a pillow is an example of a surface force. The fact that the ultimate cause of the surface force acting on the pillow is the body force acting on the weight is not of rheological importance, as the compressive force on the sample could equally well be supplied by means of a testing machine, and the observed relationship between force and deformation would be the same.

Placing a weight having a mass of 1 kg on a small pillow will cause more compression than placing it on a large pillow. From the point of view of the material, it is obviously not the total force that is important. In fact, since the deforming force acts on the upper surface of the sample, it is found that if the force is divided by the area of the surface we obtain a quantity suitable for describing the properties of the material. We call this quantity the 'stress'.

In general, then, the stress is calculated by dividing the force by the area over which it acts. In the case of a test like the one with the pillow which involves squeezing, we call this the compressive stress. A more common type of test method for elastic materials involves stretching rather than compression, as shown in Fig. 1.2. Again, the stress is the force divided by the cross-sectional area of the sample, and in this case we call it a "tensile stress". We will use the symbol σ_E for this quantity.

$$\sigma_E = \frac{\text{Stretching force}}{\text{Cross-sectional area}} \qquad \text{... (1.1)}$$

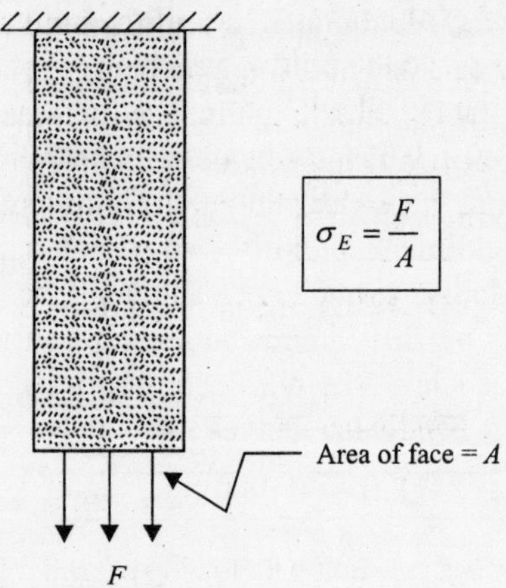

Fig. 1.2. Uniaxial (simple) extension.

Compressive and tensile stresses are the two types of 'normal stress', so called because the force is normal (perpendicular) to the surface on which it acts. In addition to normal stresses, we can have a 'shear stress'; in this case, the direction of the force is tangential to the surface on which it acts, as shown in Fig. 1.3.

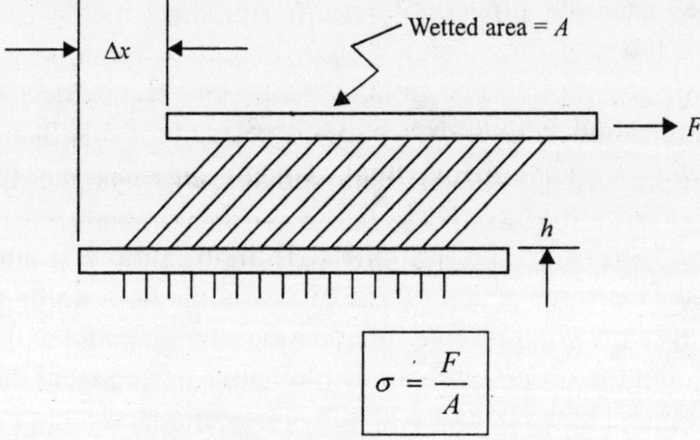

Fig. 1.3. Simple shear (2 plates, gap = *h*).

Fig. 1.3 shows the deformation called 'simple shear', in which the sample is contained between two flat plates with a fixed spacing, *h*, between them. The upper plate moves in a direction parallel to itself while the lower plate is stationary. The shear stress is the shear force divided by the tangential area. We will used the symbol σ, with no subscript, to refer to the shear stress in a simple shear deformation.

$$\sigma = \frac{\text{Shear force}}{\text{Tangential area}} \quad \text{... (1.2.)}$$

STRAIN AS A MEASURE FOR DEFORMATION

In the previous section, we stated that shear stresses and normal stresses are useful measures of the forces that act to deform a material. Now we need a quantitative measure of deformation that is rheologically significant. The description of deformation in terms of strain is more complex than the description of force in terms of stress, and there are many alternatives, rheologically significant, measures of strain. We will define here only those measures that are useful in the description of deformations commonly used to make rheological measurements, namely simple shear and simple extension. Later in this section a strain measure for small deformations that is not restricted to describing simple shear or simple extension will be defined. The thing that complicates the definition of a measure of strain is that it is necessary to refer to two states of a material element. In other words, it is not possible to specify the strain of a material element unless we specify at the same time the reference state relative to which the strain is measured. In the case of an elastic material that cannot flow, for example a cross-linked rubber, this is straightforward, because there is a unique, easily identifiable, unstrained state that a material element will always return to when ever deforming stresses are not acting.

For materials that flow, i.e., fluids, such as unique reference state does not exist. In the case of a well-controlled experiment, however, in which a simple homogeneous deformation is imposed on a sample initially at rest and free of all deforming stresses, this initial condition provides a meaningful reference state with respect to which strain can be defined. This can be used to define strain measures for simple extension and simple shear.

Strain Measures for Simple Extension

Consider the simple extension test illustrated in Fig. 1.4. Let L_o be the length of the sample prior to the application of a tensile stress and L the length after deformation has occurred. A simple measure of the deformation is the quantity $(L - L_0)$. However, this quantity is meaningful only in terms of a specific sample, whereas we desire a measure of deformation that describes the state of a material element. We can easily form such a quantity by dividing this length difference by the initial length to obtain the "linear strain" for simple extension.

$$S = (L - L_0)/L_0 \qquad \text{... (1.3)}$$

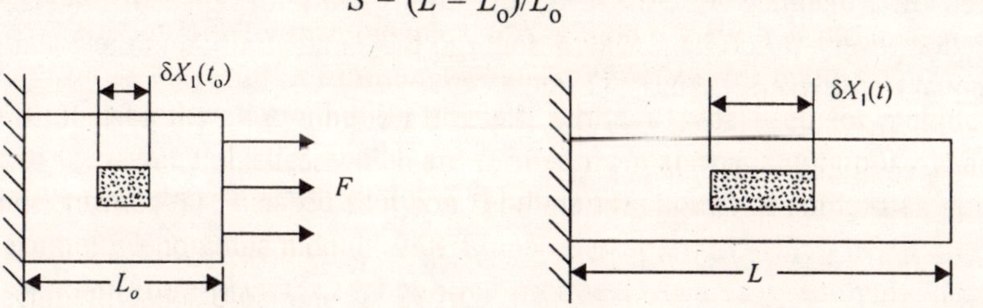

Fig.1.4. Quantities used to describe simple extension.

For a uniform deformation, every material element of the sample experiences this same strain. For example, if the initial length at time t_o of a material element measured in the direction of stretching is $\delta X_1(t_0)$, and the length at a later time, t, after deformation has occurred is $\delta X_1(t)$, the linear strain of the material element is:

$$S = [\delta X_1(t) - \delta X_1(t_0)]/\delta X_1(t_o) \qquad \text{... (1.4)}$$

This measure of deformation has some convenient features. It is independent of sample size, and it is zero in the unstressed, initial state.

However, it is not the only measure of deformation that has these properties. Another is the Hencky strain, which is defined as follows in terms of the length of a material element.

$$\varepsilon = \ln[\delta X_1(t)/\delta X_1(t_o)] \qquad \text{... (1.5)}$$

For a sample with initial length L_o undergoing uniform strain the Hencky strain can also be expressed as:

$$\varepsilon = \ln (L/L_0) \qquad \text{... (1.6)}$$

For materials that flow, e.g., molten plastics, this quantity is more useful than the linear strain. In fact, the linear and Hencky strains become equivalent in the limit of very small strains. This can be demonstrated for simple extension by noting that $\varepsilon = \ln(1 + S)$ and that the first term of the series expansion of $\ln(1 + S)$ is S.

The Hencky strain rate is also a useful quantity for describing rheological phenomena in simple extension. This is defined in equation 1.7 in terms of the length, L, of the sample.

$$\varepsilon = d \ln (L)/dt \qquad \text{... (1.7)}$$

We note that the initial length does not enter into the Hencky strain rate but does enter into the linear strain rate dS/dt.

Shear Strain

Now consider simple shear, which is the type of deformation most often used to make rheological measurements on fluids. Referring to Fig. 1.3, an obvious choice of a strain measure is the displacement of the moving plate, ΔX, divided by the distance between the plates, h.

$$\gamma = \Delta X/h \qquad \text{... (1.8)}$$

Referring to the two material particles shown in Fig. 1.5 rather than to the entire sample we can define the shear strain, γ, for the fluid element located at (x_1, x_2, x_3) as

$$\gamma\,(x_1, x_2, x_3) = \Delta X_1/\Delta X_2 \qquad \text{... (1.9)}$$

where ΔX_1 is the distance, measured in the x_1 direction, between two neighbouring material particles that are separated by a distance ΔX_2 in the x_2 direction. In the absence of edge effects, i.e., for a uniform deformation, every fluid element will undergo the same strain, and the local shear strain will everywhere be equal to the overall sample strain:

$$\gamma\,(x_1, x_2, x_3) = \Delta X/h \qquad \text{...1.10)}$$

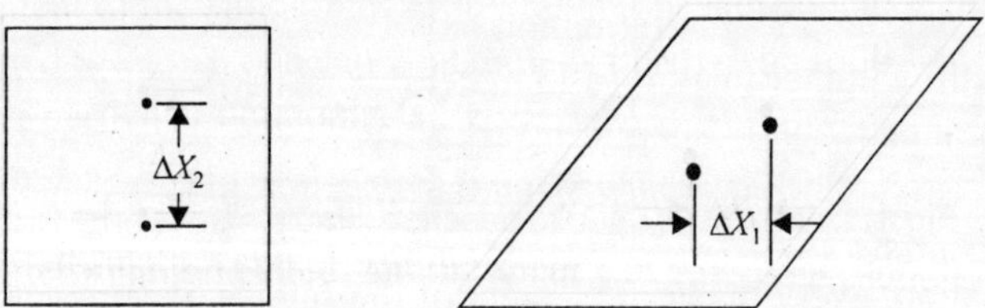

Fig. 1.5. Two material particles in simple shear.

And the shear rate is simply the rate of change of the shear strain with time:

$$\dot{\gamma} = \frac{1}{\mathrm{h}}\frac{dX}{dt} = \frac{V}{h} \qquad \text{... (1.11)}$$

where V is the velocity of the moving plate.

RHEOLOGICAL PHENOMENA

In this section we will examine the general types of rheological behaviour that can be exhibited by materials. These are elasticity, viscosity, viscoelasticity, structural time dependency, and plasticity. Although we will use simple extension and simple shear behaviour in this section to describe rheological phenomena, it is important to remember that for rheologically complex materials such as polymeric liquids, the behaviour observed in these simple deformations does not tell the whole rheological story.

Elasticity: Hooke's Law

Elasticity is a type of behaviour in which a deformed material returns to its original shape whenever a deforming stress is removed. This implies that a deforming stress is necessary to produce and maintain any deviation in shape from the original (unstressed) shape, i.e., to produce strain. The simplest type of elastic behaviour is that in which the stress required to produce a given amount of deformation is directly proportional to the strain associated with that deformation. For example, in simple extension this can be expressed as:

$$\sigma_E = E\,(L - L_0)/L_0 = ES \qquad \text{... (1.12)}$$

The constant of proportionality, E, is called Young's modulus. This relationship is called Hooke's law. The corresponding form of Hooke's law for simple shear is:

$$\sigma = G\gamma \qquad \text{... (1.13)}$$

where G is the shear modulus or modulus of rigidity. We note that in a purely elastic material like this, all the work done to deform the material is stored as elastic energy and can be recovered when the material is permitted to return to its equilibrium configuration.

Another way of describing elastic behaviour is to specify the strain that results from the application of a specific stress. For a Hookean material we have, for simple shear:

$$\gamma = J\sigma \qquad \text{... (1.14)}$$

where J is the shear compliance. Obviously, for a material following Hooke's law:

$$J = 1/G \qquad \text{... (1.15)}$$

Viscosity

Viscosity is a property of a material that involves resistance to continuous deformation. Unlike elasticity, the stress is not related to the amount of deformation but to the rate of deformation. Thus it is a property peculiar to materials that flow rather than to solid materials. We will consider first the simplest type of rheological behaviour for a material that can flow. For simple shear this type of behaviour is described by a linear relationship between the shear stress and the shear rate:

$$\sigma = \eta\dot{\gamma} \qquad \text{... (1.16)}$$

where η is the viscosity. A material that behaves in this way is called a Newtonian fluid.

For a Newtonian fluid, the viscosity is a 'material constant', in that it does not depend on the rate or amount of strain. Single phase liquids containing only low molecular weight compounds are Newtonian for all practical purposes. For multiphase systems, for example suspensions and emulsions, and for polymeric liquids, the relationship between stress and strain rate is no longer linear and cannot be described in terms of a single constant. It is still convenient, however, to present the results of a steady simple shear experiment in terms of a viscosity function $\eta(\dot{\gamma})$ defined as follows:

$$\eta(\dot{\gamma}) \equiv \sigma / \dot{\gamma} \qquad \text{... (1.17)}$$

where σ is the shear stress and $\dot{\gamma}$ is the shear rate. A typical viscosity-shear rate curve for a molten polymer is shown in Fig. 1.6. The important features of this curve are listed below.

1. At sufficiently low shear rates, the viscosity approaches a limiting constant value η_o called the zero shear viscosity.
2. The viscosity decreases monotonically as the shear rate is increased. This type of behaviour is called shear thinning. (An older terminology is 'pseudoplastic').
3. At sufficiently high shear rates the viscosity might be expected to level off again, although a high-shear-rate limiting value is not observed in melts, because viscous heating and polymer degradation usually make it impossible to carry out experiments at sufficiently high shear rates

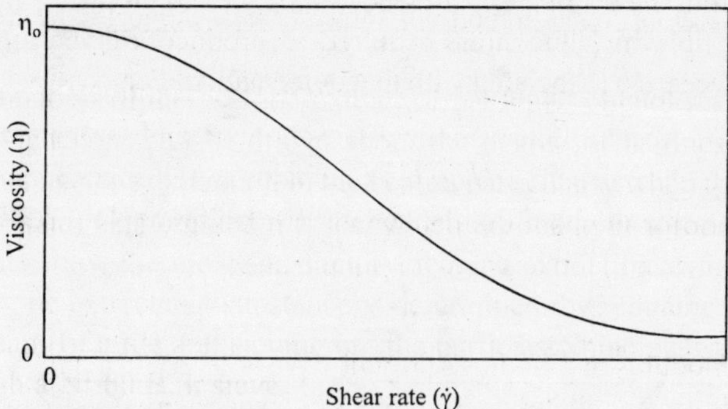

Fig. 1.6. Viscosity-shear rate curve for a typical molten polymer.

Viscoelasticity

Polymeric materials, including solutions, melts, and cross-linked elastomers, exhibit both viscous resistance to deformation and elasticity. In the case of a vulcanised (cross-linked) rubber, flow is not possible, and the material has a unique configuration that it will return to in the absence of deformating stresses. However, the viscous resistance to deformation makes itself felt by delaying the response of the rubber to a change in stress. To illustrate this point, consider the phenomenological analog of a viscoelastic rubber shown in Fig. 1.7. This mechanical assembly consists of a spring in parallel with a dashpot.

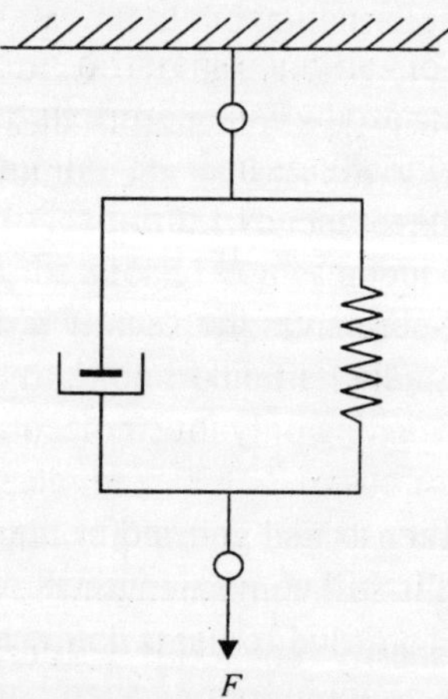

Fig. 1.7. Voigt body analog of a viscoelastic solid.

The force in the spring is assumed to be proportional to its elongation, and the force in the dashpot is assumed to be proportional to its rate of elongation. Thus, the spring is a linear elastic element, in which the force is proportional to the extension, X, and the dashpot is a linear viscous element in which the force is proportion to the rate of change of X. Note that this assembly will always return to a unique length, the rest length of the spring, when no force is acting on it. This assembly, called a Voigt body, is not intended to be a physical or quantitative model for a rubber.

However, the qualitative characteristics of its response to changes in force are similar in some ways to those exhibited by rubbers.

Consider how this assembly would respond to the sudden application of a tensile load, F. This is called a 'creep test'. The force, F, is the sum of the force in the spring, $K_e X$, and that in the dashpot, $K_v(dX/dt)$. Thus.

$$F = K_e X + K_v(dX/dt) \quad \text{... (1.18)}$$

We note that some of the work put into the assembly to deform it is dissipated in the dashpot, while the remainder is stored elastically in the spring. If X is initially zero, and the force F is suddenly applied at time $t = 0$, this differential equation can be solved to yield:

$$X(t) = F/K_e \; [1 - \exp(-K_e t/K_v)] \quad \text{... (1.19)}$$

The important point to note is that the viscous resistance to elongation introduces a time dependency into the response of the assembly, and that this time dependency is governed by the ratio (K_v/K_e), which has units of time. If we take the force to be analogous to the deforming stress in a viscoelastic material, and the elongation to be analogous to strain, we see that a viscoelastic rubber has time constant and cannot respond instantaneously to changes in stress. This is called a 'retarded' elastic response. As the time constant approaches zero, the behaviour becomes purely elastic.

Now we turn to the case of an elastic liquid. To illustrate certain qualitative features of the rheological behaviour of such a material, consider the mechanical assembly shown in Fig. 1.8. This assembly, called a Maxwell element, consists of a linear spring in series with a linear dashpot. Note first that unlike the Voigt body, this assembly has no unique reference length and will deform indefinitely under the influence of an applied force, assuming the dashpot is infinite in length. This is analogous to the behaviour of an uncross-linked polymeric material above its glass transition and melting temperatures. Such a material will flow indefinitely when subjected to deforming stresses.

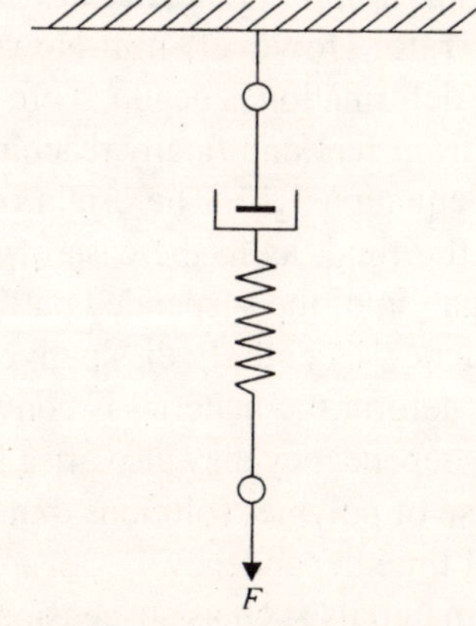

Fig. 1.8. Maxwell element analog of a viscoelastic liquid.

Now we examine the force on the Maxwell element when it is subjected to a sudden stretching by an amount X_0. The force, but not the displacement, is the same in both the spring and dashpot.

Thus:

$$F = K_e X_e = K_v (dX_v/dt) \qquad \text{... (1.20)}$$

Again we note that some of the work done is dissipated in the dashpot and the remainder is stored in the spring. The total displacement of the assembly, X_0, is the sum of X_e and X_v:

$$X_e + X_v = X_o \qquad \text{... (1.21)}$$

Thus:

$$F(t) = K_e (X_o - X_v) = K_v (dX_v/dt) \qquad \text{... (1.22)}$$

This ordinary differential equation can be solved to yield

$$F(t) = K_e X_o [\exp(-K_e t/K_v)] \qquad \text{... (1.23)}$$

Note that (K_v/K_e) is a time constant. The force thus decays or relaxes exponentially. If we take the force to be analogous to the deforming stress in an elastic liquid and the elongation to be analogous to the strain, this process is analogous to a stress relaxation experiment. As in the case of the viscoelastic rubber, we note that the combination of viscous and elastic properties endows the material with a characteristic time and makes its response time dependent. As this 'relaxation time' becomes shorter and shorter, however, it becomes more and more difficult to devise an experiment that will reveal the elastic nature of the liquid, and its behaviour appears more and more like that of a purely viscous material.

When we examine the rheological behaviour of actual polymeric materials, we find creep and relaxation behaviour that is qualitatively like those described above. In particular, the response to a sudden change in stress or strain is always time dependent, never instantaneous, and there is both elastic storage of energy and viscous dissipation. On the other hand, the creep and relaxation curves cannot be described by a single exponential function involving a single characteristic time. However, practical use can still be made of the concept of a relaxation time by describing the viscoelastic behaviour of real materials in terms of a spectrum of relaxation times.

Structural Time Dependency

In our discussion of the viscosity function, we took the shear stress to be independent of time at constant shear rate. For a Newtonian fluid this is appropriate, because the stress responds instantaneously to the imposition of a constant shear rate. However, non-Newtonian fluids may not respond instantaneously so that when the shearing deformation is begun, there is a transient period during which the shear stress varies with time, starting from zero and finally reaching a steady state value that can be used to calculate the viscosity by use of equation 1.17. The origin of this time dependency may be a flow-induced change in the structure of the fluid, as in the case of concentrated suspension of solid particles. For example, the state of aggregation of the suspended particles can be changed significantly by shearing. This 'structural time dependency' is entirely dissipative, that is to say there is no elastic energy storage, and all the work done to deform the material is converted immediately into 'heat', or more precisely, into internal energy. Time dependency may also arise from changes in the conformation and orientation of molecules, as in the case of polymer solutions, but this type of time dependency is a viscoelastic effect rather than a structural time dependency.

Restricting attention here to inelastic materials such as suspensions of rigid particles in Newtonian fluids, in which no elastic energy storage is possible, the most common type of structural time dependency is that shown in Fig. 1.9, where the shear stress decreases with time after the start up of steady shear flow. This type of behaviour, in which the viscosity decreases with time, is called 'thixotropy'. If the shearing is stopped, the viscosity will gradually increase as the particles respond to inter-particle forces in the absence of a shear stress, but the time required for the complete reformation of the equilibrium structure can be very long. There are some materials, such as corn starch suspensions, that exhibit the opposite effect, viz., the stress increases with time after the start up of steady shear. This type of behaviour is called 'rheopexy' or 'antithixotropy'. In the case of thixotropy, the shear tends to break down structure that has been established in the unsheared material, while in the case of a rheopectic material, the shearing promotes the buildup of structure.

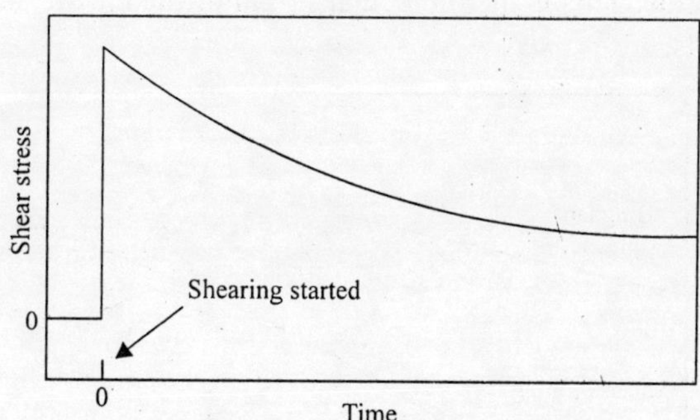

Fig. 1.9. Shear stress versus time for a material with structural time dependency.

Plasticity and Yield Stress

The structure in a concentrated suspension can be sufficiently rigid that it permits the material to withstand a certain level of deforming stress without flow is called the 'yield stress', and this type of behaviour is called 'plasticity'. Metals generally exhibit plasticity, as do semi-crystalline polymers at temperatures between their melting and glass transition temperatures. Highly filled melts are also thought to have a yield stress, although precise measurement of this property is difficult. The simplest type of plastic behaviour is that in which the excess stress, above the yield stress, is proportional to the shear rate. For simple shear flow, this type of behaviour is described by equation 1.24.

$$\sigma = \sigma_o + \eta_p \dot{\gamma} \qquad (\sigma \geq \sigma_o) \qquad \text{... (1.24)}$$

Here, σ_o is the yield stress and η_p is the plastic viscosity. A material that behaves in this way is called a 'Bingham plastic'. This is an idealised type of behaviour that is not precisely followed by any real material, but it is sometimes a useful approximation to real behaviour. Note that Equation 1.24 is not a constitutive equation, as it does not describe all the components of the stress tensor in any type of a deformation but only the shear stress in simple shear. Fig. 1.10 compares shear stress versus shear rate curves for a Bingham plastic, a Newtonian fluid and shear thinning fluid.

POLYMERIC LIQUIDS ARE NON-NEWTONIAN

It is important in applied polymer science to be able to relate physical properties, including rheological properties, to molecular structure. We will mention here only the general mechanisms by which polymeric molecules endow liquids with complex rheological behaviour.

First, we should examine the question of why polymer molecules are elastic. Our physical picture of a polymer molecule is that of a long chain with many joints allowing relative rotation of adjacent links. The presence of this large number of joints makes the molecule quite flexible and allows many different configurations of the molecule. At temperatures above the glass transition temperature a molecule will continually change its configuration due to Brownian motion, but we can describe the state of a large number of molecules in terms of statistical averages. For example, at a given temperature there will be unique average value of the end-to-end distance, R, for the molecules of a polymeric liquid that has been at rest for a sufficient length of time that it is in its equilibrium state. Deforming the liquid will alter this average length, but if the deformation is stopped, Brownian motion will tend to return the average value of R to its equilibrium value. This is the molecular origin of the elastic and relaxation phenomena that occur in polymeric liquids.

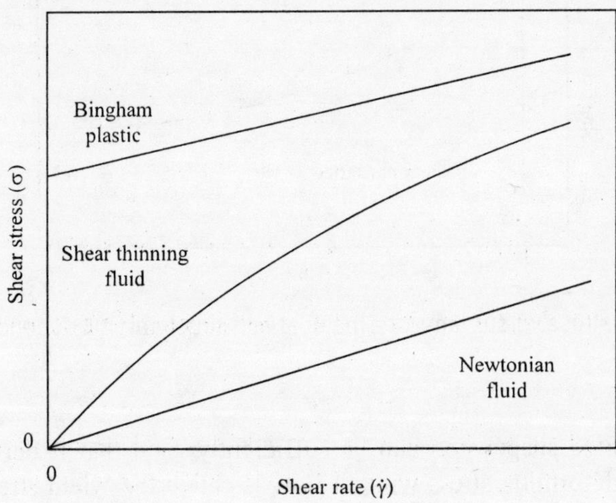

Fig. 1.10. Shear stress versus shear rate for a Bingham plastic, a Newtonian fluid, and a shear thinning fluid.

Polymer Solutions

We consider first the behaviour of a dilute solution in which the forces acting on the polymer molecule are primarily those due to the flow of the solvent. This situation is much simpler than that existing in concentrated solutions and melts, where the rheological behaviours is governed by interactions between polymer molecules. In the absence of a velocity gradient in the solvent, the polymer molecules will have an average end-to-end distance that is small, and a distribution of orientations that is random. When the solution undergoes deformation, dissolved molecule will be dragged along with it but the polymer molecule is so large that different parts of it will be exposed to different solvent velocities. In other words, the solvent moves relative to the segments of the polymer molecule and exerts forces on these segments that change the shape (conformation) of the molecule, increasing the average end-to-end distance and promoting some degree of orientation in a preferred direction. For example, in simple extension, the preferred direction is the stretch direction.

Some types of deformation have a greater ability to increase R and orient molecules than others. Deformations that can generate a high degree of stretching and orientation are said to be 'strong flows', while those that cannot are said to be 'weak'.

We have seen how the deformation of a polymeric liquid can alter the configuration and orientation of the molecules. Thus, the physical nature of the fluid is altered by the deformation, and this, in turn, alters the mechanical (rheological) properties of the fluid. This explains, for example, why the viscosity of a polymer solution depends to some extent on shear rate. At high shear rates, the shape of the polymer molecules is different from that at low shear rates, and this alters the solution's resistance to flow and thus its viscosity. Furthermore, the partial orientation of the molecules, i.e., the shift of the average orientation of the R vectors away from zero, introduces a strain-induced anisotropy that is responsible for differences in the normal stresses in steady simple shear. Finally, the tendency of Brownian motion to return the system of molecules to its most probable, i.e., equilibrium, configuration distribution explains why these fluids are elastic and have relaxation times.

Molten Plastics

In the case of concentrated polymeric liquids, including melts, the transmission of deformation stresses to a molecule is primarily (in the case of a concentrated solution) or totally (in the case of a melt) due to interactions between polymer molecules rather than interactions between the solvent and the molecules. The detailed nature of these interactions is not completely understood at this time, but it is observed that the short term response to rapid deformations of high molecular weight molten polymers is very similar to that of a cross-linked rubber. This has inspired the concept of a 'temporary network' that exists in the melt and acts like a rubbery network at shorter times but whose 'junctions' can slip over longer periods of time to permit flow. The network is sometimes said to arise from 'entanglements' in the melt. However, the modern view is that the rubbery behaviour of melts is due not to an actual looping or knotting of molecules around each other but simply to the constraints on their motion resulting from the fact that molecules cannot cut through each other.

Entanglements occur because of the high degree of spatial overlap of the molecules. The existence of overlap is readily demonstrated by considering the measured size of the polymer coils. One measure of molecular size is the 'radius of gyration', R_g. For linear polyethylene, R_g depends on the molecular weight, M as follows:

$$R_g \text{ (cm)} = 4 \times 10^{-9} \times M^{1/2}$$

For a polyethylene with a molecular weight of 10^6 g/mol, the volume of the sphere occupied by one molecule is therefore about 2.6×10^{-16} cm^3. The mass of this coil is 10^6 divided by Avogadro's number, or 1.7×10^{-18} g. The density of the coil in its occupied volume is thus less than 0.01 g/cm^3. The observed melt density of about 0.77 g/cm^3 can only be accounted for if parts of many other coils are present in the volume occupied by the coil.

A similar calculation for a polymer with a molecular weight of 10^4 g/mol gives a density of 0.1 g/cm^3, which is considerably closer to the measured bulk density. This shows that the degree of coil overlap, and therefore the entanglement density increases sharply with molecular weight.

Rubbery behaviour occurs in a melt when the molecular weight is above some critical value that arise from one polymer to another. Above this molecular weigh the number of entanglements becomes sufficient to produce strong rubber like effects.

The macroscopic effects of the strong interactions between polymer molecules in a melt include high viscosity and high elastic recoil, especially just above the melting point or, in the case of an amorphous polymer, just above the glass transition temperature. At the same time, the nature of this strong interaction can be altered temporarily by deformation so that high molecular weight melts have highly non-linear properties; for example, the viscosity is a very strong function of the shear rate.

Polymeric materials are said to have a 'memory', in that when deformation stresses are eliminated, they tend to return to a previous configuration. Cross-linked polymers have a 'perfect memory" in the sense that since their network is based on permanent chemical cross-links they always return to a unique equilibrium configuration, whereas molten polymers are said to have a 'fading memory', since the entanglement network is not permanent and is altered by flow and relaxation processes.

TENSORS

For those readers who have had little if any experience in the use of tensor notation, the very word 'tensor' probably suggests a mathematical system of impenetrable mystery. However, such readers should have no fear. There is no mystery! While we do not claim to offer here a complete course in tensor analysis, we do present in the next two brief sections everything you will need to know about tensors in order to describe the rheological properties of polymeric liquids. After a careful reading of these sections, you too can impress the uninitiated with your ability to use tensor notation to describe rheological phenomena.

The concept of a tensor was introduced into physics, and thus into rheology, because it is useful; without it, the quantitative description of many physical phenomena would be hopelessly clumsy and tedious. Because of this usefulness, most of the literature on viscoelastic behaviour makes some use of tensor notation. This literature will be inaccessible to a reader having no familiarity with tensor quantities.

We have already stated that rheology involves the relationship between deformation (strain) and force (stress) for a material. It is in the quantitative description of the quantities strain and stress that tensor notation is virtually indispensable. However, before demonstrating this, it will be useful to review briefly the concept of a vector as this is central to an understanding of tensors.

Vectors

Certain physical quantities, such as force and velocity, are best specified in terms of vectors, because a vector has a magnitude and a direction. One example is the velocity vector, v. A vector can be specified by giving its components, v_1, v_2, and v_3, referring to the velocities in the three directions, x_1, x_2, and x_3. Generally, we can refer to a typical velocity component as v_i, where i can be 1, 2, or 3.

Note that while the magnitude is a physical attribute of a vector that does not depend on the choice of a particular co-ordinate system, the components of the vector do depend on the co-ordinate system selected to describe the flow. There is a simple rule that tells how to use the components of a vector in one co-ordinate system to calculate the components of that vector in second co-ordinate system that is rotate with respect to the first.

If vectors are adequate to describe the velocity of a body and the force acting on it, why are they not sufficient for describing rheological phenomena? The answer is that rheology deals not with motion *per se*, but with deformation, and specifically with the relationship between the deformation of a fluid element and the surface forces exerted in this element by the surrounding fluid. Tensors are very useful in specifying these two types of quantities, and the specific tensors that are used to represent these quantities are the strain tensor and the stress tensor.

What is a Tensor?

Like a vector, a tensor can be represented in terms of its components, and the values of these components depend on the choice of the co-ordinate system used. Furthermore, there is a rule for using the components of a tensor in one co-ordinate system to calculate the components of the tensor in another co-ordinate system, rotated with respect to the first. The existence of this rule shows that a tensor has a basic physical significance that transcends the arbitrary choice of the co-ordinate system. However, unlike a vector, the physical significance of a tensor cannot be described in terms of a directed line segment, i.e., in terms of a magnitude and a direction.

In describing rheological behaviour associated with a particular type of deformation, we generally select a co-ordinate system that gives the components a physical significance that is easily understood. For example, in describing the stretching of a rod ('uniaxial extension') we take the x_1 direction to be the direction of stretching. Then the σ_{11} component of the stress tensor is simply the tensile stress in the sample.

Whereas a vector has three components, the tensors we will use have nine. A typical component can be written using two indices, for example σ_{ij}, where each index can take on one of the values 1, 2, or 3 corresponding to the three co-ordinate directions. To present the values of all the components of a tensor, matrix notation can be used.

$$\sigma_{ij} = \begin{bmatrix} \sigma_{11} & \sigma_{12} & \sigma_{13} \\ \sigma_{21} & \sigma_{22} & \sigma_{23} \\ \sigma_{31} & \sigma_{32} & \sigma_{33} \end{bmatrix} \quad \text{... (1.25)}$$

Can any nine numbers form the components of a tensor? No. These numbers have a specific mathematical significance, which we will find particularly suited to the description of deformation and stress. Specifically, these nine components contain all the information necessary to transform one vector into another one that has a certain prescribed relationship with the first. In mathematical language we say that the tensor 'operates on' one vector to yield a second vector, which contains information taken from both the original vector and the tensor. For example, we will see that the strain tensor, i.e., the nine components of the strain tensor, can be used to operate on the components of the vector describing the relative position of fluid particles within an undeformed fluid element, to yield the corresponding position vectors after deformation. Likewise, the stress tensor can be used to operate on the unit normal vector defining the orientation of a surface of a fluid element to yield the surface force vector acting on that element.

Since the vector operate on in both cases is an arbitrarily selected one, we see that the strain tensor actually contains a complete description of the deformation that a fluid element undergoes during some flow process, while the stress tensor contains a complete description of the state of stress acting at a point in the fluid at a particular time.

The objective here is not to solve flow problems but only to describe rheological phenomena. Thus, tensor calculus will not be required, and the reader need learn no new mathematics but only the definitions of a few quantities.

With regard to notation, we will use a bold face symbol to indicate that it is a vector. For example, the velocity vector will be represented as v. The components of a vector will be indicated by means of a subscript, for example, v_i. For the components of a tensor we will use two subscripts. For example, the components of the stress tensor will be represented by σ_{ij}. In order to minimise the number of new symbols and rules that need be learned, we will not use dyadic notation or the Einstein summation convention. These are methods of notation that simplify the writing of equations involving tensors, and are described here.

THE STRESS TENSOR

The deformations that occur in the processing and use of materials are generally more complicated than simple extension and simple shear and involve a combination of these two types of deformation. For example, consider the deflection of a rubber tyre under load or the flow of a molten plastic into a mould. First, the deformation is not uniform but varies from one place to another within the material. It thus becomes a 'field variable', i.e., a quantity that varies from one point to another and is thus a function of position. Secondly, the stress is not purely tensile, compressive or shear.

The quantitative specification of the forces acting on a solid body as a result of contact with another body is straightforward; one need only give the components of the force vector acting at the interface. However, the specification of the forces acting on the surface of a fluid element is less obvious, since the orientation of the surface is arbitrary, i.e., it depends on how one defines a fluid element. It would appear that in order to completely specify the state of stress on a fluid element one would have to give the components of the stress vector for every possible orientation of the surface. Fortunately, this is not the case, and we will see that by specifying only the components of the stress tensor, the state of stress at a point in a fluid can be completely described.

For a given arbitrary, choice of a fluid element, the surface stress vector is $t^{(n)}$, where n is the unit normal vector for the surface and specifies its orientation. The (n) superscript on the surface stress vector indicates that the components of the vector depend on the orientation of the surface. This vector does not at first seem to be a useful tool for describing the state of stress at a point in a fluid because of the arbitrariness of the choice of the surface orientation. It is possible to show, however, that if the stress vectors acting on each of three mutually perpendicular planes passing through a point in a fluid are specified, the stress vectors for any other choice of planes can be calculated by means of a simple transformation rule. It is convenient to let these planes be perpendicular to the co-ordinate axes. Thus, the unit normal vector for a surface becomes equal to one of the unit normal vectors for the co-ordinate system:

$$n = e_i$$

and the components of the force vector are given by:

$$t_j^{(e_i)} = \sigma_{ij}$$

where σ_{ij} is the stress tensor.

To understand the physical significance of the nine components of the stress tensor, consider the small cubical element of material shown in Fig. 1.11. The second subscript indicates the direction of the force and corresponds to the co-ordinate axis direction. For example, the stress component shown in Fig. 1.11 acts in the x_1 direction, and the second subscript of this component is thus1. The first subscript indicates the face on which the component acts, and this is specified by reference to the co-ordinate direction normal to this face. Thus, the force shown acts on the face normal to the x_2 direction, and the first subscript of this component of the stress is thus2. The stress component shown is thus σ_{21}.

To complete our definition of the components of σ_{ij}, we need a sign convention. Here, we will use the convention generally used in mechanics, although the reader should be aware that the opposite convention is used by some rheologists. We will take the stress to be positive when it acts in the positive x_j direction, on a face having the higher value of x_i, i.e., the face further from the origin in the x_i direction. For example, the stress component shown in Fig. 1.11 is positive if the force acts in the direction of the arrow. This is because it acts in the positive x_i direction on a face having the higher value of x_2. Fig. 1.12 shows several additional components of the stress.

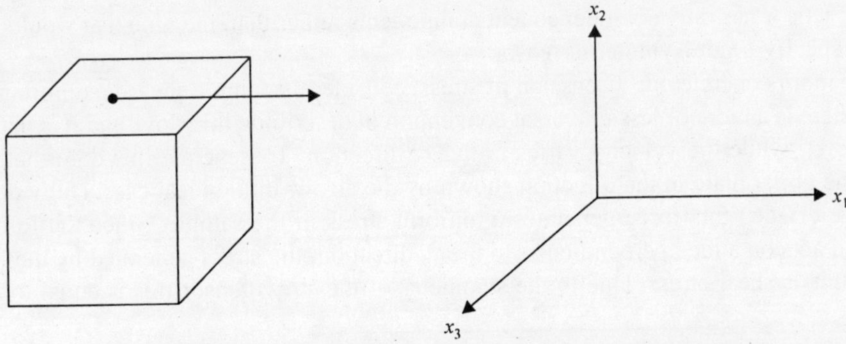

Fig. 1.11. Cubical material element with a typical stress component.

The set of nine components that is needed to specify completely the state of stress at a point in a deformable material is an example of a 'second order tensor', and the members of the set are said to be the 'components' of the tensor.

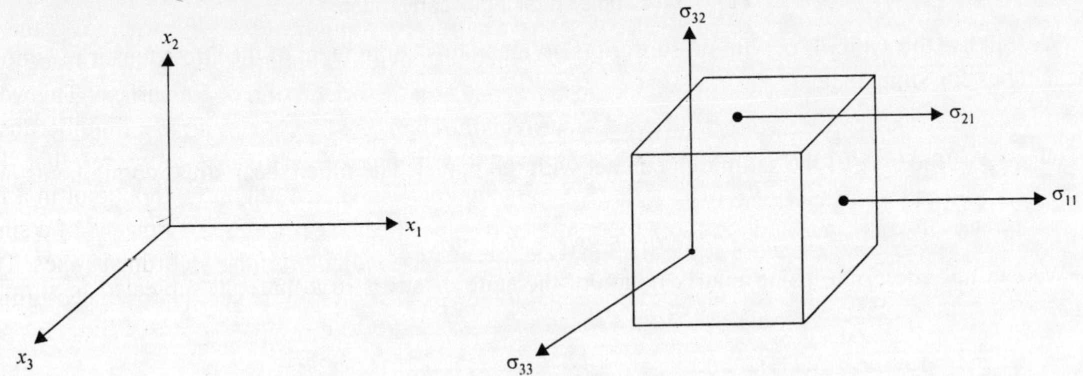

Fig. 1.12. Several additional stress components.

Since the components of the stress tensor describe the state of stress at a point in the material, the cubical element shown in Fig. 1.11 must be shrunk to an infinitesimal size. Thus, the two force vectors shown in Fig. 1.13 are acting in opposite directions at the same point. From Newton's law of action and reaction, these two forces must be equal in magnitude. They are thus both manifestations of the same component of the stress tensor, σ_{11}, and both have positive values if they act in the directions indicated. Thus, according to our sign convention, a tensile stress has a positive value.

Fig. 1.13. Equal; opposite forces at a point are represented by the same component of the stress tensor.

The principle of conservation of angular momentum can be applied to the infinitesimal material element we have been considering to show that the stress tensor has the following property:

$$\sigma_{ij} = \sigma_{ji} \quad \text{... (1.26)}$$

Thus, any two components that have the same subscripts or indexes, but in reversed order, have the same value. A tensor that has this property is said to be 'symmetric'. One result of this property is that a symmetric tensor has only six independent components rather than the nine that would be required to completely specify a non-symmetric tensor.

To make more concrete our discussion of stress, consider the simple shearing deformation shown in Fig. 1.14. There is a more or less universal convention in describing this flow, and it is that the direction of motion is x_1, while the velocity varies in the x_2 direction. To generate this deformation, a force is applied to the upper plate in the direction shown by the arrow. In the ideal case, (fully developed flow with no edge effects) this force generates a uniform stress in the sample. Since the force is in the x_1 direction and acts on a face perpendicular to the x_2 direction, the stress generated by the force F is σ_{21}. Obviously, this is shear stress. Due to the symmetry of the stress tensor this is equal to σ_{12}.

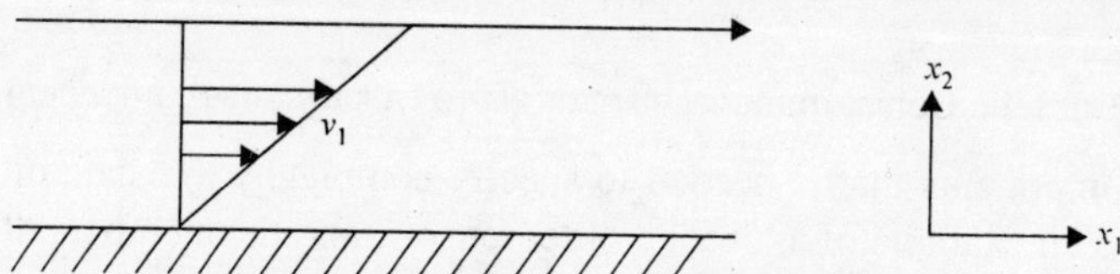

Fig. 1.14. Simple shear index convention.

We will use the symbol, σ, with no subscripts, to mean this component of the stress tensor in simple shear. Thus for simple shear,

$$\sigma_{21} = \sigma_{12} \equiv \sigma = F/A \quad \text{... (1.27)}$$

where A is the area of the sample in contact with the plates. The other shear stress components are zero:

$$\sigma_{13} = \sigma_{31} = \sigma_{23} = \sigma_{32} = 0 \quad \text{... (1.28)}$$

We can now describe, using matrix notation, the state of stress in a material subjected to simple shear:

$$\sigma_{ij} = \begin{bmatrix} \sigma_{11} & \sigma & 0 \\ \sigma & \sigma_{22} & 0 \\ 0 & 0 & \sigma_{33} \end{bmatrix} \quad \text{... (1.29)}$$

Another example of a test that is of practical interest in rheology is simple or uniaxial extension. This test is illustrated in Fig. 1.2. If we let x_1 be the direction of the applied force, the stress component resulting from this force will be σ_{11}, which is a normal stress. If it acts in the direction shown, it is a tensile stress. There are no shear stress components in this case, and the components of the stress tensor are as shown below:

$$\sigma_{ij} = \begin{bmatrix} \sigma_{11} & 0 & 0 \\ 0 & \sigma_{22} & 0 \\ 0 & 0 & \sigma_{33} \end{bmatrix} \quad \text{... (1.30)}$$

There is an additional point regarding normal stresses that should be mentioned here. While all materials are compressible to some extent, in the case of molten plastics, quite high pressures are required to produce a significant change in the volume of a sample. For this reason, for many purposes these materials can be considered to be incompressible. Now consider what happens when we subject an incompressible material to a compressive or tensile stress that is equal in all directions, i.e., an isotropic stress or 'pressure'.

The components of the stress tensor in this situation are as shown below:

$$\sigma_{ij} = \begin{bmatrix} -P & 0 & 0 \\ 0 & -P & 0 \\ 0 & 0 & -P \end{bmatrix} \qquad \text{... (1.31)}$$

A stress of this type is said to be 'isotropic'. The minus signs results from the fact that pressure is considered positive when it acts to compress a material, whereas according to our sign convention compressive stress is a negative quantity. In this situation, the sample will be totally unaffected by the forces associated with this pressure, i.e., it will not change its size or shape. Thus, such an isotropic stress field is of no rheological significance. Only when there are shear stresses acting, as in simple shear flow, or when the normal stress components are different from each other will deformation occur in an incompressible material.

Another way of saying this is that if a rheological measurement on an incompressible material is repeated at several different ambient pressures, for example by placing the rheometer in a hyperbaric chamber, the measurements at various pressures will yield exactly the same values of all rheological properties.

This means that for an incompressible material a normal component of stress has no absolute rheological significance. Only differences between two normal components are of rheological significance. For example, in the case of simple shear, it is customary to describe the state of stress from a rheological point of view by specifying the shear stress, σ, and the 'first and second normal stress differences'.

$$N_1 \equiv \sigma_{11} - \sigma_{22} \qquad \text{... (1.32)}$$

$$N_2 = \sigma_{22} - \sigma_{33} \qquad \text{... (1.33)}$$

For Newtonian fluids these two quantities are zero in simple shear, but in polymeric liquids they generally have nonzero values. One manifestation of the first normal stress difference is observed when a liquid is sheared by placing it between two flat parallel disks and rotating one of the disks. It is found that an elastic liquid exerts a normal thrust tending to separate the plates, while a Newtonian fluid exerts no normal thrust on the plates.

In the case of simple extension, there is only one rheologically significant feature of the stress field, because there are no shear stresses acting, and there is axial symmetry, so that $\sigma_{22} = \sigma_{33}$. This is the principal stretching stress, σ_E, defined in equation 1.34.

$$\sigma_E = \sigma_{11} - \sigma_{22} = \sigma_{11} - \sigma_{33} \qquad \text{... (1.34)}$$

Because the magnitude of a single normal stress component is not rheologically by itself, it is customary in rheology to define an 'extra' or 'viscous' stress as follows:

$$\tau_{ij} \equiv \sigma_{ij} + P \; (i = j) \qquad \text{... (1.35a)}$$

$$\tau_{ij} \equiv \sigma_{ij} \qquad (i \neq j) \qquad \text{... (1.35b)}$$

We see from equation 1.31 that all the components of the extra stress tensor are zero in a fluid that has been at rest for a long time. However, for materials undergoing deformation, the stress is no longer isotropic, and the '*P*' in the above equation no longer has direct relationship to the pressure as it is usually defined. However, this will not be a problem for us, as we will work only with shear stresses and normal stress differences. For these quantities, the complete specification of *P* is not essential, and it matters not whether we write our equations in terms of σ_{ij} or τ_{ij}.

A STRAIN TENSOR FOR INFINITESIMAL DEFORMATIONS

In order to describe rheological phenomena in quantitative terms, it is necessary to make use of a quantitative measure of deformation. Such a measure is called strain, and we have defined in the previous sections measures of strain suitable for specific types of deformation, namely, simple shear and simple extension. However, it is advantageous to establish a general definition of strain that is valid for all types of flow kinematics. For the present, we will define only the infinitesimal strain tensor, which will be useful in the presentation of the theory of linear viscoelasticity.

The simplest description of deformation of a fluid element is one based on the behaviour of the position vectors for the fluid particles within this element. Fig. 1.15 shows the position vector, $x_A(t_1)$, for a typical particle 'A' at a time, t_1, as well as the position vector, $x_A(t_2)$, for the same fluid particle at another time, t_2. The use of two times, or two states of the fluid element, is unavoidable in discussing strain, as strain is a measure of the difference between two states. We will take t_1 to represent a time when the fluid is in a reference configuration, with which the configuration at any other time, t_2, will be compared. Thus, we will say that a material element is in a strained state at time t_2 if it undergoes deformation in the time interval between t_1 and t_2.

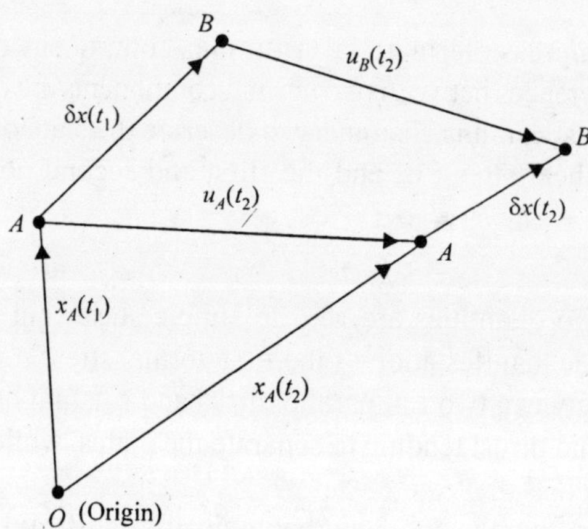

Fig. 1.15. The position vector of a fluid particle (A) at times t_1 and t_2, and the displacement vectors at time t_2 for two neighbouring fluid particles, A and B.

To describe deformation it is necessary to look at two fluid particles and the change that occurs in their relative positions between time t_1 and t_2. In Fig. 1.15, two such particles are shown, with the position of the second particle with respect to the first given by the vector $\delta x(t_1)$ at time t_1 and by $\delta x(t_2)$ at time t_2. The displacement vector, $u(t_2)$ is the position of a particles at time t_2 relative to its position at time t_1. The relative variation of the $u(t_2)$ vectors for the two fluid particles obviously provides information about the deformation of the fluid element containing these particles. For example, in a rigid displacement of the fluid element, without rotation, as shown in Fig. 1.16, the displacement vectors for the two particles are the same, and this is true for any two particles in the fluid element. This suggests that the gradient of u may be useful as a quantitative measure of strain. This quantity has nine components:

$$\frac{\partial u_i(t_2)}{\partial x_j(t_1)} \qquad \text{... (1.36)}$$

These nine quantities are the components of a tensor, and they clearly contain all the information necessary to describe the relative displacements of the two neighbouring particles. This tensor has some characteristics that make it unsuitable as a general measure of strain, but if we limit our attention to flows in which this gradient is always very small, a useful measure of strain can be defined as follows:

$$\gamma_{ij}(t_1,t_2) \equiv \frac{\partial u_i(t_2)}{\partial x_j(t_1)} + \frac{\partial u_j(t_2)}{\partial x_i(t_1)} \quad \text{... (1.37)}$$

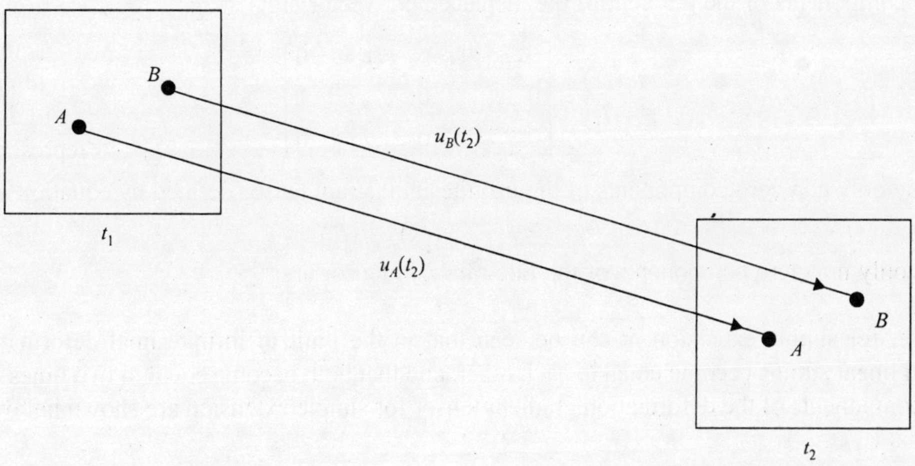

Fig. 1.16. Displacement vectors for particles *A* and *B*, of a fluid element at time t_2 and after a rigid translation of the element.

where $\gamma_{ij}(t_1, t_2)$ is the infinitesimal strain tensor at time t_2, relative to the configuration of the same fluid element at time t_1. Like the stress, this is a symmetric tensor, because if the order of the indexes i and j is reversed, the value of each component remains unchanged.

The 'strain rate tensor', also called the 'rate of deformation tensor', is defined by equation 1.38.

$$\dot{\gamma}_{ij} \equiv d\gamma_{ij}/dt \quad \text{... (1.38)}$$

Since the components v_i of the velocity vector are given by:

$$v_i = du_i/dt \quad \text{... (1.39)}$$

Equation 1.38 can be expanded as shown in equation 1.40.

$$\dot{\gamma}_{ij} = \frac{\partial v_j}{\partial x_i} + \frac{\partial v_i}{\partial x_j} \quad \text{... (1.40)}$$

This definition of a rate of deformation is not limited to use in describing deformations that are infinitesimal in total magnitude, as its definition requires only a measure of strain valid during an infinitesimal interval of time.

As an example of the use of the infinitesimal strain tensor, we note that for simple shear of magnitude γ;

$$\gamma_{12} = \gamma_{21} = \frac{\partial u_1}{\partial x_2} + \frac{\partial u_2}{\partial x_1} \quad \text{... (1.41)}$$

The variation with time of the components of the position vector for simple shear are:

$$x_1(t_2) = x_1(t_1) + x_2(t_1)\,[\gamma(t_2) - \gamma(t_1)] \quad \text{... (1.42a)}$$

$$x_2(t_2) = x_2(t_1) \quad \text{... (1.42b)}$$

$$x_2(t_2) = x_3(t_1) \quad \text{... (1.42c)}$$

The only non-zero component of the displacement vector is u_1:

$$u_1(t_2) = x_1(t_2) - x_1(t_1) = x_2(t_1)\,[\gamma(t_2) - \gamma(t_1)]$$

and the components of the gradient of the displacement vector are:

$$\frac{\partial u_i(t_2)}{\partial x_j(t_1)} = \begin{bmatrix} 0 & [\gamma(t_2) - \gamma(t_1)) & 0 \\ 0 & 0 & 0 \\ 0 & 0 & 0 \end{bmatrix} \quad \text{... (1.43)}$$

Thus, the only non-zero components of the infinitesimal strain tensor defined by equation 1.37 are:

$$\dot{\gamma}_{12} = \dot{\gamma}_{21} = \gamma(t_2) - \gamma(t_1) \equiv \dot{\gamma} \quad \text{... (1.44)}$$

and the only non-zero components of the rate of strain tensor are:

$$\gamma_{12} = \gamma_{21} \equiv \gamma \quad \text{... (1.45)}$$

Likewise, for simple extension, it can be seen that in the limit of infinitesimal deformation, the Hencky and linear strains become equal to each other, and that both become equal to two times $\partial u_1/\partial x_1$. All of the components of the deformation gradient tensor for simple extension are shown below, where $\varepsilon \equiv \varepsilon\,(t_2) - \varepsilon\,(t_1)$.

$$\frac{\partial u_i(t_2)}{\partial x_j(t_1)} = \begin{bmatrix} \varepsilon & 0 & 0 \\ 0 & -\varepsilon/2 & 0 \\ 0 & 0 & -\varepsilon/2 \end{bmatrix} \quad \text{... (1.46)}$$

Thus, the components of the infinitesimal strain tensor are:

$$\gamma_{ij} = \begin{bmatrix} 2\varepsilon & 0 & 0 \\ 0 & -\varepsilon & 0 \\ 0 & 0 & -\varepsilon \end{bmatrix} \quad \text{... (1.47)}$$

For simple extension, the components of the strain rate tensor are:

$$\dot{\gamma}_{ij} = \begin{bmatrix} 2\dot{\varepsilon} & 0 & 0 \\ 0 & -\dot{\varepsilon} & 0 \\ 0 & 0 & -\dot{\varepsilon} \end{bmatrix} \quad \text{... (1.48)}$$

NEWTONIAN FLUID

We can make use of the tensors defined above to write out the general constitutive equation for the Newtonian fluid, which is valid for low molecular weight, single phase liquids:

$$\tau_{ij} = \eta\dot{\gamma}_{ij} \quad \text{... (1.49)}$$

whereas equation 1.16 gives only the shear stress of a Newtonian fluid undergoing steady simple shear, equation 1.49 provides information about all the components of the stress tensor for any type of deformation. For example, we can use it to calculate all the stresses in steady simple shear. The velocity distribution for this deformation can be written as follows:

$$v_1 = \dot{\gamma} x_2 \quad \text{... (1.50a)}$$

$$v_2 = v_3 = 0 \quad \text{... (1.50b)}$$

and the components of the rate of deformation tensor are:

$$\dot{\gamma}_{ij} = \begin{bmatrix} 0 & \dot{\gamma} & 0 \\ \dot{\gamma} & 0 & 0 \\ 0 & 0 & 0 \end{bmatrix} \quad \text{... (1.51)}$$

Thus, the components of the extra stress tensor for a Newtonian fluid are, from equation 1.49.

$$\tau_{ij} = \begin{bmatrix} 0 & \eta\dot{\gamma} & 0 \\ \eta\dot{\gamma} & 0 & 0 \\ 0 & 0 & 0 \end{bmatrix} \quad \text{... (1.52)}$$

We note that the normal components of the extra stress are all zero in the Newtonian fluid. This means that:

$$\sigma_{11} = \sigma_{22} = \sigma_{33} \quad \text{... (1.53)}$$

and that the normal stress differences are both zero.

$$N_1 = N_2 = 0 \quad \text{... (1.54)}$$

It is important to remember that these equations for the components of the stress tensor are only valid for steady simple shear of a Newtonian fluid. Another example of the use of the constitutive equation for a Newtonian fluid is the calculation of the stresses in a deformation other than simple shear, namely simple extension. Combining equation 1.48 to 1.49 we have:

$$\tau_{ij} = \eta \begin{bmatrix} 2\dot{\varepsilon} & 0 & 0 \\ 0 & -\dot{\varepsilon} & 0 \\ 0 & 0 & -\dot{\varepsilon} \end{bmatrix} \quad \text{... (1.55)}$$

For a Newtonian fluid, we have seen that it is possible to measure the viscosity in a simple shear experiment and then use it in the general constitutive equation (1.49) to determine all the components of the extra stress tensor for any type of flow. For non-Newtonian fluids this cannot be done, i.e., it is not possible to predict a fluid's response to arbitrary deformations on the basis of rheological information obtained using only simple shear experiments.

BASIC EQUATIONS OF FLUID MECHANICS

This is the procedure that is necessary, for example, to predict the detailed flow pattern and velocity distribution for melt flow in a die or in a mould. However, in order to understand the role played by rheology in solving such problems, we will simply present the basic equations of continuum mechanics and make a few comments about them. These basic equations are the continuity equation, which is derived from the principle of conservation of mass, and Cauchy's equation, which is a generalisation of Newton's second law of motion, (F = *m*a) for the case of a deformable material.

In deriving these equations, which govern the mechanical behaviour of deformable materials, it is assumed that every point in space (specified by x_1, x_2, x_3) within a material has associated with it values for all the components of the velocity vectors, the strain tensor and the stress tensor. It is further assumed that these 'field variables' are continuous functions of position. In reality, we know that this is not precisely true, because at a sub-microscopic level we find mostly empty space punctuated by regions of concentrated mass, which are the atoms of the molecules of which the material is composed.

However, the scale of our macroscopic observation and of practical flow phenomena is so large compared to the scale at which this inhomogeneity appears, that for all practical purposes we can assume the material to be a 'continuum' and make use of the strain, velocity and stress fields to describe the deformation and flow properties of the material. This is called the 'continuum assumption'. Thus, we assume that the collection of molecules that make up the material of interest can be represented, for purposes of describing its flow and deformation, by a mathematical continuum. The equations based on this assumption are thus called the equations of continuum mechanics.

Continuity Equation

The principle of conservation of mass, when applied to a deformable continuum, tells us that the net flow of mass into any fixed volume element contained within the continuum during some interval of time is equal to the accumulation of mass within the element during that time interval. This principle is summarised by equation 1.56.

$$[\text{Net flow in}] = [\text{Flow in}] - [\text{Flow out}] = [\text{Accumulation}] \qquad \text{... (1.56)}$$

For an incompressible material, the accumulation of mass within a volume element is zero. We will not present a derivation here, but in this case it can be shown that the principle of conservation of mass can be written as shown in equation 1.57.

$$\frac{\partial v_1}{\partial x_1} + \frac{\partial v_2}{\partial x_2} + \frac{\partial v_3}{\partial x_3} = 0 \qquad \text{... (1.57)}$$

This is the 'continuity equation' for an incompressible material.

Cauchy's Equation

The second basic equation of continuum mechanics is based on Newton's second law of motion (F = *m*a). This basic principle describes the relationship between the applied force and the change of momentum of a rigid body. It was later put into a form suitable for describing the deformation of a continuum in response to deforming stresses, and the result is Cauchy's equation.

$$\rho \frac{Dv_i}{Dt} = -\rho g \frac{\partial h}{\partial x_i} + \frac{\partial \sigma_{1i}}{\partial x_1} + \frac{\partial \sigma_{2i}}{\partial x_2} + \frac{\partial \sigma_{3i}}{\partial x_3} \qquad \text{... (1.58)}$$

The quantity on the left is the density times the "substantial derivative", which is defined as follows:

$$\frac{Dv_i}{Dt} \equiv \frac{\partial v_i}{\partial t} + v_1 \frac{\partial v_i}{\partial x_1} + v_2 \frac{\partial v_i}{\partial x_2} + v_3 \frac{\partial v_i}{\partial x_3} \qquad \text{... (1.59)}$$

The substantial derivative is the quantity that tells how the velocity of a fluid element changes with time as it moves through the flow field.

Equations 1.58 and 1.59 are vector equations, and since i can have any one of the three values 1, 2 or 3, each can also be written as a set of three scalar equations.

Making use of the definition of the extra stress (Eq. 1.35), Cauchy's equation can be re-written in a form involving the pressure.

$$\rho\frac{Dv_i}{Dt} = -\rho g\frac{\partial h}{\partial x_i} - \frac{\partial P}{\partial x_i} + \frac{\partial \tau_{1i}}{\partial x_1} + \frac{\partial \tau_{2i}}{\partial x_2} + \frac{\partial \tau_{3i}}{\partial x_3} \quad \text{... (1.60)}$$

The left hand side describes the acceleration of a material element, while the right hand side accounts for the applied forces. The first term on the right represents the force due to the earth's gravitational field, which is a 'body force' and acts directly on the mass of the material element. The second term describes the net force on a material element due to pressure gradients, and the last three terms involve the extra stress tensor and result from the deformation of the fluid. It is through these last three terms that the rheological properties of the material become involved in its mechanical behaviour.

For molten polymers, the very high resistance to deformation resulting from the strong interactions between molecules makes the last three terms very large compared to the acceleration terms on the left hand side. For this reason, in flow situations involving molten plastics, the left hand side can be neglected. This is called the 'creeping flow' approximation. Taking account of the fact that Cauchy's equation represents three scalar equations, and adding in the continuity equation, we have four equations. The number of dependent variables is much greater than this, however, as it includes the three components of the velocity vector plus the six independent components of the stress tensor. Thus, no problem in continuum mechanics can be solved without an additional equation. This additional equation is the rheological constitutive equation, which relates the components of the stress tensor to the deformation.

Navier-Stokes Equation

The constitutive equation for a Newtonian fluid is given by equation 1.49. When this is combined with Cauchy's equation (1.60) and the continuity equation is used to simplify the terms, the result is the Navier-Stokes equation, which governs the flow of Newtonian fluids. If the viscosity is assumed to be a constant, i.e., independent of position, the Navier-Stokes equation can be written as follows:

$$\rho\frac{Dv_i}{Dt} = -\rho g\frac{\partial h}{\partial x_i} - \frac{\partial P}{\partial x_i} + \eta\left(\frac{\partial^2 v_i}{\partial x_1^2} + \frac{\partial^2 v_i}{\partial x_2^2} + \frac{\partial^2 v_i}{\partial x_3^2}\right) \quad \text{... (1.61)}$$

Note that this is vector equation and that by setting i equal to 1, 2 and 3, we can re-write it as three scalar equations.

In the analysis of flows involving Newtonian fluids, a dimensionless group of central importance is the Reynolds number. This is defined as follows:

$$\mathrm{Re} \equiv \frac{DV\rho}{\eta} \quad \text{... (1.62)}$$

where D and V are a characteristic length and velocity respectively. This is a measure of the relative importance of viscous and inertial (acceleration) effects. In other words, it is a measure of the relative importance of the third term on the right of equation 1.61 as compared with the term on the left. For fluids with very high viscosity, the Reynolds number is generally very low, and the inertia or acceleration term on the left hand side of equation 1.61 can be neglected. This is often referred to as the 'creeping flow' approximation. While molten polymers are not Newtonian, and equation 1.61 is therefore not appropriate, it is generally accepted that for these materials, the creeping flow approximation is valid.

CHAPTER 2

Conventional Viscometers and General Concepts

INTRODUCTION

A variety of viscometers are used in the polymers and plastics industries. These devices fall into two general categories: scientific instruments designed to make basic measurements of rheological properties and instruments that are primarily employed for industrial quality control. This chapter outlines the operating principles and basis for both general classes of instruments. It should be noted that a wide variety of rheological test methods and instruments are used in other industries. A summary of the major rheological measurement techniques for different industries is given in the last section of this chapter.

ANALYSIS OF CAPILLARY FLOW

The capillary tube viscometer is one of the simplest and most widely used instruments for rheological characterisation. Before describing its operation one must first consider the problem of capillary flow, as in Fig. 2.1, in which a fluid is forced at a steady rate from a large reservoir into a small diameter capillary tube of length L. Assume steady, isothermal flow of an incompressible fluid and L/R >>1. The co-ordinate system is defined along the r, θ, and z axes. If entrance and exit losses in the capillary tube are ignored, then

$$V_\theta = 0 \quad ; \quad V_r = 0$$
$$V_z \neq 0 \quad ; \quad V_z \neq f(\theta) \text{ (axisymmetry)} \qquad \text{... (2.1)}$$

Writing the continuity equation

$$\frac{\partial P}{\partial t} + \frac{1}{r}\frac{\partial}{\partial r}(\rho r V_r) + \frac{1}{r}\frac{\partial}{\partial \theta}(\rho V_\theta) + \frac{\partial}{\partial z}(\rho V_z) = 0 \qquad \text{... (2.2.)}$$

where density ρ is assumed to be constant.

Because the system is at steady state $(\partial P/\partial t) = 0$, since $V_r = 0$, $V_\theta = 0$, the second and third terms on the left side of equation 2.2 drop out as well, leaving:

$$\frac{\partial V_z}{\partial z} = 0 \qquad \text{... (2.3)}$$

The momentum equation for the z-component is

$$P\left[\frac{\partial V_z}{\partial t} + V_r\frac{\partial V_z}{\partial r} + \frac{V_\theta}{r}\frac{\partial V_z}{\partial \theta} + V_z\frac{\partial V_z}{\partial z}\right] = -\frac{\partial P}{\partial z} - \left[\frac{1}{r}\frac{\partial}{\partial r}(r\,\tau_{rz}) + \frac{1}{r}\frac{\partial \tau_{\theta r}}{\partial \theta} + \frac{\partial \tau_{zz}}{\partial z}\right] + \rho g_z \qquad \text{... (2.4)}$$

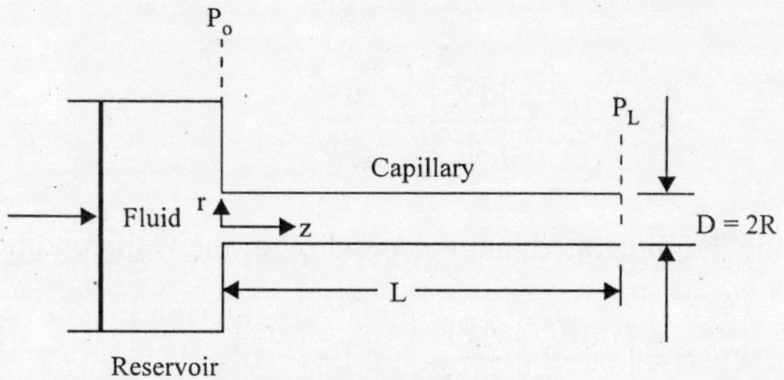

Fig. 2.1. Capillary tube flow.

From the constraints of the system imposed, the z-component momentum equation reduces to

$$\frac{dP}{dz} = -\frac{1}{r}\frac{d}{dr}(r\,\tau_{rz}) \quad \text{... (2.5)}$$

Both sides of this expression are a function of r [i.e., f(r)] and both sides are constant. It can, therefore, be assumed that $dP/dz \simeq \Delta P/L$ (note that $dP/dz < 0$), and can therefore integrate equation 2.5.

$$r\,\tau_{rz} = -\frac{\Delta P}{L}r^2 + C_1$$

$$\tau_{rz} = -\frac{\Delta P}{L}r + \frac{C_1}{r} \quad \text{... (2.6)}$$

At $r = 0$, $\tau_{zr} \neq \infty$ therefore, $C_1 = 0$ and hence:

$$\tau_{zr} = -\left(\frac{\Delta P}{2L}\right)r \quad \text{... (2.7)}$$

Equation 2.7 describes the shear stress profile for any incompressible fluid, and the profile across the capillary tube will have the form shown in Fig. 2.2.

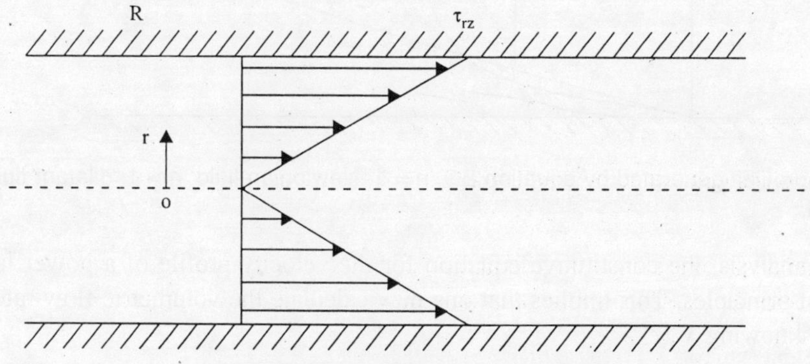

Fig. 2.2. Shear stress profile in tube flow.

Now, if the rheological properties of the fluid follow that a power law constitutive equation, i.e., $= -K|\dot{\gamma}|^{n-1}\dot{\gamma}$ where τ and τ_{zr}, $\dot{\gamma} = dV_z/dr$ and K and n are constants then

$$-K\left|\frac{dV_z}{dr}\right|^{n-1}\frac{dV_z}{dr} = -\left(\frac{\Delta P}{2L}\right)r \quad \text{... (2.8)}$$

Note that

$$\left|\frac{dV_z}{dr}\right| = -\frac{dV_z}{dr}$$

since $dV_z/dr < 0$.

This expression may now be integrated and the 'no-slip' at the wall condition may be imposed

$$V_z(r) = \left(\frac{nR}{n+1}\right)\left[\frac{\Delta PR}{2KL}\right]^{1/n}\left[1-\left(\frac{r}{R}\right)^{(n+1)/n}\right] \quad \text{... (2.9)}$$

Equation 2.9 is the velocity profile of a power law fluid in tube flow. For n = 1, K = μ; i.e., the limit of a Newtonian fluid is reached and the velocity profile of equation 2.9 reduces to the familiar parabolic velocity profile expression

$$V_z(r) = \frac{\Delta PR^2}{4\mu L}\left[1-\left(\frac{r}{R}\right)^2\right] \quad \text{... (2.10)}$$

The theoretical velocity profiles computed from Equation 2.9 have the shapes shown in Fig. 2.3: for n = 1 the parabolic shape (Newtonian, i.e., Equation 2.10); for n > 1 a dilatant fluid; for n < 1 a pseudoplastic fluid. For polymer melts and solutions, n < 1, and hence velocity profiles are typically 'plug-like'. This means that there are very high velocity gradients in the vicinity of the tube wall, as shown in Fig. 2.3.

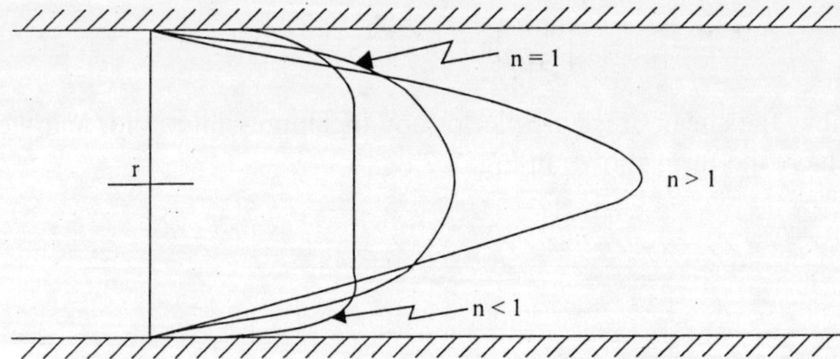

Fig. 2.3. Velocity profiles generated by equation 2.9: n = 1, Newtonian fluid; n > 1, dilatant fluid; and n < 1, pseudoplastic fluid.

In the above analysis, the constitutive equation for the velocity profile of a power law fluid was derived from first principles. This implies that one may calculate the volumetric flow rate of the fluid through the tube knowing $V_z(r)$.

$$Q = 2\pi\int_0^R V_z(r)\, r\, dr \quad \text{... (2.11)}$$

Substituting in the velocity profile expression equation 2.9 and integrating, gives

$$Q = \frac{\pi n R^3}{3n+1}\left[\frac{R\Delta P}{2KL}\right]^{1/2} \quad \text{... (2.12)}$$

Equation 2.12 can be written in a linear form as:

$$\ln Q = \ln\left(\frac{n\pi R^3}{3n+1}\right) + \frac{1}{n}\ln\left(\frac{R}{2KL}\right) + \frac{1}{n}\ln \Delta P \quad \text{... (2.13)}$$

The striking observation of Equation 2.13 is that the first two sets of expression on the right side are constant. Hence, $Q = f(\Delta P)$ and a logarithmic plot of Q vs. ΔP will give a straight line. Nature and mathematics do not always agree. Actual flow data on pseudoplastic materials tend to be non-linear, as shown in Fig. 2.4, the reason being that $n = f(dVL_z/dr)$; i.e., n tends to depend on Q. Although from a qualitative standpoint, the power law constitutive equation shows proper trends, quantitatively it is a poor predictive tool. From the standpoint of rheological characterisation, there is no substitute for experiments and therefore, a capillary tube flow experiment must be designed that is general enough to evaluate all types of time-independent fluids.

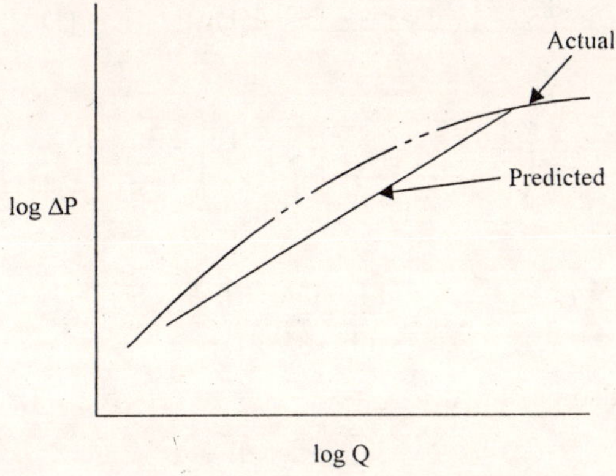

Fig. 2.4. Comparison of actual and theoretical flow curves.

CAPILLARY VISCOMETRY

Orienting the capillary tube so that the flow is vertically downward, one may write the force balance

$$\frac{\pi D^2}{4}\Delta P = \pi D L \tau_w \quad \text{... (2.14)}$$

or

$$(\tau_{rx})_{r=R} = \tau_w = \frac{D\Delta P}{4L} \quad \text{... (2.15a)}$$

Following the development of the Rabinowitsch-Mooney equation for steady, laminar flow of a time-independent fluid through a vertical tube, the flow is considered in a slice of fluid constituting an annulus between r and r + dr:

$$dQ = V\,2\pi r\,dr$$

where V is the linear velocity at r. Hence,

$$Q = \pi\int_0^R V 2r\,dr = \pi\int_0^{R^2} V d(r^2)$$

Applying the no-slip condition (V = 0, r = R), the expression becomes

$$\frac{Q}{\pi R^3} = \frac{8Q}{\pi D^3} = \frac{1}{\tau_w^3}\int_0^{T_w} \tau_{rx}^2 f(\tau_{rx}) d\tau_{rx} \quad \text{... (2.15b)}$$

where $f(\tau_{rx}) = -dV/dr$. By appropriate manipulating and replacing τ_w with DΔP/4L we obtain the Rabinowitsch-Mooney equation of shear rate at the tube wall for steady, laminar flow of a time-independent fluid:

$$3\left(\frac{8Q}{\pi D^3}\right) + \frac{D\Delta P}{4L}\frac{d(8Q/\pi D^3)}{d(D\Delta P/4L)} = \left(-\frac{dV}{dr}\right)_w \quad \text{... (2.16)}$$

Since Q = VA = 1/4 π D^2 V, equation 2.16 can be rearranged:

$$3\left(\frac{8Q}{\pi D^3}\right) = \left(-\frac{dV}{dr}\right)_w = \frac{3}{4}\left(\frac{8V}{D}\right) + \left(\frac{8V}{D}\right)\frac{d[\frac{1}{4}(8V/D)]/(8V/D)}{d(D\Delta P/4L)/(D\Delta P/4L)} \quad \text{... (2.17)}$$

or

$$-\dot{\gamma}_w = \left(-\frac{dV}{dr}\right)_w = \frac{8V}{D}\left[\frac{3}{4} + \frac{1}{4n}\right] = \frac{3n+1}{4n}\left(\frac{8V}{D}\right) \quad \text{... (2.17a)}$$

where

$$n = \frac{d \ln (D\Delta P/L)}{d \ln (8V/D)}$$

This expression as may also be stated

$$-\dot{\gamma}_w = \frac{3}{4}\Gamma + \frac{\tau_w}{4}\frac{d\Gamma}{d\tau_w} \quad \text{... (2.17b)}$$

where $\Gamma = 4Q/\pi R^3$ is the shear rate for a Newtonian fluid.

The Poiseuille equation for a Newtonian fluid is

$$\Delta P = \frac{8\mu LV}{R^2} = \frac{32\mu LV}{D^2}$$

which can also be written as

$$\ln\left(\frac{D\Delta P}{4L}\right) = \ln\left(\frac{8V}{D}\right) + \ln(\mu)$$

Because μ is constant,

$$\frac{d \ln (D\Delta P/4L)}{d \ln (8V/D)} = n = 1.0$$

Substituting this expression into Equation 2.17 shows that $(-dV/dr)_w = 8V/D$. This statement is true only for a Newtonian fluid because viscosity is a constant. In the case of a non-Newtonian n varies along the log-log plot of DΔP/4L vs. 8V/D. To apply a capillary tube viscometer in order to develop a non-Newtonian fluid's flow curve, data must be converted into a logarithmic plot of DΔP/4L vs. 8V/D, where n is determined as the slope of the curve at a particular value of the wall shear stress (i.e.,

$\tau_w = D\Delta P/4L$). The shear rate corresponding to this wall shear stress value is determined from equation 2.17. It is note that equation 2.17 is derived from laminar flow; hence, one must ensure that capillary tube flow is in this regime by checking Reynolds number. The generalised Reynold number for a power law fluid is:

$$Re_{gen} = \frac{D^n V^{2-n} \rho}{\dot{\gamma}} \quad \text{... (2.18)}$$

or

$$Re_{gen} = \frac{D^n V^{2-n} \rho}{K8^{n-1}} \quad \text{... (2.19)}$$

This expression applies to all fluids which are not time dependent. In fact, it has been shown that such non-Newtonian fluids obey the conventional Newtonian friction factor vs. Reynolds number. Therefore, laminar flow prevails for $Re_{gen} < 2100$.

Data generated in any capillary tube viscometer will require corrections for four contributions to frictional losses: (i) the head of fluid above the tube exit; (ii) kinetic energy effects; (iii) tube entrance effects; and (iv) effective slip near the tube wall. The last effect is not always present and generally can be detected as explained later.

The first three corrections can be made through application of the total mechanical energy balance. *Skelland* outlines a correction approach and provides an illustrative example. An alternative approach that generally gives good results is the *Bagley* shear stress correction method.

This method was originally developed using a variety of polyethylenes. The resulting flow curve is independent of the L/D ratio of the capillary; however, the procedure requires the assumptions of zero slip at the tube walls and time independence of the fluid material. The method uses a fictitious length of tubing $N_f R$ that is added to the actual length of the tube L, such that the measured total ΔP across L is that which would be obtained in fully developed flow over the length $(L + N_f R)$ at the flow rate used in the experiment. The shear stress at the wall for fully developed flow over the length $(L + N_f R)$ is then

$$\tau_w = R(\Delta P/2)\,(L + N_f R) \quad \text{... (2.20)}$$

If the fluid is time independent and there is no slip at the tube wall, then τ_w is a unique function of $4Q/\pi R^3 (\simeq \Gamma)$ in laminar flow. Hence

$$\tau_w = \frac{R}{2}\left(\frac{\Delta P}{L + N_f R}\right) = f\left(\frac{4Q}{\pi R^3}\right) \quad \text{... (2.21)}$$

from which

$$\frac{L}{R} = -N_f + \frac{\Delta P}{2f(4Q/\pi R^3)} \quad \text{... (2.22)}$$

The procedure involves obtaining a series of ΔP measurements made on several tubes of various L/R (or L/D) ratios, while maintaining $4Q/\pi R^3$ constant.

It is important to note that some non-Newtonian fluids display a peculiar orientation of their molecules in the vicinity of the tube walls. In an aqueous suspension, the discrete phase may actually move away from the wall, leaving a thin layer of the continuum phase in the immediate proximity of the wall. In this situation a reduction occurs in the apparent viscosity in the vicinity of the wall, known as the phenomenon of effective slip at the wall.

For a capillary tube viscometer, an effective slip coefficient can be evaluated as a function τ_w using the following relationship:

$$\frac{Q}{\pi R^3 \tau_w} = \frac{\beta}{R} + \frac{1}{\tau^4} \int_0^{\tau_w} \tau_{rx}^2 f(\tau_{rx}) d\tau_{rx} \quad \text{... (2.23)}$$

A series of capillary tube measurements is needed in order to evaluate β. For a range of tubes of various R but constant L, an plot of $Q/R^3\tau_w$ vs. τ_w can be constructed. If no slip occurs with the material (i.e., $\beta = 0$), all the curves will coincide in accordance with equation 2.23. If, however, a family of curves is obtained, then β has some value. From the family of curves, select some fixed value of τ_w and obtain values of $Q/\pi R^3\tau_w$ for each corresponding R-value curve. Now prepare a plot of $Q/\pi R^3\tau_w$ vs. 1/R. The slope of this curve provides a β-value for the selected τ_w, according to equation 2.23. By repeating this procedure for different τ_w values a relationship between τ_w and β can be established. Fig. 2.5 illustrates the method. Once a relationship between the effective slip coefficient and wall shear stress is established, the measured volumetric flow rate from the capillary tube can be corrected for slippage:

$$Q' = Q_{measured} - \beta \tau_w \pi R^2 \quad \text{... (2.24)}$$

where Q′ is the volumetric flow rate corrected for slippage and $\tau_w = D\Delta P/4L$ corresponding to $Q_{measured}$.

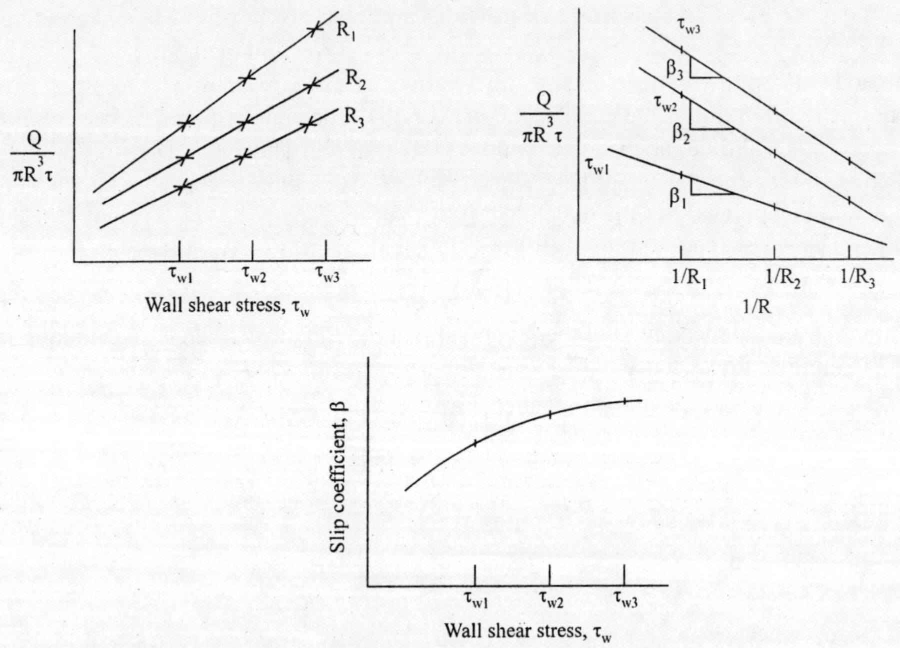

Fig. 2.5. Graphic method for estimating for effective wall slippage.

CONCENTRIC CYLINDER ROTARY AND ROTATING CYLINDER VISCOMETERS

Concentric cylinder rotary and rotating cylinder viscometers are most often applied to solution viscosity measurements and are usually limited to shear rates of < 100 s^{-1}.

The basic elements of the concentric cylinder rotary viscometer are shown in Fig. 2.6. The unit operates by applying shear to a fluid located in the annulus between the concentric cylinders. One cylinder (usually the cup) rotates while the other is fixed in space. From a series of measurements of the angular speed of the rotating cup and of the torque applied to the stationary cylinder, a flow curve for the fluid under shear can be derived.

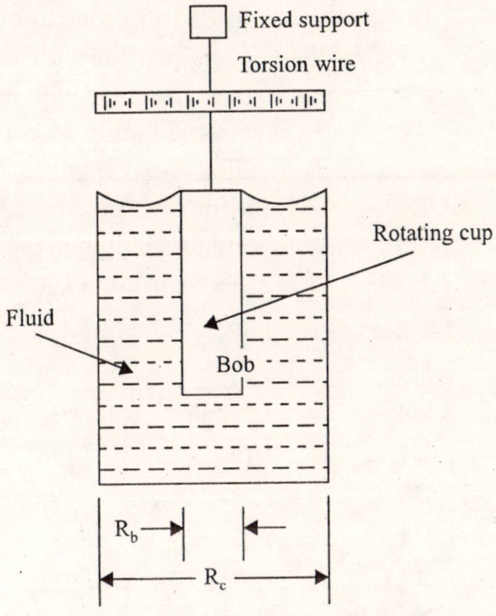

Fig. 2.6. Basic components of the concentric cylinder rotary viscometer.

The development of the constitutive equations for this type of viscometer is as follows. From a torque balance about the surface of the fixed bob while the cup revolves at a steady angular velocity, one may write:

$$\tau_b(2\pi R_b \ell)R_b = 2\pi R_b^2 \ell\, \tau_b \qquad \text{... (2.25)}$$

where T = torque, R_b = bob radius, and τ_b = shearing stress at the bob surface.

The balance equation neglects end effects at the base of the bob and assumes that shear occurs only at the cylindrical surface of the bob. In practice, however, end effects are important and can be accounted for by including an effective bob length ℓ_{eff} in equation 2.25 for the bob surface shear stress.

$$\tau_b = T/(2\pi R_b^2\, \ell) \qquad \text{... (2.26)}$$

The shear rate at the bob surface is a complex relationship, and for brevity is not included here. An approximate relationship proposed by Calderbank and Moo-Young based on the analytical derivation of Kreiger and Maron is

$$\left(\frac{dV}{dr}\right)_b = \frac{4\pi N}{1-(R_c/R_b)^2} C_R \qquad \text{... (2.27)}$$

Equation 2.27 is applicable for cup-to-bob radii ratios smaller than 1.75. The term C_R represents a series expansion of the shear rate-geometric relationship developed by *Kreiger* and *Maron* for time-independent power law fluids. Table 2.1 provides values of C_R that can be used in equation 2.27. The rheological flow curve can be developed by a logarithmic plot of τ_b vs. $(dV/dr)_b$. By employing different bob or cup sizes, the different ranges of shear rates can be studied. In the extreme, for very small annular gaps (i.e., $R_c/R_b \rightarrow 1.0$), the shear rate approaches a constant value across the annulus and can be approximated simply by

$$\frac{dV}{dr} = \frac{2\pi RN}{R_c - R_b} \qquad \text{... (2.28)}$$

Table 2.1. Values of Calderbank-Moo-Young shear rate term $(C_R)^n$.

n^H			R_c/R_b *ratio* →			
	1.070	*1.150*	*1.166*	*1.250*	*1.400*	*1.746*
0.050	2.722	4.999	5.435	–	–	–
0.100	1.708	2.617	2.801	3.735	5.184	–
0.200	1.287	1.622	1.689	2.031	2.593	3.615
0.300	1.162	1.342	1.377	1.554	1.843	2.382
0.400	1.102	1.213	1.234	1.340	1.511	1.826
0.500	1.068	1.139	1.153	1.220	1.327	1.522
0.600	1.045	1.091	1.100	1.144	1.212	1.335
0.700	1.029	1.058	1.064	1.091	1.134	1.209
0.800	1.017	1.034	1.037	1.053	1.077	1.119
0.900	1.007	1.015	1.016	1.023	1.034	1.052
1.000	1.000	1.000	1.000	1.000	1.000	1.000
1.250	–	–	0.971	–	0.941	0.910
1.500	–	–	0.952	–	0.903	0.853
1.750	–	–	0.939	–	0.877	0.814
2.000	–	–	0.929	–	0.857	0.785
2.250	–	–	0.921	–	0.842	0.763
2.500	–	–	0.915	–	0.830	0.746
2.750	–	–	0.910	–	0.820	0.732
3.000	–	–	0.906	–	0.812	0.720
3.250	–	–	0.902	–	0.806	0.711
3.500	–	–	0.899	–	0.800	0.702
3.750	–	–	0.897	–	0.795	0.695
4.000	–	–	0.894	–	0.791	0.689
Coefficient (A)	0.903	0.861	1.144	0.898	1.169	1.155
Exponent (B)	– 0.304	– 0.505	– 0.320	– 0.547	– 0.418	– 0.485
Coefficient of fit (r*)	0.946	0.963	0.891	0.974	0.928	0.957

* Each set of R_c/R_b data sets have been regressed using the least squares to a power law expression. For approximate C_R values apply above model coefficient: $C_R = A\,(n^H)_B$.

Close clearance viscometers such as this are not suitable for use with suspensions in which particle sizes are comparable to the annular gap. End effects can be corrected for by substituting an effective length ℓ_{eff} for the true bob length ℓ in equation 2.26.

The effective length can be determined by calibrating the viscometer using a Newtonian fluid of known viscosity μ and applying the following formula:

$$\ell_{eff} = \frac{T}{2\pi R_b^2 \mu (dV/dr)_b} \quad \text{... (2.29)}$$

To correct for effective slip in rotary viscometers, the correction method of Mooney can be used. The method involves obtaining measurements using three different R_c/R_b ratios. For R_1, R_2, R_3 radii, such that $R_1 < R_2 < R_3$, three radii ratios can be defined ($S_a = R_2/R_1$, $S_b = R_3/R_2$, $S_b = R_3/R_1$). The rotational speed of the cup required to obtain the same torque for each of these cases can be measured as N_{12}, N_{23}, and N_{13}. The effective slip coefficient corresponding to the stress at the surface $2\pi R^2\ell$ is

$$\beta_2 = \frac{2\pi^2 R_2^2}{T}(N_{12} + N_{23} - N_{13}) \quad \text{... (2.30)}$$

Repeating this procedure for different torque values provides a correlation between β_2 and the surface shear stress τ_2. In correcting for slippage, for a given torque value, τ_b can be computed from equation 2.26. The corresponding expression for wall shear stress at the cup surface is

$$\tau_c = \frac{T}{2\pi R_c^2 \ell} \quad \text{... (2.31)}$$

The effective slip velocities at the bob and cup surfaces are $V_{sb} = \beta_b \tau_b$ and V_{sc}, respectively. The effective velocity of the fluid at the rotating cup surface will be less than that of the cup by the amount V_{sc}. The effective velocity of the fluid at the stationary bob surface is V_{sb}.

The true shear rate corresponding to τ_b can be calculated by multiplying either equation 2.27 or 2.28 by the ratio of the actual velocity difference (i.e., from cup to bob) to the velocity difference without slip [i.e., $(V_{sc} - V_{sb})/V_{meas.}$].

A modification of the concentric cylinder viscometer is the *Brookfield viscometer*, in which the cup radius is effectively extended to infinity. Flow curves are generated from measurements of the torque needed to rotate a cylindrical rod at various speeds when immersed in the fluid of 'infinite' volume. Basically, the fluid is contained in a cup whose radius is much larger than the rod. With this arrangement, the walls of the retaining cup do not influence the shearing movement of the fluid.

Equation 2.26 defines the shearing stress at the rod surface (R_b = radius of the cylindrical rod, ℓ = length of rod immersed in the fluid). As before, an effective rod immersion length, ℓ_{eff}, can be developed to correct for end effects. The shear rate at the surface of the rotating rod for any time-dependent fluid is given by

$$\left(-\frac{dV}{dr}\right)_b = \frac{4\pi N}{n''} \quad \text{... (2.32)}$$

where n″ is the slope of a logarithmic plot of torque vs. rotational speed N, evaluated at the particular N for which T was measured . A plot of τ_b vs. $(-dV/dr)_b$ provides the flow curve. With either device a plot of DΔP/4L vs. 8/V/D, the Newtonian shear rate for laminar tube flow may also be generated.

In general, both devices allow studies over a very limited shear rate range. Commercial units are usually equipped with an electrical heating jacket and appropriate heat exchange to remove heat generated by friction in the fluid.

CONE AND PLATE VISCOMETER

The cone-and-plate viscometer is a widely used instrument for shear flow rheological properties studies. The device can be used to measure both viscosity and normal stresses. The principal features of this viscometer (Fig. 2.7) consist essentially of a flat, horizontal plate and an inverted cone, the apex of which is in near contact with the plate. The angle between the plate and cone surface is small (typically $< 2°$) The fluid sample is placed in this small gap between the cone and plate. A flow curve is developed from measurements of the torque needed to rotate the cone at different speeds. For a constant speed N, the linear velocity at r is $2\pi rN$. The gap height at r is r $\tan\phi$. Hence, the magnitude of the shear rate at r is

$$\frac{2\pi rN}{r\tan\phi} = \frac{2\pi N}{\tan\phi} \qquad \text{... (2.33)}$$

The shear rate is constant between $0 > r > R$, which means that τ_{yx} is also constant over this range. For small ϕ, $\tan\phi \simeq \phi$, and hence the magnitude of the shear rate equals approximately $2\pi N\phi$.

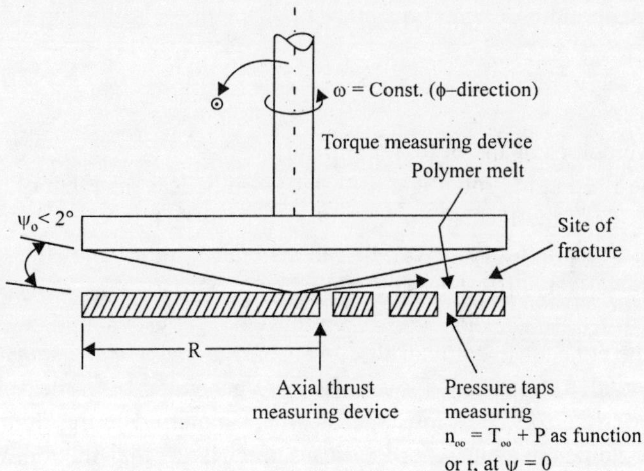

Fig. 2.7. Schematic of the cone and plate viscometer.

The following expression defines the relationship between measured torque and shear stress:

$$T = 2\pi\tau_{yx}\int_0^R r^2 dr = \frac{2}{3}\pi R^3 \tau_{yx}$$

or

$$\tau_{yx} = \frac{3T}{2\pi R^3} \qquad \text{... (2.34)}$$

Equation 2.34 enables τ_{yx} to be obtained at different measured torque values. Shear rate can be obtained from equation 2.33, and hence τ_{yx} can be plotted against shear rate for the flow curve.

SANDWICH VISCOMETER

Another type of shear flow instrument primarily used for polymer melts is the sandwich viscometer (also known as the parallel-plate viscometer). The measurement principal is illustrated in Fig. 2.8. A

Sviscous material such as an elastomer is sheared between two parallel steel plates or, as in Fig. 2.8, in sandwich between three plates. The shear rate is

$$\dot{\gamma} = \frac{V}{H} \quad \text{... (2.35)}$$

where H = interplate distance.

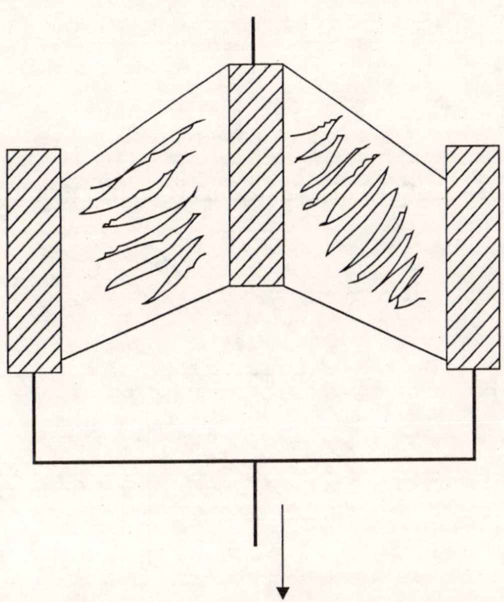

Fig. 2.8. Schematic of the sandwich viscometer.

The applied force can be determined as a function of time, which after normalisation with area provides shear stress data as a function of time:

$$\tau = \frac{F}{2A} \quad \text{... (2.36)}$$

It is possible to measure transient start-up stresses of the material with this device. After a sufficiently long time and if the material is stable, the steady-state stress can be measured, at which point the viscosity can be determined as a function of shear rate. With many materials such as elastomers, both large times and equipment strains are needed. These conditions often results in slippage. To minimise slippage contact surfaces are usually knurled.

This type of viscometer is capable of making measurements at very low shear rates on materials that usually exhibit slippage in non-pressurised rotational devices (e.g., cone-and-plate-type viscometers).

PARALLEL DISK VISCOMETER

The parallel disk viscometer (also known as a shearing disk viscometer) measures viscosity in torsional flow between a stationary and a rotating disk. Design variations are illustrated in Fig. 2.9. The operation involves the use of a serrated disk which is rotated in a sample fixed in a pressurised cavity. This system was developed for rubber applications in which both pressurisation and serrated surfaces are used to avoid slippage.

Properties are measured based on the shear rate at the outer radius of the disk, where

$$\gamma_a = \frac{a\omega}{H} \quad \text{... (2.37)}$$

Shear stress and normal stress differences are given by the following relationships:

$$\tau|_a = \frac{3T}{2\pi a^3}\left[1 + \frac{1}{3}\frac{d \ln T}{d \ln \gamma a}\right] \quad \text{... (2.38)}$$

$$P_{11} - 2P_{22} + P_{33}|_a = \frac{2F'}{\pi a^2}\left[1 + \frac{1}{2}\frac{d \ln F'}{d \ln \gamma a}\right] \quad \text{... (2.39)}$$

where F' = normal thrust. The parallel disk apparatus is limited to low shear rates.

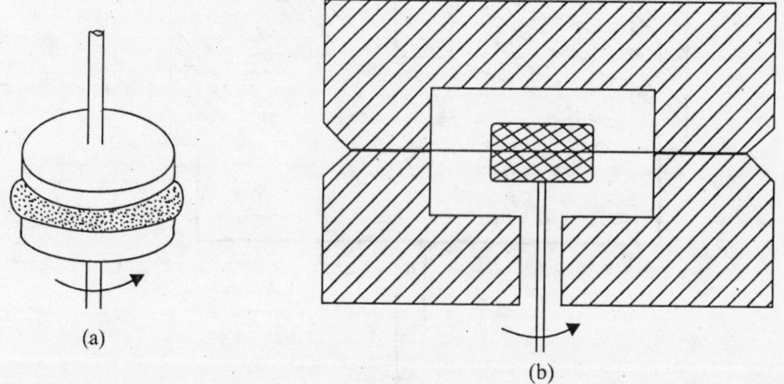

Fig. 2.9. (a) Parallel disk viscometer; (b) Mooney shearing disk viscometer.

The Mooney viscometer consists of a rotating disk in a cylindrical cavity. This instrument has become the standard quality control instrument for the rubber industry. The American Society for Testing and Materials specifies the diameter of the disk to be 19.05 mm and a thickness of 5.54 mm. A single-point viscosity measurement is made at a standardised rotor speed of 2 RPM, after a period of 4 min. and 100° or 125°C.

In the elastomers industry some manufacturers still report measurements at 8 min. and 127°C. Measurements are reported in terms of a calibrated torque measurement on a standard dial (referred to as ML-4 reading). The designation ML-4 means M, Mooney viscometer; L, large disk size (a smaller diameter disk, S, is used for vulcanisation scorch tests); and 4,4 min. of test.

ML-4 readings can be converted to torque, shear stress, and viscosity values. *Nakajima* and *Harrel* provide the following correlations:

$$T = 8.30 \times 10^{-2} D_R \quad \text{... (2.40a)}$$

$$\tau_a = 0.382 (D_R) f' \times 10^4 \quad \text{... (2.40b)}$$

where D_R is the Mooney unit dial reading, T and τ_a have units of N/m and Pa, respectively, f' is a function defined by the following relationship for shear stress at the disk's outer radius of the disk:

$$\tau|_a = \frac{T}{\pi a^3} f' \quad \text{... (2.41)}$$

$$f' = \frac{n'+3}{4}\left[1 + \frac{[(2/n')^{n'}(n'+3)hH^{n'}]}{2a^{n'+1}\{1-[a/(a+\delta)]^{2/n'}\}n'}\right]^{-1} \quad \text{... (2.42)}$$

where δ = distance between the outer disk radius and cavity, h = disk thickness, and n′ = power law exponent, ≃ (d ln T/d ln γ_a). Function f′ typically has a value of around 0.55 for the standard measurement conditions.

SLIT RHEOMETER

A slit rheometer is an instrument analogous to the capillary viscometer. The primary difference between the devices is the orifice cross-section. The device makes use of a series of flush mounted transducers located along the flow tube. These transducers measure the pressure gradients along the direction of flow, which can then be converted to wall shear stress values via

$$\tau_w = \delta_c \frac{dP}{dz}, \quad \text{... (2.43)}$$

where δ_c = half-thickness of the channel.

Han gives the following expression for the shear rate at the channel wall:

$$\dot{\gamma}_w = \frac{3Q}{2ab^2}\left[\frac{2}{3} + \frac{1}{3}\frac{d(\ln(3Q/4ab^2))}{d(\ln \tau_w)}\right] \quad \text{... (2.44)}$$

where a = half-width.

In general, the instrument is capable of operating over comparable shear rate ranges to the capillary viscometer.

MELT FLOW INDEXER

The melt indexer is widely accepted as a standard quality control instrument in the plastics industry. The device, is typically used to specify polyethylenes, polypropylene, and polystyrene. It is essentially an extrusion rheometer that is in fact similar to a capillary rheometer. The instrument employs a single capillary (diameter of 0.655 mm and L/D ratio of 12) in which extrusion of material is accomplished through pressure applied by a dead weight. The melt index is given as a measure of the weight in grams extruded for a 10-min test. The ASTM specifies a series of test conditions involving temperature and applied dead load settings. These conditions are designations A through K, and they cover the following ranges; for temperature 125° to 275°C, for applied dead load—325 to 21,600 g (or in terms of pressures: 0.5 to 29.5 atm). Specifications are selected so that melt indices lie between the values of 0.15 and 25g/ 10 min. test. The most frequently used conditions in the industry is specification E(I 90°C, 2.95 atm).

Although this is a widely used instrument, its value is that of a quality control tester which can only empirically relate to end-use processability of products. The melt index (MI) is related to the inverse of viscosity; however, no end-loss corrections have been developed nor can be MI the easily related to the *Weissenberg-Rabinowitsch* shear rate expression.

ELONGATIONAL FLOW DEVICES

Up to this point instruments based on shear flow have been discussed; however, there are numerous polymer processing operations that involve uniaxial elongational flows. An example in processing in which elongational flow occurs is in the entrance die region of an extruder. This phenomenon can be studied in a capillary die. Pressure losses through such dies can be used to determine elongational viscosity in polymer melts. For this reason a variety of laboratory test devices and techniques have been devised over the years in order to conduct rheological investigations and/or quality control testing in this type of flow. In general, measurements of elongational viscosity are complex to interpret and more difficult than measurements made in shear flow.

To make elongational viscosity measurements requires maintaining a constant elongation rate or stress while reaching steady-state conditions. These tests are typically applied to polymer melts that include high molecular weight polyolefins, styrenics, elastomers, and their compounds. A few of the most frequently used tests are briefly described below.

The most common method of measuring elongational viscosity is to stretch a filament of the material. The fluid obviously must have a sufficiently high viscosity to be maintained under controllable deformation. Materials suitable for this type of testing are elastomers and polyloefins where melt viscosities are typically 10,000 Pa/s or greater. These types of test measurements are limited to deformation rates below 5 s^{-1}. *Ballman*, *Vinogradov* and *Stevenson* describe different types of filament stretching methods. A very simple arrangement involves clamping both ends of a vertical thermostated filament, which is then stretched at a rate of dl/dt in a manner so as to maintain a constant rate of deformation. In this manner the rate of deformation ϕ is

$$\phi' = \frac{1}{\ell}\frac{dl}{dt} \quad \text{... (2.45)}$$

and

$$\ell = \ell(o)e^{Et} \quad \text{... (2.46)}$$

In other words, the length of a material undergoing stretching follows an exponential law. The test is illustrated in Fig. 2.10(a).

Another approach, illustrated in Fig. 2.10(b), involves supporting a horizontal filament on the surface of a hot, immiscible oil. The polymer filament is held at a both ends between pairs of toothed or serrated wheels that turn at linear velocity V/2. The filament is drawn at both ends through the wheels. One of these wheels contains a tension-sensing instrument so that the deformation rate is proportional to the velocity ($\phi' = (V/2)/(L/2) = V/L$).

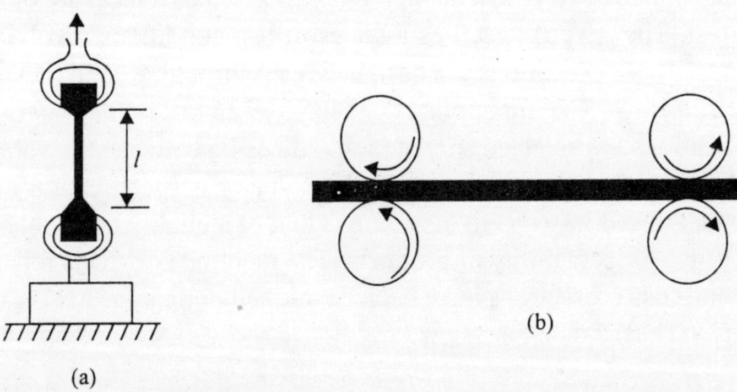

Fig. 2.10. (a) Vertical method of elongational viscosity measurement at constant elongation rate; and (b) a horizontal filament stretching test.

STRESS RELAXATION AND SMALL STRAIN MEASUREMENTS

These classes of rheological measurements are carried out on bulk polymers and are usually aimed at obtaining information on the response of the material to small strains. Strain can be applied as a step function and the stress relaxation that takes place after its application is measured.

Stress relaxation measurements are routinely made on elastomers. The test involves application of a strain to a polymer sample that is in extension, shear, or compression. The measurable parameter is the force response over time, F(t), to the applied strain. Some investigators report the material property in terms of a relaxation modulus defined by

$$E(t) = \lim_{\Delta\ell \to 0} \left[\frac{F(t) A_0}{\Delta\ell / \ell_0} \right] \quad \text{... (2.47)}$$

where A_0 = initial cross-section of sample; ℓ_0 = initial length of sample and $\Delta\ell$ = amount the sample is stretched in extension.

For measurements made under shear, a surface of the sample is subjected to a displacement in a direction normal to its thickness. The shear modulus is given by:

$$G(t) = \lim_{\Delta\ell \to 0} \left[\frac{F(t) / A}{\Delta\ell / H} \right] \quad \text{... (2.48)}$$

where H = sample thickness.

These types of tests are normally coupled with mechanical properties testing. A variety of polymers, particularly elastomers, are widely used in consumer and industry related products because of their good mechanical properties. Tensile measurements at several strain rates are often made as a quality control check of the integrity of the polymer, even though their behaviour in compound states may be quite different. The stress-strain curve provides information on both the rheological and mechanical behaviours of a polymer. Fig. 2.11 illustrates typical features of tensile stress-strain curves for polymers.

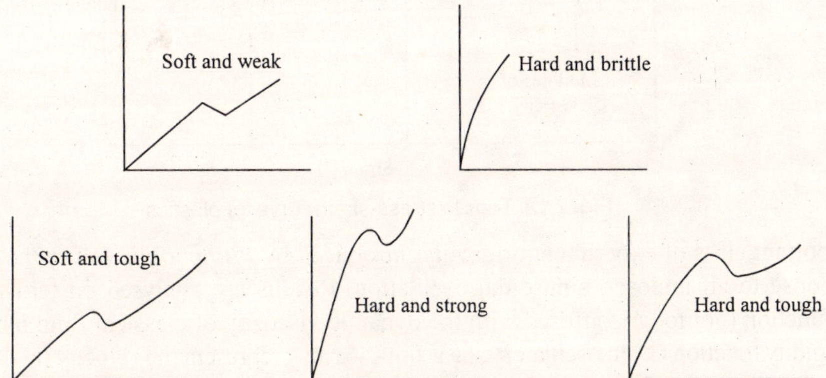

Fig. 2.11. Typical tensile stress-strain curves for polymer.

The classification of curves in Fig. 2.11 shows that polymers can be classified in general terms as rubbery, glassy, or variations in between. Compounding can greatly change the stress-strain response curve of a given polymer. Tensile testing in rubber laboratories is normally done on a tensile testing machine (such as an Instron) and performed according to ASTM standard methods (ASTM D638, ASTM D882). As a product quality control test, many elastomer manufacturers are no longer using tensile testing of neat rubber, but rely more heavily on rheometer cure response data. In product development work, however, tensile testing is very much an integral part of polymer characterisation work.

A minimal testing scheme involves obtaining a stress-strain curve at room temperature conditions and at perhaps three or more strain rates. From each curve, various quantities related to the mechanical and rheological properties of the sample can be derived. Fig. 2.12 shows the generalised tensile stress-strain curve for plastics and the various parameters of interest. The principal information obtained from these curves includes:

1. The modulus of elasticity (the initial slope of the curve).
2. The yield stress (peak stress in the early portion of the curve).
3. The stress at failure.
4. The elongation (or strain) at yield (also called the yield strength).
5. The ultimate elongation or strain at failure (also referred to as the elongation at break).
6. The energy to break, measured by the area under the stress-strain curve up to the point of failure.

Feature 6 is truly a measure of sample toughness.

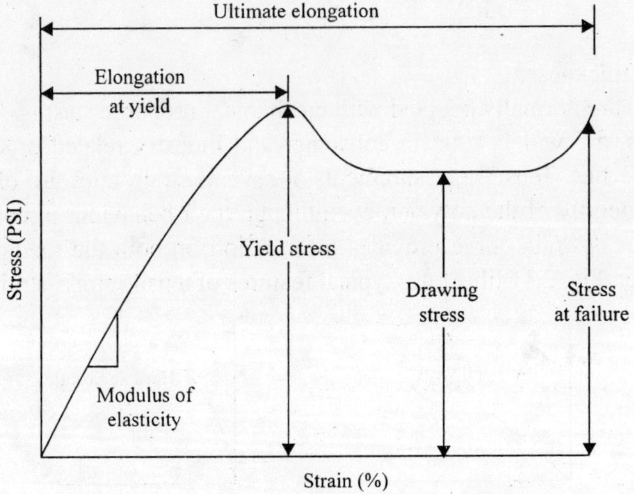

Fig. 2.12. Tensile stress-strain curve for plastics.

An important class of experiment originally introduced by *Philippoff* involves the determination of stress response to an imposed sinusoidal oscillation. Results are analysed on terms of the dynamic viscosity function (not to be confused with the dynamic viscosity of classical fluid mechanics) and the dynamic rigidity function G′. In oscillatory shear flow, we may represent the velocity field using rectangular Cartesian co-ordinates as follows:

$$V_1 = \varepsilon\, \omega\, \chi_2\, e^{i\omega t}, \quad V_2 = V_3 = 0 \qquad \text{... (2.49)}$$

where ε is small enough for second and higher order terms to be neglected and is frequency. The corresponding stress field is

$$P_{12} = \eta^*(\omega)\, \varepsilon\, \omega\, e^{i\omega t}\,; \quad P_{11} - P_{12} = P_{22} - P_{33} = 0 \qquad \text{... (2.50)}$$

where η* is the complex dynamic viscosity expressed by convention as:

$$\eta^* = \eta' - i\frac{G'}{\omega} \quad \text{... (2.51)}$$

Other functions typically defined are

$$\eta'' = \frac{G'}{\omega} \quad \text{... (2.52a)}$$

$$G'' = w\eta' \quad \text{... (2.52b)}$$

The most widely used method for measuring G′ (ω) and G″ (ω) on polymer melt systems involves placing the melt in the gap between a cone and plate and oscillating one of the elements relative to the other with a shear strain. Fig. 2.13 illustrates this technique.

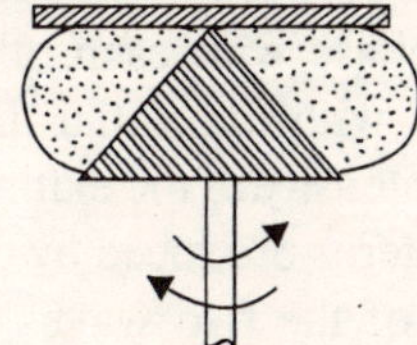

Fig. 2.13. Oscillatory shear in a cone-and-plate device.

Another arrangement, illustrated in Fig. 2.14, involves rotating two parallel eccentric disks. The polymer sample undergoes periodic sinusoidal deformation. The forces exerted on the disk are interpreted as G′ (ω) and G″ (ω).

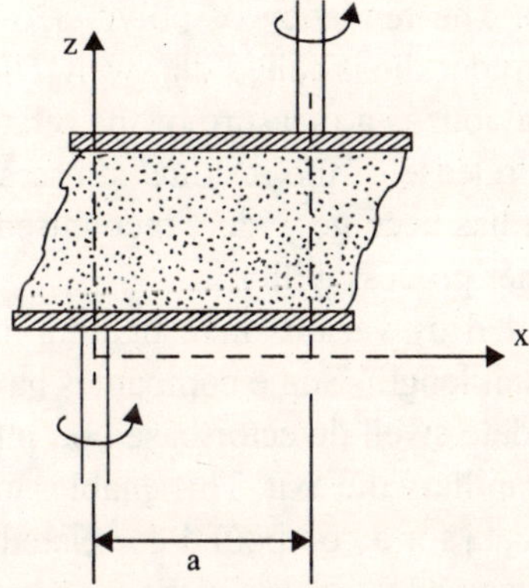

Fig. 2.14. Principle behind the orthogonal rheometer.

Many industrial research laboratories used the Weissenberg-Rheogoniometer and Rheometrics Mechanical Spectrometer for cone-and-plate sinusoidal oscillation measurements of G′ (ω) and G″ (ω).

FLOW BIREFRINGENCE MEASUREMENTS

Determination of birefringence during flow is a rheological property measurement technique applied almost exclusively to homogeneous systems of flexible polymer chains. This technique is based on a relationship between flow birefringent behaviour and stresses and is determined by the rheo-optical law established by *Philippoff*. The law relates the difference in the principal refractive indices to the differences in the principal stresses as developed during flow. The principal stresses represent the magnitudes of the principal axes of the ellipsoid of the stress tensor. The technique has been applied in purely scientific studies on flexible chain polymer melts such as polyolefin, polystyrene, and elastomers. The major difficulty in its applications is that the technique is limited to amorphous polymers. Even trace amount of crystallinity have a dramatic effect on measurements. Measurements can be performed on polymer solutions using a coaxial cylinder viscometer, whereas polymer melt measurements are normally taken on parallel-plate slit extrusion devices. These instruments seek to use flow birefringence to measure shear flow properties.

Measurement of birefringence is accomplished by studying the transmission viscometer of a polarised monochromatic light beam that is passed trough the fluid medium to some depth. Since the medium in the plane perpendicular to the transmitted wave has different refractive indices in different directions, the wave splits into two parts. Each reflected wave has a different velocity that corresponds to the two refractive indices. The waves are out of phase with each other and the beam exiting the instrument passes through a polariser crossed normal with the initial direction of polarisation. Birefringence is determined either through the reduction in intensity of the exiting beam, through the use of compensators, or through characterisation of the fringe patterns produced by the external polariser. The fringes indicate conditions under which the contributions of the two waves to the direction allowed passage by the analyser cancel.

SWELL AND SHRINKAGE TESTS

Rheological test more often used for product quality control is a measurement of the elastic recovery in polymer melts and their compounds. This is usually accomplished by measuring the swell (or shrinkage) of materials undergoing extrusion. The tendency of polymer, whether thermoplastics or rubber, to enlarge when emerging from an extruder die is called die swell. Die swell normally refers to the ratio of extruded size to die size. This behaviour is a measure of the relative elasticity in the flowing polymer stream. Die swell is caused by the release of the residual stresses when the sample emerges from the die. Measurement of this behaviour has become widely recognised in the rubber and plastics industries as an important indication of polymer processability.

Traditional measurements reported by various investigators have been made on extrudates either with micrometer or by weight per unit length. Some companies have introduced an automated capillary rheometer which employs an extrudate swell detector based on interaction with a scanning laser beam positioned immediately below the capillary die exit. This enables measurement of the running die swell. Fig. 2.15 shows running die swell data for a compound correlated against wall shear for different L/D dies. It is important to note that with some materials a distinction between running die swells and relaxed die swells should be made. Highly elastic materials will tend to swell more over time depending on their elastic recovery energy.

An optical electronic die swell measuring system can be used for continuous control and monitoring of extruders and plasticorders. The basic design features are illustrated in Fig. 2.16. An infrared, long-life diode (A) emits homogeneous infrared light through a filter (B), diffuser (C), and lens (D), then through a window (E). The light impinging on the object (F) is scattered. Only that light reaching the ambient light filter (G) is measured by the photodetector H1. Light reaching photodetector H2 is diverted through the prism (I) and is uninterrupted. This acts as a reference beam. The electronic instrumentation measures the differences between photodetector H1 and H2 and calculates the object thickness to an accuracy of 0.2% of full scale. The object to be measured must be located vertically within the measuring area. The maximum permissible deviation from vertical is 2.5°. Operation at angles > 2.5° from vertical will yield greater errors than 0.1%. If the measuring angle is > 2.5° from vertical, the measurement will be increased by the factor 1/cos a, where a is deviation from vertical.

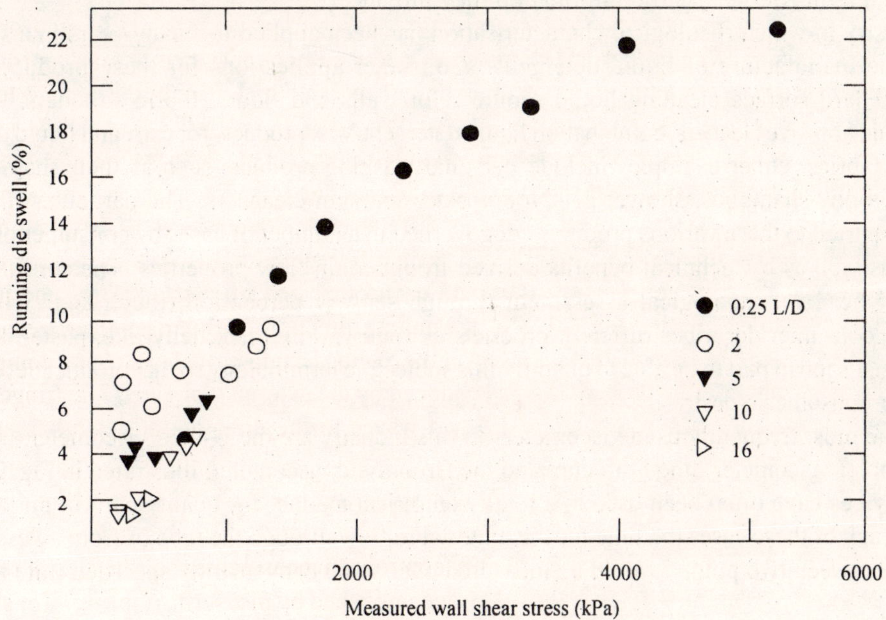

Fig. 2.15. MPT data on running die swell.

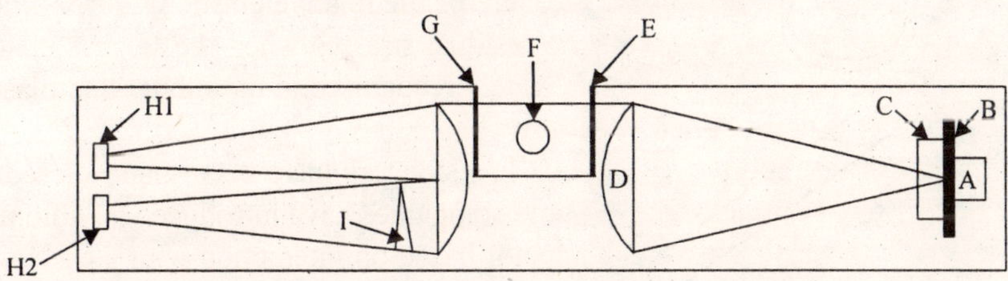

Fig. 2.16. Schematic of optical-electrical die swell tester.

While the upper collimated light serves as the measuring beam, the lower portion impacts on the second photodetector and acts as the reference beam. Thus, the temperature and age affected by drift which occurs on all semi-conductor elements is negated.

In some applications such as hose and tubing manufacturing, it is important to know both the extrudate diameter and cross-section. The device described above can provide two diameter measurements, which will give an out of round indication (eccentricity).

VISCOMETRIC TECHNIQUES FOR DIFFERENT INDUSTRIES

A variety of other instruments are employed by various industries for either product development work or product quality control. Although these methods shall be reviewed in relationship to the industries and products they are most frequently applied to, these devices are by no means restricted to those applications. One type of instrument may have become a standard tool for a particular industry largely through an evolution of preference and in some cases convenience, rather than primarily for technical reasons. Most of these techniques can be applied to any fluid characterisation; however, the techniques can greatly vary in terms of the specific information they provide.

One industry in which rheological characterisation has been applied for many years in a subjective manner is the manufacture of liquid detergents. Consumer applications for these products include dishwashing, hard surface cleaning liquid scourers for walls and floors, fabric softeners, thickened bleaches, liquid abrasive cleaners, combination liquid detergent-wax products for cars, and liquid detergents for washing fabrics. Other examples include personal washing products such as their shampoos and conditioners, body shampoos, shower gels, toothpaste, and skin cleansers. The particular rheological properties imparted to these various products are governed in a number of cases by consumer preference or attitudes as well as by technical benefits derived from certain flow properties. The consumer often makes his or her own rheological assessment through sensory perception (touch and sight) of the product. The consumer describes different products as watery, thin, thick, jelly-like, paste-like, etc. It is the rheologist's job in part to be able to quantify this subjective terminology so that product development can meet the consumer's desires.

Among the most frequently used viscometers in this industry are the U-tube viscometer, the rolling ball (or Hoeppler viscometer), the Ford cup, and the Brookfield viscometer, illustrated in Fig. 2.17. The first three devices have often been used in a totally empirical manner for quality control purposes. For example, in each of these cases the time for the material to flow through the tube or from a vessel, or for a ball to roll between two points, would be measured over some conveniently specified time interval.

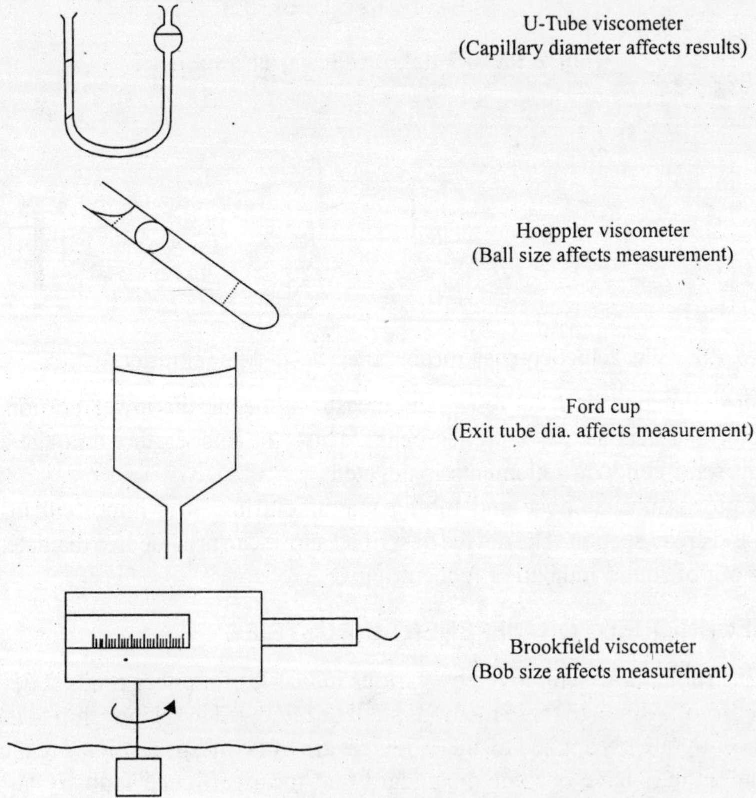

Fig. 2.17. Various viscometers employed in the detergent industry.

With some devices a scale might be included to display a reading of variables, as in the case of the Brookfield, showing readings between (for example) 10 and 90% of full-scale deflection. Many devices have been adopted, not only in this industry, on the basis of convenience and expediency rather than on a scientific or an engineering basis. A disadvantage of many of the simpler instruments is their inability to generate more than one value to characterise a fluid. Often this reading is converted into a viscosity value based on a Newtonian calibration.

Barnes describes several modifications to the various techniques used in characterising detergents. The first of these modifications concerns the Hoeppler viscometer, which normally operates using a fixed tube angle and several balls. He proposes a variable angle of rolling. By altering the angles, the average stress in the fluid can be made to vary because of the change in magnitude of the gravitational force on the ball in the tube direction.

For this arrangement, viscosity data can be fitted to a relation:

$$\eta = k_1 (\sin\alpha')\,\theta(A + Bn) \qquad \text{... (2.53)}$$

where α' = angle of the tube from the horizontal; θ = time for the ball to roll between two fixed points; and n = slope of curve generated by plotting ln (sing α') vs. n ($1/\eta$).

Note that the shear rate in this experiments is $\gamma = k_2/\theta$, and A, B, k_1 and k_2 are constants. Also, A + B = 1. Usually two experiments (i.e., tests at two angles) provide sufficient information such that a power law relation can be derived from η values.

The Ford cup provides a measure of the time for a fixed volume of fluid (e.g., 50 ml) to flow from the cup into a graduated cylinder. One problem in using this technique with such materials as detergents is the formation of foam, which can make accurate flow time determination difficult. This can be overcome by employing electroresistivity probes along the inside walls of the cup (Fig. 2.18). This provides accuratc measurement of the flow times involved. A sharp edge orifice fitted to the outlet tube provides steady calibrated flow conditions. *Barnes* gives the following expressions for viscosity and average shear rate using this technique:

$$\eta = \mu\left(\frac{1+n}{2}\right), \qquad N - s/m^2 \qquad \text{... (2.54)}$$

$$\gamma_{ij} = \frac{2(h_i - h_j)}{\theta_{ij}}\frac{R^2}{r^2}, \qquad \sec^{-1} \qquad \text{... (2.55)}$$

where i, j refer to different heights h_1, h_2, h_3, ... etc.

Measurements using a modified Ford cup are quite comparable to those obtained with concentric cylinder viscometers. Materials that are highly thixotropic generate flow times which can be much longer than predicted if steady-state concentric cylinder data are used for comparison, because the total shear in the Ford cup is small compared to the steady-state rotational data. The Ford cup is not applicable to materials in which elongational viscosity is much greater than shear flow at the same shear rates. Many detergents show anomalous elastic effects and hence the Ford cup is not suitable for viscosity measurement in all cases. Viscosity measurements in shear flow of detergent products can be made in commercial rheometers based on cone-and-plate instruments or concentric cylinder viscometers. Typical commercial units employed are the Haake Rotovisko, the Ferranti-Shirley viscometers, and the Contraves Rotary visometer. These instruments cover shear rates typically between 1 to over 1000 s^{-1}. Other instruments employed to a lesser extent are the Deer rheometer and the Weissenberg Rheogoniometer.

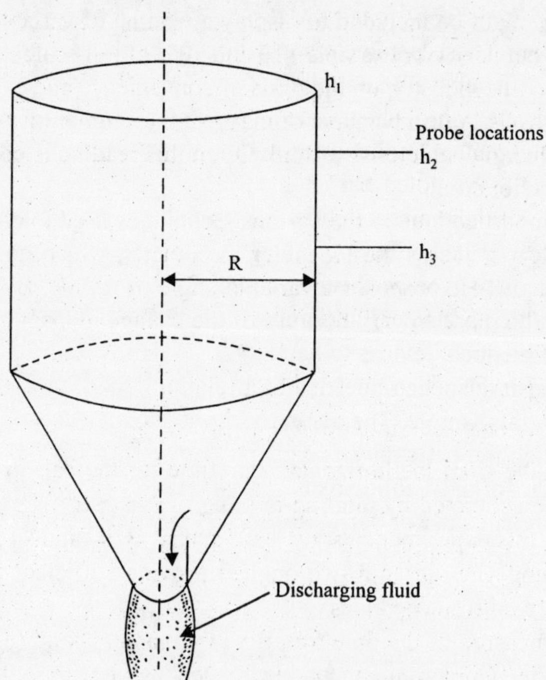

Fig. 2.18. The operation of the Ford-cup.

The Deer rheometer is designed to subject the sample to a constant stress whereby the subsequent deformation is measured. This instrument uses a frictionless air bearing so that it is capable of applying a variety of stress patterns with variable time of application and rate of application of stress. Variations of this instrument are capable of applying oscillatory stress to the fluid sample. Deformation can be studied to measure several parameters: creep, steady-state shear rate, recoverable shear, and various oscillatory parameters. Through geometry changes the instrument can cover a range of applied stresses. *Barnes* describes a modified Deer rheometer employed in the study of soap rheology. The system is illustrated in Fig. 2.19 in which the rheometer has been modified to measure the extensional properties by simple bending performed at very small deformation. Torque is applied to two vertically mounted rods of soap that are fixed at both ends. Torque is applied to the samples through the sample end retaining sleeves by magnetic knife edges mounted on radial arms. The knife edges ensure that no twisting of the rods takes place during bending. The extensional compliance is calculated from bending beam theory analysis.

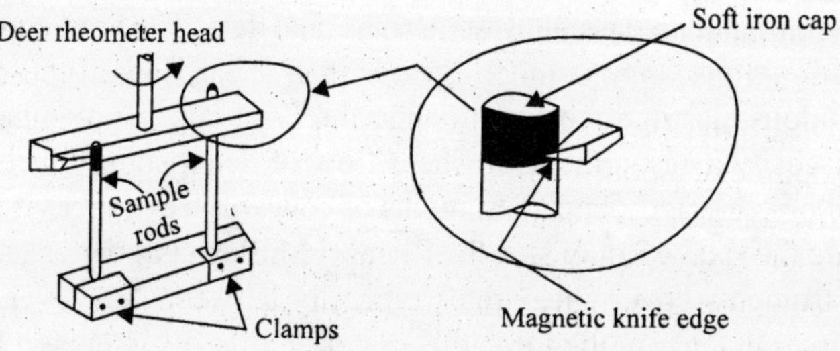

Fig. 2.19. A Deer rheometer.

The Weissenberg Rheogoniometer, as noted earlier, can be used to measure G′ and G″ as a function of frequency and deformation. These parameters can also be measured for a range of superimposed values of steady shear rate. It is most often applied in industrial laboratories at measurements over a range of shear rates and as a function of shear/time at initiation or cessation of shear. Other frequently used viscometers in the detergent manufacturing industry include capillary and orifice and orifice viscometers and penetrometers. As noted earlier in this chapter, a capillary viscometer operates by forcing the fluid through a straight circular tube. This instrument is capable of shear rates up to $10^5\ s^{-1}$. For fluids obeying power law behaviour, the Mooney-Rabinowitz correction makes possible a definition of the average viscosity-average shear rate parameters. Orifice viscometers are often used in the plastics industry to measure parameters important in pumping fluids through circular orifices. The detergents industry has applied such devices to handle very viscous semi-solid materials such as soap.

The last noteworthy device is the penetrometer, which monitors the movement of a weighted penetrator into the surface of a very viscous sample. The penetrometer is most often used for testing such materials as butter or bitumen, in which the penetrator is a sharp, truncated pin. Within the detergents industry, this device has proven useful for semi-solid soaps.

A variety of other instruments and/or variations of devices described above are available that are applied in lubrication technology. Lubricating oils are manufactured from the distillation of crude oil. Primary distillation is performed at atmospheric pressure to remove the lighter gasoline, kerosene, and gas oil fractions, and the lubricating oils remain in the residue that boils above about 370°C. The residue must be distilled under high-vacuum conditions, resulting in several lubricating-oil fractions and an asphaltic bitumen. The second-stage residue from those crudes low in asphalt is also a source of lubricating oils. These fractions are referred to as residual oils to distinguish them from distillate oils, which are boiled off in a high-vacuum still. Lubricating-oil fractions can constitute up to 5% of the total crude, depending on the crude source. A further refinement step in manufacturing the lubricating oils is usually performed with acid or solvents to remove a variety of unwanted constituents such as sulphur compounds. Hydrotreating at moderate temperatures and pressures saturates the unsaturated hydrocarbons and remove sulphur bound molecules. Paraffinic constituents separate out as wax at low temperatures. These are removed by chilling and filtering or by a solvent treatment process. Polar compounds and dark colouring costituents are removed from the oil in a final treatment stage using active clays. The refinery stream in these processes are referred to as base oils in the petroleum and allied industries. Base oils are characterised by the origin of the crude, the specific refining process used, viscosity, their viscosity index (VI), and various other properties. The VI is a measure of the variation of viscosity with temperature, and, in general, the higher the VI the smaller the variation.

An important class of products is mineral oils, examples of which are paraffins and aromatics. Oil additives are usually supplied as concentrates in mineral oil and are readily added during blending operation. Additives can have a variety of purposes for the end-use product, for example, oxidation inhibitors, additives to neutralise acids, surface-active agents to disperse and maintain other additives in suspension, ultrahigh-pressure lubricating additives. Polymers are also added to oils to modify rheological properties at extreme temperature conditions (e.g., for motor oils, to increase pouring ability at very low temperatures). A general example of a pour-point depressant is an additive that inhibits the formation of a wax-crystal network at low temperatures. This type of product extends the liquid property range of the lubricant. In lubrication technology, the Navier-Stokes equations for incompressible viscosity and proper application of these equations is usually sufficient to estimate oil film thickness and the friction. These equations are usually solved isothermally for computation simplicity and do not account for viscous heating effects. Hence, it is necessary to fully define the viscosity-temperature dependence of these products in order to correct engineering formulations from the Navier-Stokes equations. This is done, as noted earlier, through the use of the VI; standardised procedures are described in "Kinematic Viscosity", ISO Method 3104, IP Method 71, ASTM Method D445 and "Industrial Liquid Lubricants", ISO Viscosity Classification–ISO 3448.

The capillary viscometer is the standard instrument used for these measurements. Temperature conditions are standardised in these test methods so that oils can be compared on a relative basis (e.g., 40° and 100°C). In the ISO Viscosity Classification system, a 10-grade oil has a viscosity between 9.0 and 11.0 cSt at 40°C. In total there are six grades for each decade in viscosity.

Many commercial oils contain polymeric additives. Adding small amounts of polymer to an oil reduces the rate of decrease in viscosity with increasing temperature. Examples of the types of polymers used are polyisobutene, polyalkylmethacrylates, and co-polymers such as styrene-butadiene and styrene-isoprene. Molecular weights of these additives range from about 50,000 to several hundred thousand. In an automobile engine, motor oil is exposed to effective shear rates ranging from 10^4 to 10 s^{-1} over wide temperature ranges. Considerably lower shear rates are involved during engine starting. In starting an engine, the cranking speed must be high enough to enable a cylinder to fire. Furthermore, the power the engine delivers when it does fire and the assistance of the starter motor are disengaged during this process. Both these actions depend on the viscosity of the motor oil. One important industry test which is a low-temperature, high-shear test method is the Cold-Cranking Simulator (ASTM Method D2602). The US Society of Automotive Engineers (SAE) provides a classification of engine oil viscosity based on this test method.

CHAPTER 3

Viscometric Techniques and Analysis of Steady-State Flows

INTRODUCTION

Polymer-processing operations like extrusion, injection moulding, pultrusion, roll coating, blow moulding, and others are too complicated to model rigorously. Fortunately, most processing operations can be broken down into a set of simpler flow processes for which analytical or numerical solutions can be obtained using various simplifying approximations and constraints. In this way, it is possible to understand how process variables such as screw speed and temperature in an extruder affect performance variables such as volumetric output.

All polymer-processing operations involve the flow of polymer solutions or melts under a pressure gradient, shear deformation, or both. Since concentrated polymer solutions and polymer melts are non-Newtonian fluids, it is necessary to know the exact relationship between stress and strain—the constitutive equation—in order to analyse flow through even the simplest geometry.

POISEUILLE FLOW AND THE CAPILLARY VISCOMETER

Constitutive equations are expressions that define the stress rate of the deformation relationship at every point in the fluid. In practice it is relatively easy to measure macroscopic variables; for example, volume flow rate can be measured more easily than the velocity components at every point in the fluid. Consequently, a method of writing the functional dependence implied by a general constitutive equation is sought in terms of readily measurable macroscopic variables.

For a laminar steady-state flow through a circular tube or capillary of radius R, the volume flow rate Q is related to the velocity $v_z(r)$ by

$$Q = \int_0^R 2\pi r\, v_z(r) dr = \pi R^2 \langle V \rangle \qquad ...(3.1)$$

where $\langle V \rangle$ denotes the mass average velocity. Integration by parts and application of the no-slip condition at the wall (i.e., $v_z = 0$ at r = R), yields:

$$Q = -\int_0^R \pi r^2 \frac{dv_z}{dr}\, dr \qquad ...(3.2)$$

The only non-zero component of the rate of deformation tensor is $dv_z/dr = \Delta_{12}$. The shear rate-stress relationship can be expressed in the functional notation as

$$-\Delta_{12} = f(\tau_{12}) \quad \text{or} \quad \dot{\gamma} = f(\tau) \qquad ...(3.3)$$

where τ is the shear stress component. From the dynamic equations for this flow it can be shown that the shear stress is linear across the tube radius [i.e., $\tau = C_o r/2 = (\Delta P/L)(r/2)$]. As already discussed, the wall shear stress is simply $\tau_w = R\Delta P/2L$.

By simple algebraic manipulation it can be shown that:

$$\frac{4Q}{\pi R^3} = \phi(\tau_w) = \frac{4}{\tau_w^3}\int_0^{\tau_w} \tau^2 f(\tau)\, d\tau \qquad ...(3.4)$$

where $\phi(\tau_w)$ is defined by this expression.

Taking $d\phi/d\tau_w$ and using Leibniz' rule, one can develop an expression solving for $f(\tau_w)$, which is the well-known Weissenberg-Rabinowitsch-Mooney equation:

$$f(\tau_w) = \frac{3}{4}\phi + \frac{1}{4}\tau_w \frac{d\phi}{d\tau_w} \qquad ...(3.5)$$

Because ϕ and τ_w are measurable in terms of macroscopic variables, equation (3.5) allows one to find $f(\tau_w)$ over some range of τ_w. This functional dependence is identical with $f(\tau_w)$, however, so that with Equation (3.5), the shear behaviour of the fluid is determined. The usual method is to plot ϕ vs. τ_w, and from this obtain $d\phi/d\tau_w$. Then the right side of equation (3.4) is plotted against τ_w to give $f(\tau_w)$. From the resulting graph of $f(\tau_w)$ vs. τ, which is just Δ_{12} vs. τ_{12}, one can obtain and plot the viscosity η as a function of shear rate through the definition in equation (2.9) in chapter 2.

This analysis of flow in a capillary is based upon the assumption that simple shear flow exists. This is achieved in steady-state, laminar, isothermal flow in a tube of constant cross-section, as long as one does not consider regions near the entrance and exit of the tube. In these 'end' regions the flow is changing from (or to) its previous (or future) distribution outside the tube. The length of an "end" region is generally a function of tube diameter and some dynamic parameters. For example, for a Newtonian fluid, the 'entrance length', the length of tube required to achieve the fully developed simple shear flow, depends upon the Reynold's number.

One method of minimising the effect of the entrance length is to use a viscometer tube so long that the pressure drop over the entrance region is very small compared to the drop over the entire tube. Generally, this means that if L is the total tube length, then L_e/L must be small, perhaps on the order of 0.01. Often this restriction may demand tubes too long for practical purposes.

A method for eliminating the influence of end effects exists which uses two tubes identical except in length. The shorter tube must be longer than the entrance length, so that both tubes have a finite region in which the flow is in simple shear. Consider two such tubes attached to the bottoms of identical pressurised reservoirs, as illustrated in Fig. 3.1. Suppose that the pressure drops across each tube are adjusted so that the volume flow rates are identical and the reservoirs contain identical fluids. Let ΔP_L and ΔP_S be the total pressure drops measured across the long and short tubes, respectively. If the volume flow rate is the same in both tubes, one would then expect the entrance and exit lengths and the pressure drops across these lengths to be identical in both tubes. Then

$$\Delta P_L = \Delta P_{en} + \Delta P_{ex} + \Delta P_{fdL} \qquad ...(3.6)$$

$$\Delta P_S = \Delta P_{en} + \Delta P_{ex} + \Delta P_{fdS} \qquad ...(3.7)$$

where ΔP_{en} and ΔP_{ex} are the pressure drops across the entrance and exit regions and ΔP_{fdL} is the pressure drop across the long (L) or short (S) tube that arises strictly from the fully developed flow. Since the flow rates are the same, the pressure gradients in the fully developed sections are identical, and

$$\Delta P_{fdL}/L_L = \Delta P_{fdS}/L_S \qquad ...(3.8)$$

where L_L and L_S are the lengths over which the flow is fully developed in the large and small tubes. Combining the last three expressions, we obtain:

$$\frac{\Delta P_{fdL}}{L_L} = \frac{\Delta P_L - \Delta P_S}{L_L - L_S} \qquad ...(3.9)$$

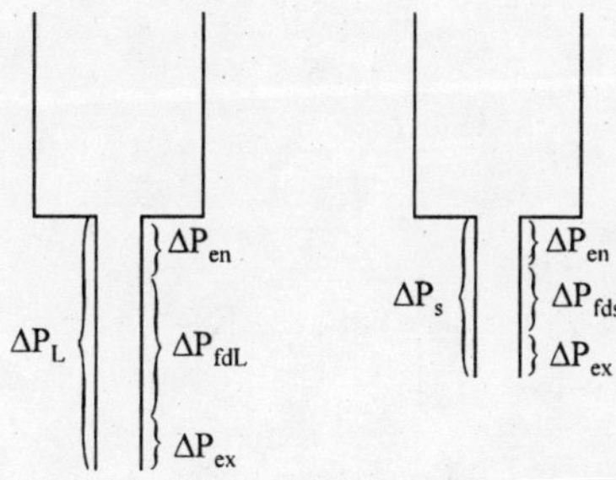

Fig. 3.1. Definition sketch for the treatment of capillary flow data.

This is the pressure gradient free of entrance and exit effects, although either viscometer tube may have involved appreciable end effects. It should be noted that the tube lengths should be sufficiently different so that the pressure difference $\Delta P_L - \Delta P_S$ is not subject to large error.

The shear stress can be expressed by a power law model:

$$\tau_w = K'\phi^n \qquad ...(3.10)$$

and from equation (3.4) one can write

$$\tau = \left(\frac{4n}{3n+1}\dot{\gamma}\right)^n K' \qquad ...(3.11)$$

This expression can be stated as the standard form for the power law

$$K = K'\left(\frac{4n}{3n+1}\right)^n \qquad ...(3.12)$$

VISCOUS HEAT GENERATION

The generation of heat through viscous dissipation can lead to significant temperature variations across the shear fields. Because fluid properties such as viscosity are strongly temperature dependent the shear stress-shear rate relation is considerably altered by non-isothermal effects. Hence, a shear stress-shear rate curve obtained under non-isothermal conditions does not reflect the basic fluid response independently of any temperature-dependent effects. The challenge then is to take data subject to viscous heating effects and separate that part of the response due to non-isothermal behaviour from that part of the response due to non-Newtonian behaviour.

Analyses reviewed thus far are applicable to non-isothermal flows; however, these analyses do not allow a separation of the temperature-dependent behaviour from the overall response of the fluid. This can be done only if a model for the fluid (or constitutive equation) is assumed, and if the temperature variation of the fluid parameters is specified. This results in a parameter fitting procedure in which data obtained at different shear rates and temperatures are manipulated until a consistent set of fluid constants that are independent of shear rate and temperature is obtained.

In Poiseuille flow, if heating is important, one must solve the energy equation for laminar flow in cylindrical co-ordinates. With appropriate boundary conditions the problem to be solved is the so-called energy equation:

$$\rho \hat{C}_P v_z \frac{\partial T}{\partial z} = \frac{1}{r}\frac{\partial}{\partial r}\left(rk\frac{\partial T}{\partial r}\right) + \Phi \qquad \text{...(3.13)}$$

$$T = T_o \text{ at } z \le 0$$

$$\frac{\partial T}{\partial r} = 0 \text{ at } r = 0$$

$$\left.\begin{array}{r} -k\dfrac{\partial T}{\partial r} = q \\ \text{or } T = T_o \end{array}\right\} \text{ at } r = R$$

where T_o is the reservoir temperature of the fluid, and k, ρ, and $\hat{C}_P$ are the fluid thermal conductivity, density, and heat capacity, respectively. The third boundary condition is arbitrary, in that it depends upon actual operating conditions. Typical conditions used state that the heat flux across the tube wall may be a constant q (including zero, the adiabatic condition), or the temperature of the wall may be specified as a constant.

Equation (3.13) is approximate in that small radial velocities arising from the effect of viscosity variation on the flow field are neglected, so that terms such as $v_r\partial T/\partial r$ do not appear. In addition, it is assumed that axial conduction of heat is unimportant, so that the term $\partial(k\partial T/\partial z)/\partial z$ does not appear. The fluid is assumed incompressible, and ρ is taken as a constant. These assumptions are valid for moderate temperature variations.

If the velocity profile is assumed to be perturbed by the varying temperature field, then the energy equation must be solved in conjunction with the dynamic equation for this flow, given by

$$0 = -\frac{\partial P}{\partial z} + \frac{1}{r}\frac{\partial}{\partial r}(r\tau) \qquad \text{...(3.14)}$$

subject to the boundary conditions

$$vz = 0 \quad \text{at } r = R$$
$$\partial v_z/\partial r = 0 \quad \text{at } r = 0$$
$$P = P_1 \quad \text{at } z = 0$$

Again, radial velocities are ignored.

If a model relating τ to $\dot{\gamma}$ is introduced, and if a functional form for the temperature dependence of the parameters of this model is introduced, the resulting set of equations constitutes a non-linear coupled boundary value problem which cannot be analytically solved.

If fluid properties are assumed to be constant, in which case the isothermal velocity profile is known, solutions exist from which the temperature profile may be calculated as a function of axial position. *Bird* gives solutions for the case of the power law fluid, with either the isothermal or adiabatic wall. *Siegel* give solutions for the Newtonian fluid with boundary conditions specifying constant prescribed flux at the tube wall. The simplest case is that of *Brinkman* for the Newtonian fluid with temperature-independent properties, with either isothermal or adiabatic wall. None of the studies accounts for the alteration of the flowfield arising from viscosity variation across the radius of the capillary. As such they may not be used to correct viscometry data. Their primary use is in the estimation of the temperature rise to be expected. If the estimate of the maximum temperature rise is small, say < 1°C, then one might judge that the results need not be corrected at all for heating effects. If the estimated temperature rise is large, then one is faced with the problem of rejecting the data or accepting the results as subject to significant error.

In experiments with ethylene propylene diene monomer (EPDM) rubbers the wall of a capillary behaves in a nearly adiabatic manner under normal operating conditions. Hence, estimates of temperature rise based upon the adiabatic theories are useful. Fig. 3.2 shows *Bird's* solutions for the power law fluid, plotted as a dimensionless wall temperature which is a function of a dimensionless axial variable, with n as a parameter. The dimensionless variables are defined as

$$\Theta = (T - T_o)\frac{4k\,K^{1/n}}{D^2 \tau_w^{(n+1)/n}}\left(\frac{3n+1}{2n}\right)^2 \quad ...(3.15)$$

and

$$Z = \frac{4}{Pe}\frac{n+1}{3n+1}\frac{Z}{D} \quad ...(3.16)$$

τ_w is the wall shear stress and Pe is the Peclet number $(= D\langle V\rangle \rho \hat{C}_P / k)$.

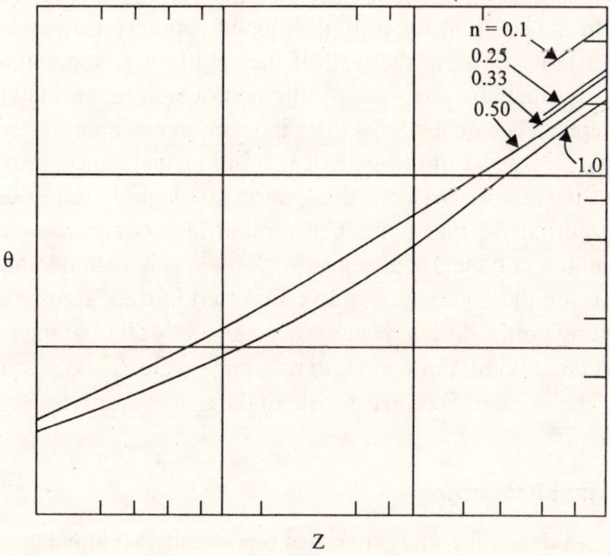

Fig. 3.2. Dimensionless wall temperature rise down the length of a capillary, due to viscous heating.

It is interesting to examine the manner in which variables such as $\langle V\rangle$ and D affect the maximum temperature rise. This is shown simply by re-writing equation (3.15) as follows:

$$T - T_o = \frac{D^2 K}{4k}\dot{\gamma}^{n+1}\left(\frac{2n}{3n+1}\right)^2 \Theta\left(n, \frac{z}{D}, Pe\right) \quad ...(3.17)$$

where τ_w has been replaced by $K\dot{\gamma}^n$. If interest is at conditions under high shear rates, then

$$\dot{\gamma} = \frac{3n+1}{4n}\left(\frac{8\langle V\rangle}{D}\right)$$

and thus a choice must be made between increasing $\dot{\gamma}$ by increasing $\langle V\rangle$ or by decreasing D.

As an example, if we consider 4 × $\dot{\gamma}$ then the coefficient of Θ is decreased by a factor of $(1/4)^2(4)^{n+1}$. For n = 1 the factor is unchanged, while for n = 1/2 the factor is halved. On the other hand, if $\dot{\gamma}$ is increased by increasing $\langle V\rangle$ by 4, at constant D, then the coefficient of Θ increases directly as $\dot{\gamma}^{n+1} = (4)^{n+1}$. For n = 1 and 1/2 this factor is 16 and 8, respectively. It becomes clear that the heating effect is minimised if $\dot{\gamma}$ is increased by using smaller diameter capillaries. In fact Θ is a function of D and $\langle V\rangle$ by way of the Peclet number. However, the Peclet number dependency is weak under usual experimental conditions, and hence the above observations apply. Of course, it has been assumed that K and n are unchanged by an increase in $\dot{\gamma}$. While this is not rigorous, in the sense that no fluid truly possesses power law behaviour, the approximation does not appear to be bad.

Capillary experiments involving highly viscous materials should always be accompanied by an estimate of temperature rise due to viscous heating. In general, many applications of viscosity data simply do not warrant the effort required to ascertain the correct for non-isothermal effects. On the other hand, one can find studies in the literature that draw conclusions based on the assumption that the high shear rate viscosity of high viscous polymer melts is free of viscous heating perturbations.

Middleman provides a nomograph for estimation of temperature rise due to viscous dissipation. The nomograph is based upon the solution for fully developed capillary flow of a power law fluid. This solution provides the temperature rise at the wall of the capillary, at some axial position z, under the assumption that the wall is adiabatic. To simplify the nomograph, approximations were made which essentially remove any dependence on the power law index n. In examples tested, the nomograph yields a temperature rise within a factor of 2 of the analytical solution, and hence, provides a rapid estimate of the order of magnitude of the effect of viscous dissipation in a highly viscous capillary flow. All scales are in centimetres per gram per second units. Conversion factors are given below. A short table of thermal properties of common polymers is also given. Note especially that k must be in centimetres per gram per second units. To use the nomograph, move from left to right across the chart. For a given set of data, begin by connecting points on the $\dot{\gamma}$ and τ scales, and locate the intersection of this line with reference scale 1. Connect that point, through D, to reference scale 2, and so on across the other scales to the temperature scale. The value of Θ is an estimate of the temperature rise at the wall of the capillary, at the given value of z/D.

Typical Values of Thermal Properties

The following are typical values of thermal properties reported by Middleman:

Properties	$\left(\frac{g/cm}{s^3/°C}\right)$	$k/\rho\hat{C}_P\left(\frac{cm^2}{s}\right)$
Benzene	1.5×10^4	8.5×10^{-4}
Water	6.3×10^4	1.5×10^{-3}
Polybutadiene/styrene co-polymer	2.1×10^4	1.2×10^{-3}

Polybutene	1.2×10^4	7.4×10^{-4}
Polyethylene	3.3×10^4	1.3×10^{-3}
Polyethylene terephthalate	1.2×10^4	5.1×10^{-4}
Polystyrene	1.2×10^4	6.2×10^{-4}

Conversion Factors

The conversion factors are given below:

To obtain k in $g/cm/s^3$ °C, multiply k in Cal/s/cm/°C by 4.2×10^7; k in BTU/h/ft/°F by 1.7×10^5.

To obtain k/ρCp in cm^2/sec, multiply $k/\rho \hat{C}_p$ in ft^2/hr by 0.26.

To obtain $k/\rho C_p$ in $g/cm/s^2$ ($dyne/cm^2$), multiply in lb . ft_f^2 by 480; in lb . $in._f^2$ by 6.9×10^4.

To obtain D in centimetres, multiply D in in. by 2.54; D in mil by 2.54×10^{-3}.

Note: It should suffice to use nominal values for $\dot{\gamma}$ and τ, i.e., $\dot{\gamma} = 8 \langle V \rangle / D$ and $\tau = D\Delta P/4L$. For polymer melts, $\dot{\gamma}$ would be underestimated but τ would be overestimated (assuming no corrections are made to ΔP) and the errors would tend to be compensatory.

We now direct attention to the case of Couette flow. To assess viscous dissipation one must solve the dynamic and energy equations together:

$$0 = \frac{d}{dr}(r^2 \tau) \qquad \text{...(3.18)}$$

$$0 = \frac{1}{r}\frac{d}{dr}\left(rk\frac{dT}{dr}\right) + \Theta \qquad \text{...(3.19)}$$

subject to boundary conditions

$$v_\theta = 0 \text{ and } \quad T = T_o \quad \text{at} \quad r = R_o$$

$$\left.\begin{array}{l} v_\theta = V = 2\pi\Omega R \quad \text{and } T = T_o \\ \qquad \text{or } \dfrac{\partial T}{\partial r} = 0 \end{array}\right\} \text{at } r = R$$

Note T and v_θ are functions only of radial position, and hence may be described by ordinary differential equations. Because of this it is possible to introduce some realistic complications into the analysis which cannot be readily handled in capillary flows that are described by partial differential equations.

The technique outlined below follows that used by Turian and Bird. It is necessary to assume a constitutive equation and a temperature dependence of fluid properties. The power law is a useful model for the rheological behaviour; the fluid properties are taken to be functions of temperature expressible in a power series about some reference temperature at which properties are assumed known. Shear stress is defined by:

$$\tau = K\dot{\gamma}^n \qquad \text{...(3.20)}$$

and K is taken as expansion formula

$$K = K_o[1 - K_1(T - T_o) + K_2(T - T_o)^2 - \ldots] \qquad \text{...(3.21)}$$

For moderate temperature changes the thermal conductivity may be taken as constant.

Equations (3.18) and (3.19) may be solved by a perturbation method. The results outlined below depend on whether one assumes an isothermal inner wall or an adiabatic inner wall.

For an isothermal wall, the maximum temperature rise is

$$\frac{\theta_{max}}{B_R} = \frac{(T - T_o)_{max}}{T_o B_R}$$

$$= \frac{n^2 C_o^{1+(1/n)}}{4}\left[1 - \frac{n}{2} f(s,n)\left(1 - \ln\frac{n}{2} f(s,n)\right)\right] \quad ...(3.22)$$

The shear rate at the moving wall is

$$\dot{\gamma}_R = \dot{\gamma}_o\left\{1 - \frac{\beta B_R s n^2 C_o^{1+(2/n)}}{16}\left[1 - S^{-4/n} - n(1 - s^{-2/n}) f(3,n)\right]\right\} \quad ...(3.23)$$

For the case of an adiabatic wall, the maximum temperature is

$$\frac{\theta_{max}}{B_R} = \frac{n^2 C_o^{1+(1/n)}}{4}\left[1 - S^{-2/n} - \frac{2}{n} S^{-2/n} \ln S\right] \quad ...(3.24)$$

The shear rate at the moving wall is

$$\dot{\gamma}_R = \dot{\gamma}_0\left\{1 - \frac{\beta B_R s n^2 C_o^{1+(2/n)}}{16}\left[1 - S^{-4/n} - n(1 - S^{-2/n}) f(S,n)\right]\right\} \quad ...(3.25)$$

In the above relationships, $f(s, n) = (1 - S^{-2/n})/\ln S$, $C_0 = S^{2-n}(\gamma_0/\Omega)^n$ and B_R is the dimensionless Brinkman number $\left(B_R = \frac{R^{1-n} K_o V^{1+n}}{K T_o}\right)$. Finally, $\beta = K_1 T_o$ is a measure of the temperature sensitivity of K. If βB_R is not small compared to unity, then the solutions above are inaccurate. In that case more terms in the perturbation series are required.

Expressions for $\dot{\gamma}_R$ are in the form of a correction factor multiplied by the isothermal shear rate. To apply the correction factor it is necessary to have at hand values of fluid properties such as β, K_o, and n. This poses a dilemma (i.e., these values are not known *a priori*). This suggests that correction for non-isothermal effects must be based on a trial and error technique. The data are treated as if free of heating effects, and K_o and n are estimated. The maximum temperature may then be estimated and a second set of data obtained at this temperature level. From estimates of K_o at two temperatures an estimate of β may be obtained. From β, K_o, and n an approximate correction factor is obtained and new values of β, K_o, and n are calculated. The procedure is repeated until satisfactory convergence is obtained.

CHAPTER 4

Normal Stress Measurements

INTRODUCTION

Some of the techniques of viscometry were discussed in previous chapters. Many fluids, when subjected to a simple shear flow, develop not only shear stresses but also normal stresses. These normal stresses manifest themselves in the so-called 'normal stress phenomena' which commonly occur in viscometric experiments.

These stresses also materialise in many important flows. For example, when a rod (such as the shaft of a stirrer) is rotated about its axis perpendicular to the free surface of a Newtonian liquid, the liquid surface is depressed in the neighbourhood of the rod, as a consequence of centrifugal forces that accompany the induced rotational flow. Some fluids, however, are observed to develop an elevated surface at the rod. This 'rod climbing' effect is perhaps the best known normal stress phenomenon, and is called the '*Weissenberg effect*'.

When certain non-Newtonian fluids (e.g., polymer melts) are ejected from an orifice or tube, the resulting jet is commonly observed to expand to a diameter much larger than its initial ejection diameter. This effect is often referred to industrially as 'die swell' or 'extrudate swelling'.

This chapter is concerned with information of fundamental rheological significance from the measurement of normal stress phenomena. In contrast to the measurement of shear stresses, the measurement of normal stresses includes the isotropic stress $p\delta$ in addition to the dynamic stress τ. In order to generate information specifically about the dynamic stress components, it is necessary to 'remove' the influence of p from the measurements. This is most commonly done by expressing results in terms of stress differences since, for example

$$T_{11} - T_{22} = -p + \tau_{11} - (-p + \tau_{22}) = \tau_{11} - \tau_{22} \qquad \text{... (4.1)}$$

and the isotropic component does not appear. When this is not possible it may be necessary to impose some assumptions upon p.

The fundamental meaning of the isotropic component p is still not clear for a viscoelastic material. While the classical definition is

$$p = p_m = -\frac{1}{3}(T_{11} + T_{22} + T_{33}) \qquad \text{... (4.2)}$$

for such fluids, so that p is by definition 'mean normal stress', it is not obvious that this is rigorous.

As in the case of viscometry, the flows studied approximate simple shear flows. In contrast to the techniques of viscometry, many of the methods used for normal stress measurement are incompletely developed, in the sense that the effects of perturbations which cause deviations from a simple shear flow, such as exit effects in capillary flows, are not fully understood. Techniques for removing the influence of such effects from data are subject to assumptions not yet theoretically justified.

ANALYSIS OF THE CAPILLARY JET

When a fluid discharges from a tube as a free jet it exerts an axial thrust upon the tube. This thrust can be calculated as the difference between the flux of axial momentum leaving the tube and the axial stresses that exist in the fluid as it is ejected. The thrust F can be expressed as follows:

$$F = \int_0^R \rho v_z^2 \, 2\pi r \, dr - \int_0^R T_{zz} \, 2\pi r \, dr \quad \text{... (4.3)}$$

This expression is subject to the assumption that surface tension and gravity do not contribute significantly to the axial momentum.

It is possible to invert this relationship via a technique very similar to that used in developing the *Weissenberg*-Rabinowitsch-Mooney equation and thereby solve for T_{zz} at the wall. The resultant expression is:

$$(T_{zz})_R = \rho\langle V\rangle^2\left[\frac{3n+1}{n} - 2\int_0^1\left(\frac{v_z}{\langle V\rangle}\right)\frac{r}{R}\,d\left(\frac{r}{R}\right)\right]$$

$$-\frac{F}{\pi R^2}\left(1+\frac{1}{2n}\,\frac{d\ln F}{d\ln 8\langle V\rangle/D}\right) \quad \text{... (4.4)}$$

The shear rate at the tube wall can be obtained from the *Weissenberg*-Rabinowitsch-Mooney expression. It is therefore possible to obtain T_{zz} and the corresponding shear rate. This result is subject to the assumption that the velocity and stress distributions at the tube exit correspond to those for fully developed flow. Recall that n is the slope of the log τ vs. log $\dot{\gamma}$ flow curve at the conditions of interest.

Introducting the components of τ in place of the components of T,

$$(T_{zz})_R = -\,p(R, L) + (\tau_{zz})_R \quad \text{... (4.5)}$$

$(T_{zz})_R$ is measured, using equation 4.4, at the exit (Z = L). Hence, the isotropic stress component p(R, L) at the tube exit must be known in order to find $(\tau_{zz})_R$.

It can be assumed that p(R, L) denotes atmospheric pressure at the tube exit. Examination of the dynamic equations, however, shows this to be false. The radial component of the equations of motion gives

$$0 = -\frac{\partial p}{\partial r} + \frac{\partial \tau_{rr}}{\partial r} + \frac{\tau_{rr} - \tau_{\theta\theta}}{r} \quad \text{... (4.6)}$$

Integration from r = 0 to some arbitrary r gives

$$p(r,z) = p(0,z) + \tau_{rr} + \int_0^1 \frac{\tau_{rr} - \tau_{\theta\theta}}{r}\,dr \quad \text{... (4.7)}$$

This is subject to the reasonable assumption that $\tau_{rr} = 0$ along the centre line of the tube, where the deformation rate vanishes. Hence,

$$(T_{11})_R = -\,p(0,L) + (\tau_{11} - \tau_{22})_R - \int_0^R \frac{\tau_{22} - \tau_{33}}{r}\,dr \qquad ...(4.8)$$

In the absence of dynamic normal stresses it follows that the only isotropic stress would be ambient press P_o and $p(0,L) = p(R,L) = P_o$ would be true.

The axial stress T_{11} may also be determined from a jet expansion experiment. Defining the final radius and velocity of the jet as R_j and V_j and assuming that the liquid is free of stress (relative to the ambient pressure) at this point, the flux of momentum across any position beyond the region where these final values are achieved is $\pi \rho R_j^2 V_j^2$. Because momentum is conserved in the jet, however, this quantity must just equal the thrust exerted on the tube. Hence, in equation 4.4, one can replace F by $\pi \rho R_j^2 V_j^2$.

$$(T_{11})_R = \rho\langle V\rangle^2\left[\frac{3n+1}{n} - 2\int_0^1\left(\frac{v}{\langle V\rangle}\right)^2 \frac{r}{R}\,d\left(\frac{r}{R}\right)\right.$$

$$\left. - \frac{1}{n\alpha^2}\left(1 + n - \frac{d\ln\alpha}{d\ln 8\langle V\rangle/D}\right)\right] \qquad ...(4.9)$$

where $\alpha = R_j/R$. Equation 4.8 is applicable regardless of whether one measures thrust or expansion.

The stress T_{11} is associated with the normal stress in equilibrium with the fully developed shear flow within the capillary. There is, however, a viscous contribution to T_{11} which arises from the relaxation of the velocity profile subsequent to ejection of the liquid as a free jet. In Newtonian fluids, for which there is no normal stress except that developed by profile relaxation, this viscous stress can be large enough to cause the jet to swell. For Reynolds numbers larger than about 200, however, the viscous contribution is negligible.

In non-Newtonian fluids it is not possible to separate the viscous effect from the primary normal stress effect. It is usually assumed that the viscous stresses are negligible under the experimental conditions which are peculiar to a study of jets of polymeric fluids. The validity of this assumption can be checked by plotting values of T_{11} obtained with different diameter tubes against shear rate. Since the viscous effect would not be a function of shear rate, but rather would be a function of a viscous parameter, such as a non-Newtonian Reynolds number, data taken in different diameter tubes would fall on a single curve, when plotted as T_{11} vs. $\dot{\gamma}$ only if free of viscous effects.

At very high flow rates the ratio R_j/R approaches a limiting value which depends only upon the slope n of the log τ vs. log $\dot{\gamma}$ curve for the particular fluid. For the power law fluid this limiting value is given by

$$\alpha_\infty = R_j / R = \sqrt{(2n+1)/(3n-1)} \qquad ...(4.10)$$

and lies in the range $0.866 \le \alpha \le 1$ for $1 \ge n \rangle 0$.

A characteristic relaxation time for a fluid can be defined as the time required for the stress in the stress in the fluid to adjust to a step change in shear rate. Fig. 4.1 gives a possible quantitative definition for a relaxation time. One would expect that the normal stresses are in equilibrium with the shear field in the tube if the residence time $L/\langle V\rangle$ is large than the relaxation time λ. The criterion for fully developed normal stresses is

$$\frac{L_n/\langle V\rangle}{\lambda} = k_n \quad \text{or} \quad \frac{L_n}{D} = k_n \frac{\langle V\rangle\lambda}{D} \qquad \text{... (4.11)}$$

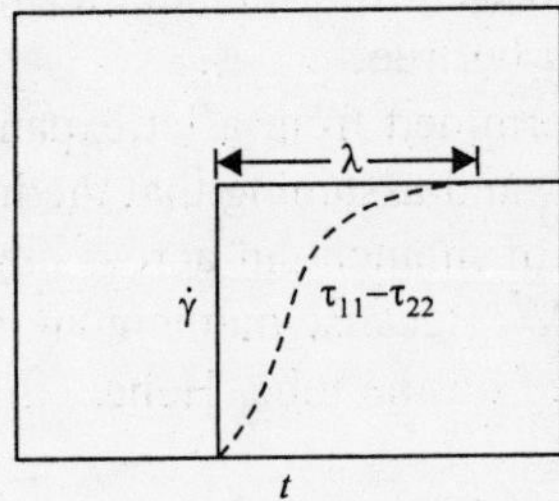

Fig. 4.1. Concept of a relaxation time, based on the hypothetical response of a viscoelastic fluid to a step change in shear rate.

The parameter k_n must be established through experiment (but one would expect it to be of the order of unity). Estimates of λ are on the order of magnitude of 10^{-2}s for fluids that show significant expansion. For example, typical values of $\langle V\rangle$ and D are $\langle V\rangle = 500$ cm/s and $D = 0.1$ cm; a typical estimate for L_n/D is 50.

Since shear rates in capillary flows are generally high (around 10^4/s), and since fluids that exhibit significant swelling are fairly viscous, heat generation within the capillary can be appreciable. In addition to producing a temperature different than the one that the fluid is presumed to have on the basis of the reservoir temperature, the temperature field alters the velocity profile. The velocity profile appears in equation 4.4 and it is clear that viscous heating effects could complicate the proper interpretation of normal stress measurement.

Under normal conditions jet experiments involve shear rates in the range 10^{-4} to 10^5/s. To extend the range over which useful normal stress data may be obtained in capillary flow, a pitot tube method is suitable to lower shear rates. Using Fig. 4.2 as a reference, the pressure P_i would be an 'impact' pressure, involving the axial normal stress in addition to the dynamic pressure $1/2\ \rho v_1^2(r)$ or

$$P_i = -T_{11}(r) + \frac{1}{2}\rho v_1^2(r) \qquad \text{... (4.12)}$$

The wall pressure P_w would be just $-T_{22}(R)$.

From the dynamic equations for laminar Poiseuille flow, one finds for the radial direction

$$T_{22}(R) = -p(0,z) - \int_0^R (\tau_{22} - \tau_{33}) \frac{dr}{r} \qquad \text{... (4.13)}$$

and, for the axial direction, equation 4.8 with R and L replaced by r and z we obtain:

$$T_{11}(r) = -p(0,z) + (\tau_{11} - \tau_{22}) - \int_0^r (\tau_{22} - \tau_{33}) \frac{dr}{r} \qquad \text{... (4.14)}$$

The pressure difference $P_i - P_w$ becomes:

$$P_i - P_w = -(\tau_{11} - \tau_{22}) + \frac{1}{2}\rho v_1^2(r) - \int_r^R (\tau_{22} - \tau_{33}) \frac{dr}{r} \qquad \text{... (4.15)}$$

Experiments shows that the normal stress difference $\tau_{22} - \tau_{33}$ is considerably smaller than the normal stress difference $\tau_{11} - \tau_{22}$. It is therefore normal practice to assume that $P_i - P_w$ allows the determination of the normal stress difference $\tau_{11} - \tau_{22}$.

This assumes that the velocity profile is known. If, however, viscous heating is significant the velocity profile will be perturbed, and the calculation of $\tau_{11} - \tau_{22}$ will be in error. At low shear rates, viscous heating effects are likely to be minimal. A thermocouple attached to the pitot tube would serve to indicate the presence of a significant temperature variation across the capillary.

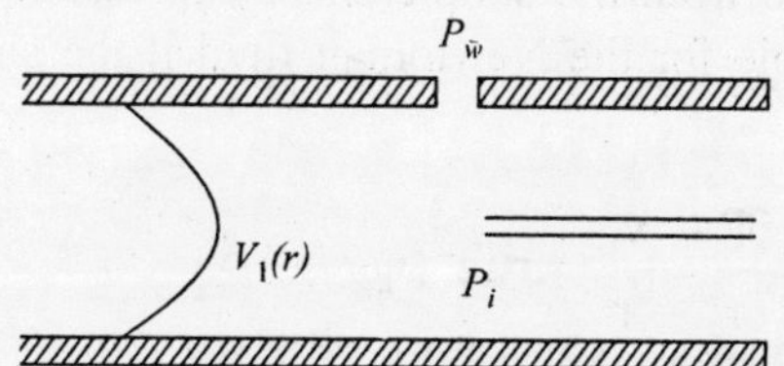

Fig. 4.2. Definition sketch for normal measurement using a pitot tube in laminar poiseuille flow.

Note also that equation 4.15 indicates that a pitot tube may not be used directly to measure the velocity profile in a fluid that exhibits normal stress phenomena. Any pitot tube measurement is subject to the criticism that the probe distorts the flow, and hence does not measure the velocity profile which would exist if the probe were not present.

The success of pitot tube measurements in Newtonian flows lends some confidence to its use in more complex flows. However, with a viscoelastic flow in the neighbourhood of an obstruction, the flow my be different from a Newtonian flow which is kinematically similar at some distance from the obstruction. The difference arises primarily from a markedly different behaviour of the boundary layer in viscoelastic stagnation flows. Evidence based on markedly reduced heat transfer coefficients from heated cylinders placed in a stream of a viscoelastic fluid supports this view.

NORMAL STRESSES IN THE CONE AND PLATE CONFIGURATION

In discussing the cone-and-plate system as a viscometer it was stated that the velocity field is

$$v_\phi = v_1 = r\,\Omega\,\psi/\psi_o \quad \text{... (4.16)}$$

and the shear rate $\dot{\gamma} = \Omega/\psi_o$. The r-component of the dynamic equations, in spherical coordinates, is reducible to the form

$$-\rho\frac{v_1^2}{r} = -\frac{\partial P}{\partial r} + \frac{\partial \tau_{33}}{\partial r} - \frac{\tau_{11} + \tau_{22} - 2\tau_{33}}{r} \quad \text{... (4.17)}$$

Since shear rate is independent of r, and since τ is a function only of shear rate, terms such as $\partial\tau_{33}/\partial r$ are identically zero. The stress measured by pressure taps in the plate ($\psi = 0$) is

$$T_{22}(r,0) = -\,p(r,0) + \tau_{22} \quad \text{... (4.18)}$$

If T_{22} is introduced into the dynamic equation, the result is

$$\frac{\partial T_{22}}{\partial r} = -\left(\frac{\rho v_1^2}{r} - \frac{X(\dot{\gamma})}{r}\right) \quad \text{... (4.19)}$$

where $X = \tau_{11} + \tau_{22} - 2\tau_{33}$. If the isotropic component of T is taken as the mean normal stress, then $X = -3\tau_{33}$. If equation 4.19 holds at the plate surface ($\psi = 0$), then the measured stress distribution becomes:

$$\frac{\partial T_{22}}{\partial r} = -\frac{3\tau_{33}}{r} \quad \text{or} \quad \frac{\partial T_{22}}{\partial \text{ in } r} = -3\tau_{33} = \text{Constant} \qquad \text{... (4.20)}$$

Deviations between this prediction and experimental results may be ascribed to inertial effects that cause the velocity profile to deviate from the simple shear expression assumed above. A first-order estimate of inertial effects is possible for the Newtonian fluid if the simple shear velocity profile is inserted in equation 4.17 so that

$$\frac{1}{\rho}\frac{\partial P}{\partial r} = \frac{v_1^2}{r} = r\,\Omega^2(\psi / \psi_o)^2 \qquad \text{... (4.21)}$$

Integration yields

$$(P - P_o) = \frac{1}{2}P\Omega^2(\psi / \psi_o)^2 r^2 + \rho C_N(\psi) \qquad \text{... (4.22)}$$

where Po is some arbitrary pressure and C_N is a function of ψ, but not r. For $\psi = 0$, the pressure along the plate is independent of radial position, according to this result.

Experimental measurements are not in agreement with this prediction; instead they indicate a variation of p with r. A possible interpretation of this result seems to have no physical foundation, but does load to a prediction consistent with the experiment. It is claimed that the stress measured at the plate is really the ψ-average of the stress distribution between the cone and plate.

A detailed evaluation of the dynamic equations for a non-Newtonian fluid, coupled with the assumption that the stress measured along the plate surface, gives the result

$$\overline{T}_{22} + P_o - \tau_{22} = -3\tau_{33}\left(\frac{1}{3} - \ln\frac{R}{r}\right) - \frac{1}{6}\,\rho\,\Omega\,R^2\left(\frac{r^2}{R^2} - \frac{3}{5}\right) \qquad \text{... (4.23)}$$

τ_{22} is not averaged because it is a function only of shear rate and so is independent of ψ, and Po is taken to be the ambient pressure.

A plot of the measured quantity

$$\overline{T}_{22} + P_o + \frac{1}{6}\rho\Omega R^2\left(\frac{r^2}{R^2} - \frac{3}{5}\right)$$

against ln (R/r) should be linear, with a slope equal to 3t33 and a zero intercept ata radial position ro such that

$$\ln(R / r_o) = \frac{1}{3}(1 - \tau_{22} / \tau_{33}) \qquad \text{... (4.24)}$$

Hence, both τ_{33} and τ_{22} may be obtained directly from such an experiment, if the inertial correction is known.

The inertial correction may be determined experimentally be measuring the stress distribution along the plate for a Newtonian fluid. If inertial effects perturb the stress field without causing a significant deviation of the velocity field from that assumed in equation 4.21, then a non-Newtonian fluid will behave, with respect to inertial forces, like a Newtonian fluid, as the shear field will be uniform throughout the fluid. Thus, the inertial correction at a particular shear rate can be obtained from the measured stress distribution of a Newtonian fluid whose viscosity is the same as the apparent viscosity of the non-Newtonian fluid at that shear rate.

Because the inertial stresses depend upon Ω, but not upon ψ_o, data taken with different cone angles should be a function only of Ω/ψ_o when the proper inertial corrections have been made. Thus, one should always take data with more than a single cone angel whenever possible.

If the total vertical force exerted against the cone is measured, it is possible to determine the normal stress difference T_{22} directly, without a measurement of the stress distribution in the system. For the small cone angles usually employed, this force may be calculated from.

$$F = -\pi R^2\left[(-p(R) + \tau_{33}) + \frac{1}{2}(\tau_{22} - \tau_{11})\right] \quad \text{... (4.25)}$$

If the system is in equilibrium with the atmosphere on its outer boundary, then

$$-p(R) + \tau_{33} = T_{33}(R) = 0$$

and it follows that

$$\tau_{11} - \tau_{22} = 2F/\pi R^2 \quad \text{... (4.26)}$$

Inertial effects may be introduced as before. The most common procedure is to correct F on the basis of experiments with Newtonian fluids of appropriate viscosities. Then $\tau_{11} - \tau_{22}$ is calculated from F′, a corrected force from which the inertial effect has been removed.

PARALLEL PLATE CONFIGURATION

Another system useful for the measurement of normal stresses at low shear rates is the parallel plate torsional configuration, most easily achieved in the space between the bottom of two concentric cups. Fig. 4.3 shows the geometry for this system.

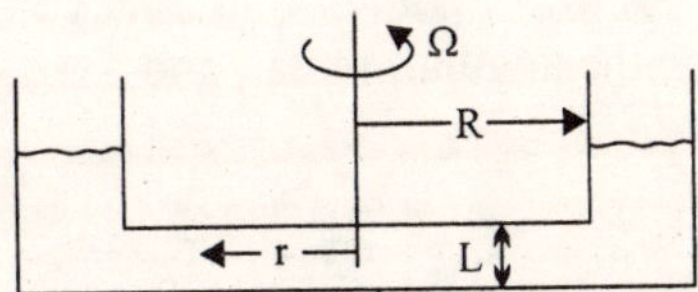

Fig. 4.3. Definition sketch for parallel plate torsional shear flow.

If inertial effects are unimportant, and if edge effects at the radial boundary can be ignored, then the velocity profile may be shown to be

$$v_1 = \Omega\, rz/L \quad \text{... (4.27)}$$

The only non-zero components of the rate of deformation tensor are

$$\dot{\gamma} = \Delta_{12} = \Delta_{21} = \Omega\, r / L \quad \text{... (4.28)}$$

The shear rate varies radially in this system.

The governing expression for this system is

$$T_{22}(r) + p(0) = \tau_{22} - \tau_{33} + \int_0^r \frac{\tau_{11} - \tau_{33}}{r}\, dr \quad \text{... (4.29)}$$

subject to the assumption that T_{22} and T_{33} are zero at r = 0, as the shear rate vanishes at that point.

Because T_{22} may be measured as a function of r, the right side of equation 4.29 may be obtained. However, further information about the individual stress differences requires arbitrary assumptions about their relative values or their dependencies on shear rate.

A more useful result is obtained from the determination of the total vertical force exerted against the upper plate. The force F is given by

$$F = -2\pi \int_0^R T_{22}(r)\, r\, dr \qquad \text{... (4.30)}$$

If $\dot{\gamma}$ is substituted for r through Equation 4.28 and F is differentiated with respect to $\dot{\gamma}_R$ the result is

$$-3\tau_{22}(\dot{\gamma}_R) = \frac{2F}{\pi R^2}\left(1 + \frac{1}{2}\frac{d \ln F}{d \ln \dot{\gamma}_R}\right) \qquad \text{... (4.31)}$$

where $\dot{\gamma}_R = \Omega R/L$.

As in the case of the cone-and-plate system, F must be replaced by F′ where F′ includes a correction for inertial effects. Inertial stresses are far more significant in the parallel-plate system than in the cone-and-plate, at comparable shear rates. Because the shear rate varies spatially in the parallel-plate system, the inertial correction cannot be obtained exactly from experiments with a Newtonian fluid; however, the Newtonian stress estimate will considerably reduce the error due to inertial effects. Any residual anomalous behaviour can be detected by plotting τ_{22} against the plate separation L, at constant shear rate. If the curve is not flat, it is suggested that the τ_{22} curve be extrapolated to zero plate separation, and that value of τ_{22} used as a corrected stress.

CONCENTRIC CYLINDER (COUETTE) SYSTEM

This system may also be used to measure normal stresses. The primary measurement is the difference in radial stresses exerted at the two cylindrical surfaces. As illustrated in Fig. 4.4, this measurement is usually made by monometers.

If 1, 2, and 3 denote, respectively, the angular, radial, and axial directions, then the radial dynamic equation may be written as

$$-\rho\frac{v_1^2}{r} = -\frac{\partial P}{\partial r} + \frac{\partial \tau_{22}}{\partial r} + \frac{\tau_{22} - \tau_{11}}{r} \qquad \text{... (4.32)}$$

subject to the usually assumption of a laminar simple shear flow. If the inertial term is neglected, Equation 4.32 may be integrated between the inner and outer radii to give

$$\Delta(p - \tau_{22}) = -\Delta T_{22} = \int_{SR}^{R} \frac{\tau_{22} - \tau_{11}}{r}\, dr \qquad \text{... (4.33)}$$

Since $r^2\tau$ = constant, τ may be substituted for the variable of integration, and the resulting equation may be differentiated with respect to the shear rate at the inner wall γ_i. This leads to (after differentiation and use of the Leibniz rule):

$$\frac{d\Delta T_{22}}{d\dot{\gamma}_i} = \frac{1}{2}\left[\left(\frac{\tau_{22} - \tau_{11}}{\tau}\right)\frac{d\tau_o}{d\gamma_i} - \left(\frac{\tau_{22} - \tau_{11}}{\tau}\right)_i \frac{d\tau_i}{d\gamma_i}\right] \qquad \text{... (4.34)}$$

where quantities subscripted 0 and i are be evaluated at the conditions of the outer and inner walls, respectively.

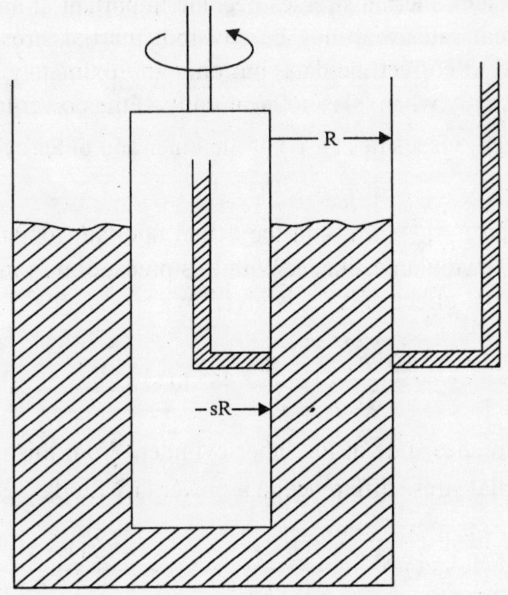

Fig. 4.4. Definition sketch for normal stress determination in cylindrical couette flow.

For a fixed geometry, since $\tau_o = s^2\tau_i$, it follows that

$$\frac{d\tau_o}{d\dot{\gamma}_i} = s^2 \frac{d\tau_i}{d\dot{\gamma}_i} \quad \text{or} \quad \frac{d \ln \tau_o}{d \ln \dot{\gamma}_i} = \frac{d \ln \tau_i}{d \ln \dot{\gamma}_i} = n \qquad \text{... (4.35)}$$

where n, as defined here, is the slope of the logarithmic plot of τ vs. $\dot{\gamma}$.

Equation 4.34 can be rearranged to the form

$$\frac{2}{n}\frac{d\Delta T_{22}}{d \ln \dot{\gamma}_i} = (\tau_{22} - \tau_{11})_0 - (\tau_{22} - \tau_{11})_i \qquad \text{... (4.36)}$$

The left side of this equation may be measured as a function of shear rate $\dot{\gamma}_i$, with $\dot{\gamma}_i$ calculated from equation 4.33. As s approaches unity, the shear rates at the inner and outer walls approach each other, and the difference on the right side of equation 4.36 becomes subject to considerable error.

A useful approximation is possible if 's' is not close to unity. Let $\tau_o = s^2\tau_i$, and suppose the viscosity behaviour is approximate by the power law model.

$$\dot{\gamma}_o = s^{2/n}\dot{\gamma}_i \qquad \text{... (4.37)}$$

As a further approximation

$$(\tau_{22} - \tau_{11}) = A\dot{\gamma}^{2n} \qquad \text{... (4.38)}$$

Also, $(\tau_{22} - \tau_{11})_0 = s^4(\tau_{22} - \tau_{11})_i$, follows as an approximation. If 's' = 1/2, the normal stress difference at the outer wall is only 1/6 of the value at the inner wall. Hence, for 's' ⟨1/2, a good approximation would be

$$\frac{2}{n}\frac{d\Delta T_{22}}{d \ln \dot{\gamma}_i} = -(\tau_{22} - \tau_{11}) \qquad \text{... (4.39)}$$

As in the parallel-plate system, inertial stresses become important at high shear rates, and limit to some extent the range of shear rates that may be covered. Inertial stresses based on studies with newtonian fluids may be used to correct the data, but only approximately, since the shear rate varies across the annular gap significantly when 's' is not near unity. If the power low model is assumed it can be seen that the ratio of apparent viscosities ($\tau/\dot{\gamma}$) at the inner and outer cylinder is

$$\eta/\eta_o = s^{2(1/n - 1)} \quad \text{... (4.40)}$$

If s = 1/2 and n = 1/2, then η_i/η_o = 1/4, and the actual non-Newtonian inertial stresses might be poorly approximated by any Newtonian estimate. With this precaution in mind, it may be noted that a first-order estimate of the inertial effect of ΔT_{22} is given by

$$\mathfrak{F} = \frac{\rho s^2 R^2}{8}(1 - s^4 + 4s^2 \ln s)\dot{\gamma}_i^2 \quad \text{... (4.41)}$$

where $\dot{\gamma}_i$ is the Newtonian shear rate at the inner cylinder. From this expression it is possible to obtain an estimate for the inertial stress difference in a power law fluid,

$$\mathfrak{F} = \frac{\rho s^{2/n} R^2 n^2}{8}\left[s^{2/n}(1 - s^2) - \frac{2s^{2/n}(1 - s^{2(1-1/n)})}{1 - 1/n} + \frac{s^{2/n}(1 - s^{2(1-1/n)})}{1 - 2/n}\right]\dot{\gamma}_{in}^2 \quad \text{... (4.42)}$$

where $\dot{\gamma}_{in}$ is the power law shear rate.

If $\mathfrak{F}$ is estimated, the foregoing equations are unchanged so long as ΔT_{22} is replaced by $\Delta T_{22} + \mathfrak{F}$. If inertial stresses are effectively removed by this method, then data for $\tau_{22} - \tau_{11}$ should fall on a single curve, when plotted against shear rate, with no dependence upon s or R.

ELONGATIONAL FLOWS

Simple elongation can be defined as a flow with which another material property, the elongational viscosity, can be measured. One method of achieving a simple elongation is illustrated in Fig. 4.5. The upper end of a cylindrical sample is clamped and stationary. The lower end is clamped and moves with the velocity U. If end effects in the region of the clamps can be ignored then the deformation will be uniform, in the sense that the shrinking of the radius will not depend upon axial position, and the axial velocity will be uniform across the radius. Hence, v_z is not a function of r, nor is v_r a function of z.

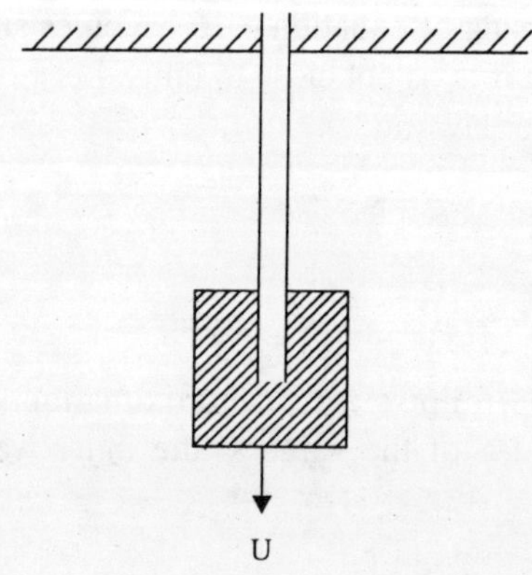

Fig. 4.5. Definition sketch for a simple elongation flow.

The axial velocity and radial velocity must satisfy the continuity equation:

$$\frac{\partial v_z}{\partial z}+\frac{1}{r}\frac{\partial}{\partial r}(rv_z)=0 \quad \text{... (4.43)}$$

In order to satisfy continuity, then it must be true that

$$\frac{\partial v_z}{\partial z}=\text{Constant}=-\frac{1}{r}\frac{\partial}{\partial r}(rv_r) \quad \text{... (4.44)}$$

A solution which satisfies the boundary conditions $v_z = 0$ at z = 0, $v_z = U$ at z = L(t), and $v_r = 0$ at r = 0, is

$$v_z = Uz/L, \quad v_r = -Ur/2L \quad \text{... (4.45)}$$

The length of the sample increases from its initial value L_o according to

$$L(t) = L_o + \int_0^t U(t)dt \quad \text{... (4.46)}$$

Because the volume of the sample must remain constant, it follows that the area of the cross-section decreases with time, and is given by

$$A(t)/A_o = L_o/L(t) = 1/\alpha_1 \quad \text{... (4.47)}$$

where α_1 is the 'principal extension ratio'.

The strain rate is

$$\dot{\varepsilon}=\frac{\partial vz}{\partial z}=\frac{U}{L}=\left(\frac{1}{L}\right)\frac{dL}{dt}=\left(\frac{1}{\alpha_1}\right)\frac{d\alpha_1}{dt} \quad \text{... (4.48)}$$

In the usual elongation experiment the velocity U is held constant, for example, $U = U_o$. The $\dot{\varepsilon}$ becomes

$$\dot{\varepsilon} = U_o / L = U_o / \alpha_1 L_o \quad \text{... (4.49)}$$

and it is clear that the strain rate continually decreases as elongation proceeds.

Because material functions may depend on $\dot{\varepsilon}$, one must be careful in interpreting results from such an experiment.

A quantitative example of behaviour under constant speed extension can be obtained by examining a simple linear viscoelastic model of the form

$$\tau + \lambda\frac{d\tau}{dt} = \eta_o\Delta \quad \text{... (4.50)}$$

In the limit of vanishing relaxation time λ, one obtains the Newtonian fluid.

For the axial direction, Δ_{11} is $2\dot{\varepsilon}$, and hence is given by $\Delta_{11} = 2U_0/\alpha_1 L_o$. Since $L = L_o + U_o t$, it follows that $\alpha_1 = 1 + U_o t/L_o$. If the time variable is replaced by α_1 in the constitutive equation above and dimensionless variables are introduced, the resulting differential equation may be written as

$$\tau^* = \Lambda\frac{d\tau^*}{d\alpha_1} = \frac{1}{\alpha_1} \quad \text{... (4.51)}$$

where

$$\tau^* = \tau_{11}L_o/2\eta_o U_o \quad \text{and} \quad \Lambda = \lambda U_o/L_o$$

The initial condition is simply $\tau^* = 0$ at $\alpha_1 = 1$. The solution for this model may be represented as τ^* vs. α_1 with Λ as a parameter. Fig. 4.6 shows a family of solutions. In the limit of vanishing Λ, the Newtonian solution, $\tau^* = 1/\alpha_1$, is approached.

A constant rate of strain experiment may be achieved by an elongation programme given by

$$U = \beta L_o e^{\beta t} \qquad \text{... (4.52)}$$

where β is some constant. For this choice of U it is found that L is

$$L = L_o e^{\beta t} \qquad \text{... (4.53)}$$

and

$$\dot{\varepsilon} = U / L = \beta \qquad \text{... (4.54)}$$

Hence, by suitable choice of β, the desired value of strain may be achieved.

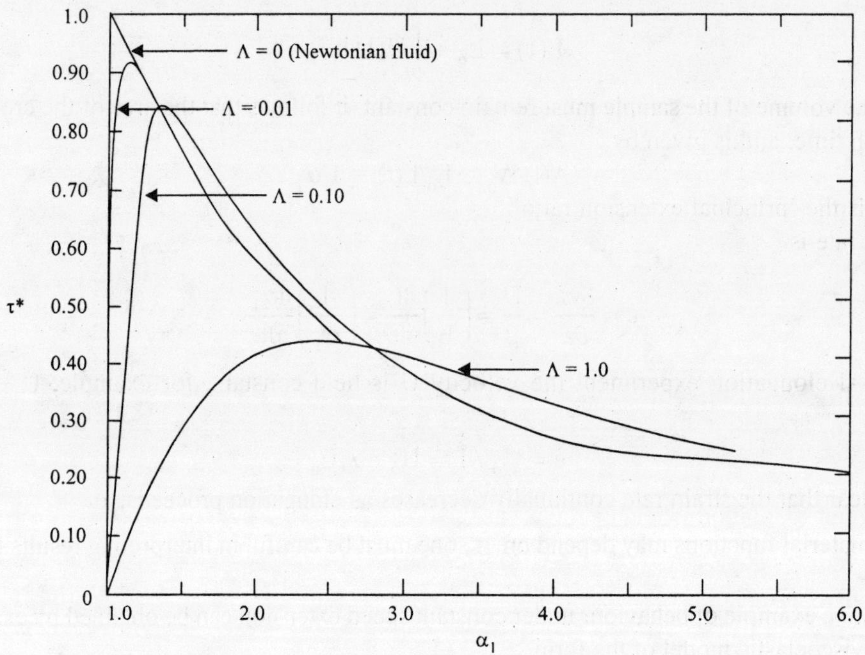

Fig. 4.6. Comparison of Newtonian and viscoelastic elongational flows, at constant speed, plotted as reduced stress vs. principal extension ratio.

DYNAMIC MEASUREMENTS

The flows discussed up to this point have been those maintained under steady-state (time-independent) conditions. It is possible to carry out tests under dynamic conditions in which an external variable, such as the pressure drop across a capillary or the angular velocity of the one cylinder of a Couette instrument, undergoes some programmed time variation. One then measures some features of the transient response of the fluid and infers therefrom the properties of the fluid under examination.

Two flows that particularly lend themselves to dynamic measurements are the oscillatory flow and the relaxational flow. In the former experiment the system has imposed upon it sinusoidal external conditions of known frequency and amplitude. The frequency, amplitude, and phase of the response of the fluid are measured. In a relaxational flow external conditions, such as force or strain, undergo a rapid change from one steady state to a second steady state. The response of the fluid as it approaches a new equilibrium state is then measured.

In both types of flows, by appropriate measurements, it is possible in principle to determine the viscosity and the normal stress coefficients for the fluid. If the experiments are properly designed and interpreted these coefficients should be identical to those measured in some steady-state configuration. It has been more common, however, to define new coefficients that are peculiar to the particular type of dynamic test undertaken. It is important, then, to investigate the possibility that these new coefficients might be related to the viscosity and normal stress coefficients, or otherwise to establish these coefficients properties of the material.

Dynamic experiments are of interest for a number of reasons. Historically, because polymeric fluids are recognised to exhibit time effects such as stress relaxation, it has been natural to examine such effects directly through stress relaxation experiments. Furthermore, a very severe test of the applicability of a proposed constitutive equation is its ability to predict both steady-state and dynamic behaviour. Such tests are often useful in differentiating among models which make similar predictions for steady-state flows, but differ significantly in predicted dynamic response. Finally, it is often possible to achieve a time scale in a dynamic experiment which is not easily accessible to a steady-state experiment. Hence, dynamic experiments are useful for extending the range of time scale over which a particular material is examined.

First, consider the problem of oscillatory flow. Consider a fluid filling the region between long concentric cylinders. Suppose the outer cylinder oscillates with frequency $\omega(s^{-1})$ and amplitude ϑ_o. The angular velocity ω, the shear rate $\dot{\gamma}$, and the shear stress τ will exhibit sinusoidal behaviour of frequency $\tilde{\omega}$. In general these quantities will not be in phase with the input oscillation. If the inner cylinder is free to respond, for example, its angular velocity might be

$$\omega_i = \Omega_i \cos(\tilde{\omega}t + \delta) \qquad \text{... (4.55)}$$

while the input oscillation would be

$$\omega_o = \Omega_o \cos\tilde{\omega}t \qquad \text{... (4.56)}$$

A convenient representation of such oscillating quantities is in terms of complex exponentials. Equation 4.55 may be written

$$\omega_i = \text{Re}\{\Omega_i e^{i\tilde{\omega}t}\} \qquad \text{... (4.57)}$$

where $\Omega_i = \Omega_{iR} + i\,\Omega_{iI}$ is a complex amplitude and Re { } means "the real part of { }". In terms of these quantities it may be shown that

$$\Omega_i = \sqrt{\Omega_{iR}^2 + \Omega_{iI}^2} \qquad \text{... (4.58)}$$

and

$$\delta = \tan^{-1}(\Omega_{iI}/\Omega_{iR}) \qquad \text{... (4.59)}$$

Similarly, the shear stress and shear rate are

$$\tau = \text{Re}\{Te^{i\tilde{\omega}t}\} \qquad \text{... (4.60)}$$

and

$$\dot{\gamma} = \text{Re}\{\dot{\Gamma}e^{i\tilde{\omega}t}\} \qquad \text{... (4.61)}$$

By analogy with earlier definitions one might define a viscosity coefficient as $\eta = \tau/\gamma$. Because τ and $\dot{\gamma}$ are not necessarily in phase with one another such as definition leads to a time-dependent viscosity which then does not really represent a material property. Instead, it is common to define a complex viscosity coefficient as

$$\eta^* = T/\dot{\Gamma} = \eta' - i\eta'' \qquad \text{... (4.62)}$$

The parameter η' is the 'dynamic viscosity' and is a function of frequency $\tilde{\omega}$ in essentially the same way that the steady shear viscosity η is a function of shear rate. The parameter η'' is found to be a measure of elastic response of the material, and can be shown to be related to the steady shear normal stress coefficient ψ_{12}. A common method of operation is to have the bob suspended from a thin torsion wire. The outer cylinder, or cup, oscillates and, after initial transients damp out, a sinusoidal motion is induced in the annular body of fluid concentric with the cup and bob. A sinusoidal torque is exerted upon the bob which is opposed by the elasticity of the suspending wire. At equilibrium the bob oscillates with an amplitude and phase which depend upon the fluid properties, the amplitude and frequency of the cup motion, the moment of inertia of the cup, and the torsion constant of the wire.

The cup motion may be arbitrarily taken to be strictly cosinusoidal, so that ϑ_o (angular deflection from equilibrium) is strictly a real quantity. The bob response is, in general, out of phase with the cup and so ϑ_1 is complex quantity: $\vartheta_i = \vartheta_R + i\,\vartheta_1$. The primary measurements consist of the amplitude ratio $q = |\vartheta_i|/\vartheta_o$ and phase difference $\delta = \tan^{-1}(\vartheta_1/\vartheta_R)$. From these two quantities, at any particular frequency $\tilde{\omega}$, the coefficients η' and η'' may be determined. By performing experiments over a range of frequencies, the frequency dependence of η' and η'' may be determined. This analysis laeds to the following:

$$\eta' = -\frac{Gq\sin\delta}{1+q^2-2q\cos\delta} \qquad \text{... (4.63)}$$

$$\eta'' = \frac{G(q\cos\delta - 1)}{1+q^2-2q\cos\delta} \qquad \text{... (4.64)}$$

where

$$G = \frac{I\tilde{\omega}^2 - k}{4\pi L\tilde{\omega}}\frac{R_o^2 - R^2}{R_o^2 R^2} + \frac{\tilde{\omega}\rho}{8}\frac{(R_o^2 - R^2)^2}{R_o^2}$$

It is also possible to perform an experiment in which the cup is fixed and the upper end of the torsion wire is twisted periodically through an angle $\psi = \psi_o \cos\tilde{\omega}t$. The bob then performs periodic oscillations of frequency but of different amplitude and phase, so that $\vartheta_i/\psi_o = qe^{-i\delta}$. Again, measurement of the amplitude ratio q and the phase angle δ permits calculation of η' and η''. For this case

$$\eta' = \frac{R_o^2 - R^2}{4\pi LR_o^2R^2}\left[I\tilde{\omega}^2 - k + \frac{k\cos\delta}{q}\right] \qquad \text{... (4.65)}$$

$$\eta'' = \frac{R_o^2 - R^2}{4\pi LR_o^2R^2}\frac{k\sin\delta}{q} \qquad \text{... (4.66)}$$

subject to the approximation $\rho\tilde{\omega}R_0^2/|\eta^*| << 1$.

It is common practice among those who investigate solid-like materials to define material coefficients based upon the shear strain γ rather than the shear rate $\dot{\gamma}$. In a sinusoidal experiment γ would be written as

$$\gamma = \text{Re}\{\Gamma e^{i\tilde{\omega}t}\} \quad \text{... (4.67)}$$

However, because $\dot{\gamma}$ = dγ/dt, γ could also be written in the form

$$\gamma = \text{Re}\{(\dot{\Gamma} / i\tilde{\omega})e^{i\tilde{\omega}t}\} \quad \text{... (4.68)}$$

A 'complex shear modulus' is defined as G* = T/Γ . Since Γ = i $\dot{\Gamma}$ it follows that G* = i$\tilde{\omega}\eta$*. G* is usually written as G′ + iG″, where G′ is called the 'storage modulus' or 'dynamic rigidity', and G″ is called the 'loss modulus'. As noted above, G′=$\tilde{\omega}\eta$″. It can be seen also that G″ = $\tilde{\omega}\eta$′.

A 'complex compliance' J* may also be defined as J* = J′ – iJ″ = Γ/T. By simple manipulation it can be shown that J′ = G″/(G′^{2} + G″^{2}) and J″ = G″/(G′^{2} + G″^{2}); J′ and J″ are called the storage and loss compliances, respectively.

CREEP AND RELAXATION

It is important to note that given material may show different classes of the mechanical response under conditions involving different time scales. Consider, for example, a sheet of polymer suddenly subjected to a constant shear stress τ by the action of a force pulling on one plate bonded to the polymer; as in Fig. 4.7, limiting responses are possible. If the material is a true elastic solid, then the shear strain γ immediately attains and remains at a constant value as long as the stress is maintained. If the stress is removed the strain relaxes instantaneously and the solid returns to its initial unstressed state. If the material, however, is a fluid, then the shear strain continuously increases with time as long as the stress is maintained. If the stress is suddenly removed the strain does not relax, but instead remains constant. An irreversible deformation, flow, has occurred. Fig 4.8 illustrates these responses.

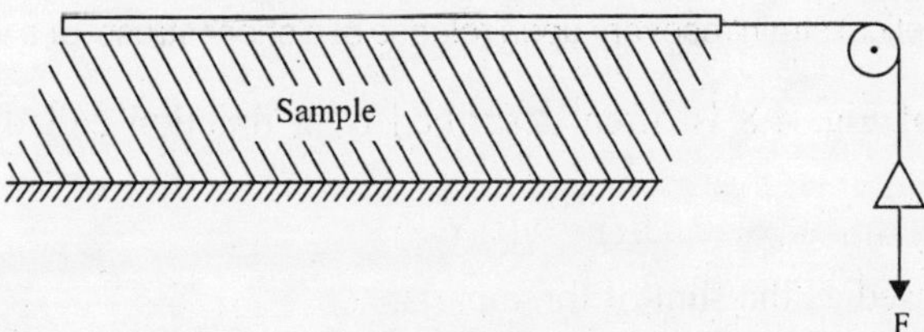

Fig. 4.7. Definition sketch for a shear creep experiment at constant shear stress.

In most polymeric materials neither of these simple responses occurs. Instead, a combination of elastic and viscous behaviour gives rise to the strain history indicated in Fig. 4.9. One associates such behaviour with viscoelastic materials.

If a material is observed over a very short time scale compared to the time λ of Fig. 4.8 it is likely that the observer would conclude that the material is an elastic solid. In contrast, if the time λ is very short compared to the response time of suitable measuring instruments, one is likely to conclude that the material under examination is a fluid. For many polymeric materials, at least in certain ranges of temperature and molecular weight, the complete response curve is accessible to measurement, and so viscoelastic parameters may be measured.

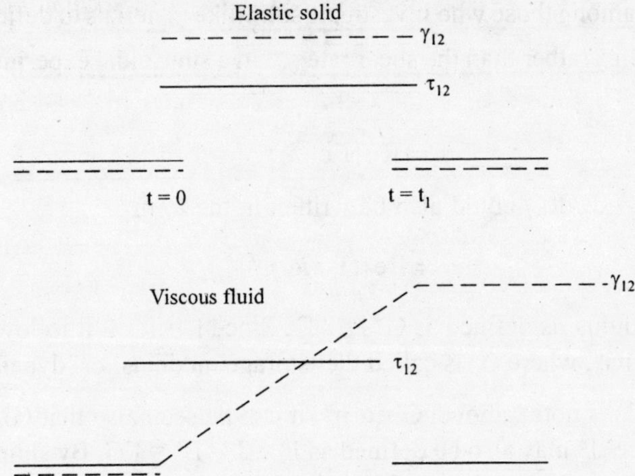

Fig. 4.8. Hypothetical shear strain response to a constant shear stress imposed for a finite time. The perfectly elastic solid shows instantaneous and complete recovery of strain upon release of stress. The purely viscous fluid shows no recovery of strain.

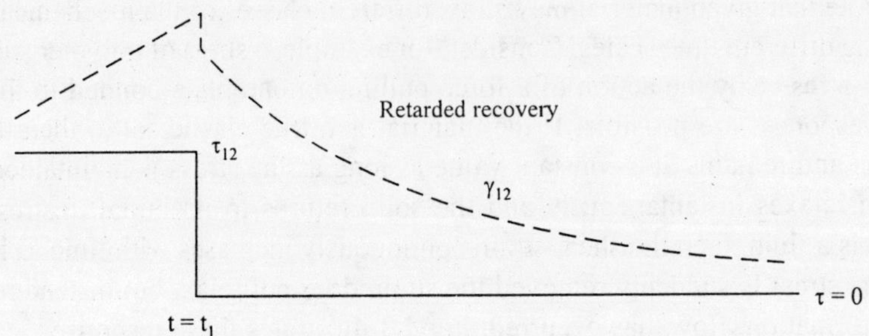

Fig. 4.9. Hypothetical strain recovery upon release of a shear stress for a viscoelastic fluid.

The response curve of Fig. 4.8 is often described by a function called the 'creep compliance', defined as

$$J(t) = \gamma(t)/\tau \qquad \text{... (4.69)}$$

and is typically expressed as the sum of three parts:

$$J(t) = J_o + J\psi(t) + t/\eta \qquad \text{... (4.70)}$$

J_o is the 'ordinary elastic compliance' and represents the initial elastic response to stress. The term t/η represents the long time flow response and $\psi(t)$ is called the 'retarded elasticity function', and is defined in such a way that it approaches unity as time increases indefinitely. J is referred to as the steady-state shear compliance, and like J_o and η, is a material constant for a given shear stress. The retarded elasticity function, while not a material constant, is nevertheless a characteristic property of a material. An experiment giving rise to a response such as is shown in Fig. 4.8, is referred to as a 'creep experiment'.

If the material is in elastic equilibrium by virtue to some previous creep behaviour followed by steady flow, and if the stress causing this flow is suddenly removed, then retarded elastic recovery occurs, as shown in Fig. 4.9. The strain relaxation history is then the reverse of the creep curve.

Defining $\gamma(0)$ as the strain at the instant of stress removal

$$\gamma(0) - \gamma(t) = \tau[J_o + J\psi(t)] = \tau[J(t) - t/\eta] \quad \text{... (4.71)}$$

where τ represents the constant stress that led to $\gamma(0)$. Note that in strain recovery there is no flow mechanism because the external stress no longer acts on the material. An obvious test of consistency of any creep experiment is the prediction of the recovery curve from measurements, under creep, of J_o, J, and $\psi(t)$.

The viscosity of the material can be obtained from the long time portion of the creep curve. Because it is practically impossible to measure viscosity of such rubber-like materials by simple techniques such as capillary flow or cone-and-plate shearing, which must be restricted to relatively low viscosities, the creep experiment provides an important extension of viscometry methods to the region of extreme viscosities. Because the time scale of the creep experiment is so low, these viscosities are "zero-shear" viscosities and hence the polymer is essentially Newtonian in its flow behaviour under these conditions.

Another instrument used to measure viscoelastic properties is the torsional pendulum. A circular disk hung from a torsion wire is in contact with the material to be studied. The disk is displaced through a small angle about its axis, the torsion in the wire tends to restore the disk to its equilibrium position, but the inertia of the disk causes it to overshoot the equilibrium position. Thus, the disk performs torsional oscillations of fixed frequency and damped amplitude. The frequency and damping factor of the oscillations are measured, and may be related to instrument constants and properties of the material.

The creep experiments and their corresponding dynamic coefficients are related to shear behaviour. Elastic materials are commonly studied in tension as well as in shear, and it is common to study viscoelastic materials in tensile experiments. Thus, one may study elongation (creep) under constant tensile load or tensile stress relaxation at constant tensile strain.

A simple relaxation experiment involves the sudden stretching of a strip of a material to a fixed elongation. The tendency for creep is counteracted by continually reducing the force maintaining the elongation in such a way that the initial elongation is not altered. From the resulting data of stress vs. time it is possible to calculate a relaxation modulus.

As in the case of shear creep, tensile creep experiments may be used to measure the viscosity of semi-solid materials. Viscosities as high as 1015 P can be measured. Tensile creep is similar to the simple elongation experiment. The difference lies in the fact that in elongational flows the strain is programmed and the measured response is the stress (force), while in the creep experiment the force is usually fixed and the measured response is the strain.

CHAPTER 5

Rheometric Techniques for Polymer Melts

INTRODUCTION

The relationship between stress and shear rate, and therefore the dependence of apparent viscosity upon shear rate, can be determined over a wide temperature range by a variety of techniques that utilise some of the simple pressure or shear geometries. These include capillary and Couette rheometry, which are based upon simple pressure flow through a capillary and simple shear flow through two rotating, concentric cylinders, respectively. Other common methods include cone and plate and parallel-plate rheometry, which can also give information concerning normal stresses through measurements by force transducers mounted in the direction normal to the plane of shear. Slit rheometers, which can be used to measure exit pressures in pressure flow through a rectangular channel, can be used to measure normal stress under typical processing conditions.

PRACTICAL CONSIDERATIONS OF RHEOLOGY

The response of thermoplastics during processing can be categorised by three classes of property:

1. Deformation processes, which are necessary to form the product.
2. Heat and heat transfer, which are necessary to achieve the plastic state.
3. Chemical and physical change, which may be deliberately induced or adventitious.

The subject of rheology is concerned with the study of deformation and flow: of all the responses of materials it is the one that is most readily observed. We have all squeezed toothpaste, kneaded bread dough, or tried to wipe glue from our fingers. As noted in earlier chapters, rheology is one way of describing those sensations. The fundamentals of rheology are drawn from mechanics and provide the support for this study, which is mainly concerned with how materials systems differ from the ideals of classical mechanics. One approach to rheology is to broaden the field of classical mechanics by defining more generalised materials, the properties of which are derived logically from an equation of state. The advantage of this approach is that it offers full predictive capacity in any deformation history once the equation of state is described by appropriate experiments. At the other end of the spectrum lies the view that the interaction between complex materials and complex histories provides a series of unique situations that can only be studied in their own environment. An intermediate course is to evaluate the response of a material system in controlled experiments which are qualitatively similar to elements of its processing history and to discover from such experiments what the properties of the material appear to be.

Three material states are relevant to polymer processing: granular, the form in which materials are fed to the process; melt, the form in which they are usually shaped; and solid, the form of the final product, but also one in which some shaping may take place. This chapter deals with the rheology of melts, where most of the deformation occurs. However, in illustrating material property data some discussion on granular and solid response are included where these properties are relevant to processing.

Rheology is concerned with the relationship between stress, strain and time. In reviewing some of the principles presented in earlier chapters, there are three simple deformations:

1. In simple shear the stress is applied tangentially (Fig. 5.1):

Stress (N/m²) $\sigma_s = F/A$ ...(5.1)

Strain (unity) $\gamma = x/h$ ...(5.2)

Rate of strain (s^{-1}) $\dot{\gamma} = \frac{1}{h}\frac{dx}{dt} = \frac{v}{h}$ ...(5.3)

2. In simple extension the stress is applied normal to the surface of the material, as shown in Fig. 5.2.

Stress (N/m²) $\sigma_E = F/A$ ...(5.4)

Strain (unity) $\in = \int_{\ell_0}^{\ell_1} \frac{dl}{\ell}$ ...(5.5)

Rate of strain (sec^{-1}) $\dot{\in} = v/\ell$ ...(5.6)

3. In bulk deformation the stress is applied normal to all faces. The stress is the applied pressure, P, and the strain is the change in volume per unit volume, $\delta V/V$.

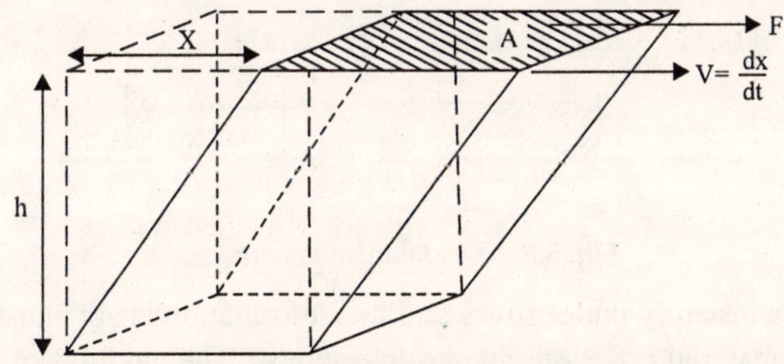

Fig. 5.1. Simple shear: area A and distance h remain constant during deformation.

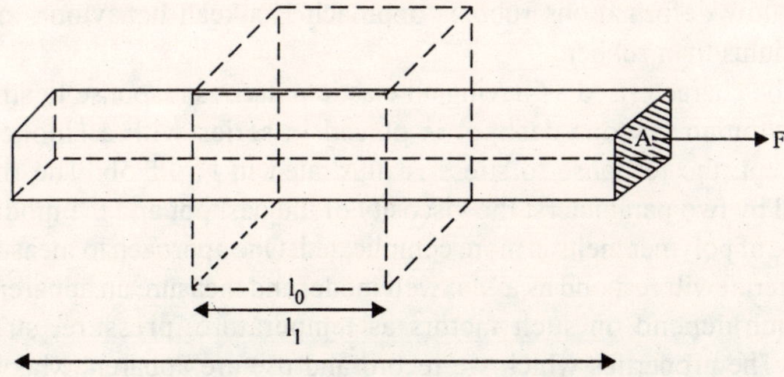

Fig. 5.2. Simple extension: cross-sectional area A and sample length l both vary during deformation.

In the case of polymers, the response to a simple deformation process may be complex. Thus, in a simple shearing flow not only is there a shear stress, but also a pull along the lines of flow if usually described as the normal stress. The practical deformations of polymer processing are themselves usually complex flows compounded of shear, extension, and bulk deformations. One solution to this complexity is to introduce tensor notation which is the starting point of many rheological treatises. For the purposes of this discussion, it is sufficient to recognise and remember that those complexities exist; with that appreciation it is possible to seek simplifications in the practical response of real materials.

There are three types of response to an applied stress: viscous flow, elastic deformation, and rupture.

In viscous flow a material continues to deform as long as the stress is applied and the energy put in to main the flow is dissipated as heat. Viscosity is defined as the ratio of stress to rate of strain (units of Newton second per square metre).

If the only response of a material is to flow with a viscosity that is independent of stress level, the material behaves as a Newtonian liquid. At very low stress levels many polymer melts approach Newtonian behaviour, which is illustrated in Fig. 5.3.

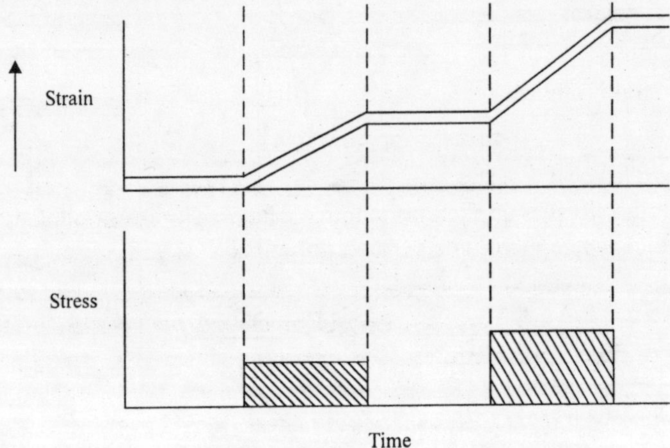

Fig. 5.3. Newtonian behaviour.

If the material deforms instantly under stress and the deformation is spontaneously reversed when the stress is removed, the material possesses an elastic response. The modulus of such a material is the ratio of stress to recoverable deformation (units of Newtons per square metre). If all the deformation is reversible and the deformation is proportional to the applied stress the material is said to have a Hookean response (Fig. 5.4). At low deformations rubbers approach Hookean behaviour, and many polymer melts have a lower modulus than rubber.

Polymer melts may be characterised as having an elastico-viscous response to stress. The Maxwell model in which a Newtonian viscous dashpot is placed in series with a Hookean elastic spring (Fig. 5.5a) is one concept; the response to stress is illustrated in Fig. 5.5b. The rheology of such a material is characterised by two parameters: the viscosity of the dashpot and the modulus of the spring. The rheological response of polymer melts is more complicated. One approach to measuring that rheology is to assume that the material will respond as a Maxwell model and measure an apparent viscosity and an apparent modulus which depend on such factors as temperature, pressure, stress, geometry of deformation, and time. The properties which we record and use are apparent Maxwell viscosity and apparent Maxwell modulus.

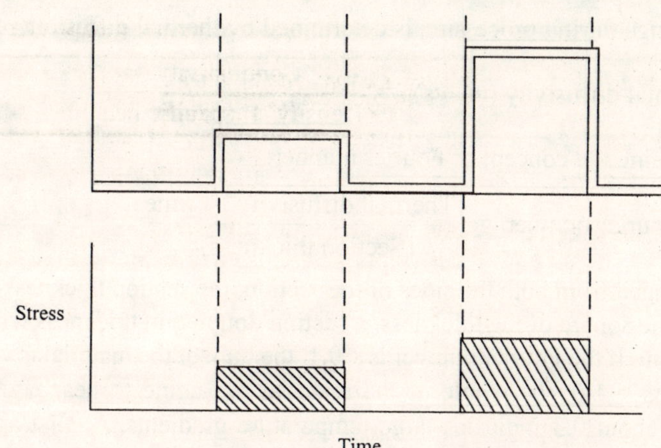

Fig. 5.4. Hookean behaviour.

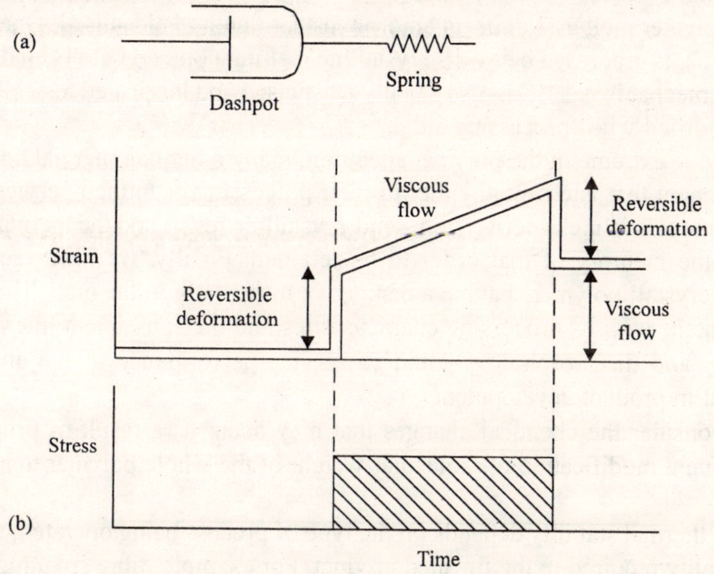

Fig. 5.5. (a) Maxwell model; (b) its response to stress.

Polymer melts are commonly very brittle. Rupture plays an important role in determining the maximum rate at which a deformation process proceeds. In the practical response of polymer processing it is frequently the interactions between viscosity, elasticity, and rupture phenomena that determine the success of the operation.

In considering the bulk compressibility of materials the thermodynamic properties have already been acknowledged, and in viscous flow, where the work input is dissipated as heat, density and specific heat at constant pressure are immediately invoked. For a proper understanding of polymer processing we require a full appreciation of the pressure:volume:temperature relationships, or the material together with the heat transfer characteristics. It is important to note that polymers are excellent insulating materials. All those properties may be measured and handled by the techniques used for conventional materials.

The rate of heat exchange during processing is determined by thermal diffusion:

$$\text{Thermal diffusivity } (m^2/s) = \frac{\text{Conductivity}}{\text{Density} \times \text{specific heat}} \qquad ...(5.7)$$

which is readily applied using the concept of Fourier number:

$$\text{Fourier number} = \frac{\text{Thermal diffusivity} \times \text{time}}{(\text{Section thickness})^2} \qquad ...(5.8)$$

(If heat is being exchanged from both the sides of the section, the section thickness is halved). Since heat transfer depends on the square of the thickness, a section doubled in thickness will take four times to reach thermal equilibrium. If the Fourier number is < 0.1; the bulk of the material is very little affected by its environment; if it is > 1.0, the whole of the material has come to near equilibrium with its surroundings. A value of about 0.3 indicates large temperature gradients. At first approximation the thermal diffusivity of polymer melts 10^{-7} m^2/s. From this result it may be readily deduced that a section 2 mm thick, cooled from both sides, will approach thermal equilibrium in 10 s, while a 20 mm section will require about 15 min.

The concept of a polymer melt as a uniform amorphous continuum is idealised. One bulk polymer, polyvinyl chloride (PVC), is processed below its crystalline melting point and has been demonstrated to have a particulate flow mechanism. PVC is also usually plasticised and lubricated to achieve easier flow and may be further modified with 'processing aids'.

While PVC is near one extreme of the polymer spectrum, many other commercial polymers include some additional component that is essential either to their processing or to their service performance. Many commercial thermoplastic melts derive from semi-crystalline resins and the thermal history of the process may allow some memory of that order to be retained. Finally, we have recently seen the introduction of 'liquid crystal' polymers having a distinctive mesophase in the melt. The interaction of the morphology of a 'melt' with its processing characteristics and the subsequent interaction between that processing history and the morphology—and so service performance of the end product is a subject of great interest in product development.

It is important to consider the chemical changes that may occur as a result of processing, which may range from significant modification of molecular weight of the whole polymer to specific surface conditioning.

The requirement of thermal stability depends on the type of process being operated, the scale of the process, and on the quality required in the finished product. For example, fibre spinning, polymers are deliberately degraded in order to reduce the molecular weight to obtain a very low viscosity polymer melt having a narrow molecular weight distribution. More often the requirement is for good thermal stability. The more sensitive processes are continuous extrusion processes and (especially) processes requiring a stretching of the melt—film blowing and blow moulding. The least sensitive processes are fully constrained, intermittent flows such as injection moulding.

The typical time scale for a processing operation is on the order 3 to 5 min from feeding granules to solidified product. This mean time masks a residence time distribution such that some 5% of the material will have a residence time on the order of 10 min, and a very small amount will have a very long residence time. The problem of residence time distribution is accentuated in processes such as sheet or film extrusion in which a pocket of degraded material may be disturbed from a dead space and locally lubricates the flow. Several commercial processes run at a high level of reclaim (blow moulding is typically 50%) so that a simple residence time of 5 min becomes greatly extended when reclaim is taken into consideration. The problem of residence time distribution is also accentuated if the degradation mechanism is autocatalytic. Because of residence time distribution around protrusions, sometimes accentuated by catalysis from the metal parts, mechanical and appearance defects due to weld lines are common in easily degraded polymers.

The scale of a processing operation also alters the requirement for thermal stability. During extrusion from a 25 mm diameter extruder, it is possible to maintain the melt temperature to within 5°C of the set barrel temperature and within 10°C of the melting point. In larger machines running fast (more typical of commercial practice), heat generation and low thermal conductivity result in a mean melt temperature usually at least 30°C above the melting point—local hot spots may be up to 20°C higher. Heat generation is a greater problem when working with polymers of high molecular weight. Running the extruder cold accentuates heat generation (and hence hot spots); for any material prone to degradation there is an optimum machine temperature.

RHEOLOGY AND POLYMER STRUCTURE

The rheology of a polymer strongly depends on the chemical and morphological structure of the material. If we consider the balance between service properties and ease of processing, a qualitative inverse relationship may be developed, as shown in Fig. 5.6. That field shows three areas: along the line, the region of common experience; above the line, the target of a successful product; below the line, the disaster region. Movement toward the target may be achieved by altering the structure which may itself be modified by processing history. It is important to gain appreciation of the structural features that influence flow.

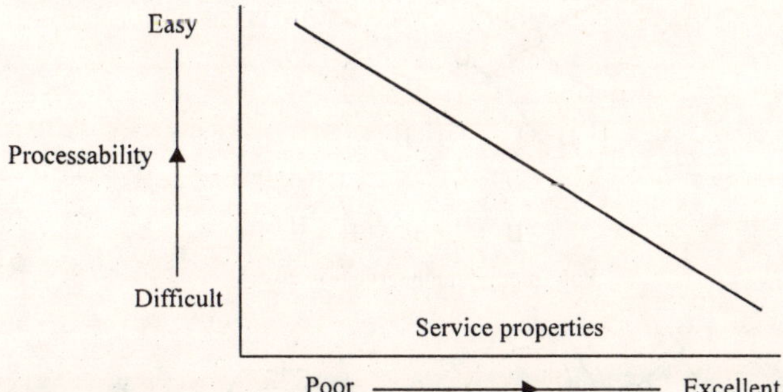

Fig. 5.6. Relationship between service properties and ease of processing.

Part of this appreciation of how chemical structure influence chain flexibility, and thus rheology, may be obtained through molecular visualisation of polymers, as shown in Fig. 5.7. This means the molecule of polyethylene is easily recognised by highly flexible, with full freedom to rotate about each of the bonds. By comparison, polyparahydroxybenzoic acid has very little freedom of movement because of the rigidity of the aromatic links, and is intractable as a melt requiring powder metallurgy to shape.

In addition to the stiffening of the main chain, the molecule may be stiffened by adding side groups as in poly(methyl methacrylate), which greatly restricts the freedom of the molecule.

To determine the influence of chain stiffness and conformation on rheology quantitative description of molecular size is first needed. For a polymer of molecular weight M the length of the molecule L, is obtained from

$$L = M(L_0/M_0) \qquad ...(5.9)$$

where L_0 is the length and M_0 is the molecular weight of the repeat unit.

The space that a chain occupies, as distinct from its molecular volume, is obtained first from a consideration of the number of molecules, N, in 1 kg of polymer:

$$N = \frac{1000 \times 6.02 \times 10^{23}}{M}, \quad \text{molecules / kg} \qquad ...(5.10)$$

(6.02×10^{23} being Avogadro's number), in which the volume of one molecule, V, is given by

$$V = (\rho N)^{-1} \qquad ...(5.11)$$

where ρ is the density.

The effective diameter of the chain, D,

$$D = \left[\frac{4}{\pi}(L_0 / M_0)^{-1}(\rho \times 6.02 \times 10^{26})^{-1}\right]^{1/2} \qquad ...(5.12)$$

and $V = \pi D^2 L/4$.

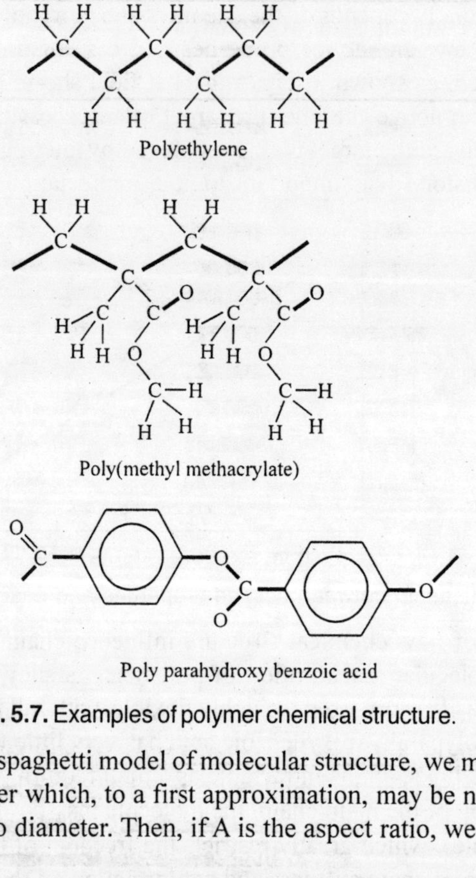

Polyethylene

Poly(methyl methacrylate)

Poly parahydroxy benzoic acid

Fig. 5.7. Examples of polymer chemical structure.

Using the so-called polymer spaghetti model of molecular structure, we may anticipate a dependence on molecular length and diameter which, to a first approximation, may be normalised by considering a constant aspect ratio of length to diameter. Then, if A is the aspect ratio, we may deduce the molecular weight

$$M = AD(L_0/M_0)^{-1} \qquad ...(5.13)$$

Consider molecules of aspect ratio 1000 as typical of molecules having commercially useful properties. This aspect ratio is based on weight average molecular weight, M_w, for linear polymers of narrow molecular weight distribution ($M_w/M_n = 2$) (which is naturally obtained in the polymerisation of many materials).

Taking a simplified view of the rheology, in the form of three parameters: η_o, the viscosity at a shear stress of 10^3 N/m^2(Ns/m^2); G_o, the modulus at a shear stress of 10^3 N/m^2 (N/m^2); and $\sigma_{1/2}$, and shear stress at which $\eta/\eta_o = 1/2$. $\sigma_{1/2}$ is a description of non-Newtonian behaviour, and from these terms we may derive the parameters

$$\tau_o = \eta_o/G_o \quad ; \gamma_C = \sigma_{1/2}/G_o \qquad ...(5.14)$$

The former is a characteristic of natural time, the latter is an elastic shear strain or orientation function.

Melt rheology depends greatly on temperature. As already discussed it is thus necessary to make any comparison of rheologies on the basis of comparable temperatures. The choice of a common fixed temperature is not appropriate, because at any given temperature some polymers decompose while others are not molten. The choice of the temperature at which the melt is most commonly processed must be arbitrary; polyethylene, for example, is commonly processed at between 150° and 300°C. For this comparison we select a temperature 200°C above the glass transition temperature, T_g, a temperature that lies within the processing range of many commercial polymers and one at which the molecules may be considered to experience a common state of excitation.

The shear modulus, G_o, varies by more than an order of magnitude for different polymers. That difference is readily reconciled with what we know of the chemical structure. Taking polyethylene as a standard we may attribute the chain stiffness to the flexibility of the backbone. Other molecules [polypropylene, poly(4-methyl pentene-1), polystyrene] have the same backbone, but because of the bulky side groups, have a larger cross-sectional area. Because the stress can only be supported through the backbone we may anticipate an approximate inverse relationship between D^2 and G_o. We also see molecules, e.g., polyethylene terephthalate, which are much stiffer than polyethylene. Here, qualitative inspection of the chemical structure reveals the presence in the backbone of inflexible aromatic units that naturally stiffen the molecule.

The viscosity of melts of polymers of the same aspect ratio at 200°C above T_g likewise varies by an order of magnitude, but that variation is in direct correlation with the variation in modulus, so that the ratio η_o/G_o is approximately constant. From this we can deduce that under these conditions viscosity depends only on the stiffness of the chain.

Further, the non-Newtonian character of the melt, as defined by the shear stress at which the viscosity has decreased to one half its low shear value, also varies directly with chain stiffness, implying a positive correlation between non-Newtonian flow and orientation.

The polymer that seems least consistent with the others, Nylon 66, is one that differs in molecular interaction by possessing strong hydrogen bonding. This almost certainly raises T_g and thus depresses viscosity at T_g + 200°C, and also has the effect of increasing non-Newtonian flow. A second polymer that falls outside the theory of the others is poly(methyl methacrylate), which appears to have a surprisingly high viscosity, possibly reflecting the stearic hindrance of the bulky side groups.

There are two common ways in which the viscosity of a polymer is reduced: plasticisation and the incorporation of a more flexible co-monomer unit in the backbone. Both courses reduce the glass transition temperature and hence the viscosity at a given processing temperature. The chain stiffness, and thus viscosity, is further reduced by increasing the flexibility of the backbone by the use of the co-monomer, or by increasing the effective cross-sectional area of the molecule by the incorporation of plasticiser. Finally, the increased chain flexibility leads to an enhancement of non-Newtonian behaviour and consequently a further reduction in viscosity under any given high shear condition.

Molecular weight is the single most important parameter in determining the viscosity of a polymer. A factor of two in molecular weight produces a tenfold change in the viscosity at a given shear stress in the range of molecular weight, M, of interest for thermoplastics. This relationship can be expressed as

$$\eta = kM^{3.5} \qquad ...(5.15)$$

Equation (5.15) is true for a narrow range of molecular weights; a more appropriate relationship is

$$\eta = k_1M + k_2M^3 + k_3M^5 + . . . \qquad ...(5.16)$$

In general, non-Newtonian flow is more pronounced in polymers of higher molecular weight. This feature may be qualitatively attributed to the argument that the greater the average number of entanglements per chain, the greater the probability of there being a region of intense entanglement which is broken down under lower average stress.

Molecular weight may be defined in many ways, of which weight average M_w and number average M_n are the most common:

$$\overline{M}_w = \Sigma M_i N_i^2 / \Sigma M_i N_i \qquad ...(5.17a)$$

$$\overline{M}_n = \Sigma M_i N_i / \Sigma N_i \qquad ...(5.17b)$$

The question as to which of these controls the viscosity is open. The answer can in part be gained by making physical blends of different molecular weights. For flexible polymers such as poly(4-methyl pentene-1), the longer chains appear to play the dominant role; for more rigid molecules such as poly(methyl methacrylate), it is the shorter chains that play the controlling role.

Polymers of lower molecular weight have lower viscosity and so are easier to shape. All other things being equal, polymers of higher molecular weight will have superior strength properties, thus the optimum polymer for any application requires a balance to be struck between processibility and performance. The expectation that service performance can always be enhanced by increasing molecular weight is confused by the fact that in polymer processing, all else is not equal, especially with respect to such features as the degree of orientation induced during processing.

Orientation will tend to be higher when processing polymers of higher molecular weight and, while that orientation may enhance strength in the flow direction, this gain is commonly at the expense of a loss of properties in the transverse direction. There may thus be an optimum molecular weight to obtain the best possible service properties.

In addition to molecular weight, molecular weight distribution (MWD) is also an important contributor to rheology. MWD is usually represented by the polydispersity index as the ratio of weight to number average molecular weight.

When a sample has a narrow MWD, all the chains may be assumed to have a similar degree of entanglement so that they are uniformly stressed during flow. In a sample of broad molecular weight distribution, the longer chain molecules appear to form a protective network around their shorter brethren, as illustrated in Fig. 5.8.

On the assumption that it is the entanglements that resist deformation, the long chains in the polymer of broad molecular weight distribution accept a disproportionately large amount of the total stress. This above average stress causes above average elastic response in those molecules, and this response is retarded by viscous resistance as the smaller molecules conform.

Thus, in a sample of broad MWD there is usually more elastic response than in a sample of similar viscosity with narrow MWD, but that elastic response is delayed (Fig. 5.9).

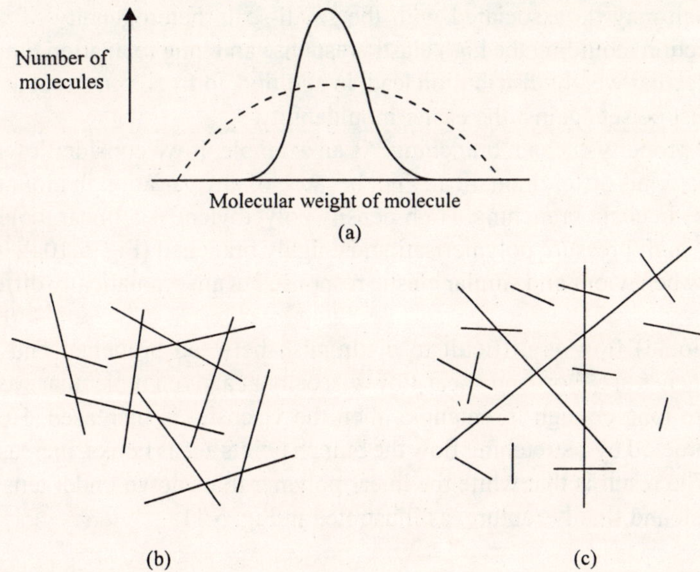

Fig. 5.8. (a) Molecular weight distribution, broad MWD, narrow MWD; (b) narrow MWD, about 2 entanglements per chain: (c) broad MWD, long chain = 4 entanglements, short chain = 0 to 1 entanglements.

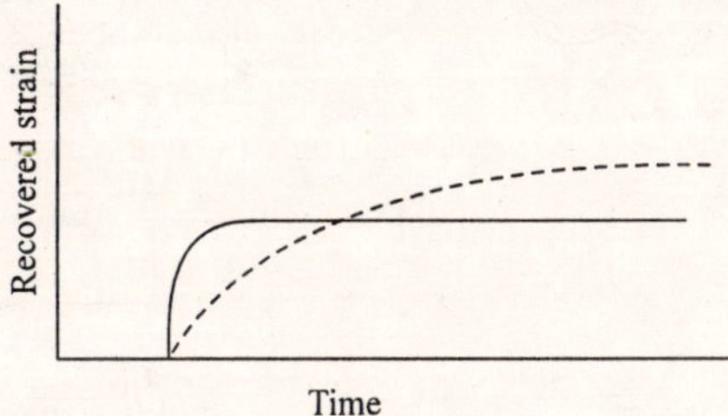

Fig. 5.9. Effect of MWD on strain recovery after steady flow–broad MWD, narrow MWD.

If it is stress that orients the molecules and thus reduces entanglements, then another consequence of the uneven stress distribution is that the entanglements in the longer chains start to disentangle at a lower average stress level.

Thus, if a small high molecular weight tail is polymerised on top of a sample having low molecular weight the resulting flow curve shows two distinct transitions of non-Newtonian response: the first occurs at a low average stress when that average stress is largely taken by the few long molecules, placing them under high local stress; the second takes place at the same stress level as the unmodified low molecular weight polymer. The more general response for polymers of broad, rather than bimodal, MWD shows a more gradual transition.

Note that while the two polymers illustrated have approximately the same viscosity under melt flow index conditions, the viscosity at very low shear stress may be several orders of magnitude greater for the polymer of broad molecular weight distribution, while at high shear that polymer has significantly easier flow. In selecting materials for such applications as blow moulding, high shear processibility and low stress form stability are commonly enhanced by broadening the MWD. However, these advantages are usually offset by poorer strength properties associated with the low molecular weight tail, and a loss of surface finish which may be associated with the small-scale heterogeneity of such materials. In processes such as injection moulding the high elastic response and long relaxation times associated with polymers of wide molecular weight distribution lead, in addition, to frozen-in strain and the possibilities of warping, which must be set against the easier mouldability.

Another important property is chain branching. As an example, if we consider low- and high-density polyethylene, these materials differ dramatically not because of any variation in monomer or MWD, but because of a difference in chain branching. High-density polyethylene is a linear molecule; low-density polyethylene made by high-pressure polymerisation, is highly branched (Fig. 5.10a). The two polymers have similar shear flow behaviour and similar elastic response but are dramatically different in stretching flows.

In shearing (rotational) flow is difficult to distinguish between branched and linear molecules (Fig 5.10b). If the branches are short and shear flow viscosity at a given molecular weight is depressed, but if the branches are long enough to entangle, then the viscosity is increased. However, when the molecules are highly ordered by a stretching flow the branch points act as hooks, increasing the resistance to flow (Fig. 5.10c). The result is that while the linear polymer thins down under tension, the branched polymer tends to stiffen and finally rupture, as illustrated in Fig. 5.11a.

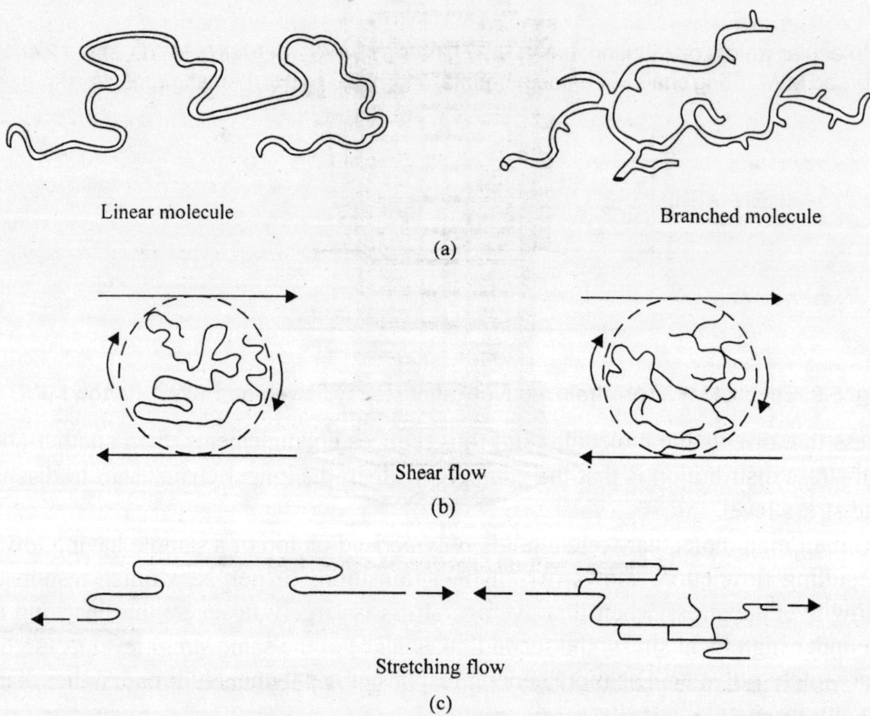

Fig. 5.10. (a) Branched and linear molecules; (b) shear flow; (c) stretching flow.

Polymer melts that are superficially similar in other rheological tests may be 100-fold different in their resistance to high stress extensional flows. The tendency for the elongational viscosity of branched polymers to increase with stress means that local stress concentrations are less critical in establishing stretching flows (Fig. 5.11b).

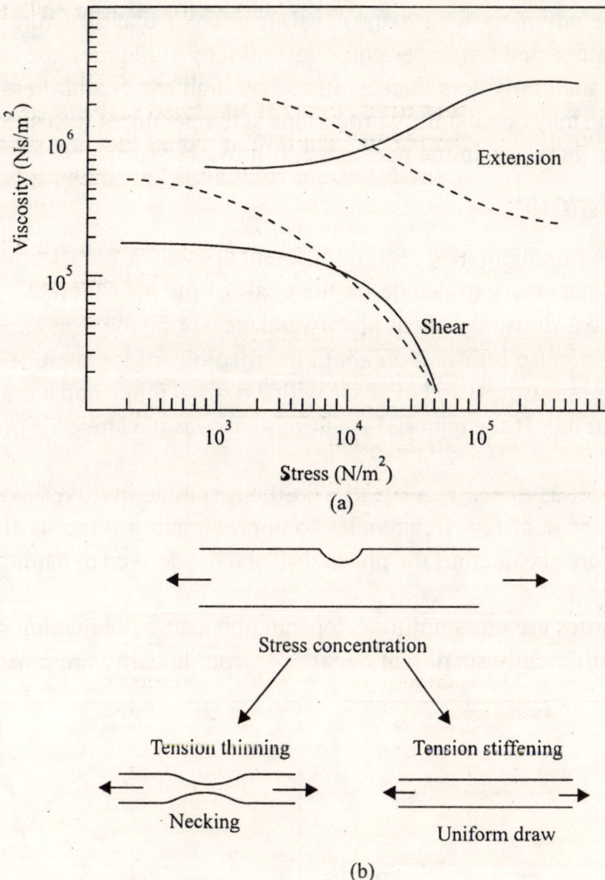

Fig. 5.11. (a) Shear flow and stretching flow behaviour of linear (——) and branched (- - - -) polyethylene melt flow index 0.3 at 150°C; (b) effects of stress concentrations on stretching flows.

A further effect of branching, as evidenced by linear and branched polyethylene, is the dependence of viscosity on temperature. The viscosity of branched polyethylene is twice as sensitive to temperature as that of linear polyethylene. It is clear that it is not the short branches that are responsible for this change, since the temperature sensitivity of polypropylene, which could be considered a short-branched molecule with the same backbone, is the same as that of linear polyethylene.

Polymer melts that are superficially similar in other rheological tests may be 100-fold different in their resistance to high stress extensional flows. The tendency for the elongational viscosity of branched polymers to increase with stress means that local stress concentrations that are less critical in polymer melts exhibit an apparently liquid crystal mesophase, allowing them to be anisotropic at rest. Such polymers may be induced to develop a liquid crystalline state by the action of mechanical history under conditions of temperature and composition where that state does not spontaneously form.

Another situation involves block co-polymers of the ABA type, in which surface tension forces drive the compatible ends to associate in star-shaped accretions. This structure is exploited in thermoplastic elastomers. The low stress rheology of such polymers is modified by mechanical history.

Melts from crystalline polymers tend to possess memory of that structure, re-crystallising in the same pattern after being subjected to temperatures several tens of degrees above their nominal melting points. Crystallisation in such polymers is accentuated by high stress and pressure. In polypropylene, for example, crystallisation may occur during processing at temperatures well above the nominal melting point, leading to dramatic increases in the resistance to flow.

RHEOMETRIC APPLICATIONS

Different categories of measurement may require different approaches to rheometry. Before embarking on any measurement it is necessary to decide on the goals of the measurement.

Subjecting a polymer to a sinusoidal shear history and measuring the stress response is established as a useful way of obtaining precise information about the rheology of the melt. If the material is entirely elastic the stress response is sinusoidal and in phase with the strain input, and has an amplitude dependent on the stiffness of the material. If the material is entirely viscous the stress response is exactly $\pi/2$ out-of-phase.

No melt is entirely viscous or entirely elastic, so the response involves a phase shift (Fig. 5.12). It is common for the response at low frequencies to approximate to viscous flow, while that at high frequency is usually far more elastic; thus the phase shift and the derived dynamic viscosity and modulus are frequency dependent.

Generally those properties are not amplitude dependent because the maximum strain and strain rate in the measurement are sufficiently small that departures from linearity are rare.

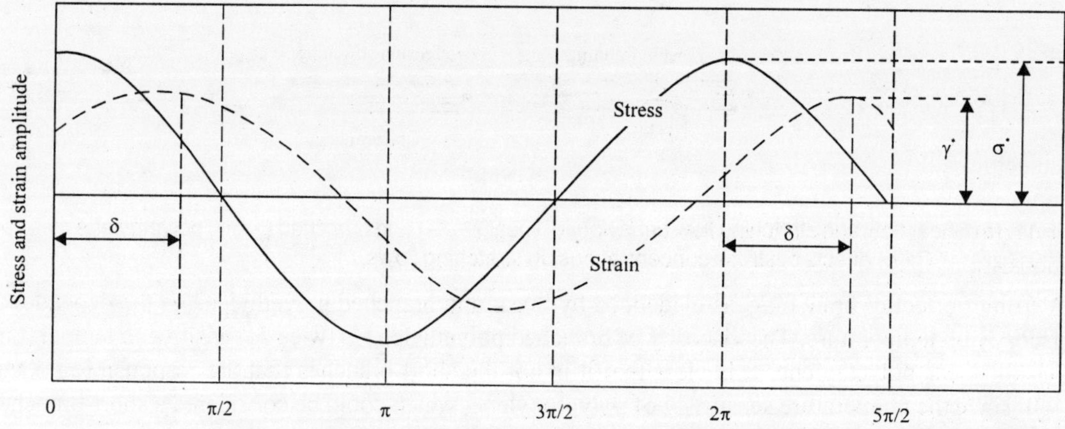

Fig. 5.12. Measurement of stress response, σ^* where γ^* is the strain amplitude and δ is the phase lag.

This class of experiments provides precise data characterising the rheology of the melt. There are, however, several ways of representing such data. The differences in interpretation rest on whether stress or strain is assumed to be the controlling mechanism. Most equipment actually operates under imposed strain, but the response would be identical if stress were the driving mechanism. Most instruments operate under the principle that strain is assumed to be the controlling variable, thus yielding the dynamic viscosity:

$$\eta' = \frac{\sigma^*}{\omega\gamma^*}\sin\delta \qquad \text{...(5.18)}$$

and the modulus:

$$G' = \frac{\sigma^*}{\gamma^*} \cos\delta \qquad ...(5.19)$$

where σ^* and γ^* are the stress and strain amplitudes, δ is the phase lag, and ω is the angular velocity.

If the dynamic viscosity and modulus are independent of frequency we can deduce that the material is behaving similar to a Voigt element (Fig. 5.13a). Because these parameters depend on frequency we may alternatively describe them as the apparent Voigt viscosity and modulus of the material. An alternative representation of the same data may be made using the stress as the controlling variable giving

$$\eta_M = \sigma^*/\omega\gamma^* \sin\delta \qquad ...(5.20a)$$

$$G_M = \sigma^*/\gamma^* \cos\delta \qquad ...(5.20b)$$

where, if η_M and G_M are independent of frequency, the material is responding like a Maxwell model (Fig. 5.13b) and the properties η_M and G_M may be described as the apparent Maxwell parameters.

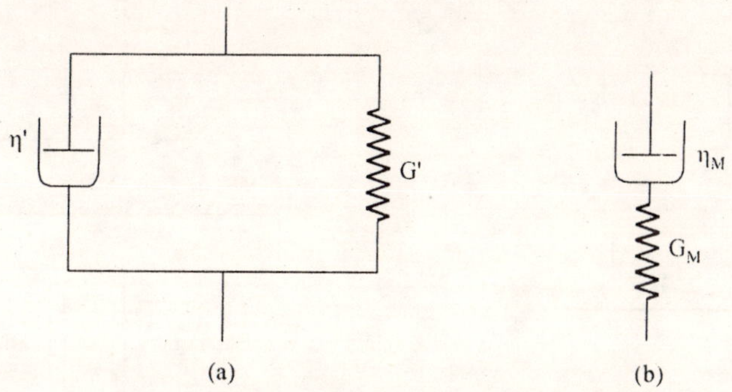

Fig. 5.13. (a) Voigt model; (b) Maxwell model.

In selecting the most useful method of representation one may choose between that which is most commonly used—the apparent Voigt parameters—and that which most closely approximates to the behaviour of the system that is being measured. The Voigt element is a solid in that it will not deform continuously under a sustained stress. The Maxwell element is a liquid, and, at sufficiently low frequency, most polymer melts approximate to a Maxwell model. Steady-flow rheological experiments on polymer melts implicitly use the Maxwell rather than the Voigt interpretation.

Most commercially available oscillatory shear rheometers operate in the range 10^{-3} to 10^{3} rad/s. The lower limitation is one of time scale of measurement and the upper limitation is the complication arising from the inertial effects. Crystal excitation has been used to study the behaviour of liquids at very high frequencies. Most dynamic measurements operate with strain amplitudes of the order 0.1.

The principal advantages of dynamic measurements for the study of the rheology of melts are:

1. The techniques have been rigorously analysed and may be used to obtain precise data in the linear region over a very wide range of frequency.
2. They provide direct measurement of elastic as well as viscous properties, and for materials of low viscosity they may be the only practical way of comparing elastic response.
3. Through the analogy between inverse frequency and time, dynamic methods yield information relevant to the very short time scale response of melts.

Disadvantages in using this type of measurement are:

1. The uncertain correlation between the rheological response in the linear elastico-viscous region and the evident non-linear response apparent in most polymer processing.

2. The relatively high cost of the equipment.

As already discussed, rigorous treatment has been given to the more common types of rheometers. We will now discuss practical limitations of these instruments. The first of these is the cone-and-plate rheometer. There are two major variants, depending on whether the apparatus is designed to operate under constant stress or constant rate. Both are illustrated in Fig. 5.14.

The constant-rate instrument, in which one face is rotated at a series of predetermined rates and the torque is measured on the opposite face, is preferred in most commercial instruments. If the instrument is being used as a viscometer only, this method of operation has an advantage over the constant stress variant in that when the system reaches steady flow, the torque output is a constant reading.

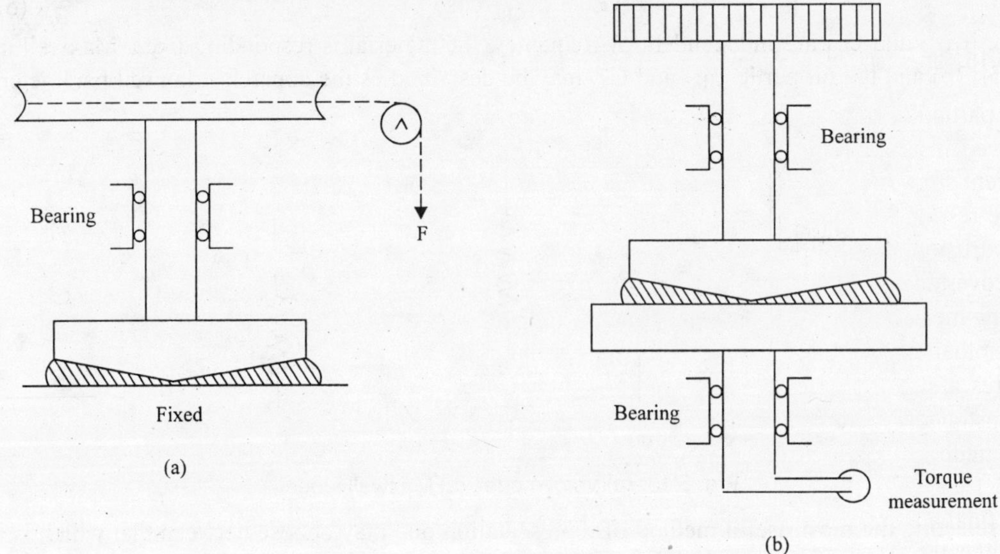

Fig. 5.14. Cone-and-plate rheometers: (a) Constant-stress type; (b) constant-rate type.

Cone-and-plate instruments may be used to determine the elastic response of melts. Several measurements can be made, including stress and strain transients to and from steady flow, normal stresses, and strain recovery.

When an elastic liquid is subjected to simple shear flow, as well as the shear stress there is a pull along the lines of flow. In a cone-and-plate rheometer pull strangulates the flow, causing a pressure profile across the face of the instrument and a normal thrust forcing the faces apart. That thrust is most conveniently measured by instruments in which the strain rate is imposed, which accounts for the common preference for that class of instrument.

The measurement of the stress transients during start-up and cessation of steady shear flow offer opportunity for deducing the elastic response. Such transients can be of particular value for comparative measurements, but absolute measurements demand very accurate setting of the apparatus. Many observers report an effect usually described as the stress overshoot phenomenon when, on start-up of flow at a constant rate, the stresses go through a maximum before reaching its equilibrium condition (Fig. 5.15). The phenomenon can be extremely reproducible and some workers deduce from it features of the viscoelastic response.

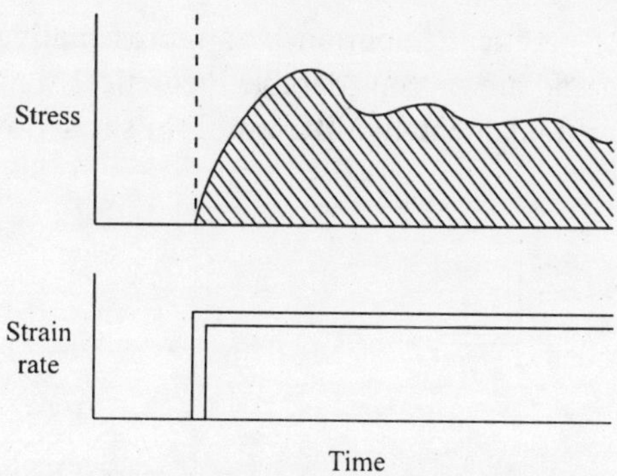

Fig. 5.15. Stress overshoot phenomenon: stress passes through a maximum and subsequently decays.

A particular advantage of the constant stress cone-and-plate is the ease with which the stress may be removed instantaneously, allowing the observer to record strain recovery as a function of time for different stress histories. The ratio of shear stress to recovered shear makes a simple comment on the elastic response which may be related to practical observations of frozen-in strain in mouldings and postextrusion swelling, in that all represent a desire of the material to revert to an earlier state. The use of recoverable shear as the prime measure of elastic response has the additional advantage that while it is being measured, it is the only thing that is happening.

Another advantage of the use of constant stress compared to an experiment at constant rate is that it requires a shorter time to reach equilibrium response. This difference in time scale is most clearly demonstrated for a Maxwell model in which, under constant stress, steady flow is attained instantaneously, while under constant strain rate the approach to equilibrium conditions is very slow. This is illustrated in Fig. 5.16.

A final advantage of the constant stress method of operation is that most rheological phenomena, in particular the reaching of a defined recoverable strain and the onset of non-Newtonian flow, tend to occur at a near-constant shear stress rather than constant shear rate for a range of temperatures and molecular weights. Comparative testing of elastic phenomena is thus uncomplicated by viscosity. Further, comparisons of viscosity in shear thinning systems at constant shear stress are more discriminatory than comparisons at constant shear rate in non-Newtonian systems.

The advantages of cone-and-plate rheometry are:

1. It gives a precisely defined flow at low shear stress.
2. It can provide data on both elasticity and viscosity in a variety of experiments.
3. As a measure of rheology under low stress it is especially sensitive to the structural characteristics of the melt, the structure of which may break down under higher stress.

The limitation of this class of rheometry for melts is associated with the very large ratio of surface area to volume of the instrument, with its attendant problems of slip that limit its usefulness to relatively low stress measurements. Even it if were possible to overcome the problems associated with the polymer-metal interface, the rate of heat generation would still preclude effective operation of the instrument at higher stress. The second class of instruments to discuss are elongational devices. As already described, the principle involves the use of simple instruments that have provided useful information on stiff melts. More sophisticated machines can supply high-quality data on a broader range of materials. Direct elongational flow measurements have a particular advantage over shear measurements in that the measurement does not involve an interface with the wall of the instrument, so that slip is not a problem.

Such measurements are of direct practical importance as representative of the important drawing operations of practical processing and are of considerable theoretical significance as representing the class of rotational flows which are fundamentally different from shear flows.

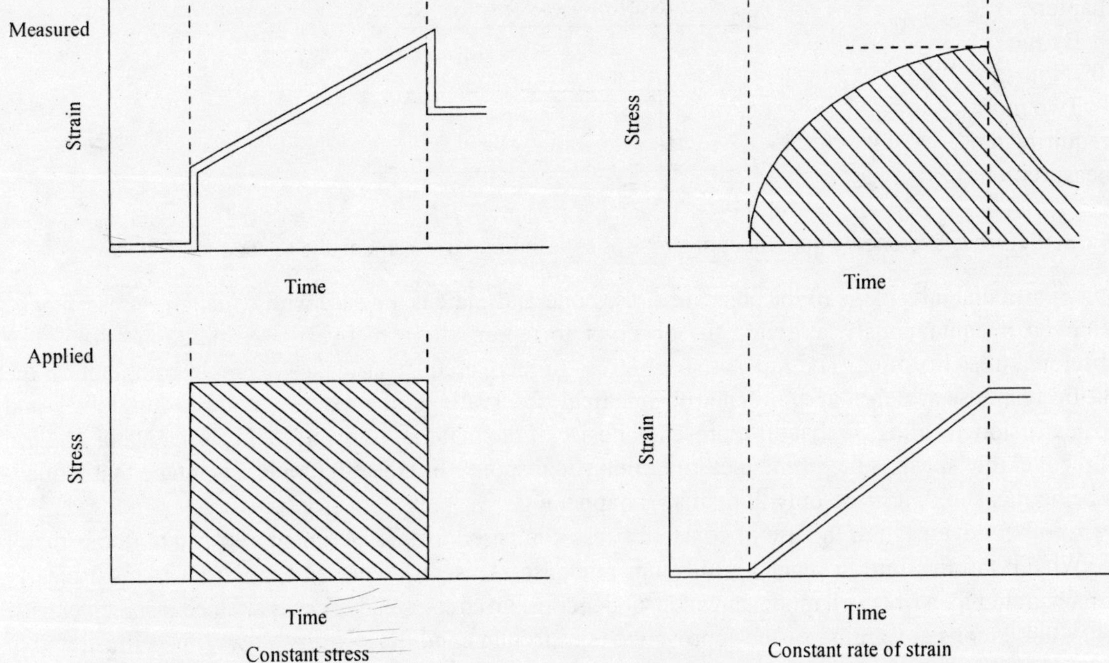

Fig. 5.16. Deformation of Maxwell model: under constant stress steady flow is attained instantaneously.

The advantages of constant stress or constant rate as the controlling mode of deformation apply to extension as well as shear. The reduction of the time transient available when making measurements at constant stress is a special advantage in elongation flow when making measurements on melts whose viscosity decreases with increasing tensile stress. In such experiments the small residual unrecovered strain from a brief crpee and recovery experiment (Fig. 5.17) may be interpreted to give an apparent rate of viscous deformation in circumstances in which waiting for steady flow conditions to be achieved would allow the sample to neck.

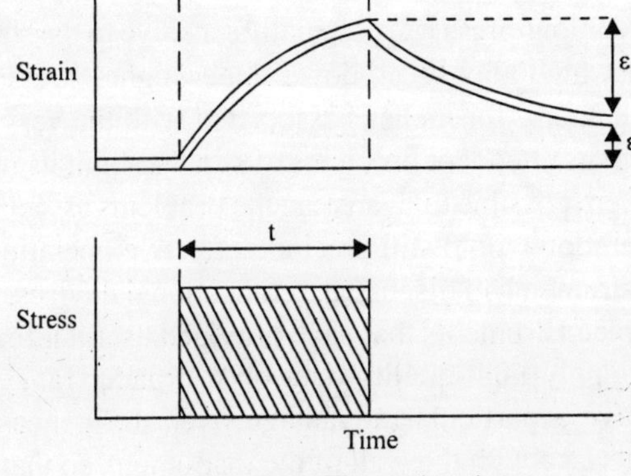

Fig. 5.17. Elongational state rate.

The most important argument in favour of making direct measurements of elongational flow is that only by making such measurements can one satisfy oneself of the significant differences between shear and extensional flows and between different materials in such flows.

The subject of measuring apparent rheological properties via capillary flow was discussed in earlier chapters. Additional comments on practical limitations of this instrument follow.

By the selection of appropriate equipment it is possible to measure shear stresses in the range 10^3 to 10^6 N/m^2 and shear rates in the range 0.001 to 500,000 s^{-1}.

Two modes of operation have been used: controlled pressure by dead weight loading or gas pressure (requiring a measurement of the volumetric flow rate), and controlled volume displacement (requiring a measurement of pressure) (Fig. 5.18).

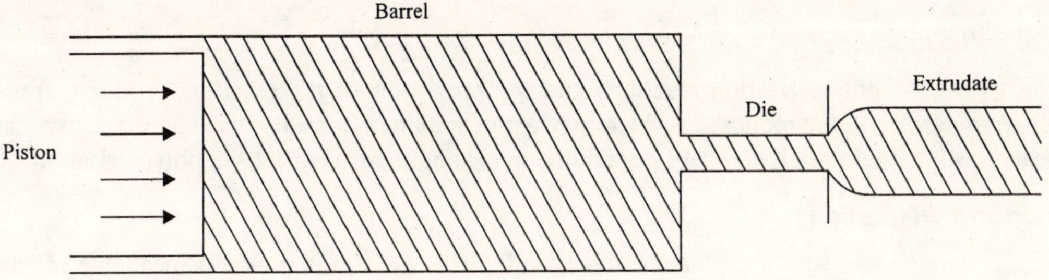

Fig. 5.18. Capillary extrusion rheometer.

What is required for the measurement is the relationship between flow rate and pressure drop through the die, and because pressure drops may be associated with the flow in the extruder barrel (and, more particularly, frictional losses between the piston and the barrel wall), it is highly desirable to make the measurement of pressure drop directly in the melt just above the die rather than relying on inferring pressure drop from the force required to drive the piston. The choice of barrel size is determined by the following considerations:

1. The available charge.
2. The heat-up period to an equilibrium temperature (especially when the polymer is subject to degradation).
3. Engineering tolerances.

The larger the reservoir size, the more convenient it will be to engineer. The larger volume flow rates available from larger pots allow the same shear rate range to be covered with larger radius dies, which themselves will have better tolerances so that viscosity (proportional to R) may be more accurately determined. (In general, dies of radius 0.25 to 2 mm are found to be most convenient). Unfortunately, larger pots require longer heating-up times and larger charges and it is this factor that limits the size appropriate to laboratory measurements.

Interpretation of capillary rheometry data using the Poiseuille equation yields an apparent rather than a true measure of viscosity at the operating temperature. A series of corrections is appropriate to derive the true viscosity, but only one of these, the ends (or Bagley) correction, is recommended.

It is further recommended that the long die used in such measurement should have a length-to-radius ratio of between 20 and 50 and preferably 32.

The principal sources of error when measuring high viscosity by capillary flow include:

1. Reservoir and friction losses.
2. End pressure drop (Bagley correction).

2. Disproportionation is the transfer of a hydrogen atom of one macro-radical to another to form two macromolecules, one of them with a double bond at its end:

$$R{-}CH_2{\cdots}CHX{\cdots}CH_2{\cdots}CHX\bullet + \bullet CHX{-}CH_2{\cdots}CHX{-}CH_2{-}R' \longrightarrow$$
$$\longrightarrow R{-}CH_2{-}CHX{\cdots}CH{=}CHX + CH_2X{-}CH_2{\cdots}CHX{-}CH_2{-}R'$$

Chemical substances capable of reacting with free radicals to terminate the reaction chain are called polymerisation inhibitors (examples are hydroquinone and trinitrobenzene). Chemical compounds which are only chain-transfer agents, i.e., do not affect the rate of polymerisation but determine the molecular mass of polymer, are called polymerisation regulators, or modifiers. These substances are added to the reaction mixture in strictly rated amounts.

Purity of initial monomer

The rates of radical polymerisation reactions and the properties of the resulting products are very sensitive to impurities in the monomer, which may act as initiators or inhibitors. Therefore to obtain a homogeneous high-quality polymer very pure monomers should be taken for the polymerisation.

Inhibition and retardation

A retarder is defined as a substance that can react with a radical to form products incapable of adding monomer. If the retarder is very effective, no polymer may be formed; this condition is sometimes called inhibition and the substance an inhibitor. The distinction is merely one of degree. The two phenomena are illustrated in Fig. 5.1, where the rate of polymerisation is displayed in idealised fashion. The action of a retarder is two-fold: It both reduces the concentration of radicals and shortens their average lifetime and thus the length of the polymer chains.

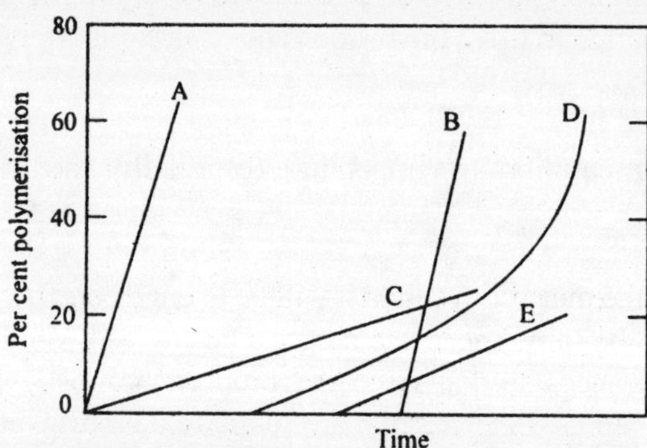

Fig. 5.1. Idealised time-conversion curves for typical inhibition and retardation effects: curve A, normal polymerisation; curve B, simple inhibition; curve C, simple retardation; curve D, non-ideal inhibition; and curve E, inhibition followed by retardation.

In the simplest case, the retarder may be a free radical, such as triphenylmethyl or diphenyl-picryl-hydrazyl, which is too unreactive to initiate a polymer chain. The mechanism of retardation is, simply the combination or disproportionation of radicals. If the retarder is a molecule, the chemistry of retardation is more complex, but the product of the reaction must be a radical too unreactive to initiate a chain. Inhibitors are useful in determining initiation rates, since their reaction with radicals is so rapid that the decomposition of inhibitor is independent of its concentration but gives directly the rate of generation of radicals. As a result, the length of the induction period before polymerisation starts is directly proportional to the number of inhibitor molecules initially present. This number then represents the number of radicals produced during the time of the induction period. For this analysis to be valid it must be known that one molecule of inhibitor is exactly equivalent to one radical. There are a number of reasons why this may not be so, and each case must be investigated separately.

Configuration of Monomer Units in Vinyl Polymer Chains

The addition of a vinyl monomer to a free radical can take place in either of two ways:

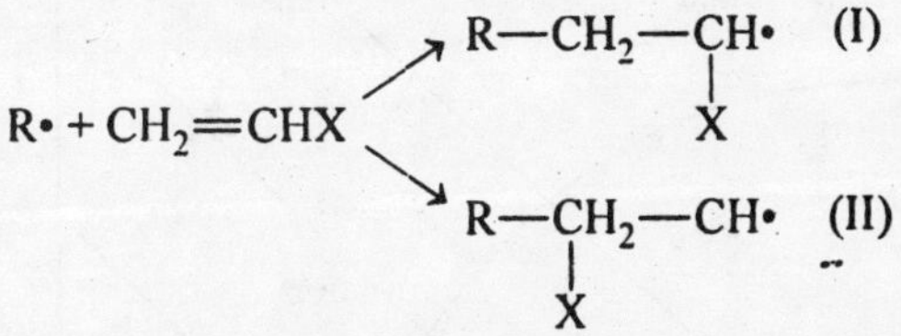

$$R\cdot + CH_2{=}CHX \nearrow R{-}CH_2{-}\underset{\substack{|\\X}}{CH}\cdot \quad (I)$$

$$R\cdot + CH_2{=}CHX \searrow R{-}\underset{\substack{|\\X}}{CH_2}{-}CH\cdot \quad (II)$$

the reaction leading to the more stable product being favoured. Since the unpaired electron can participate in resonance with the substituent X in structure I but not in II, reaction I is favoured. Steric factors also favour reaction I.

The occurrence of reaction I exclusively would lead to a head-to-tail configuration in which the substituents occur on alternative carbon atoms:

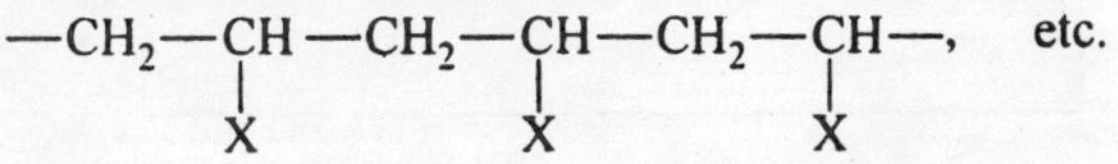

$$-CH_2-\underset{\substack{|\\X}}{CH}-CH_2-\underset{\substack{|\\X}}{CH}-CH_2-\underset{\substack{|\\X}}{CH}-, \quad \text{etc.}$$

Alternative possibilities are a head-to-head, tail-to-tail configuration:

$$-CH_2-\underset{\substack{|\\X}}{CH}-\underset{\substack{|\\X}}{CH}-CH_2-CH_2-\underset{\substack{|\\X}}{CH}-\underset{\substack{|\\X}}{CH}-CH_2, \quad \text{etc.}$$

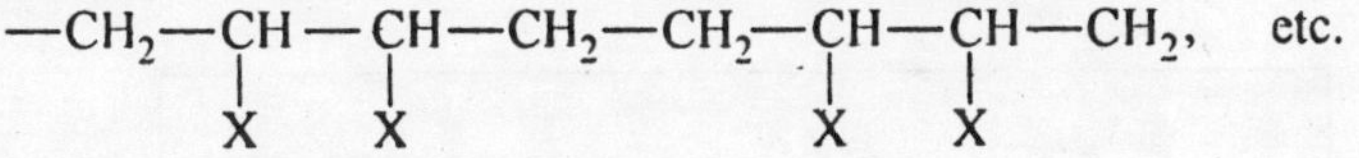

or a random structure containing both arrangements. The possibility of obtaining a regular head-to-head, tail-to-tail configuration exclusively by chain polymerisation is remote, and it appears that only an occasional monomer unit enters the chain in a reverse manner to provide an isolated head-to-head, tail-to-tail linkage.

Laboratory Methods in Vinyl Polymerisation

Small-scale laboratory polymerisations, carried to low conversion with pure monomer and minor amounts of initiator and other agents, are of great importance in obtaining kinetic data for substantiating the kinetics and mechanisms of polymerisations. Laboratory manuals are available describing the methods and experiments that are most useful. The experimental data of interest include the overall rate of polymerisation, the rate of initiation (and transfer where present), and the molecular weight of the polymer.

Rate of polymerisation

A chemical method or isolation and weighing of polymers can be used to measure the rate of polymerisation, but it is customary to work in an evacuated and sealed system to avoid extraneous effects due to the presence of oxygen.

of the solidified extrudate within 5 mm of the leading edge. The two measurements of swell ratio from an orifice die and from a die of length-to-radius ratio 32 approximate to the maximum and minimum values of swell ratio.

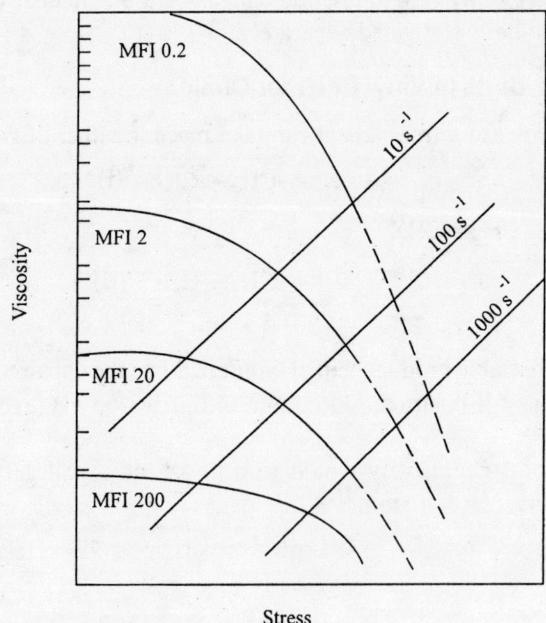

Fig. 5.21. Apparent viscosity vs. wall shear stress: low-density polyethylene at 170°C.

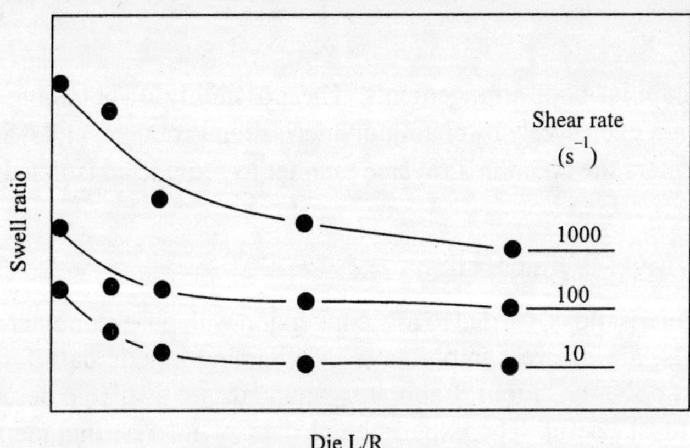

Fig. 5.22. Effect of die length-to-radius ratio on post-extrusion swelling: polypropylene melt flow index 230/2 = 15 at 210°C.

There are several accessories that can be adapted to a capillary rheometer to provide even more detailed information on the rheological properties of a melt. The first of these involves stretching flow, which is an obvious accessory to an extruder with an instrumented haul-off. Measurements of in-line tension, drawing profile, and rate can be of great value in comparing the stretchability of polymers. Simply clipping a weight on a sample and allowing it to draw until it solidifies is a simple method of comparing highly viscous materials in stretching flows. The main drawback of such measurements is the difficulty of interpreting the flows, with velocity gradients and complex history and temperature gradients, to yield fundamental data. The only parameters that are easily defined are the draw ratio and maximum stress to which the material has been subjected (force/area in thinnest section). The measurement is thus most suitable for studying phenomena such as rupture in which maximum stress is the most important factor, or for comparative views of the extensibility of materials.

The melt from a capillary rheometer may also be extruded past a chamber where the material is subjected to a well-defined shear history, as illustrated in Fig. 5.23. Such a section may be used for preparing melts of different shear history so that the influence of this important, though often neglected, parameter may be evaluated in separate experiments.

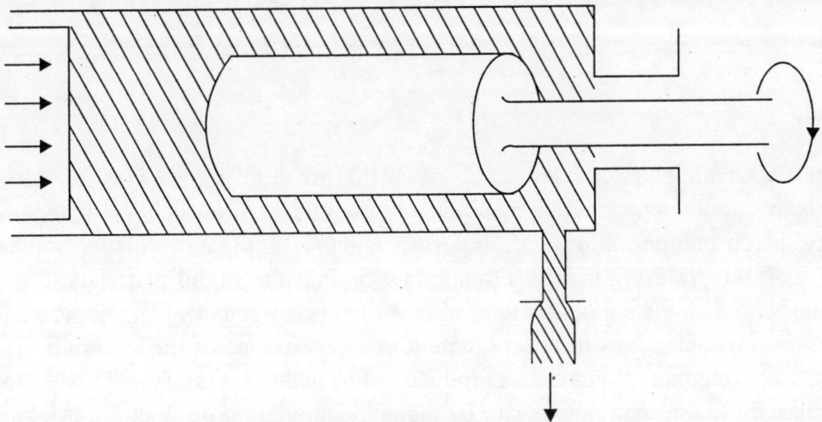

Fig. 5.23. Illustrates pre-shearing.

In addition to mechanical measurements of stress and strain, it is possible to make direct measurement of the streamlines and of stress birefringence during flow. The measurement of stress birefrigence requires sophisticated equipment, experimentation, and interpretation, but the information yielded is comprehensive. The introduction of markers allows a qualitative judgment to be made of the streamline pattern.

The melt flow index test is a quality control test used to assess the fluidity of a melt under standard conditions. As a single point test, melt flow rate is extremely sensitive in distinguishing between members of the same polymerisation family. Under ideal conditions it can give reproducible values to within 3%, discriminating variations of 1% in molecular weight (on polymers whose molecular weight is on the order of 100,000). This sensitivity does not necessarily distinguish polymers that have been made by different routes in which the viscosity of polymers having the same melt flow rate may vary by more than an order of magnitude under other conditions.

Of all the members of the family of capillary flow systems, the melt flow indexer is the most widely used. It is also capable of much more versatile use than the measurement of a single value. Flow rate may be determined as a function of applied load so that a qualitative indication of pseudoplasticity is readily obtained by comparing the ratios of flow rate at two different loads. The extrudate is available for elasticity determination by means of the swelling ratio and may also be subjected to simple drawing experiments. The measurement of melt flow rate and swelling ratio before and after processing, especially at a low temperature, can be an especially useful indication of changes that occur during processing.

CHAPTER 6

Torque Rheometers and Processability Testing

INTRODUCTION

Torque rheometers with miniaturised internal mixers (MIM) are multi-purpose instruments that are well suited for formulating multicomponent polymer systems, and studying flow behaviour, thermal sensitivity, shear sensitivity, batch compounding, etc. The instrument is applicable to thermoplastics, rubber (compounding, cure, scorch tests), thermoset materials, and also to liquid materials.

The rheometer with a single screw extruder allows the measurement of rheological properties and extrusion processing characteristics to differentiate lot-to-lot variance of the materials. The rheometer also enables the process engineer to simulate a production line in the laboratory and to develop processing guidelines. Because all the measurements can be made by simulating production, it is much easier to solve problems that the engineer may face.

The torque rheometer with a twin screw extruder is considered a scaled-down continuous compounder. It allows the compounding engineer to develop polymer compounds and alloys. It also permits the formulation engineer to ensure his formulation is the optimum.

Because the instruments are very versatile, low viscosity (liquid) and high viscosity materials, high temperature engineering thermoplastics, rubber, and thermoset materials can be tested. Therefore, they are widely used in quality and control and research and development areas in the polymer industries.

A torque rheometer is an instrument which is widely used in studying formulations, developing compounds, and characterising the polymer flow behaviour by measuring viscosity-related torque caused by the resistance of the material to the shearing action of the plasticating process.

Torque is the effectiveness of a force to produce rotation. It is defined as the product of the force and the perpendicular distance from its line of action to the instantaneous centre of rotation.

A microprocessor-controlled torque rheometer is a very unique system, consisting of two basic units: electromechanical drive unit, and a microprocessor unit. Its advantages are :

1. The microprocessor-based torque rheometer can control two pieces of equipment simultaneously. An example of this would be a feeder, or postextrusion system. It allows the user to regulate feed rate or downstream equipment.
2. The system can drive equipment, control and monitor parameters, and has the additional feature of data acquisition capabilities.
3. The unit displays all parameters graphically or numerically on-line during testing.

4. The rheometer also has the capability to plot any parameter against another in the off-line mode.
5. One can make his own programme using either on-line or off-line during testing.
6. The unit has a safety cut-off system when the torque or totalised torque exceeds maximum values.
7. All data can be printed on a printer on-line or off-line.
8. The rheometer will monitor and record several pressures and melt temperatures, as well as torque, totalised torque, and speed(s).
9. The rheometer is able to programme several speeds and temperatures for studying shear sensitivity or temperature sensitivity.
10. The system allows one to magnify specific portions of the test data in order to better differentiate small differences between data.
11. Systems have automatic calibration capabilities for the torque load cell and pressure transducers in any range.
12. Every parameter measured is recorded for later reference.

Fig. 6.1. shows a block diagram of a typical system.

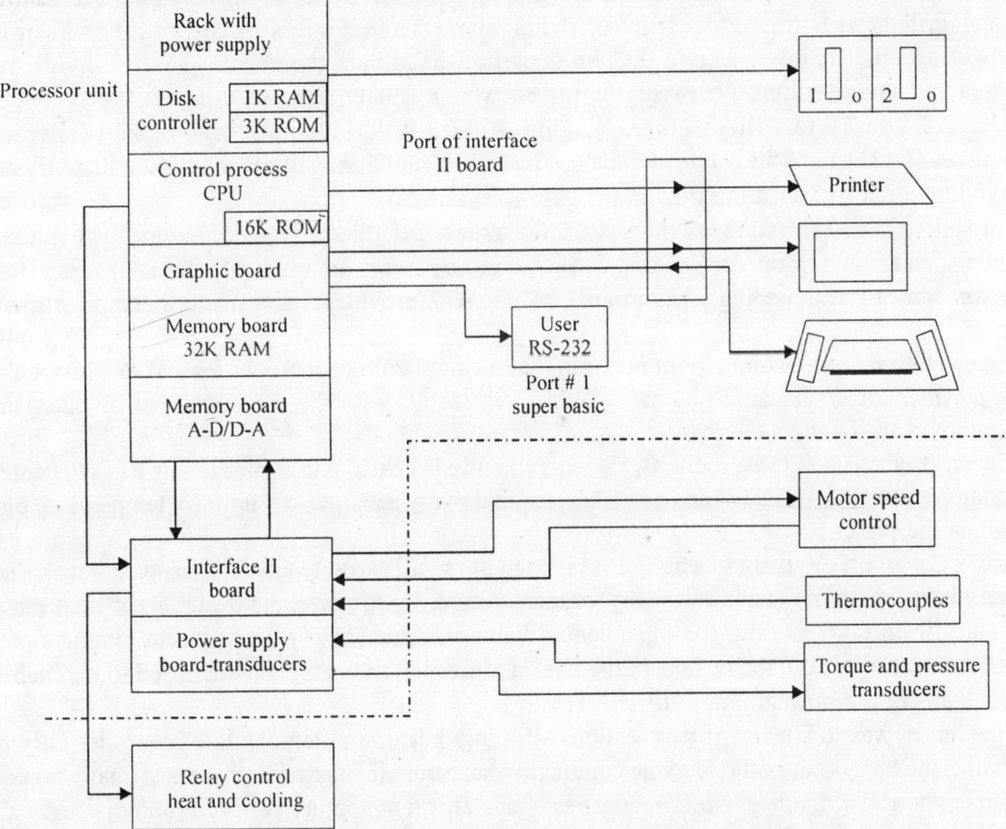

Fig. 6.1. Schematic block diagram of microprocessor-controlled torque rheometer.

MINIATURISED INTERNAL MIXER (MIM)

Most polymeric products are not pure polymers, but mixtures of the basic polymer with a variety of additives, such as pigments, lubricants, stabilisers, antioxidants, flame retardants, antiblock agents, cross-linking agents, fillers, reinforcement agents, plasticisers, UV absorbants, and foaming agents. All these additives must be incorporated into the polymer prior to fabrication. Some of the additives take a significant portion of the mixture; others only minute amounts. Some are compatible; others are not.

Depending on the quality of resin and additives and homogenisation of the mixtures, the quality of the final product will be varied. Therefore, developing a quality resin and additives when met with desired physical and mechanical properties of the product and quality control associated with them plays an important role in the plastic and rubber industries.

The mixer consists of a mixing chamber, with a spiral lobed rotor in each chamber. A totally enclosed mixing chamber contains two fluted mixing rotors that revolve in opposite directions and at different speeds to achieve a shear action similar to a two-roll mill.

In the chamber, the rotors rotate in order to effect a shearing action on the material mostly by shearing the material repeatedly against the walls of the mixing chamber. The rotors have chevrons (helical projections), which perform additional mixing functions by churning the material and moving it back and forth through the mixing chamber. The mixture is fed to the mixing chamber through a vertical chute with a ram. The lower face of the ram is part of the mixing chamber.

There is a small clearance between the rotors, which usually rotate at different speeds (e.g., gear ratio 3:2, 2:3, and 7:6 for different mixers) at the chamber wall. In these clearances, dispersive mixing takes place. The shape of the rotors and the motion of the ram during operation ensure that all particles undergo high intensive shearing flow in the gaps (clearances).

Normally, roller rotors are used for thermoplastics and thermosets, cam rotors for rubber and elastomers, and sigma rotors for liquid materials. Banbury rotors are used with a miniaturised Banbury mixer for rubber compounding and formulation. Fig. 6.2 provides a schematic diagram of the MIM mixer.

Temperature is the most important factor in measuring rheological properties and/or flow behaviour of the polymer mixes, because one can get different results with the same material by changing the temperature even by a few degrees.

The mixer consists of three sections, and each section is heated and controlled by its own heater and temperature controller. It is designed as such to maintain very accurate and uniform temperature profiles throughout the mixer.

Since mechanical dissipation heat is developed in the small gap between rotors and chamber, the heat conducts to the centre bowl and raises the set temperature. In this case, the heater at the centre bowl is automatically shut off, and the cooling solenoid valve is automatically energised to circulate cold air through the centre section; the heaters in the back and front sections remain energised to maintain their own set temperature through the test.

If the mixer was to consist of two sections with one heater and temperature controller only at the bowl, when the heat is shut off due to the conduction heat from the material, there is virtually no heating at the time; therefore, the back section becomes cold. This temperature control problem causes a non-uniform temperature profile throughout the mixer. Additionally, heating takes longer than that of the mixer with three heaters.

The loading chute is also designed for quick loading, and the chute can be removed from the mixer after the material is fed into the mixer. If the cold chute is not removed, the chute will draw the heat out of the mixer and the set temperature will be disturbed. This will not allow the measurement of flow behaviour at the desired temperature. Additionally, if the chute cannot be removed after loading, the chute will be hot; it will cause problems in loading for the test to follow.

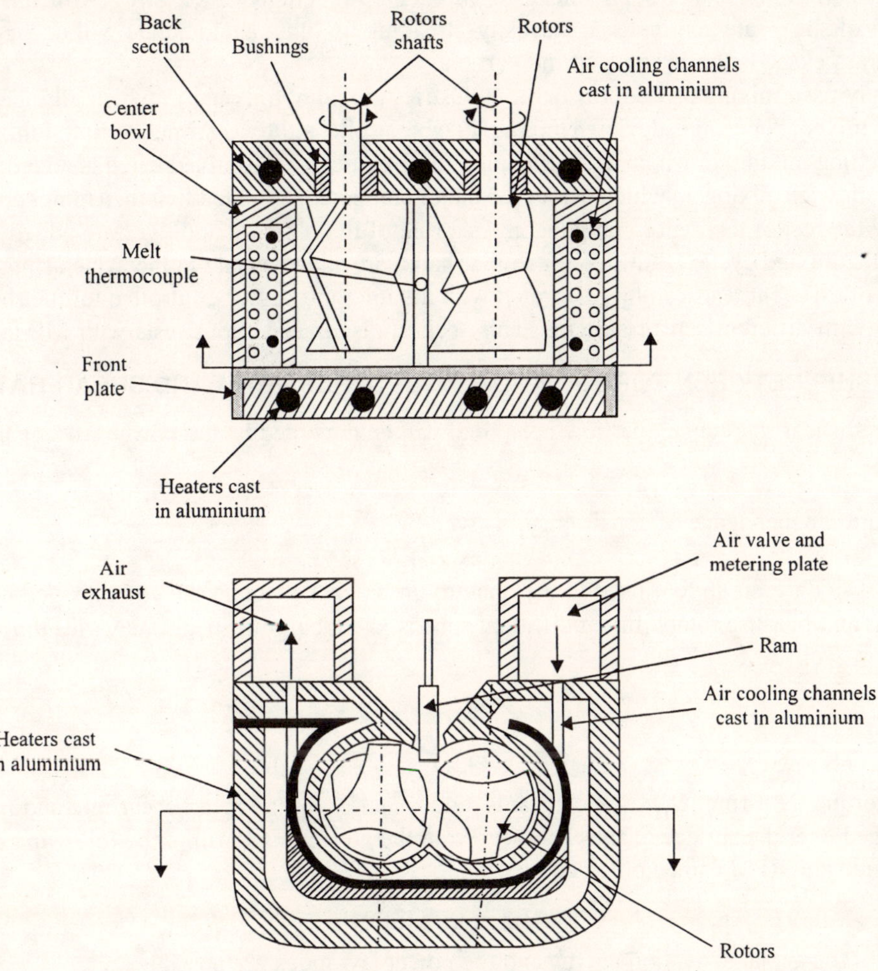

Fig. 6.2. Schematic diagram of MIM.

It is well known that plastic materials with two or more components being processed should be well mixed, so that the compounded material would provide the best physical properties desired for the final product.

There are distinctive types of mixing processes. The first is the spreading of particles over positions in space (distributive mixing) and the second is shearing and spreading of available energy of a system between the particles themselves (dispersive mixing). In other words, distributive mixing is used for any operation employed to increase the randomness of the spatial distribution of particles without reducing their sizes. This mixing depends on the flow and the total strain, which is the product of shear rate and residence time or time duration. Therefore, the more random the arrangement of the flow pattern, the higher the shear rate; the longer the residence time, the better the mixing.

A dispersive mixing process is similar to that of a simple mixing process, except that the nature and magnitude of forces required to rupture the particles to an ultimate size must be considered.

A more important function of the intensive mixing is incorporation of the pigments, fillers, and other minor components into the matrix polymer. This mixing is a function of shear stress, which is calculated as a product of shear rate and material viscosity. Breaking up of an agglomerate will occur only when the shear stress exceeds the strength of the particle.

Commonly used mixing procedures in the industries are dump mixing, in which all ingredients are added to the mixer at once; upside down mixing, in which solid additives are added first, followed by the polymer; seeding mixing, in which a small amount of previously well-mixed batch is added to the new batch; and sandwich mixing, in which part of the major component is added first then minor components, followed by the rest of the major component. As a rule of thumb, hard additives are added as early as possible in the mixing cycle. Dilutes, on the other hand, are added as late in the cycle as possible.

A series of different tests can be performed with the microprocessor-controlled torque rheometer in conjunction with different sensor systems. This section discusses various tests with MIMs.

RELATIONSHIP BETWEEN TORQUE AND SHEAR STRESS/RPM AND SHEAR RATE

Over a limited shear rate range, the flow behaviour can be governed by the power law equation :

$$\tau = m\dot{\gamma}^{n} \quad ...(6.1)$$

The temperature dependence of consistency is given by :

$$m = m_0 \exp(\Delta E/RT) \quad ...(6.2)$$

where exp(ΔE/RT) refers to 'e' raised to the quantity in brackets. Due to the complex geometry of the mixer rotors, and because rotors turn at different speeds, the local wall shear rate varies throughout the chamber; however, we can write;

$$\bar{\dot{\gamma}} = C_1 N \quad ...(6.3)$$

$$\bar{\tau} = C_2 \hat{T} \quad ...(6.4)$$

where $\hat{T}$ = torque, N = rpm, C_1, C_2 = constants, and $\bar{\dot{\gamma}}$ and $\bar{\tau}$ are the mean shear rate and mean shear stress averaged over the surface of the chamber. From the above relationship, the following expression between torque and RPM can be obtained :

$$\hat{T} = k \exp(\Delta E / RT) N^{n} \quad ...(6.5)$$

Calculation of the actual activation energy and the power law index are possible.

Shear Sensitivity

Shear sensitivity of the test material can be obtained from data generated as illustrated in Figs. 6.3. and 6.4. In order to obtain the data, the power law constitutive equation is adopted.

Since torque is proportional to shear stress and RPM is proportional to shear rate, as discussed earlier, the following relationship between torque and RPM may be obtained :

$$\log(\hat{T}) = \log m' + n \log N \quad ...(6.6)$$

Therefore,

$$n = \frac{d \ln \hat{T}}{d \ln N} \quad ...(6.7)$$

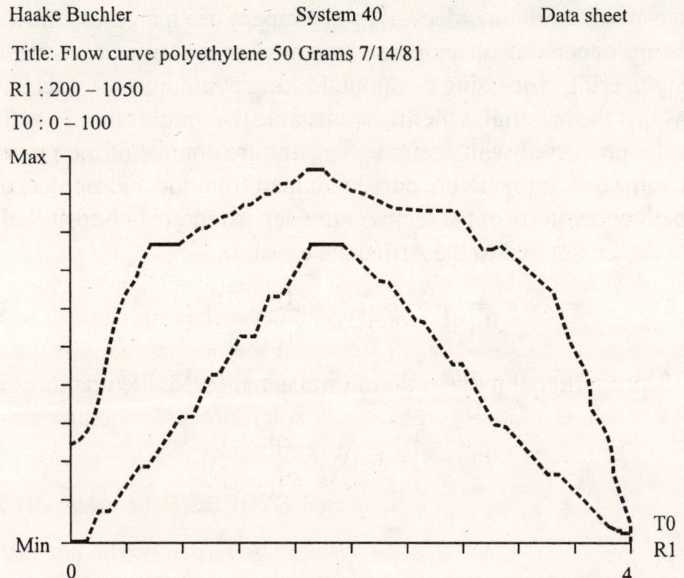

Fig. 6.3. Data generated by using speed programming to test the shear sensitivity of the polymer.

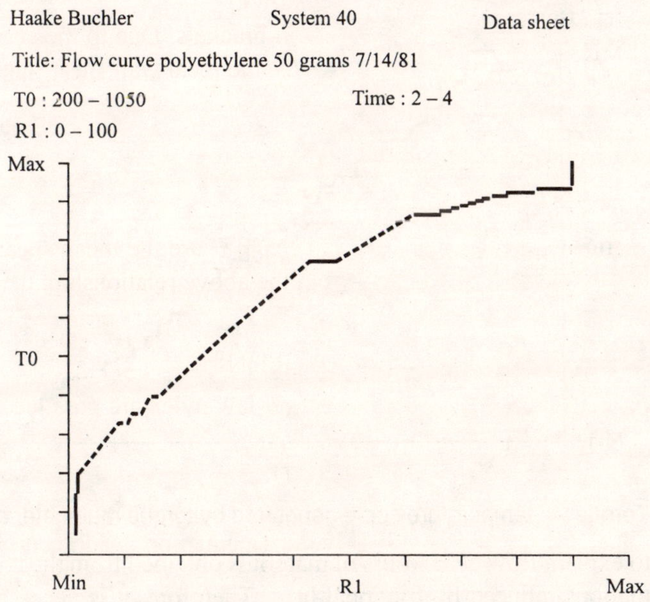

Fig. 6.4. Plot of torque vs. RPM.

Fig. 6.3 illustrates the study of polyethylene using speed programming: (0 to 80 RPM) ramp time with 2 min up and down. The data obtained are presented as a plot of torque, RPM vs. time curve. Fig. 6.4 shows a flow curve of the polyethylene, re-plotted (torque vs. RPM) from the data shown in Fig. 6.3. This test is used to obtain the flow behaviour of the material as a function of speed.

Generally, the viscosity of a polymer decreases as temperature increases, and vice versa. Properties of the material also change depending on temperature. Knowing the temperature dependence of viscosity is important for the engineering processing compounds. Large variations of viscosity for a certain range of temperatures means that the material is thermally unstable (i.e., requires large activation energy). This kind of material must be processed with accurate temperature control of the processing equipment.

Fig. 6.5 shows the torque vs. temperature curve obtained from the microprocessor-controlled torque rheometer. In this experiment a study of the temperature dependency of viscosity-related torque on PVC was made. These data can be applied to the Arrhenius equation

$$\ln(\eta) = \ln(A') + \frac{\Delta E}{R}\left(\frac{1}{T}\right) \qquad ...(6.8)$$

Because viscosity is proportional to the ratio of torque and RPM, the relationship

$$\ln\left(\frac{\hat{T}}{N}\right) = \ln A' + \frac{\Delta E}{R}\left(\frac{1}{T}\right) \qquad ...(6.9)$$

may be obtained.

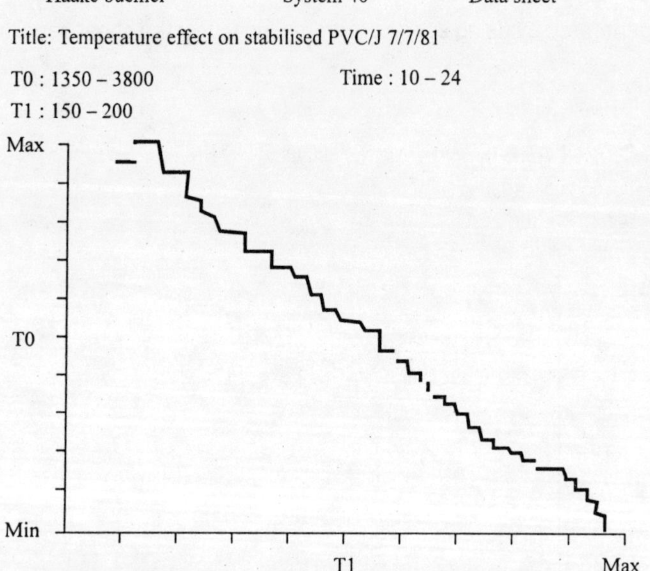

Fig. 6.5. Torque vs. temperature curve generated by torque rheometer and MIM.

It is often difficult to explain flow behaviours of materials obtained from the torque rheometer due to melt temperature fluctuations induced by frictional heat. Therefore, it is necessary to compensate for the temperature effect of the material. Temperature-compensated torque can be obtained from an integrated form of Arrhenius equation:

$$\ln\left(\frac{\hat{T}}{\hat{T}'}\right) = \frac{\Delta E}{4{,}576}\left(\frac{1}{T} - \frac{1}{T'}\right) \qquad ...(6.10)$$

where $\hat{T}$ = measured torque at temperature T (degrees Kelvin) and $\hat{T}'$ = calculated torque at reference temperature T′ (degrees Kelvin).

Induction time is defined as the time at which a testing material begins to oxidise. Induction time of a material can be obtained from torque rheometer data. It can be seen as decreasing or increasing torque on the torque curves, depending on structural breakdown or network formation due to oxidation at constant temperature, respectively.

These data can be obtained using the temperature-compensated torque described above. Fig. 6.6 shows data obtained from the torque rheometer utilising temperature-compensated torque software.

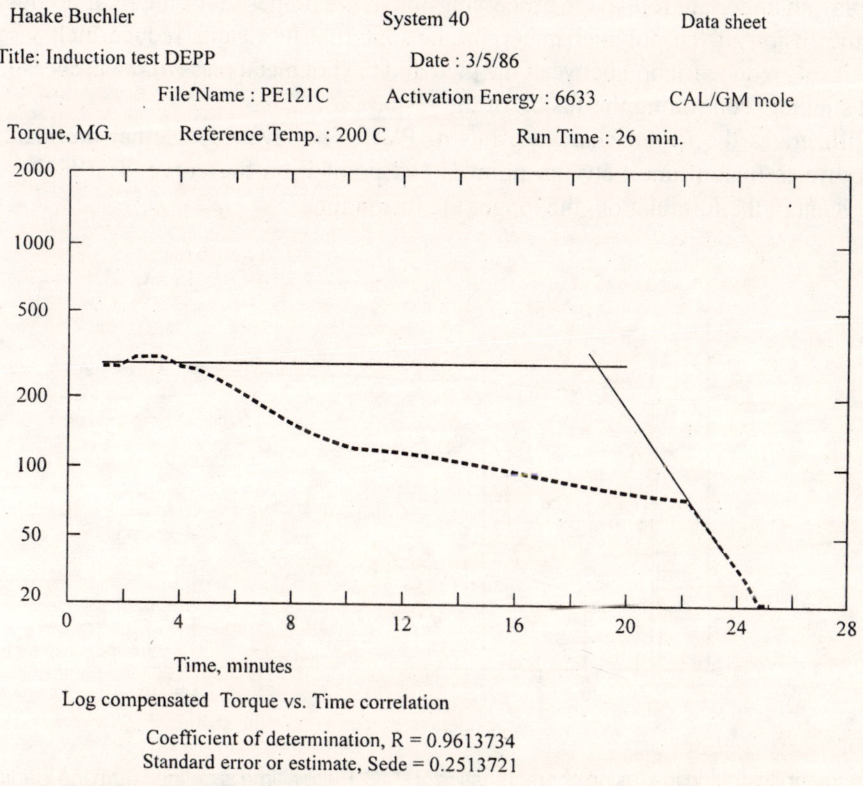

Fig. 6.6. Typical data showing effect of temperature on degradation of polymer.

Induction time of the polymer at a temperature can be obtained by extrapolating the stable torque and sharp decreasing torque regions as shown in Fig. 6.6. The data also can be obtained using torque curves directly obtained from the rheometer.

The unit work is defined as the work energy required to process the unit volume or unit mass of material. This can be calculated from the totalised torque, which is obtained from the rheometer directly. Totalised torque is defined as the energy required to process a certain material for a certain period of time at given conditions, and simply as the area under the torque curve. The totalised torque can be converted to work energy:

$$W_u = \frac{W_t}{V_b}$$

$$= \frac{2\pi N \int_{t_1}^{t_2} \hat{T}(t)dt}{V_b} = \frac{61.588 \times N \times TTQ}{V_b} \qquad ...(6.11)$$

where a unit of unit work = [J/cm^3], W_t = total work energy, $\hat{T}(t)$ = torque, N = RPM, V_b = charged sample volume, and TTQ = totalised torque.

Fusion Characteristics

Lubricants play an important role in the processing and in the properties of the final product. Lubricants also affect the fusion of the polymer materials; i.e., internal lubricants reduce melt viscosity, while external lubricants reduce friction between the melt and the hot metal parts of the processing equipment and prevent sticking, controlling the fusion of the resin.

Fig. 6.7 illustrates the fusion characteristics of PVC. The level of external lubricant used in the formulation affects fusion time between point L and point F on the curve. The higher the level of external lubricant in the formulation, the longer the fusion time.

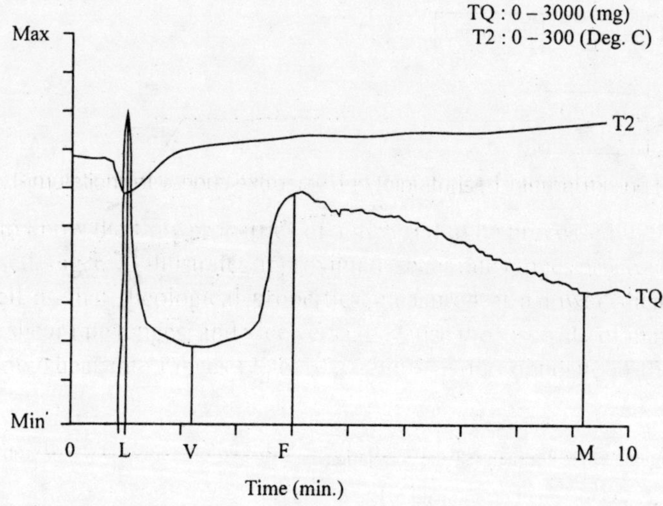

Fig. 6.7. Experiment illustrating the fusion characteristics of PVC. Fusion time is defined from the loading peak (point L) to the fusion peak (point F).

If an unnecessarily high level of external lubricant is used in the formulation, it will take a longer period of time to melt the material processing, which results in the reduction of production, an increase in energy consumption, and poor products. Meanwhile, if too low a level of external lubricant is used, the material will melt too early in the processing equipment, which may result in degradation in the final product. Therefore, selecting the optimum amount of external lubricant is a necessity for the improvement of processing and for good quality of products.

Selection of Blowing Agent for Foam Products

Foam products can be produced by adding chemical blowing agents to the formulation. The blowing agents are inorganic or organic materials and decompose under heat to yield gaseous products.

The selection of a blowing agent is very important in foam products. If the proper blowing agent is not chosen, it may react with other ingredients used in the formulation, which results in retarding the blowing agent from being activated. The level of blowing agent used in the formulation is also important. If the level of the agent used is too high, too many unnecessary cells form in the foam product and it will collapse easily. If the level is too low, the product will not have enough cells, and so it will decrease the cushionability of the product.

Figs. 6.8 and 6.9 illustrate evolution of gas from the activation of blowing agents used in rubber. Fig. 6.8 gives information on the temperatures at which the two blowing agents begin to activate. Fig. 6.9 provides an idea of the torque levels at which they activate.

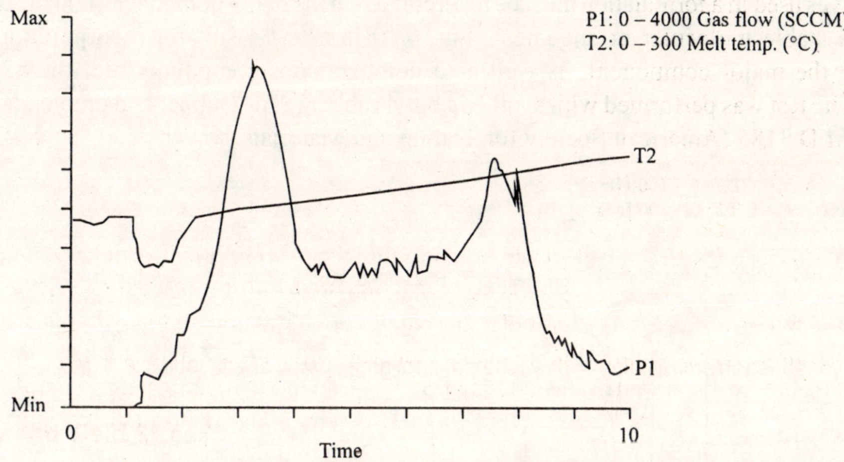

Fig. 6.8. Experiment in which activation of blowing agent used in an EPDM formulation was studied as a function of temperature.

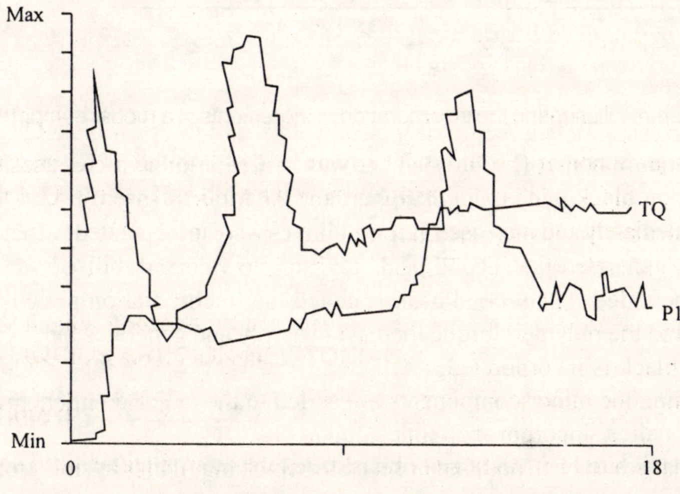

Fig. 6.9. Torque level when blowing agents in an EPDM compound are activated.

One can study the suitability of the blowing agent chosen by the processor for formulation and determine if it reacts to any other ingredients in the formulation processing.

Formulation

Most polymers are used with additives such as fillers, lubricants, flame retardants, stabilisers, colour pigments, plasticisers, impact modifiers, etc., to improve processibility, physical and mechanical properties, uniformity, flexibility, and to reduce production cost.

Additive Incorporation and Compounding

All of the additives used in a formulation must be incorporated in the major component, and the components should be in a stable molecular arrangement. Fig. 6.10 is a test result for incorporation of minor components to the major component, as well as a homogeneous compound after the additives are incorporated. The test was performed with synthetic butyl rubber (SBR) rubber and preblended additives based on ASTM D 3185 (American Society for Testing and Materials.

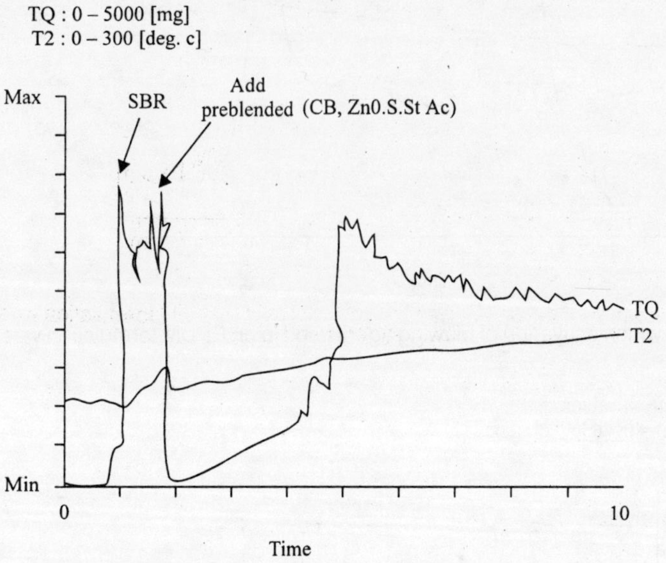

Fig. 6.10. Experiment illustrating the incorporation of ingredients in a rubber compound.

Ethylene propylene diene monomer (EPDM) rubber was loaded into the mixer and mixed for 30 s. Preblended additives (carbon black, zinc oxide, sulphur, and stearic acid) were added at 30 s. Torque values dropped sharply immediately and increased as the additives were incorporated; when the ingredients were fully incorporated, it generated the second peak, and torque values stabilised when the material was homogeneously compounded. The second peak is called the 'incorporation peak'.

If hard fillers are added to the polymer, torque increases sharply and generates the second peak. This can be seen when carbon black is incorporated.

The interval from the time the minor components are added to the major component to the point of the 'incorporation peak' is called 'incorporation time'.

Cross-linkable thermoplastics, rubber, and thermoset materials change their physical properties, usually from liquid to solid (cure) via chemical reaction, heat, or catalyst during processing.

It is important for process engineers to know how long it takes to cure, how high a torque is generated, and how long it will be in stable flow. This information would help the engineers predict how the material should be processed to provide the best processability and properties. Fig. 6.11 illustrates the test results of the cure characteristics of rubber. Once the rubber is introduced to the heated mixer, it generates a sharp increase in torque called the 'loading peak'. After it is loaded the material begins to soften, which results in decreasing torque. It reaches minimum torque (maximum flow) and starts to scorch prior to the curing reaction. When it cures, the torque value increases and torque decreases after the curing reaction ends. The cycle time is considered to be the time from loading to cure peak.

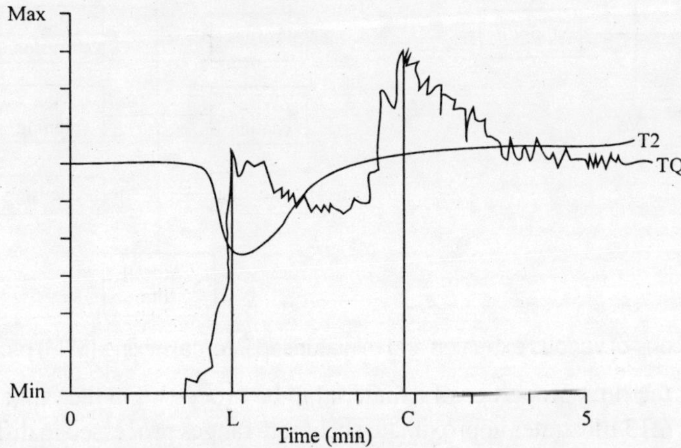

Fig. 6.11. Experiment shows scorch and cure characteristics of a rubber.

SINGLE SCREW EXTRUSION

Solids conveying, melting, mixing, and pumping are the major functions of polymer processing extruders. The single screw extruder is the machine most widely used to perform these functions.

Applications of Single Screw Extruders

Production simulations, rheological property measurements, and processing characteristics are discussed in this section. The applications mentioned above that use a single screw extruder are widely acknowledged in polymer industries and research organisations because the production line can be closely simulated in the laboratories.

Simulation of the extrusion process in the laboratory is one of the most important applications of the torque rheometer in conjunction with single screw extruders. Fig. 6.12 illustrates simulations of widely used extrusion processes in the industries.

It is very important for a process engineer to know the rheological properties of a material, since the properties dominate the flow of the material in extrusion processes and also dominate physical and mechanical properties of the extrudates.

Therefore, it is also important to measure the properties utilising a similar miniaturised extruder in the laboratory, so that a process engineer knows the flow properties in the system by simulating the production line. This will enable engineers to see lot-to-lot variances of materials for better quality control of the resins.

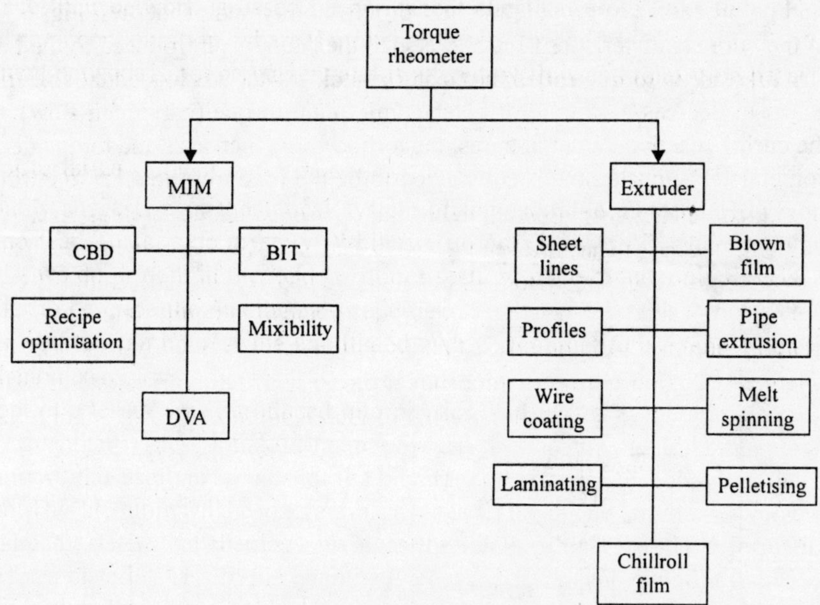

Fig. 6.12. Simulations of various extrusion and miniaturised internal mixing (MIM) processes.

It is desirable to know the flow properties of a material to be processed in the range of shear rates of equipment to be used. Fig. 6.13 illustrates approximate shear rate ranges processed in different processing equipment, as well as that rheological properties measured at a lower shear rate ranges cannot be applicable to high shear rate ranges, and vice versa, because the viscosity of material A is lower than that of material B at low shear rate ranges (1 to 10/s), but it is the opposite at the high shear rate ranges (> 100/s).

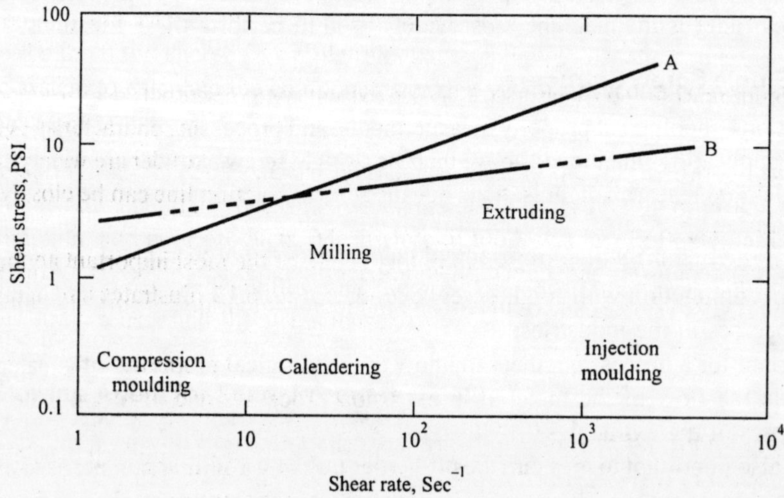

Fig. 6.13. Approximate shear rate ranges for different processing equipment.

In order to obtain viscosity data, a slit die with three pressure transducers along the die can be used. Because the slope of pressure drop vs. length of die (L) curve in the fully developed region is used to calculate shear stress, entrance effect correction is not necessary in slit rheometry.

The viscosity of materials can be calculated as follows:

Apparent shear rate (Γ_w):

$$\Gamma_w = \frac{6Q}{wh^2} \qquad ...(6.12)$$

True shear rate ($\dot{\gamma}_w^*$) with Rabinowitch corrections is:

$$\dot{\gamma}^* = \frac{\Gamma_w}{3}\left(2 + \frac{d \ln \Gamma_w}{d \ln \tau_w{}^*}\right) \qquad ...(6.13)$$

True shear stress (τ_w*) is:

$$\tau_w{}^* = \frac{h}{2L}\,(\Delta P) \qquad ...(6.14)$$

where h = opening of slot (cm), w = width of slot (cm), L = length of slit die (cm), Q = volume flow rate (cm^3/s), ΔP = pressure drop ($dyne/cm^2$).

Even though they are the same polymer materials, often they show lot-to-lot variances in processing and in product quality. The reasons the variance occurs are difficult to detect in the production line.

The processing characteristics can be easily tested in the laboratory by simulating the production line and analysing the data microscopically. A miniaturised single screw extruder in conjunction with a torque rheometer can characterise the processing by measuring total shear energy (mechanical energy), specific energy, residence time, shear rate, and specific output.

Total Shear Energy (TE)

Total shear energy (TE) is the energy introduced to the polymer material by the motor drive during processing. This is calculated from the measured torque multiplied by screw speed (N).

$$TE = Torque \times N \times 9.807 \times 10^{-3} \qquad ...(6.15)$$

These data provide information about total mechanical energy required to process the material.

Specific Energy (SE)

Specific energy (SE) is defined as the energy required to process a unit mass of material. This is calculated from TE divided by the total mass flow rate.

$$SE = \frac{0.0167\ TE}{\dot{m}} \qquad ...(6.16)$$

where $\dot{m}$ = total mass flow rate. These data provide information about viscous dissipation heat built-up in the system based on screw speeds.

Residence Time (t′)

Residence time is the time required for a material to reside in the extruder before it comes out of the die.

It can be obtained from volume of screw (V) divided by volumetric flow rate (Q), or length of screw (L) divided by angular velocity of screw (πDN):

$$t' = \frac{V}{Q} \quad \text{or} \quad t' = \frac{L}{\pi DN} \qquad ...(6.17)$$

Maintaining uniform residence time is important in achieving a quality product. If it is not maintained, the material will sometimes degrade because of excess exposure to heat.

Shear Rate ($\dot{\gamma}$)

Shear rate in the metering section of the screw can be obtained as follows :

$$\dot{\gamma} = \frac{\pi DN}{h} \qquad ...(6.18)$$

where D = diameter of screw, N = RPM, and h = gap between screw and barrel.

Specific Output

Specific output (SO) is defined as mass flow rate per unit RPM of screw and is calculated from total mass flow rate ($\dot{m}$) divided by RPM (N) used.

$$SO = \frac{\dot{m}}{N} \qquad ...(6.19)$$

This information characterises the uniformity of solids-conveying melting, and pumping mechanisms of the screw being used.

The following data have been obtained using fibre-graded polypropylene from different lots. Sample A is standard control material, and B and C are from different batches. The test was performed using an instrumented single screw extruder (L/D = 25, D = 3/4 in.), with a two-stage screw and capillary die (L/D = 40), in conjunction with a microprocessor-controlled torque rheometer. The temperature profile on the extruder barrel was 220°C at zone 1, 200°C at zone 2, 190°C at zone 3, and 190°C at the die. The melt temperature was measured with an exposed thermocouple, and entrance pressure was directly measured at the entrance to the capillary.

This test provides a microscopic view of the processing characteristics of the polymer, showing differences that would normally not be seen individually in the overall macroscopic view seen in the production line. Fig. 6.14 shows normalised output (grams of material extruded per screw revolution) vs. RPM.

Normalised output tends to decrease as screw speed increases, even though overall output increases as screw speed increases because of melt limitations of the material in the system. Fig. 6.15 illustrates specific energy (energy needed to process 1 g of material) vs. RPM. The specific energy increases steadily as screw speed increases because the overall viscosity of the material in the system increases due to decreasing residence time as screw speed increases.

Fig. 6.16 shows a plot of viscosity vs. shear rate. The viscosities of samples B and C show much similarity, but the specific output and energy still show large differences.

The test results lead to the conclusion that the same material from different lots show different processing characteristics.

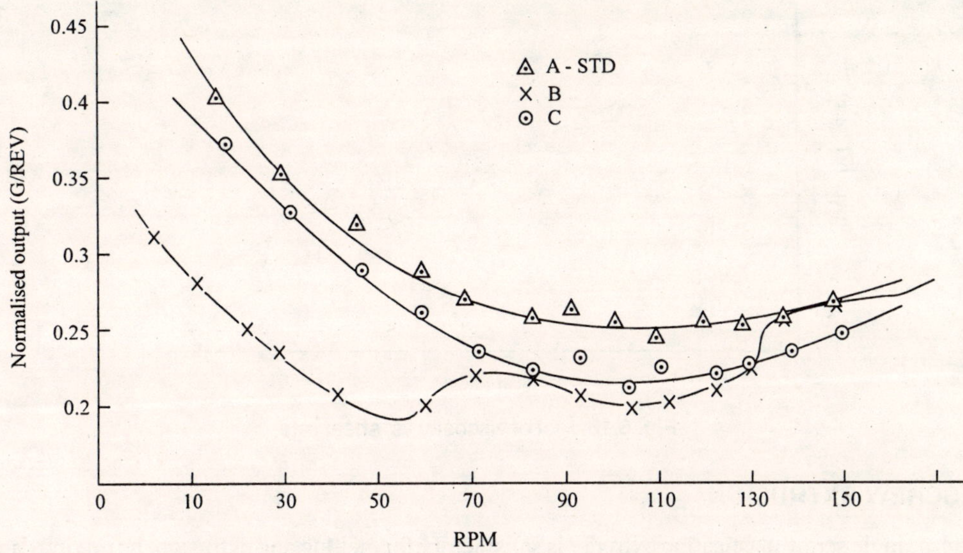

Fig. 6.14. Normalised output as a function of screw RPM.

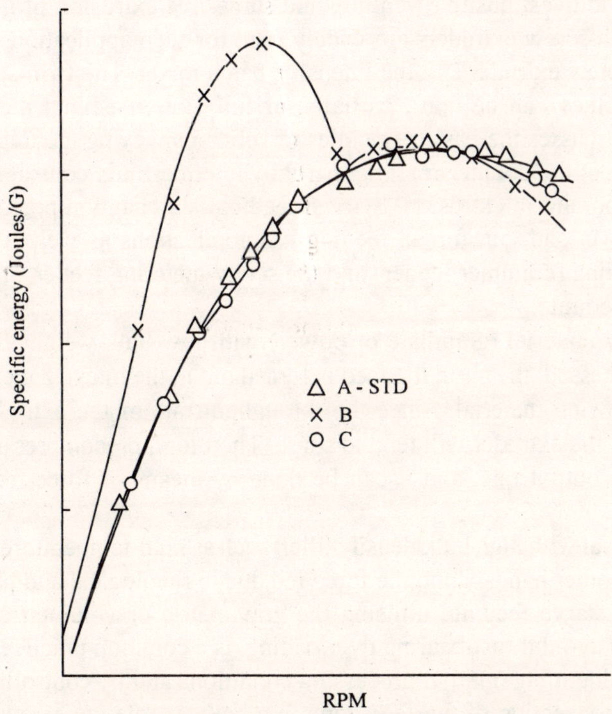

Fig. 6.15. Specific energy as a function of screw RPM.

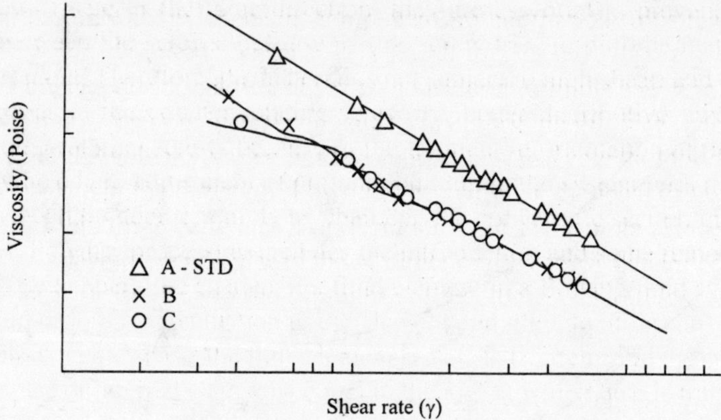

Fig. 6.16. Plot of viscosity vs. shear rate.

TWIN SCREW EXTRUSION

Although a single screw plasticating extruder is satisfactory for melting and extrusion, its mixing capabilities are limited. Tougher specialised thermoplastic materials developed for new applications must be processed at conditions requiring very accurate temperature control, uniform residence time distribution, better dispersion of fillers and additives, positive venting, and surge-free extrusion of homogeneous melt or high-quality products. Twin-screw extruders are widely used for such applications, and they overcome limitations of the single-screw extruder and the intensive batch mixer. The twin-screw unit comes in a variety of configurations; they can be non-, partially, or fully intermeshing; they can be conical or cylindrical. This section discusses the basic principles of solids-conveying melting, mixing, degassing, energy, and residence time of the counter-rotating, partially intermeshing, conical twin-screw extruder.

Material feeding is important in extrusion. Starved feeding is a common practice in order to avoid drive and thrust-bearing overload, but forced feeding has applications as well. Two types of feeding devices, i.e., the force feeding (crammer feeder) and the screw-metering feeder, are commonly used to regulate and control the feed rate.

When low bulk density material (granulate or power) with low bulk weight fillers such as fumed silica or wood dust are processed, the air in the feed is forced out in the intake zone, and can impede the smooth intake of the inflowing material. Since the throughput rate of the extruder is limited by the volumetric-conveying rate, the extruder will tend to surge. Therefore, precompressing the feed material is required to increase the output rate, and it can be done by means of force-feeding devices at the throat.

High bulk density material with high bulk density filler, such as high-temperature engineering plastics with carbon blacks, on the other hand, cannot be force fed due to rheological and physical properties of the material. In this case, starve feeding, utilising the gravimetric or volumetric metering feeder to control the feed rate and to avoid thrust-bearing overloading, is a common practice. The desired output rate can be obtained by maintaining optimal processing conditions and by controlling the feed rate.

Fig. 6.17 shows the cross-section of the feed zone in a counter-rotating-screw system. In order to feed as much material as possible, utilising the largest conveying volume, the screws turn outward on top and inward at the bottom. To obtain more material conveying, a different screw geometry just underneath the hopper often uses a greater pitch and multiple thread starts. The positive displacement of such a filling zone is greater than that of the rest of the extruder. Polymer enters the screw in the intake zone, and is compressed and conveyed toward the downstream into the melting zone, where transition from solids to melt occurs.

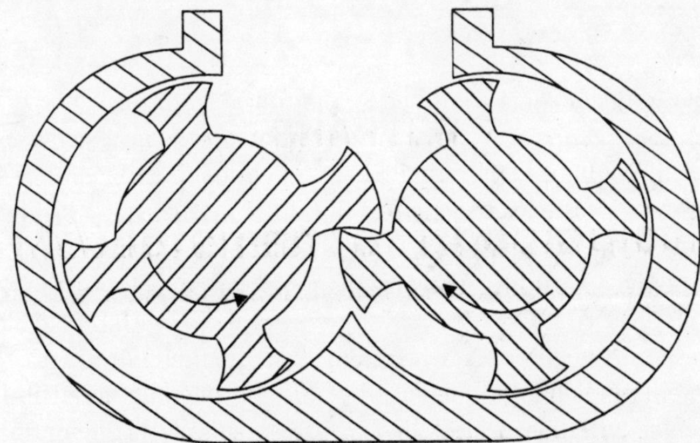

Fig. 6.17. Cross-section of the feed zone in a counter-rotating screw system.

Meanwhile, at the edge of the solids bed, a melt pool is created that draws in solid particles and softens them, and the plasticated mass is taken in by a calender gap. Inflowing unmelted polymer goes into the melt stream by kneading action in the zone in which the melting begins to accelerate the melting process. Melt can be conveyed while the trapped air can leak back. The melt will eventually reach a point at which pressure buildup commences as a result of the flow of melt back through the various leakage gaps.

In a single screw extruder, the solids bed steadily diminishes along the melting zone. In a twin screw extruder, however, a complete melting sequence can be found in each chamber where the solids bed and melt interface move from the diverging side to the converging side of the screws while the chambers move through the extruder. Exteusive study has been carried out in the melting mechanism in the twin screws.

Fig. 6.18 shows unwound chambers to illustrate the melting process. The melting mechanism at low backpressure is also different from that at high backpressure. At low backpressure, the melting process takes place at the barrel wall so that heat transfer plays an important role. With high backpressure, however, the fully filled length of the extruder reaches back to the melting zone, and considerable leakage flows take place. In this case, melting starts at the solids/melt interface. Therefore, headpressure affects the melting inside the system.

It is well known that compounds with two or more components should be mixed well to provide optimum properties in the final product. There are two distinct types of mixing processes: the spreading of particles over positions in space (distributive mixing), and the sharing or spreading of the available energy of a system between the particles themselves (dispersive mixing).

Distributive mixing is used for any operation employed to increase the randomness of the spatial distribution of particles without reducing their sizes. This mixing depends on the orientation of flow and the total strain, which is the product of shear rate and residence time. Therefore, the more random the arrangement of flow pattern, the higher the shear rate; the longer the residence time, the better the mixing.

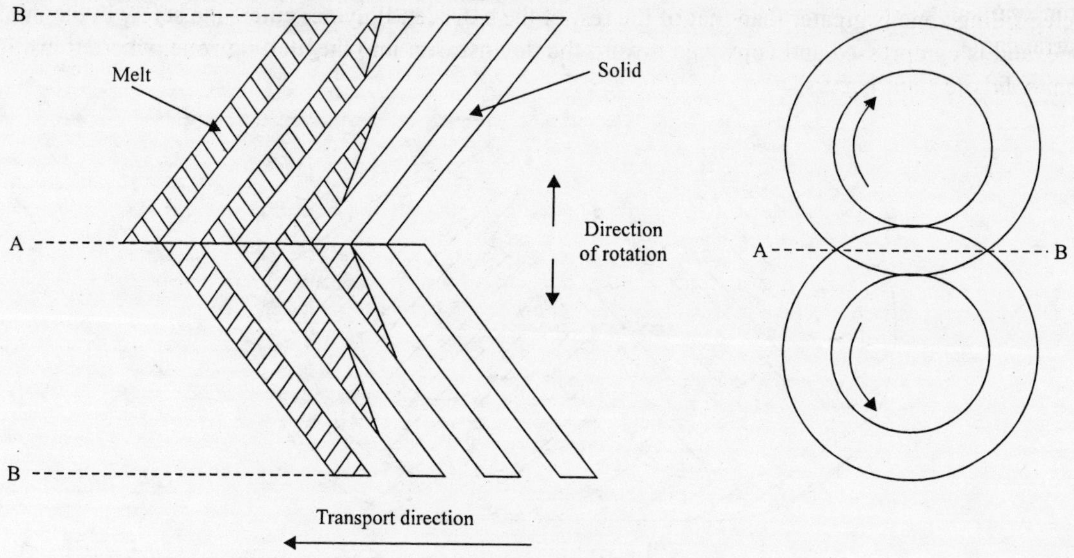

Fig. 6.18. Unwound chamber illustrating the melting process.

Dispersive mixing is similar to that of a simple mixing process, except that the nature and magnitude of forces required to rupture the particles to an ultimate size must be considered. An important application of such intensive mixing is the incorporation of pigments, fillers, and other minor components into the matrix polymer. This mixing is a function of shear stress, which is the product of shear rate and material viscosity. Breaking up of an agglomerate will occur only when shear stress exceeds the strength of the particle.

Fig. 6.19 illustrates the various leakage flows, the magnitude of these leakage flows is dependent on the level of pressure differences between the chambers. But because there is communication between chambers all the way to the die, the leakage flows become a function of the die pressure.

The leakage flows between chambers, although reducing the output rate, can bring about much intensive mixing, which is of great importance in the twin screw extruder. Idealised theoretical output (Q_{th}) from a twin screw extruder can be calculated as:

$$Q_{th} = 2mNV \qquad ...(6.20)$$

where m = number of thread starts per screw, N = screw speed, and V = chamber volume. In practice, the output rate (Q_{pr}) is less than theoretical throughput because of the presence of leakage flows within the extruder.

$$Q_{pr} = dQ_{th} \qquad ...(6.21)$$

where the empirical constant d is between 1 and 0.

Such interaction between the chambers generates high shear at points at which the screws intermesh, since the material encounters highest shear rates in this region of the machine. Fig. 6.20 shows the different flow patterns of co-rotating and counter-rotating screws. When the screws are rotating in opposite directions, material is forced and milled in the calender gap and is subjected to extremely high shear. These high shear rates contribute to dispersive mixing. Fig. 6.21 shows pressures buildup in co- and counterrotating screws. Pressure buildup allows this material to compress and expand so that vigorous mixing can be achieved. Meanwhile, at the intermeshing point, material is also being exchanged from one screw to the other. In this way, distributive mixing is also being affected because of the rearrangement of flow patterns.

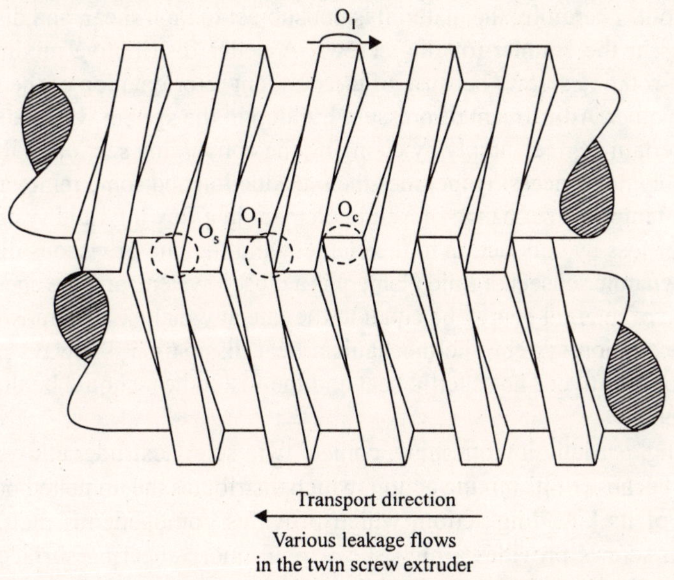

Fig. 6.19. Various leakage flows in partially intermeshed twin screws.

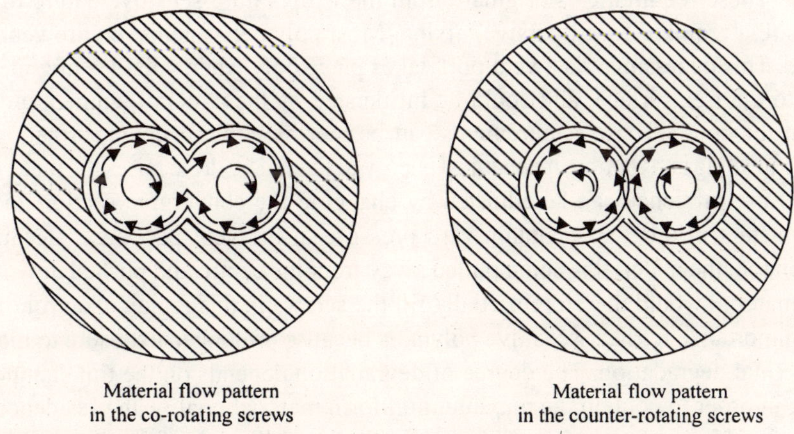

Fig. 6.20. Chamber-to-chamber material flow pattern in co-rotating (left) and counter-rotating (right) twin screws.

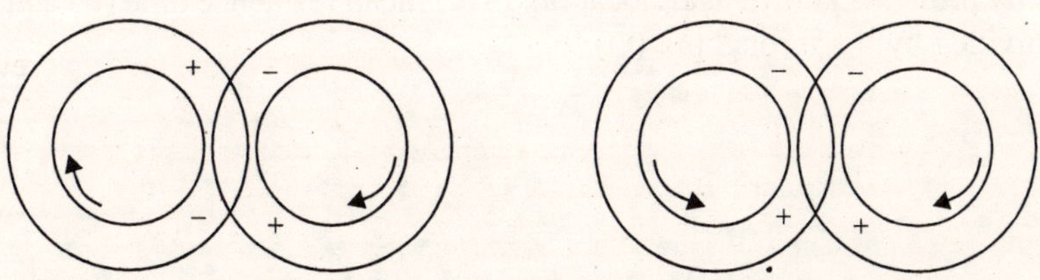

Fig. 6.21. Pressure buildup in a co-rotating (left) and a counter-rotating (right) extruder.

When the screws rotate in the same direction, they are essential in preventing the material from entering the gap between the screws, because as one screw tries to pull the material into the gap, the other screw pushes it out. Therefore, the material is not subject to high shear, and dispersive mixing will not be as thorough as in the counter-rotating screws. A better distributive mixing, however, can be accomplished with co-rotating screws because of the frequent re-orientation of the material as it passes from one screw to the other. Adjustment of pressure buildup in the system with the conical twin screws is also versatile to a certain degree, simply by changing the converging section in the barrel, depending on the applications. Polymer processing includes the introduction and some removal of mechanical and thermal energies. The temperature change of a fluid element in a flowing fluid system is determined by the sum of heat gain or loss by conduction to the element and the rate of viscous dissipation within the element. In a thermodynamic sense, the fluid element is a closed system, and the energy balance requires that its rate of increase of internal energy be equal to the rate at which heat is transferred to it and work is done on it. In the extrusion system, the mechanical heat dissipation is always positive; conduction heat may be positive or negative. The specific heat and the fusion heat should be the average value of a multicomponent system.

The counter-rotating, partially intermeshing, conical twin screw extruder allows the best utilisation of mechanical energy because of its mixing action, which distributes the unmelted polymer into the melt stream, and because of its kneading action, which provides homogeneous melt. Because the flow streamline in the twin screws provides a high shear input, more shearing surface is generated, high pressure is developed, which can be transmitted through the system, and very effective heat transport can be achieved. Practical considerations in mixing equipment usually require maintaining a relatively low temperature. These requirements originate from the temperature-sensitive nature of polymers and the large shear stress required for dispersive mixing. Most polymer systems require venting in order to remove volatiles. This degassing, via a vent port, takes place only when volatiles migrate from the core of the material to the free surface, and could be influenced by the material temperature, the pressure, and the material thickness. These requirements can be met by conical, counter-rotating twin screws. Due to intensive kneading and mixing, mechanical energy delivered by drive raises the material temperature of the melt. Pressure influence can be adjusted by changing the converging section and by changing processing parameters in order to maintain zero pressure in the vent port area. Because the screws rotate outward on top, the material is being pulled away from the centre and pushed down into the barrel along the wall; material would therefore partially fill the screws in this region. Control of the residence time is of great importance for heat-sensitive polymers because prolonged exposure to high temperature may result in thermal degradation. The degree of degradation depends on the time/temperature history of the polymers; in order to contain the maximum uniform material quality, the residence time must be short and uniform. This can be achieved by constantly controlling the feed rate and by maintaining accurate processing control. Residence time distribution can be determined from the velocity profile directly, if the velocity profile is known at all locations. This mean residence time (t′) equals the volume of the system (V) divided by the throughput (Q).

CHAPTER 7

On-Line Measurement of Rheological Properties

INTRODUCTION

There are three types of application of on-line rheometers: (i) process monitoring; (ii) quality control; and (iii) automatic process control. The first type arises primarily in the development of a new material or process, when it is useful to be able to monitor the effects of changes in formulation or operating parameters on product characteristics. In a quality control application, a slow and labour-intensive laboratory test procedure is replaced by a direct indication of product quality. The most sophisticated type of application is in automatic process control, where the rheometer is used as a sensor providing an input signal to the controller.

The rheological properties of an in-process material may be of interest for its own sake, or as a measure of some other attribute such as molecular weight. For some resins, such as 'controlled rheology' types or high-flow injection moulding materials, the viscosity is the primary property that distinguishes different grades, so if the resin manufacturing process could be controlled to produce a preset viscosity, quality control would be automatic. More often, the rheological property is of interest as a measure of some other attribute. For example, the storage and loss moduli might serve as a measure of the molecular weight distribution, or the viscosity might be used as a measure of composition or the state of dispersion of a compound or blend.

TYPES OF ON-LINE RHEOMETERS FOR MELTS

It is common practice to distinguish between 'on-line' and 'in-line' instruments. An on-line rheometer makes use of a side stream and a gear pump to carry a sample of melt to the actual sensor, whereas an 'in-line' rheometer is one that senses the properties of the melt as it passes a particular point in the process. An 'off-line' measurement is one made in a laboratory.

It is more difficult to design a rheometer for in-line use, but such an instrument has the advantage of a short signal delay, and it avoids the use of gear pumps. On the other hand, the sample cannot be brought to a prescribed temperature prior to the measurement. Since fluctations in temperature are a normal part of any industrial process, this means that a temperature compensation procedure must be used to reduce the measured property to a standard temperature. If the material being processed obeys time-temperature superposition, this is quite straightforward, once the shift factor is known as a function of temperature.

On-Line Capillary and Slit Rheometers for Melts

As is the case in the laboratory, the simplest and most common type of process rheometer is the capillary rheometer. A side stream is taken from the main process line and forced through the capillary by a gear pump. The capillary is thermostated to control the melt temperature. The flow rate is calculated from the speed of the gear pump, and the pressure drop is measured directly by means of pressure transducers. By cycling between several pump speeds, the apparent viscosity at several shear rates can be determined.

If the rheometer exits to the atmosphere, only a single vented pressure transducer is required. However, in this type of installation, all the material that enters the rheometer flows out as scrap at a rate of about 0.5 to 1 kg/h. If this is to be avoided, a second gear pump is required to return the capillary efflux back to the main process stream. A second pressure transducer is then required, and the speed of the second gear pump must be carefully controlled to maintain a reasonable and constant level of pressure in the capillary.

Clearly it is not possible to carry out all the measurements necessary to determine the entrance corrections and the Rabinowitch correction, and therefore on-line capillary rheometers are generally used to indicate either the apparent viscosity or to simulate a melt index measurement. The melt index test is simulated as follows. The pressure drop is set at 298.2 kPa, which is the nominal value for the melt index test, and a die having the prescribed L/D is used. Then the motor driving the gear pump is regulated to maintain the pressure drop at the set level, and the resulting pump speed is noted. The simulated melt index value is calculated as follows:

$$MI = 10KSV\rho(D_m/D_c)^3 \qquad ...(7.1)$$

where:
- MI = melt index in g/10 min
- V = pump displacement, cc/rev.
- S = pump speed, RPM
- K = empirical calibration factor to account for all the differences between a melt indexer and the capillary rheometer
- ρ = melt density
- D_m = standard die diameter for melt index test
- D_c = capillary diameter

In practice, the product $KV\rho$ is determined experimentally for the particular resin of interest. On-line rheometers based on slit flow have also been used. Flow in a slit and the equations necessary to calculate quantities of rheological significance are presented there. The advantage of the slit over the capillary is that it is possible to mount pressure transducers flush with the wall. This has been of special interest because, it has been proposed that the first normal stress difference can be determined from wall pressures measured in a slit. Both the hole pressure method and the exit pressure method are described there, and commercial on-line sensors based on both these methods are available. However, as explained questions remain about the reliability of these two methods, and there have been no published reports of their successful use for the on-line measurement of N_1. Another pressure-flow device that can, in principle, be used to measure N_1 is the curved slit rheometer developed by *Geiger*. As in the case of the previously mentioned slit rheometers, however, it is not clear whether this device can form the basis of a useful industrial sensor.

The Rheometrics Melt Flow Monitor is a commercial slit rheometer that was developed not to measure N_1 but to eliminate certain disadvantages of other on-line instruments.

The problems addressed were:

1. Long signal delay due to long side stream.
2. Waste stream produced by rheometer efflux.
3. The need to mount major components near processing machinery.

This rheometer is equipped with two gear pumps, each driven by its own stepper motor, with the speed of the downstream motor regulated to provide a preset pressure at the exit of the capillary. The entire unit is designed to be installed just above a special adapter plate that can be placed at the end of an extruder or incorporated into the barrel assembly so that melt is sampled before it reaches the end of the extruder.

Rotational On-line Rheometers for Melts

In a rotational on-line rheometer, a side stream from a process line carries melt into the rheometer test chamber in which a rotating element is mounted, and means are provided to measure the torque on either the rotating or stationary element. The flow through the rheometer can be continuous, although the through-flow will have an effect on the measured torque, and more reliable results are obtained by periodically stopping the flow to make a measurement. Although the response of such an instrument is relatively slow, it represents a vast improvement over periodic laboratory testing. One challenge in the design of a rotational process rheometer is to provide a method for measuring the torque without interference from a static or dynamic seal. This problem is usually solved by use of a torsion tube, which transmits torque while simultaneously providing a static seal.

In the Rheoprocessor developed by *Kepes* the rotating member is a bicone, and the speed or displacement arising from the application of a constant or time-varying torque is monitored. A seal is not necessary as the rheometer chamber drains to the atmosphere by gravity and is thus not under pressure. A four-way valve allows the periodic stopping of the flow for measurement or the use of a calibration standard.

The Rheometrics On-line Rheometer is a concentric cylinder device designed for the measurement of linear viscoelastic properties. A gear pump provides a controlled flow of melt into the rheometer. The outer cylinder is made to oscillate sinusoidally, and the torque on the inner cylinder is measured by means of a torque tube assembly that provides a perfect seal. The dynamic seal used for the shaft driving the outer cylinder introduces no error, since the torque is not measured on this shaft. *Starita* and *Rohn* have described this rheometer and *Zeichner* and *Macosko* have discussed its application for monitoring the molecular weight distribution of polypropylene.

In another concentric cylinder process rheometer described by *Heinz* the concentric cylinder sensor is a self-contained unit that is inserted into the melt stream, either in an on-line or an in-line installation. The sensor consists of three concentric, thin-walled cylinders. The second (middle) cylinder is driven in an oscillatory mode, and a shear stress is generated on both sides due to the presence of the two stationary cylinders. The torque on the drive shaft is transmitted to an external displacement sensor by means of a torque tube. Bellows are used to provide seals for the moving parts.

Khachatryan have described a concentric cylinder rheometer for use with fibre-forming resins. Two thin-walled concentric cylinders are suspended in a chamber through which the polymer is made to flow. The inner cylinder is rotated by a motor, and the torque transmitted to the outer cylinder is monitored. The published report does not explain how the outer cylinder is suspended nor how the torque arm is sealed where it passes through the wall of the chamber. Another concentric cylinder on-line melt rheometer has been described by *Menges*.

In-Line Melt Rheometers

All of the on-line instruments described above require the use of a side stream and at least one gear pump. In some installations, where it is important to minimise the signal delay or when the use of a gear pump is not practical, an in-line installation is advantageous a 'shear stress transducer' that has been developed for use in a sliding plate rheometer. It has been proposed that such a device could be used as the basis for an in-line rheometer. One possible configuration is shown in Fig. 7.1. A rotating drum (A) generates a nearly uniform shearing deformation in the small gap (B), and the resulting shear stress is monitored by the shear stress transducer (C). The motion of the rotor also generates a drag flow that draws melt into the gap to aid the process of sample renewal. A rheometer of this type has been used to control reactive extrusion processes.

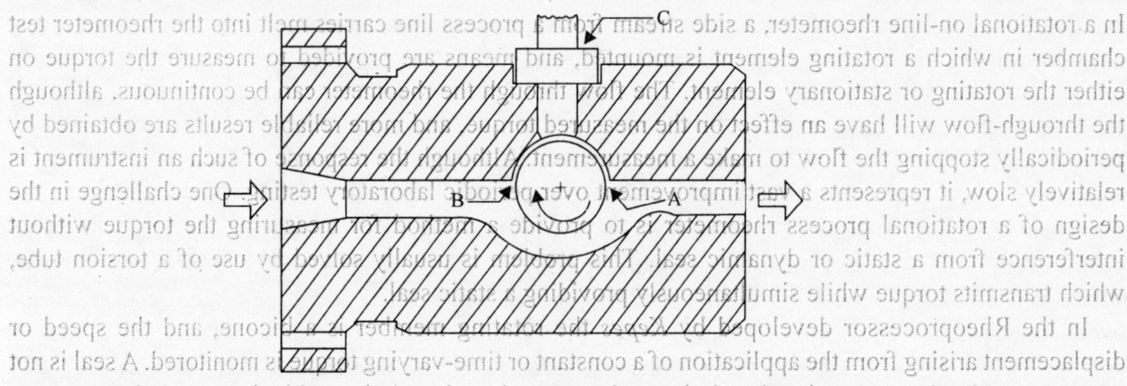

Fig. 7.1. An in-line rotational rheometer that makes use of a shear stress transducer. A = rotary drum driven by constant speed motor or servo-motor; B = gap in which concentric-cylinder drag flow is generated; and C = shear stress transducer that senses the shear stress at the outer wall of the concentric-cylinder drag flow zone.

A single 'flight' machined onto the drum could be used to ensure a clean sweep of the shearing gap as often as desired to ensure complete sample renewal and to avoid the build-up of a degraded layer of material at the outer wall. Using a servo-motor to drive the drum, any desired shear history can be generated, including steady shear, oscillatory shear and step strain. Thus, not only can the viscosity be monitored but also the storage and loss moduli and even non-linear viscoelastic properties. By using a cam-like drum having two or more diameters over various sections of its perimeter, the viscosity at two or more shear rates could be measured once each revolution by simply driving the drum at a constant rotational speed. Another possible application of a shear stress transducer for in-line measurements is the measurement of the shear stress on the wall of an extruder barrel. Because the flow field is neither simple nor under complete control, this could not provide information about any well-defined rheological property. However, if a special 'rheometric module' were inserted into the screw, and if the screw speed were accurately monitored, it would be possible to make a measurement of the viscosity.

Revesz has described a device for measuring the shear stress exerted by the melt on the tip of an extruder screw. He suggests a method for estimating the viscosity of the melt as it enters the die, although the principal purpose of the device is to provide information about melt flow rate.

SPECIFIC APPLICATIONS OF PROCESS RHEOMETERS

Some types of processes in which process rheometers have potential applications are: (i) polymerisation; (ii) extrusion and melt forming; (iii) compounding; (iv) blending; and (v) reactive extrusion. The advantages to be gained by the use of such sensors include improvements in product quality, reduction of waste, and increases in production rates.

Possible manipulated variables in process control applications are:

1. Feed rates of reactants or additives.
2. Process temperature.
3. Extruder speed.
4. Geometry of die or other mechanical components.

Feed rate can be controlled by a liquid pump or by a loss-in-weight solids feeder, and this is the obvious choice of manipulated variable in reactive extrusion operations. For example, it may be desirable to accomplish a controlled degree of degradation or of cross-linking to produce an extrudate having a specified viscosity or elasticity. Temperature is not truly a possible manipulated variable, as it is actually changed by manipulation of either the thermal or mechanical energy input to the process. Processing machinery usually has a large thermal inertia, and this leads to a slow response to a change in the temperature set point if thermal energy input is used as the manipulated variable. Changing screw speed alters the rate of mechanical energy input, and this is a much faster way of changing the temperature. However, this also changes the flow rate, pressure, and strain history of the melt. Changing machine geometry, for example by means of a hydraulically variable die gap, can change the rheological properties of a shear sensitive polymer such as LDPE. An obvious area of application is in the polymerisation process itself, although there have been no published reports of such uses. If the product of a polymerisation process is itself a melt, then a process rheometer can be used to monitor this product directly. However, several modern polymerisation processes produce a powder rather than a melt, and in order to use rheology as a tool for process control it is necessary to provide a mechanism for automatically sampling this stream and converting powder to melt. The Rheometrics ROR-QC addresses this problem by use of pneumatic transport. *Starita* and *Macosko* have discussed some ways in which variations in resin characteristics can cause problems in melt forming operations such as blow moulding. If a process rheometer could provide continuous information regarding the nature of the raw material, operating conditions could be altered to compensate for such changes.

Variations in viscosity can cause serious difficulties in an injection moulding operation, especially when large parts are being produced, and it would be useful to be able to monitor melt consistency on a continuous basis. At present, nozzle pressure is sometimes used as a rough indication of viscosity.

The use of compounding operations to produce sophisticated multicomponent materials is growing rapidly, and there is a widely recognised need for methods to ensure that the product has the correct formulation and is properly mixed. Process rheometers show promise in this area, specifically to monitor the state of dispersion of a masterbatch, a compound or a blend.

Considerable attention has been focused on the use of process rheometers for the closed-loop control of reactive extrusion processes. If a rheological property is sensitive to the extent of reaction, then a process rheometer can be used to provide a signal for the control of the feed rate of one of the reactants. Several reports have been published on the use of rheometers to control the peroxide-initiated degradation of polypropylene. This is a process that is used to produce a range of lower molecular weight resin grades starting from a single high-molecular weight grade using an extruder. Both rotational and capillary on-line rheometers have been evaluated for this application. In addition, *Pabedinskas* used the die pressure drop as a measure of the amount of degradation.

SECTION II

Properties and Testing of Polymers

SECTION II

Properties and Testing of Polymers

CHAPTER 8

Identification of Polymers

INTRODUCTION

There is no simple way of identifying polymers plastic materials used for an end-product just by having a feel by touch, odour or taste. There are a large varieties of polymers and in each polymer group, there are a number of grades with varying properties. In addition, there are copolymers, blends, alloys, elastomers and reinforced plastics. Also available are end-products using combination of more than one polymer as in the case of co-extruded multilayer films, sheets, bottles and tubes of laminates using various films for a composite structure. Quite often an end-product is manufactured using varying quantity of reprocessed material with the virgin material.

Also, many polymers undergo significant change in their structure on processing making the identification of basic polymer quite difficult. Therefore, identifying any plastic material is a very complex process. This chapter gives general methods commonly employed in the plastic industry for spot testing of end-products.

SETTING-UP OF INHOUSE IDENTIFICATION FACILITIES

Identification of most of the plastics can be easily done following the simple identification techniques. For carrying out this, a small laboratory can be set up with minimum expenses and tests can be carried out by any qualified chemist. A detailed specification of the laboratory, instruments and glasswares, chemicals and other accessories needed is presented alongwith simple ways of preparing various solutions.

Laboratory

Space required for setting up the laboratory is around 30 × 20 square feet. The room should be well ventilated and lit. There should be minimum of two exhaust fans fixed at the top of the wall. Working table of 3' height, 3' width and 10' length is to be fixed. This may be made of either wood or concrete with ceramic tiles fixed on top of it. Two such tables can be fixed parallel to each other with necessary electrical and gas connection fixed in between. There should be 15 amps and 5 amps electrical sockets two each. Three gas taps are to be fixed for connection to burners. Taps are to be connected to LPG gas cylinders which are kept away at a safe distance. Two gas cylinders are required one in usage and one for replacement. There should be two to three water taps with sinks and one or two fire extinguishers in the laboratory. There should be enough shelves under the table for keeping the laboratory wares and chemicals. Laboratory wares and chemicals are to be kept separately in the cupboards. A fuming cupboard should be fixed adjacent to one of the walls. One exhaust fan and sliding door which can move up and down are to be provided. Inside the fuming cupboard two 15 amps sockets are to be fixed, one for muffle furnace and one for any other heating requirement.

Equipments

One air circulating oven which can reach 200°C with ± 2°C accuracy is required for drying of glass wares and any other material. Corrosive chemicals and easily inflammable materials should not be dried inside. One muffle furnace with temperature range room temperature to 1200°C with suitable thermocouple is needed for burning the plastics to find out the fillers present in them. One water distillation unit for distilling water is needed.

Two balances one with 0–150 gm range and 0.1 mg accuracy and one with 1 kg range are required. In the first case electronic balance is preferred.

Glasswares and Other Accessories

Glasswares and other accessories commonly used in any chemical laboratory only are needed for identifying plastic materials. These are available with all the scientific instrument suppliers. All the wares needed are listed below :

1. Beakers, 50, 100, 250, and 1000 ml
2. Combustion tubes, approx. 8 × 70 mm
3. Crucibles (platinum or nickel) approx. 3–3.5 cm in diameter
4. Decicators
5. Glass funnels, approx. 4 and 7 cm in diameter
6. Glass rods
7. Graduated cylinder, 10 ml, 100 ml, 500 ml
8. Pipets, 1 ml, 10 ml
9. Porcelain crucible, approx. 3–3.5 cm in diameter
10. Porcelain cups, approx. 5 cm in diameter
11. Porcelain mortar with pestle, approx. 10 cm in diameter
12. Porcelain plate
13. Safety glasses
14. Specific gravity bottles
15. Test tubes, small with approx. 7 mm diameter and medium with approx. 15 mm diameter
16. Watch glasses
17. Corks or rubber stoppers to fit the test tubes
18. Crucible tongs
19. Filter paper
20. Knife
21. Oil bath (silicon oil is preferred)
22. Spatula
23. Stands (metal)
24. Test tube racks
25. Test tube tongs
26. Thermometers

27. Tweezers
28. Wire mesh (for heating purpose).

Optional accessories

1. Distillation flask, reflux condenser
2. Heating mandles
3. Melting point apparatus
4. Mill for grinding plastic samples
5. Shore hardness testers
6. Sohxlet apparatus.

Chemicals

Chemicals needed for carrying out identification tests are listed below. All these chemicals are available with chemical suppliers. Chemicals are normally available in 500 ml bottles. As for indicator requirements, it is generally sufficient to order 1-5 gm. Chemicals should be stored in unambiguously identified bottles.

It must be kept in mind that many organic solvents are flammable and should therefore be stored away from flame. Concentrated acids, bases and other corrosive chemicals need special care during handling since they might cause injuries.

All solvents and chemicals are available in several degree of purity, for example commercial, analytical grade etc. As far as possible only analytically pure reagents should be used. Solvents that have turned yellow or dark during storage should be distilled before use.

Solvents

Acetone, Benzene, Benzyl alcohol, Carbon tetrachloride, Chloroform, *m*-cresol, Cyclohexanone, Di-ethyl ether, Ethanol, Ethyl acetate, Ethylene glycol, Isopropyl alcohol(propanol), *n*-Hexane, Methanol, Methylene chloride, Nitrobenzene, Petroleum ether, Pyridine, Tetrahydrofuran, Toluene, *p*-xylene.

Organic reagents

Acetic anhydride, Alizarin red, Aniline, Anthrone, Benzidine, Chromotropic acid, 2,6-dibromoquinone-4-chloramide, *p*-Dimethyl amino benzaldehyde, Diphenyl amine, Formalin, Hydroquine, Hydroxyl amine hydrochloride, Malachite green, Maleic anhydride, Methylene blue, Marpholine, Alpha-Naphthol, Beta-Naphthol, *o*-Nitrobenzaldehyde, 4-Nitrophenyldiazoniumfluoroborate (Cochlight, England), Phenyl hydrazine, Phenyllithium, Phthalic anhydride, Resorcinol, Thymol.

Inorganic chemicals

Ammonium molybdate, Ammonium sulphate, Barium hydroxide solution, (approx. 0.2 N), Bismuth nitrate, Borax, Calcium chloride, Copper acetate, Copper sulphate, Iodine, Iodine monochloride, Iron chloride (1–5) N soln in water, Iron sulphate, Lanthanum nitrate, Lead acetate as 2N solution(26.7 g in 100 g of water), Magnesium chloride and anhydride for density determination, Mercury, Mercury oxide, Mercury chloride, Nickel sulphate, Phosphoric acid, Potassium chloride, Potassium dichromate, Potassium hydroxide, Potassium iodide, Potassium thiocyanate, Silver nitrate, 2% solution, Sodium (to be immersed in kerosene), Sodium acetate, Sodium carbonate, (anhydrous), Sodium hydroxide, Sodium hypochlorite or chloride of lime solution, Sodium nitrate, Sodium nitroprusside, Sodium peroxide, Sodium thiosulphate, Sulphur, Wijs solution (or) Iodine monochloride, Zinc chloride, (anhydrous) for density determination, Zirconium nitrate.

Acids and bases

Acetic acid, Hydrochloric acid, Nitric acid, Sulphuric acid, Ammonia solution.

Miscellaneous

Activated charcoal, Congo red paper, Copper wire, Cotton floak, Fine sand, Glass wool, Lead acetate paper(keep in a closed bottle), Litmus paper (red and blue), pH paper, Silver coin (for sulphur identification).

IDENTIFICATION OF PLASTICS BY SIMPLE METHODS

Physical Identification

Visual appearance

Raw polymers are usually transparent or translucent, whitish or pale brown. Thermoplastics are generally produced as powders or granules; uncompounded thermosetting resins are in the form of either high viscous liquids or low melting temperature solids and compounded stocks are generally coloured and opaque.

Method of fabrication

Injection moulded articles will have gate marks and flash marks. In such cases the material may be thermoplastic or rarely injection grade thermosetting material. If the products have only mould split lines without any gate marks, the material may be thermoset. Sometimes even thermoplastic materials like PTFE and UHMHDPE are compression moulded and sintered and those articles will not have any gate marks. Cast nylons also do not have any marks.

Penetration to hot rod and cutting with a knife

A hot soldering iron rod is kept over the material. If the material is thermoplastic, it will soften and melt and if it is a thermoset it will char directly without melting. When cut with a sharp knife, thermoplastics can be easily pared off, whereas, thermosets will be difficult to pare off. Also thermosets will become flaky or powdery. Thermoplastics like PMMA and Polystyrene(PS) are also difficult to pare.

Floatation test

If the material floats in water it will be polyolefins, usually polyethylene and polypropylene. But filled polyethylene and polypropylene will sink in water.

Colour

Phenolic resins are invariably dark in colour whereas UF and MF resins are available in light colours. PTFE is milky white in colour. Virgin Acrylics, PC and some grades of PVC are clear and transparent.

Odour

Plasticised PVC will have the smell of plasticisers (example: DOP). In cellulose nitrate camphor is added as plasticiser and will have the smell of camphor.

Burning test

The observation that may be made with the aid of flame are: (i) how the material burns in flame; (ii) whether the material burns after the removal from the flame; and (iii) what kind of odour is given off.

Bending test

Some plastics will exhibit particular characteristics in the manner in which they respond to bending. It is important that the specimens are of approximately the same dimension for getting better comparative results.

Film tear test

Many plastics are available as thin films and because of the method of production, may exhibit specific characteristics in the manner in which they tear. Cut a piece of film to about 50 mm × 12 mm. Then make a short cut at one end with a pair of scissors. Pull and attempt to tear the film the rest of the way. Repeat the test in a direction at right angles.

Chemical Indentification

Pyrolysis test

A few milligrams of polymeric sample is pyrolysed in a 50 mm × 10 mm ignition tube and the pyrolysed vapour is tested with a moistened indicator paper. Acid vapours turn blue litmus to red and alkaline (basic) vapours turn red litmus to blue.

Acid vapours may come from carbohydrate polymers and their derivatives. High acid vapours often indicate the presence of chlorine. eg. PVC, PVDC or chlorinated rubber.

Alkaline vapours indicate the presence of Nitrogen. Examples of polymers containing nitrogen are polyamides (Nylons), polyurethanes, urea formaldehyde and melamineformaldehyde. Neutral vapours are evolved from hydrocarbon polymers, silicones and some polyesters.

Solubility test

Solubility tests form a basis for the identification of some main types of plastic materials. However, in many cases solubility varies considerably for different samples of the same resin and it is difficult to interpret the results.

Softening and melting points

The temperature range of softening and melting of polymers as well as their densities can also be used to identify them. The determination is carried out in the usual apparatus, melting point apparatus, the Kofler microscope or the kofler hot bench observation with a microscope using polarised light. For the accurate determination of melting points differential scanning calorimetry can be used.

Detection of elements

Preparation of sodium fusion extract

A small piece of clean sodium (about 0.02 g) is placed in an ignition tube warmed and small quantity (about 0.1 g) of the sample to be tested is added. The tube is gently heated until reaction subsides and then strongly until the glass is red-hot. The tube is then plunged into distilled water (10 ml) contained in a china dish. The filtrate is divided into four equal portions which are used for the detection of elements.

Nitrogen

Two ml of the sodium fusion extract is boiled with 3 drops of freshly prepared aqueous ferrous sulphate and then cooled, after acidification with acid, a drop of 0.5N aqueous ferric chloride is added. A blue precipitate confirms the presence of nitrogen.

Chlorine and Bromine

Two ml of the sodium fusion extract is acidified with 5N nitric acid, the solution is boiled for 2 min. 0.01N aqueous silver nitrate is then added. A white precipitate which is soluble in ammonia confirms the presence of chlorine whereas a yellow precipitate which is insoluble in ammonia confirms the presence of bromine.

Fluorine

An aqueous solution of zirconium nitrate (0.1%) and alizarin red (0.1%) is prepared Filter paper is immersed in this solution and then allowed to dry. When required, a small piece of the paper is moistened with aqueous acetic acid (50%). The solution to be tested is neutralised with 5N hydrochloric acid and a drop placed on the moistened test paper. A red spot turns yellow indicates the presence of fluorine.

A 0.5 g of the sample is pyrolysed in a Bunsen flame and after cooling, few ml of concentrated sulphuric acid is added. The presence of fluorine is indicated by a characteristic non-wettability of the wall of the test tube.

Sulphur

Two ml of sodium fusion extract is added to three drops of freshly prepared aqueous sodium nitro prusside (approx. 5%). A violet colour indicates the presence of sulphur.

Detection of phosphorus

One gm of the sample is heated with 3 ml of concentrated nitric acid and few drops of the concentrated sulphuric acid, cooled, diluted with water and some drops of molybdate solution (preparation given below) is added and heated. A yellow precipitate indicates the presence of phosphorus.

Preparation of ammonium molybdate solution

Dissolve 30 g ammonium molybdate in hot water, cool the solution and add 100 ml of water. Pour a thin stream of this solution into a solution of 10 g ammonium sulphate in 100 ml of approximately 55% nitric acid and mix. Let stand for one day, filter through a thick acid-proof filter and store well-sealed in the dark.

CONFIRMATION TESTS

Tests for Polyolefins

A piece of dry sample is pyrolysed in a tube closed with a filter paper which is drenched with a solution of 0.5 g yellow mercury oxide in sulphuric acid (1.5 ml conc. sulphuric acid added to 8 ml. water). If the vapour gives a golden yellow spot, indicates polyisobutylene, butyl rubber and polypropylene (the latter only after a few minutes). Polyethylene does not react. Natural and nitrile rubber, as well as polybutadiene yield a brown spot.

Test for Chlorine Containing Polymers

Heat a small amount of polymer in pyridine and make a solution. Add few drops of 5% alcoholic (methanol) NaOH to a portion of boiling solution and a portion of solution in cool condition. Observe the change of colour taking place in the solutions.

Test for Caprolactum in Nylon 6

About 0.5 gm of sample is heated in 5.0 ml. of distilled water and allowed to boil for 10 to 15 min. After cooling, 2-3 drops of con. H_2SO_4 is added to 0.5 ml of the above solution followed by addition of 2 ml.

of potassium iodo bismuthate (a solution of 5 gm. of basic bismuth nitrate and 25 gm. of potassium iodide in 10 ml. of 2% H_2SO_4). Precipitation of an orange red complex indicates the presence of caprolactum.

Test for Adipic acid in Nylon 6,6

About 0.2 gm of sample is heated in low flame in a test tube and the vapour coming out of the tube is passed over a filter paper moistened with a saturated solution of *o*-nitrobenzaldehyde in 2N aqueous NaOH. A deep mauve(violet) colour confirms the presence of adipic acid. A yellowish green colour develops in the case of polyethylene terephthalate and polybutylene terephthalate confirming the presence of terephthalic acid.

Test for Polycarbonate

1. About 1.0 gm of sample is pyrolysed in an ignition tube which is plugged with cotton. The cotton is removed and immersed in 1% methanolic solution of *p*-dimethylamino benzaldehyde and then one drop of 5 N-Hydrochloric acid is added. A dark blue colour appears in the case of polycarbonates. A red colour that does not change to blue is observed in the case of polyamides.
2. The cotton floak in which the pyrolysed vapours of the sample absorbed is treated with dilute (1:1) HCI. An intense red colour which is unaffected by methanol indicates the presence of polycarbonate.

Test for PMMA

About 1 gm of sample is heated in an ignition tube and the pyrolysate is collected in a test tube wrapped with a wet filter paper. To the distillate, 1 ml of conc. HNO_3 is added and heated just to boiling. After cooling, 5 ml. of water is added and then sodium nitrite (0.1 gm) is added. A blueish green colour confirms the presence of polymethylmethacrylate.

Test for Polyacetals

Polyacetals produce formaldehyde on heating. A small amount of sample is heated with 2 ml. conc. sulphuric acid and a few crystals of chromotropic acid for about 10 min at 60-70°C. A strong violet colour indicates formaldehyde.

Tests for PET and PBT

PET and PBT are soluble in nitrobenzene. A small sample is pyrolysed in a glass tube covered with filter paper. The filter paper is drenched with a saturated solution of *o*-nitroben zaldehyde in dilute sodium hydroxide. A blue-green colour which is stable against dilute hydrochloric acid, indicates terephthalic acid. Then PET and PBT are differentiated based on melting points.

Test for Polyurethane

1. About 0.5 gm of sample is dissolved in 10 ml. of glacial acetic acid and 0.1 gm. of *p*-dimethylamino benzaldehyde is added. The solution turning yellow after several minutes shows the presence of polyurethane.
2. Reaction with *p*-nitrophenyl diazonium fluroborate.

Filter paper is moistened with 1% of *p*-nitro-phenyl diazonium fluroborate. The filter paper is inserted in the fumes produced from the pyrolysing of the sample.

The di-isocynate used in the manufacture of polyurethane can be identified according to the colour that develops in the paper as given below:

Reddish Brown	-	Toluene di-isocyanate or Dimeric toluene di-isocyanate (TDI)
Violet	-	Napthalene di-isocynate (NDI)
Yellow	-	Methylene di-isocynate (MDI)
Beige		Phenylene di-isocynate (PDI)

3. Colour reaction to differentiate polyester urethanes from polyether urethanes.

A sample of about 5.0 gm is finely powdered and treated with few drops of the sodium hydroxide solution (2N methanolic solution with little phenolphthalein) and then with a few drops of a saturated methanolic solution of hydroxylamine hydrochloride solution. The mixture must be alkaline. For highly cross linked polymer, the mixture is heated for 20-30 secs. The mixture is then acidified with 1N hydrochloric acid and each drop of the acid is followed by an addition 3% ferric chloride reagent until a colour develops.

In the presence of a polyester-urethane, the mixture turns violet. In the case of polyester-urethane based on castor oil or dimerised fatty acid, the mixture turns brown or violet brown, polyether-urethanes do not give the characteristic colour reaction.

Test for Cellulose in Cellulosics (Molisch Reaction)

Sample is dissolved in acetone, reacted with 2 to 3 drops of 2% ethanolic solution of 1-napthol and a few drops of conc. H_2SO_4 is allowed to flow down the wall of the test tube. Red to reddish brown ring at the interface indicates cellulose.

Test for acetates and propionates

The sample (about 0.2 gm) and few ml of Hydrochloric acid are heated for 10 minutes and dilute ammonia is added to make alkaline. One ml of this solution is reacted with 1-2 drops of a 5% lanthanom nitrate solution and 1 drop of 0.1N Iodine solution.

A fast colour change to blue confirms the presence of acetates and a brown colour indicates the presence of propionate. For differentiation between cellulose acetate and cellulose acetate-butyrate, it is usually sufficient to examine the vapour produced by dry heating of the sample. The acetate smells like acetic acid, the acetate-butryate has the smell of both acetic acid and butyric acid (like rancid butter).

Test for cellulose ethers

The important cellulose ethers are methyl cellulose, ethyl cellulose and benzyl cellulose.

Detection of methyl cellulose

The sample is treated with a 10% tannic acid solution. A flaky (flocculent) precipitate confirms the presence of methyl cellulose. A sample is treated with iodine - potassium iodide solution. A colour change from violet brown to brown, which disappears on the addition of strong NaOH confirms the presence of methyl cellulose.

Reaction to heating and burning

Cellulose ethers melt and char after heating and continuously burn after ignition. Nature of flame and smell of the cellulose ethers are as follows:

Methyl cellulose	-	Brilliant yellow flame	
Ethyl cellulose	-	Less brilliant flame accompanied by dripping	Odour of burnt paper
Benzyl cellulose smells like bitter almonds	-	Brilliant sooty flame	

Detection of ethyl cellulose

The sample is treated with one drop of dichromate solution (1.0 g of potassium dichromate $K_2 Cr_2 O_7$ + 60 ml water + 7.5 ml of con.H_2SO_4) and then heated to 100°C. The mouth of the tube is covered with a filter paper moistened with a mixture of equal volumes of 20% aqueous morpholine solution and 5% aqueous sodium nitroprusside. A blue colour indicates the presence of ethyl cellulose.

Tests for phenol formaldehyde (PF), urea formaldehyde (UF) and melamine formaldehyde (MF)

A small amount of sample is boiled in water in presence of H_2SO_4 and filtered. Few drops of 5% aqueous chromotropic acid solution (1,8-dihydroxynaphthalene-3, 6-disulphonic acid) and excess of conc.H_2SO_4 are added to the filterate and heated at 100°C for few minutes. In the presence of formaldehyde, the solution turns violet/dark violet. Poly(vinyl formal), polyoxymethylenes, PF, UF and MF resins contain formaldehyde.

1. *Test for phenol in PF:*
   (a) *Million's reagent:* About 10 gm of Hg is dissolved in 10 ml. of HNO_3 by gentle heating and then diluted with 15 ml. of distilled water.

   The sample is heated to boiling for two minutes with 1 ml. of clear million's reagent. Red colour indicates the presence of phenol in the sample.
2. *Test for urea in UF:* About 0.5 gm. of sample is refluxed for 30 min. with 20% acetic acid (50 ml.). The mixture is cooled and filtered. To the filterate added a solution of (2 ml.) xanthydrol in methanol (1% solution) and boiled for 1-2 minutes. White bulky precipitate confirms the presence of urea.
3. *Test for melamine in MF:* About 0.5 gm of sample is refluxed with 80% acetic acid (25 ml.) for 30 min., cooled and filtered. Then the filterate is evaporated to dryness and added 2 ml. of water, cooled and filtered. To the filterate one drop of saturated aqueous picric acid is added. Yellow precipitate indicates the presence of melamine.

Test for epoxy resin

Foucry test

About 0.25 gm. of sample is dissolved in 98% H_2SO_4 by slight warming. After cooling 1 ml. of 63% HNO_3 is added. After shaking, the mixture is poured into 100 ml. of 5% aqueous NaOH. A bright red or orange red colour indicates the presence of epoxy resin.

Test for alkyd resins

Oil modified polyesters are called as alkyd resins. Since in most of the alkyd resins, phthalic anhydride is used as the major polybasic constituent the presence of phthalic anhydride is identified by the following test.

Test for phthalate

The sample (0.1 gm) is carefully melted in a test tube with 0.3 gm of crystalline phenol and one drop of concentrated sulphuric acid. After cooling, the melt is dissolved in water (10-20 ml) and the solution is made alkaline with 5% NaOH. A red colour indicates the phthalate which confirms the alkyd resins.

IDENTIFICATION OF PLASTICS MATERIALS

Thermoplastics

ABS

ABS is an amorphous terepolymer with a specific gravity of 1.04. It is soluble in toluene and ethylene dichloride. It burns with a yellow flame and an acrid odour, and continues to burn after the removal of the flame source with black smoke with soot and drippings. Infrared spectroscopy is the best method for the positive identification of ABS. The same technique can also be used to further identify a percentage of each polymer in the terepolymer composition.

Acetal

Acetal is available as a homopolymer and copolymer. Both types burn with a blue flame and give off a toxic formaldehyde odour. They are both crystalline with a specific gravity of 1.41. The homopolymer can be distinguished from the copolymer by a Fisher-John melting point test. The copolymer acetal has a lower melting point than the homopolymer. Both types are soluble in hexafluoroacetone sesquihydrate. However, due to the extreme toxicity of this solvent, it is rarely used for identification purposes. A differential thermal analysis (DTA) can also be used to distinguish a homopolymer.

Acrylic

Acrylic is an amorphous polymer with a specific gravity of 1.18. Acrylic is one of the few transparent plastics used in outdoor applications. It burns with a blue flame with a yellow tip, gives off fruity odour, and does not produce smoke while it burns. Acrylic is soluble in acetone, benzene, and toluene. Infrared spectroscopy is the best means of identification of this polymer.

Cellulose acetate

Cellulose acetate is an amorphous material with a specific gravity of 1.30. It burns very slowly with a yellow flame, drips, gives off an odour of vinegar, and has black smoke with soot. It is soluble in acetone, furfuryl alcohol, and acetic acid. Cellulose acetate is positively identified by infrared spectroscopy.

Cellulose acetate butyrate

This polymer burns with a yellow flame with a blue tip, drips, and gives off an odour of rancid butter. The specific gravity is 1.24 and is soluble in acetone and trichloromethane. Cellulose acetate butyrate is also one of the few weatherable transparent plastics.

Cellulose propionate

The cellulose propionate is also an amorphous material like other celluloids with a specific gravity of 1.20. It burns with dark yellow flame and an odour of burnt sugar. The material drips while burning and the drippings also burn like other celluloids. Acetone, carbon tetrachloride, and trichloromethane are the most common solvents.

Fluorocarbons (FEP, CTFE, PTFE, PVF)

Fluorocarbon plastics do not actually burn when exposed to a flame. They can be easily identified by a copper wire test that indicates the presence of fluorine by a bright green-coloured flame. The fluoroplastics have a very high melting point. PTFE has a waxy surface and a specific gravity of 2.15. Fluoroplastics are practically impossible to dissolve in any chemical.

Nylons

There are many types of nylons with a specific gravity ranging from 1.04–1.17. All types burn with blue flames and yellow tips and give off a burnt wool or burnt hair odour. Nylon self-extinguishes on removal of the flame. Phenol, *m*-cresol, and formic acid are the most common solvents. Different types of nylons can be identified by the Fisher-Johns melting point test (ASTM D 789). Solubility and specific gravity tests are also used to differentiate between the types of nylons. However, infrared spectroscopy is the best method for positive identification.

Polycarbonate

Polycarbonate is one of the toughest transparent thermoplastics. It has a specific gravity of 1.2, burns with a yellow or orange flame with the odour of phenol, and gives off black smoke with soot in the air. Polycarbonate is soluble in methylene dichloride and ethylene dichloride. Infrared spectroscopy is the best method of identification for polycarbonate.

Thermoplastic polyester

Thermoplastic polyester burns with a yellow flame with a blue edge, and as it burns it drips and gives off black smoke with soot. The characteristic odour is one of burning rubber.

PVC

This self-extinguishing amorphous plastic has a specific gravity of 1.2–1.7. It burns with a yellow flame and a green tip, and gives off the odour of hydrochloric acid and white smoke. Tetrahydrofuran and methyl ethyl ketone are common solvents for PVC. PVC can be easily identified by the copper wire test which indicates the presence of chlorine by a bright green flame. There are many different types of vinyl polymers such as a polyvinyl acetate, polyvinylidene chloride, and ethylene vinyl acetate. The type of vinyl can be positively identified by infrared spectrometric techniques.

Polyethylene

Polyethylene is one of the very few crystalline plastics that will float on water. The specific gravity ranges from 0.91–0.96. It burns quickly with a blue flame with a yellow tip. Polyethylene drips while it burns and gives off paraffin odour similar to a burning candle. Polyethylene is impervious to most common solvents. However, it can be dissolved in hot toluene or hot benzene. Infrared spectroscopy is used to confirm the identity of polyethylene.

Polypropylene

This polyolefin family polymer has a specific gravity of 0.88 and floats on water. It burns with a blue flame with a yellow tip and gives off an acrid odour similar to diesel fumes. The material is soluble in hot toluene. Polypropylene is positively identified by infrared spectrometer.

Polystyrene

Polystyrene is an amorphous plastic with a specific gravity of 1.09. It burns with a yellow flame and gives off the odour of illuminating gas or marigolds. The material burns and drips, creating a dense black smoke with soot in the air, also dripping continues to burn. Polystyrene is soluble is acetone, benzene, toluene, and ether. Polystyrene is often modified with rubber to improve its impact strength. These different types of polystyrenes can be identified by infrared spectroscopy.

Polyphenylene oxide (PPO)

PPO has a specific gravity of 1.06. It burns with a yellow-orange flame without dripping and gives off an odour of phenol or burning gas. It exhibits self-extinguishing characteristics. PPO is soluble in toluene and dichloroethylene. Like other polymers, PPO can be identified by infrared spectroscopy.

Polysulphone

This self-extinguishing amorphous thermoplastic has a specific gravity of 1.24. It burns with an orange flame, give off the pungent odour of sulphur, and produces black smoke with soot in the air while it burns. Polysulphone is soluble in methylene chloride.

Polyurethane (thermoplastic)

This generally easy-ignite plastic burns with a yellow flame with a faint odour of apple. It produces black smoke. Thermoplastic polyurethane is soluble in tetrahydrofuran and dimethyl formamide. The specific gravity of this material is 1.2.

Thermosetting Plastics

Diallyl phthalate (DAP)

The specific gravity of this compound ranges from 1.30–1.85 depending upon the type of filler. It burns with a yellow flame, produces black smoke, and gives off the faint odour of phenol. It is self-extinguishing in nature and is difficult to ignite.

Epoxy

The specific gravity of epoxy compounds range from 1.10–2.10. It burns with a yellow flame and gives off a black smoke that gives off pungent amine odour. Different types of epoxies can be characterised by infrared spectroscopy and thermal analysis.

Phenol formaldehyde

Phenol formaldehyde burns with a yellow flame and gives off the odour of phenol. Depending upon the type of filler, it may or may not be self-extinguishing. Phenolic compounds are only available in dark colours. The specific gravity ranges from 1.30–1.90. Phenol formaldehyde is soluble in acetone and acetic acid.

Urea formaldehyde

Urea formaldehyde burns with a yellow flame with a greenish blue edge. It gives off strong odour of formaldehyde and the material usually swells and cracks as it continues to burn.

Melamine formaldehyde

This self-extinguishing thermoset has a specific gravity of 1.47–1.8. It burns with a yellow flame with a blue edge and gives off a fish-like odour. The material swells, cracks, and turns white at the edges of a burns section. Melamine formaldehyde is soluble in acetone and cyclohexanone.

Polyesters

Polyesters burn with a yellow flame with blue edges and produce black smoke with soot in the air. They continue to burn without dripping after the removal of the ignition source and give off a sour cinnamon odour. The specific gravity ranges from 1.30–1.50.

Silicones

Silicones burn with a bright yellow flame without odours and continue to burn after the removal of the flame source. The specific gravity ranges from 1.05–2.82. The positive identification is made by an infrared spectrometer.

IDENTIFICATION OF POLYMERS BY INSTRUMENTAL METHODS

In recent years sophisticated instrumental techniques are playing an important role in identification and characterisation of polymers. More recently, advances in computer technology have been combined with analytical instruments to give analytical speed, resolution, simplicity and minimal sample requirements, unimagined a few years ago. Here the more important instrumental methods used for plastic analysis are described briefly and Table 8.1 highlights instrumental techniques of polymer and additive separation, identification and analysis.

Table 8.1. Instrumental techniques of polymer and additive separation, identification and analysis.

LC/GPC	*GC*	*IR*	*TA*	*X-ray/OES*	*NMR*	*AA/Wet chemistry*	*Microscopy*
Antioxidants Phenols Thioesters Phosphites Plasticisers Lubricants Polymer M.W. distribution	Residual monomers Non-polymeric compounds; Oils Plasticisers	Polymer composition Additives: Qualitative Quantitative Phosphates	Fillers: Carbon black Glass Lubricants: Molybdenum disulphide Polymer M.W. (degradation) PE cross-linking	Fillers: Talc/mica Flame retardants Alumina-trihydrate Antimony trioxide Stabilisers: Organotin	Polyesters Silicones Phenols Mineral oil	Lubricants Flame retardants Halogenated compounds Catalyst residues	Contaminants Surface films Continuity Crystallinity

Infrared Spectroscopy

For the past last two decades infrared spectroscopy(IR) has proved to be a valuable method in studying polymers. With infrared analysis and its versatility in sampling techniques, the structural identification of many polymeric materials is made easy and quick. Not only one is able to elucidate the polymeric structure with this instrumental technique, but also obtain valuable information concerning the characteristics of a specific material.

The advent of FTIR has brought an increased interest in infrared spectroscopy as a characterisation technique. The increased speed as higher signal to noise of FTIR compared to dispersive infrared has lead to a substantially greater number of applications in polymer research. Also FTIR instrument has allowed the digitised spectra to be treated by sophisticated data processing techniques and has improved the quality of infrared spectra and reduced the time required for analysis. FTIR has served to rekindle our interest in applying the IR technique to complex multicomponent polymer system such as blends and alloys. There are two kinds of fundamental vibrations for molecules, stretching, in which the distance between two atoms increases or decreases, but the atoms remain at the same bond axis and bending (or deformation), in which the position of the atom changes relative to the original bond axis. The various stretching and bending vibrations of a bond occur at certain quantised frequencies. When infrared light of the same frequency is incident on the molecule, energy is absorbed and the amplitude of that vibration is increased. When the molecule reverts from the excited state to the original ground state, the absorbed energy is released as heat.

Sample preparation

According to the nature of the investigations, and the properties of the compound under examination, a suitable sample preparation technique is adopted. Analysis can be conducted with capillary films, solutions, films, Nujol suspensions(mulls) and in potassium bromide pellets.

Capillary films

Liquids are most conveniently studied as capillary films. A drop of a liquid is placed between two KBr of NaCl plates which are fixed in the mounting and inserted into the spectrophotometer. In the reference beam of the instrument is placed a potassium bromide or sodium chloride plate twice as thick as that used in the sample cell.

Solutions

Spectral examinations of solutions are primarily used in quantitative analysis and are largely concerned not with the polymer themselves but with the secondary compounds, additives and monomers present in polymers.

Films

One of the most frequent techniques of sample preparation in the spectral analysis of polymers is the preparation of films which are pressed or cast from solutions. Polymer solutions are spread over NaCl or KBr plates and the spectra measured after removal of the solvent under an infrared lamp. The films may also be cast from solutions into glass plates or mirror surfaces from which, on removal of the solvent, the films are stripped and inserted directly into the beam.

Pellets

In this technique, the polymer is ground with KBr, then the powder is placed in a suitable mould. The sample is evacuated and pressed under a pressure of about 10 tonnes/cm^2.

Mulls

Preparation of samples by the mulling technique, which is a suspension in paraffin oil, is usually adopted. A fine powder of the sample is suspended in paraffin oil or other immersion liquid and placed between two sodium chloride plates.

Identification

By following the Fig. 8.1 and Fig 8.2 one can inspect the unknown spectrum for the presence (P) or absence (A) of the specific band listed in each block. If the band is present the (P) branch of the diagram is followed. If the band is absent, one proceeds along the (A) path.

This matching process is continued until one locates a box with a class of polymer or specific polymer name. A comparison is then made between the unknown and the materials listed in the final block.

The number in parentheses in the final block refers to the spectrum number in this study. This method of sample identification may assist in obtaining a positive match but it is even more helpful in identifying the type or class of polymeric material.

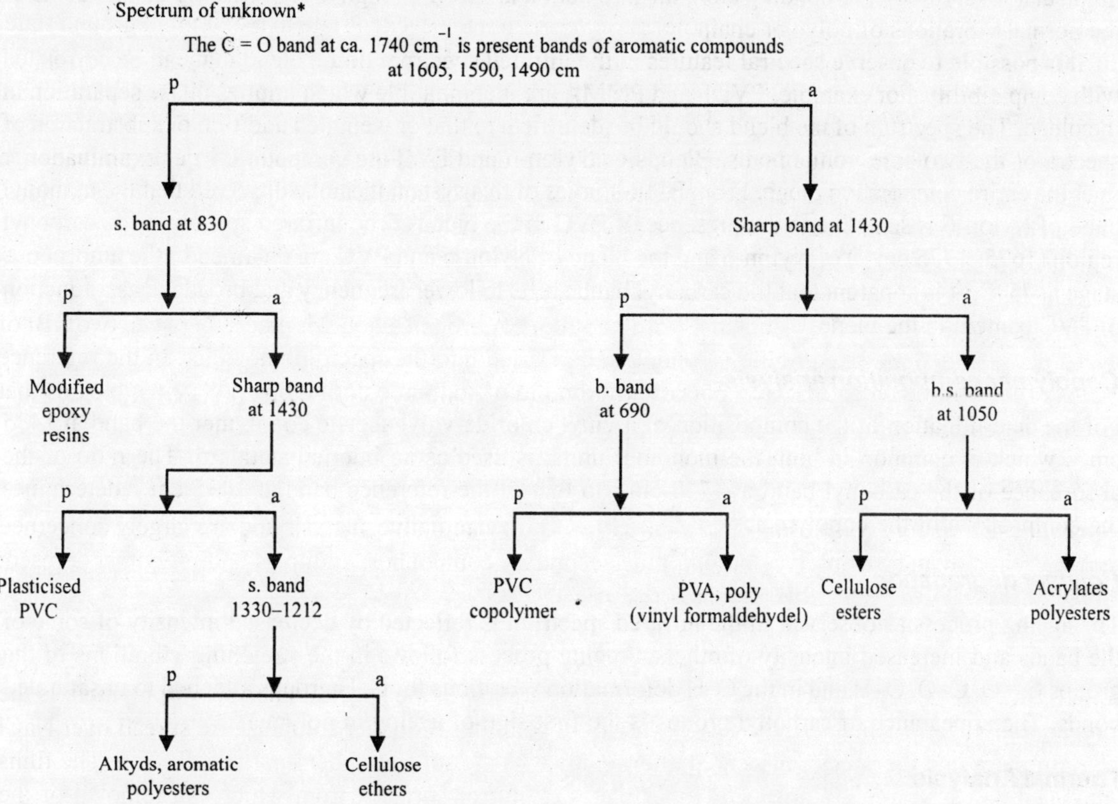

Fig. 8.1. Polymer identification diagram from infrared spectra.

Monomer content studies

For better product performance the evaluation of monomer content in plastics is essential. For example, a quantitative measure of the number of unreacted isocyanate group in the polyurethane foam is obtained through the use of NCO stretching bond at 2270 cm^{-1}. The area A (NCO) of the band is calculated with respect to a base line drawn between the points at 2334 and 2150 cm^{1}. Similarly the CH stretching band is integrated to give A (CH) using a base line drawn between 3012 and 2550 cm^{-1}. The ratio of A (NCO)/ A (CH) is directly related to the amount of unreacted isocyanates in the polyurethane foam.

Crystallisation of polymers

The kinetics of the crystallisation process can be evaluated using FTIR techniques. A particular example comes from the study of polymer blends of PVDF/PMMA where crystallisation of the alpha and beta forms have been followed during heating of the blend sample which is quenched from the melt and crystallised by heating at 2 k/min rate in the spectrometer. When blend has 70% (weight percentage) PVDF, the beta crystal form is obtained directly but for other compositions the alpha form is dominant.

Compatibility of polymers

FTIR is a valuable tool to analyse the crystallinity and compatibility nature of components in multi-component blend systems. For example, in two polymers A and B which are completely miscible on a molecular level (dipole, H-bonding etc.), the interaction is strong enough to significantly disturb some of the normal vibrations of polymer chain.

It is possible to observe spectral features in the infrared spectrum of the blend that can be correlated with compatibility. For example, PVDF and PMMA are incompatible which implies phase separation in the blend. The spectrum of the blend should be identifical to that of weighted addition or substraction of spectra of the two pure components. Blends of Nylon-6 and PVC are compatible in amorphous stage over the entire composition range. From FTIR studies of this system it can be observed that the carbonyl band of Nylon-6 is sensitive to the presence of PVC in the blends. The infrared spectra in the carbonyl region (1675–1775 cm^{-1}) of Nylon-6 and the blend of Nylon-6 and PVC are obtained in the amorphous stage at 75°C. It is apparent that the carbonyl band shifts to lower frequency and broadens as a function of PVC content in the blend.

Copolymer composition analysis

For the determination of the composition of a vinyl chloride-vinyl acetate copolymer the band at 1435 cm^{-1}, which is common in both the monomer units, is used as an internal standard. The ratio of the absorbance of the carbonyl band at 1730 cm^{-1} to that of the reference band at 1435 cm^{-1} determines the composition of the copolymer.

Polymer degradation

The ageing process as observed in the infrared spectrum is reflected by decreased intensity of some of the bands and increased intensity of others. Ageing process follows in the stretching vibrations of the groups C = O, C–O, O–H and in the C–H deformation vibrations for C–H groups attached to unsaturated bonds. The appearance of carbonyl groups is the first sign of ageing in polymers.

Thermal Analysis

Thermal analysis comprises a group of instrumental analytical techniques which are particularly powerful tools for the identification and characterisation of polymer materials. Essentially a thermal analytical method is one which follows the change in property in a sample whilst that sample is subjected to a predetermined temperature regime. The collective term thermo analytical methods comprises of different types of measurements. Differential thermal analysis (DTA) and Differential scanning calorimetry (DSC) are of importance, both measure the amount of heat taken up from or emitted to the surroundings per unit time with changing/constant temperature. In this manner, transition temperatures, heat capacities melt enthalpies, specific heat etc., can be measured and from this information further indications regarding phase transitions crystallisation processes etc., can be derived. Change in weight of polymers as a function of temperature can be determined by employing thermogravimetric analysis (TGA) which thus permits the analysis of desorption, degradation and decomposition processes.

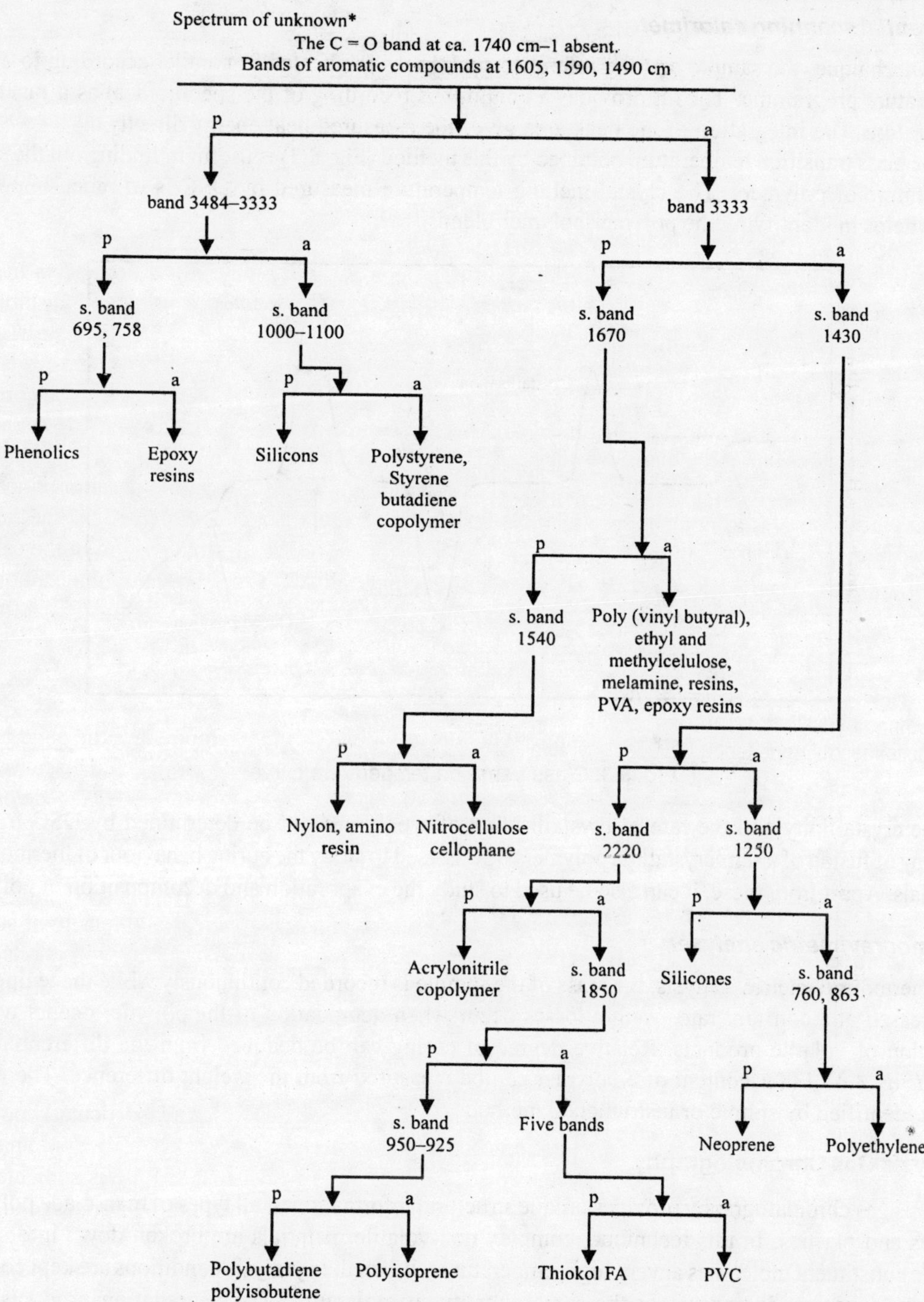

Fig. 8.2. Polymer identification diagram for infrared spectra.

Differential scanning calorimetry

In this technique, the sample and the reference substances are heated in parallel according to a linear temperature programme. Thus it provides a continuous recording of the specific heat as a function of temperature. The integration of the peak area gives the measured heat energy directly.

The glass transition temperature obtained by this method (Fig. 8.3) is useful in finding out the service temperature of polymers. The crystal melting temperature measured by DSC is of much importance which helps in identifying the polymer/polymer blends.

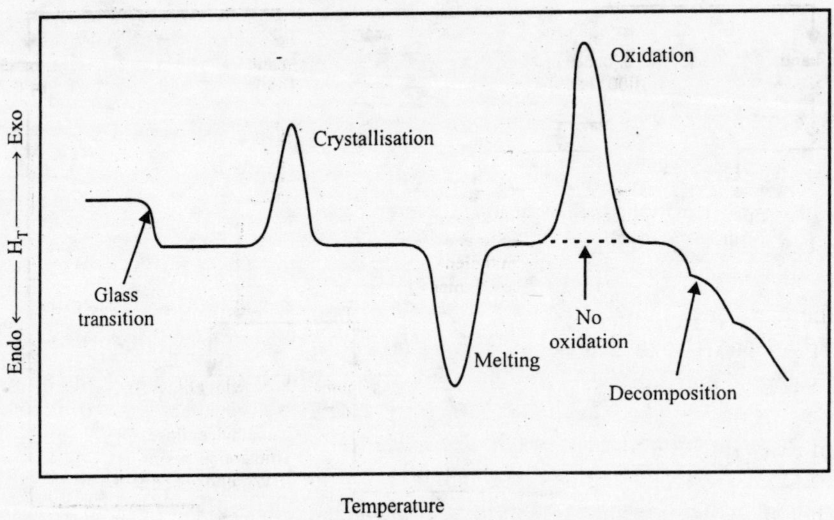

Fig. 8.3. Glass transition temperature curve.

The crystallinity and the rate of crystallisation of a polymer can be determined by DSC from the enthalpy of fusion of a semicrystalline polymer. DSC is used to study the curing behaviour of thermosetting materials. Apart from these, it can also be used to study the evaporation and decomposition of polymers.

Thermogravimetric analyser

With thermogravimetric analysis, the mass of the sample is recorded continuously while the temperature is increased at a constant rate. Weight losses occur when degradation of the polymer occurs with the formation of volatile products. Relative degree of curing can be deduced from the differential TGA curve (Fig. 8.4). Filler content of a polymer can be measured from the weight difference. The residue can be identified by simple or instrumental methods.

Pyrolysis Gas Chromatography

Pyrolysis gas chromatography provides unique structural information of all types of man-made polymers, rubbers and plastics. In this technique, complex non-volatile materials are broken down into smaller volatile constituent molecules at very high temperatures. When the pyrolysis conditions are kept constant, the sample always degrades into the same constituent molecules. The degradation products when introduced into a gaschromatograph, the chromatogram obtained remains same throughout the analysis representing the degradation products of a particular polymer.

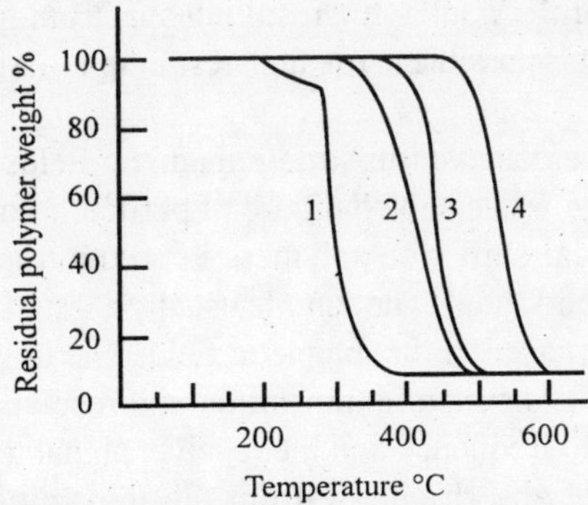

Fig. 8.4. Thermogravimetric analysis curve.

The instrument consists of a pyrolyser, where sample is heated at high temperatures upto 900°C. The pyrolysed gaseous product is carried by a carrier gas like Helium or Nitrogen into the gas chromatograph. Chromatograph consists of fused silica capillary columns, where the gaseous product is separated into different fractions. These fractions are then passed into detectors for identification. Detectors like thermal conductivity detectors (TCD) and flame ionisation detectors (FID) detect most of the pyrolysed products. Also there are some specific detectors for detecting compounds of phosphorus, nitrogen or other elements. Also the pyrolysates are detected by coupling GC with fourier transform infrared spectrometer or mass spectrometer.

In the case of analysing volatile matter like vinyl chloride monomer in PVC, head space sampler is used. In head space sampler the volatile matter is releaved from the sample by heating to relatively low temperatures. Then the volatile matter is automatically injected into GC and analysed.

Applications

Gas chromatography is successfully used in identifying many of the polymers including acrylics, rubbers, polyesters, polyolefins, vinyls and fluorocarbon polymers. It is possible to distinguish between copolymers and blends of similar composition by means of stepwise pyrolysis procedure. The composition of a copolymer can be determined from the relative intensities of the characteristic peaks. The microstructural characterisation of polyolefins is done from the relative peak intensities of pyrolysates. This technique is a most powerful tool to study the thermal degradation mechanisms of polymers especially those containing polar groups such as polyamide. Also it can be used to identify the various additives added to a polymer. Qualitative and quantitative determination of monomer in a polymer such as vinylchloride, styrene, methymethacrylate etc. can be made by Gas chromatograph analysis.

Nuclear Magnetic Resonance Spectroscopy (NMR)

NMR spectroscopy is now, one of the most useful instrumental technique for characterising polymeric materials. It helps in understanding the detailed chain structure of polymers and copolymers and of the morphology and transitions in the solid state. NMR also provides means for identifying intermediate structures, formed during polymerisation reactions so permitting more detailed reaction mechanisms to be proposed.

NMR spectroscopy like other spectroscopic methods is concerned with the detection of the absorption of electromagnetic radiation by energy transitions between energy levels. Since the energy levels are influenced by the environment of the nuclei, the resulting spectrum gives more or less direct evidence regarding the chemical structure. It is possible to use high resolution fourier transform NMR for quantitative analysis of certain polymers.

Nuclear magnetic resonance is observed in strong magnetic fields and is restricted to nuclei with a nuclear spin not equal to zero. Two effects in NMR are of particular importance as regards the structure analysis of polymers i.e., chemical shift (for polymers in solution and at high temperature) and line broadening (for polymers in the solid state). The sample usually a liquid or a solid in solution, is contained in a precision glass tube. This is placed in the magnetic field. The unit which supports the sample tube is known as the probe. The probe consists of a transmitter and receive coils which surround the sample and act to couple the incident radio frequency and the receiver signal to the sample. When the resonance condition is satisfied, the sample absorb energy from the transmitter and the change in the coil is detected by the receiver coil and is then amplified as the NMR signal. The higher the spinning frequency of the sample, the higher the intensity and hence higher resolution of the signal and the smaller the spinning side band signals which result from sample spinning.

Applications of NMR to polymers

NMR is used for the determination of stereochemical configurations of the polymer backbone. The existence of the configurational sequences (activity) gives rise to distinguishable magnetic environments for the nuclei in the polymer units and hence different NMR absorption frequencies.

NMR appears to be a suitable technique for characterising copolymers. The information that may be derived ranges from determination of the overall co-monomer compositions in the copolymer to detailed descriptions of the arrangement of monomer sequences in the chain and determination of configurational sequences in the chain.

Structural information that may be derived from NMR studies of polymers in the bulk state depends on the form of the sample. For oriented polymers, information can be deduced from the anisotropy evident in the spectra. For the majority of bulk polymers which are unoriented, information can be deduced from the resolution of the spectra and the typical line shape in the spectra.

NMR can be used to study the crystallinity of polymers. The technique relies on the fact that molecular chain motion and the consequent line widths differ in the crystalline and amorphous regions.

CHAPTER 9

Testing of Polymers—Basic Concepts

INTRODUCTION

Not too long ago, the concept of testing was merely an afterthought of the procurement process. But now, with the advent of science and technology, the concept of testing is an integral part of research and development, product design, and manufacturing. The question that is often asked is 'why test?' The answer is simple. Times have changed. The manner in which we do things today is different. The emphasis is on automation, high production, and cost reduction. There is a growing demand for intricately shaped, high-tolerance parts. Consumer awareness, a subject totally ignored by the manufacturers once upon a time, is now a major area of concern. Along with the requirements, our priorities have also changed. While designing a machine or a product, the first order of priority in most cases is safety and health. Manufacturers and suppliers are now required to meet a variety of standards and specifications. Obviously, relying merely on past experience and quality of workmanship is simply not enough. The following are some of the major reasons for testing :

1. To prove design concepts.
2. To provide a basis for reliability.
3. Safety.
4. Protection against product liability suits.
5. Quality control.
6. To meet standards and specifications.
7. To verify the manufacturing process.
8. To evaluate competitors' products.
9. To establish a history for new materials.

A true picture of the testing of these materials must therefore contain not only a description of existing methods, but also some indication of the directions in which changes are expected or needed. Such a picture, with its emphasis on limitations and obstacles may present a pessimistic view, and indeed many phases of polymer plastics testing are still unsatisfactory. Frequently test results from one laboratory will not agree with those from another as closely as would be desired; the technique of predicting the behaviour of a finished article from tests on the material from which it was made is insufficiently developed; and much of the data which would be useful for theoretical interpretations of the behaviour of plastics and elastomers is lacking.

In the last two decades, just about every manufacturer has turned to plastics to achieve cost reduction, automation, and high yield. The lack of history along with the explosive growth and diversity of polymeric materials has forced the plastics industry into placing extra emphasis on testing, and developing a wide variety of testing procedures. Through the painstaking efforts of various standards organisations, material suppliers, and mainly the numerous committees of the American Society for Testing and Materials (ASTM) over 2000 different test methods have been developed.

The need for developing standard test methods specifically designed for plastic materials originated for two main reason. Initially, the properties of plastic materials were determined by duplicating the test methods developed for testing metals and other similar materials. The Izod impact test, for example, was derived from the manual for testing metals. Because of the drastically different nature of plastic materials, the test methods often had to be modified.

As a result, a large number of non-standard tests were written by various parties. As many as eight to ten distinct and separate test methods were written to determine the same property. Such practice created total chaos among the developers of the raw materials, suppliers, design engineers, and ultimate end-users. It became increasingly difficult to keep up with various test methods as well as to comprehend the real meaning of reported test values.

The standardisation of test methods acceptable to everyone solved the problem of communication between developers, designers, and end-users, allowing them to speak a common language when comparing the test data and resuls.

In spite of the standardisation of various test methods, we still face the problem of comprehension and interpretation of the test data by an average person in the plastics industry. This is due to the complex nature of the test procedures and the number of tests and testing organisations. The key in overcoming this problem is to develop a thorough understanding of what the various tests mean and the significance of the result to the application being considered. Unfortunately, the plastics industry has placed more emphasis on how and not enough on why, which obviously is more important from the standpoint of comprehension of the test results and understanding the true meaning of the values. The lack of understanding of the real meaning of heat deflection temperature, which is often interpreted as the temperature at which a plastic material will sustain static or dynamic load for a long period, is one such classic example of misinterpretation.

In the chapters to follow, we concentrate on the significance, interpretation, and limitations of physical property data and test procedures. Finally, a word of caution: it is extremely important to understand that the majority of physical property tests are subject to rather large errors.

As a general rule, the error of testing should be considered ± 5 per cent. Some tests are more precise than others. Such testing errors come about from three major areas: one from the basic test itself, the second from the operators conducting the tests, and the third from the variations in the test specimens. While evaluating the test data and making decisions based on test data, one must consider the error factor to make certain that a valid difference in the test data exists.

STANDARDISATION IN VARIOUS COUNTRIES

Standardisation of a test means that many of its users have agreed on all of the critical requirements and that these have been written up by a standardising group. Furthermore, the method will have been tried in an inter-laboratory programme so that its repeatability within an laboratory, and reproducibility between laboratories, are known. Each developed country has a national body with overall responsibility for such standardisation. In the USA this is the American National Standards Institute (ANSI), which coordinates the various groups that actually establish standards. Committee D-11 of ASTM has the principal responsibility for standards which affect producer-consumer relationships in the American rubber industry. Specifications for rubber products are written by many other groups such as Underwriters Laboratories, AMS Aeronautical Board, National Hospital Association, and International Electro-technical commission. The predominant international standards group is the international organisation for standardisation (ISO), of which technical committee 45 is concerned with rubber products. Activities of this letter group have included many physical tests on rubber which are performed in a standard manner worldwide.

USA

Here the ASTM specifications are most generally adopted but in addition to these one can find others. They are on a high level and growing very rapidly since the outbreak of the war. A close cooperation between the government agencies (National Bureau of Standards) and industry exists.

Great Britain

The British Standards Specifications are also on a high level especially regarding rubbers and the thermosetting types of plastics. In many respects they deviate considerably from the ASTM specifications.

Germany

Germany had, before the war, developed a system of classification and of standardising, far in advance of all other countries of the world. The VDE specifications in 1939 were already the 22nd edition of a steadily growing series of standards. One of the striking features in Germany was the very intensive collaboration along the whole line, starting from the manufacturer of raw materials, via the maker of mixtures, the moulder and the fabricator to the final consumer. For all these groups the system was really alive and it was not considered as an unavoidable burden. The trade ordered its primary and final materials according to the specifications and a systematic control, carried out by the 'Materialprüfungsamt' (Bureau of testing materials) in the producing plants, furnished the guarantee that all materials were constantly maintained up to standard. Practically all moulders pressed "their number" in all their mouldings, giving the guarantee to the consumer that they were under control of the central organisation.

India

Indian Standards Institute, (Bureau of Indian Standards), our national organisation is actively pursuing standardisation of all kinds of materials and products related to polymers since 1947.

Other Countries

About Russia no data is available. (From the smaller European countries Switzerland has a system of its own, although a close analogy to the German testing methods can be traced. In our opinion it is much to be regretted that such small countries go their own way, making the international mutual understanding still more difficult. In this respect the situation in France and in Holland is much more to be welcomed, since practically the system of one of the larger units (Germany) has been adopted. It is clear from the above survey that there is a very intensive need in the world for better international cooperation.

SPECIFICATION AND STANDARDS

A specification is a detailed description of requirements, dimensions, materials, etc. A standard is something established for use as a rule or a basis of comparison in measuring or judging capacity, quantity, content, extent, value, quality, etc.

A specification for a plastic material involves defining particular requirements in terms of density, tensile strength, thermal conductivity, and other related properties. The specification also relates standard test methods to be used to determine such properties. Thus, standard methods of test and evaluation commonly provide the bases of meausrement required in the specification for needed or desired properties.

As discussed earlier, the ultimate purpose of a standard is to develop a common language, so that there can be no confusion or communication problems among developers, designers, fabricators, end-users, and other concerned parties. The benefits of standards are innumerable. Standardisation has provided the industry with benefits such as improved efficiency, mass production, superior quality goods through uniformity, and new challenges. Standardisation has opened the doors for international trade, technical exchanges, and establishment of common markets. One can only imagine the confusion the industry would suffer without the specific definition of fundamental units of distance, mass, and time, and without the standards of weights and measures fixed by the government.

Standards originate from a variety of sources. The majority of standards originate from industry. The industry standards are generally established by voluntary organisations that make every effort to see that the standards are freely adopted and represent a general agreement. Some of the most common voluntary standards organisations are the American Society for Testing and Materials, the National Sanitation Foundation, the Underwriters Laboratories, the National Electrical Manufacturers Association, and the Society of Automotive Engineers. Quite often, the industry standards do not provide adequate information or are not suitable for certain applications, in which case, private companies are forced into developing their own company standards. The company standards are generally adapted from modified industry standards.

The Federal Government is yet another major source of standardisation activities. The standards and specifications related to plastics are developed by the U.S. Department of Defence and the General Services Administration under the common heading of Military Standards and Federal Standards, respectively.

After the Second World War, there was a tremendous increase in international trade. The International Standards Organisation (ISO) was established for the sole purpose of international standardisation. ISO consists of the national standards bodies of over 50 countries from around the world. The standardisation work of ISO is conducted by technical committees established by agreement of five or more countries. ISO's Technical Committee 61 on plastics is among the most productive of all ISO committees.

PURPOSE OF SPECIFICATIONS

There are many reasons for writing specifications, but the major reason is to help the purchasing department, purchase equipment, materials, and products on an equal basis. The specifications, generally written by engineering departments, allow the purchasing agent to meet his requirements and ensure that the material received at different times is within the specified limits. The specification is intended to ensure batch-to-batch uniformity, as well as remove confusion between the purchaser and supplier, as we all known that more often than not what is provided by the supplier is not what is expected by the purchaser.

PRECISION, ACCURACY, AND VALIDITY

A test is valid if the results are actually a measure of the desired property. If a valid wear test for tyre tread rubber existed, for example, there would be little need for expensive fleet testing for wear resistance. Even though an apparatus and method may be valid, however, the accuracy can be impaired by either instrumental or procedural defects. An oven aging test, for example, is inaccurate if the temperature measurement is incorrect. A test is precise if results are closely reproducible. A low temperature brittle point test, for example, may be both valid and accurate if results on many specimens are averaged, yet have a wide scatter of results, and hence be imprecise, on certain types of materials.

Problems of precision, accuracy, and validity each exist in varying degrees in all physical tests on rubber. Many weaknesses have been found in rubber testing and much of it is non-productive. Examples of non-valid testing include unnecessary specifications by the consumer. High tensile strength may be specified routinely when, in fact, it is totally unnecessary to the proposed use. In this case it is not a valid test. Another example cited is use of the 'peanut' or 'angle' specimen for tear testing. Since results are unduly influenced by stock modulus, they do not truly represent tear resistance. While tests may be valid in a limited application, such as the simple hysteresis tests (rebound, Yerzley Oscillograph), they may become invalid because of faulty interpretation of data.

'The inadequacy of the simple device lies in the acceptance of the data as an unqualified measure of hysteresis, regardless of the service conditions for which the stock is intended'. The validity of a method for a particular application should be indicated in its scope. A further caution is that tests should be chosen which can be interpreted. Some tests measure such complex combinations of basic factors that it is nearly impossible to understand the results.

Inter-laboratory tests on rubber usually give a wide scatter of data. Invariably some laboratories report results which are so far from the mean of the others that some instrumental or procedural problem is indicated. In other words, the data is inaccurate to a degree unsuspected by personnel of that laboratory. Surprisingly often the problem is temperature measurement or calibration of force measuring equipment. These problems can be avoided by testing standard materials periodically or by continual reciprocal testing with other laboratories.

Precision or reproducibility problems may be caused by variability in test material improper sampling, inadequate control of test conditions, or instrument problems such as friction or electronic drift. Certainly the precision of a test should be known, either from previous records or by testing many specimens, so that confidence levels of test results can be established statistically.

SPECIMEN PREPARATION

Most physical tests on vulcanisates are performed on specially prepared samples. Some variations in properties must be expected even on supposedly identical samples because of non-uniformities in raw materials and processing techniques. For this reason a control stock is often included in each group of experimental stocks. Dispersion of test results on the control must then be considered in determining a degree of confidence in evaluations of results on the experimental stock.

Some standard methods for stock mixing, sample curing, and specimen forming are given in ASTM methods D3100 through D3109. Many properties of rubber important to good physical testing practice are implicit in these directions. For example, different aeging periods after mixing, after remilling, and after curing are required for compounds based on different polymers. In general, no tests should be made until at least 16 hours after vulcanising a sample because significant post-vulcanisation changes in structure of the material occur during this period. Of course, in control tests such delays are intolerable, so the change in properties between actual testing time and 16 hours after curing should be ascertained and taken into consideration.

Some test results are influenced by flow of rubber in the mould, so a good general practice is to use no specimens cut from material near a mould edge. Also the grain in rubber caused by the action of the mill in the final pass may affect certain properties. Tensile tests, for example, are always made along the grain. Specimens are sometimes required from a finished product rather than from laboratory prepared samples. In such cases they are often buffed to obtain uniform thickness, to remove fabric or fabric impressions, or to remove glaze or skin coats from the rubber. Abrasive grinding wheels of about 30 grit, 5-inch diameter, running at 2500 to 3500 rpm are satisfactory for this purpose. The grinder should have a slow feed so that very little material is removed at each cut, to avoid both high heat and coarse abrasive pattern at the surface.

STANDARD TEMPERATURE

ASTM D1349 gives a list of standard test temperatures which agrees with the ISO list. An advantage of using only the standard temperature is that many unnecessary temperature changes in environmental chambers are eliminated. A standard humidity is also provided although it is doubtful that humidity has much effect on physical properties of rubber. Properties of fabrics or rubber-fabric composites, however, are very sensitive to moisture.

BREAK-IN

For some tests a rubber specimen should be broken-in before recording any results. This means that it should be deformed several times to the same extent that it is in the recorded test. This usually gives a softer material, but one on which results are reproducible. Upon retesting the specimen properties gradually return toward the initial values. The rebound test is an example of one which requires a break-in.

NON-DESTRUCTIVE TESTING

Non-destructive testing of polymer products is a growing field which is at the present time mostly limited to specific items. With their proven capabilities, however, more general tests are likely to appear. These methods include holography, ultrasonics, infrared radiation, microwaves, X-rays, and gages based on nuclear, capacitive, optical, or magnetic principles.

Ultrasonics have been used for thickness measurements, including that of a coating of one material on another, and for locating voids. Infrared radiation permits non-contacting surface temperature measurements to be made on stationary or moving objects. Micro-waves can be used for non-contacting thickness measurements without radiation hazard. Both the loss tangent and the dielectric constant of the material can be measured independently. X-rays, including both xeroradiography and fluoroscopy, are sensitive to flaws of various kinds – especially to inclusion of foreign material that is more opaque to X-rays than is the test material.

Any test instrument which produces an electrical signal that can be calibrated in terms of the measured property level is a potential candidate for tie-in with a computer. This may be an on-line tie-in, in which the coded results are fed directly to the computer for immediate processing, or the results may be stored and introduced to the computer at a later time. Depending atleast partly on the speed at which data must be logged, storage can be by punched card, paper tape, or magnetic tape.

An on-line (real time) computerised system of four tensile testers as used at Dupont has been described by Stanton. This used a dedicated computer (for that purpose only) which was programmed to accommodate nine tests performed on the tensile testing machine: stress-strain, Finch tear, ASTM Die B tear (Winkleman), ASTM Die C tear (Grains), flexural modulus of elasticity, compression-deflection, paper tensile, peel adhesion, and trouser tear. Four of these programmes had alternate routines, making

13 tests in all. The computer not only processed the data and printed out the calculated results but also controlled a segment of the electronic circuit in each unit. This application illustrates the principal reasons for computerising physical tests: saving employee's time, accurate lodging and calculating of large amounts of data, consistent test control, and rapid operation.

BASIC SPECIFICATION FORMAT

Many guidelines and directives have been written for writing specifications. The specifications for materials should include the following:

1. A descriptive title and designation.
2. A brief but all inclusive statement of scope.
3. Applicable documents.
4. A classification system.
5. Definition of related terms.
6. Materials and manufacturing requirements.
7. Physical (property) requirements.
8. A sampling procedure.
9. Specimen preparation and conditioning requirements.
10. Reference to, or descriptions of, methods of test acceptable for determining conformance.
11. Inspection requirements.
12. Instructions for retest and rejection.
13. Packaging and marking requirements.

Classification System

Since plastic materials are seldom supplied without the addition of certain additives and fillers, a classification system must be used to avoid confusion. For example, the specification for acetal resin covers the two types of acetal resins classified according to melt flow. Type I is the material for injection moulding and Type II is the material for extrusion. The two types are subdivided into classes according to applications. Class A is natural colour only, containing an antioxidant for general purpose use. Class B is black, containing carbon black pigment for weather-resistant uses. Class C is natural, containing a light stabiliser for weather-resistant use. Table 9.1 lists the detailed requirements for acetal resin.

Table 9.1. Detailed requirement for acetal resin injection moulding and extrusion materials.

Properties	*Type I*	*Type II*
Specific gravity, 23/23°C (73.4/73.4°F):		
min	1.39	1.39
max	1.44	1.44
Melting point, min, °C	158	158
Flow rate, g/10 min:		
load 1050 g:		
min	3.5	—
max	—	2.0
or		
load 2160 g:		
min	7.5	—
max	—	4.3

Requirements

Physical requirements should include specific quantitative information according to the classification. Table 9.2 lists the detailed requirements for moulded specimen properties.

Table 9.2. Detailed requirements for moulded specimen properties.

Properties	*Class A*	*Type I Class B*	*Class C*	*Type II Class A*
Tensile yield strength, 23°C (73.4°F), min				
MPa	55.2	55.2	55.2	55.2
psi	8000	8000	8000	8000
Elongation, 23°C (73.4°F), min, per cent	10	10	10	30
Modulus (tensile) of elasticity, 23°C (73.4°F), min				
MPa	2415	2415	2415	2415
psi	350 000	350 000	350 000	350 000
Deflection temperature under load, 1.82 MPa (264 psi), °C, min	100	100	100	100
Whether resistance, retention of original elongation, min, per cent	–	65	50	–

Sampling and Conditioning

In order to reproduce the test results time after time, sampling and conditioning procedures must be religiously followed. The procedures for sampling and conditioning must be clearly specified.

Test Methods

The specifications, without adequate information regarding test methods, are useless. The standard and generally accepted test methods must be specified. If no standard test methods exist, a detailed description and requirements of the test should be an integral part of the specification.

Standardisation in Various Countries

It will appear that in various countries a reasonable status has been reached in standardising test methods. It is however very regrettable that up to now hardly any cooperation between various parts of the world exists, resulting in the fact, that now 'various technical languages are talked' and nobody can understand anybody else. The chief reason is that most testing methods are arbitrary, measuring no pure physical constants. The values obtained being dependent on the dimensions of the test pieces, the speed of testing, the size of the apparatus etc, etc.

The astonishing lack of cooperation already appears in such a simple thing as the form and the dimensions of test pieces. To mention one example only we refer to the impact specimen for moulded plastics, for which the size of the notch has a radius at the base of 0.010 inch (0.25 mm) in the United States. In Great Britain the standard is 0.040 inch (1 mm) and in Germany no final dimensions have been settled up to now. Experiments have shown however that for rigid materials the results are strongly influenced by the form and size of the notch, showing more clearly how necessary standardising is in this field.

FACTORS AFFECTING TEST RESULTS

A major probelm faced by the quality control for personnel in an attempt to standardise the testing methods for plastics is accridated to the wide variety of materials included under that name. Some of these materials are rigid and brittle, some are soft and rubberlike; some absorb water rapidly and approach equilibrium in a few hours or days depending on their dimensions; while others require years under similar conditions to approach saturation; and still others are scarcely affected by water at all.

Clearly no single set of test methods would be completely suitable for all these material. Methods originally developed for other materials, e.g., rubbers, metals and textiles, which showed promise of being useful in evaluating the particular types of synthetic materials under consideration were adopted for testing plastics as well standardisation of tests for any given property involves the selection from among these methods of one or more which are applicable to a wide range of plastic materials. Where practicable, these methods have been expanded by the addition of extra procedures to extend the range of their applicability.

In case, where the required procedures differ too widely, separate test methods have been been established to measure a similar property for separate groups of materials. As a rule, the use of different test methods for different materails does not cause any practical difficulty, because the end uses of these materials are generally so different that direct comparisons of their properties are rare. It is essential, however, to recognise, in any atttempt at comparison, that two different methods may measure related properties, but not identical ones.

Conditions which Affect Test Results

For any single type of these materials there are a number of factors which must be carefully controlled if consistent results are to be obtained. The relative importance of each depends on the material. Some of the factors are: (i) moisture content; (ii) temperature of specimen; (iii) size and skin dimensions of the specimen; (iv) testing speed; (v) specimen preparations.

Moisture content

Moisture exerts a decided effect on the mechanical properties of most plastic materials, and consequently on the serviceability of articles made from them. For example the effect of average monthly indoor relative humidity on the breakage of telephone equipment that telephone breakage is usually due to impact forces resulting from accidental dropping.

The exposure of dry phenolic test specimens to an atmosphere of 90% relative humidity for 12 days will cause an increase in their Izod impact strength of about 30% and, since these materials absorb water very slowly, a longer exposure would be expected to produce a considerably greater change. Some of the cellulosic plastics and vinyl butyral plastics are much more rapidly affected. Other materials. e.g., the vinyl chloride plastics and polystyrene are substantially unaffected by moisture.

The speed with which some plastics absorb moisture makes it imperative to test them in the same atmosphere to which they were conditioned; since otherwise their properties would change during the test. Table 9.3 shows the effect of time of exposure to 50% relative humidity on the flexural strength of specimens of a cellulose acetate plastic which had previously been kept over water for one week. The first value under 'flexural strength' is the average result for the control, which had not been exposed over water. The second is the result for a set of bars that were tested under water, and represents the correct value for throughly wet specimens. The remainder of the data are for bars on which tests were finished after the times shown in the third column.

Table 9.3. The effect of time of exposure to 50% relative humidity on flexural strength of a cellulose acetate plastic.

Conditioning time at			*Flexural strength*		*% of original*
50%	*100%*	*50%*	*kg/mm²*	*psi*	*strength*
7 days	0	0	6.16	8750	100
7 days	7 days	0	2.93	4160	47.5
7 days	7 days	1 1/2 min	3.04	4310	49.3
7 days	7 days	1 h	3.46	4930	56.4
7 days	7 days	24 h	4.62	6560	75.0
7 days	7 days	64 h	5.25	7480	85.4
7 days	7 days	192 h	6.02	8550	97.8
7 days	7 days	312 h	6.17	8770	100

Materials whose properties change thus rapidly with changing ambient relative humidity are usually conditioned at 50 ± 2% relative humidity, and are tested in the conditioning atmosphere. Those whose properties change more slowly require too long to approach equilibrium values under these conditions to warrant the practice except for special studies. It has been shown, however, that at elevated temperatures they can be dried out very quickly. It is customary, therefore, to dry them in an oven at 50ºC (122ºF) for 48 hours followed by 16 hours in a desiccator over calcium chloride. Their low rates of water absorption generally make it possible to test them at the prevailing humidity without any appreciable change of properties during the test.

Temperature of the specimen

Although the temperature affects the properties of all plastics it is only with the thermoplastics that the effect is so large as to constitute a serious problem in testing. Values for strength properties of some of the commoner thermoplastics at two or more temperatures have been given and explained in Fig. 9.1.

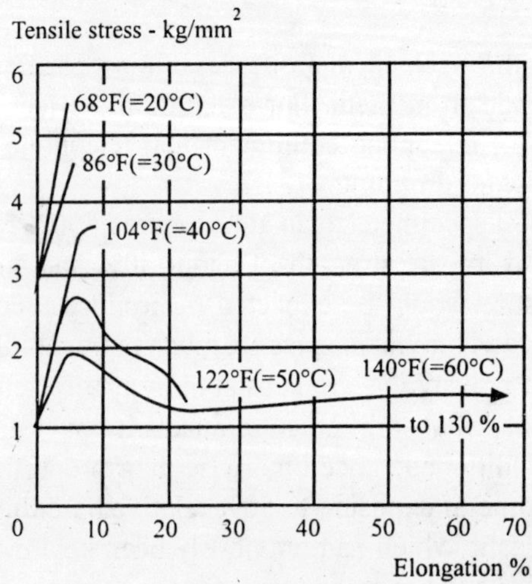

Fig. 9.1. Effect of temperature on the tensile stress strain properties of methyl methacrylate.

In this figure stress strain curves given for a cast methyl methacrylate resin, at 6 temperatures between + 20°C and + 60°C (+ 68°F + 140°F). In this temperature range the material has changed from a very hard brittle plastic at the lower temperature to a very soft one with properties approaching those of an elastomer at the higher. Correspondingly great changes in strength with temperature change are shown by other thermoplastics.

It is customary, therefore, to test thermoplastics and, frequently, thermosetting materials, in a room maintained at a constant temperature. The standard laboratory atmosphere is defined as having a relative humidity of 50 ± 2%, and a temperature of 25 ± 1°C (77° ± 1.8°F).

For tests at temperatures for which a controlled room is not available, the testing machine, or the parts that come in contact with the specimen are enclosed in an insulating container and air at a suitable temperature is circulated through the enclosure. Units for supplying the conditioned air are supplied by several manufacturers. The flexural tool of a testing machine is used for low temperature testing. The enclosures and servicing units have been specified by the ASTM; and methods for testing for tensile and compressive strength and for impact strength at high and low temperatures have been standardised.

When tested at high temperatures, specimens lose moisture during the test and during the time required to reach equilibrium at the test temperature. For this reason all specimens intended for testing at high and low temperatures are conditioned by drying and are moved directly from desiccators to the atmosphere in which they are to be tested.

Size and skin dimension

Test results are expressed in terms of specimens of unit dimensions, eg., pounds per square inch or kilograms per square centimeter for tensile strength, which means that specimens of any size would give test results reducible to the same figure. It is well established however, that the dimensions of the test piece frequently exert a large influence on the test results. One of the causes is the existence of microscopic defects in all test specimens. The defects cause concentrations of stress in the test piece as a whole. The larger the piece the greater will be the number of these defects, and the greater the probability of the existence of more severe defects. A second cause is the difference which is frequently observed between the outer and inner parts of a test specimen or a finished article and obviously the smaller the specimen the greater will be the contribution of the surface layers to the test results.

This skin effect is especially pronounced in plastics. The difference in properties between the skin and the remainder of the piece may be caused by overheating of the outer layers to insure good moulding of the inner ones, by high orientation of the skin in injection moulding of materials with chain molecules; by plasticiser loss due to evaporator from the surface in plasticised materials; or by a flow of resin to the surface of an object during the moulding operation.

The size effect is illustrated in Fig. 9.2. The data represents sheet thickness in mm versus tensile strength in kg/mm^2 on sheets of various thicknesses cut from a single block of cellulose acetate plastic. It is probable that an additional factor beside size effect is involved in the rapid decrease of tensile strength with increasing thickness, since these sheets are made by a process requiring the use of solvents which are subsequently removed by 'seasoning' in hot air, and the difficulty of thorough removal of solvents increases disproportionately with increasing thickness. This added factor may very well be considered as one of the causes of the size effect in this type of plastic, though it is doubtless not the major cause in this case.

Means of eliminating or correcting for the size effect in plastics have not been standardised except in so far as the specifying of the dimensions of a test piece may be said to do so. Care in moulding, seasoning or polymerisation processes can reduce its magnitude, wherever these processes are contributing factors; but until the effect has been more thoroughly studied for these materials, it will be necessary to continue to supplement laboratory tests with tests on finished articles when the dimensions of articles and test pieces differ widely.

Testing speed

Since high polymeric materials possess viscosity as well as elasticity it is to be expected that their mechanical properties would vary with testing speed. The equipment consists of a massive wheel which rotates at a controlled speed. Projecting from the periphery of the wheel is a tool with an edge for impacting. When the wheel is operating at the desired speed the specimen, a thin strip of sheet material mounted in a vise as a cantilever beam is thrust into path of the impacting edge and is struck on one face. If the speed of the wheel is sufficiently low the specimen bends easily out of the way and allows the tool to brush past it. When, however, the speed of the wheel is so high that the viscosity of the specimen prevents its bending away rapidly enough, it is shattered like a brittle material. The test consists of finding the speed at which shattering just occurs.

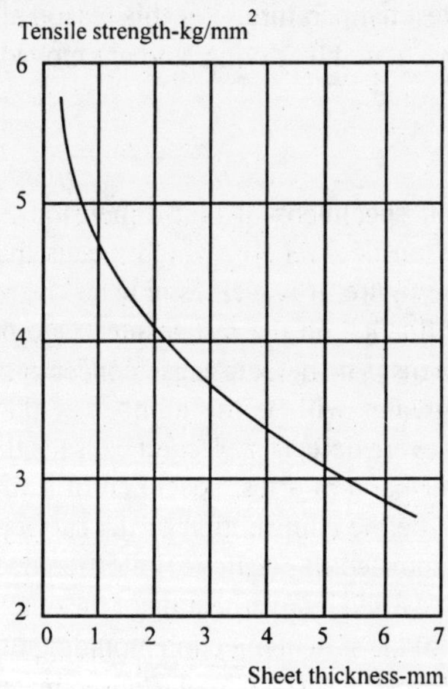

Fig. 9.3. Effect of test specimen size on the tensile strength of a cellulose acetate plastic.

This type of test has found its greatest application with some of the vinyl copolymers which are sensitive to testing speed variations. A modification of it has been used for determining the brittle point of plastics and elastomeric materials.

Table 9.4 illustrates of the effect of viscosity at lower testing speeds for a cellulose acetate plastic. The data shows the effect of testing speed on tensile strength and the time required to break identical specimens under constant tensile load. The effect of speed on yield point, elongation and reduction of area of methyl methacrylate plastics also shows a similar effect of speed on strength, and the effect of time of compressive loading on strain. These effects of speed are enormously greater than corresponding effects, for example in the testing of steel.

Table 9.4. Tensile test 25°C (77°F) 50% relative humidity.

Testing speed. No load crosshead separation		*Tensile strength*	
cm per min.	*inches per min.*	*kg.mm²*	*p.s.i.*
0.13	0.052	3.24	4500
1.6	0.63	3.55	5050
2.9	1.13	3.70	5250
4.8	1.88	4.10	5800
Crepe test 25°C (77°F) 50% Relative humidity.			
Tensile stress		*Time to break*	
kg mm²	*p.s.i.*	*hours*	
1.89	2695	100	
1.77	2505	300	
1.62	2305	700	
1.42	2008	6500	

They have necessitated the standardisation of testing speeds for determining mechanical properties and the establishment of tests with widely differing experimental time scales for measuring resistance to any single type of stress. Obviously the load that a plastic part can carry continuously is a very small fraction of its ultimate strength in a short time test.

Specimen preparation

The effects of the variables which have been so far discussed are susceptible to measurement and control, and while much of this work remains to be done, at least, the direction that the work should take is clear. For materials in the form of unoriented sheets, and for end products stamped or machined from them, these are the principal variables. A test piece cut from a sheet, and a flat airplane window from the same sheet are basically alike. But if the sheet is formed by heat and pressure, the forming may introduce stresses, local orientations, local changes in thickness, 'unmoulding', curing, or plasticiser loss that give it different properties from those predictable from the original test piece. When a test piece and a finished article are made from the same sample of moulding powder they experience such widely different and complex combinations of temperature, pressure and flow distribution that the specimen can hardly be said to represent any part of the finished products. The differences in properties between parts of a single moulded piece can be demonstrated using impact bars cut from various sections of an injection moulded telephone base of a cellulosic plastic. The great differences in properties have lead the Germans to develop a method of testing very tiny specimens [10 mm × 15 mm by the available thickness from 1.5–4 mm (0.4 × 0.6 × 0.06–0.16 inch], which can be cut from the objects.

In the study of the effect of moulding conditions on the physical properties of test specimens, some manufacturers used injection moulded impact on tensile specimens from several different compositions of ethyl cellulose, cellulose acetate, and cellulose acetate butyrate. This can be carried on one machine at temperatures ranging from the lowest primary cylinder temperature at which complete bars can be consistently moulded to the highest temperature that could be used without causing exessive flashing or discoloration. Other moulding conditions such as die temperature and time of moulding cycle were kept sufficiently constant to avoid their influencing the results. These observers found that with increasing temperature the impact strength of several compositions of each of these three materials was at least doubled and generally was increased by a factor of 3 or 4. Under similar conditions the tensile strength was reduced by about 50% and the elongation was generally increased and in some cases doubled.

Impact bars injection moulded at low temperatures are generally highly oriented in a lengthwise direction. When the molecular chains are relatively inflexible, as with cellulosic plastics, orientation reduces the impact strength, but materials with more flexible chains, such as polystyrene, or methyl methacrylate, shows increased impact strength with the increasing orientation that reduced injection moulding temperatures produce. Moulding conditions are generally well established in many laboratories to the extent that each laboratory can mould test specimens that will give reproducible results, but tests on individual materials moulded at different laboratories show very poor agreement. This problem is receiving increasing attention, but predictions about its ultimate solution are unsafe.

The compression moulding of test specimens from thermosetting materials has been somewhat better controlled and standard moulds have been specified for some specimens. However, the mechanical properties for these materials can be considerably modified by the type and size of mould, the temperature, and the moulding cycle, and their behaviour is further complicated by their use with fillers such as wood flour, canvas or mica, which may be damaged, oriented or unevenly distributed during the moulding process. Thus, service testing on finished parts made from the thermosetting materials is also a necessary part of a testing programme.

CHAPTER 10

Physical Tests of Polymers

INTRODUCTION

The physical properties of materials are of critical importance for the design, production and quality control of products. Consequently, it is not surprising that a whole spectrum of test methods have been devised to measure these properties. Whilst many features of testing materials are common to all materials, the particular characteristics and uses of each group of materials, such as metals, ceramics, polymers, etc., have provided good reasonss why each group has developed its own procedures. That is to say that there are also bad reasons, such as insularity, and that there is not room for greater cooperation and hence unification of methods.

The physical properties of a plastic composition vary greatly with the test conditions. For example, consider the effect of temperature. The tensile strength of a plastic composition at 100°C may be as little as 10 to 20% of the same composition tested at 28°C. The rate of application of a load also has an effect. The load required to produce a given elongation in a plastic compound under conditions of rapid loading may be as much as three times that required when tested at a low rate of loading.

MECHANICAL PROPERTIES OF PLASTICS

The mechanical testing of plastics is difficult. The number and variety of the materials makes it difficult to define tests which can be widely applied to give results which can be used for comparative purposes. For a given plastic the results vary with such factors as the method of making the plastic, the method of fabrication of the test piece and the history of the sample. The mechanical properties of a plastic vary considerably with temperature. The problem is further complicated by the time dependence of the changes, which occur in plastics under mechanical test.

Tensile Strength and Modulus

Much useful information can be obtained about a plastic by studying the plot obtained from a load-deformation test. The equipment used for such a test on a plastic will be similar in design to that used for other materials. There are many test devices with fairly universal application to all materials. The greater requirement for constant loading speeds, when testing plastics, gives rise to most modifications from standard metal-testing equipment. The results of such a test are usually expressed in terms of a stress-strain diagram.

The stress-strain diagram differs from plastic to plastic, but conforms to some extent to that shown in Fig. 10.1. Five parameters are usually obtained from such a diagram. These are :

1. *Young's modulus of elasticity:* This is the ratio of stress to strain in the elastic region in which Hooke's law holds, either closely or approximately. It is the slope of the initial steep portion of the plot.
2. *Tensile strength:* This is the maximum stress the material can withstand without breaking.
3. *Yield stress:* This is the stress at which the first sign of non-elastic deformation occurs.
4. *Elongation at break:* This is always quoted as a percentage of the original length.
5. *Energy to break per unit volume:* This is the area under the graph, and gives a rough measure of the toughness of the material.

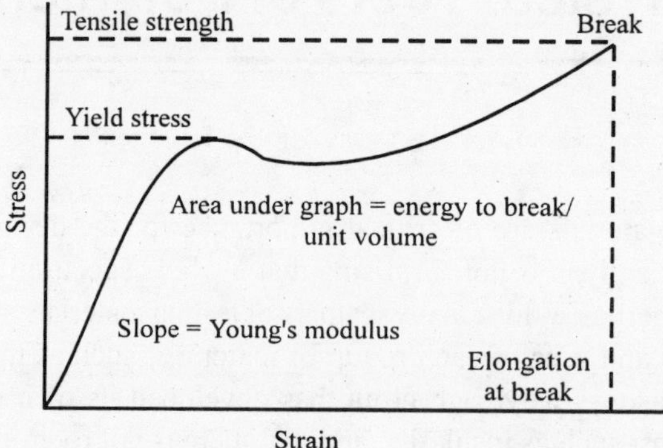

Fig. 10.1. Typical stress-strain diagram for a plastic under tensile test.

The following terms are often used in describing the mechanical properties of plastics :

1. Hard and soft, indicating high and low values of Young's modulus.
2. Strong and weak, indicating high and low yield points.
3. Brittle, indicating a material which breaks before the yield point is reached.
4. Tough, indicating a material with a high energy to break per unit volume.

Stress-strain curves for these various types of material are shown in Fig. 10.2. The tensile properties of a plastic cannot be predicted with any accuracy from a knowledge of its molecular structure. Nevertheless, certain generalisations can be made. Highly cross linked materials and thermoplastics below their glass transition points are hard and brittle. They can be strong, the stress at break being high even though this occurs before the yield point is reached. A thermoplastic above its glass transition point is softer, as extension is possible due to chain straightening and molecular slipping, this type of material will yield. The position of the yield point is determined by the relative time dependencies of the chain-straightening and molecular-slipping processes. If the rate of slipping is low, compared with that of chain straightening, considerable elastic extension can occur before the non-elastic effect of slipping becomes apparent. If the rates are similar, permanent set occurs at low stresses. The appearance of permanent set will also depend upon the rate of loading. If the plastic is crystalline it will be harder than the completely amorphous material. The presence of the crystallites increases the yield stress and the Young's modulus, but molecular slipping can still occur in the amorphous regions. In a lightly cross linked plastic, chain straightening can occur but molecular slipping cannot. Such a material above its glass transition point is soft, but also strong, allowing large elastic extensions.

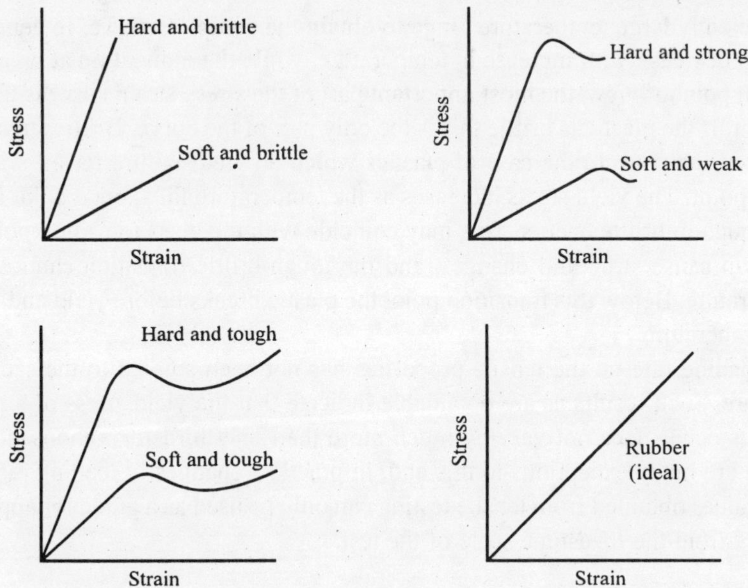

Fig. 10.2. Generalised stress-strain diagrams for various plastic types.

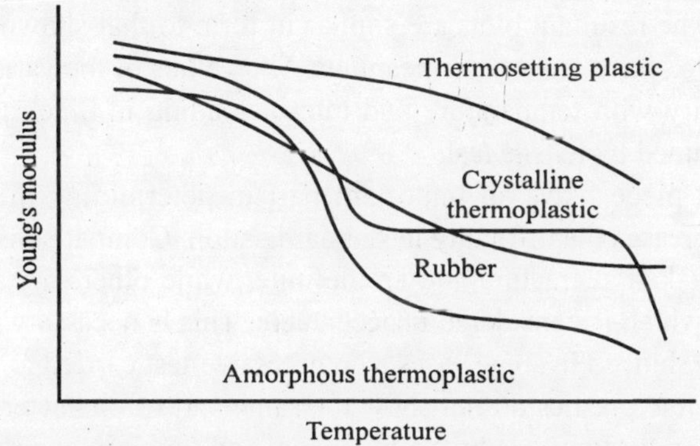

Fig. 10.3. Variation of Young's modulus with temperature for different plastic types.

A general pattern for the effect of temperature on Young's modulus is shown in Fig. 10.3. For a highly cross-linked material the value is high over the whole temperature range, falling only slowly as the temperature increases. For an amorphous material below its glass transition point the value is similar, but above the glass transition point the value falls, due to the onset of chain straightening. It falls slowly at first and then much more rapidly, as temperatures are reached at which molecular slipping becomes important. For a crystalline plastic the fall in value above the glass transition point is less marked, as the crystallite regions limit the chain-straightening effect.

The fall off in value at a higher temperature is much more marked, as molecular slipping will only assume importance when the crystalline melting point is reached. For a lightly cross-linked material the effect is similar to that for an amorphous thermoplastic, except that the final fall in Young's modulus, due to molecular slipping, does not occur. It must be realised that the characteristics shown in Fig. 10.3 are generalisations and many variations from this pattern of behaviour are found. It is often not possible to test over a sufficiently large temperature range to obtain the complete curve. In general, yield stress and tensile strength decrease with increase in temperature, while the elongation at break increases.

From a practical point of view, the most important part of the stress-strain curve is the initial portion up to the yield point. If the plastic is brittle this is the only part of the curve. Brittle stress is the stress at break for this type of material. In the case of plastics which do yield failure for all practical purposes occurs at the yield point. The yield stress increases as the temperature falls, but a point is reached when a transition from tough to brittle occurs. This may coincide with the glass transition point, but the high stresses at break can cause structural changes, and the tough-brittle transition cannot be assumed to occur at this temperature. Below this transition point the plastic breaks before yield and the brittle stress also varies with temperature.

The effect of loading rate on the tensile properties has not been studied to the same extent as the effect of temperature. Such results as are obtainable indicate that the yield stress of a plastic in which viscoelastic changes occur does not vary by much more than one-third for a thousandfold change in loading rate. There are many exceptions to this and, in practice, changes in loading rate may be much larger. Therefore, values obtained from tensile testing can only be used as a guide for applications which do not differ widely from the load-time scale of the test.

Compression and Shear

The general characteristics of the stress-strain curves for plastics under compressive test are the same as those for tensile tests. The resultant plots are similar in form to that shown in Fig. 10.1. The terms hard, soft, brittle, tough, etc., have analogous meanings. The values of the various parameters obtained from compressive tests vary with temperature and rate of loading in much the same way as do the equivalent parameters obtained by tensile test.

The nature of the test piece plays an important part in determining the values obtained from compressive tests. This increases the difficulty in standardisation. Comparisons between values are not always valid as some are taken at fracture and are definite, while others are arbitrary, being taken at some degree of distortion which is considered unacceptable. This is necessary, as many plastics do not rupture under compression. Indeed in some cases a compressive test carried to the limit will reduce the test piece to a flat disc. A few plastics do not show the same sort of characteristics under tension and compression. For example, polystyrene is brittle in tension and tough in compression.

Compressive tests on plastics are carried out with equipment similar in design to that used for other materials. Many test devices are available. The requirement is the application of a compressive load to a suitable test piece between two parallel plates, the rate of loading to be steady. Test pieces are moulded or machined from moulded samples.

Several methods have been used to apply shear stress to plastics. The best method is probably the application of a torque on a cylindrical tube. This has the obvious limitation that many plastics cannot be fabricated in this form. The most common method of finding the shear strength of a plastic is to measure the maximum load when a hole is punched in a standard sheet by a standard punch. Apart from the problems involved in standardising the test sheet, results form this type of test are suspect, as the stress applied is by no means pure shear. There is also evidence that some samples yield before the maximum load for puncture is reached. In such cases the yield stress would be a more accurate working value for the shear strength.

Flexural Properties

The normal test for flexural strength is to support a bar of the plastic at either end and load it steadily at the centre. The distortion is measured as the load is increased, and the process continued until fracture occurs. The flexural strength is the maximum stress in the outer fibre of the test piece at the moment of crack or break. It can only be determined for those plastics which do break.

Flexural properties vary with specimen dimensions, temperature, humidity and loading rate. Values can only be compared if they are obtained by identical tests. Quoted values should only be used as a guide to behaviour under conditions approximating to those of the test.

Impact Resistance

The ability of a plastic to withstand impact depends upon its toughness. An approximate measure of the impact resistance can be obtained from the area under the stress-strain curve, provided it applies to a rate of loading equivalent to the impact. Measured impact strength is a measure of the work done in breaking a test piece. This is not always applicable to other samples, as the measured value varies considerably with the nature of the test piece and the way in which it is broken. The most common method used for measuring impact strength is the Izod notchedbar test. A notch is machined in a rectangular test piece which is supported at either end. A free-swinging pendulum impacts on test piece at the notch and breaks it. The energy absorbed is found from the reduction in swing of the pendulum. The difficulty of making standard test pieces and allowing for energy losses to the supports and the kinetic energy of the fragments, makes the reproducibility doubtful. While the notch test is widely used, it is not very realistic. The presence of the notch alters the characteristics of a plastic, and also means that the impact stress is a complicated combination of tensile and shear. Impact strengths, determined in this way, give a good indication of the ability of a plastic to withstand impacts when fabricated in such a way that it has a notch. Extension of these results to unnotched pieces can only be very approximate.

An alternative method of measuring impact strengths is to use a drop-weight test. A clamped test piece is subjected to a sudden stress, when a known weight falling from a known height impacts on the clamps at one end. In such a test the stress is purely tensile, and there is no need for a notch. There is still the problem of reproducibility of test pieces, but many different test piece shapes can be used, so that the contribution of test piece design can be evaluated to some extent. This type of test is time consuming as there is no guarantee of a break every time, and a number of tests are required for each result. The values obtained are much more realistic than those obtained from Izod tests, but are still only valid for general comparison. One other test method is worthy of mention. In this a moving flywheel impacts on a clamp at one end of a sample which is rigidly fixed at the other. The sample breaks and the energy absorbed in the break is determined by the reduction in energy of the flywheel. This has the advantages of the drop-weight test, and gives more rapid results.

There is considerable variation in impact strength with temperature for a given plastic. The changes are those to be expected from the variation of tensile properties.

From a purely practical point of view, it would seem that flexural impact tests are better than tensile impact tests. Flexural impact tests in which a rectangular sheet of plastic is supported at either end and a linear impact applied between the supports, correspond more closely to impacts likely to be met in practice.

An alternative is to support a sheet on a circular support and apply a point impact at the centre. Such results can only be used to give a general guide to impact resistance, but they are probably more realistic for assessing performance in fabricated articles.

The value of impact strengths is considerably enhanced, if they are used in conjunction with those obtained by other test methods and the results of tensile tests.

Crepe

While tensile properties are studied at various rates of loading the time scale involved is always short, and the results give little information concerning the changes which may occur in a plastic under continual stress for long periods. In practical applications, these periods vary from hours to years. The changes in plastics may be large, and this slow deformation is one of the most important limiting factors in their application to load-bearing units.

The standard method of studying this phenomenon is to apply a fixed load to a standard test piece and examine the strain in it as a function of time. The time scale of the test may be as long as several years, depending on the nature of the plastic. A typical strain-time curve is shown in Fig. 10.4. The immediate elastic extension and recovery are those expected for the particular stress and can be calculated from the value of Young's modulus for the plastic.

The strain increases at a decreasing rate, over a period of time, and this is known as crepe. The reverse process is recovery, which occurs after the stress has been removed and the immediate elastic recovery has taken place. The specimen does not return to its original state, there being a permanent set. If the stress is not removed, the crepe may tend to zero, or more likely it will settle to a constant value, when it is called secondary crepe. In some cases the crepe rate increases again, and the test piece ultimately fractures. This is known as tertiary crepe. The crepe rate usually increases markedly with increase in temperature and tests must be carried out a constant temperature.

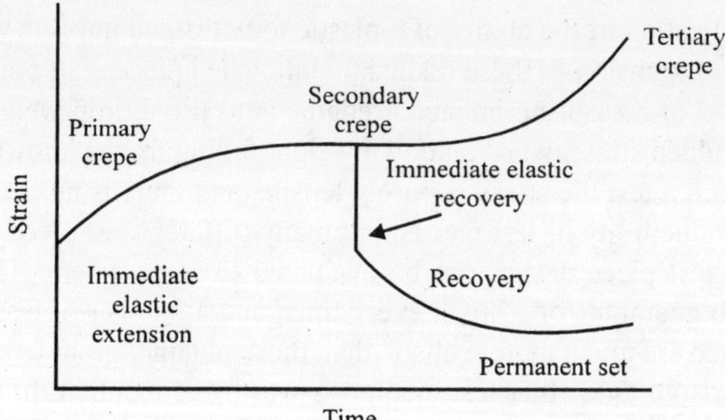

Fig. 10.4. Strain-time curve for a typical plastic under constant stress.

The crepe characteristics of plastic depend upon its structure. The greater the freedom of the molecules to move, relative to one another, the more extensive the crepe. Thus thermosetting plastics usually crepe less than thermoplastics; crystalline plastics less than amorphous ones and thermoplastics below their glass transition points less than those above them. Because crepe is the result of a viscous movement in the plastic, the effect of a small stress for a long period can be just as great as that of a larger stress for a shorter period. A plastic will crepe to some extent if it is subjected to any stress, no matter how small, for a continuous period. This considerable, practical limitation must be accepted. However, plastics particularly thermosetting ones, can be used with fillers, metal inserts or in laminates for the manufacture of load-bearing units.

Other deformations can occur in plastics over a long time period. These are not directly connected with stress, but may alter the apparent crepe characteristics. They are due to such changes as :

1. The formation of further cross links in a thermosetting plastic which will cause shrinking.
2. The absorption or desorption of water.
3. A slow crystallisation in super-cooled thermoplastics which causes shrinking.
4. A gradual relieving of stresses locked in during fabrication, which can cause warping, particularly at elevated temperatures.

Hardness

The term hardness cannot be precisely defined, as it comprises many characteristics of a plastic. The only way in which an estimate can be made is to carry out a number of standard tests, each designed to study one of these characteristics. The more important ones which are studied are :

1. Indentation under load and after the release of the load.
2. Resistance to abrasion in terms of loss of weight or change in optical properties. The optical properties usually studied are transmission and reflection. The ability of a surface to retain its gloss is called the mar resistance.
3. Scratch resistance.

The most commonly quoted hardness values are obtained by indentation tests. These should be used only for rough comparison. Thermosetting plastics are generally harder than thermoplastics, and in the latter category, amorphous plastics below their glass transition points are harder than others. The hardness shows a large temperature dependence.

Effect of Vibrational Stress

Like other materials, plastics are subject to fatigue under repeated cycles of stress, even if the maximum value of the stress is below the yield point of the plastic. Fatigue is a gradual deterioration in the mechanical properties due to the formation of tiny cracks. The end result is failure. The extent of the fatigue depends upon the frequency, amplitude, mode and extent of the cyclic stress and on the temperature. In testing for fatigue the number of cycles needed to produce failure at a given stress value is determined. This number increases as the stress value decreases. Some plastics have a fatigue limit. This is the stress value below which no amount of cycling will produce failure. It is not possible to generalise concerning the fatigue properties of a plastic in terms of its structure. A unit which is likely to be subjected to vibrational stresses must be made from a plastic which has been tested for fatigue. These tests need not be carried to failure, but must include at least as many cycles as are likely to be met in practice. It is also important to define failure correctly, as fatigue is a progressive effect and a unit may become too weak to function properly, well before fracture occurs. There are many equipments used for fatigue tests. Some use cycles of tension and compression, others flexural stress cycles and others torsional stress cycles. Correlation between the results of the various tests is difficult, and it is best to use results from the type of test which approximates most closely to operational conditions.

Damping

The damping capacity of a plastic is a measure of the amount of mechanical energy which is converted into heat, when the material is subjected to alternating or reversing stresses. Perfectly elastic materials have no damping capacity. Energy is absorbed and stored as potential energy when the material is stressed, and then converted into kinetic energy as the material springs back into its original shape, when the stress is removed. Non-elastic materials, on the other hand, retain their new shape when the stress is removed, all the energy absorbed having been converted into heat. Steel springs and rubber bands correspond fairly closely to perfectly elastic materials, while viscous liquids are non-elastic. Even in steel springs a small amount of energy is converted into heat. When a plastic is distorted, elastic and viscous mechanisms are involved. Thus, mechanical energy will be converted into heat. The extent to which a plastic deviates from the perfectly elastic varies with the molecular structure of the plastic and the nature of the stress. The deviation is usually quite appreciable, and the stress-strain curve for a typical plastic, subjected to cycles of tension and compression, is shown in Fig. 10.5. The area of the hysteresis loop is a measure of the amount of mechanical energy converted into heat per cycle.

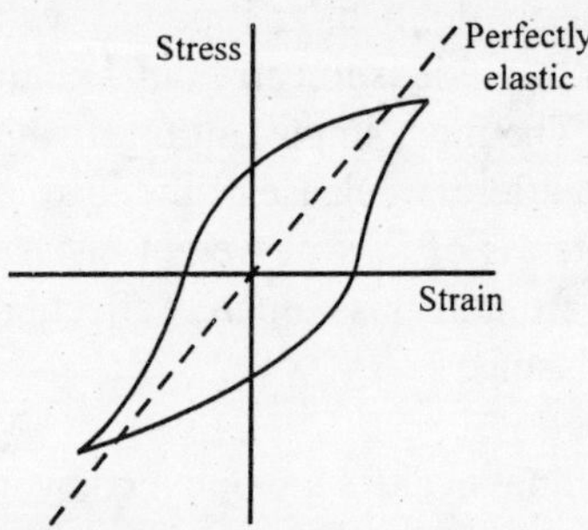

Fig. 10.5. Stress-strain diagram for a plastic under cyclic stress.

The property of high energy absorption is made use of in the damping of vibrations. Vibrational energy produced in one part of a structure is absorbed by a plastic and converted into heat, which is then dissipated in another part of the structure. It is necessary to ensure that the heat produced in the plastic can be dissipated, otherwise the resultant temperature rise may cause deterioration of the plastic. This temperature limitation affects the choice of the plastic for a particular damping function. A thermoplastic with a higher damping capacity may be less satisfactory than a thermosetting plastic with a lower damping capacity, but better heat-resistant properties.

Even rubbers show this hysteresis effect as their distortion involves viscous mechanisms. The term resilience is used in this context, the resilience being a measure of the mechanical energy which is recovered. Thus, the higher the hysteresis, the lower is the resilience. The development of heat in rubbers during cycles of stress is an important factor in the design of tyres and other items.

Dynamic testing can be carried out to determine the damping capacity of plastics. Results of such tests can only be used if the loading conditions of the test are similar to those to be met in practice. The variation of damping capacity with the mode of loading is considerable. Damping capacity also varies with temperature. Thus, any test must be carried out under conditions, such that the heat dissipation is similar to that in practice.

Stress Cracking

Stress cracking is the apparently brittle failure of a plastic under stresses considerably less than those which would be expected to produce such failure. The effect of the failure can vary from crazing of the surface with a series of fine cracks, to complete rupture. The determining factors in the onset of stress cracking would seem to be, the nature and structure of the plastic, its history, the nature of the stress and the presence of.some external initiating agent.

Environmental stress cracking is caused by a surface effect of a substance which is not a solvent for the plastic, and does not appear to modify it chemically. It is thought that the mechanism involves the formation of a strongly absorbed film on the surface of the plastic. This type of stress cracking seems to be limited to polythene. The range of initiating agents is wide, including surface active agents, oils and greases. Solvent stress cracking is caused by the softening and swelling of the plastic surface by solvents or solvent vapours. It occurs in a number of plastics including polystyrene, acrylics and polycarbonates. Thermal stress cracking is heat initiated, and is due to structural changes within the plastic. Stress cracking can also result from the degradation of the surface caused by oxidation.

The elimination of stress cracking is mainly a design problem involving elimination of built in stresses during fabrication limitation of applied stresses to an acceptable level, selection of the most suitable grade of the plastic and, in some cases, the use of surface coatings. The most important precaution for the user is to ensure that plastics subject to solvent stress cracking are not brought into contact with solvents or their vapours, while under stress.

ELECTRICAL PROPERTIES OF PLASTICS

The main use of plastics in electrical engineering is for insulation of all sorts. Plastics are good insulators, but there are considerable differences in the electrical properties of various types. Before a plastic is selected for use as an insulator, or for some other purpose, which makes use of its electrical properties, it is essential that these are known. The advantages of plastics lie in their cheapness and ease of fabrication, compared with the non-polymeric materials of comparable electrical performance. If a flexible insulation is required, there are few suitable alternatives. For rigid insulators ceramics may be more suitable, especially if high temperatures are to be encountered.

Polarisation

All molecules consist of a number of positive and negative charges. The chemical bonding determines the orientation of these charges relative to one another. Individual bonds may, or may not, have a permanent charge separation resulting in a dipole. Even if the bond has no polar nature, the positive negative charge distribution is flexible and if an electric field is applied a charge separation can result. Polarisation of this type is known as distortion polarisation, and disappears as soon as the field which induces it is removed. This type of polarisation occurs in molecules which have permanent polar bonds, as well as those which do not. The dipole caused by the distortion has no fixed orientation relative to the molecular geometry, but will line up with the inducing field, whatever its direction, without movement of the molecules. This independence of molecular motion means that the extent of the polarisation is not altered by temperature change. The high flexibility of the charge pattern allows the polarisation to follow the distorting field very rapidly, even if it is alternating at very high frequencies. This applies to all frequencies in the range below the visible region.

Polarisation of plastics containing molecules with polar bonds is enhanced by the effect of these permanent dipoles. In addition to the distortion polarisation, there is another effect greater in magnitude, known as orientation polarisation. In materials with small molecules, this is due to the rotation of the whole molecule to align the dipole with the field. This only occurs if the molecule is free to rotate. Thus polar molecules held in crystals do not show this effect, but the liquid produced when the crystal melts, does. Increasing the temperature of such a liquid, increases the thermal agitation and reduces the extent of the orientation polarisation, as fewer molecules are aligned. If alternating fields are applied the molecules must rotate to maintain alignment. This is possible at low frequencies, but as the frequency increases the molecules lag behind the field and, ultimately, do not align at all. Thus the degree of orientation polarisation falls as the applied frequency is increased. Even at high frequencies distortion polarisation still occurs.

Orientation polarisation occurs in plastics, but the process is somewhat different. Even in the solid state, alignment can occur. In a polymer molecule with polar bonds, there will be a large number of dipoles spaced along the chain. These can be aligned with the field. The alignment involves the rotational transitions, which cease to be random under the influence of the applied field. This cannot occur in crystalline regions, amorphous plastics below their glass transition points, or highly cross-linked thermosetting plastics. In a crystalline thermoplastic there may be one direction of the applied field for which the dipoles are aligned but this would be a very special case. In an amorphous thermoplastic below its glass transition point, or a highly cross-linked thermosetting plastic, orientation polarisation can occur, if there are polar bonds in side groups off the main chain or network, which can rotate despite the rigidity of the main skeleton. The effect of increasing the temperature of a plastic is to increase the freedom for rotational transitions, and hence the extent of the orientation polarisation. The dipoles will follow an alternating field at low frequencies. As the frequency is increased the rotational transitions cannot occur fast enough to give complete alignment. The degree of alignment falls as the frequency is increased, finally reaching zero. This decrease in the degree of alignment usually occurs over the radio-frequency range.

Dielectric Constant

The dielectric constant of a plastic is defined in the usual way, as the ratio of the capacity of a condenser with plastic between the plates, to that of the same condenser with air between the plates. It is a measure of the ability of a plastic to store electrostatic energy when subjected to an electric field. Dielectric constant or permitivity is important in the study of capacitive effects. It is also important in the calculation of the electrical stress distribution in composite insulators.

The increase in the capacity of condenser when the air between the plates is replaced by a plastic is due to the polarisation in the plastic. The magnitude of the dielectric constant depends on the ease of polarisation of the plastic. Materials with no polar bonding have low dielectric constants. These are temperature and frequency independent, as they are due only to distortion polarisation. This also applies to plastics with polar bonds which are not free to align with the field. Plastics with alignable polar bonds have much higher dielectric constants, which are temperature dependent. At higher frequencies the value will fall when the distortion polarisation remains, but the orientation polarisation decreases. Whenever dielectric constants are quoted for plastics, the frequency to which the value applies should be stated. The value of the corresponding temperature should also be stated, but this is often omitted, if normal temperatures are being considered.

Plastics with low dielectric constants are used in transmission and insulation applications, especially where high frequencies are involved. The high dielectric constant plastics have applications in the manufacture of capacitors. Even in this field it is often necessary to use non-polar plastics, in order to avoid the temperature dependence of the dielectric constant of the polar materials.

Dielectric Strength

The dielectric strength of a plastic is defined as the voltage gradient at which the plastic fails as a dielectric, by electrical breakdown. This occurs when the voltage across an insulator reaches such a value that there is a sudden rush of current. The failure usually occurs at a point rather than over the whole area. A few electrons are torn free by the electrical force and a small current flows. The temperature is raised by this current flow easing the freeing of more electrons so that a cascade occurs. The final failure may be a complete puncture due to thermal decomposition, which will spread as arcing continues. Another form of failure is the formation of a carbon conducting path through the plastic due to the decomposition.

It is not possible to relate dielectric strength with structure and the values for most plastics are fairly similar.

The value for the dielectric strength for a plastic shows considerable variation, depending upon the method of determination. If a thorough investigation is required, it is necessary to study dielectric strength as a function of four inter-related variables. These are voltage, time, temperature and thickness. In order to standardise testing, three voltage-time systems have been suggested. The first involves the increase of the voltage from zero to breakdown value at a uniform and rapid rate. The resultant value is known as the short time dielectric strength. The second method gives the 'minute value' by determining the value of the voltage, which will lead to breakdown within a minute, when suddenly applied. The third method gives the step by step value, the applied voltage being increased stepwise and maintained at each value for a finite time. Tests of this sort show that the longer the duration of electrical stress, the lower is the voltage at which failure occurs. The values obtained from such tests are useful in comparing plastics, but other tests such as the electrical equivalent of fatigue and crepe tests are necessary to give the overall picture.

As the thickness of a sample increases so does the voltage necessary to cause breakdown. The relationship is not linear, and the voltage gradient at breakdown decreases as the thickness of the sample increases. It is important that tests on insulators should be carried out on samples of the thickness which is to be used in practice. This variation of breakdown voltage gradient with thickness accounts for the remarkably good insulating properties of thin films of plastics compared with their performance in bulk.

The variation of dielectric strength with temperature is complex. Values usually fall as temperatures are increased.

Power Factor

A direct current involves the passage of electrons through the conductor. This requires the presence of free electrons. An alternating current can be transmitted by a suitable time variable displacement of charge. Thus, a material can carry an alternating current, even if there are no free electrons, provided charge displacements can occur in it. An insulator in which polarisation can occur is capable of carrying an alternating current, and power will be dissipated in the insulator, if there is a current component in phase with the applied voltage.

$$\text{Power} = \text{Voltage} \times \text{Current} \times \text{Power Factor},$$

where the power factor is the cosine of the phase angel between voltage and current as shown in Fig. 10.6. The ratio of the in phase and 90° out of phase components of the current is called the dissipation factor. The power is dissipated as heat in the dielectric.

In a plastic in which orientation polarisation occurs the extent of the in-phase current depends on the frequency. At very low frequencies the dipoles can be considered to be aligned with the field at all times. When the voltage is at its maximum the dipoles are not moving and the current is zero. When the voltage is zero, the dipoles are changing direction at maximum speed, and the current is a maximum. This is shown in Fig. 10.7(a) using one simple dipole to indicate the effect. The result is a current which is 90° ahead of the applied voltage. The plastic is purely capacitive and the power factor is zero. As the frequency is increased, so the alignment of the dipoles lags behind the field as shown in Fig. 10.7(b). There is now a component of the orientation polarisation current in phase with the voltage and, therefore, a finite power factor. The power factor and the energy dissipated increase with increased frequency, until it becomes impossible for the dipoles to attain complete alignment. Further increase in frequency still increases the lag, but the extent of the movement of the dipoles gets smaller. This means a reduction of current and, at higher frequencies, this becomes more important than the increase in lag, so that the energy dissipated falls.

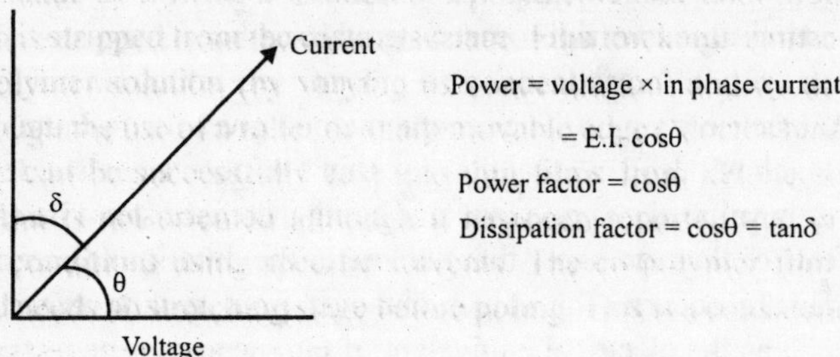

Fig. 10.6. Vector representation of voltage and current in a plastic.

Distortion polarisation maintains accurate alignment with the field up to extremely high frequencies so that the current due to this effect, always leads the field by 90°. For plastics in which this is the only type of polarisation, the power loss is very small. This power loss is not due to the polymeric material, but to the presence of impurities, moisture and other trace constituents which introduce polar molecules.

Power loss can be considered energetically rather than electrically. There is no frictional opposition to distortion polarisation, and so no energy will be dissipated as heat when this occurs. At low frequencies the work done against friction by the dipoles as they move is small, and little power is dissipated. As the frequency is increased the rate of dipole movement increases and so does the work done against friction. At some frequency a stage is reached at which the dipoles no longer completely align. The extent of the work done against friction then falls, as the extent of dipole movement diminishes. Thus the change in power loss with frequency follows the pattern of increase followed by decrease.

The thermal energy produced in a plastic in an alternating field will cause a temperature rise, unless it is dissipated. As rise in temperature in a polar plastic will alter the polarisation and affect the electrical properties. It may also affect the mechanical properties, particularly if the glass transition point or crystalline melting point is reached. It is important to design components in which power is dissipated in such a way that no temperature rise occurs. Dielectric heating of plastics has many applications in fabrication, particularly if uniform heating throughout a mass of plastic is important.

Insulation Resistance

The insulation resistance of a plastic is a measure of the resistance it provides when two terminals are placed in contact with it. It is the ratio of the applied D/C voltage on the electrodes to the total current between them. The insulation resistance is a composite of the volume resistance and the surface resistance. The former represents the extent to which current can pass through the bulk of the material, and is mainly dependent on the nature of the plastic. It can be determined by applying a voltage of about 500 volts across plate electrodes which are in contact with a disc of the plastic under test. A ring guard is used to eliminate surface currents. The small current flowing between the electrodes varies somewhat with time and is conveniently measured 1 minute after the application of the voltage. The volume resistivity is the resistance between electrodes covering opposite faces of a centimetre cube.

Surface resistance is not so much a property of the plastic but of the nature of the surface, its cleanliness and purity. Surface resistivity is the ratio of the potential gradient parallel to the current along the surface, to the current per unit width of surface. It can be measured, but values are not usually quoted, depending as they do on the nature of the surface rather than the plastic. The most important factor to be taken into account is the presence of moisture, which will considerably lower the surface resistance.

This is particularly marked if a continuous film of moisture can form. Thus plastics which are not readily 'wet' by water are better in this respect than others. The insulation resistance of most plastics decreases rapidly with increase in temperature.

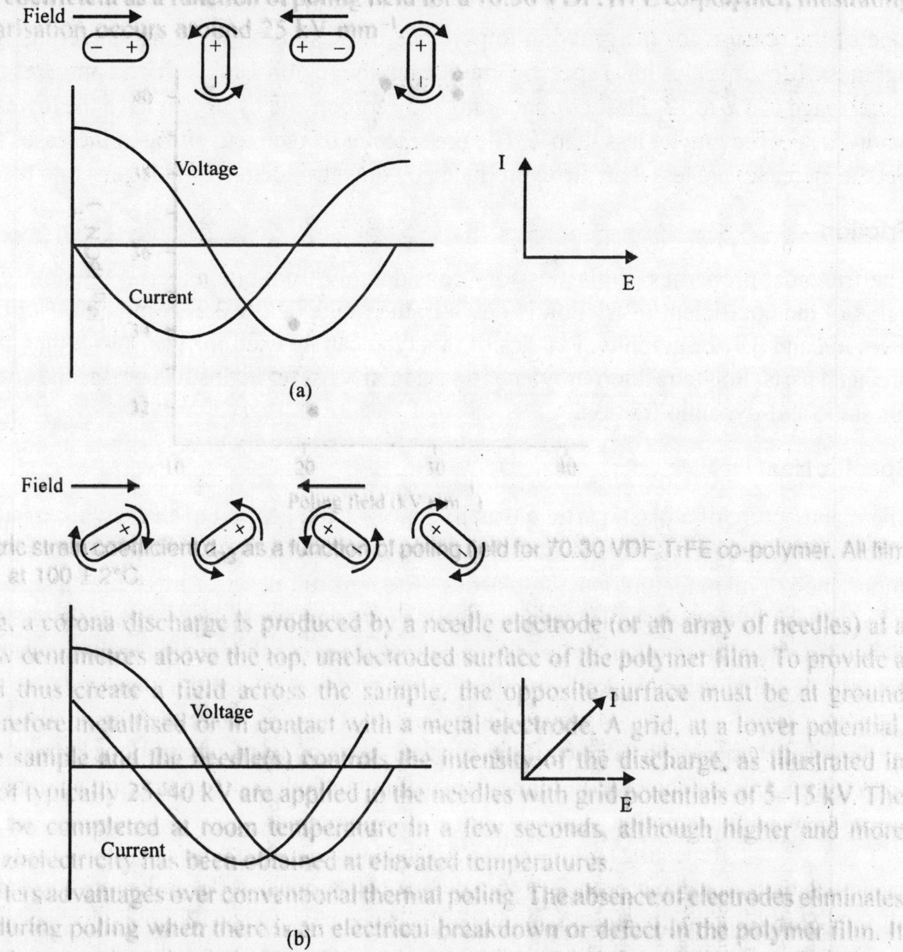

Fig. 10.7. Current due to a simple dipole following an alternating field. (a) very low frequencies; (b) higher frequencies.

Arc Resistance

If an electric arc passes over the surface of a plastic, there is a tendency for tracking to occur. The arc initially passes through the air, but the surface of the plastic decomposes along the line of the arc. The result of the decomposition is a carbon track which can conduct the current. This causes further decomposition and increases the size of the current carrying track until the current path is completely along the surface. The arc resistance of organic plastics is low, because of the carbon present when decomposition occurs. Tracking will not occur in inorganic plastics, unless the decomposition gives rise to alternative conducting products. The value of the arc resistance depends on the ease of decomposition of the plastic and the nature of the decomposition products.

PHYSICAL PROPERTIES

Specific Gravity

One of the reasons for the growing importance of plastics as structural materials is their comparative lightness. Most plastics have specific gravities in the region 1 to 2 which compares favourably with the metal range of 1.6 to 12. Plastics containing only carbon and hydrogen have the lowest values, polythene having a specific gravity less then 1. The presence of oxygen and nitrogen increases the specific gravity, but the increases are less than those produced by the introduction of halogens into the polymer molecule.

Friction

The frictional properties of plastics vary considerably from one material to another. In some cases the value of the coefficient of friction is down in the range below 0.2, comparable with values of 0.05 for diamond and 0.1 for graphite. Plastics of this type can be used for bearings either alone, or as coatings on metal parts. Polytetrafluoroethylene is particularly useful in this respect, as the coefficient is the same for static and dynamic friction.

Specific Heat

The manufacture of a plastic article usually involves the heating of the raw material, and the higher its specific heat the greater is the cost of the heating process. Thus, a knowledge of specific heats is of importance to manufacturers using plastics. The specific heats of most plastics lie in the range 0.2 to 0.4 which is higher than the values for metals. Precise determination of specific heats of plastics may be difficult, as heat can be taken up in bringing about chemical changes in the material. Approximate values can be calculated using the known values of the atomic heats of the elements.

$$\text{Atomic Heat} = \text{Specific Heat} \times \text{Atomic Weight.}$$

For most elements the value can be taken as 6.3, the important exceptions being carbon 1.8, hydrogen 2.3, oxygen 4.0 and fluorine 5.0. The use of this method to determine the specific heat of nylon 6 is given below as an example:

Repeating unit $-\!\!(CH_2)_5CONH-$.

Sum of atomic heats $= (6 \times 1.8) + (11 \times 2.3) + 4.0 + 6.3 = 46.4$

Sum of atomic weights $= (6 \times 12) + 11 + 16 + 14 = 113.$

Thus the specific heat of the repeating unit and hence of the plastic is

$$\frac{46.4}{113} = 0.41$$

Thermal Conductivity

The thermal conductivity of plastic is low, and they are usually considered as being heat insulators. The value of the thermal conductivity of most plastics lies in the range 2 to 10 $\times 10^{-4}$ cal cm/cm^2 °C sec. Some other values for comparison are: metals greater than 800 $\times$ 10^{-4}, asbestos 18 $\times$ 10^{-4}, wood 10 $\times$ 10^{-4} and cork 1 $\times$ 10^{-4}. Plastics are widely used for heat insulation. It must always be remembered that a low thermal conductivity will give rise to large temperature gradients in a material. Care is needed in the design of heat insulation units to ensure that temperatures are not produced which can cause decomposition. Another disadvantage of a low thermal conductivity is the difficulty in raising a mass of the material to a uniform temperature. This may be necessary during the fabrication of large items, and elaborate heating equipment may be needed.

Coefficient of Expansion

The coefficients of expansion of plastics are high, those for linear expansion being in the region of ten times as great as the values for metals. This can be made use of in a few applications, such as those requiring a shrink-fit process. In most cases the large expansions and contractions of plastics presents considerable problems, due to the thermal strains set up as temperatures within a mass of the material vary. There are also fabrication problems, particularly in the manufacture of moulded items, where allowance for large shrinkage must be made in mould design. This means that plastic items cannot be moulded to such close tolerance as can metal items.

Useful Temperature Limits

One of the main disadvantages of plastics is the limited temperature range in which they can be used. This applies particularly to the upper limit, which is much lower than that for many other types of material. The limits at either end of the useful range may be set by transitions, or by the gradual changes in properties with temperature. If the upper limit is due to a transition it is fixed, but if it is due to the gradual change of properties it will vary with the application. For instance, a plastic above its glass transition point will soften as the temperature is increased, and the upper limit is lower for load-bearing units than for others. Similar considerations apply to the lower useful limit. The problem is further complicated by the variation in behaviour with the thermal loading rate. A plastic may fail at lower sustained temperatures than that at which it will fail if the temperature is rapidly raised.

Many test procedures have been proposed for determining the upper useful temperature limit. It is important that any results from such tests should only be applied on design , if the conditions to be met closely resemble those of the test. One value which is often quoted is the heat-distortion temperature. This is obtained by supporting a standard sample at either end and loading it at the centre. The temperature is raised at a uniform rate. The heat-distortion temperature is the temperature at which the deflexion reaches a given value. Another temperature often quoted for plastics is the Vicat softening point. A standard flat-ended needle carrying a standard load is placed on the surface of a suitable sample, which is heated at a fixed rate. The softening point is the temperature at which the penetration reaches a given value. Softening points are usually higher than heat-distortion temperatures, and are not usually obtainable for thermosetting plastics.

Except for certain applications in the field of refrigeration the lower useful temperature limit is less important. The requirement is normally that no failure should occur at the lowest temperatures encountered. Highly cross-linked plastics and thermoplastics, whose glass transition point is above normal temperatures, usually meet this requirement. The only change in properties is a gradual increase in brittleness as the temperature falls. If the glass transition point is below normal temperatures, the requirement may not be met. The lower useful temperature limit is normally set at about 10°C above the glass transition point.

Flammability

The burning characteristics of the plastics are extremely important if high temperatures are to be met. The most important is the temperature at which ignition occurs. The second most important factor is the rate at which burning proceeds after the plastic has been ignited. In all cases oxygen is needed for the combustion, although the presence of oxygen in the polymer molecule in the form of nitro or nitrate groups could eliminate this necessity. In practice, cellulose nitrate is the only plastic of this type, and the non-propellant grades still require additional atmospheric oxygen. Nevertheless, it is easily ignited and burns fiercely. At the other end of the scale, some of the thermosetting plastics are extremely difficult to ignite and then char rather than burn. Most plastics need a large amount of heat for ignition and burn slowly. The rate of burning depends on the availability of oxygen, and the rate of consumption of the material depends on the geometry of the burning piece.

Light Transmission and Colour

A few plastics, such as polymethyl methacrylate, have a higher percentage transmission than glass. They are often used as glass substitutes where the shattering effect of glass is to be eliminated, or the ease of moulding of the plastic is an advantage. Most plastics scatter a greater percentage of the incident light than glass, but this percentage is low enough for the plastic to be classed as transparent, or at least translucent. Some of these can be made in highly transparent film form, and are used in packaging and similar applications. One or two of the highly crystalline plastics scatter so much of the incident light that they are opaque in bulk. The presence of plasticiser or filler may also produce an opaque material.

The extent of transmission and scattering is independent of wavelength, and plastics have no colour of their own. In some cases the presence of impurities gives some absorption of the blue end of the spectrum and the plastic has a yellowish tinge. Colouring of plastics is done by the addition of dyestuffs or pigments. All colours are usually possible, but in one or two cases, such as phenol formaldehyde, the yellowing is so marked that only darker colours are possible. Depending on the nature of the colouring matter, the extent of light scattering may be unchanged or increased, and many plastics are available in both coloured transparent and coloured opaque forms.

Most transparent plastics transmit in the UV region as well as the visible. For some applications this is a disadvantage and can be overcome by the inclusion of a material which absorbs UV. Several complex organic substances can be used for this purpose without altering the transmission characteristics in the visible region. They also minimise the UV initiated photo decomposition, which can occur in some plastics. The products of this decomposition can give the plastic a yellowish tinge or reduce its transparency. The decomposition may also cause considerable reduction in the mechanical strength.

Refractive Index

The refractive indices of transparent plastics lie in the range 1.35 to 1.7. This is the range in which the values for most types of glass are found. The use of plastics in optical devices, such as lenses, is limited by their low scratch resistance. If plastic lenses are required with any degree of accuracy, they must be ground. Lenses can be moulded but, as yet, high-quality products cannot be made in this way. The dispersion effect of plastics is quite low. There is obviously scope for the use of plastics in optical devices, but at present this is limited to the fields in which the non-shatter properties are of prime importance.

Effect of High-Energy Radiation

High-energy radiation can cause changes in the chemical structure of a plastic. The radiation may produce cross linking, or it may produce cleavage of the molecules. Polythene is affected at the C–H bond, and when this occurs on adjacent chains, cross links are formed. The polythene becomes stronger and more rigid. In PTFE, on the other hand, it is the C–C bond which is attacked and the chains are broken down. The PTFE becomes brittle and loses strength. The amount of radiation needed to produce changes varies from plastic to plastic. Polystyrene requires about fifty times as much radiation as polythene for cross linking. Polythene can be cross linked by doses between 10 and 50 Mrad. Polymethyl methacrylate and PTFE are degraded by less than a tenth of these doses. The breaking and forming of bonds in highly cross-linked materials does not have such a marked effect on the properties.

The effect of the radiation also depends upon outside factors, including the size, shape and temperature of the sample, and the nature of the surrounding atmosphere. The radiation produces reactive units in the polymer, which may continue to react, and produce changes in the material, after the irradiation periods. Plastics which have been subjected to reactor radiation will, themselves, become radioactive. For hydrocarbon systems the induced radioactivity is small, but fillers and other constituents may become highly radioactive.

Controlled radiation is used in the preparation of special grades of cross-linked polythene. It is also used for other cross-linking processes, such as the vulcanisation of some synthetic rubbers, and the curing of some polyester resins.

Water Absorption

One or two plastics are water soluble and their applications are often based upon this property. Most plastics are insoluble in water, but will absorb it to some extent. The absorption of water produces changes in properties, which can reduce the effectiveness of the plastic. There is a marked deterioration in the electrical properties, and the mechanical strength may be reduced. The absorption of water may also lead to dimensional changes. The presence of water in the raw plastic, before fabrication, can lead to difficulties in the fabrication process and flaws in the final product.

The standard method for determining water absorption is to dry and weigh a thin disc of the plastic and then re-weigh it after complete immersion in water for a standard time. The increase in weight is an indication of the water absorption of the plastic. The conditions of the test can be varied. The parameter usually varied are the thickness of the sample, the duration of immersion and the temperature of the water. The test can also be used to indicate the extent of any dimensional changes which may occur. The results of tests of this type must only be used as a rough guide to operational performance. The absorption of water by a fabricated item will depend on the nature of the exposure (complete immersion, partial immersion, wet atmosphere, dry atmosphere, etc.), the size and shape of the part and many other factors. Thus, additional tests should be designed, in which the operating conditions are reproduced.

Resistance to Solvents

In order to dissolve a solid, the solvent must overcome the inter-molecular forces. Solvents attack plastics by penetrating between the molecules and forcing them apart. The ease with which this can occur will vary with the plastic as well as the solvent. Thermoplastics, in which the molecules are tightly bound and inflexible, will be less prone to attack than those with loose packing and flexible chains. The inter-molecular forces in a thermoplastic vary with the length of the molecules. Thus, the effect of a given solvent on a particular plastic varies with molecular weight, the longer chains being less prone to attack. Changes in temperature can bring about large changes in the structure of the plastic, which will affect its resistance to solvents. If the plastic contains a cross-linked structure there are no molecules to separate and the plastic does not dissolve. The solvent merely penetrates the solid, causing it to swell. The amount of swelling will depend upon the degree of cross linking. Lightly cross-linked materials may swell as much as 50 per cent, while highly cross-linked structures hardly swell at all. Even in the case of thermoplastics, the solution process may be very slow, and considerable swelling can occur before any appreciable solution takes place.

The standard solubility test is to immerse a suitable sample in the solvent, at a known temperature, for a fixed time. The change in weight and alteration in dimensions and appearance are noted. Mechanical strength tests on treated and untreated samples are sometimes carried out. The results of such tests are available for most plastics and a wide range of solvents. Nevertheless there are so many variables that data should only be used if they apply to the particular grade of plastic and conditions of attack, which are under consideration.

Resistance to Chemicals

In general, the resistance of plastics to chemical attack is high. The test procedure is the same as that used to determine the resistance to solvents. A wide range of acids and alkalis of varying concentrations, as well as solutions of a number of salts, are used in the tests. Data from these tests should only be applied if the grade of plastic, and the conditions of attack, correspond to those of the test. The grade of plastic can make a considerable difference, particularly if fillers or plasticisers are included.

It is difficult to generalise concerning the chemical resistance of plastics. Materials which absorb water are more prone to attack by water-soluble reagents. Unsaturated materials are more prone to attack by strong oxidising acids. If the polymer chain contains linkages which can be hydrolysed easily, it is likely to be attacked by hydrolysing agents. Such generalisations can be dangerous as there are many exceptions.

OPTICAL PROPERTIES

Measurement of the optical properties of plastics, is naturally limited to the types and forms which are transparent, and the determination of the majority of these properties differ little from the methods in use for glasses.

Haze

The haze of plastics has been subject to many purchase specifications, and a simplified method for measuring this property has been standardised Fig. 10.8 shows the positions in the instrument. A collimated beam passes from the aperture, B, to the aperture, A, and falls on a photoelectric cell equipped with a filter to make its response correspond with that of the human eye. The photoelectric cell is connected to a microammeter. Total light transmission is the ratio of the microammeter reading with the specimen in position A to the reading with no specimen in the instrument. With the specimen in position B, the only light that can reach the cell is that transmitted without deviation by haze particles or by surface irregularity of the specimen. The ratio of the microammeter reading with the specimen in position B to the reading with no specimen is the parallel light transmission.

The haze, H, is given by the expression

$$\frac{T_1 - T_2}{T_1} = H$$

where T_1 is the total light transmission and T_2 is the parallel light transmission. Current work indicates that this method will eventually be replaced by a method using an integrating sphere.

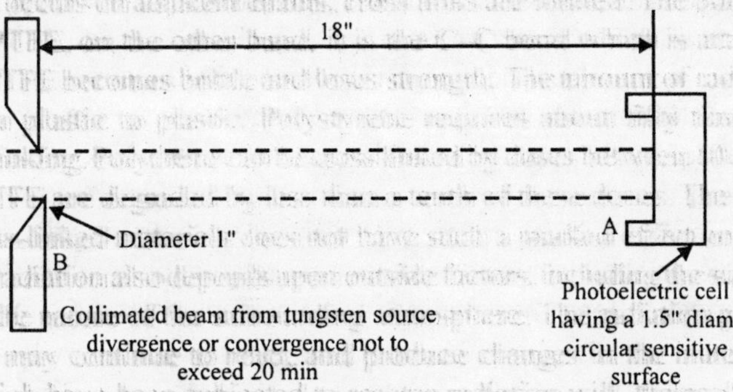

Fig. 10.8. Path of light in haze meter.

Birefringence

When monochromatic polarised light falls on an isotropic specimen and then strikes an analyser that is crossed with the polariser, both the specimen and the field appear dark, since the specimen has no more differential effect on the two components of the polarised light that pass through it than air that separates the polariser and the analyser. If, however, the specimen is anisotropic in the plane perpendicular to the direction of propagation of the light; due either to structure, or stress, and is not oriented with one of its axes of anisotropy in the plane of polarisation; then the polarised light beam is broken into two perpendicular components which have different velocities in the specimen and thus arrive at the analyser out of phase. Their components which pass through the analyser are again in the some plane and can interfere, to produce darkness or varying degrees of light depending on the phase difference or retardation. If the specimen is oriented at an angle of 45 degrees with the plane of polarisation the two components in the specimen have equal amplitudes, and the effect shows its maximum distinctness. When white light is used to illuminate the specimen, it may show colours which are characteristic of the amount of retardation, for phase differences of a very few wave lengths, due to destructive interference of certain wave lengths. At higher retardations, this destructive interference is so frequent that colour disappears and the specimen appears white. Within the elastic limit the retardation $n\lambda$ due to stress above is given by the expression.

$$n\lambda = CST$$

where S is the stress, T is the thickness and C is the stress optical constant of the material. A second constant is sometimes included to permit use of the customary units for these quantities. The letter n is an integer and λ is the wave length of the light under consideration.

Transparent plastics have stress optical constants that vary from 1 for some cast methyl methacrylates to 100 for some specially prepared photoelastic compositions. It is very misleading, therefore, to attempt to judge the residual stress conditions of an unknown plastic specimen by examination of the colour pattern in polarised light. Exact interpretation is further complicated by the mixture of chain orientation and residual stresses, both of which contribute to the retardation. If the stress optical coefficient is known however, a good estimate of the anisotropy of the specimen can be made by observing the colour for slight degrees of anisotropy or for very thin films. Where the retardation is so high as to produce no colour or non distinctive colours, a compensating device is necessary. Quartz compensators with high retardation (5 to 20 orders) are very expensive. Fortunately they can be made inexpensively from plastics. Recently a method has been developed for making a compensating plastic wedge in which the material is heated and stretched until it has a high orientation. Then one end is bevelled with a shaper or other suitable machine to give a uniform spacing of the retardation lines. Only the first few orders are visible in white light. When the wedge is crossed with the direction of orientation of a test specimen, the position of the coloured lines on the wedge indicates the approximate number of wave lengths of retardation caused by the specimen and the position of the black line at their centre indicates it exactly.

Some of the monofilments such as nylon where the asymmetry is very high show enormous retardations. It is possible, in these samples to determine retardation by direct measurement of their difference in index of refraction for light polarised in two perpendicular directions.

CHAPTER 11

Chemical Analysis of Polymers

INTRODUCTION

The principal obstacle in analysing polymers is that one has to deal with substances which have very high molecular weights, cannot be crystallised, etc. If one always had to do with pure plastics on their own, these analyses would not be so difficult to carry out; such cases do in fact occur, but not often. In most cases, the material to be examined is a mixture. This may only be a plastic to which a plasticiser has been added, or it may contain a filler and possibly a dyestuff. It may also be a mixture of plastics, which can contain additional plasticisers, fillers, and dyes.

PREPARATION FOR ANALYSIS

In the analysis of a polymer the first operation may consist in crushing the material. In a few cases a mortar can be used. In many cases a file or rasp is used. In others, as for example for rubber or rubbery materials, a rapidly rotating but if possible, a not too fine abrasion machine helps, or turning on a lathe with a sharp tool. Rock-cutters and mincing machincs are also useful. If a small laboratory crusher is available, it can be used. During each pulverisation care must be taken that the material being crushed does not become heated, otherwise changes may occur.

Having finally obtained the powdered material, the plasticiser present can be extracted from the plastic with a suitable solvent, such as ether, chloroform or methyl alcohol, and quantitatively estimated. For this purpose, a small *Soxhlet apparatus* is most useful.

The examination of the plasticisers so obtained is discussed in detail in the end of this chapter. In most cases no great difficulty is involved. Most plasticisers give saponification values which already, yield extensive information. In many cases, a boiling point determination suffices.

The second problem is the removal of the filler from the plastic. In the majority of cases it is the filler in the substance being treated which is not soluble in any solvent, and may thus be separated from the plastic, the later being dissolved and the filler being removed by filtration. The filtered residue can be well washed, dried and weighed. If preferred, this operation may also be carried out in a *Soxhlet apparatus*, and a small pressure filter may be used for the filtration of sticky or viscous solutions. It must, of course, be seen that the residue left behind on the filter is in fact the pure filler, and that it does not contain any plastic which does not dissolve in the solvent used. In this case, the extraction must be repeated using a second solvent in which the undissolved plastic is soluble.

There are also cases in which the separation of the filler by dissolution of the plastic is not successful. In such cases a chemical treatment of the plastic (e.g., saponification) may be helpful, but in such cases caution is required so that the filler also is not changed. There are cases in which the filler cannot be separated. Thus, for example, there are hardened phenol resins which often contain ground phenol resins as the filler. Here the filler is chemically identical with the plastic, but must however be recognised as filler.

If the plasticisers and fillers are separated from the true artificial resins and plastics, the analysis can begin properly. This is not to say that a fairly practised plastics expert would not often be in a position to determine straight away to which class the material belongs, merely by inspection or at least by a few simple tests. But the experienced man is also quite well aware that, with plastics, one can very easily be deceived and that it is always safer if a systematic analysis is carried out. On the other hand he will not neglect to make as many preliminary tests as possible on the plastic, in order to get the greatest possible number of clues. For this reason, all preliminary tests, which are of use to the experienced plastics analyst and which are also indispensible for the detailed chemical analysis, will be mentioned first.

PRELIMINARY EXAMINATION

A large number of plastics consist only of the elements carbon and hydrogen, others contain oxygen in addition. A further group of plastics contains nitrogen besides carbon and hydrogen, and also oxygen in certain cases. Again, others contain chlorine, or phosphorus, or sulphur, perhaps even some of them simultaneously. It appears therefore that it is important to know the group into which a particular plastic must be placed. The first object, after having separated the polymer from the filler is therefore to test for the elements mentioned.

Nitrogen

The well-known test for the detection of nitrogen is that of *Lassaigne*. The procedure is as follows. A small specimen, preferably powdered, of the plastic is ignited in a small ignition tube together with a small piece of sodium (or better with a drop of a liquid alloy of sodium and potassium). The resulting melt is extracted with a little water and filtered if necessary. The filtrate is heated after adding several drops of ferrous sulphate solution and acidifying with nitric acid. If a blue precipitate or a blue-green coloration forms, the presence of nitrogen is indicated.

This test can often fail to work when only a small percentage of nitrogen is present in the substance. Therefore in order to be more certain whether nitrogen is present in the substance or not, the *Epprecht-Hornung* method is used, the sensitivity of which is so great that it can readily be carried out in a small melting-point tube. The procedure is to mix a few milligrams of the substance to be tested with about ten times as much calcium oxide, or more, in a small test-tube, and ignite the mixture over a small *Bunsen flame*. At, or in, the opening of the tube is held a small strip of filter paper which is moistened with dilute hydrochloric acid.

Ammonia vapour, resulting from the ignition of the nitrogen containing substance with calcium oxide, issues from the tube in this way and is absorbed by the hydrochloric acid. The ammonia in the filter paper strip can be detected with any of the sensitive ammonia reagents. For this it is recommend *Riegler's* reagent (diazotised *p*-nitraniline) a drop of which is placed on a watch glass, and the filter paper dipped into it, the solution being buffered with some powdered calcium oxide. If ammonia is present, a distinct red colour appears, or in some cases a red precipitate, whilst in the absence of ammonia at the most a brown coloration appears, which cannot however lead to confusion.

With this method the nitrogen in a small specimen of rubber, the nitrogen-content of which was determined as 0.04% by the *Dumas* method, could be detected in that a marked orange colour appeared on immersing the filter paper strips.

The preparation of *Riegler's* reagent is carried out in the following way. One gram of *p*-nitraniline is dissolved in 20 ml of hot water containing 2 ml of concentrated hydrochloric acid. The clear solution is diluted, by shaking, with 160 ml of cold water, and the cold solution diazotised with 0.5 g of sodium nitrite dissolved in 20 ml of water. A precipitate which appears at the start dissolves after brief shaking. The solution keeps for a long time when kept in the dark. If necessary the solution can be filtered, before use, from any precipitate which may possibly be present.

Chlorine

A small sample of the plastic is placed in an ordinary test-tube with a small crystal of silver nitrate, and treated with 1-2 ml of concentrated nitric acid and a drop of sulphuric acid (the reagents must themselves be proved to be halogen free). The mixture is heated over a small flame to steady boiling. Should the nitric acid distil off before the plastic decomposes, a further portion of the acid is added. As soon as all is clearly decomposed, the mixture is diluted with distilled water. In the presence of chlorine (or halogen), a precipitate of silver halide is formed which as is known dissolves in ammonia as a complex salt and can be reprecipitated by adding nitric acid.

Alternatively the *Beilstein* test can be used, according to which the substance is placed on a copper wire and held at the edge of a *Bunsen flame*, the presence of chlorine being indicated by a slight green coloration of the flame.

Sulphur

In this case also, a small sample of the substance is ignited with sodium in the same way as in the *Lassaigne* test given for nitrogen, and the resulting melt extracted with water. Whatever the form in which the sulphur is present in the test, it is reduced in the melt to sulphide, and as such can be detected with any sulphide reagent, such as the violet coloration with sodium nitroprusside, the blackening of a bright silver coin, or the precipitation of black lead sulphide.

Phosphorus

This determination is carried out exactly as that described for chlorine, only with the difference that silver nitrate is not added during the test, as this could possibly interfere. The solution from the halogen determination should not be used for the phosphorus determination also. After the sample has been decomposed with nitric acid in the presence of sulphuric acid, distilled water is added to dilute the solution, and several drops of ammonium molybdate reagent are added, after which the whole is heated for about a minute. In the presence of phosphorus a yellow precipitate of ammonium phosphomolybdate forms, at least after cooling. Caution is required since under certain circumstances a white precipitate of molybdic acid can be formed.

The ammonium molybdate reagent is prepared as follows: 100 g of ammonium sulphate are covered with one litre of nitric acid, specific gravity 1.35–1.36, in a 3 litre flask and dissolved by stirring. Likewise 300 g of ammonium molybdate are dissolved in hot water, cooled to room temperature, and diluted to one litre, and the solution poured in a thin stream into the ammonium sulphate solution with stirring. Finally the whole mixture is allowed to stand for at least 48 hours at room temperature, and filtered through an acid-proof thick filter, the final solution being kept well closed in the dark.

Having now carried out these reactions on the unknown plastic, classification is essentially more easy. There are still some other equally useful preliminary tests which one can carry out, in order to narrow down the possibilities still further. The description of these tests will be directly appended. It is, however, to be recommended that these tests be conducted on the lines of the analysis scheme which is described later on to avoid unnecessary work.

Saponification Number

As the saponification rates of various plastics (generally esters) are markedly different, one must differentiate between saponifications carried out solely with the purpose of classifying the plastic, and those carried out to examine accurately the saponification products. For the classification, it is important that the saponification is always carried out under exactly equal conditions, whilst for the saponification with the additional examination of the products as complete a saponification as possible is required, a longer saponification time and a stronger alkaline solution also being necessary.

In many cases it is found convenient to draw the so-called saponification curve, i.e., the same quantity of the unknown plastic is always taken, and it is decomposed using the same amount of solvent and alkali, varying only the saponification time. By comparison of such a curve with the known curve of a known plastic, there is the possibility that additional help may be obtained for the plastic analysis.

For our purposes, i.e., the classification of plastics we are content with saponifications which are always carried out under equal conditions. Convenient conditions are as follows: about 3 g of the plastic are weighed accurately on an analytical balance and placed in a 250 ml *Erlenmeyer flask* with a plane-ground glass top. To the plastic is added a mixture of 25 ml of alcohol and 50 ml of benzene and dissolved if possible. Exactly 25 ml of 2N-alcoholic potassium hydroxide are run in from a pipette. Such a solution is conveniently obtained by preparing a saturated aqueous solution of potassium hydroxide in a platinum or paraffin-coated glass vessel and allowing it to stand for some time. Ultimately the potassium carbonate present comes out of solution and separates on the surface. By weighing and titrating several drops of this alkaline solution, the potassium hydroxide content can be determined accurately. Now, enough of this alkali is weighed in a measuring flask so that on filling to the graduation mark with 95% alcohol, an exactly 2N-alcoholic potassium hydroxide solution is obtained, the titre of which is controlled preferably by phenolphthalein.

The *Erlenmeyer flask* containing the substance to be saponified, the solvent and the alkali, is now connected with a reflux condenser which is closed with a soda-lime tube in order to prevent admission of carbon dioxide to the saponification solution. The flask is placed in a water-bath and the contents brought to the boil. After exactly one hour it is taken from the water-bath and the condenser well washed through, and after adding several drops of phenolphthalein the excess of alkali is back-titrated with aqueous 1N-hydrochloric acid, shaking well. The saponification number of the substance is then calculated from the following formula :

Weight of substance = A g.
Quantity of alcoholic potassium hydroxide added = 25 ml
Titre of alcoholic KOH = T
Quantity of HCl used in the back-titration = a ml
Titre of HCl = t.

$$\text{Saponification number} = \frac{(25T - at)\,56.11}{A}$$

Phenols

Although phenols are practically always perceptible by their characteristic odour, some rapid, simple phenol reactions will be given here:

1. A small piece of the plastic to be examined is melted in a test-tube with solid sodium hydroxide. The resulting melt is extracted with boiling water, acidified with sulphuric acid and filtered. The clear filtrate is made just weakly alkaline with ammonia and then treated with sodium hypochlorite. If on warming a blue colour appears, phenol is present.
2. The sample can also be tested for phenol with *Millon's* reagent, or with *Folin's* reagent.

Methyl Alcohol

This test is used to determine which ester and acid, possibly present, may exist as. The saponification products of the plastics serve for this detection. If the alcohol used for the saponification is free from methyl alcohol the solution which is obtained during the saponification (determination of saponification value), can be used directly. After setting up a small distilling tube, one proceeds by distilling carefully 1 ml of liquid from the *Erlenmeyer* flask into a test-tube, and treating this distillate with 4 ml of 20% sulphuric acid. Cooling the mixture by immersing the test-tube in cold water, it is treated gradually, with continuous stirring, with 1 g of finely powdered potassium permanganate. After the violet colour of the permanganate has disappeared, the solution is filtered through a small dry filter into another test-tube; the filtrate, which has generally a light red colour, is heated gently for about 30 seconds and the solution, which is now colourless, is cooled. To the cold solution a freshly prepared solution of 0.05 g of morphine sulphate (or morphine chloride) in 2.5 ml of concentrated sulphuric acid is added, and mixed by stirring with a clean glass rod. In the presence of methyl alcohol, the solution shows a violet to a deep reddish-violet colour after 20 minutes at the most.

Ethyl Alcohol

The methyl ester is met in plastics almost as frequently as the ethyl ester. To confirm the presence of the latter, the saponification solution is tested for ethyl alcohol, the procedure being that a certain quantity of solvent is distilled off with the aid of a small fractionating tube; it must of course be seen that a fraction is obtained in which the ethyl alcohol can be present. It is also clear that in this case the use of ethyl alcohol as the solvent for the saponification is not permitted, and that methyl alcohol must be used.

From the one fraction which must contain the ethyl alcohol (it boils at 78.5°C) a few millilitres are taken, some iodine solution (not an ethyl alcoholic solution of iodine) added, preferably an iodine-potassium iodide solution, together with some sodium carbonate and the whole warmed gently. When ethyl alcohol is present, the penetrating smell of iodoform is observed and yellow crystals of iodoform separate, which are easily identified under the microscope.

In this reaction it must be seen that no acetone is present, as it also gives the iodoform reaction. Acetone may be recognised by treating a small sample with a few drops of a freshly prepared sodium nitroprusside solution and either sodium or potassium hydroxide, when the solution assumes a red colour, which changes to purple red or violet on acidification with acetic acid.

Phthalic Acid

Phthalic acid occurs not only in a large number of plasticisers but also in the so-called phthalic resins (Glyptals, Alkydresins etc.). In most cases it is recognised quite easily by the fact that it crystallises in the form of the potassium salt after saponification with potassium hydroxide. These crystals can be filtered off, and from them the free acid can be prepared. The following tests for phthalic acid are however rapidly carried out.

About 1 g of the plastic to be tested is mixed with 3 g of pure resorcinol and the mixture heated for five minutes but not above 130°. A blank test should always be carried out simultaneously, as by heating to a higher temperature also without phthalic acid a green fluorescence can be seen, probably owing to a thermal decomposition of the resorcinol to a dicarboxylic acid. After cooling the reaction products are tritrated with cold water, the solution is filtered, and the filtrate makes weakly alkaline after diluting with about 50 ml of water. If phthalic acid is present in the plastic, the appearance of the characteristic fluorescence of fluorescein is observed.

In all cases however, it must be borne in mind that other organic compounds can also give a fluorescence with resorcinol. The following are known to behave thus: tricarballylic acid, citric acid, tartaric acid, nitrophthalic acid, hemipinic acid, trimellitic acid, dihydroxy-maleic acid, succinic acid, succinamide, malic acid, saccharin, asparagine and naphthalic acid.

Possible difficulties can be avoided by using the *Toeldte test* for phthalic acid. The procedure is to filter off the potassium phthalate which is obtained in the alcoholic saponification, and to mix a portion of it with twice as much powdered thymol and one drop of sulphuric acid. Heat in an oil bath at 150°C for ten minutes, cool, and pour in some 50% alcohol. Make weakly alkaline with dilute sodium hydroxide. An intense blue colour appears, which is due to the formation of thymolphthalein thus indicating the presence of phthalic acid.

In most cases dry distillation of plastics also indicates the presence of phthalic acid in the plastic, as a white sublimate of phthalic acid or of phthalic anhydride separates in the test-tube.

Colophony Resins

A small sample of the finely powdered resin is boiled with about three times its weight of acetic anhydride and the resulting solution poured off into a small watch glass. Here a drop of sulphuric acid, specific gravity 1.53, (63% H_2SO_4) is allowed to flow slowly into the side of the solution. In the presence of colophony, a bluish violet colour is formed, changing to brown.

Other Resins

The above described test can be exploited further when carried out as follows. The small sample of powdered resin is boiled in acetone and treated as above. If colophony is present a luminous red colour remains after the fading of the violet. If a bluish green colour arises, fading into green, the presence of a vinyl resin is indicated. Coumarone resins give a permanent cherry-red colour, whilst cyclohexanone resins develop a marked pinkness which changes to a dirty orange colour. Shellac gives no colouration at all, on the whole.

Nitro-groups

This test is important for proving the presence of nitrocellulose. Its use allows the nitro-groups in nitrocellulose or in its saponification product to be detected. A drop of diphenylamine reagent, which is prepared by dissolving 6 g of pure diphenylamine in 100 ml of concentrated sulphuric acid, is added to a sample of the plastic being examined and if a dark blue colour forms, the presence of nitrocellulose is indicated. As already mentioned, this test can also be performed with the saponification solution, in which case several drops of the latter solution are taken in a test-tube, together with as many drops of the reagent, and a deep blue colour arises if nitrate ions are present in the saponification solution.

Aldehydes

This test is naturally not carried out in cases where dry distillation of the plastic has already established, with certainty, the presence of aldehydes, in most cases formaldehyde. If however one is not quite sure the following directions are followed; use being made of fuchsine-sulphurous acid reagent which can easily be prepared as follows. 1 g of fuchsine is powdered finely and dissolved in 1 litre of water. To this solution, 20 ml of 38% sodium bisulphite solution are added and after about ten minutes 20 ml of hydrochloric acid (sp. gr., 1.18) are also added. After standing for about two hours the solution is colourless and ready for use. The solution should be stored in the dark, well stoppered, and out of contact with air.

The detection of the aldehyde is now carried out by boiling a sample of the plastic to be examined with 25% sulphuric acid and distilling off the resulting, or liberated, aldehyde. 2 ml of the fuchsine-sulphurous acid reagent are placed in each of two test-tubes and to one of them, 1 ml of distillate is added; to the other an equal quantity of distilled water is added. Both tubes must then be stoppered. If after some time, 30 minutes at the most, a violet colour appears in the test-tube, whilst the blank test remains colourless, aldehyde is present in the plastic.

Furfural

In this test use is made of the fact that furfural develops a red colour with xylidene (also with aniline acetate) in a non-alcoholic solution. The reaction with xylidene is specific for furfural. It must be seen that the test is not carried out in bright daylight since the resulting colour is strongly sensitive to light.

A sample of the plastic is heated with some 25% sulphuric acid and the aldehyde distilled off. 2 ml of the distillate, which should contain not more than 0.01 to 0.05 mg of furfural, is treated with a drop of phenolphthalein solution and a few drops of strong sodium hydroxide solution. 0.5 ml of freshly distilled aniline is now added, or better xylidene, and the whole treated with 4 ml of glacial acetic acid and 10 ml of water.

Allow to stand in the dark for 15 minutes. If a red colour is observed, furfural is present. In this test, it is not essential that the aldehyde be distilled off first; the sulphuric acid solution can also be used.

Coumarone

About 0.1 g of the plastic is dissolved in about 7 ml of chloroform. To this, or to the clear solution decanted from the insoluble portion, add about 1 ml of an approximately 10% by weight chloroform-bromide solution and shake strongly. After some time, coumarone is present, a red colour forms which darkens to deep red in a few days.

Aniline

A finely powdered sample of the plastic is boiled with 25% aqueous sodium hydroxide and extracted with ether. If the plastic contains aniline, the latter is now found in the ether and can be removed from it with dilute hydrochloric acid. The aqueous solution is now warmed a little in order to drive off any ether dissolved in it. Then some drops of hypochlorite solution are added. After a dirty red colour has developed, a dilute solution of phenol and ammonia is added, and the solution becomes a blue colour in the presence of aniline. Sufficient ammonia must be added to make the solution weakly alkaline. Addition of acid makes the solution red.

With this test, it is strongly recommended that a blank test be carried out simultaneously, that is with pure dilute hydrochloric acid, together with a second test containing a trace of aniline. The reason for this is that a colour also develops with phenol alone, from which however, it differs in appearance.

Another test for aniline results from the fact that a very dilute solution (1:10000 being suitable) remains colourless on addition of calcium chlorohypochlorite, or at the most becomes weakly coloured, and on adding a few drops of ammonium sulphide solution it develops a pink colour.

Glycerol

A sample of the plastic is saponified and after cooling, exhaustively extracted with ether. The aqueous portion is then acidified and again extracted with ether. The aqueous solution is subsequently neutralised with sodium hydroxide, evaporated on the water-bath, and the residue completely dried in a drying oven. The residue is then ground up with a little absolute alcohol and the alcohol solution filtered into a small hard-glass tube. Here the alcohol is completely distilled off and the resulting residue treated with a little finely powdered potassium bisulphate. The tube is now heated in a small *Bunsen flame*, whilst over, or in, the opening of the tube is held a small strip of filter paper which is moistened with a freshly prepared 1% aqueous solution of sodium nitroprusside. If glycerol is present (5 μg are sufficient) it is converted by the bisulphate into acrolein which colours the impregnated filter paper a deep gentian blue. On subsequent treatment of the filter paper with dilute sodium hydroxide the colour changes to yellowish red.

Carbohydrate (Cellulose)

A small sample of the plastic is treated with about an equal quantity of oxalic acid and several drops of dilute sulphuric acid (1:3) in a small crucible, and heated on an asbestos plate over a small flame. As soon as the content of the crucible begins to turn brown, the crucible is covered with a perforated watch glass, over the opening of which is placed a filter paper which is impregnated with a saturated solution of *o*-dianisidine in glacial acetic acid. If carbohydrates are present, furfural is formed by the reaction in the crucible, the vapour of which colours the impregnated paper violet (10-20 μg of cellulose are sufficient).

Acetic Acid

A drop of the saponified solution of the plastic, or the steam-distillate of the same solution acidified with sulphuric acid, is mixed with one drop of 5% lanthanum nitrate solution and one drop of 0.01 N-iodine solution. One drop of 1N-ammonia solution is added and the whole allowed to stand for several minutes. If acetic acid is present, a blue to brown ring forms round the ammonia drop.

In the following section quantitative analysis of individual polymers which are importance to plastics analyst are discussed. Of course it would mean far too much work if each of these tests is applied to every single plastic to be examined: and even then it would only be in the fewest cases that a really conclusive and certain result would be obtained. However, in order to achieve the desired purpose i.e., the identification of plastics as quickly as possible with the fewest possible number of tests one proceeds in practice according to the following scheme. As far as possible however, as has already been mentioned earlier, plasticisers and fillers should be removed from the substance before the examination proper is begun. Only if this is successfully carried out, can one reckon on obtaining the information required, with the help of the scheme. It is, of course, important that the scheme be followed step by step and only continued with when each single step is clear.

QUANTITATIVE ANALYSIS OF INDIVIDUAL POLYMERS

Cellulose Ethers

Methylcellulose

Behaviour in the flame: Flame surrounded by yellow-green mantle. Burns readily once started, producing a slightly sweet odour. *Behaviour on dry distillation:* Decomposes with charring. *Saponification number:* Below 20.

Methylcellulose differs from other cellulose derivatives considered here in that it is soluble in cold water. On warming to over 40°C, it separates out again in the form of a gel. If one has unswollen methylcellulose in the form of a powder, a simple test is inspection under the microscope. By swelling in an iodine-potassium iodide solution, methylcellulose is coloured an intense blue. Under the microscope a single filament of it is easily recognised since it appears as if the single filament contained a colourless capillary boring. In this respect methylcellulose strongly resembles cellulose glycollic ester, which, displays a more violet colour.

The degree of esterification of methylcellulose can best be estimated by determining the number of methoxyl groups, the *Pregl* modification of the *Zeisel* method being suitable. With this method, ethoxyl groups can be determined. Both the gravimetric and the volumetric methods used are based on the fact that on boiling substances which contain a methoxyl or ethoxyl group with hydriodic acid, these groups are split off as methyl- or ethyl iodide. The determination is best carried out in the form of a microchemical method. A qualitative test results from the fact that methylcellulose forms methyl acetate on boiling with acetic anhydride and concentrated sulphuric acid, which is easily recognised by its characteristic smell. On the other hand however, the methyl acetate can also be determined by fractionating and determining its boiling point.

On heating methylcellulose with concentrated sulphuric acid and acetic anhydride, making alkaline and boiling again, free methyl alcohol is formed which can be distilled off. In conclusion, methylcellulose, like all other cellulose esters, gives the *Molisch* reaction for carbohydrates.

Ethylcellulose

Behaviour in the flame: See methylcellulose. *Behaviour on dry distillation:* See methylcellulose. *Saponification number:* Below 20. *Specific gravity:* 1.14. *Refractive index:* 1.47. *Melting point:* 100–135°C (softening point).

Ethylcellulose has no such characteristic properties as methylcellulose and is therefore somewhat more difficult to characterise. Here the determination of the number of ethoxyl groups is of interest. Accordingly an ethoxyl content of 55% corresponds to tri-ethylcellulose. An ethoxyl content of 23.68% corresponds to one ethoxyl group per $C_6H_{10}O_5$.

Corresponding to methylcellulose, ethylcellulose gives ethyl acetate on boiling with concentrated sulphuric acid and acetic anhydride, the acetate being recognised by its smell or boiling point. Also, as for methylcellulose, ethylcellulose can be saponified and the ethyl alcohol formed in this way can be detected.

To ascertain the alkyl of the ether, the excellent method due to *Furter* is used. The alkyl iodide split off on treatment with hydriodic acid is collected in a supension of silver 3:5 - dinitrobenzoate and the resulting ester of dinitrobenzoic acid is isolated and its melting point determined. If ethylcellulose is present, the ethyl ester of 3:5 dinitrobenzoic acid, melting point 92–93°C, is obtained, whilst in the presence of methylcellulose, the corresponding methyl ester is obtained, which melts at 100–111°C.

Benzylcellulose

Behaviour in the flame: Almost incombustible, burns, however, slowly in an open flame. *Behaviour on dry distillation:* Chars. *Saponification number:* Below 20. *Specific gravity:* 1.22. *Softening point:* 95–139°C.

Benzylcellulose differs from the other ether celluloses primarily by its essentially difficult combustibility. In most cases also, benzylcellulose is coloured a faint yellow.

The content of benzyl groups is quantitatively estimated by heating an accurately weighed quantity of benzylcellulose with a mixture of acetic anhydride and concentrated sulphuric acid in equal parts, under reflux for an hour. The acetic acid and benzyl acetate, formed during the reaction, are then driven off by super-heated steam. The acetic acid in the distillate is exactly neutralised with sodium hydroxide and the benzyl acetate saponified, after addition of an accurately measured quantity of 1 N-sodium hydroxide, by boiling under reflux for one hour. Here again it must also be seen that no carbon dioxide is absorbed by the solution being saponified : this can easily be achieved by closing the condenser with a soda-lime tube. After an hour the excess of alkali is back-titrated, after the addition of several drops of phenolphthalein, with 1 N-hydrochloric acid, and the benzyl content of the benzylcellulose calculated by the following formula :

A = weight of benzylcellulose

B = quantity of 1 N-NaOH added

C = quantity of 1 N-HCl used in the back-titration

$$\%C_6H_5O = 9.3\frac{B - C}{A}$$

The benzyl residue can also be detected, purely qualitatively, by distilling off the benzyl acetate which is formed when the benzylcellulose is boiled with acetic anhydride and concentrated sulphuric acid, saponifying a part of it and oxidising with potassium permanganate. If benzyl alcohol is present, it is oxidised to benzoic acid and can be detected as such.

Cellulose Esters

Cellulose acetate

Behaviour in the flame: Burns only with difficulty and develops a smell of acetic acid. *Behaviour in ultra-violet light:* Pale blue. *Behaviour on dry distillation:* Sharp acrid odour, evolved gases react acid to litmus. *Saponification number:* Over 200. *Specific gravity:* 1.27–1.38. *Refractive index:* 1.47–1.51. *Softening point:* 50–110°C.

Cellulose acetate is readily soluble in ketones and esters and is softened somewhat by alcohols. It is scarcely affected by hydrocarbons. The triacetate is notable by its solubility in chloroform.

To determine the acetyl content, an accurately weighed portion of the plastic is dissolved in acetone, or better in pyridine, and then saponified with alkali at about 55°C. After half an hour the excess of alkali is back-titrated with acid using phenolphthalein. It is important to see that the sample is well dissolved during the saponification in order to avoid a low result. The hydrocellulose remaining after the saponification is characterised after being washed—it dissolves in cuprammonium solution—or it can be weighed after filtering, washing and drying and acetic acid is qualitatively tested.

Some recommend saponification with 89% phosphoric acid for a safer determination of the acetyl content of cellulose acetate. The liberated acetic acid is distilled off and acidometrically titrated.

Cellulose acetobutyrate

Behaviour in the flame: Continues to burn after removal from the Bunsen flame, producing sparks and a smell of acetic acid and rancid butter. *Behaviour on dry distillation:* Sharp acrid odour, the gases evolved react acid to litmus. *Saponification number:* Over 200. *Specific gravity:* 1.20–1.22. *Refractive index:* 1.47–1.48. *Softening point:* 50–110°C.

These mixed esters are not as easily saponified as cellulose acetate so that the saponification time is suitably increased to two hours. A higher temperature is also to be recommended. The pyridine, which is again used as solvent, is then distilled off under vacuum and the residue extracted with water, the precipitated hydrocellulose being removed by filtration. The clear filtrate is then transferred to a long-necked fractionating flask and after adding absolute alcohol and some sulphuric acid, it is heated on the water-bath till gently boiling. For the first hour care is taken that no distillate is allowed to come over. After this, the contents of the flask are subjected to fractional distillation, when first the ethyl acetate boiling at 77°C distils with the excess of ethyl alcohol, whilst the ethyl butyrate formed, boils at 120°C. Both fractions are collected separately and saponified after adding an accurately measured quantity of 1 N-alkali and finally the excess of alkali is back-titrated with acid.

The acetyl content of the first distillate is calculated from the formula,

$$\%CH_3CO = 4.3\frac{B-C}{A}$$

Where A g signifies the weight of plastic taken, B ml the quantity of 1 N-sodium hydroxide added, and C ml the quantity of 1 N-hydrochloric acid used in the back-titration.

The butyl content of the second distillation is obtained from:

$$\%\ C_3H_7CO = 7.1\ \frac{B-C}{A}$$

In order to be certain that no loss has occurred, the first saponification of cellulose acetobutyrates can also be carried out quantitatively; in this case calculated on the same weight of sample, the same quantity of alkali must be consumed as in the saponification of the separated ethyl esters in the above method of determination.

Nitrocellulose

Behaviour in the flame: Burns quickly with a bright flame, also continues to burn quickly on removal from the Bunsen burner. *Behaviour on dry distillation:* Explodes if the nitro-content is high. Nitrous gases evolved in all cases. The decomposition vapour reacts acid to congored. *Behaviour in ultra-violet light:* Yellow-brown. *Saponification number:* Over 200. *Specific gravity:* 1.35–1.60. *Refractive index:* 1.50–1.46. *Softening point:* 75–90°C (decomposes in 30 minutes at 130°C).

Nitrocellulose is certainly one of the oldest plastics, and today is still extraordinarily widely distributed. The plastic was introduced into daily life under the name of celluloid, in which the smell of camphor can easily be recognised especially on warming by friction. The main disadvantage of nitrocellulose is its low flash point and therefore the resultant danger of fire.

Qualitatively nitrocellulose can easily be recognised with the help of diphenylamine-sulphuric acid reagent which gives a strong blue colour in contact with nitrocellulose itself, or its saponified solution.

Quantitatively the nitro-group content is of particular interest, and for its determination one of the following methods preferably are used.

The plasticiser present is removed from the finely powdered nitrocellulose by means of a chloroform extraction, and a determined quantity of the plasticiser-free, well-dried (caution!) sample is weighed into the tube of a *Lunge nitrometer*, and 5 ml of pure sulphuric acid (60° Beaumê) is added. Previously the nitrometer was filled with mercury with the compensating tube attached. After connecting the cup and sulphuric acid to the nitrometer, the whole is well shaken. The nitrometer is allowed to stand, and the mercury given time to separate itself clearly, after which atmospheric pressure is resorted and the gas volume read. After correcting for temperature and pressure, to 0° C and 760 mm of mercury respectively, the nitro-group content is obtained by multiplying the corrected volume by the factor 3.286, thus giving the result as a percentage with regard to the portion weighed.

More recently the *Nitron*-reagent has been used with advantage. The procedure is to weigh out accurately about 1 g of plasticiser-free nitrocellulose, and to make use of the fact that nitrocellulose is insoluble in benzene, and to saponify at 50 to 60°C with 10 ml of a 30% sodium hydroxide solution containing 1 ml of 30% hydrogen peroxide, on the water-bath. Hydrocellulose is precipitated, removed by filtration, washed, dried and weighed, while the filtrate is boiled to destroy any excess of hydrogen peroxide present. Then the filtrate is made slightly acid with acetic acid and made up exactly to 50 ml in a measuring flask. From this stock solution, exactly 5 ml are removed in a pipette, a few more drops of acetic acid added and after warming to 60 to 70°C, 10 ml of a 10% nitron acetate solution are added. After standing for 24 hours, as much as possible of the clear supernatant liquid is poured off, and the remaining needles of nitron nitrate filtered off by suction on to a weighed filter crucible, previously dried at 105°C, and washed with a little ice-cold water. The needles are dried for three hours at 105°C and subsequently weighed. If A g represents the weight of plasticiser-free nitrocellulose, and B g the amount of nitron-nitrate weighed, then

$$\% \ NO_2 = 1.227 \ B/A$$

Naturally the nitro-groups can also be determined by carrying out a *Kjeldahl* nitrogen determination on the nitrocellulose and calculating the nitro-group content from the nitrogen content. Here, however, no foreign nitrogen must be allowed to be present.

If one wishes to determine the hydrocellulose in nitrocellulose without saponification as above, there is still the following possibility. The nitrocellulose may be treated at 100°C with a concentrated solution of ferrous chloride after which hydrochloric acid is added. After the development of oxides of nitrogen has ceased, the hydrocellulose remaining is washed neutral with water, washed repeatedly with alcohol, dried and finally weighed.

Polyvinyl Esters

Polyvinyl acetate

Behaviour in the flame: Burns smoothly and steadily, smells of lower fatty acids. *Behaviour on dry distillation:* Evolved gases react acid to litmus and smell according to the acid with which the polyvinyl alcohol is esterified (acetic acid, butyric acid, etc.). A sublimate (maleic acid) can appear in the case of polyvinyl maleate.

Behaviour in ultra-violet light: Brilliant bluish white. *Specific gravity:* polyvinyl acetate 1.15–1.19. *Refractive index:* Acetate, 1.473. *Softening point:* 130–160°C. *Saponification number:* Over 200.

As usual, for an accurate analysis the plasticiser and filler are, in this case also, removed by extraction of the finely powdered or broken up material. So long as there is no mixed polymer present, no difficulty arises in the analysis. The plastic is first saponified and the acid with which the polyvinyl alcohol was esterified is determined in the saponification solution. If the acid is now known and it is recognised that only one acid is present, the quantity of alkali used up by the acid present can be found by back-titration of the excess of alkali, and this leads to a direct determination of the quantity of acid present. The calculation is exactly like that given for cellulose acetate assuming that polyvinyl acetate is being treated.

In conclusion the degree of esterification can also be determined just as for cellulose esters, a certain weighed quantity of polyvinyl ester being boiled with acetic anhydride, and the acetic acid remaining behind after the esterification, determined by titration. From the amount of acetic acid used up, the proportion of hydroxyl groups which were not esterified in the original ester can be calculated.

In polyvinyl esters, maleic acid can also be found after the saponification, (as polyvinyl maleate readily finds an application in the preparation of tubing and similar materials), on its own or in a mixture with polyvinyl acetate. If it has been ascertained by qualitative analysis that only acetic and maleic acids are present in the saponification product these can both be titrated side by side, preferably potentiometrically or if a potentiometer is not available, by the addition of two different indicators. The consumption of acid in the presence of phenolphthalein is noted first, and again the consumption using thymol blue. These titrations are made possible by the fact that maleic acid is a very strong acid.

If A g of substance were weighed out for the saponification, C ml of 1 N-acid were used with phenolphthalein present, and D ml was the total acid required to change the thymol blue, whilst B ml of 1 N-alkali were used for the saponification, the contents of maleic and acetic acids are calculated in the following way:

$$\% \text{ Maleic acid} = 11.6\,\frac{B-D}{A}$$

$$\% \text{ Acetic acid} = 6\,\frac{2D-C-B}{A}$$

These titrations can naturally also be used for other strong and weak acids, as for example, for propionic or butyric acid in place of acetic acid. The formula must naturally be corrected accordingly.

The method can also be used for phthalic acid in place of maleic acid, only the factor 16.6 must be used instead of 11.6. In all such titrations it is of paramount importance to ensure the absence of other acids including also carbonic acid, as these acids also use up alkali and then completely falsify the results obtained. To determine the polyvinyl alcohol it is precipitated quantitatively from the saponification solution with alcohol, filtered and washed neutral with alcohol. After careful drying the polyvinyl alcohol can be weighed.

Polyvinyl chloride

Behaviour in the flame: The plastic extinguishes itself on removal from the *Bunsen flame*. A green zone arises in the flame of the plastic when the latter is held in the *Bunsen flame*. A pungent odour is given off, but not that of burning rubber.

Behaviour on dry distillation: Chars on strong heating, and the gases evolved react acid to congo, and give a positive chlorine test with silver nitrate.

Behaviour in ultra-violet light: Brownish-blue. *Specific gravity:* 1.2–1.6. *Softening point:* Plasticised 65°C. *Saponification number:* Under 20 with a normal saponification.

The chlorine content in polyvinyl chloride is of particular interest. As however, polyvinyl chloride is very difficult to saponify there is the risk that not all the chlorine will be detected, when it is determined in the saponification products, even after a long saponification. It is better to carry out the chlorine determination on the polyvinyl chloride direct, a certain quantity being weighed out accurately, and decomposed in a *Kjeldahl* flask with fuming nitric acid to which some concentrated sulphuric acid and silver nitrate (crystals) have been added. The decomposition can also take place with fuming nitric acid in a *Carius* pressure tube. In both cases, after cooling, distilled water is added and the resulting silver chloride precipitate filtered on to a weighed glass filter, washed with a little dilute (1:1) nitric acid, dried and weighed. Also, however, the amount of silver nitrate added can be weighed accurately, and the excess present after the decomposition can be back-titrated according to the *Volhard* method, or better potentiometrically. From the difference, the chloride content is calculated.

Let the amount of polyvinyl chlorine weighed be about 0.100 g = A g, and the amount of pure silver nitrate be about 0.5 g = B g. In the back-titration C ml 0.1 N, NaCl solution are used.

The chlorine content of the polyvinyl chloride is then calculated from the formula :

$$\%\ \text{Chlorine} = \frac{20.86\ B - 0.3545\ C}{A}$$

or using the weights of 0.1 and 0.5 g given above for polyvinyl chloride and silver nitrate respectively,

$$\%\ \text{Chlorine} = 104.3 - 3.545\ C.$$

The data given for the weights should in some degree be adhered to, as in this way a suitable quantity of silver is always present, and also the quantity of sodium chloride solution required for the back-titration is not too large.

Once again reference is made to the fact that it is advantageous to remove, if possible, any plasticiser and other added substance before the analysis. With polyvinyl chloride this is best done by finely powdering the resin, drying well, and then extracting thoroughly with ether. The material freed from the ether solution is finally boiled repeatedly with a mixture of methyl alcohol and water (1:1) and the liquid decanted off. By this means the emulsifier and stabiliser are successfully removed. Finally the resin can be dissolved in cyclohexanone at 50—60°C and after diluting with acetone filtered off from the filler, using a pressure filter. After vapourising and quantitatively driving off all solvent from the cyclohexanone solution, the chlorine determination can be carried out. If any filler has been present, this can also be tested for chlorine. If the test is positive it can safely be deduced that an after-chlorinated polyvinyl chloride is under examination.

Polyvinyl chloride-acetate

Mixed polymers of this kind are very wide-spread, and do not offer any great difficulty in analysis. The polyvinyl acetate portion is readily saponified. The determination is therefore so conducted that a sample of the substance is saponified with alcoholic potash as usual, and the alkali portion used up by the free acids determined by back-titration with hydrochloric acid of the alkali excess which remains. Here, of course it must be remembered that beside the acetic acid a certain portion of hydrochloric acid is set free, the presence of which can easily simulate large amounts of polyvinyl acetate. For this reason, the titration with HCl is continued until thymol blue, which is added as a second indicator, changes. The titration is therefore carried out in the following way; first the excess of alkali is back-titrated with hydrochloric acid using phenolphthalein, then thymol blue is added and the titration continued until the additional hydrochloric acid liberates acetic acid from the sodium acetate present, and the indicator changes.

The difference between the phenolphthalein and thymol blue titres corresponds with the actual acetic content. If A g of plastic are weighed out, B ml of 1 N-KOH are added for the saponification, C ml of 1 N-HCl are required for the back-titration using phenolphthalein, and an additional D ml of 1 N-HCl are required to liberate the acetic acid.

The following equations result:

$$\% \ CH_3CO\sim = 4.3 \ D/A$$

$$\% \text{ Chlorine freed by the saponification} = 3{:}5 \ \frac{B - C - D}{A}$$

Naturally one can also back-titrate with sulphuric acid using phenolphthalein after the saponification and then ascertaining the chlorine content in the solution by the *Volhard* method. Finally the hydrochloric acid, split off during the saponification, can be calculated from the chlorine content and the acetate by deduction.

Other mixed polymers sometimes occur. For example, there is the mixed polymer of polyvinyl chloride and acrylic ester. Here the portion of the ester component is quantitatively determined by a *Zeisel*, alkoxyl determination.

Generally speaking, with mixed polymers, the chlorine atom is much more easily split off than with pure polyvinyl chloride. As the firmness of attachment of the chlorine atom is a relatively important factor and information about it is of interest, a simple method of characterisation has been worked out. This firmness of attachment is determined most simply by measuring the temperature at which hydrochloric acid is split off from the polyvinyl chloride or its mixed polymer, by placing a small quantity of the substance in a test-tube and hanging this in an oil-bath the temperature of which is slowly raised. A moistened congo paper is held at the open end of the tube and the appearance of a blue colour watched for. As soon as this occurs the temperature of the oil-bath is read and hence the temperature at which hydrochloric acid begins to split off from the polymer is found. Reference, however, must be made to the fact that it is advisable to test the congo paper for chloride ions immediately, as other strong acids which may be split off from the polymer also colour the congo paper blue. For this reason it is often advisable to carry out the test so that the test-tube which contains thc assay is closed with a doubly bored cork which contains two glass tubes, a very slow stream of nitrogen or carbon dioxide being introduced into one of them, the other leading to a silver nitrate solution slightly acidified with nitric acid. In this way the splitting off of hydrochloric acid from the polymer is observed directly by the appearance of silver chloride.

Polystyrene

Behaviour in the flame: Burns with a yellowish white flame, and gives off a sweetish, bloomy smell.

Behaviour on dry distillation: Depolymerises to a colourless liquid with the appearance of a smell of hyacinths. The smell of benzene can perhaps also be observed. The liquid formed by the depolymerisation boils, on redistillation, at 146°C.

Behaviour in ultra-violet light: Blue-violet. *Specific gravity:* 1.05–1.07. *Refractive index:* 1.59–1.60. *Softening point:* 65–75°C. *Saponification number:* Under 20.

On heating above 300°C, polystyrene depolymerises at least partially. The resulting monomer, styrene, has a boiling point of 146°C and can be recognised by means of a boiling point determination. Characteristic of styrene also is its property of adding on bromine to form a dibromide which melts at 74–75°C. As can easily be seen from the formula

$$C_6C_5.CH = CH_2,$$

styrene is readily oxidised to benzoic acid by dilute nitric acid or by a sulphuric acid-dichromate mixture. The benzoic acid can itself easily be recognised by colour reactions. Best is the test in which the benzoic acid is sublimed from the dry residue obtained after oxidation and treated with one drop of 0.3%

hydrogen peroxide solution and one drop of 3% ferric chloride solution. Slowly (or more rapidly on warming in a water-bath) the benzoic acid is oxidised to salicylic acid, and the well-known violet colour for salicylic acid appears.

A quantitative method of determining the polystyrene in a plastic is extremely difficult. In the end one can only determine the portion which is depolymerisable and convert it to the dibromide.

As it is known that an addition of diphenyl chloride to polystyrene increases its strength and hardness and at the same time lowering its inflammability, the presence of diphenyl chloride in polystyrene must be reckoned with.

Polymethacrylic and Polyacrylic Esters

Behaviour in the flame: The plastic burns, and continues to do so after removal from the *Bunsen flame*. Strong sweetish smell appears.

Behaviour on dry distillation: Depolymerises above 250°C to monomeric acrylic and methacrylic esters. Sweetish smell appears. *Behaviour in ultra-violet light:* Methacrylic-ester, bright blue; acrylic ester, dull bluish or grey-white.

Specific gravity: methyl methacrylate 1.18–1.19. *Refractive index:* methyl methacrylate 1.49. *Softening point:* 60–70°C. *Saponification number:* over 200° (the methyl ester of polymethacrylic acid is saponified only with difficulty).

Polymethacrylic esters are uncommonly widespread today and are relatively easily recognised. Again after removing the plasticiser etc. by suitable treatment, the product is depolymerised by heating to about 300°C. The monomer can now be characterised either by its boiling point or by saponification and characterising the product by the alcohol which is split off. A safe differentiation between methacrylates and acrylates is that the saponified solution of methacrylic esters gives off the characteristic smell of isobutyric acid after acidification with dilute hydrochloric acid and warming with zinc dust, since methacrylic acid in contrast to acrylic acid, can be reduced to isobutyric acid.

The following boiling points of several esters of methacrylic and acrylic acid may be useful for a closer characterisation:

Methacrylic acid —	methyl ester	100°C
	ethyl ester	117°C
	n-propyl ester	141°C
	isobutyl ester	155°C
	n-butyl ester	163°C
Acrylic acid —	methyl ester	80.5°C
	ethyl ester	99.5°C

Phenol-formaldehyde Condensation Products

Behaviour in the flame: Phenol resins are incombustible, and they retain their shape: a smell of phenol and formaldehyde is perceptible.

Behaviour on dry distillation: Phenol resins decompose on dry distillation, phenol, formaldehyde and sometimes ammonia also, are liberated.

Behaviour in ultra-violet light: Green without a filler - with a filler, blue-violet.

Saponification number: Under 20 for unmodified phenol resins

Specific gravity: Alcohol soluble, 1.21–1.27
Moulded, 1.3–2.0
Oil soluble, 1.05–1.23

Refractive index: 1.70–1.47

Softening point: Hardened resins do not soften.

With phenolic resins, one must differentiate between the so-called modified and unmodified resins, as has already been seen from the analysis table. Unmodified phenol resins consist solely of phenol and aldehyde, apart from any plasticiser, filler, accelerator etc. present. Such unmodified phenol resins occur in various degrees of ripeness (degrees of hardness) i.e., either they are soluble in various solvents, or else they are insoluble and thermally deformable, or they are insoluble and not thermally deformable.

It is of interest to know how much free phenol or free formaldehyde is present in such resins as this knowledge enables one to determine the quality of the phenol resin.

To determine the free phenol (also cresol, xylenol etc.) content of hardened phenol resins, they are powdered very finely in which operation the most important factor is that a warming of the resin must not take place. Then the powder obtained is extracted in a *Soxhlet* apparatus with carbon dioxide-free water, the condenser of the *Soxhlet* being closed with a soda-lime tube. The aqueous extract is transferred to a measuring flask and made up to exactly 100 ml. From this stock solution, exactly 5 ml are removed in a pipette and added to 10 ml of 0.1 N-bromide-bromate solution, which is in a 100 ml stoppered flask. The solution is now diluted to 30 ml and treated with 5 ml of dilute (1:1) hydrochloric acid. The mixture so obtained is then shaken and allowed to stand for half-an-hour after which 5 ml of 10% aqueous potassium iodide solution are added and the iodine liberated is back-titrated with 0.05 N-sodium thiosulphate solution using starch as indicator. The difference between the quantity of thiosulphate used in this case, and that which is used for a parallel blank run, represents the quantity of free phenol which is present in the aliquot portion (1/20) used in the estimation. Representing this difference by B (ml) and the quantity of phenol resin weighed out for extraction by A (g), the content of free phenol in the sample is calculated from :

$$\text{\% free phenol} = 3.133 \frac{B}{A}$$

To determine the free formaldehyde content the gravimetric method using dimedon as precipitating agent can be used, in which case, 50 ml of the above aqueous extract are taken from the 100 ml measuring flask using a pipette, and treated with 5 ml of 5% alcoholic dimedon solution. A few drops of bromophenol blue are added, and the solution acidified carefully with dilute sulphuric acid until the indicator just turns yellow. It is best to add some acetate buffer solution, the mixture allowed to stand for several hours, filtered, and the precipitate washed with a little cold water.

After drying to constant weight at 100°C the precipitate is weighed. If B g is the weight of precipitate obtained from a weight A g of resin, the free aldehyde content is given by

$$\text{\% free formaldehyde} = 20.4 \frac{B}{A}$$

Here it should be noted specially that a considerable free formaldehyde content can give rise to a false result in the free phenol determination since formaldehyde can be oxidised under certain circumstances by the bromide-bromate solution.

In connection with the quality of phenol resins it is also of interest to know the free ammonia content. For this purpose, about 20 g of finely powdered resin is placed in a distillation apparatus just as is used for the *Kjeldahl* nitrogen determination, an excess of sodium hydroxide added, and the ammonia distilled off into an accurately measured quantity of 0.1 N-hydrochloric or sulphuric acid, the end of the condenser always being immersed in the acid. By back-titrating the acid excess after completing the distillation, the free ammonia content in the resin is found. If B ml of 0.1 N-acid corresponds to the acid neutralised during the distillation of ammonia, the free ammonia content is calculated froms:

$$\% \text{ free ammonia} = 0.17 \frac{B}{A}$$

where A g is the weight of phenol resin taken.

To determine the total nitrogen, the *Kjeldahl* method is used, as described under the heading aminoplasts.

Further the determination of the acetone extract of hardened phenol resins is of material interest. From this valuable conclusions can be drawn about the degree of hardness of resins. Thick objects show a materially higher acetone extraction in the centre, particularly when the resin is not temperature-treated before compression. Compressed masses, which are made in a form which can be heated by high freqeuncy currents show the most satisfactory acetone extraction values.

The content of acetone extractable substance is determined, by reducing a sample of the substance to fine particles by means of a file and sieving out the particles which pass through a Nr 30 B.S. sieve but do not pass through a Nr 60 B.S. sieve.

Somewhat more than 5 g of this powder are dried for two days in a vacuum dessicator, and then weighed accurately and extracted with acetone in a *Soxhlet* apparatus. After six hours extraction, the residue is dried well and weighed. The loss of weight which corresponds with the weight extracted is calculated as a percentage of the dried starting material.

A good phenol resin should contain less than 1.5% free phenol, less than 0.05% free formaldehyde, less than 0.1% free ammonia, and an acetone extract of less than 6%.

Modified phenol resins are best analysed by first saponifying them and examining the saponification solution and residue separately. Resinic and fatty acids present are best determined by the methods given for the analysis of alkydals.

Aminoplasts

The expression aminoplast is used in different countries to denote different resins; nevertheless, one can, under that name, class together all those resins whose structure consists of units containing amino-groups. This collective group naturally comprises, therefore, resins which are relatively different. Fundamentally this group can be divided into two classes, the first arising by condensation of amino-substances with formaldehyde and the second comes about by polymerisation.

Proteinoplasts, urea and thiourea resins, sulphonamide-formaldehyde, aniline-formaldehyde, and melamine-formaldehyde resins are counted in the first class, whilst nylon belongs to the second class.

Proteinoplasts

Behaviour in the flame: Distinct smell of burning hair or milk. The plastic burns but extinguishes itself on removal from the Bunsen flame. *Behaviour on dry distillation:* The plastics decomposes and chars. Alkaline reacting gases are evolved which smell of burning hair, and perhaps of formaldehyde also.

Behaviour in ultra-violet light: White. *Saponification number:* Less than 20. *Specific gravity:* 1.33–1.35. *Softening point:* Above 100°C.

Aniline-formaldehyde

Behaviour in the flame: Burns feebly. Smell of formaldehyde. *Behaviour on dry distillation:* Smell of formaldehyde and aniline. *Specific gravity:* 1.22–1.25. *Saponification number:* below 20. *Softening point:* about 130°C

Urea-resins

Behaviour in the flame: Does not burn, the article retains its shape. Smell of formaldehyde. *Behaviour on dry distillation:* Smell of ammonia and formaldehyde. Vapour reacts alkaline. *Behaviour in ultra-violet light:* Bluish white. *Saponification number:* Below 20. *Specific gravity:* 1.45–1.50. *Refractive index:* 1.90–1.54. *Softening point:* Chars above 150°C.

Melamine-formaldehyde resin

Behaviour in the flame: Strong fishy odour. As for the other characteristics, as urea resin.

Thiourea resin

Behaviour in the flame: No flame results, articles retain their shape. Smell of hydrogen sulphide is detectable. *Behaviour on dry distillation:* Chars above 175°C. Ammonium sulphide is formed and as such can be recognised and identified. *Behaviour in ultra-violet light:* Bluish white. *Saponification number:* Below 20. *Specific gravity:* 1.48. *Refractive index:* 1.66.

Sulphonamide-formaldehyde resins

Behaviour in the flame: Scarcely burns. Smell of formaldehyde. *Behaviour on dry distillation:* Distinct characteristic smell of formaldehyde. *Behaviour in ultra-violet light:* Pale blue, similar to Glyptal.

Nylon

The name was first used for all synthetic materials, which can be used for making fibres. From the chemical point of view, such plastics, which are formed from aliphatic dicarboxylic acids and aliphatic diamines by condensation, are classed as nylon. The melting points of some such polyamides are:

Polyhexamethylene-adipamide	248°C
Polydecamethylene-adipamide	230°C
Polydecamethylene-sebacamide	195°C

In general, nylon only softens to a small degree below the melting point, but it turns brown on melting if air is accessible.

Behaviour in the flame: Burns only with difficulty. *Specific gravity:* 1.08. *Refractive index:* 1.53.

Nylon is marked by its solubility in phenol, cresol, formic and acetic acids, as well as mineral acids. Thus, as nylon is hydrolysed by strong mineral acid, the hydrolysate can be tested for the amine and acid used. However, a *Kjeldahl* nitrogen determination readily gives information on the qualitative synthesis.

Analysis of aminoplasts

The presence of nitrogen being a characteristic of amino-plastics, the determination of nitrogen is of foremost importance in their analysis. Nitrogen can be determined most rapidly, safely and also most accurately, by the *Dumas* method. It is however certain that a combustion furnace for the volumetric determination is not available in every laboratory, so that the *Kjeldahl* method is also used, especially if a substance with a large nitrogen content is being treated. For this reason the method will be discussed briefly here.

A micro—or at least a semimicro—method is best used. The substance under examination (about 0.1 g) is weighed in a small weighing glass and introduced into a long-necked, hard-glass, round-bottomed flask of about 50 ml capacity. About 5 ml of concentrated nitrogen-free sulphuric acid and a little copper sulphate, the latter catalysing the decomposition, are placed in the flask which is then heated on a wire gauze over a small flame, slowly at first, then more strongly. Towards the end of decomposition some potassium sulphate is added to the substance to make a shorter decomposition time possible. As soon as the contents of the flask are colourless, they are diluted cautiously with distilled water to about 20 ml and the flask connected directly with the steam-distillation apparatus. After heating the steam developer till the water inside boils, and also after 50 ml of 0.1 N-hydrochloric or sulphuric acid have been placed into the distillate receiver, 10 ml of an approximately 40% sodium hydroxide solution are allowed to flow into the distillation apparatus through the tube which admits the steam, the tube then being fastened quickly to the steam developer. From the start of the introduction of steam it must be seen that the end of the condenser is immersed in the acid already present in the receiver. If, only after connecting with the distillation apparatus, the water in the steam-developer is boiled, there is a danger that air will be driven through it, which the ammonia (which should distil over) will take up, this leading eventually to failure of the ammonia to combine with the acid. After 30 ml of distillate have come over the receiver is lowered so that the condenser tube is no longer immersed, several additional millilitres are distilled off and the condenser tube rinsed through, as well as that portion of the tube which had been immersed, with distilled water. Phenolphthalein is now added to the receiver and the excess of acid is back-titrated with 0.1 N-sodium hydroxide solution. If A g of substance were weighed, B ml of acid were used (in the receiver) and C ml of alkali were used for the back-titration, the substance contains a nitrogen content given by the following formula :

$$\%\ \text{nitrogen} = 1.4\ \frac{B-C}{A}$$

When the nitrogen content of the amino-plastic is known, one can conclude the nature of the plastic, or if the plastic being treated is known one can conclude its purity. The following table gives some idea of the nitrogen contents of particular aminoplasts :

Aminoplast	*Nitrogen content*
Aniline-formaldehyde	about 10%
Proteinoplast	10–15 %
Pure polyacrylonitrile	26 %
Urea-resins	about 30 %
Thiourea-resins	about 30 %
Melamine-formaldehyde	over 30 %

If, besides, nitrogen, sulphur and perhaps phosphorus also, are present in the plastic being examined, they can likewise be determined quantitatively, the sulphur by treating a certain weight with fuming nitric acid in a *Kjeldahl* flask, and also by decomposing the substance in a *Carius* pressure tube, after which the material is treated with barium nitrate or chloride. In this way, the sulphur is oxidised to sulphate and with the barium salt is converted to barium sulphate which is filtered off after diluting with water, and ignited. When igniting the substance together with a filter paper (using an ash-free filter paper) the residue from the ignition which is mostly pale yellow, must be treated with a drop of sulphuric acid and re-ignited after fuming the sulphuric acid, as the filter usually reduces part of the barium sulphate to sulphide, which can easily lead to false results in the analysis. The determination of phosphorus, occurring for example in proteinoplasts, is referred to in the following section on plasticisers.

Chlorinated Plastics

Chlorinated rubber

Behaviour in the flame: Burns feebly, producing a green zone in the flame. Smell of burning rubber is given off. *Behaviour on dry distillation:* Smells of burning rubber and hydrochloric acid. The latter can be detected with silver nitrate, it also colours congo paper blue.

Behaviour in ultra-violet light: Very faintly bluish. *Softening point:* About 90°C. *Saponification number:* Low, variable. *Specific gravity:* 1.64. *Refractive index:* 1.56.

Chlorinated diphenyl

Behaviour on dry distillation: Gases react acid to congo, chlorine detectable. *Softening point:* Low melting to a liquid of low viscosity. *Specific gravity:* 1.57–1.80. *Refractive index:* 1.61–1.70.

Chlorinated naphthalene

Behaviour on dry distillation: Can be distilled and sublimed. *Softening point:* Liquid at 136°C.

Chloroprene

Behaviour in the flame: Green overlapped by a yellow zone. Burns poorly, smells similar to burning rubber. *Behaviour on dry distillation:* Decomposes, smells of rubber. Hydrochloric acid is split off.

Chloroprene can be vulcanised on heating, rapidly on addition of metal oxides (MgO, ZnO) which can be detected in a vulcanised sample.

Analytically, the determination of the chlorine content is of primary interest, and the section on polyvinyl chloride can be referred to. The firmness of attachment of the chlorine is also of interest, and its determination can be carried out in the way which has been described.

Natural and Synthetic Rubber

The smell on burning serves first of all to differentiate natural rubber from artificial rubber-like substances. Yet this alone is not sufficient, as other substances also smell of burning rubber. On oxidative degradation of natural rubber, acetic acid is formed from the isoprene units present in the rubber. Most artificial rubber-like substances contain inorganic catalysts which can be detected in their ash. They are also differentiated by the vulcanisability of natural rubber. A further difference is revealed particularly in the behaviour towards a given solvent; thus, for example, natural rubber is not affected by acetone whilst Buna swells considerably. Conversely benzene or petrol show almost no effect on Buna whilst rubber is considerably affected.

Buna N differs from rubber by its nitrogen content of about 6%, since Buna N or Perbunan, as it is also known, is formed by the polymerisation of butadiene with about 25% acrylonitrile. Buna S, on the other hand, contains styrene and can be detected as such because it is a butadiene-styrene copolymer.

Plasticisers

It is perhaps somewhat exaggerating to say that almost every substance has found a use somewhere as a plasticiser, or that their use has been suggested, yet one can take it that almost 20,000 plasticisers are commonly used.

Considering the properties required for a good plasticiser the number given seems almost improbable. A plasticiser should be a substance with a low vapour pressure, if possible a liquid with a small viscosity dependence on temperature; further, if possible, it should be chemically inactive and insoluble in water. There is yet this point, that a plasticiser which is excellent for a certain plastic, may be of no use for another because for one reason or another it is incompatible with this plastic, causing it to become turbid or even to make it sweat.

By far the most widely occurring as plasticisers are esters. This fact is of use to the analyst; esters can be saponified, thereby obtaining a chemically active substance, the corresponding acid, which can be characterised.

In the first place, one must be careful that as much as possible of the plasticiser can be extracted as pure as possible from the plastic. To purify the plasticiser the extract obtained is evaporated carefully, and the oily residue put in a short-path high-vacuum distillation apparatus. In this one attempts to fractionally distil it, which with practice is successfully done without great difficulty. In most cases, a main fraction is obtained which quantitatively surpasses all others and which is identical with the plasticiser itself. Then a boiling point determination is attempted on the pure distillate under normal pressure. In most cases, however, this leads to decomposition of the substance.

Whether this boiling point determination is successful or not, a saponification of the plasticiser is now carried out in all cases. A certain weight of the plastic is heated for several hours with alcoholic sodium or potassium hydroxide, carbon dioxide being excluded, and afterwards the excess of alkali is back-titrated. If A g of substance were weighed out, B ml 1 N-alkali being present, and C ml of 1 N-acid were used in the back-titration, the equivalent weight of the plasticiser is calculated from the formula,

$$\text{Equivalent weight} = 1000\,\frac{A}{B-C}$$

If, after decolorising the phenolphthalein in the back-titration, methyl orange is added and the titration continued, it can easily be determined whether one is dealing with a *mono-*, *di-* or *poly* basic acid, as in the later case, supposing that both carboxyl groups are not far distanced from one another, a further quantity of acid is used to change the methyl orange. Assuming that the desired result, namely the identity of the plasticiser, is not yet known, a further step which can be taken is an attempt to characterise the acid present in the saponification solution. Phthalic acid and phosphoric acid are tested for first, as these form the basis of the most-used plasticisers. Also testing the resulting alcohols can be useful, and is generally necessary. To do this, the alcohol is distilled off from the alkaline solution, or, should it be a higher alcohol which is being treated, it is extracted with ether. Having extracted the alcohol the aqueous extract remaining is acidified and the acid in it extracted, again using ether. Acids can now be recognised directly from their melting points; should this lead to difficulties, they can be converted to their phenacyl esters or their *p*-nitrobenzyl-esters, which can be re-crystallised and characterised by their melting points.

The alcohols, for their part, are esterified with acetic acid or benzoic acid, if they cannot be recognised by their boiling points, and the resulting esters characterised by their boiling or melting points, or also by saponifying them quantitatively again and determining the molecular weights of the esters by the formula already mentioned above. As the acid is now known, conclusions about the alcohol can be drawn. This method can also naturally be used for acids.

If a phosphoric acid ester is being treated, which can easily be determined by the detection of phosphorus, one proceeds by carrying out a quantitative phosphate determination on the saponfication product, in which determination the saponfication solution is evaporated to dryness, and the residue dissolved in water making up the solution to a certain volume (according to the quantity of plasticiser used) from which an aliquot portion is removed by means of a pipette for the determination.

Particularly suitable for the determination is the precipitation method using ammonium molybdate reagent which is prepared. 20 ml of the phosphate solution are slightly acidified with nitric acid, several drops of sulphuric acid are added, and then heated to boiling on an asbestos-wire gauze over a small flame. It is essential to see that the glass above the solution does not become overheated. By means of a pipette, about 20 ml of the ammonium molybdate reagent are added to the phosphate solution when boiling, in such a way that the liquid flows directly into the solution without touching the glass. After the reagent has been added, the solution is again brought to the boil, allowed to boil for about thirty seconds and then left standing for several hours. The cold liquid is now filtered through a previously weighed glass filter and after drying, the ammonium phosphomolybdate obtained is weighed.

By multiplying the weight of ammonium phosphomolybdate obtained, by the factor 0.0378, the phosphoric acid contained in the aliquot portion is obtained in grams of P_2O_5.

For an accurate determination it is recommended that the precipitate obtained be dissolved in as little ammonia as possible and the solution treated with an excess of acidified magnesia mixture and heated to boiling. A little phenolphthalein is now added, and the solution treated with dilute ammonia until a faint red colour forms. After cooling it is treated with a fifth part of the total volume of ammonia (concentrated) and filtered after ten minutes. The precipitate is washed with 2.5% ammonia and after drying it is ignited at a bright red heat. After cooling in a desiccator, the precipitate of magnesium pyrophosphate $Mg_2P_2O_7$ thus obtained is weighed, and the P_2O_5 content calculated by multiplying by the factor 0.638.

CHAPTER 12

Chemical Analysis of Additives in Polymers

INTRODUCTION

The wide range of additives used now-a-days in polymer technology, makes positive identification difficult by all but the most sophisticated analytical techniques. The presence of several types of additives in a single polymer formulation, e.g. plasticisers, ultra-violet stabilisers, slip agents and possibly two antioxidants, one for processing and one for service, may all be present in a single formulation. Depending on processing history and age of the polymer, the possibility that additive decomposition products may also be present to complicate the analytical problem in hand.

The latter type of additive decomposition should be distinguished from that occurring during analytical processing. For example, a particular type of polymer additive may undergo partial thermal degradation during extrusion operations involved in its manufacture and then during analysis may degrade by another route under the influence of light.

To summarise then, the determination of additives in polymers presents the analyst with some difficult problems. Only small concentrations are present, complex mixtures may be involved and, moreover, frequently the mixture is of compounds of completely unknown type.

DIRECT SPECTROSCOPY OF POLYMER FILMS

The difficulties involved in the extraction of additives from polymers have lead to a search for analytical techniques not involving a prior solvent separation of an additive extract. Of all the techniques tried, only those based on spectroscopy can claim any measure of success.

In situ spectroscopic techniques are not likely to be of value, then, in the analysis of samples of unknown composition. If known amounts of additive can be incorporated into additive-free polymer, however, these techniques are likely to be extremely useful in the study of solvent extraction procedures and the study of additive ageing processes (i.e. the effects of heat, light, sterilisation, radiation, etc.), since the rate of disappearance or of decay can be measured directly by the decrease in absorbance of the sample at a suitable wavelength.

An example is given below of a method based on direct polymer film infra-red spectroscopy for the determination of the ultra-violet absorber Cyasorb UV 531 (2-hydroxy-4-*n*-octoxybenzophenone) at a concentrations of 0.1 to 1% in unpigmented high density polyethylene. Antioxidants such as Polygard and Santonox R do not interefere in this procedure.

Apparatus

Double-beam spectrometer covering the 15–17 μ region (e.g. Grubb Parsons GS2A).
Hydraulic press with heated and water-cooled platens.
Stainless steel moulding plates (6 in. × 6 in. × 1/6 in.).
Shims 0.06 cm thick (circular 1 in. diameter or rectangular 1 in. long).
Aluminium foil.
Clear plastic rule calibrated in millimetres.
Dial gauge calibrated in 0.01 mm divisions.

Procedure

Preparation of sample film

Cover two stainless steel moulding plates with aluminium foil and place up to six 0.06 cm thick shims on one of these. Place approximately 0.2 g of polymer sample in the centre of each shim and carefully place the second moulding plate on top. Position the two plates in the press and apply contact pressure. Switch on the heating supply and set the thermostat to 120°C. When the temperature reaches 120°C increase the pressure to 3000 lb/in^2, switch off the heating supply and water-cool to room temperature. Carefully strip off the aluminium foil from the polymer films and push out the films from the shims. Check the thickness of each film by means of the dial gauge. Six readings on each film should not vary by more than 0.03 mm. Reject the film which has air bubbles, is uneven or is wedge-shaped. Shape the film to fit the spectroscopic sample holder and gently scrape one of the surfaces with a fine emery board to produce a series of fine parallel lines. This reduces the incidence of interference fringes.

Recording the infra-red spectrum

Place the film in the sample holder and position in the infra-red instrument so that the beam passes through the film at right angles to the scratch marks.

Record the infra-red spectrum from 15.5 to 16.5 μ in accord with the spectrometer operating instructions using a scanning speed of 1/2 μ per minute.

Before removing the film from the instrument mark the position of the infra-red beam. Remove the sample from the holder and measure the thickness to the nearest 0.01 mm by means of the dial gauge at six points within the marked area. Calculate the mean of these six measurements.

Measurement of Absorbance

Remove the chart from the spectrometer and with a sharp pencil rule a base-line from approximately 15.8 μ to approximately 16.2 μ. With a ruler measure I and I_0 to the nearest 0.1 mm at the wavelength of the peak maximum (Fig. 12.1).

Calculate the absorbance at 15.94 μ and hence the absorbance per unit thickness by means of the following expression :

$$\text{Absorbance per unit thickness} = \frac{\log_{10} I_0 / I}{\text{Film thickness (in cm)}}$$

Calibration

Prepare duplicate films from the standard sheets containing 0.1, 0.3, 0.5 and 1.0 wt. %. UV 531 as above. Record the infra-red spectrum as described above and calculate the absorbance per unit thickness as described under measurement of absorbance. Construct a calibration curve by plotting absorbance per unit thickness against percentage weight UV 531 for each standard film. Use this calibration curve to obtain the UV 531 content of the polyethylene sample.

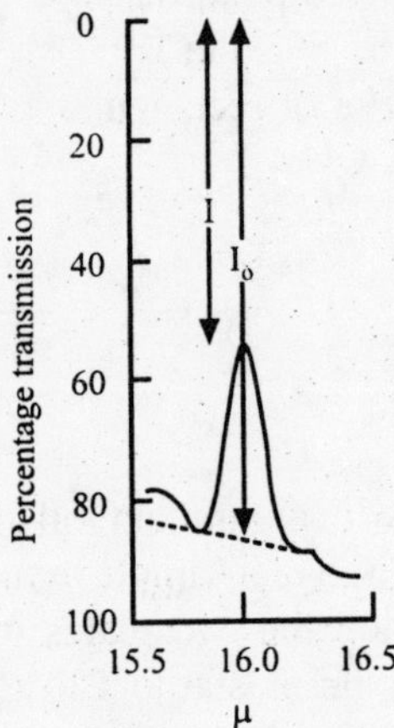

Fig. 12.1. Determination of UV 531 in high-density polyethylene by direct film infra-red spectroscopy.

PRELIMINARY SOLVENT EXTRACTION OF ADDITIVES FROM POLYMER

As *in situ* methods of analysis for additives are not frequently applicable, it is necessary in most instances to use various methods, usually involving a solvent extraction procedure, for the preliminary isolation of an extract of the polymer in which is recovered a maximum amount of the additive and a minimum amount of polymer.

Solvent Extraction Procedures

One of the most difficult types of polymer to deal with, because of its insolubility, is the polyolefin variety. The British Standard method is favoured but it involves dissolution of the polymer in boiling toluene under reflux, followed by precipitation of the high molecular weight fraction with ethanol. The filtrate then contains the additives (provided these are toluene- and ethanol-soluble), plus an amount of low molecular weight polyethylene wax. The extraction and extraction time required by the antioxidants Ionol (2,6-di-t-butyl-*p*-cresol) and Santonox (4,4'-thio-bis-(6-t-butyl-*m*-cresol)) from polyethylene, at room temperature varies linearly with polymer density and particle size, and also with the nature of the extraction solvent. It is concluded that if the polyethylene is powdered to 50 mesh, 3 hr shaking in a wrist action shaker is sufficient to recover 98% of either of the above additives from polymer of any density, i.e. between high and low density polyethylene. In support of these findings some correlation has been found between the density of polyethylene and its permeability to solvents. Some extracting solvents with comments are given in Table 12.1. Chloroform is a useful solvent for the quantitative extraction of additives from polyolefins. The experimental technique is illustrated below by means of an example involving the determination of Tinuvin 326 in natural or pigmented polypropylene.

Determination of Tinuvin 326 in polypropylene

The Tinuvin 326 is extracted from the polymer with chloroform under reflux conditions. The concentration of Tinuvin 326 in the chloroform extract is determined by measuring the ultra-violet absorption peak at 355 mμ and by reference to a prepared calibration curve.

Table 12.1. Solvents for antioxidant extraction.

Polymer type	*Substances extracted*	*Extracting solvent*	*Comments*
Polythene	Cresols Phenolic antioxidants	Chloroform	Heat at 50°C for 3 hrs in closed container
Polythene	Cresols	Hexane	Heat at 50°C for 24 hrs
Polythene	Antioxidants	Ether	in the dark at room temperature
Polythene	Phenolic antioxidants	Chloroform	
Polythene	Antioxidants	Toluene	Reflux to dissolve polymer and ppte with methanol
Polythene	Antioxidants	Water	At 70 under nitrogen
Polythene	Phenolic antioxidants	Compares carbon disulphide with iso-octane	
Polyolefins	2,6-Di-t-butyl-*p*-cresol	Cyclohexane	Reflux for 30 min
PVC	Diphenyl thiourea 2-Phenylindole dicyandiamide	Methanol or ether	
PVC	Stabilisers Lubricants Plasticisers	Ether	
Rubbers	Amine and phenolic antioxidants	Ethanol/HCl	Reflux, then steam-distil amines from extract
Rubbers	Phenyl salicylate Resorcinol benzoate	Ether	
Rubbers	Antioxidants	Acetone	
Rubbers	Ketone-amine condensates Phenols 2-Mercaptobenzimidazole	Acetone	
General	*p*-Phenylenediamine derivatives	95% Methanol or ethanol	For 16 hrs in an extraction cup

Apparatus

Ultra-violet spectrophotometer, recording type. Sample cells, quartz—1 cm path length, water bath, and miscellaneous glassware.

Reagents

(i) Chloroform AR; and (ii) Tinuvin 326.

Procedure

Calibration

Dilute the 0.20% w/v chloroform solution of Tinuvin 326 to obtain solutions of 0.001, 0.0008, 0.0004, 0.0002% w/v.

Record the ultra-violet spectrum of each of these solutions against a chloroform blank from 300 mμ to 380 mμ using 1 cm cells. Measure the absorbance of the absorption peak at 355 mμ as shown in Fig. 12.2. Plot a calibration curve of absorbance against concentration of Tinuvin 326 in chloroform.

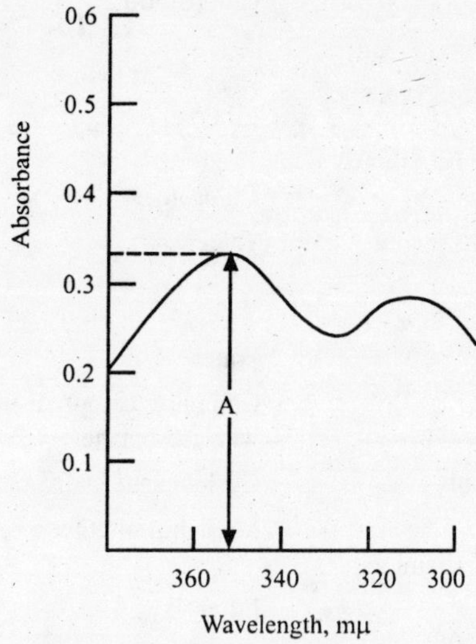

Fig. 12.2. Determination of Tinuvin 326 polypropylene by ultra-violet spectroscopy.

Polymer extraction

Weigh accurately into a 250 ml round-bottomed flask a portion of the polypropylene sample, containing about 5 mg Tinuvin 326. Add 50 ml of chloroform, place the flask under a water-filled reflux condenser, bring the mixture gently to the boil on a waterbath and heat for 30 min. Allow the solution to cool and decant into a 100 ml volumetric flask after filtering through a No. 42 Whatman filter paper. Transfer any polymer collected in the filter paper to the round-bottomed flask and add 40 ml of chloroform. Repeat the extraction procedure. Filter the cooled solution into the volumetric flask containing the first extract. Wash the residue with a little more chloroform, filter into the volumetric flask and make up to the 100 ml mark with chloroform. Shake the solution well and then pipette 5 ml into a 50 ml volumetric flask and make up to the mark with chloroform.

Record the ultra-violet spectrum of the diluted chloroform extract from 300 mμ to 380 mμ against a chloroform blank using 1 cm cells.

Measure the absorbance of the Tinuvin 326 absorption peak at 355 mμ (Fig. 12.2). Determine the amount of Tinuvin 326 present in the diluted extract by reference to a prepared calibration curve.

$$\%\ \text{w / w Tinuvin 326 in polymer} = \frac{(\%\ \text{w / w Tinuvin 326 in diluted extract}) \times 1000}{\text{Weight of original polymer}}$$

DETERMINATION OF PHENOLIC ANTIOXIDANTS

A very popular method of estimating antioxidants in polymer extracts is by coupling or oxidising them to form coloured products and measuring the resulting absorbance in the visible region of the spectrum. This technique is not particularly specific for individual antioxidants, but is specific for phenolic antioxidants in general (also amine antioxidants), and hence can often be applied without interference from other types of polymer additives.

A British Standard method for estimating the total phenolic antioxidant content of polyethylene involves a preliminary extraction of the polymer with hot toluene to extract the additive, followed by addition of ethanol to precipitate any dissolved polymer, then coupling the extract with diazotised sulphanilic acid to produce a colour which can be compared with standards by visible spectrophotometry. Di-cresylol propane and Santonox R (4,4'-thio-bis-(3-methyl-6-tert-butyl phenol)), in particular, are mentioned as additives that can be determined by this technique.

DETERMINATION OF AMINE ANTIOXIDANTS

Procedure for the colorimetric determination of amine antioxidants is given below:

Apparatus

All of the absorption measurements are made on a Cary Model II spectrophotometer using quartz cells having a light path of 1.000 cm. The slit widths used are shown below:

Slit widths used in the determination of amine antioxidants

Wavelength *mμ*	*Slit width* *mm*	*Wavelength* *mμ*	*Slit width* *mm*
350	0.054	550	0.006
375	0.029	575	0.006
400	0.017	600	0.008
425	0.012	625	0.012
450	0.009	650	0.018
475	0.007	675	0.027
500	0.006	700	0.043
525	0.006	—	—

Reagents

The reagents used are:

Methanol—Pure synthetic methanol (99.85 %).

Hydrochloric acid—Pure laboratory reagent grade, specific gravity 1.188 (37.2% hydrochloric acid by weight).

p-Nitroaniline—Micro-analytical reagent.

Sodium nitrite—Reagent grade.

Coupling agent—2.800 g of *p*-nitroaniline is dissolved in 32 ml of hot concentrated hydrochloric acid and the solution is diluted with distilled water to 250 ml. After cooling to room temperature the volume of liquid is adjusted to exactly 250 ml. A second solution is made containing 1.44 g of sodium nitrite in exactly 250 ml of distilled water. Both of the above solutions are reputed to be stable indefinitely. Twenty-five ml of each of these solutions are pipetted into separate 100 ml beakers and are chilled in ice to below 10°C. The contents are mixed by combining the solutions and pouring back and forth from one beaker to the other. Pure nitrogen is bubbled through the mixture as it is allowed to warm to room temperature. Finally, 10 mg of urea (or 1 ml of 0.1 g/10 ml solution) is added to destroy any excess nitrous acid. The reagent becomes cloudy after an hour or so but is suitable for use for several hours. Fresh reagent should be made every day.

Methanol-hydrochloric acid solvent—Three volumes of methanol are placed in a glass stoppered bottle and one volume of concentrated hydrochloric acid is added gradually with gentle mixing. The mixture is allowed to come to room temperature before carrying out the analysis.

Procedure 'A'

The sample to be analysed must be very thinly sheeted or powdered. Weigh a 1.000 ± 0.0005 g sample. Wrap with extraction cloth which has been previously extracted to remove sizing, etc. Place in an Underwriter's extraction cup and extract for 16 hrs with 95% ethanol or methanol. Transfer the alcohol extract to a 100 ml volumetric flask. Cool to room temperature and bring to the mark with the extraction solvent. Transfer a 10-ml aliquot to a 100 ml volumetric flask. Add 15 ml methanol-hydrochloric acid solution and 1 ml of coupling agent. Place in dark for 1 1/2 hr and then bring to the mark with methanol-hydrochloric acid. Determine the absorption spectrum from 700 mμ to 350 mμ using a Cary Spectrophotometer or other spectrophotometer (Fig. 12.3).

If a red colour is formed immediately upon coupling and the colour then fades to an amber or brown, it is likely that one of the antioxidants containing phenyl-betanaphthylamine (PBNA) is present. In this case the alternate procedure 'B' must be used.

Alternate Procedure 'B' for PBNA

Extract is indicated in the regular procedure. Take a 10 ml aliquot from the 100 ml volumetric flask. Bring almost to the mark with methanol-hydrochloric acid mixture and add 1 ml of coupling agent. Bring exactly to the mark, mix and immediately determine the absorbance. The absorbance will rise to a maximum in 2 to 5 min. A British standard procedure describes a method for the determination of Nonox CI (N, N'-di-2-naphthyl-*p*-phenylenediamine) in low density polyethylene. In this method the antioxidant is separated from the polymer by bringing the polyethylene into solution in toluene, followed by precipitation of the polymer with ethanol. The mixture is then filtered to remove polymer, the antioxidant remaining in the filtrate. The antioxidant content of the filtrate is determined colorimetrically by oxidation with hydrogen peroxide, in the presence of sulphuric acid. This reagent produces a green colour with Nonox CI, which gradually reaches a maximum intensity. The colour is evaluated at 430 mμ when the maximum depth of colour is reached.

A modification to this procedure for the determination of Nonox CI in high density polyethylene which, compared to low density polymer, has only a small solubility in toluene. The procedure is extended to the determination of oxidised Nonox CI, and a further product called degraded Nonox CI. This latter material is produced only in polymers containing acidic forms of carbon black and is believed to be produced by decomposition of oxidised Nonox CI by carbon black acidity. These procedures can be applied to study the effect of twelve different acidic and alkaline carbon blacks on Nonox CI degradation, occurring during polyethylene extrusion.

DETERMINATION OF PLASTICISERS

The quantitative separation of plasticisers from the other ingredients is the first and most important step in the analysis of plasticised polyvinyl chloride compositions. The most effective and convenient method of separation is by extraction with a suitable solvent, using the Soxhlet apparatus. The efficacy of this procedure depends mainly on the choice of solvent. The ideal solvent would not dissolve any of the polyvinyl chloride, but would remove all the plasticiser, and all, or none, of the other ingredients of the composition.

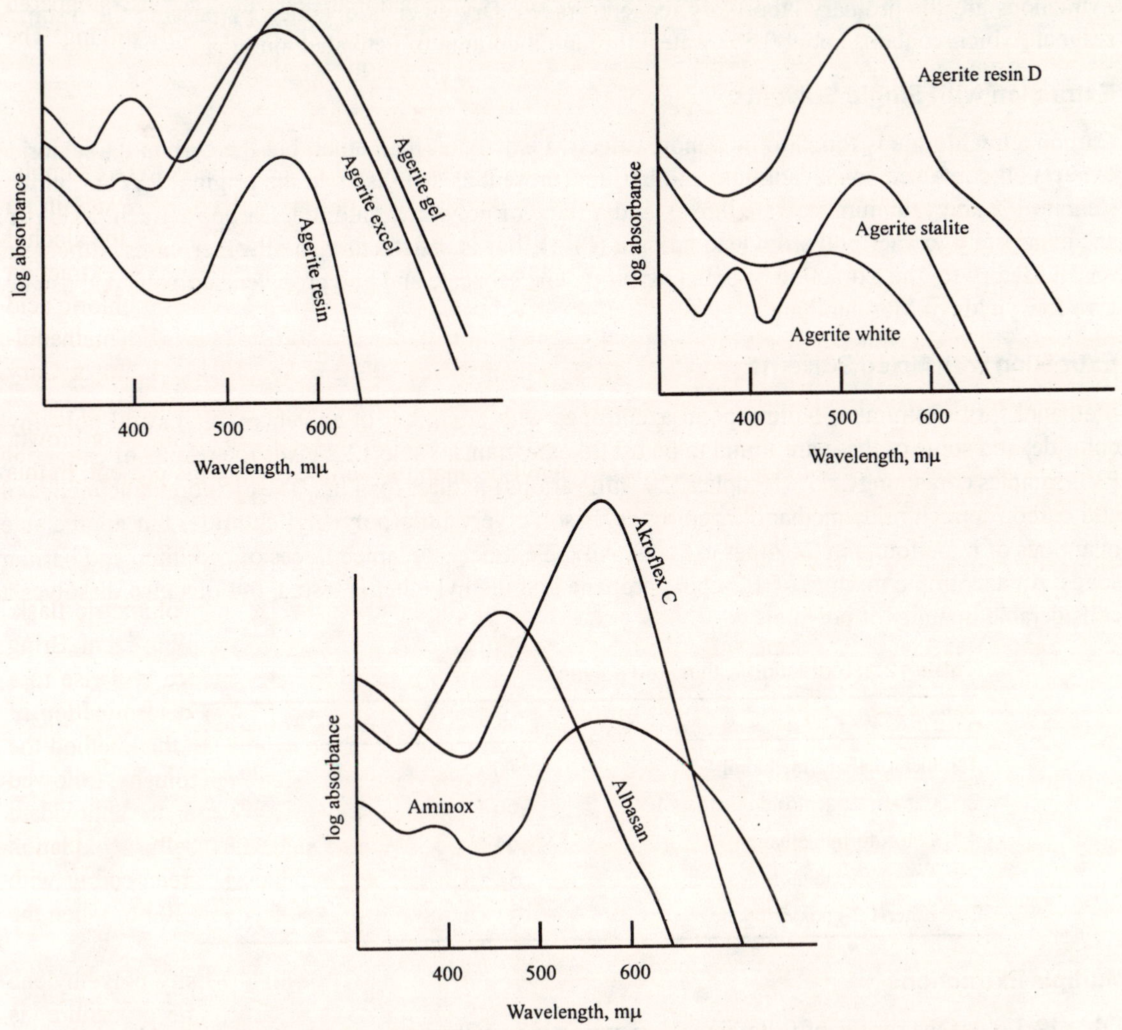

Fig. 12.3. Visible spectra of amine antioxidants diazotised *p*-nitroaniline method.

The polyvinyl chloride compositions are prepared by mixing the weighed ingredients by hand, then blending on a hot two-roll open mill. They are removed from the mill as 0.05 cm sheets. Each sheet of plasticised polyvinyl chloride are cut into pieces 1 cm × 0.3 cm. In every extraction the weight of sample is 2 g, and the volume of solvent 50 ml. An all-glass jacketed Soxhlet apparatus is used with Whatman 19 mm × 90 mm extraction thimbles.

When each extraction is completed the solvent is distilled off, and the last traces are removed by heating in the water-bath for 1 hr and in the oven at 105°C for 3 hrs. After cooling in air, the flask and extract are weighed, then heated in the oven for a further 30 min, cooled and re-weighed. If the weights agreed to within 2 mg the first value is taken, otherwise the flask is heated for another 30 min and weighed once more.

Polymer dissolved in the extracts are detected by adding methanol to a solution of the extract in tetrahydrofuran: any polyvinyl chloride could be seen as a white precipitate. The solvents used for the extractions are the ordinary laboratory reagent grades. Dry ether is obtained by passing the ordinary reagent (which contains about 0.3% water) through a column of activated alumina.

Extraction with Single Solvents

Carbon tetrachloride is found to be a more effective extractant than ether, but the carbon tetrachloride extracts all contained some cadmium and barium (present as stabilisers in the original PVC as barium stearate 1% and cadmium stearate 0.5%), and a trace of polyvinyl chloride. Carbon tetrachloride may not completely extract polypropylene adipate (PPA), but is satisfactory in all other cases. Ether was unsatisfactory for the extraction of both polypropylene sebacate and polypropylene adipate, and usually give low yields of Mesamoll and TTP.

Extraction with Mixed Solvents

Methanol forms minimum boiling-point azeotropes with a number of solvents which swell polyvinyl chloride, and some of these are found to be useful extractants. Table 12.2 gives the results of extracting PVC samples containing 28.5% Reoplex 220 with various mixtures for 4 hrs. The dichlorethane/methanol and carbon tetrachloride/methanol azeotropes dissolve very little polyvinyl chloride, but appreciable quantities of it are found in the other extracts. All the extracts contained traces of cadmium and barium soaps. An azeotropic mixture of 1,2-epoxypropane and diethyl ether is tested, but this also dissolves a considerable quantity of polymer.

Table 12.2. Extraction with mixed solvents.

Solvent mixture	*B.p.*	*Extract, %*
1:1, dichlorethane/methanol	60°C	30.2
2:1, Carbon tetrachloride/methanol	56°C	30.0
4:1, Chloroform/methanol	54°C	31.8
7:1, Acetone/methanol	56°C	39.2
7:6, Ether/1,2-epoxypropane	33°C	35.7

Multiple Extractions

Table 12.3 gives the results of following 4 hrs extractions with ether by extractions with other solvents. It can be seen that extraction for 4 hrs with carbon tetrachloride/methanol azeotrope gives better results than extraction for 15 hrs with methanol.

Table 12.3. Multiple extractions.

		Extracts, %	
Plasticiser	*Concn. %*	*4 hr ether, 5 hr MeOH*	*4 hr ether, 4 hr CCl_4 MeOH*
Bisoflex 791	28.5	27.7	28.7
Tritolyl phosphate	28.5	25.8	28.9
Mesamoll	28.5	25.5	28.1
Bisoflex 79S	28.5	28.1	29.0
Reoplex 220	28.5	20.4	29.3
Hexaplas PPA	23.5	11.0	17.4

If a composition containing a polyester plasticiser and a monomeric plasticiser is extracted first with ether, then with methanol or carbon tetrachloride/methanol, the ether extract will contain nearly all the monomeric plasticiser, and some of the polyester, and the second extract will contain almost pure polyester. Table 12.4 gives the results of extracting a composition containing Reoplex 220 (19.0%), Bisoflex 791 (7.5%), and Bisoflex 79S (7.0%), using various procedures.

Table 12.4. Extraction of mixed plasticisers.

Solvent	*Time, hr*	*Extract, %*	
Diethyl ether	8	28.6	
Carbon tetrachloride	4	33.4	
Diethyl ether	4	27.3	} 27.8
Methanol	15	0.5	
Diethyl ether	4	27.2	} 33.2
Carbon tetrachloride/methanol	4	6.0	

Effects of Extraction Time and Plasticiser Concentration

The extraction of plasticisers from polyvinyl chloride compositions must involve diffusion processes, and the rate of extraction should depend, to some extent, on the initial plasticiser concentration. This is found to be true for the extraction of Mesamoll by ether. Compositions made from I.C.I. Corvic D65/1, lead stearate (0.1%), and Mesamoll (10–50%) are extracted with ether for 1–8 hr, and the plasticiser contents m (% w/w), at times t (hours), are calculated from the weights of the extracts. Fig. 12.4 shows the curves obtained by plotting log (m/m_0) against t, for different values of m_0, the initial plasticiser content. In every case, except when m_0 = 10%, there is initially a very rapid loss of plasticiser, followed by a period when m/m_0 decreases exponentially. When m_0 = 10%, the rate of decrease of m/m_0 is very small at all stages of the extraction.

Improvement of Extractions

The aim in the separation of plasticisers from polyvinyl chloride compositions is to remove all the plasticiser, but in practice it is satisfactory if m is reduced to about 0.5%. It is found that the residual plasticiser content after extraction with ether is frequently very much more than 0.5%. Extractions could be improved by increasing t, $(S/V)^2$, or D.

Increasing t is useful only if $(S/V)^2D$ is reasonably large, or m is reduced to a low value in the early stages of the extraction. This can be seen if the curves of Fig. 12.4 are extrapolated to values of m/m_0 equivalent to m = 0.5%. For m_0 = 50% the time required to reduce m to 0.5% is about 12 hr; for m_0 = 40%, 15 hr; for m_0 = 30%, 24 hr; for m_0 = 20%, 36 hr; for m_0 = 10%, several days (it was found experimentally that after 64 hr m was 4.8%).

The surface-to-volume ratio should have a considerable effect on the rate of extraction, and the sample being extracted should always be reduced to as fine a state of sub-division as possible. Samples can be cut in strips, cooled so that they became rigid, and ground them up in a pencil sharpener. This method could be used for compositions plasticised with tritolyl phosphate, but is not suitable for the type of composition likely to be encountered now-a-days. Samples received in sheet form are usually just cut into strips, and S/V will depend mainly on the thickness of the sheet, but wherever possible the material should be grated, using one of the tools now available. For sheet material cut into strips with scissors S/V will generally be less than 200 cm^{-1}, and may be as little as 20 cm^{-1}, but for rasped material S/V should be at least 200 cm^{-1}.

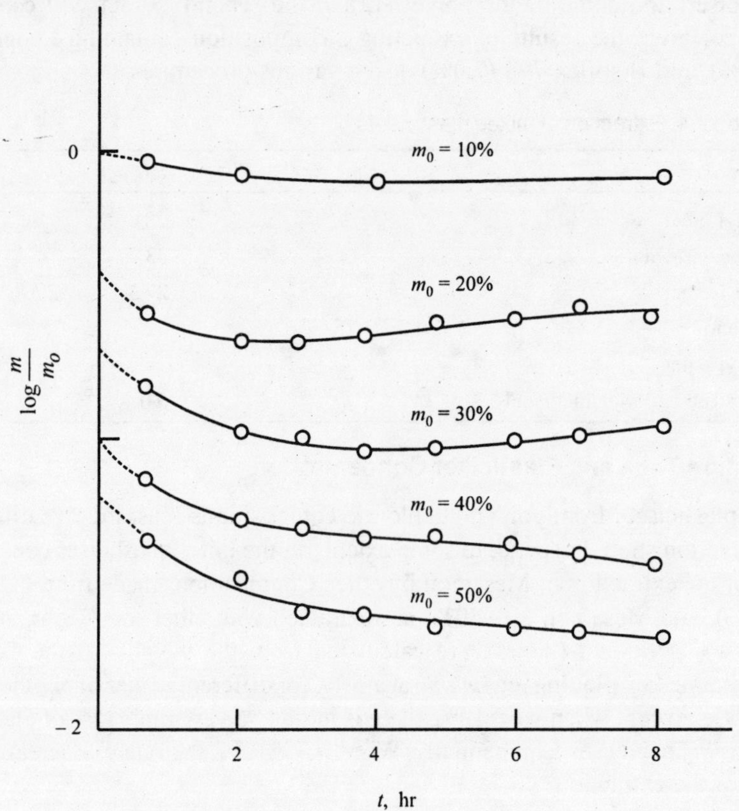

Fig. 12.4. Effect of extraction time (t) and plasticiser concentration (m_o) on ether extraction of Mesamoll from PVC.

The diffusion coefficient can be increased by increasing the temperature at which extraction takes place, or by using a solvent which swells the sample more than it is swollen by diethyl ether. The first method is not practicable, but the second has been shown to be effective. The main difficulty is in finding a solvent which swells, but does not dissolve, polyvinyl chloride; the use of a constant boiling mixture of carbon tetrachloride and methanol seems to be a solution of this difficulty. This mixture can be used for the extraction of residual plasticiser from a great number of commercial polyvinyl chloride compositions, after first extracting for 8 hr with ether.

The only difficulties encountered are with compositions containing blends of butadiene/acrylonitrile copolymer and low molecular weight polyvinyl chloride, and by those containing vinyl chloride/vinyl acetate copolymer: the carbon tetrachloride/methanol mixture extracts appreciable amounts of polymer from such compositions.

The most common method of plasticiser is to use ether to extract the plasticisers to be determined. If a quantitative analysis is required, the sample should be thin (0.1 mm or less) and should be extracted for several hours (usually 10 hr) to ensure that extraction is complete. If extraction is to be followed by spectrometry, care must be taken to eliminate all solvent by drying for 2–3 hr at 80°C. If the analysis is to be done by chromatography, drying is unnecessary since the solvent is much more volatile than the plasticiser and will be well separated.

The spectrum reveals immediately the chemical type of a single plasticiser or of the principal one if there are several. Comparison with authentic spectra often leads to the unequivocal identification of the principal plasticiser if its spectrum is available in a collection, and under good conditions homologous or isomeric plastcisers may be distinguished. For example, di-*n*-octyl and *bis*-2-ethylhexyl phthalates. The chemical type of secondary plasticisers (phthalates, phosphates, esters of di-acids, etc.) may be established with a degree of certainty depending on the analyst's experience and the presence of absorption peaks that do not interfere with other substances extracted from the original polymer.

The same extraction technique may be followed by gas-phase chromatography on a special, sufficiently rapid, column. A single plasticiser is identified by its retention data with, however, an uncertainty which may be practically eliminated by the use of a chromatogram obtained on a column of different polarity. Fig. 12.5 shows on the one hand the poor resolution of three plasticisers: (i) diethyl phthalate; (ii) dimethyl sebacate; and (iii) tributyl phosphate on a non-polar column; and on the other hand, the good separation of these substances on a polar column. The two chromatograms are recorded in a period of 1/2-1 hr, depending on the nature of the plasticiser, the spectrogram being obtained in 10 min. These times are negligible compared to the time of extraction.

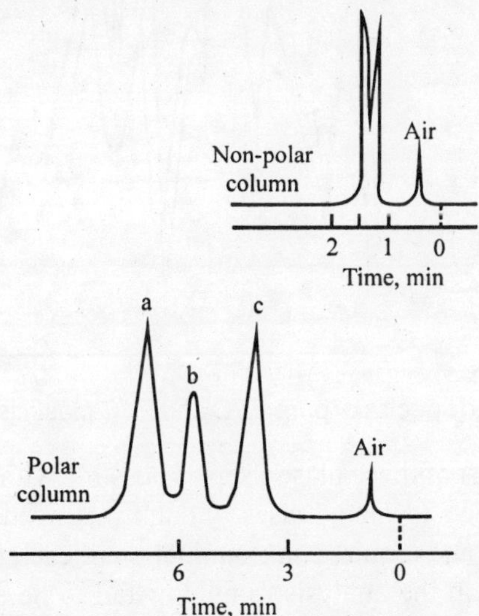

Fig. 12.5. Gas chromatograms showing the poor resolution of three plasticisers. (a) diethyl phthalate, (b) dimethyl sebacate and (c) tributyl phosphate—on a non-polar column; and the good separation of these substances on a polar column.

Gas-phase chromatography has the advantage of providing the identification, with practically equal ease, of secondary plasticisers with a concentration of possibly only 1% of that of the principal plasticiser. At the same time traces of acid esters or heavy alcohols may easily be detected in commercial plasticisers, as well as symmetrical esters that are usually found in unsymmetric plasticisers.

The direct analysis of plasticisers in a polymer is of considerable interest as it would eliminate the preliminary extraction. Effectively this is possible by infra-red spectrometry in certain favourable cases in which one can establish the chemical type of the main plasticiser, but to draw the maximum of information from spectrometry it would be necessary to apply differential spectrometry techniques or electrical analysis of spectrograms. Fig. 12.6 compares the spectra of pure polyvinyl chloride and PVC plasticised with 5% dibutyl phthalate. The presence of a phthalate can be detected or, if its identity is known, its concentration determined by means of the peak at 1725 cm^{-1}.

On the other hand, gas-phase chromatography, because of the physical separation it effects, furnishes both a qualitative and a quantitative analysis of polymer-plasticiser mixtures with almost the same ease as the analysis of plasticisers alone. It suffices to submit the sample, prepared as for the pyrolysis of plastics, to a controlled pyrolysis in order to disengage the vapourised plasticisers. The polymer is partially degraded, but its pyrolysis products are in all the cases much lighter than the plasticisers and in no way prevented their separation and identification. Fig. 12.7 shows the separation thus obtained of four plasticisers: (i) dibutyl succinate; (ii) tributyl phosphate; (iii) dimethyl sebacate; and (iv) diethyl phthalate and the pyrolysis products of polyvinyl chloride. The latter are elutriated during the first minute of operation.

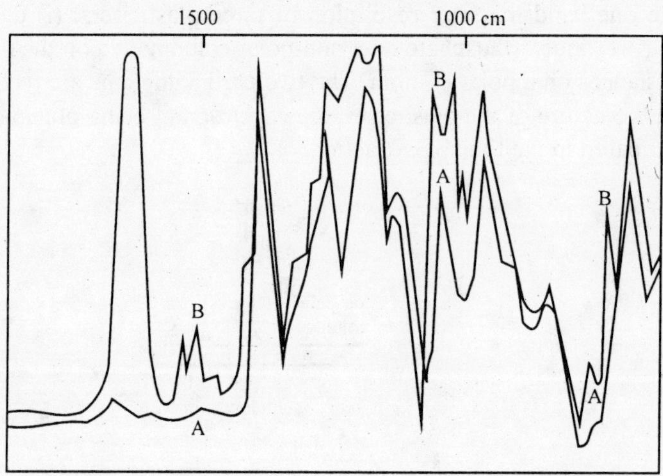

Fig. 12.6. Comparison of infra-red spectra of pure PVC and PVC plasticised with 5% dibutyl phthalate.

The quantitative analysis of most of these plasticisers is possible with a relative standard deviation of 5%. The technique can be extended to many cases, though this needs care because of the possible thermal degradation of plasticisers and requires an examination of each particular case. It is also always possible for plasticisers to remain in the analysis sample or fail to be elutriated from the column and escape detection. For example tricresyl phosphates and heavier products, mainly the so-called non-migrating plasticisers (polyesters, polyacrylonitrile), are quite unsuitable for determination by gas-phase chromatography. Non-migrating plasticisers, which find increasing use in industry, also present a problem in infra-red spectrometry. They cannot be extracted by solvent and must thus be regarded as polymer mixtures with all their difficulties.

DETERMINATION OF ULTRA-VIOLET ABSORBERS

Among the numerous additives commonly used in plastic materials, the ultra-violet absorbers are increasing in importance because they are often used in food packaging materials to protect the plastic materials as well as the foodstuff packaged from the actinic action of ultra-violet radiation. Actinic effects may cause discoloration of both the plastic material and the foodstuff, and may on occasion also change the taste and loss of vitamins in the food.

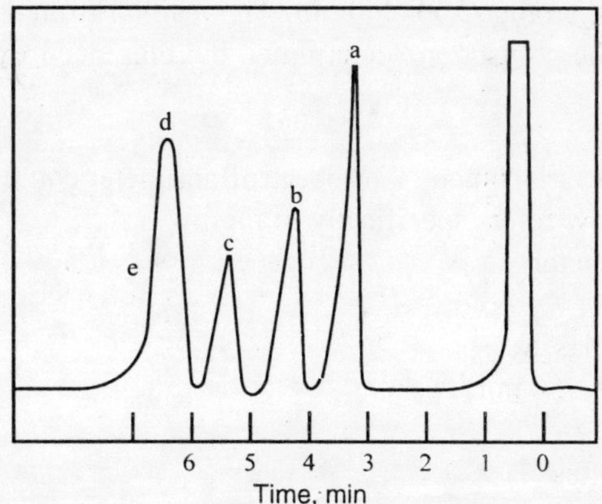

Fig. 12.7. Chromatogram showing separation of (a) dibutyl succinate, (b) tributyl phosphate, (c) dimethyl sebacate and (d) diethyl phthalate, (1) pyrolysis products of PVC.

The ultra-violet absorbers can be divided in different groups :

1. Benzophenone derivatives.
2. Salicylic acid esters.
3. Resorcinol esters.
4. Benzotriazole compounds.
5. Coumarine derivatives.

By their nature, many of these types of compounds are amenable to analysis by fluorimetric analysis, thus Uvitex OB has an intense ultra-violet absorption at a wavelength of 378 mμ, which is high enough to bc outside the region where many potentially interfering substances present in the polymer extract would be excited to fluoresce. This is illustrated in the following fluorimetric procedure for the determination of down to 10 ppm Uvitex OB in polystyrene. Antioxidants such as (2,6-di-tert-butyl-*p*-cresol), (1,3,5-tri-methyl-2, 4, 6-tri (3, 5-di-t-butyl-4-hydroxybenzyl)benzene), polygard (tris(nonylated phenyl)phosphite) do not interfere in this procedure.

Some ultra-violet absorbers for use in plastic materials are given below :

1. 2-hydroxy-4-methoxy-benzophenone
2. 2,4-dihydroxy-benzophenone
3. 2-hydroxy-4-methoxy-4'-methyl benzophenone
4. 2,4,5-trihydroxy butyrophenone
5. 4-dodecyloxy-2-hydroxy benzophenone
6. 2-hydroxy-4-n-octoxybenzophenone
7. 2,2'dihydroxy-4-methoxy benzephenone
8. 2,2'dihydroxy-4,4'-dimethoxy benzophenone
9. *p*-tert-butylphenylsalicilate
10. Resorcinol mono benzoate
11. Hydroxyphenylbenzotriazole
12. 7-diethylamino-4-methyl coumarin.

Method

0.2 g of the polystyrene sample is shaken with chloroform to achieve solution. The sample is excited by ultra-violet radiation of wavelength 370 mμ from a mercury vapour lamp and the fluorescence spectrum of the sample recorded over the range 400–440 mμ. The reading from the fluorimeter is noted and the *Uvitex OB* concentration in the polystyrene determined by reference to a prepared calibration graph.

Apparatus

Spectrometer with fluorimetric attachment. Any spectrofluorimeter could be used for the determination. The procedure described below refers specifically to the:
Optica CF4NR spectrofluorimeter
1 cm silica fluorimetric cells
25 ml and 100 ml graduated flasks
Ultra-violet lamp (wavelength 350 mμ)
Chance OXI optical filter
5 ml and 1 ml graduated pipettes
Mechanical shaker.

Reagents

Chloroform: Spectroscopic or analar grades are suitable. *Sample of polystyrene* free of *Uvitex OB*. *Uvitex OB* standard solution made by dissolving 0.0100 g of *Uvitex OB* in 100 ml of chloroform.

Calibration

Into each of seven 100 ml volumetric flasks weigh 0.800 g of additive free polystyrene then make additions of *Uvitex OB* standard solution as indicated below:

Standard no.	*Chloroform (ml)*	*Uvitex OB standard solution (ml)*	*ppm Uvitex in standard*	*ppm Uvitex in polymer*
1	100	Nil	Nil	Nil
2	95	5	5.0	625
3	98	2	2.0	250
4	99	1	1.0	125
5	99.5	0.5	0.5	62.5
6	99.8	0.2	0.2	25.0
7	99.9	0.1	0.1	12.5

Setting up the fluorimeter

Mount the mercury vapour source on the Optica CF4NR and connect as described in the manufacturer's manual. Set the controls of the instrument as follows :

Console controls:

Double beam mains control	ON
Fluorimeter mains control	ON
Mode switch	SB
% switch	100 %
Time constant	Normal
Gain control	5

Dynode supply:

Control button 3

Filter:

Mount Chance OXI filter between mercury vapour source and sample compartment.

Recorder:

Set to read % transmission.

Record the fluorescence spectrum of each solution, prepared as above from 400 to 440 mμ at a suitable slit width. If the 435 mμ peak is less than 15 divisions, or greater than 80 divisions in height, re-run the trace on a higher or lower slit width, respectively.

Wash out the fluorimeter cell thoroughly with chloroform and check that the final washings show no fluorescence at 435 mμ before re-filling the cell with the next solution.

Repeat the above procedure on the various solutions at various slit width settings. Measure the height of the peaks at 435 mμ for the set of calibration solutions at each slit width setting used.

Plot a calibration curve of peak height versus the corresponding concentrations of *Uvitex OB* in the standard solutions. The fluorimeter must be re-calibrated whenever the mercury lamp is changed.

Analysis of Polystyrene

Fill the fluorimeter cell with standard No. 1 (i.e., chloroform and polystyrene blank) and place the cell in the fluorimeter beam. Switch on the rotating mirrors, set the monochromator to 400 mμ and, with the recorder chart running, set the pen on zero by means of the "set zero" control.

Set the monochromator to 435 mμ. Fill the fluorimeter cell with a 1.0 ppm *Uvitex* standard (Standard No. 4). Set slit width to a suitable value and adjust 100% control until the recorder pen gives the same reading as in the current calibration curve.

Weigh accurately 0.2 g of polymer into a 25 ml volumetric flask, add 15 ml chloroform and shake until dissolved, make up to volume with more chloroform. This solution is now ready for analysis.

Calculations

Determine the concentration (C ppm) of *Uvitex OB* in the solution examined by reference to the peak height vs. *Uvitex* concentration graph. Calculate the concentration of *Uvitex* in the polymer as follows:

$$\text{ppm } Uvitex\ OB \text{ polystyrene} = C \times 125 \text{ ppm}$$

DETERMINATION OF POLYGARD (TRIS(NONYLATED PHENYL)-PHOSPHITE) IN STYRENE-BUTADIENE RUBBERS AND LATEXES

The widely used method for the determination of this stabiliser is based on the determination of phosphorus. This involves a tedious preliminary digestion with nitric and perchloric acids. A colourimetric method for determining polygard based on hydrolysis to nonyl phenol, followed by coupling with *p*-nitrobenzene-diazonium fluoroborate and colorimetric estimation at 550 mμ has been described.

Various other phenolic antioxidants produced dyes under these conditions. There is method for the determination of polygard in SBR which utilises the bathochromic shift in the spectrum of phenols resulting from the formation of phenolate ions in alkaline solution.

Polygard in iso-octane has an ultra-violet spectrum with a peak at 273 mμ in neutral solution. By adding a strong base (tetrabutylammonium hydroxide) the polygard is hydrolysed and the peak is shifted to 296 mμ. The difference in absorbance at 299 mμ between the neutral and alkaline solutions is directly proportional to the amount of polygard present. By use of this bathochromic shift, interference of non-phenolic impurities is eliminated and a background correction factor is not required.

The latex is flocculated by addition of acid. Polygard is extracted by agitation and boiling with iso-octane. Excess acid, water, and carbon disulphide resulting from the decomposition of dithiocarbamates are distilled over. A portion of the extract is made alkaline and the absorbance of the alkaline solution is determined using the neutral extract as reference.

The method is accurate, simple, rapid (elapsed time less than 1.5 hr for polymer, less than 2.0 hr for latex), specific for phenolic compounds, and insensitive to interference except xanthates. Results are repeatable and agree well with those obtained with the perchloric acid method (Table 12.5). The method is unsatisfactory for oil-extended stock or mixtures of Polygard with other phenolic-type antioxidants.

Table 12.5. Determination of polygard. Comparison of ultraviolet with perchloric acid methods.

SBR latex	*% Polygard perchloric acid (Phosphorus)*	*Ultra-violet*
Sample 1	1.31 1.30	1.18 1.30
Sample 2	1.26 1.28	1.24 1.36
Sample 3	1.23 1.24	1.36 1.32
Sheet rubber		
Type 1019	1.25 1.28	1.04 1.06
Type 1503	1.62 1.63	1.53 1.60
Type 1018	1.59 1.56	1.57 1.58
Type 1022	1.27 1.31	1.16 1.14

DETERMINATION OF ORGANIC PEROXIDES

Organic peroxides can occur in small amounts in some types of polymers such as polystyrenes as a result of the fact that a peroxide has been used as a polymerisation catalyst in polymer manufacture. Also, stable organic peroxides such as dicumyl peroxide have been used as synergists, in conjunction with bromine and or phosphorus-containing additives, to impart fire resistance to cellular expanded polystyrene and other types of plastics.

Polarography of twenty-three commercially available organic peroxides have been studied. Polagrams can be recorded, using. Sargent Model XXI polarograph on solutions of these compounds in an electrolyte consisting of a 0.3 M lithium chloride dissolved in 1:1 benzene:methanol, using ethyl cellulose as a maximum suppressor.

The compounds fell into six different groups, based on their structures and behaviour in the polarographic cell. The first group of eight compounds, have reduction waves at or near zero voltage. Included are diaroyl and diacyl peroxides and peroxy acids. The diaroyl peroxides (benzoyl and bis(2,4-dichlorobenzoyl) peroxides) showed linear relationships between diffusion current and concentration in the range 9×10^{-2} to 4×10^{-4} M, and the diacyl peroxides (acetyl, lauroyl, and succinic acid peroxides) between 1×10^{-2} and 1×10^{-4} M. The succinic acid peroxide showed a second reduction wave, at –1.44 V, which are attributed to the free acid group; a polarogram of succinic acid showed a reduction wave at approximately the same potentials.

Peroxyacetic and peroxybenzoic acids give reduction waves at 0.00 voltage in the presence of ethylcellulose and methylene blue, but they do not demonstrate a linear relationship between diffusion current and concentration. The acids evidently reacted slowly with the methanol in the solvent, as a continuous decrease in diffusion current are observed with increased time of contact. Peroxyacetic acid showed an additional wave at a half-wave potential of –1.41 V, probably because of the presence of acetic acid, whose half-wave potential are observed to be 1.44 V. Bis (1-hydroxyheptyl) peroxide give two reduction waves at 0.00 and –1.20 V. A linear relationship existed between the concentration (1×10^{-2} to 1.3×10^{-4} M) and diffusion current at half-wave potential of –1.20 V, but not at 0.00 voltage.

Methyl ethyl ketone peroxide in dimethyl phthalate showed three half-wave potentials, one of which (–1.82 V) is attributed to the phthalate ester, The first reduction wave (0.60 V) is observed only when the concentration of peroxide is below 0.01 M, and the relationship between diffusion current and concentration are non-linear, since diffusion current showed a maximum at a concentration of 2.1×10^{-3} M. The second reduction wave (half-wave potential, –1.26 V) demonstrated a linear relationship between diffusion current and the above concentrations. The samples of phenylcyclohexane hydroperoxide and di-tert-butyl perphthalate showed, for both reduction waves, a linear relationship between diffusion current and concentration in the range of 10^{-2} to 10^{-4} M.

Group three consisted of two peroxy esters (tert-butyl perbenzoate and tert-butyl peracetate) which give a single reduction wave at about –1.0 V. Both compounds showed a linear relationship between diffusion current and concentration in the range of 10^{-2} to 10^{-4} M.

The fourth group of seven hydroperoxides also showed a single reduction wave, and at a slightly more negative potential than the third group. The group consisted of diisopropylphenyl, tert-butyl-isopropylphenyl, *p*-menthane, cumene, pinane, and tert-butyl hydroperoxides, and hydrogen peroxide. Five of the more complex members of the group reduced in the range of -1.02 ± 0.02 V. All gave a linear relationship between diffusion current and concentration in the range 10^{-2} to 10^{-4} M.

In a class by itself (Group 5) was the transannular peroxide ascaridole, which reduced at –1.22 V. It showed a linear relationship between diffusion current and concentration in the range 9.2×10^{-2} to 1.6×10^{-3}. M. Two peroxides, di-tert-butyl peroxide and 1-phenylmethyl-tert-butyl peroxide, are not reduced in the voltage span of 0.00 to –2.00 V.

Determination of *p*-Tert-butyl Perbenzoate in Polystyrene

Apparatus

Cathode-ray polarograph, with dropping mercury electrode, 10 ml polarograph cells and thermostatted (25°C) water-bath.

Agla micrometer syringe capable of delivery 0.01 cc with an accuracy of 0.0002 ml.

Volumetric glassware, pipettes 50 ml, 25 ml, 5 ml; volumetric flasks 100 ml.

Centrifuge 'Super-medium' to take 250 ml centrifuge bottles.

Reagents

Toluene, re-distilled pure grade, rendered peroxide-free by shaking 1 l of the solvent with 10-20 ml of ferrous sulphate solution—prepare by mixing 60 g ferrous sulphate, 6 ml concentrated sulphuric acid and 110 ml distilled water—followed by re-distillation. Store in a well-stoppered brown glass bottle.

Base electrolyte (0.6 M) remove peroxides from 1 l of methyl alcohol as described above. Weigh out 2.544 g lithium chloride (Analar) and make up to 100 ml with methyl alcohol. *Para-tert-butyl perbenzoate.*

Standard addition solution. Prepare by diluting a suitable weight of *p*-tert-butyl perbenzoate with toluene. Nitrogen white spot, oxygen content less than 25 ppm.

Mercury, pure for polarographic analysis, use trebly distilled mercury, the manufacturers should be requested to supply this mercury in stone containers. Over a period of time, mercury picks up an impurity from polyethylene storage bottles which might interfere in polarography.

Procedure

Weigh out 5.0 ± 0.01 g of sample and transfer to a 250 ml Pyrex glass centrifuge bottle. Measure 50 ml of toluene into the bottle. Drop a polythene coated magnetic stirrer rotor into the bottle and stopper with a cork (avoid rubber bungs). Stand the bottle on a magnetic stirrer and leave for several hours until the sample has completely dissolved or dispersed in the solvent. Accurately pipette into the gently stirred contents of the bottle 50 ml of 0.6 M lithium chloride reagent.

Insert the bottle in a centrifuge and spin at 700–900 g until insolubles have completely settled to the bottom of the bottle leaving an absolutely clear upper phase containing all the peroxide present in the original sample.

Dilute 50 ml toluene with 50 ml of lithium chloride solution to serve as a blank. Pipette 5 ml of the sample solution and 5 ml of the blank solution into two polarographic cells and immerse these in the constant temperature tank of the cathode-ray polarograph (thermostatted at 25°C). Carry out the degassing operations described below on the sample and on the blank solutions immediately before carrying out all polarographic measurements.

Connect a supply of oxygen-free nitrogen to the polarographic cell. Lower the dropping mercury electrode-system over this cell, so that the outer glass sleeve of the electrode dips 1–2 mm into the water tank (providing a water seal to prevent the ingress of atmospheric oxygen). Immerse the anode connection in the side arm of the polarographic cell. Pass nitrogen for 15 min to completely displace oxygen from the cell solution, then switch off the nitrogen. Leave the glass sleeve in position to prevent re-entry of atmospheric oxygen into the cell solution and carry out the polarographic measurements described below within 1 to 3 min of stopping the gas purge.

The analytical conditions with the Southern Analytical Davis Differential A1660 cathode-ray polarograph (with single cell operation) are as follows:

Cathode	dropping mercury
Reference anode	mercury pool
Circuit	forward sweep (with derivative units control switch off)
Sensitivity control	select a suitable current scale-factor and keep this constant throughout the analysis. Alter the instrument sensitivity by means of the shunt scale-factor control.
Start potential	–0.7 V

By means of the 'Y'-shift control, adjust the light spot to the origin of axes on the left-hand side of the graticule on the cathode-ray tube. Repeat this operation at different shunt scale-factor settings, until the polarographic wave is visible on the graticule.

The analytical conditions with the Southern Analytical K1000 polarograph are as follows:

Cathode	Dropping mercury
Reference anode	mercury pool
Circuit	cathodic direct
Start potential	–0.7 V

Adjust the polarograph to its maximum sensitivity setting. Adjust the 'X'-shift control and the 'Y'-shift controls until the light spot commences its sweep at the origin of axes at the left-hand side of the graticule on the cathode-ray tube. Repeat this operation at different sensitivity settings until the polarographic wave is visible on the graticule.

Take the readings on the freshly degassed solutions as follows. Adjust the polarograph to the para-tert-butyl perbenzoate start potential (–0.7 V) and obtain the wave as described above for the polystyrene sample solution. Read off from the graticule the maximum height of the peak (at about –0.9 V). Raise the dropping mercury electrode from the cell and deliver as quickly as possible into the sample solution by means of a horizontally held Agla syringe a suitable 'standard addition' of a toluene solution of para-tert-butyl perbenzoate (ensure that the weight of peroxide present in the 'standard addition' is similar to that already present in the cell solution). Limit the volume of 'standard addition' solution to less than 0.05 ml in order to avoid dilution errors. Lower the electrode into the sample cell and again pass nitrogen for 2 min. Immediately read the new peak height at –0.9 V. Similarly, determine the peak height at –0.9 V on the degassed polystyrene-free reagent blank solution.

Calculations

Para-tert-butyl perbenzoate (ppm w/w)

$$= \frac{100 \times M \times 10^6}{5 \times W} \times \left[\frac{(h_1 S_1 - h_3 S_3)}{(h_2 S_2 - h_1 S_1)} \right]$$

where W = weight (g) of polystyrene sample taken for analysis (assuming 5 ml portion taken for polarography),
h_1 = peak height (graticule divisions) of sample solution before standard addition,
h_2 = peak height (graticule divisions) of sample solution after standard addition,
h_3 = peak height (graticule divisions) of polymer-free reagent blank solution,
S_1, S_2 and S_3 are the corresponding instrument sensitivity settings (the product of h and S are known as peak currents in micro-amps),
M = weight (g) of *para*-tert-butyl perbenzoate in volume of 'standard addition' solution added to cell solution.

Gas chromatography has also been used to determine certain types of organic peroxides. For example, procedures involving chromatography of heptane solutions of peroxide (Table 12.6) on phthalate/diatomaceous earth or silicone/diatomaceous earth columns and using helium as the carrier gas. No doubt, this type of procedure could be easily adapted to the examination of solvent extracts of polymers.

Gas chromatographic procedure for the determination of di-tert-butyl peroxide is based on the thermal decomposition of the peroxide in benzene solution into acetone and ethane when the solution is injected into the gas chromatographic column at 310°C. The technique is calibrated against standard solutions of pure di-tert-butyl peroxide of known concentration.

Certain types of peroxides used in polymer formulations are extremely stable and unreactive. This applies to substances such as dicumyl peroxide used as an ingredient of some self-extinguishing grades of polymers.

This substance cannot be determined by polarography and will not react with many of the reagents normally used for determining organic peroxides.

Table 12.6. Gas chromatography. Retention times for organic peroxides.

Compound	*Column*[a]	*Temp.°C*	*Helium*[b] *pressure psi*	*Retention time,min.*
tert-Butyl hydroperoxide	2 m-A	80	20	22.7
tert-Pentyl hydroperoxide	1 m-A	80	15	19.9
tert-Butyl peracetate	1 m-A	100	20	6.5
tert-Butyl peroxyisobutyrate	1 m-A	100	20	15.3
Di-tert-butyl peroxide	2 m-A	80	20	8.1
Di-tert-pentyl peroxide	1 m-A	80	15	15.4
2,5-Dimethyl-2,5-di-(tert-butyl peroxy)3-hexane	1 m-O	138	15	4.9
2,5-Dimethyl-2,5-di(tert-butyl peroxy)hexane	1 m-O	138	15	6.9
n-Heptane	2 m-A	80	20	6.1
n-Dodecane	1 m-O	138	15	3.1
n-Pentane	1 m-A	100	20	0.4
n-Nonane	1 m-A	100	20	4.9

[a] Length (metres) and type of column used.
[b] Inlet pressure.
A, didecyl phthalate on diatomaceous earth.
O, silicon grease on diatomaceous earth.

Recently, a method has been developed for determining dicumyl peroxide in polystyrene, which is not subject to interference by other organic peroxides or additives that may be present in the polymer. The dicumyl peroxide is extracted from the polymer with acetone and then separated from any other additives present by thin-layer chromatography on silica gel. The gel in the area of the plate containing dicumyl peroxide is then isolated and digested with potassium iodide in glacial acetic acid followed by titration of the liberated iodine by titration with very dilute sodium thiosulphate solution. This procedure has a precision of ± 12% of the determined value with polymers containing 0.25 to 0.5% dicumyl peroxide. It is a rather time-consuming procedure but has the advantage of avoiding all risk of interference from other types of peroxides present in the sample.

DETERMINATION OF STYRENE, ACRYLONITRILE AND METHACRYLONITRILE MONOMERS

Probably the best procedures available for the determination of styrene monomer in polymers are those based on gas chromatography. Alternative methods are, however, available based on ultra-violet spectroscopy and on polarography.

Direct Ultra-violet Spectroscopic Method for Styrene

In this method, the polystyrene sample (0.5 g) is dissolved in 50 ml of chloroform or another suitable spectroscopic solvent and the ultra-violet spectrum recorded in the region 280–310 mμ against polymer-free solvent in the reference cell. If the polymer is incompletely soluble in the solvent (e.g. due to the presence of gel or pigments), it is dissolved in 30 ml of solvent and the suspension filtered rapidly under light vacuum through an asbestos pad to remove insolubles.

It is essential to ensure that negligible solvent evaporation occurs during filtration. The filtrate is then made up to 50 ml and the ultra-violet spectrum recorded as already described. Calibration is performed by applying the technique to solutions prepared by including various known amounts of pure styrene monomer in 1% w/v solutions of styrene-free polystyrene.

Conventional and high impact polystyrenes contain various non-polymer additives (e.g. lubricants) which result in widely different and unknown background absorptions at the wavelength maximum at which styrene monomer is evaluated (292 mμ). The influence of the background absorptions on the evaluations of the optical density due to styrene monomer is overcome by the use of an appropriate baseline technique, claimed to make the method virtually independent of absorptions due to polymer additives. In this technique, a straight line is drawn on the recorded spectrum across the absorption peak at 292 mμ in such a way that the baseline is tangential to the absorption curve at a point close to the absorption minima occurring at 288 mμ and 295–300 mμ (Fig. 12.8). A vertical line is drawn from the tip of the styrene absorption peak at 292 mμ to intersect the baseline and the height of this line is then a measure of the optical density due to the true styrene monomer content of the test solution.

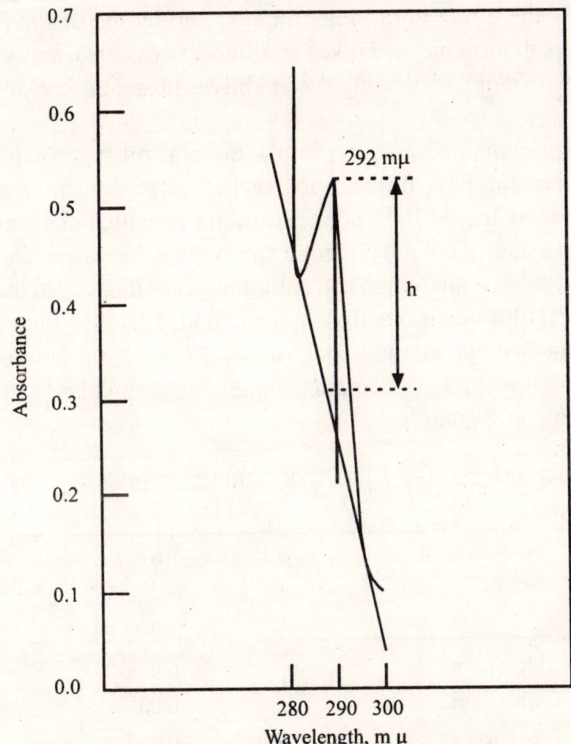

Fig. 12.8. Typical absorption curve of a polystyrene containing styrene monomer ultra-violet.

This baseline correction technique can obviously be applied to the determination of styrene monomer in polystyrene only if any other ultra-violet absorbing constituents in the polymer extract (e.g. lubricants, antioxidants) absorb linearly in the wavelength range 288–300 mμ. If the polymer extract contains polymer constituents other than styrene with non-linear absorptions in this region, then incorrect styrene monomer contents will be obtained. An obvious technique for removing such non-volatile ultra-violet absorbing compounds is by distillation of the extract followed by ultra-violet spectroscopic analysis of the distillate for styrene monomer.

Distillation/Ultra-violet Spectroscopic Method for Styrene

The polystyrene sample (0.5 g) is dissolved in chloroform or ethylene dichloride (20 ml) in a stoppered flask and the solution is poured into an excess of methyl alcohol (110 ml) to reprecipitate dissolved polymer. The polymer is filtered off and washed with methanol (120 ml) and the combined filtrate and washings gently distilled to provide 200 ml of distillate containing only styrene monomer and any other distillable component of the original polystyrene sample. Non-volatile polymer components (viz. stabilisers, lubricants and low molecular weight polymer) remain in the distillation residue.

The optical density of the distillate is measured either at 292 mμ or by the baseline method against the distillate obtained in a polymer-free blank distillation. Calibration is performed by applying the distillation procedure to solutions of known weights of pure styrene monomer in the appropriate quantities of methyl alcohol and the chlorinated solvent.

Table 12.7 shows results obtained for styrene monomer determinations carried out on two samples of pigmented polystyrene by the direct ultra-violet method and by the distillation modification of this method. It is seen that the distillation method gives results that are consistently some 0.1% higher than those obtained by direct spectroscopy, indicating that additives present in the polystyrene are interfering in the latter method of analysis.

Both polystyrene samples contained an ester and a mineral oil type of lubricant together with a phenolic antioxidant. The lubricants have little absorption in the 280–300 mμ region and do not interfere in either method of analysis at the 5–10% concentrations at which they are used in polystyrene formulations. The absorption spectrum of the phenolic antioxidant, however, shows a sharply decreasing non-linear absorbance in the 280–300 mμ region and contributes significantly to the background absorption of the test solution in the direct ulra-violet spectroscopic method. This invalidates the baseline correction procedure and leads to erroneous styrene monomer values. In the distillation procedure, however, the test solution used for spectroscopy does not contain the phenolic antioxidant and there is no interference in the determination of styrene monomer.

Table 12.7. Comparison of direct ultra-violet and distillation/ultra-violet methods for the determination of styrene monomer.

Method	*Solvent*	*Styrene monomer, % w/w Polystyrene sample*	
		No. 1	*No. 2*
Direct UV method	Chloroform	< 0.05	0.13 —
	Carbon tetrachloride	< 0.05	0.13, 0.14
	Ethyl acetate	< 0.05	0.16, 0.18
			0.14, 0.12
			0.14, 0.14
			0.15 —
Distillation/UV method	Ethylene dichloride/ methanol	0.16, 0.18	0.27, 0.29
		0.16, 0.18	0.26, 0.29
		0.20	0.29 —

Table 12.8 shows the results obtained by applying the two methods discussed above to synthetic blends of pure styrene monomer in polystyrene both in the presence and in the absence of 0.5% w/w

added phenolic antioxidant. The direct ultra-violet method gives correct styrene monomer contents only when the polymer does not contain the phenolic antioxidant. In its presence, determined styrene monomer contents are appreciably lower than the true values. Also, as expected, when the phenolic antioxidant is removed from the test solution by distillation, correct styrene monomer contents are obtained.

Similar effects to these can also be observed with polystyrene containing other ultra-violet absorbing additives. Thus, the influence of a mixture of 0.4% w/w tris-(nonylated phenyl) phosphite (Polygard) and 0.2% w/w 2,6-di-tert-butyl-*p*-cresol (Ionol CP) on the determination of styrene monomer is shown in Table 12.9.

Table 12.8. Influence of phenolic antioxidant[a] on the determination of styrene monomer by direct ultraviolet and by distillation/ultraviolet methods.

	Styrene found, % w/w			
	Direct UV method[b]		*Distillation/UV method*	
Styrene added to polystyrene % w/w	*No phenolic antioxidant[a] addition*	*0.5% phenolic antioxidant[a] added on polymer*	*No phenolic antioxidant[a] addition*	*0.5% phenolic antioxidant[a] addition on polymer*
0.11	0.11, 0.11	0.04, 0.08	0.12	0.12
0.22	0.22	0.11	—	—
0.27	0.26	0.18	0.29	0.26
0.41	0.41, 0.40	0.30, 0.27	0.40	0.40

[a] Wingstay T.
[b] Chloroform used as a sample solvent.

Table 12.9. Influence of Polygard/Ionol CP mixture on determination of styrene monomer by direct ultra-violet and by distillation/ultra-violet methods.

	Styrene found, % w/w			
	Direct UV method[b]		*Distillation/UV method*	
Styrene added to polystyrene % w/w	*No additive*	*0.4% w/w polygard and 0.2% w/w Ionol CP added on polymer*	*No additive*	*0.4% w/w polygard and 0.2% w/w Ionol CP added on polymer*
0.10	0.10, 0.10	— [b]	0.11	0.11 0.09
0.26	0.25, 0.26	0.16 0.16	0.26	0.27 0.25
0.41	0.42, 0.41	0.29, 0.29	0.41	0.40 0.42

[a] Chloroform used as sample solvent.
[b] Not measurable due to strong interference by additives.

Direct ultra-violet spectroscopic methods are reliable only in the case of polystyrene samples that do not contain ultra-violet absorbing antioxidants or any other type of strongly ultra-violet absorbing additive. The more lengthy distillation/ultra-violet spectroscopic method, however, will give correct results in the presence of such additives unless the additive is both sufficiently volatile to distil and absorbs in the same region of the spectrum as styrene monomer. Frequently the nature of the additives present in a polystyrene sample submitted for monomer analysis is unknown and, to overcome as far as possible any interference from this source it is recommended that the distillation step is included in the analysis.

Polarographic Method for Acrylonitrile

Residual amounts of styrene and acrylonitrile monomers usually remain in manufactured batches of styrene-acrylonitrile copolymers. As these copolymers have a potential use in the food-packaging field, it is necessary to ensure that the content of both of these monomers in the finished copolymer is below a stipulated level.

Styrene monomer can be determined in styrene-acrylonitrile copolymers by ultra-violet spectroscopy. A satisfactory spectroscopic method has not been described, however, for determining low concentrations of acrylonitrile monomer. The procedure for determining acrylonitrile monomer involves reaction with an excess of a standard solution of dodecyl mercaptan and subsequent determination of the unreacted mercaptan by titration with standard potassium bromate-potassium iodide solution in acidic medium. However, this procedure cannot be used for determining less than 300 or 400 ppm of acrylonitrile in polymer.

A polarographic method for determining down to 30 ppm of residual acrylonitrile in styrene-acrylonitrile copolymers has been developed by improving its sensitivity to a lower detection limit of 2 ppm acrylonitrile in polymer. It is possible, using the same base electrolyte, to determine styrene monomer in amounts down to 20 ppm in styrene-acrylonitrile copolymers. Methacrylonitrile can also be determined if acrylonitrile is absent.

Both styrene and acrylonitrile can also be determined in acrylonitrile-butadiene-styrene terpolymers, using the same procedure. Tetrabutylammonium iodide (0.1 M) in aqueous dimethyl formamide can be used as the base electrolyte and polarography is carried out using a Southern Instruments Davis differential cathode ray polarograph.

Figs. 12.9 and 12.10 show polarograms of single solutions of acrylonitrile and styrene monomers in the base electrolyte, and of a mixture of the two monomers. It is seen that the waves for the two monomers are well resolved from each other.

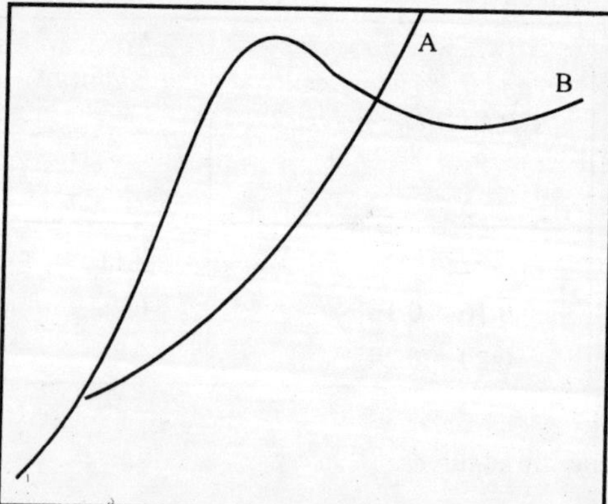

Fig. 12.9. Cathode-ray polarogram of a synthetic solution of acrylonitrile in tetrabutylammonium iodide-dimethyl-formamide base electrolyte. Curve A, base-electrolyte blank solution. Curve B, 9 ppm of acrylonitrile in base electrolyte. Start potential, −1.7 V.

The method is described below:

Apparatus

Cathode-ray polarograph—K 1000 or Davis Differential Model. Complete with stand for dropping mercury electrode, polarographic cells (10 ml) and thermostatted (at 25°C) water-bath. Agla micrometer syringe—capable of delivering 0.01 ml with an accuracy of 0.0002 ml. Calibrated glassware—pipettes, 25- and 5-ml capacities; 50-ml calibrated flasks.

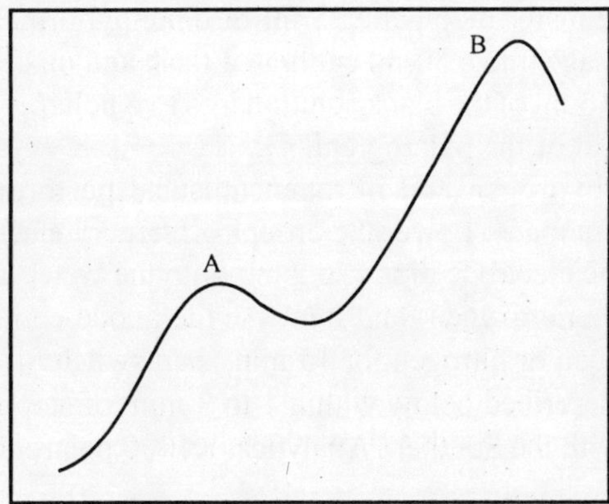

Fig. 12.10. Cathode-ray polarogram of a synthetic solution of 11.2 ppm of styrene and 9.3 ppm of acrylonitrile monomer in tetrabutylammonium iodide-dimethyl-formamide base electrolyte. Wave A, acrylonitrile wave. Wave B, styrene wave. Start potential −1.7 V.

Reagents

N,N'-Dimethylformamide: This material is usually sufficiently pure, as supplied, for polarographic analysis. If the reagent is impure, high blank readings are obtained during polarography. The dimethylformamide blank value may be reduced by suitable purification of the reagent as described below.

First dry the solvent (1l) by standing it over 50 g of anhydrous potassium carbonate for 3 days. Distill the product through an 80-cm Fenské column. Discard the first 200 ml of distillate and collect in a dry flask the fraction boiling between 151.5° and 153°C. Tetrabutylammonium iodide base electrolyte, 0.2 m in dimethylformamide-pure solid.

Weigh 7.39 ± 0.01 g of solid anhydrous tetrabutylammonium iodide into a beaker and dissolve it in 100 ml of a dimethylformamide–water (9 + 1 v/v) mixture. (Use dimethylformamide that has been freshly purged with hydrogen or nitrogen). Prepare this reagent freshly each day. With impure batches of tetrabutylammonium iodide a high reagent blank value is obtained in the potential range used in the polarographic analysis. In this instance purify the solid base electrolyte as described below.

Dissolve 10 g of tetrabutylammonium iodide in 100 ml of a methanol-acetone (1 + 3 v/v) mixture and filter it through paper. Leave this solution to evaporate to about half its original volume in a clean laboratory atmosphere. Add 10 ml of water to reprecipitate the salt in the form of fine crystals. Recover the crystals by filtration, dry them *in vacuo* and store them in an evacuated desiccator.

Acrylonitrile and styrene monomers: Re-distill the monomers immediately before use.

Mercury—for polarographic analysis: Use trebly distilled mercury. Request the manufacturers to supply this mercury in stone containers as it has been shown that, over a period of time, mercury picks up an impurity that might interfere in polarography from polyethylene storage bottles.

Hydrogen or nitrogen: extremely low oxygen content.

Procedure

Purge the stock bottle of dimethylformamide with oxygen-free nitrogen or hydrogen for 30 to 40 min to sweep out dissolved oxygen. Keep the bottle well stoppered and repurge each time it is opened. Use freshly purged dimethylformamide solvent throughout the analysis.

Weigh 1.25 ± 0.01 g of styrene-acrylonitrile sample into a 50-ml calibrated flask. Transfer by pipette 25 ml of anhydrous dimethylformamide to the flask, stopper the flask and swirl it to dissolve the polymer. Dilute the solution to the mark with the 0.2 m tetrabutylammonium iodide reagent and mix it well. For the blank solution, transfer by pipette 25 ml of dimethylformamide and 25 ml of the 0.2 m tetrabutyl-ammonium iodide reagent to a 50-ml calibrated flask and mix them well. Transfer by pipette 5 ml of the sample solution and 5 ml of the blank solution into two polarographic cells and immerse these in the constant temperature bath of the polarograph.

Purging of the cell with hydrogen and nitrogen must be performed immediately before any polarographic measurements are made. Lower the dropping mercury electrode system over this cell, so that the outer glass sleeve of the electrode dips 1 to 2 mm into the water tank (providing a water seal to prevent the ingress of atmospheric oxygen) and immerse the anode connection in the side arm of the polarographic cell. Pass hydrogen or nitrogen for 15 min, then switch off the gas supply. Carry out the polarographic measurements described below within 1 to 3 min. of stopping the purge with gas.

The analytical conditions with the Southern Analytical K1000 polarograph are as follows:

Cathode	dropping mercury
Reference anode	mercury pool
Circuit	cathodic direct
Start potential	styrene –2.0 V
	acrylonitrile –1.7 V

Adjust the polarograph to its minimum sensitivity setting (i.e. × 25). Adjust the 'X'-shift and the 'Y'-shift controls until the light spot commences its sweep at the origin of axes at the left-hand side of the graticule on the cathode-ray tube. Repeat this operation at decreasing sensitivity settings until the polarographic wave is visible on the graticule.

The analytical conditions with the Southern Analytical Davis Differential cathode-ray polarograph (with single-cell operation) are as follows:

Cathode	dropping mercury
Reference electrode	mercury pool
Circuit	forward sweep (with derivative units control switched off)
Sensitivity control	select a suitable current scale factor and keep this constant throughout the analysis. After the instrument sensitivity by means of the shunt scale-factor control
Start potential	styrene –2.0 V
	acrylonitrile –1.7 V

By means of the 'Y'-shift control, adjust the light spot to the origin of axes on the left-hand side of the graticule on the cathode-ray tube. Repeat this operation at different shunt scale-factor settings until the polarographic wave is visible on the graticule.

CHAPTER 13

Applications of Gas Chromatography

INTRODUCTION

Gas chromatography is particularly useful for the characterisation of more volatile substances such as mixtures of fatty acids or of alcohols, esters or hydrocarbons which are used in polymer formulations. It has the advantage that it provides more information regarding the carbon number distribution of each of these types of compounds than can be obtained by infrared spectroscopy alone. Similarly, if the polymer is known or suspected to contain dicarboxylic acids, polyol esters, polyhydric alcohols or alcohol esters of phthalic, sebacic, adipic or azelaic acids, then, frequently, information on the type of compound present can be obtained by comparing under the same conditions the retention time of the unknown with those of known authentic substances. Some gas chromatographic methods for the separation of amine and phenolic type antioxidants are listed in Table 13.1.

Gas liquid chromatography is concerned with such matters as the examination of monomers, of the products obtained on depolymerisation of polymers, of the products obtained on chemical attack of polymers and plasticisers, and of solvents and infra-red examination of separated gas liquid chromatographic fractions.

Gas chromatography, and in particular gas liquid chromatography, has probably had a greater influence than any other single factor, on the methods of analysis employed in the plastics industry in the past decade. The subject of gas liquid chromatography in relation to the plastics industry is so wide that it is probably best to describe some of the implications of the procedure on a pretty broad basis.

To begin with, the analytical chemist in the plastics industry are required to be very well informed about the gas liquid chromatographic detection and determination of a whole host of impurities in substances that are starting points for polymer production, e.g. in ethylene, propylene, butadiene, acrylonitrile, styrene, chloro-fluoro hydrocarbons, chlorinated hydrocarbons, fluorinated hydrocarbons, methyl methacrylate, methyl acrylate, ethyl acrylate, ethyl methacrylate, vinyl chloride, vinylidene chloride, vinyl acetate and so on.

Moreover, a knowledge of the behaviour of the lower saturated and unsaturated hydrocarbons will also be needed and there are cases where information on substances such as terpenes is useful in solving problems. Again gas liquid chromatographic examination of the products of depolymerisation of polymers, whether carried out externally or internally over the chromatographic column, will often be extremely informative.

Table 13.1. Separation of antioxidants—gas liquid chromatographic techniques.

Substances separated	*Stationary phase*	*Column temp. °C*	*Other details*
2,6-Di-t-butyl-*p*-cresol, 2-(2-Hydroxy-5-methyl-phenyl) benzotriazole	25% LAC-2R/446 (adipate ester) + 2% H_3PO_4 on chromosorb	135	H_2 carrier gas, F.I.D. Error ± 1%
2,6-Di-t-butyl-*p*-cresol, (I) 2,6-Di-t-butyl phenol	10% Apiezon N on celite 545	164	H_2 carrier, F.I.D. 10^{-3} M (I) in presence of others can be detected
2,4,6-tri-t-butyl phenol			
Diphenylamine			
2,6-Di-t-butyl-*p*-cresol	Apiezon		F.I.D.
Phenyl-2-naphthylamine			
Halogenated bis-phenol	10% DC-710 Silicone oil on chromoport 80–100 mesh	225-250	12 in. glass column 1/4 in. o.d. Carrier: 130 ml He/min
Low b.p. phenols	Capillary column coated with 10% xylenol phosphate	125	F.I.D.
Phenol and 5-t-butyl derivatives	Silicone oil 550-carbowax 400 (3:2)	200	Mean deviation 0.4%
Phenols and cresols	5% w/w of various phosphate esters of phenols	110	120 cm × 4.5 mm column, Pye-Argon Chromatograph
Ionox 330	(a) 20% DC-710 Silicone oil on chromosorb.	200-300 in. 10 min.	(a) 12 × 3/16 in. column
	(b) 2% SE.30 Silicone gum on chromosorb mesh		(b) 12 × 1/16 in. stainless steel column.
Low molecular weight phenols	Silicone-coated capillary column		Converted to trimethylsilyl esters before chromatography.
2,6-Di-4-methyl phenol	20% SE.30 on HMDS-treated 60 mesh chromosorb W	200	E.C. detector

HMDS, hexamethyldisilazane. E.C., electron capture. F.I.D., flame ionisation detector.

Certain polymers, and in particular acrylic polymers will respond to chemical attack by such chemical agents as hydriodic acid, and gas liquid chromatographic test of the resulting iodides will often be extremely effective. In our case, perhaps the greatest all round advances have resulted from our study of solvents from plastic adhesives. We have shown that certain solvents can be identified by recognition of their gas liquid chromatographic patterns, whilst in other cases mixtures of solvents may be separated by gas liquid chromatography on two columns of quite different polar character; the individual solvents may be recognised by the figures they yield for the relative retention times versus a standard substance on the two different columns.

All this work has led us to gas liquid chromatographic separation of solvent mixtures, and, in its turn, to the infra-red identification of the separated substances both in the vapourised and liquid conditions. This work has had a far reaching effect on other aspects of the analysis of plastics.

Then again it was the practice, some years ago, to obtain information about phthalate plasticisers by isolation of the corresponding alcohols and direct infra-red examination of those alcohols. More recently we have shown that gas liquid chromatographic separation of those alcohols is even more informative, either as a final step or only as a preliminary step to infra-red identification.

Finally, it should be realised that a great deal of old fashioned gas analysis has been superseded by modern gas solid and gas liquid chromatographic procedures.

Although for particular purposes very sensitive detectors such as argon detectors have been used in connection with the analysis of plastics etc. e.g., the argon detector in the determination of acetylene, allene and methyl acetylene in propylene, almost all the methods described have been worked out with simple conventional equipment.

APPARATUS FOR GAS LIQUID CHROMATOGRAPHY

An outline of the apparatus generally used is given in Fig. 13.1. The columns Fig. 13.2 used for separation work are conventional U type columns usually 6 ft or 12 ft in length with screw caps for the introduction of samples.

The column support is usually Celite. The stationary phase varies with the particular separation being effected, but usually consists of a high proportion, e.g. 30% of a high boiling ester such as dinonyl phthalate admixed with the Celite. The apparatus works under reduced pressure and katharometer detection is employed. If calibration is required, it is usually carried out within the apparatus and with appropriate internal standards.

Infra-red Spectroscopy

The infra-red absorption spectrum is a unique property of the molecule of a substance, and for this reason infra-red examination has proved to be among the more powerful methods of identification of organic materials. The art of infra-red spectroscopy has now advanced to a stage at which it is possible to obtain, on relatively cheap and commercially available equipment, a reliable infra-red spectrum of an unknown within a few minutes; the infra-red spectrum can therefore be considered alongside the more conventional tests used in characterisation, such as melting point, refractive index, and preparation of derivatives, as a general identification method.

In the field of synthetic polymers however, the more conventional tests are of restricted value and hence examination by the infra-red method is generally acknowledged to be the most important single test available for polymer identification.

The infra-red method is based both upon the assignment of absorption bands to particular structural groups, and upon matching the spectra of unknowns against those of well characterised standard substances prepared for examination in a definite manner. Thus for the proper use of the infra-red method a collection of such spectra is of the greatest importance.

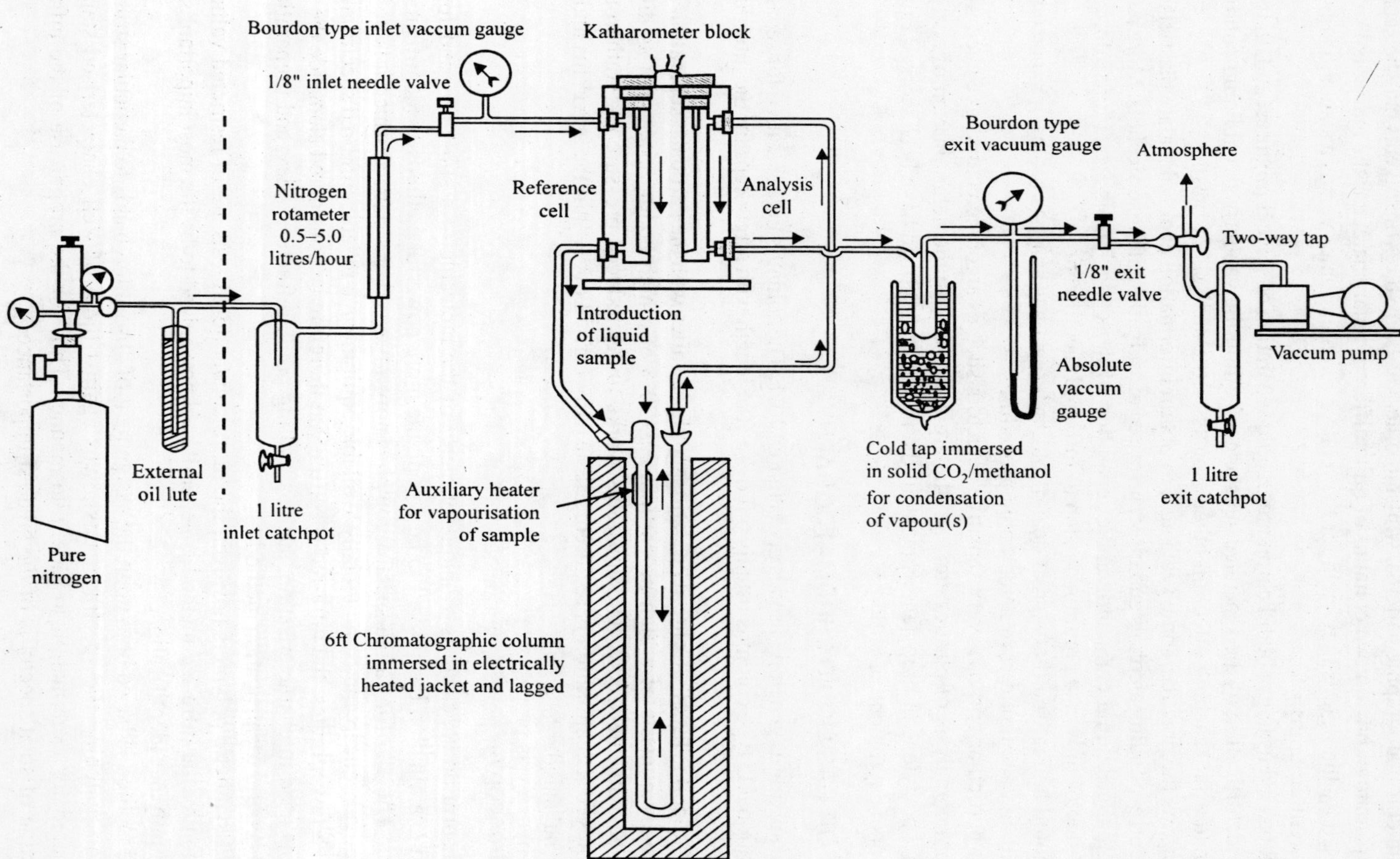

Fig. 13.1. Apparatus for gas liquid chromatography.

Fig. 13.2. Column for gas liquid chromatography.

The programmed temperature gas chromatography between 200° and 300°C on a silicone elastomer coated column can elute phenolic anti-oxidants of very high molecular weight and boiling point. This method is applied to the determination of Ionox 330 in various polymers. The conditions are given below:

Chromatograph	Aerograph Model 600-C with flame ionisation detector.
Column	1/8 in. × 18 in. stainless steel with 5% SE-30, Silicone Gum Rubber, methyl on 80/90 Anakrom ABS (acrylonitrile-butadienestyrene).
Temperatures	
Injection port	330°C
Column	290°C
Detector	290°C
Carrier gas	65 ml/min helium
Retention time for Ionox 330	8.5 min

Calibration standards representing 0.1 g/l. to 0.8 g/l. Ionox 330 are analysed with an average repeatability of ± 2.0%. The calibration curve obtained as a plot of peak area versus concentration was linear and passed through the origin, indicating that no Ionox 330 is lost through decomposition in the hot injection port.

Sometimes a component of a polymer extract is insufficiently volatile to elute on a gas chromatographic column, yet the hydrolysis products of this compound are volatile enough to be gas chromatographed. Thus, a surface coating on polystyrene beads, which has been isolated by shaking the polymer with methanol, shown by preliminary infrared measurements to consist of a mixture of a metal salt of a fatty acid and a primary aliphatic amide. The metal salt is non-volatile and can not be chromatographed and the amide decomposes to a nitrile on the column. The methanol surface extract of the polymer is evaporated to dryness, weighed, and its nitrogen content determined. The residue is hydrolysed by refluxing for 6 hour with 30% sulphuric acid and free fatty acids then extracted with benzene. Gas chromatography of this extract shows that it consisted almost entirely of three groups of carboxylic acids, each with its own carbon number distribution as follows:

1. C_2—C_{15} mainly palmitic acid 40%
2. C_{16}—C_{17} Mainly stearic acid 30%
3. C_{21} Erucic acid 25%

The original polymer surface coating therefore probably consists of a mixture of a metal stearate, erucic acid and palmitamide. This assumption leads to a nitrogen content of 2.3% whereas the nitrogen content found was 2.1%. This type of hydrolysis technique can be applied to other types of compounds which might be difficult to chromatograph in their original form. Thus, either acid or alkaline hydrolysis will convert alkanolamine/fatty acid condensates to the fatty acid and the free alkanolamine. Plasticiser esters are converted to the acid and alcohol which are often easier to identify than the original ester.

Other types of compounds, which might be amenable to this approach, include aliphatic amides, sugar esters, and other types of esters such as esters of phosphorus acid, e.g. Polygard (tris-2:3-dibromopropyl phosphite).

CHARACTERISATION AND DETERMINATION OF ADDITIVES

The attraction of gas chromatography lies in its ability to simultaneously separate, identify and estimate sub-milligram quantities of complex mixtures. There are certain drawbacks to the technique. Retention times are no more specific for gas chromatography than are R_f values for thin-layer and paper chromatography. Day-to-day reproducibility of retention times is not good for some types of instruments, especially when operating at high column temperatures. Since many polymer additives are not very volatile, low stationary-phase loadings have to be used in order to reduce retention times to a sensible value, which means that large areas of uncoated solid support appear which in turn lead to bonding with compounds such as phenolic and amine antioxidants. This results in of peaks and in lengthening of retention times. There are measures which can be taken to meet these difficulties. Thus, relative retention times are often more reproducible than unadjusted retention times. Non-volatile sample components can be converted into more volatile derivatives such as tri-methyl-silyl ethers. The solid support can also be treated with hexamethyldisilazane, for example, to reduce the number of active sites on the column material available for bonding. The formation of derivatives before chromatography does, however, in some applications lead to sample loss and can often lead to the appearance of spurious peaks on the chromatogram. Nevertheless, despite these limitations gas chromatography has found numerous applications in the identification, characterisation and estimation of additives in polymers and some of these are discussed below in further detail.

Roberts and *Swank* describe a procedure for the determination of Ionol (2,6-di-tert-butyl-*p*-cresol) and Tinuvin P (hydroxyphenylbenzotriazole) in polystyrene. In this spectra is obtained by the above technique with spectra of standard vapours. Generally speaking, the additive constituents of polymers have a boiling point which is appreciably higher than 200°C and hence cannot be handled in infrared gas cells. For such substances, infrared examination as a liquid or a solid, as discussed below, is more relevant. Obviously, in order to avoid volatilisation during polymer processing and subsequent life, most types of additives used in polymer formulations are fairly high melting point solids. Volatile constituents are, however, sometimes encountered, viz. expanding agents, plasticisers, lubricants, adhesives, solvents, monomers and degradation products of additives or of the polymer itself, and infrared gas cell techniques can be of value in the examination of gas chromatographic fractions containing these types of substances.

For the collection of volatile and liquid fractions emerging from a gas chromatographic column and their subsequent identification by infrared spectroscopy, the techniques to various polymeric materials encountered in the plastics industry are applied. They incorporated several of the techniques developed by *Anderson* into their procedures, but, unlike him, found that his technique could not be applied effectively to substances with a boiling point much in excess of 120°C. Moreover, they found that many of the standard infrared spectra of substances encountered in the plastics industry were recorded only for liquid compounds and not for the corresponding vapours. All the fraction-collecting apparatus used in conjunction with gas-chromatographic columns operated under reduced pressure and with katharometer detection is of two types:

1. Apparatus involving a column thermostatically controlled within the range 0° to 130°C and a katharometer at room temperature. This apparatus is used for separating mixtures when the highest-boiling constituent boils below 160°C at atmospheric pressure.

2. Apparatus involving a column and katharometer contained within the same thermostatically controlled chamber and maintained at some temperature in the range 100° to 200°C. This apparatus is used for the separation of mixtures with constituents boiling up to 250°C, which give trouble in apparatus of type 1 owing to condensation in the katharometer.

The trap shown in Figs. 13.3 and 13.4 has been successfully used with apparatus of type 1 for the condensation of substances boiling in the range–100° to +160°C. The packing used is small Dixon rings held in place by a small loose-fitting piece of glass rod. The efficiency of this type of trap varies between 85 and 95%, depending on the chemical class of substance. The tap is connected to the exit of the katharometer by means of the minimum length of stainless-steel tubing (internal diameter 1 mm) brazed into a metal cup. This in turn is cemented to a glass socket with Araldite cement. The exit ends of the four traps are connected via rubber tubing and separate stopcocks to a vacuum manifold. The traps are partly immersed in liquid nitrogen contained in 1-pint Thermos flasks.

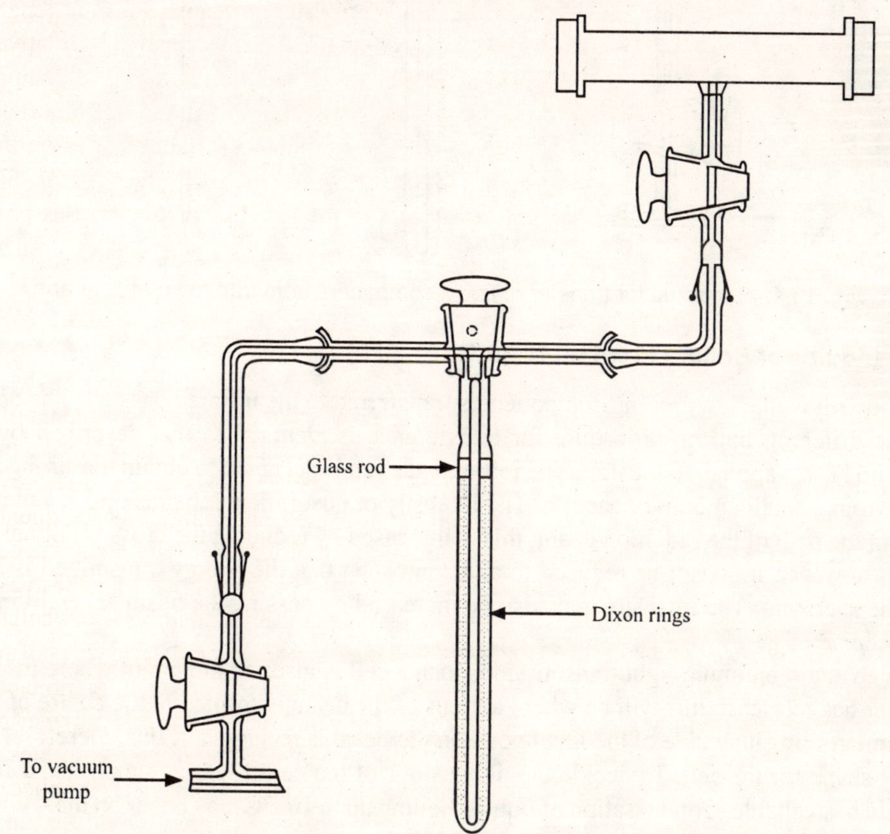

Fig. 13.3. Apparatus for transfer of volatile component from trap to infra-red gas cell.

A preliminary chromatogram indicates the complexity of the mixture. On a second run, the traps are switched in by following the recorder trace and any overlap between peaks is allowed to go to waste. The delay time between the recorder signal and the component reaching the trap is negligible.

Having condensed the desired component in a trap it has to be decided whether to record the spectrum in the vapour state (method 1 below) or the liquid phase (method 2 below). *Anderson's method* is retained for identifying all components that, from the gas-chromatographic evidence, would appear to boil below 60°C. Small amounts of such substances in liquid form are readily lost by vapourisation in attempting the transfer to a small liquid infrared cell.

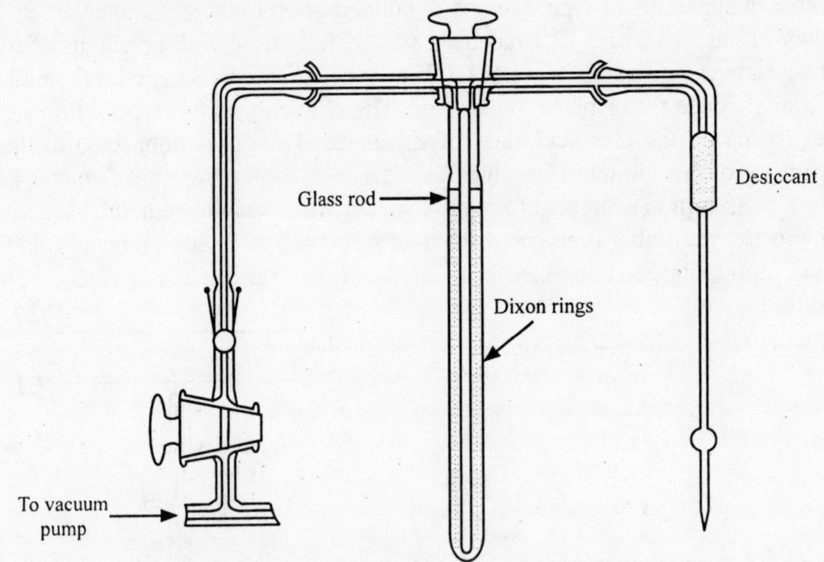

Fig. 13.4. Apparatus for transfer of liquid component from trap to tip of side-arm.

Method 1, Liquids of Boiling Point up to 60°C

Fig. 13.3 shows how the low-boiling component is transferred to the infrared gas cell. The design of the apparatus is different, but the procedure for transference is identical to that described by *Anderson* except that no heat is applied to the trap. The infrared gas cell is designed to obtain the strongest possible absorption from a small amount of sample. The intensity of absorption depends only on the number of molecules in the path of the radiation beam; this is increased by reducing the cross-sectional area of the cell, which, however, must not be reduced to an extent such that the energy transmitted is insufficient to record the spectrum. The intensity may also be increased by passing the beam several times through the same volume of gas.

In order to attain optimum light transmission, such a cell is inserted at a point where the area of the spectrometer beam is least; this will be where a focus of the beam is formed in the centre of the cell. At a few centimetres on either side of the focal position the beam is rectangular; this, therefore, is the most economical shape for the cell. To decide the dimensions of the cell it was assumed that 1 ml of gas at N.T.P. would be available. Condensation of liquids boiling up to 100°C may occur on the walls of the cell at pressures above 80 mm; the cell must therefore be of capacity about 9 ml.

The image to be found in the centre of the cell should normally have the length and width of the entrance slit of the spectrometer. Knowing the *f* number of the instrument it is possible to calculate the size of the rectangular aperture required to admit the radiation cone in terms of the length of the cell. For the instrument used (Hilger H800, aperture *f* 11) the dimensions for a 9-ml cell are: length, 10 cm; height, 1.5 cm; width, 0.6 cm (Fig. 13.3). The body consists of rectangular brass tube (1 mm thick) cut from wave-guide tubing and has the internal dimensions stated above. A flat brass flange is brazed on each end of the tube, and a short side-arm (stainless-steel capillary tubing) is brazed in one side of the body. A glass capillary tap that will just slide over the side-arm is cemented in place by filling the space between the two tubes with Araldite resin; by this means, the volume of the side-arm is rendered negligible. Two rock-salt windows are attached to the flanges with Picene wax. Although the cell is designed for an *f* 11 system, it will give acceptable performance, although with some loss of energy, on an *f* 7 spectrometer.

Method 2, Liquids Boiling Above 60°C

If the gas-chromatographic evidence indicates that the substance boils above 60°C, the fraction condensed in the trap is treated as shown in Fig. 13.4. The trap and contents are held in liquid nitrogen while the apparatus is completely evacuated. After testing the system for leaks, the source of vacuum is cut off. The flask containing the liquid nitrogen is removed from the trap and placed over the side-arm so that the bulb is just immersed in the refrigerant. As the trap attains room temperature, the small amount of liquid in it distils through the desiccant and is condensed round the sides of the bulb of the side-arm. The liquid is encouraged into the capillary tip by lowering the vacuum flask containing the liquid nitrogen until only the last 1/4 in. of the tip is immersed. The bulb is then warmed with the fingers. During this distillation the trap is not warmed, as this may cause some of the liquid to distil in the wrong direction. The apparatus may be left for 1 or 2 hr in order to achieve complete distillation of higher-boiling liquids. Having obtained the liquid in the capillary tip, this tip is broken off at the constriction just below the bulb. Fig. 13.5 shows a piece of apparatus which may be used to transfer the liquid to the micro infrared cell. It has a rubber test and a very fine internal capillary. The devices as a detachable tip, which is discarded after use and permits close control over the transfer operation. In the design of a small liquid cell, first consideration must be given to reducing the cross-sectional area, the precise area required for a particular spectrometer being found by trial and error. The radiation beam of the instrument is obstructed at the cell-mounting position, to find the least dimensions at which workable energy transmission is retained. A cross-section of 5 mm × 1.5 mm is suitable for the Hilger H800 spectrometer, although the resulting cell could be used satisfactorily on other instruments.

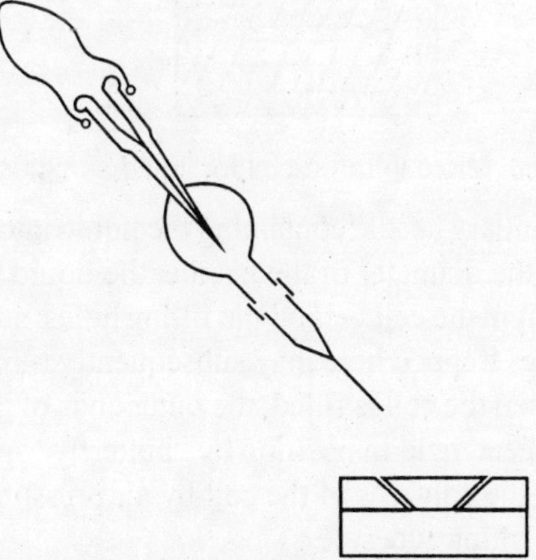

Fig. 13.5. Pipette for transferring liquid to micro infrared cell.

Most of the samples placed in a normal infrared liquid cell is used in the filling tubes rather than in the area exposed to the beam. Consideration was therefore given to a cell in which the filling tubes need not be occupied with liquid, and a suitable design is shown in Fig. 13.6. The two rock-salt plates are about

20 mm × 10 mm × 3 mm in size. Two holes are drilled in one plate with a No. 86 drill; these holes taper inwards, as shown, being 7 mm apart on the inside and 12 mm apart on the outside of the plate. A gold-foil washer (25 μ thick) is cut, the inner space being 1.5 mm wide and approximately 7 mm in length (just sufficient to clear the edges of the filling holes); the width of this washer is about 2 mm. The cell is first stuck together with gold amalgam. The washer is immersed in mercury for a short period and withdrawn after a layer of amalgam has formed on its surface. The cell is then carefully assembled in a small clamp and left under pressure for 24 hr for the amalgam to set. The space between the plates outside the washer is then filled with Araldite 700 resin and a reasonable surplus of resin is allowed to bridge the edges of the plates on all four sides. The resin was found to be necessary, as the gold amalgam is a poor adhesive and the cell tends to open up and channel; the amalgam joint acts as a barrier to prevent the resin from contaminating the cell window.

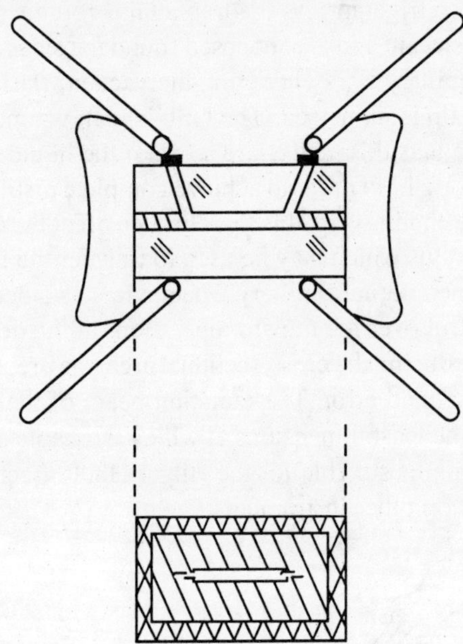

Fig. 13.6. Micro infrared cell for liquid samples.

The cell is filled by inserting a capillary needle containing the liquid into one filling hole. Provided that the thickness of the cell is less than the diameter of the needle, the liquid will run in to fill the cell area. It is essential that no free space be left in the cell behind the filling holes; such space will not immediately fill by capillary action and air bubbles trapped here may subsequently run into the useful part of the cell and are most difficult to remove. When the cell is filled, the outer ends of the filling tubes are covered by two small squares of nitrile-rubber sheet, held in position by "butterfly" spring clips. This method forms an excellent seal without displacing the contents of the cell by air pressure, as may occur with stoppers that are pushed or screwed into the filling tubes.

The amount of liquid required to fill the cell by the method described above (as opposed to the theoretical capacity of the cell calculated from the dimensions) is determined experimentally by weighing the cell on a microbalance before and after filling. For six fillings with α-dodecane (density 0.766) the weights recorded were 0.53, 0.36, 0.27, 0.29, 0.67 and 0.31 mg. This compares with the calculated capacity of 0.20 mg of n-dodecane (7 × 1.5 × 0.025 × 0.766). With this cell, infrared spectra have been obtained from 0.5 to 1.0 μl portions of liquid samples, but practice is required in the transfer of these small volumes. On the other hand, 2 to 3 μl of liquid present no difficulties.

Method 3

This is an extremely simple, but effective, method that has been used in conjunction with a *Griffin* and *George* mark II gas chromatograph. The back plate of the oven is removed and the copper pipe (1/8 in. Internal diameter) from the katharometer is cut, leaving approximately 1/2 in. protruding beyond the oven wall. An Agla syringe barrel containing a tiny piece of cotton-wool (2 to 3 mg) is connected into the exit-gas stream by sleeving one end of the barrel over the copper pipe with silicone-rubber tubing. This entails running a preliminary chromatogram and 'breaking into' the gas stream at the appropriate point in the chromatogram on a second run. The piece of cotton-wool, although not particularly efficient, has the effect of stopping 'vapour fog' from passing straight through the barrel. The syringe barrel is cooled during this operation by packing solid carbon dioxide round the outside. Having wetted the cotton-wool with the unidentified component, the syringe barrel is removed from the gas stream and a specially adapted nozzle is inserted. This nozzle consists of a fine piece of hypodermic tubing sleeved and cemented into the normal Agla glass nozzle and filed flush with the glass at one end. This has the effect of cutting down the 'dead' volume of the syringe to the minimum. The syringe is held nozzle uppermost, the piston is inserted and the air expelled. The sample is then expressed from the cotton-wool by pressure of the piston directly into the micro infrared liquid cell. Unfortunately, it has been found that 10 to 12 μl of sample are needed in order to express sufficient liquid to fill the cell. However, if insufficient liquid is available, a drop of carbon disulphide is placed on the cotton-wool from a second syringe and the carbon disulphide solution of the sample is expressed into the cell.

This method proved valuable in the examination of a sample of acrylic sheet known from elemental analysis to contain phosphorus. Dry vacuum distillation of the polymer and solvent extraction both yielded a liquid, which, on direct infrared examination, proved to be an impure material containing phosphate or phosphite. The vacuum distillate at 200°C was submitted to gas-liquid chromatographic examination on a 6-ft (1/4 in. diameter) column packed with 30% w/w of silicone E301 on Celite maintained at 130°C. A sample of the main component was taken by method 3 and the infrared spectrum obtained was readily identified as that of triethyl phosphate.

Method 4

In a modification of method 3, a heated outlet pipe is fitted to the back of the katharometer and condensed the vapour in the trap shown in Fig. 13.7. The outlet pipe consists of a stainless-steel tube (1 mm internal diameter) approximately 9 in. long and terminates in a brass plug tapered to fit a B7 cone. The opposite end is joined to the copper exit pipe of the katharometer with silicone-rubber tubing. The pipe is heated directly by passing approximately 0.5 V at 10 A through it from a variable transformer. The lagging is of asbestos string and fire clay cement covered by a second layer of asbestos string.

The trap is connected into the gas stream at the appropriate point in the chromatogram. The lower part of the trap is immersed in liquid nitrogen and the fraction condensed. The subsequent procedure is similar to that described for method 2, the small amount of substance being vacuum-distilled into the tip, which is then broken off, and the liquid is transferred to the micro infrared cell.

Mass spectrometric examination of fractions separated by gas chromatography has been used by a number of investigators, for identification and quantitative analysis. This technique could undoubtedly by used in the solution of additive identification problems.

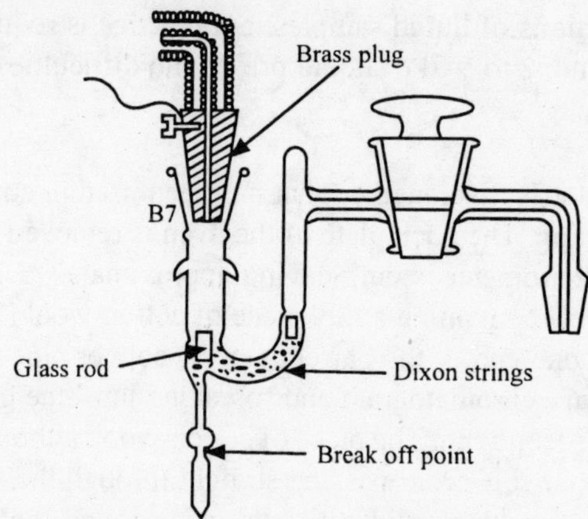

Fig. 13.7. Trap used for collection of fractions emerging from gas chromatography column.

IDENTIFICATION OF ADDITIVES BY COMBINATION OF GAS CHROMATOGRAPHY AND INFRARED SPECTROSCOPY

Some of the limitations encountered in applying gas chromatography to the unequivocal identification of polymer additives have already been discussed. In addition to these limitations there are others. Retention times are purely relative and not always exactly reproducible; even when the chemical class of the sample constituents is known, identification by comparison of the retention time of the unknown with those of standards requires exact reproduction of column operating conditions.

Difficulty can also arise when non-symmetrical peaks are produced; the peak-maximum retention time of a component then depends on concentration. Peaks can also overlap, and certain combinations of substances are often difficult to separate. Chromatographically, a single peak is no criterion of purity, since more than one substance may be present; the components present could often be resolved if their presence was suspected and alternative column operating conditions were selected.

Confirmation of homogeneity of fractions is therefore required together with unequivocal identification and accurate quantitative determination of the product. The analytical method used should involve some property of the molecule other than boiling-point. Often, only fractional milligram amounts of material can be recovered from a column. Of the few techniques that are applicable to these small quantities of material, mass and infrared spectroscopy are particularly suited.

The use of fraction collecting techniques in conjunction with gas chromatography is now well established and is an attractive proposition in additive identification problems, despite the fact that little published work has yet appeared on the application of this technique to polymer additives.

In this technique, the separated compound as it emerges from the gas chromatographic column is swept by the carrier gas through a cold trap where it condenses. The material in the trap is then either transferred to an infrared gas cell for examination in the vapour phase, is transferred as a liquid to a suitable micro cell or may be condensed on a cold surface as a solid for examination by conventional spectroscopic techniques. In the past few years there have been improvements in commercially available equipment which permit infrared spectroscopic examination of gas chromatographic fractions of less than 1 mg in quantity. Beam condensers to focus the energy beam of infrared spectrometers in an area of 1 mm × 5 mm or smaller enable the use of smaller infrared cells and potassium bromide pressed plates than before. Infrared cells have been developed which require only a milligram or less of liquid to fill them.

Cells have also been described for infrared spectroscopy of fractions while still vapourised in the gas stream emerging from the gas chromatographic column and also for obtaining the vapour spectra of fractions after trapping and revapourising them.

Trapping of gas chromatographic fractions in dry-ice cooled glass or metal capillary tubes is fairly efficient for high boiling compounds, decreasing in efficiency with lower boiling substances. Presumably, with the lower boiling substances, losses are due to some of the gas chromatographic fraction being swept through the trap as a fog or as frozen particles. When helium is used as the carrier gas it is possible that small amounts of helium impurities, notably water, can be trapped out together with the gas chromatographic fraction. In these circumstances it is necessary to carefully dry the carrier gas entering the gas chromatograph in order to eliminate any interference effects in the subsequent infrared spectroscopy of trapped fractions.

Various others have been described techniques for collecting individual gas chromatographic fractions and transferring these as liquids to micro infrared cells. These include the examination of 0.4 μl of a liquid sample in the infrared without the use of beam condensers. This fraction trapping technique is efficient enough to permit quantitative collection of fractions. *Blake*. describe a similar technique for collecting and obtaining the infrared spectra of gas chromatographic fractions in the 10–100 μg range. The equipment is simple, inexpensive, and spectra may be scanned with a simple table top infrared spectrometer without beam condensers using a cell made by drilling a 1 mm diameter hole in a cube of a sodium chloride.

A specular reflectance system for the infrared analysis of micro-sized samples is shown in Fig. 13.8. The advantages and limitations of this technique can be compared with other micro infrared techniques. Samples are mounted on small metal mirrors which reflect the light beam back through the sample. A transmission spectrum is thereby obtained but the effective path length is twice that of the actual sample thickness and a given absorption band consequently has twice the absorbance obtained by conventional transmission measurements.

This system was applied successfully to gas chromatographic fractions, and is claimed to be particularly useful for the examination of non-volatile liquids such as, for example, dioctyl phthalate. Crystalline solids are easily deposited and handled on the metal mirrors, but crystallinity effects on the spectrum must be carefully watched out for. Its limitations include a 'stray-light' artefact, polymorphism effects and difficulties in obtaining sample uniformity.

CHARACTERISATION AND DETERMINATION OF MONOMERS AND VOLATILE IMPURITIES

Polymers often contain substances of medium volatility such as residual monomers, residual polymerisation solvents and expanding agents. In addition, when polymers are heated they may release volatiles as a result of the thermal degradation of either the polymers themselves or their additives or catalyst residues. These volatiles can have an important bearing on such properties as processability, the tendency to form voids and, in the case of foodstuff-packaging grades, the possible tendency to impart taste or odour to the packed commodity.

One way of identifying non-polymeric constituents of polymers is to extract the polymer with a low-boiling-point solvent, remove the solvent from the extract by evaporation or distillation and analyse the residue. This procedure is, of course, inapplicable to the analysis of extracted polymer constituents which are volatile enough to be lost during the solvent-removal stage. Alternatively, an extract or a solution of the polymer may be examined directly for volatile constituents by gas chromatography, in which case losses of volatiles are less likely to occur.

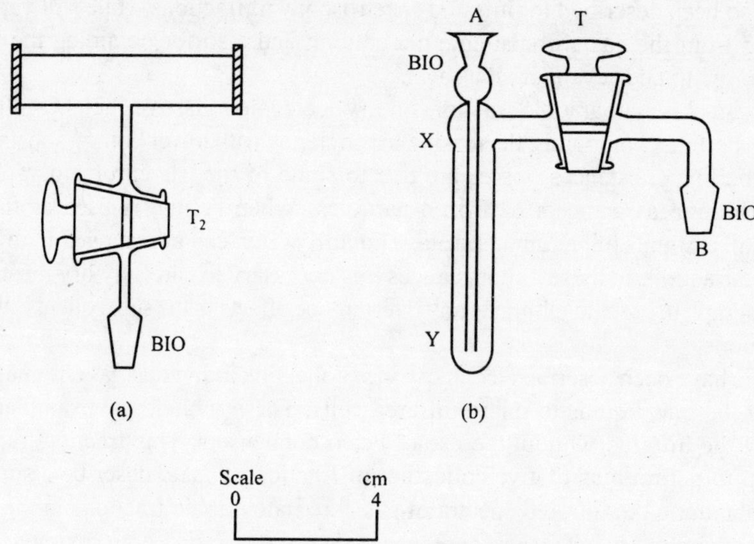

Fig. 13.8. Trap for collecting products separated by gas chromatography.

In such a procedure, however, the large excess of solvent used for extraction or solution might interfere with the interpretation of the chromatogram, obscuring some of the peaks of interest. Trace impurities in the solvent may also interfere with the chromatogram. Of course, none of these procedures is suitable for studying the nature of volatile breakdown products which are produced only upon heating a polymer.

A simple and inexpensive apparatus has been described for liberating both existing volatiles and those produced by thermal degradation from polymers by heating at temperatures up to 300°C, in the absence of solvents, prior to their examination by gas chromatography. The technique avoids the disadvantages resulting from the use of extraction or solution procedure.

The apparatus illustrated in Fig. 13.9 consists of a glass ignition tube, supported as shown in a Wade 1/4-in. diameter brass coupling nut, covered with a silicone rubber spectrum and sealed with a Wade 1/4-in. brass stop-end body. The stop-end body. The stop-end body has two 1-mm diameter holes drilled through the cap. The whole unit is placed in a slot in a cylindrical copper block (3 in. long × 2 in. dia.) which is heated by two (240 V, 85 W) cartridge heaters and controlled at temperatures up to 300°C from a variable transformer.

The temperature is measured with a thermocouple capable of accurately measuring temperatures in the 100–300°C temperature range with a maximum error of ± 5%. The thermocouple is inserted in the slot adjacent to the ignition tube; it has been shown that under these conditions the thermocouple records the true temperature of the contents of the tube. The provision of a slot in the copper block enables more than one ignition tube to be heated simultaneously if required.

A sample of the polymer (0.25–0.50 g) is placed in an ignition tube and sealed with Wade fittings and a septum, as described. If necessary, the tube is then purged with a suitable gas by inserting two hypodermic needles through the septum via the holes in the cap of the stop-end body and passing the gas into the tube through one hole and allowing it to vent through the other. After purging, the two needles are removed simultaneously and the tube is then heated in the copper block under the required conditions of time and temperature. A sample (1–2 ml) of the head-space gas is withdrawn from the ignition tube into a Hamilton gas-tight hypodermic syringe via the septum and injected into a gas chromatograph. It is advisable to fill the syringe with the gas used initially in the ignition tube and to inject this into the tube before withdrawing the sample. This facilitates sampling by preventing the creation of a partial vacuum in the ignition tube or the syringe, or both. It also minimises any undesirable entry of air into the ignition tube.

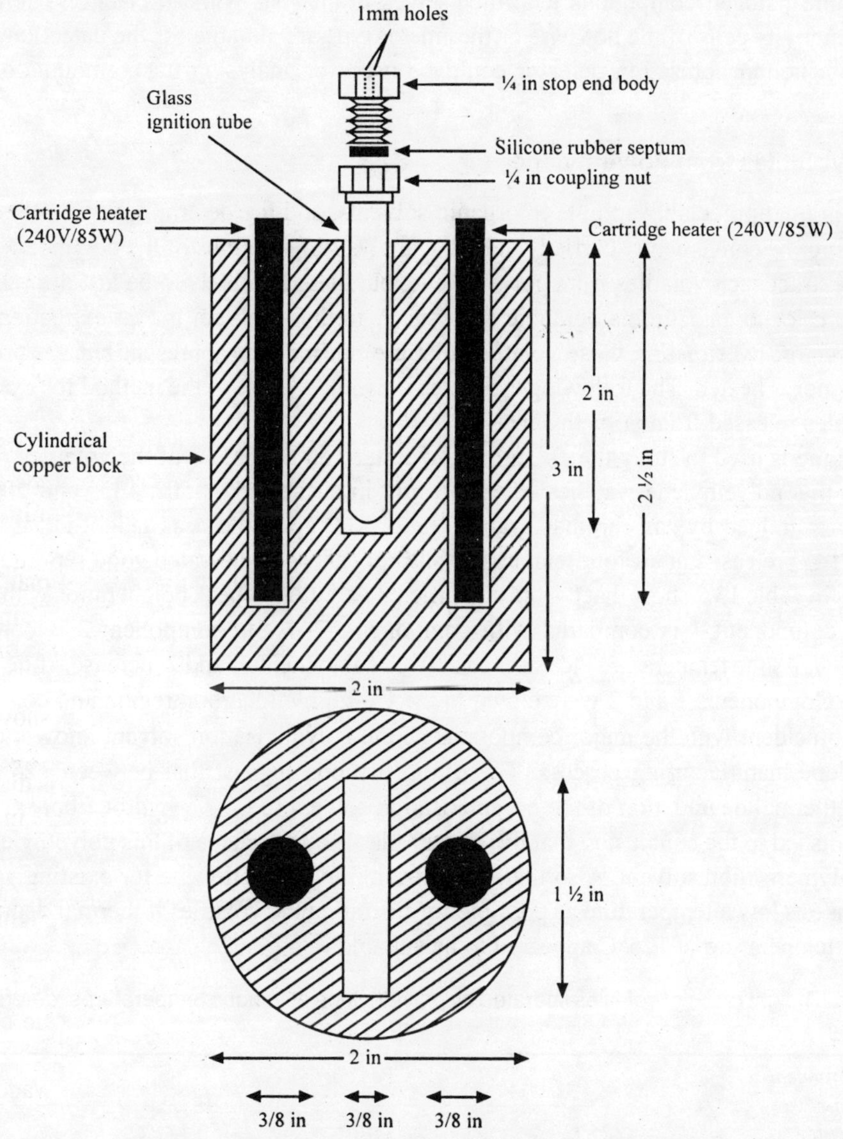

Fig. 13.9. Apparatus for liberating volatiles from polymers.

With the apparatus, a polymer may be heated under any desired gas and, while this may frequently be the carrier gas used with the gas chromatograph, it is also possible to carry out studies in oxidising or reducing atmospheres. A polymer may also be heated to any temperature and samples of the head-space gas may be withdrawn at intermediate temperatures and times to determine under what conditions any particular volatile is liberated.

By using gas chromatographic detectors of suitable sensitivities and selectivities, it is possible to examine polymers for the presence or formation of volatiles at both the percentage and the parts-per-million levels. For example, traces of organic volatiles may be examined with an ionisation detector, and traces of organic halogen compounds lend themselves to analysis with an electron capture detector. Thermal conductivity cells of the hot-wire or thermistor type are suitable for the detection of inorganic volatiles, and a helium ionisation detector could be used for analysing trace amounts of permanent gases.

Analysis of Volatiles from Polyethylene

Polyethylene is not appreciably soluble in organic solvents, and the determination of existing volatile impurities cannot be conveniently carried out by the analysis of solutions of the polymer. Many organic solvents will extract such volatiles from polyethylene, but these are likely to be lost during the solvent-removal stage prior to gas chromatographic analysis. A further problem is that extraction procedures cannot be used for investigating those volatiles that are not inherently present but are produced only when the polymer is heated. The following examples illustrate the use of the method for examining both types of volatiles released from polyethylene.

The technique is used to study the effect of temperature on the nature of the volatiles released, and to investigate this polyethylene was heated for 15 min in air at 125°, 150°, 135° and 200°C and the volatiles were examined by gas chromatography. Table 13.2 gives the peak height ratios of the major components (1–7) released at all four temperatures. Replicate runs indicated good reproducibility.

The results in Table 13.2 show that the only significant variation of peak height ratios with temperature occurs when component 1 is compared with components 2–7 and component 2 is compared with components 4–7. As the temperature increases to 200°C, components 1 and 2 increase while components 3–7 decrease. Components 1 and 2 were eluted in the C_2–C_4 hydrocarbon region and components 3–7 were eluted coincident with the major compounds of the polymerisation solvent known to be used in this polyethylene manufacturing process. These observations suggest that between 125° and 200°C there is some thermal degradation of the polymerisation solvent to C_2–C_4 hydrocarbons.

These results led to the conclusion that the major volatile components of this polyethylene were due to residual polymerisation solvent. Also that, when examining polyethylene for existing volatiles, it is necessary to use as low a temperature as possible for liberating the volatiles if thermal degradation is to be avoided. A temperature of 125°C appeared to be suitable.

Table 13.2. Peak height ratios of volatiles liberated by polyethylene at various temperatures, determined by gas chromatography.

Component number ratio	*Peak height ratio*			
	125°C	*150°C*	*175°C*	*200°C*
1:2	4.4	5.4	3.1	3.1
1:3	0.3	0.3	0.3	1.7
1:4	2.1	3.2	2.4	14.6

(Contd ...)

Component number ratio	*Peak height ratio*			
	125°C	*150°C*	*175°C*	*200°C*
1:5	0.7	1.0	0.8	4.7
1:6	0.5	0.8	0.6	3.5
1:7	0.9	1.2	1.0	5.9
2:3	0.06	0.06	0.08	0.06
2:4	0.5	0.6	0.8	4.6
2:5	0.2	0.2	0.3	1.5
2:6	0.1	0.1	0.2	1.1
2:7	0.2	0.2	0.3	1.9
3:4	8.1	9.6	9.1	8.3
3:5	2.9	3.1	3.1	2.7
3:6	2.1	2.3	2.3	2.0
3:7	3.4	3.7	3.8	3.4
4:5	0.4	0.3	0.3	0.3
4:6	0.3	0.2	0.3	0.2
4:7	0.4	0.4	0.4	0.4
5:6	0.7	0.7	0.7	0.8
5:7	1.2	1.2	1.2	1.3
6:7	1.7	1.6	1.6	1.7

ANALYSIS TO VOLATILES FROM POLYSTYRENE

In Fig. 13.10 (a–e) are shown gas chromatograms of volatiles obtained when polystyrenes from several different manufacturers were heated at 200°C for 15 min under helium. All the samples liberated the same range of aromatic hydrocarbons, these differing only in their relative concentrations. However, the non-aromatic hydrocarbon material, eluted from the gas chromatographic column prior to ethyl benzene, shows marked differences from sample to sample. The results do show the value of the technique for detecting differences in the volatiles liberated by polystyrenes produced by different manufacturers.

Quantitative Determination of Volatiles

Gas chromatography has been used for the quantitative determination of low levels of aromatic hydrocarbons in polystyrene, and also for the determination of expanding agents such as normal and isopentane in expandable grades of polystyrene.

These gas chromatographic analyses are carried out on solutions of the polymers in the presence of internal standards. To avoid interferences in the analysis, it is essential for the solvent and the internal standard to have retention times different from those of the volatile compounds being determined in the polymer. Application of the volatiles apparatus to a polystyrene provides a rapid means of determining the retention time of the volatile compounds present in the polymer, enabling a suitable solvent and internal standard to be selected for the subsequent quantitative analysis by solution procedures.

The following examples illustrate the use of the technique during the analysis of polystyrene for residual aromatic hydrocarbons and expanding agents. Fig. 13.11 shows a chromatogram of the volatiles liberated by a polystyrene after heating at 200°C for 15 min under helium.

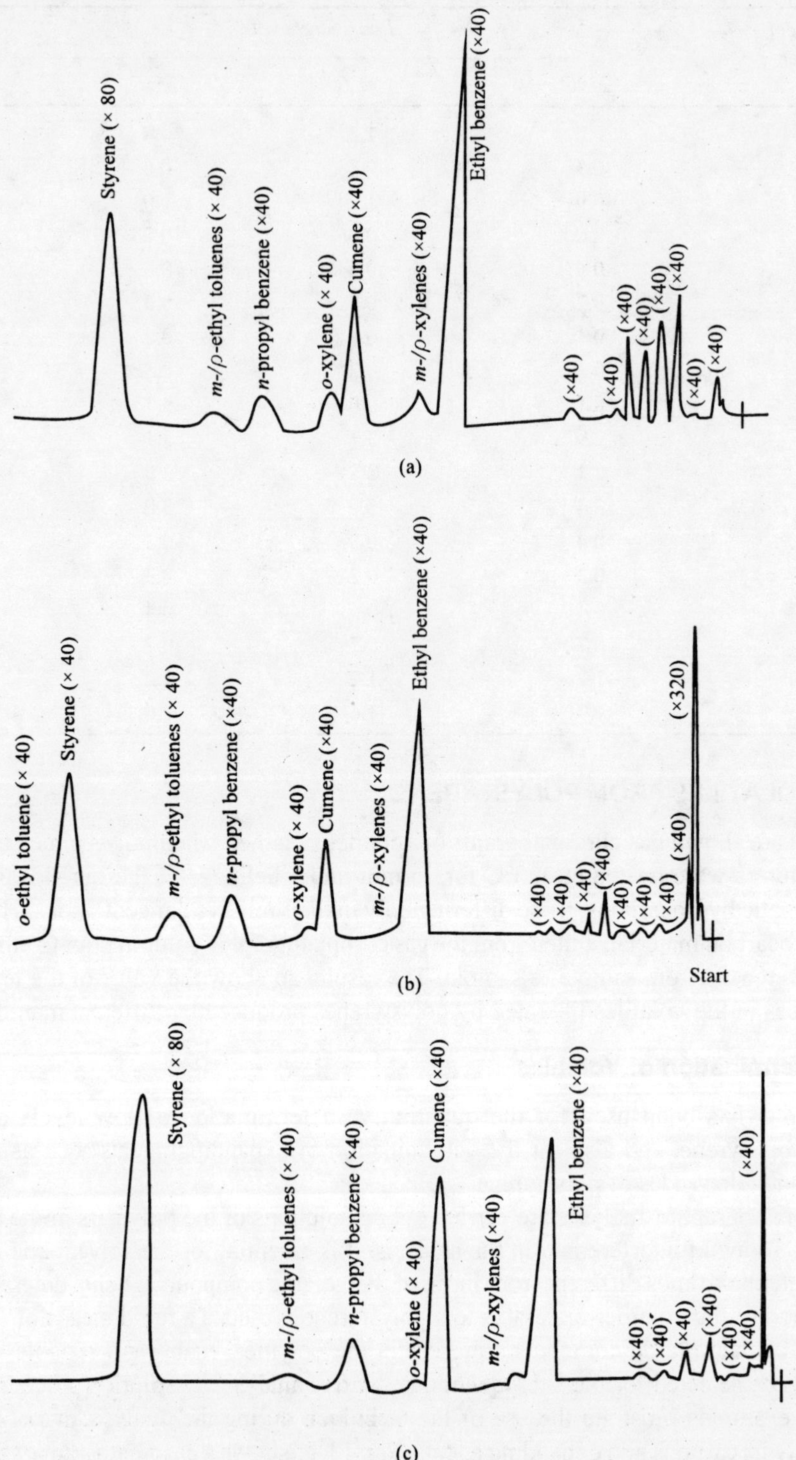

Fig. 13.10. Gas chromatograms of volatiles liberated from different polystyrenes at 200°C for 15 min in helium.

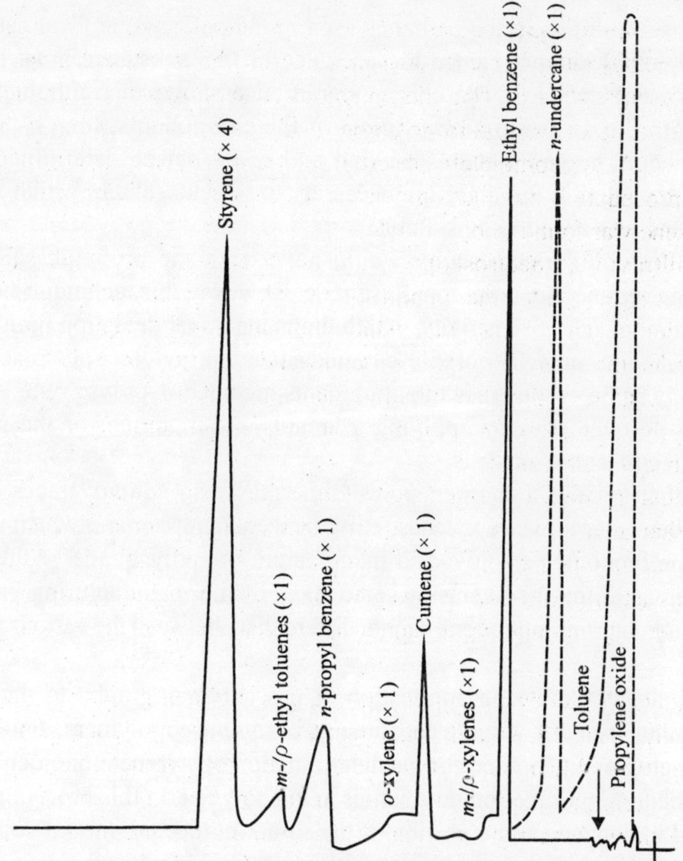

Fig. 13.11. Gas chromatogram of aromatic hydrocarbons liberated from polystyrene at 200°C for 15 min In helium.

Superimposed on the chromatogram are the solvent and internal standard peaks which would be present if the analysis were carried out on a solution of the polymer in propylene oxide in the presence of n-undecane. The quantitative determination of the aromatic hydrocarbons in polystyrene would not be interfered with by either the propylene oxide or by the n-undecane internal standard.

Mixtures of isopentane and n-pentane are commonly used as expanding agents in expandable grades of polystyrene. An available procedure for doing this involved the gas chromatographic analysis of a solution of the polymer in propylene oxide in the presence of 2:2-dimethyl butane as an internal standard. This method is entirely satisfactory for analysing grades of expandable polystyrene in which it is known that isopentane and n-pentane are the only expanding agents present. However, if other types of expanding agent have been used in the polymer formulation then it is possible that, under the selected conditions, their retention times might coincide with those of propylene oxide or 2:2-dimethyl butane, with the result that the analysis would be invalidated.

Fig. 13.12. shows a chromatogram of the volatiles liberated by an expanded polystyrene after heating at 200°C for 15 min under helium. Superimposed on the chromatogram is the solvent peak which would be present if the analysis were carried out on a solution of the polymer in propylene oxide. In addition to n-pentane and isopentane, the sample contains lower concentrations of several C_6 and C_7 hydrocarbons, all of which originate as impurities in the pentanes used in the polystyrene formulation. One of these hydrocarbons is 2:2-dimethyl butane. Hence the presence of this substance must be allowed for by using an alternative internal standard. The chromatogram also shows that although propylene oxide would not interfere with the determination of most of the components, then *n*-hexane and one of the C_7 hydrocarbons would be completely obscured and could not be determined. For a complete analysis by a solution procedure it was therefore necessary to select a solvent which is eluted later than propylene oxide. Benzene was found to be suitable.

Some years ago, ultraviolet spectroscopy would have been the probable selected method for determining residual free styrene monomer in polystyrene. However, this technique suffers from several disadvantages. In addition to lack of sensitivity, which limits the lower detection limit to about 200 ppm styrene in polymer under the most favourable circumstances, ultraviolet spectroscopic methods are subject to interference by some of the types of antioxidants included in polystyrene formulations. Such interferences can only be overcome by applying a lengthy pre-treatment of the sample to remove antioxidants prior to spectroscopic analysis.

In addition to residual styrene monomer, polystyrene may also contain traces of other aromatic hydrocarbons such as benzene, toluene, xylenes, ethyl benzene and cumene which originate either as impurities in the styrene monomer employed to manufacture the polystyrene or they may have been used in small quantities as dilution solvents at some stage of the manufacturing process. Ultraviolet spectroscopic methods for determining styrene cannot differentiate between the various volatile substances present in polystyrene.

It has been investigated to study the application of gas chromatography to the determination of styrene monomer and other aromatic volatile constituents in styrene copolymers, emulsions and lattices. In this case a gas chromatographic method for the determination of styrene monomer in polystyrene has been discussed for the determination of other volatiles in the polymer. In this procedure a known weight of polymer is dissolved in benzene and a portion of the solution injected on to a column of diisodecyl phthalate on Chromosorb W support. Argon is used as the carrier gas with an argon ionisation detector (with radium source).

Two methods, both based on gas chromatography, for the determination of styrene monomer and other volatiles in polystyrene. In one method an orthodichlorobenzene solution of the polymer is distilled to isolate volatiles as a concentrate in the distillate. The orthodichlorobenzene used to dissolve the polymer contains a known amount of toluene which is used as an internal standard. The distillate is chromatographed on a polyethylene glycol column using helium as a carrier gas and a katharometer detector. This method is claimed to determine styrene and ethyl benzene in polystyrene. In their second method, Pfab and Noffz dissolve the polymer in methylene dichloride containing a known amount of 1-phenyl butane as internal standard.

The polymer is then reprecipitated by the addition of excess methyl alcohol. The filtrate is chromatographed as described above, except that a flame ionisation detector is used instead of a kathorometer in order to increase the overall sensitivity of the procedure. This method is claimed to be capable of determining styrene monomer, ethyl benzene, cumene and xylenes.

More recently, a gas chromatographic method for the determination of various volatiles in polystyrene, styrene-acrylonitrile copolymers, styrene-acrylonitrile-butadiene termpolymers and other copolymers is used. In this procedure, the polymer is dissolved in dimethyl formamide containing a known amount of toluene as internal standard. A portion of this solution is injected into two columns in series comprising 20% Tween 81 on Chromosorb W, followed by 10% Resoflex-446 on Chromosorb W. Using a hydrogen flame ionisation detector, it is claimed that less than 10 ppm of various monomer and other volatile impurities can be determined in the polymer by this procedure. It has been investigated that the polymer present in the solution injected into the gas chromatographic column deposits on the injection block and is removed by reaming after every fifty sample injections.

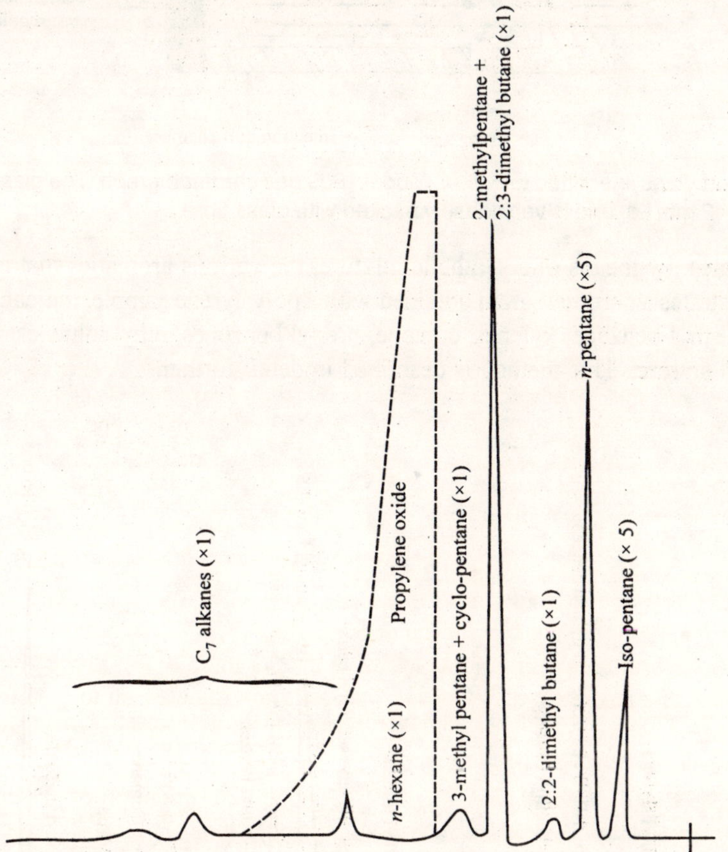

Fig. 13.12. Gas chromatogram of expanding agents liberated from expandable polystyrene at 200°C for 15 min in helium.

Recently few scientists have extended the gas chromatographic technique to the determination in polystyrene of styrene and a wide range of other aromatic volatiles in amounts down to the 10 ppm level. In this method a weighed portion of the sample is dissolved in propylene oxide containing a known concentration of pure n-undecane as an internal standard. After allowing any insolubles to settle an approximately measured volume of the solution is injected into the chromatographic column which contains 10% Chromosorb 15–20 M supported on 60–70 BS Celite. Helium is used as carrier gas and a hydrogen flame ionisation detector is employed.

In Fig. 13.13 is shown a device which is connected to the injection port of the gas chromatograph in order to prevent the deposition of polymeric material in the injection port of the chromatograph with consequent blockages. When a solution of polystyrene is injected into the liner, polymer is retained by the glass fibre and volatile components are swept on to the chromatographic column by the carrier gas.

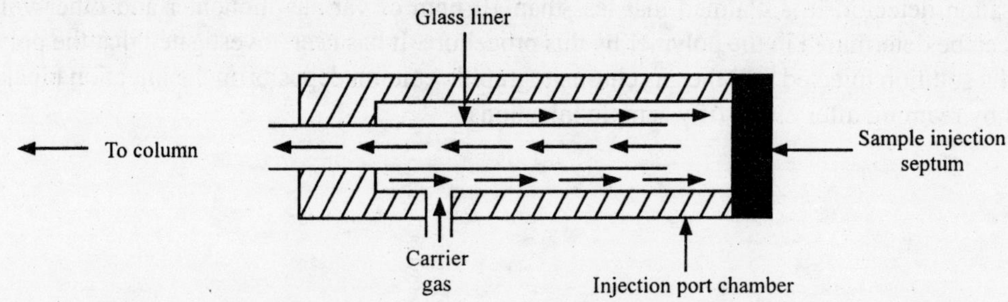

Fig. 13.13. Injection port glass liner fitted to F. & M. Model 1609 gas chromatograph. The glass liner measures 60 mm × 40 mm o.d. × 2 mm i.d. and is very loosely packed with glass fibre.

Fig. 13.14 illustrates, by means of a synthetic mixture, the various aromatics that can be resolved, whilst Fig. 13.15 illustrates a chromatogram obtained with a polystyrene sample, indicating the presence of benzene, toluene, ethyl benzene, xylenes, cumene, propyl benzenes, ethyl toluenes, butyl benzenes, styrene and α-methyl styrene. This method is described in detail further.

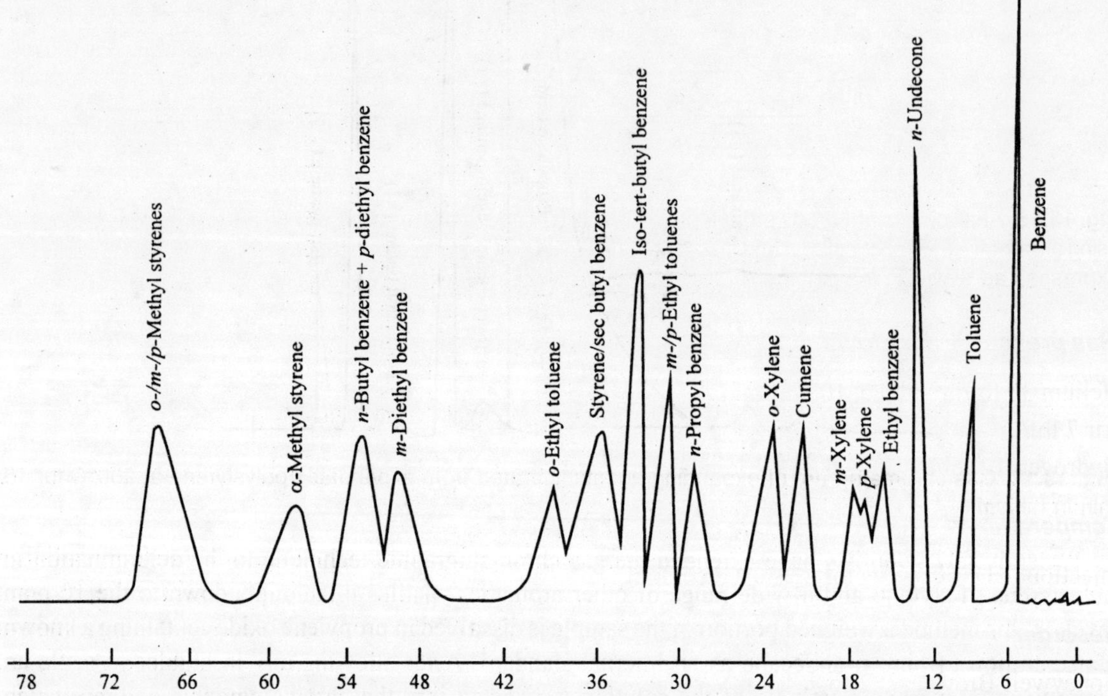

Fig. 13.14. Gas chromatogram of synthetic blend of hydrocarbons likely to occur in polystyrene on a Carbowax 15–20M column at 80°.

Column

Copper tube (15 ft × 1/4 in. o.d. × 3/16 in. i.d.) packed with 10% w/w Carbowax 15–20 M supported on 60–72 B.S. mesh Celite.

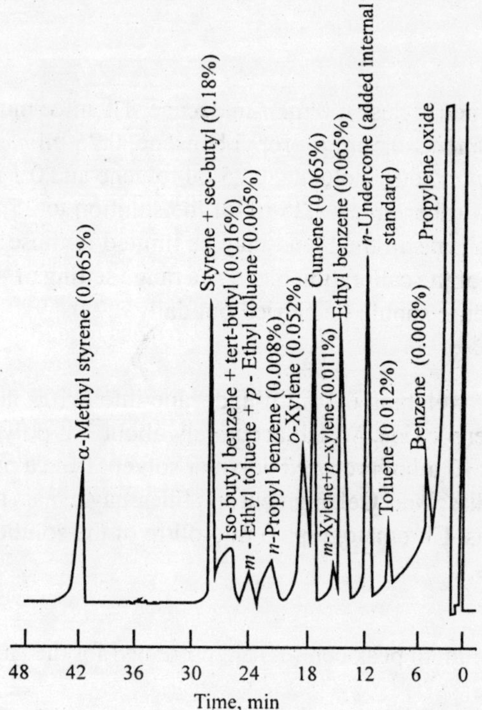

Fig. 13.15. Analysis of anhydrous propylene oxide solution of polystyrene for residual aromatic hydrocarbons—using a Carbowax 15–20M column and glass liner in injection port. Internal standard n-undecane. Concentrations are expressed as % w/w in the polymer.

Gas pressures and flows

Helium 30 lbf/in^2 gauge, 100 ml/min.
Air 7 lbf/in^2 gauge, 650 ml/min.
Hydrogen 12 lbf/in^2 gauge, 75 ml/min.

Temperatures

Injection – 150°C; detector – 150°C; column – 80°C; and flame – 200°C.

Recorder

Honeywell-Brown, –0.2 to + 1.0 mV range, 1 sec. response, 15 in. (381 mm)/hr chart speed.

Reagents

Carbowax 15–20 M and 60–72 B.S. mesh Celite Styrene and Propylene oxide. *Ortho*-, *meta*- and *para*-xylenes, cumene, *n*-propyl benzene, isobutyl benzene, α-methyl styrene and mixed *meta*/*para*-diethyl benzenes, Tert-, sec- and *n*-butyl benzenes, *ortho*-methyl styrene, mixed *meta*/*para*-methyl styrenes, *meta*- and *para*-ethyl toluenes, *Ortho*-ethyl toluene.

Analytical procedure

Calibration

Weigh in turn into a 10 ml volumetric flask 1.0 ml *n*-undecane, 1.0 ml α-methyl styrene, 1.0 ml styrene, 0.5 ml ethyl benzene, 0.5 ml cumene, 0.5 ml *n*-propyl benzene, 0.25 ml *meta*- or *para*-xylene, 0.25 ml *ortho*-xylene, 0.25 ml *meta*- or *para*-ethyl toluene, 0.25 ml toluene and 0.1 ml benzene. Dilute to 10 ml with propylene oxide and then further dilute 0.25 ml of the solution to 25 ml with propylene oxide. If necessary, the aromatic hydrocarbons in the blend may be limited to those it is specifically required to determine. Chromatograph 5 µl of the calibration blend at a range setting of × 10 and suitable attenuation settings. This calibration procedure should be carried out daily.

Analysis

Prepare the polymer solvent by weighing 0.1 ml n-undecane into a 100 ml volumetric flask and then diluting to 100 ml with propylene oxide. Weigh accurately about 1 g polymer into a 25-ml stoppered measuring cylinder, add exactly 10 ml of prepared polymer solvent from a pipette, seal with a serum cap and shake until the polymer has dissolved. Gel, pigment and filler may remain as an undissolved suspension without detriment to the analysis. Chromatograph 5 microlitre of the solution at a range setting of × 10 and suitable attenuation settings.

Relative retention distances

Retention distances (injection point to peak centres) are corrected for the gas hold-up of the column and are expressed relative to styrene:

Benzene	0.17	*tert*-Butyl benzene	0.92
Toluene	0.29	*sec*-Butyl benzene	1.00
n-Undecane	0.40	Styrene	1.00
Ethyl benzene	0.47	*ortho*-Ethyl toluene	1.05
meta-Xylene	0.50	*meta*-Diethyl benzene	1.35
para-Xylene	0.50	*para*-Diethyl benzene	1.44
Cumene	0.62	*n*-Butyl benzene	1.44
ortho-Xylene	0.66	α-Methyl styrene	1.60
n-Propyl benzene	0.77	*ortho*-Methyl styrene	1.86
meta-Ethyl toluene	0.84	*meta*-Methyl styrene	1.86
para-Ethyl toluene	0.84	*para*-Methyl styrene	1.86
iso-butyl benzene	0.92		

Calculation

On the calibration and analysis chromatograms, measure the peak heights of the n-undecane and the aromatic hydrocarbons allowing for any attenuation factors. From the calibration chromatogram, determine the response fraction for each component as follows :

A = weight of component in calibration blend,

B = weight of n-undecane in calibration blend,

C = peak height of component on calibration chromatograph,

D = peak height of *n*-undecane on calibration chromatograph.

$$\text{Component response factor, } F = \frac{A}{B} \times \frac{D}{C}$$

From the analysis chromatogram, determine the concentration of each component as follows :

f = component response factor,

g = peak height of component on analysis chromatogram,

h = peak height of n-undecane on analysis chromatogram,

j = % w/w n-undecane in polymer solvent,

k = weight (g) of polymer taken for analysis.

$$\text{Component \% w / w} = \frac{10 \times G \times F \times J}{H \times K}$$

Gas chromatography is eminently suitable for the determination of volatile expanding agents present in expanded cellular grades of polystyrene. Normal and isopentane are commonly used for this purpose. The method quoted below enables these and wide range of other volatiles to be determined in polystyrene with a reasonable degree of accuracy.

The sample is chromatographed as a solution in methylene dichloride in the presence of 2:4-dimethyl pentane as an internal standard. The concentrations of the components are determined from peak height ratios and response factors.

Operating conditions

Instrument:	F & M Model 1609 or equivalent instrument with injection liner (Fig. 13.13).
Column:	Copper tube (10 ft × 1/4 in. o.d. × 3/16 in. i.d.) packed with 25% w/w di-*n*-butyl phthalate on 44–60 Celite.
Gases:Helium	30 psig, rotameter at 4.0
Air	10 psig, rotameter at 2.0
Hydrogen	7 1/2 psig, rotameter at 3.0
Temperatures:	°C
Injcction	165
Column	50
Detector	125
Flame	170
Recorder:	1 mV, 24 in./hr.

To a 100 ml volumetric flask add 95 ml of a suitable solvent. Seal with a serum cap and weigh. Through the serum cap, add in turn 0.25 ml methyl cyclopentane, 1 ml 2:4-dimethyl pentane, 0.25 ml *n*-hexane, 0.5 ml 3-methyl pentane, 1 ml 2-methyl pentane, 0.5 ml 2:2=dimethyl butane 0.5 ml *n*-pentane and 0.5 ml isopentane, re-weighing after each addition. Include in this calibration mixture only those substances it is required to determine in the polystyrene sample. Chromatograph 5 μl at a range of × 100 and suitable attenuations.

Weigh 1 g polymer sample into a 25 ml measuring cylinder, add 10 ml of a 1% v/v solution of 2:4-dimethyl butane in solvent, seal with a serum cap and shake to dissolve. Chromatograph 5 μl at a range of × 100 and suitable attenuations.

Measure the peak heights of all the components, allowing for any attenuation factors, on both the calibration and analysis chromatograms.

From the calibration chromatogram, calculate response factors as follows:
P = peak height of component,
H = peak height of 2:4-dimethyl pentane,
A = weight of component in calibration mixture,
B = weight of 2:4-dimethyl pentane in calibration mixture.

$$\text{Response factor, } F = \frac{A}{B} \times \frac{H}{P}$$

From the analysis chromatogram, calculate the concentration of each component as follows:
W = % w/w 2:4-dimethyl pentane in solvent.
S = weight of sample taken,
C = peak height of component,
D = peak height of 2:4-dimethyl pentane,
F = response factor for component.

$$\text{Component (\% w / w)} = \frac{C}{D} \times \frac{10 \times W}{S} \times F.$$

The following are uncorrected retention distances (injection point to peak apex) and the corresponding relative values (2:4-dimethyl pentane = 1.00).

Another method is used in the determination of residual monomers, non-polymerisable polymer volatiles and additives in styrene based resins by a procedure involving chromatography of a solution or dispersion of the sample in dimethyl formamide on a tandem column made up of 3 ft of 20% Tween 81 followed by 10 ft of 10% Resoflex 446 at a column temperature of 120°C. *Stevens* describes the determination of acrolein and furfural monomers in phenol/acrolein and phenol/furfural resins and also the determination of phenol and formaldehyde in phenol formaldehyde resins.

Phenol is estimated on a silicone SF-96 column, using *m*-cresol as internal standard, whilst formaldehyde is estimated on a sucrose octa-acetate column, using *n*-butanol as the internal standard. *Priori* and *Panetti* describe the determination of free phenol, cresols and xylenols in phenoplasts by first separating the phenolics chemically and then completing the analysis gas chromatographically on a silicone oil column.

	DR	
Component	*mm*	*rel.*
Isopentane	21 1/2	0.30
n-Pentane	25 1/2	0.36
2:2-Dimethyl butane	33 1/2	0.46
2-Methyl pentane + 2:3-dimethyl butane	44	0.61
3-Methyl pentane + cyclopentane	50 1/2	0.70
n-Hexane	57 1/2	0.80
2:4-Dimethyl pentane	72	1.00
Methyl cyclopentane	84 1/2	1.17

Note 1. The following solvents are recommended:

Present in sample	*Solvent*
n-pentane	propylene oxide
isopentane	
2:2-dimethyl butane	
isobutane	
cyclopentane	methylene dichloride
isopentane	
n-pentane	
2:2-dimethyl butane	
2-methyl pentane	
2:3-dimethyl butane	
3-methyl pentane	
cyclopentane	
n-hexane	
2:4-dimethyl pentane	
methyl cyclopentane	
iso-octane	benzene

Note 2. The following internal standards are recommended when paraffins up to C_5 only are present in the polymer.

Present in sample	*Internal standard*
isobutane and/or 2:2-dimethyl butane	*n*-pentane
n-and/or isopentane	2:2-dimethyl butane
iso-octane	methyl cyclopentane

Gas chromatography has been successfully applied to the determination of monomers other than styrene and other types of volatiles in various types of polymers, and rubbers. *Tweet* and *Milelr* determined monomers in ethyl acrylate/styrene copolymer emulsions by distilling off the monomers in the presence of toluene and then examining on a silicone grease column using toluene as the internal standard.

Brodsky and *Nelson* discussed the determination of monomers, including styrene, in latex samples. These procedures involve the direct injection of the latex into the gas chromatographic column. *Haslam* identified mixed solvents, and impurities in plastics and in plastic adhesives by a combination of gas chromatography and infrared spectroscopy, and chemical analysis.

CHAPTER 14

Application of Chromatography for Analysis of Unknown Additives

INTRODUCTION

A full knowledge of all the types of additive present in the polymer helps in the selection of tests methods, however, due allowances can be made for other types of additive constituents present or of any decomposition products of additives present. Whilst this information might be to hand if an analyst is examining materials of known origin, this would not always be so. In such cases, it is mandatory that the first step must be to completely identify the additives present, before any consideration can be given to the problem of selecting or devising a method of quantitative analysis for any constituent present in the polymer.

The problem resolves itself into the preparation of a total solvent extract of the polymer in which all additives are completely recovered, followed by separation of the mixture into purge single components by a suitable form of chromatography and, finally, by identification of each separated pure component by suitable means, usually involving visible infrared, ultraviolet, mass or nuclear magnetic resonance spectroscopy and perhaps micro-analysis for elements present.

Only after this stage can the details of the quantitative determination of particular polymer constituents be considered.

It is advisable when commencing the analysis of a polymer for unknown additives to determine first its content of various non-metallic and metallic elements. Any element found to be present must be accounted for in the subsequent examination for, and identification of, additives. Hence, elemental analysis reduces the possibility of overlooking any additive which contains elements other than carbon, hydrogen and oxygen.

The analytical methods used to determine elements should be sufficiently sensitive to determine about 10 ppm of an element in the polymer, i.e., should be able to detect in a polymer, a substance present at 0.01% and containing down to 10% of the element in question.

This requirement is met for almost all the important elements by use of optical emission spectrography and X-ray fluorescence spectrometry. Using these two techniques, all metals and non-metals down to an atomic number of 15 (phosphorus) can be determined at the required concentrations. Nitrogen is determinable at these levels by micro-Kjeldahl digestion techniques.

Three forms of chromatography are worthy of serious consideration for the separation of additive mixtures, viz. column, paper and thin-layer chromatography. Of these, thin-layer chromatography is by far the most useful general technique.

COLUMN CHROMATOGRAPHY

Thin-layer chromatography often provides enough of each of the individual polymer additives in a sufficiently high state of purity to enable them to be identified by spectroscopic methods. However, if insufficient sample is provided by this technique then larger quantities in the 50–500 mg range can be obtained by scaling up the separation on to a column. Column chromatography is more time-consuming than thin-layer chromatography, therefore the latter technique should be used in preliminary experiments aimed at finding suitable adsorbents and development solvents for achieving a satisfactory separation. These conditions can then usually be translated to a column without difficulty. Similarly, the thin-layer technique is useful as a rapid method of monitoring the purity of fractions obtained in column separations.

Successful separations can often be achieved on columns of activated silica gel by successive elution of the sample on the column with a range of solvents of increasing polarity.

A hexane solution of the sample is transferred to the top of the column of gel which has been filled with the same solvent. The column is then successively eluted with *n*-hexane, mixtures of *n*-hexane and benzene (or toluene), neat benzene (or toluene), mixtures of benzene (or toluene) and absolute ethanol (or acetone), and finally absolute ethanol (or acetone). Only solvents of suitable purity should be used in this work, e.g., spectroscopic or chromatographic grades.

This process provides a series of fractions in the collector from which particular groups of fractions can be used for further evaluation. In some cases, of course, some of the fractions will consist of the low molecular weight component of the original polymer.

One of the difficulties of column chromatography is the problem of identifying the fractions in which the separated compounds are concentrated. This can be achieved by the laborious process of examining all the fractions, for example by infrared or ultraviolet spectroscopy or by evaporating to dryness and weighing the residues; or by the less laborious process of monitoring the effluent as it leaves the chromatographic column so that solute-containing fractions from the fraction collector can be picked out from the fractions which do not contain any substances. Several types of effluent monitors are available, based on the measurement of the ultraviolet absorption, conductivity, etc. These have the disadvantage of being too specific for dealing with mixtures of compounds of unknown type. For example, compounds which do not either absorb in the ultraviolet or ionize would be missed using these detectors. The most useful general purpose monitors are those based on the measurement of refractive index and on thermal effects.

This instrument operates on the principle that as each separated compound moves down the column it is accompanied by heat of adsorption and desorption due to interaction between solute molecules and the stationary phases. These heat pockets (i.e. separated compounds) are detected by a thermistor at the column outlet and recorded on a strip chart which can be operated in conjunction with a fraction collector. Thus, separated fractions can be readily located and bulked if necessary for further examination.

This apparatus has been used to separate up to 200 mg of anti-oxidants from each other using a silica gel column. Santonox R (4,4′-thio-bis-(3-methyl-6-tert-butyl phenol) and Ionol CP were easily separated on this scale using benzene as the development solvent. Taponol CA and diluauryl thiodipropionate were well resolved using cyclohexane : ethyl acetate (9:1 v/v).

Separation of Mixture of Additives

Adsorbents other than silica gel can be used for the separation of mixtures of polymer additives : column chromatography on neutral alumina for the separation of antioxidants, plasticisers, etc., in rubber extracts is another technique used. The separated compounds can be detected by monitoring the effluent with a LKB 254 mμ ultraviolet detector Fig. 14.1. In this procedure a carbon tetrachloride solution of the sample is applied to an alumina column wetted with the same solvent and the column is successively eluted with carbon tetrachloride, mixtures of carbon tetrachloride and benzene, benzene, mixtures of benzene and absolute ethanol, and finally, ethanol. Separations were carried out on a scale to provide enough to each separated compound for the preparation of infrared and ultraviolet spectra.

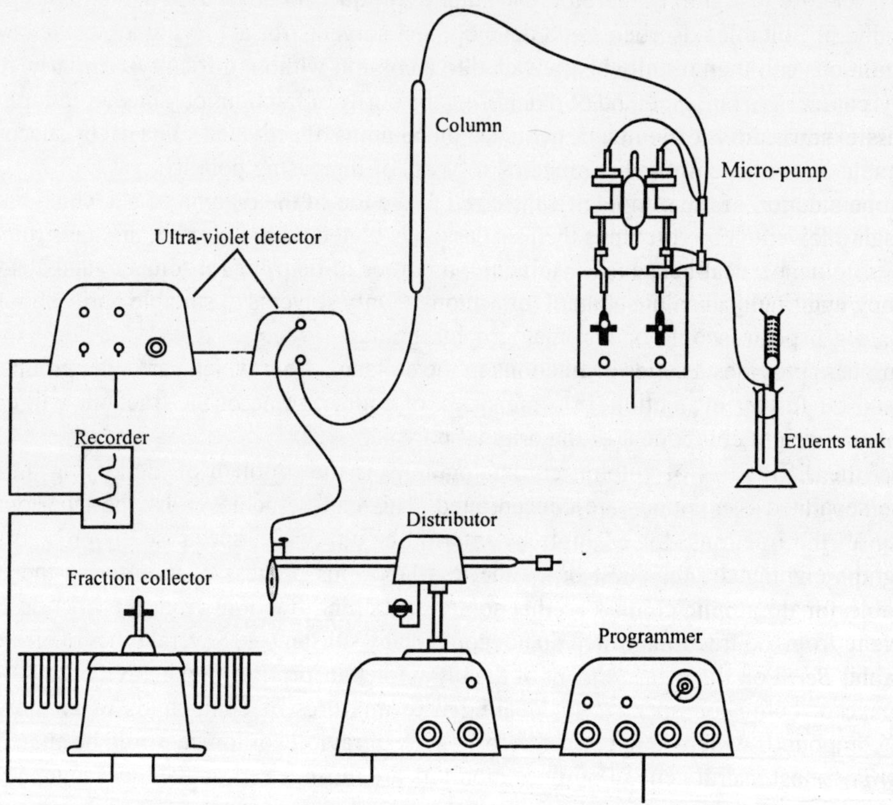

Fig. 14.1. Experimental system employed in column chromatography of rubber extracts.

A synthetic rubber absorbent is used to separate antioxidants. The stationary phase is applied to the column as a suspension in light petroleum and separation carried out with a mobile phase of 21 % v/v acetone in water. An alumina column to separate phenolic antioxidants with chloroform followed by 10% v/v water in ethanol is also used.

A separatory analytical scheme based on column chromatography in which aliquots of the sample solution were chromatographed on seven alumina columns, each with a different mobile phase is also reported. From the position of the components on the various columns after a suitable elution time and from the colours obtained with specific detecting agents, almost unambiguous identification of antioxidants can be obtained.

Identification of Antioxidants in Rubber Vulcanisates

First, the sample extract are chromatographed on an alumina column. Separated compounds are detected on the column by ultraviolet light and/or by extruding the moist column from the tube and streaking a narrow band down the side of the column with various chromogenic reagents such as sulphuric acid, 1% ammonium vanadate in sulphuric acid, 1%, potassium dichromate in sulphuric acid or 1:3 v/v nitric:sulphuric acid. Discussed here are, the behaviour on the column of certain individual antioxidants such as phenyl-alpha-naphthylamine, phenyl-beta-naphthylamine, Agerite White, Neozone HF, Flectol H, and Agerite Stalite. These compounds are generally easily eluted from an alumina column with 1:99 ethanol:benzene and hence could be easily separated from other less mobile constituents of the original mixture.

Infrared and ultraviolet spectroscopy can be utilised for the examination of individual components of vulcanisate extracts, mainly antioxidants and accelerators, after separation from all other compounds present by column chromatography. After chromatography, portions of the eluate are evaporated to dryness and examined by infrared spectroscopy. Mixtures of the following types of compounds are in many cases successfully dealt with; thiazoles, dithiocarbamates, thiurams, xanthates, guanidines, aldehyde amine compounds and various types of amine and phenolic antioxidants.

Although infrared methods offer a reasonable solution to the problem of the qualitative analysis of vulcanisates for accelerators and antioxidants, it is unlikely that they could be made quantitative. Ultraviolet spectroscopy would be more amenable to quantitative analysis and more sensitive for the determination of accelerators and antioxidants.

The chromatographic behaviour on silica gel/Celite packed columns of 32 accelerators and antioxidants of the type used in vulcanised rubber formulations has been explored. Methylene chloride is used as sample solvent. The separated compounds by viewing the developed column in ultraviolet light, by application of various chromogenic streaking reagents to the extruded chromatographic column and by examination by ultraviolet spectroscopy. Silica gel/Celite mixtures as adsorbent have certain advantages over alumina. Thus, it has a weak adsorptive power for accelerators and antioxidants, which permits the chromatography of labile compounds without decomposition. Also, silica gel/Celite is claimed to be more suitable for the application of a wide variety of streak reagents, and is more amenable to the quantitative recovery of adsorbates from the column. Table 14.1 lists the types of compounds discussed by *Parker* and *Berriman*.

Table 14.1. Compounds studied by *Parker* and *Berriman*.

Accelerator or antioxidant	*Reputed composition*
Group A	
Vulcafor DOTG	di-*o*-tolyl guanidine
Vulcafor DPG	diphenyl guanidine
Vulcafor TPG	triphenyl guanidine
Vulcafor TC	thiocarbanilide
Vulcafor MBT	mercaptobenzthiazole
Vulcafor MBTS	dibenzthiazolyl-2-disulphide
Vulcafor TMT	tetramethylthiuram disulphide
Vulcafor MS	tetramethylthiuram monosulphide

(Contd...)

Accelerator or antioxidant	Reputed composition
Vulcafor TET	tetraethylthiuram disulphide
Santocure	2-benzthiazolyl-N'-cyclohexyl sulphenamide
Neozone A (PAN)	phenyl α-naphthylamine
Neozone D (PBN)	phenyl β-naphthylamine
Agerite white	sym-di-β-naphthyl-p-phenyenediamine
MTD	m-toluylene diamine
DPPD	sym-diphenyl-p-phenylenediamine
Group B	
ZMC	zinc dimethyldithiocarbamate
Vulcafor ZDC	zinc diethyldithiocarbamate
Vulcafor ZNBC	zinc di-n-butyldithiocarbamate
Vulcafor MA	formaldehyde-aniline condensation product
Vulcafor RN	acetaldehyde-aniline condensation product
Nonox NS	phenol-aldehyde-ketone
Nonox S	aldol-naphthylamine condensation product
Flectol H	polymerised trimethyl dihydroquinoline
Perflectol	(found Flectol H and DPPD)
Neozone C	(found m-toluylene diamine (MTD) and phenyl α-naphthylamine)
Neozone HF	(found phenyl β-naphthylamine and symdi-phenyl-p-phenylenediamine (DPPD))
Ureka white	(found MBT and quanidine derivative)
Vulcafor DAU	(found TET and MBTS)
Vulcafor DAW	(found MBT and quanidine derivative)
Vulcafor DHC	(found MBT and ZDC)
Vulcafor F	(found MBTS and guanidine derivative)
Vulcafor FN	(found MBTS and guanidine derivative)

A number of colour reactions for accelerators and antioxidants also are shown in Table 14.2. Many of these are suitable for application as streak reagents. Table 14.3 lists the reagents used for the detection of the single Group A compounds of Table 14.1.

The limiting sensitivities of the various streak reagents used for detecting the compounds listed in Table 14.1 were obtained by developing increasing quantities of each of the compounds on 1-cm columns by the procedures described below. In most cases the smallest detectable quantity corresponded to 0.05 mg on a 1-cm column under the conditions quoted.

Table 14.2. Some colorimetric reagents.

Reagent	Compounds reacting
$Bi(NO_3)_3$-1% $NaOH$–HNO_3	MBT, thiuram
Aq. $Bi(NO_3)_3$ in acetone	MBT
$AuCl_3$	DPG
Cu oleate in $CHCl_3$	dithiocarbamates
Ditto, after Na_2SO_3	thiuram sulphide

(Contd...)

Reagent	*Compounds reacting*
Co-oleate in benzene	DPG, DOTG, *o*-tolyldiguanide dithiocarbamates, MBT, TMT
$CuSO_4$ aq. + acetone, etc.	dithiocarbamates, thiurams, etc.
Phenolphthalein	DPG
HCl and phenol or α-naphthol	diazoaminobenzene
Diazotized *p*-nitraniline	aromatic amines
NaOH and diazotized *p*-nitraniline	aromatic amines
Diazotized sulphanilic acid	aldol naphthylamine
Acetic acid or HCl and *p*-dimethyl-amino benzaldehyde	1-naphthyliminoaldol PBN
p-Phenylenediamine, Br_2 and NH_3	1-naphthyliminoaldol, thiuram
p-Phenylenediamine and $FeCl_3$	primary and various amines
$FeCl_3$ or $CuCl_2$	aldol-naphthylamine
Aq. NaOCl + phenol (3%)	aldehyde-aniline condensation products
$SnCl_4$ and amyl nitrite in benzene	diarylamine and naphthylarylamines
SnCl4 and benzotrichloride in ethylene dichloride	diarylamine-ketone condensation products
$SnCl_4$ benzoyl peroxide in benzene	aryl substituted *p*-phenylene diamines
$SnCl_4$ and bromine in ethylene dichloride	aniline-acetone condensation produts, etc.
H_2SO_4 and trace HNO_3	diphenyl and dinaphthyl *p*-phenylenediamines
H_2SO_4	
H_2SO_4 + SeO_3	
H_2SO_4 + $K_2S_2O_2$	
Conc. HNO_3	reactions of 40 commercial antioxidants investigated
Arsenic acid in H_2SO_4	
Amm. molybdate in H_3SO_4	
10% H_2O_2 in H_2SO_4	
H_2SO_4	
HNO_3	
$(NH_4)_2S_2O_8$ in H_2SO_4	
0.5 % MoO_3 in H_2SO_4	reactions of 8 commercial anti-oxidants investigated
0.5% SeO_2 in H_2SO_4	
Acetic acid	
Acetic acid + bromine	
1 % ammonium vanadate in conc. H_2SO_4	
1 % potassium dichromate in conc. H_2SO_4	reactions with 9 commercial anti-oxidants recorded
Nitric acid/sulphuric acid 1:3	

For qualitative work columns of silica gel–Celite mixture 1 cm in diameter and 15 cm long are employed. 1.0-mg quantities of most of the compounds form well-defined zones on such columns and in most cases very strong streak tests are obtained with this quantity. Methylene chloride readily dissolves all the compounds concerned and its developing power is not unduly great. However, it is not suitable for placing very weakly adsorbed compounds such as the phenyl-naphthylamines. Light petroleum or carbon tetrachloride is suitable for these compounds.

Table 14.3. Colours of streaks obtained with various reagents.

Compound	*30% w/v Sodium hypochlorite*	*5% w/v Aqueous $CuSO_{4.5}H_2O$*	*5% w/v $Bi(NO_2)_3$ in 0.5 N nitric acid*	*Bismuth nitrate in 0.5 N nitric acid after reduction*	*5% w/v Aqueous $Pb(C_2H_3O_2)_2$ $_3H_2O$*	*Aqueous lead acetate after reduction*	*1% w/v $(NH_4)VO_3$ in 60% w/w sulphuric acid*	*Mixture of conc. HNO_3 (1 vol.) and conc. H_2SO_4 (3 vols.)*	*0.5% w/v Selenium dioxide in conc. sulphuric acid*
Vulcafor DOTG	dark reddish-brown	nil	nil	nil	nil	nil	nil	nil	nil
Vulcafor DPG	dark reddish-brown	nil	nil	nil	nil	nil	nil	nil	nil
Vulcafor TPG	reddish-brown	nil	nil	nil	nil	nil	nil	nil	nil
Vulcafor TC	pale orange on standing	light brown	yellow	yellow	nil	nil	nil	pale violet fades rapidly	nil
Vulcafor MBT	nil	faint yellow	bright chrome-yellow	bright chrome-yellow	lemon-yellow	lemon-yellow	faint green	nil	faint yellow
Vulcafor MBTS	nil	nil	nil	bright chrome-yellow	nil	lemon-yellow	nil	nil	nil
Vulcafor TMT	nil	bright yellow-green	pale lemon-yellow	pale lemon-yellow	nil	nil	v. pale green to faint blue	nil	nil
Vulcafor MS	nil	strong yellow	pale yellow	pale yellow	nil	nil	v. pale green to faint blue	nil	nil
Vulcafor TET	nil	bright yellow-green	pale lemon-yellow	pale lemon-yellow	nil	nil	v. pale green to faint blue	nil	nil
Santocure	nil	nil	nil	bright chrome-yellow	nil	lemon-yellow	faint green	nil	nil
PAN	light orange to orange-yellow	nil	nil	nil	nil	nil	prussian blue	dark olive-green	blue on standing
PBN	orange	nil	nil	nil	nil	nil	dark brown	green rapidly turning brown	pale greenish-yellow
Agerite white	orange-pink	nil	pale green or pale blue	nil	nil	nil	dark greenish-blue	mauve	deep blue
MTD	orange-brown	yellow-green	nil	–	nil	–	pink-brown on standing	faint orange	nil
DPPD	pale orange-yellow	nil	light blue	–	nil	–	crimson	magenta	purple

Table 14.4. Adsorption series of some simple accelerators and antioxidants from four binary developers.

(These data apply when the compounds are placed on the column in methylene chloride: some of these compounds giving diffuse zones under these conditions give tight zones when placed in carbon tetrachloride, e.g. Agerite white).

Benzene/light petroleum		*Ethyl ether/light petroleum*		*Ethyl acetate/benzene*		*Ethyl acetate/carbon tetrachloride*	
Strongly adsorbed	{ DOTG DPG TPG MDT }	100%	{ DOTG[a] DPG[a] }	strongly adsorbed	{ DOTG DPG }	strongly adsorbed	{ DOTG DPG }
		100%	MTD	40%	MTD	50%	MTD
		50%	TPG[b]	20%	TP[b]	50%	TPG[b]
100% (slightly developed)	{ TC MS }	40% 30%	{ MS TMT } { TMT TC }	1 1/2%	{ TC MS }	5–8%	{ MS TC TMT }
100%	{ TMT MBT TET[b] }	20%	MBT	1/2%	{ TMT MBT }	2%	MBT
100%	{ TET[b] MBTS }	10%	TET		{ MBT MBTS TET }	2%	TET
		10%	{ Agerite white[b] DPPD[b] }	1/2%	TET	1%	{ Santocure MBTS }
100%	Santocure	10%	{ MBTS Santocure }	0%	Santocure		
30%	{ Agerite white[b] DPPD[b] }			0%	Agerite white[c]	1%	{ Agerite white[b] DPPD[b] }
				eluted	DPPD		
15%	{ PBN PAN }	1%	PBN	eluted	PBN	0%	{ PBN PAN (mainly eluted with 25 ml) }
		1%	PAN	eluted	PAN		

[a] Diffuse zone at top of column.
[b] Formation of a very diffuse zone covering a large proportion of the column.
[c] Mainly eluted but a pink colour remains on column.
Compounds whose zones overlap are bracketted.

All chlorohydrocarbons are washed and dried before use as the development of traces of chlorine produced serious decomposition of many solutes. Table 14.4 presents the adsorption series obtained with the four binary solvent mixtures which are most fully investigated.

The concentration of the more polar constituent of the development solvent, which produce a convenient development of the zone, is shown against each compound. Where compounds are not completely separated by one binary developer, a good separation can often be obtained by the use of a different binary solvent. For example, using mixtures of ethyl acetate and benzene and zone of TMT overlaps that of MBT, but with ethyl ether in light petroleum the TMT can be cleanly separated from the MBT. With the above chromatographic data as a guide, it was possible to devise a series of chromatographic treatments for obtaining optimum separation of each of the compounds in Table 14.11 from others giving interfering streak reactions.

Paraffin wax, mineral jelly and liquid paraffin, all possible constituents of polymer and rubber extracts, are eluted from the column even with light petroleum alone and can thus be separated from accelerators and antioxidants. Any impurities they contain produce little or no colour on the column and do not therefore interfere with the detection of the accelerators and antioxidants. Stearic acid can be detected by streaking the column with bromocresol green indicator solution; the acid changed the colour of the streak from greenish-blue to yellow-green. Stearic acid is completely eluted with 20% ethyl alcohol in light petroleum or 20% ethyl ether in light petroleum, and can be recovered quantitatively. With the weaker developers, stearic acid forms wide diffuse zones but does not interfere with the identification of the other compounds. Another column chromatographic procedure for the identification and semi-quantitative determination of plasticisers in PVC that can be used is as follows:

The plasticiser is first Soxhlet extracted from 1 to 2 g of PVC sample using anhydrous diethyl ether. Ether is then evaporated from the extract and residual traces of PVC precipitated by the addition of 2 ml absolute ethanol. Following filtration of any polymer, the ethanol is finally evaporated off to provide a PVC free plasticiser extract.

Chromatographic separation of the mixture is achieved on a column comprising equal parts by weight of Celite 545 (100–200 mesh) and silica gel (100–200 mesh), both dried immediately before use for 24 hr at 100–110°C. A mixture of 0.5 g polymer extract and 10 ml carbon tetrachloride is then applied to a carbon tetrachloride wetted column of the Celite-silica gel mixture. Various elution solvents were used. The column eluate is collected in fractions which are evaporated and weighed. It is seen from Table 14.5 that good separations are achieved by this method for various plasticiser mixtures.

Selected fractions isolated on the chromatographic column are used for the preparation of infrared spectra which enable the various components to be identified.

PAPER CHROMATOGRAPHY

Applications of paper chromatography has find a definite place in the examination of polymers for antioxidants. Since most antioxidants are highly polar, they cannot be efficiently separated on normal paper except by the use of highly polar mobile phases. Consequently reversed-paper chromatography or acetylated papers are frequently used to reduce the effects of 'tailing'.

The detecting spray reagents generally used for antioxidants are either diazotized amines which form coloured products with amines and phenols, or are oxidising agents, since the oxidation products of antioxidants are generally highly coloured. Sometimes the sample solution is treated with the colouring reagent first, and the coloured products are then chromatographed, but multiple spots can be obtained from a single antioxidant in this way. A comprehensive scheme for the systematic identification of antioxidants and accelerators by paper chromatographic methods is given in Table 14.6.

Table 14.5. Column chromatography of PVC plasticisers elution sequence for plasticiser mixture.

Eluant	*Fraction*	*Plasticiser found*
Carbon tetrachloride	1	Cereclor (chlorinated hydrocarbon)
	2	—
	3	—
	4	Mesamoll (alkyl ester of a sulphonic acid)
Carbon tetrachloride/di-isopropyl ether (2%)	5	—
	6	Tritolyl phosphate
	7	Di-*n*-butyl phthalate
	8	—
Carbon tetrachloride/di-isopropyl ether (5%)	9	Di-*n*-butyl sebacate
	10	—
	11	—
	12	—
Carbon tetrachloride/acetone (2%)	13–16	—
Carbon tetrachloride/acetone (5%)	17	Diethylene glycol
	18	dibenzoate
	19	Abrac 'A' (epoxidised vegetable oil)
	20	—
Carbon tetrachloride/acetone (7.5%)	21	—
	22	Polypropylene sebacate
	23	Polypropylene sebacate (trace)
	24	—

Table 14.6. Separation of antioxidants by paper chromatography.

Substances separated	*Stationary phase*	*Mobile phase*	*Derivative or treatment*	*Detection*	*Comments*
Gallate's gallic acid butylated hydroxy-anisole	7% liquid paraffin on paper	Light petroleum	—	Ammoniacal Silver nitrate	Descending run–4 hr
Antioxidants in food and fats	Paper	Acetic acid-water (1:4)	—	0.2% $Fe_2(SO_4)_3$–0.1% $K_3Fe(CN)_6$ (1:1)	Blue spots
Amine antioxidants	Paper	Acetic acid-water acetone (3:6:1)	React with 3-methyl-benzo-thiazolin-2-one hydrazone HCl/$FeCl_3$ before chromatography	Products coloured	
Antioxidants	Paper	Not given	Heated under reflux with HCl	Sulphanilic acid-sodium nitrite or ninhydrin	

(*Contd...*)

Substances separated	*Stationary phase*	*Mobile phase*	*Derivative or treatment*	*Detection*	*Comments*
Aromatic amines and phenothiazine anti-oxidants	Dipropylene glycol on paper	Cyclohexane saturated with dipropylene glycol		UV light or *p*-nitrobenzenediazonium fluoroborate	112 μg detected
Antioxidants	Whatman acetylated paper No. AC82	Ethanol-benzene-acetylacetone (10:10:1)	Antioxidants extracted from accelerators with ethanol	Potassium *p*-diazobenzene sulphonate	Ascending against the grain–5 hr
Antioxidants	Acetylated Whatman No. 1	Not reported	Extract into ethanol add 4 M NH_4OH, 20% $SrCl_2$ and filter		
Catechols	What No. 1 impregnated with formamide +H_3PO_4, dimethylformamide or liquid paraffin	(a) Isopropylether (b) Chloroform (c) Heptane (d) Heptane-benzene (1:1) (e) 80% methanol			
Antioxidants and accelerators	Paper		Coupled with *p*-diazobenzene sulphonic acid, or alkali	Products coloured	
Urea-based stabilisers	Paper	Propanol-methanol-water (2:1:1)		*p*-Dimethylamine benzaldehyde	
Basic antioxidants	Acetylated Whatman No. 1	96% Ethanol-benzene (1:1)		4% Benzoyl peroxide in benzene	
Phenolic antioxidants	Acetylated Whatman No. 1	Butyl acetate-pyridine-methanol-water (1:5:1:3)		Tollen's reagent Millon's reagent.	
Basic antioxidants	Schleicher and Schüll 2043b/45ac	96% Ethanol-benzene (1:1)		1% Diazobenzene sulphonic acid (DBS) in 25% aqueous acetic acid or 20 mg. of DBS in 5 ml of 0.1 M NaOH +5 ml of 96% ethanol	
Phenolic antioxidants	Schleicher and Schüll 2043b/45ac	Butyl acetate-pyridine-methanol-water (1:5:1:3)		Tollen's reagent, Millon's reagent, phosphomolybdic acid, vanicillin or potassium ferricyanide	

In the part of this scheme relating to antioxidants acetylated paper and two solvent systems, one for basic and one for phenolic acidic constituents are used. Identification is based mainly on the R_f value of each constituent and on the colours produced by various spay reagents (Tables 14.7 and 14.8). Circular paper chromatography with satisfactory results is also used with the same solvent systems.

Another mėthod employs different solvent systems. The sample extract is evaporated to dryness at 80°C and dissolves the residue in 96% ethanol. Controlled additions of strontium chloride and ammonia solutions are made to precipitate out fatty acid and other impurities which are then removed by filtration and the clear filtrate examined for antioxidants by paper chromatography.

Table 14.7. Paper chromatographic separation of amine antioxidants (Acetylated paper).

Systematic name	*Colour of reaction product*	*Identification limits in micrograms before/after chromatogr. separation*		*Rf values*
1. Phenyl-x-naphthylamine	Light yellow	5	10	0.64
2. Phenyl-β-naphthylamine	Blue-grey	5	20	0.64
3. Diphenyl-p-phenylenediamine	Yellow-orange	< 1	2	0.56
4. Phenyl-cyclohexyl-p-phenylenediamine	Yellow	< 1	10	0.73
5. Di-β-naphthyl-p-phenylene-diamine	Pink	1	5	0.55 (Tailing)
6. p-Isoprop-oxydiphenyl-amine	Yellow-brown	—	—	0.73
7. p,p'-Dimethoxydiphenylamine	Brown-pink	—	—	0.68
8. p-(p-Tolyl-sulphonylamino)-diphenylamine	Brown-red	< 1	5	0.65
9. p-(pTolyl-sulphonylamino)-phenyl-p-tolyl-amine	Red	1	5	0.65
10. Mono- and diheptyldiphenyl-amine	Green	—	—	0.81; 0.91
11. 2,4-Diamino-diphenylamine	Brown	< 1	10	0.38
12. p,p'-Diaminodiphenylmethane	Red-brown	2	40	0.50
13. Diphenyl-ethylenediamine	Red-brown	10	80	0.65
14. Di-o-tolyl-ethylene-diamine	Red-brown	10	80	0.67

A procedure based on paper chromatography for the identification of nitrogen-containing antioxidants in synthetic aviation turbine oil formulations which with minor modification can be applied to the analysis of plastics. The most successful solvent system for chromatography is dipropylene glycol as the stationary phase and cyclohexane saturated with dipropylene glycol as the mobile phase.

Whatman paper is used for separation. A combination of methods is used to locate the separated antioxidants on the chromatogram. The chromatogram is first examined under ultraviolet radiation, when the compounds are detected either by their intense fluorescence or by absorption when they appear as dark spots on the paper. The chromatogram is then sprayed with a 0.05% solution of p-nitrobenzenediazonium fluoborate in acetone to reveal those spots which form azo dyes. It was possible to detect as little as 2 μg of each antioxidant, and in some cases even less than 1 μg is detected. Paper chromatogram is shown in Fig. 14.2.

Table 14.8. paper chromatographic separation of phenolic antioxidants.

Acetylated paper, Mobile phase,
(1) Butyl acetate-pyridine-methanol-water 1:5:1:3.
(2) Isopropanol-methanol-water 3:3:3.
Spray reagents:
(1) Tollen't reagent.
(2) Million's reagent.

Systematic name	*Trade name*	*Identification in micrograms before/after chromatogr. separation (Tollen's reagent)*		*Rf values*	
				Mobile phase 1	*Mobile phase 2*
2,6-Di-tert-butyl-4-methyl phenol	Ionol Deenax	10	80	0.60	0.44
2,2'-Methylene-bis-(4-methyl-6-tert-butyl phenol)	2246	1	10	0.66	0.55
2,5-Di-tert-amyl-hydroquinone	Santovar A	1	40	0.69	0.53 T
2,5-Di-tert-butyl-hydroquinone	Santovar O	1	10	0.71	0.55 T
Hydroquinone-monobenzylether	Agerite Alba	10	10	0.66	0.28 T
4,4'-dihydroxybiphenyl	Alternungs-chutz-mittel D.O.D.	1	2	0.77	0.43 T
4,4'-Thio-bis(6-tert-butyl-3-methyl phenol)	Santowhite Crystals	10	40	0.71	0.63 T
4,4'-Thio-bis-(2,5-di-tert-amyl phenol)	Santowhite L	10	40	0.58	0.70 T
4,4'-Butylidene-bis (6-tert-butyl-3-methyl phenol)	Santowhite Powder	10	40	0.69	0.73 T

Note: T means tailing of the spot.

The use of acetone as a location reagent solvent is advantageous, since it evaporates off the paper very quickly, thus a chromatogram is not wetted and is easy to handle. Other reagents used in paper chromatography are as follows:

1. A solution of bromine in carbon tetrachloride detects phenothiazine as a blue-grey spot, 3,7-dioctylphenothiazine as a red-brown spot, and di-2-pyridylamine as a transient orange spot.
2. If the antioxidant solutions are spotted directly onto an untreated paper, and then sprayed with a solution of tetracyanoethylene in benzene, coloured spots are formed similar to the azo dyes obtained with *p*-nitrobenzenediazonium fluoborate. However, after solvent development of the anti-oxidants on paper impregnated with dipropylene glycol, only phenothiazine was detected as a blue-green spot when sprayed with tetracyanoethylene reagent.

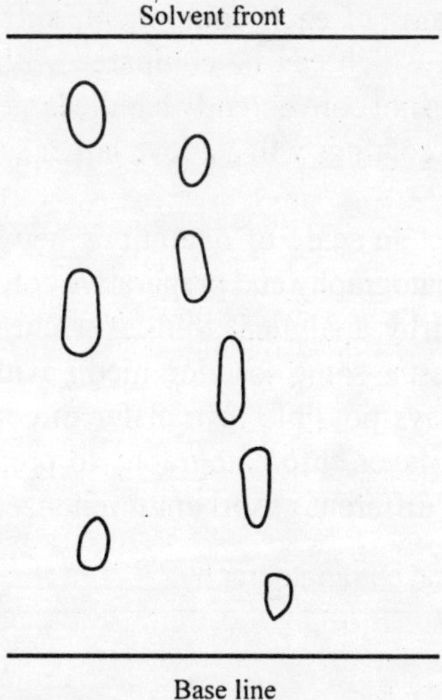

Fig. 14.2. Paper chromatography of nitrogen containing anti-oxidants. Separated antioxidants visualised with *p*-nitrobenzene diazonium fluoride.

The R_f values of the antioxidants and the effects observed under ultraviolet radiation and on spraying with *p*-nitrobenzenediazonium fluoborate are given in Table 14.9.

Table 14.9. The paper chromatographic detection and identification of nitrogen containing antioxidants.

Antioxidant	*Observation under ultraviolet radiation*	*Colour of azo dye*	*Rf values (23° C)*
Diphenylamine	Nil	Orange-red	0.55
Di-(p-octylphenyl)amine	Absorbs	Yellow	0.86
Phenothiazine	Fluoresces*	Pink	0.19
3,7-Dioctylphenothiazine	Absorbs	Brown-yellow	0.82
N-phenyl-α-naphthylamine	Fluoresces	Mauve	0.57
N-phenyl-β-naphthylamine	Fluoresces	Mauve	0.43
Di-2-naphthylamine	Fluoresces	Mauve	0.34
Di-2-pyridylamine	Fluoresces	Nil	0.22

* At low concentration phenothiazine is best detected as a dark spot against a white paper background. Paper chromatograms of the pure antioxidants are shown in Fig. 14.2.

THIN-LAYER CHROMATOGRAPHY AND SPECTROSCOPY

The identification of mixtures of unknown additives in solvent extracts of polymers presents some difficult problems. The solvent extract is usually available in only fairly small quantities, often consists of a complex mixture requiring preliminary separation into pure components before identifications can be attempted, and is frequently a mixture of compounds of completely unknown type.

Thin-layer chromatography using plates coated with 250 μ adsorbent is an excellent technique for efficiently separating quantities up to 20 mg of total additive mixtures into their individual components.

This technique provides a few milligrams of each component, sufficient to prepare a recognisable infrared or ultraviolet and mass spectra which can be compared with the spectra of authentic known compounds. However, the technique does not conveniently handle larger quantities, although preparative thin-layer chromatography using thicker coatings will achieve larger scale separations with some loss of resolution.

Table 14.10 illustrates the differences in scale of operations between normal analytical thin-layer chromatography, preparative layer chromatography and preparative column chromatography. Separations which have been successfully achieved by analytical thin-layer chromatography may be transferred directly to the preparative layer, because the same sorption media with the same grain size are used in each technique. However, it is not always possible to transfer directly separations which have been successfully achieved by analytical thin-layer chromatography to preparative column chromatography with equal success because appreciably different adsorbent grain sizes are used in the two techniques.

Table 14.10. Comparison of layer and column chromatography.

Ingredients	*Analytical TLC*	*Preparative layer chromatography (laboratory scale)*	*Preparative column chromatography (laboratory scale)*
Quantity of substance mixture applied	several micrograms	0.1–50 g	1–250 g
Average grain size of adsorbent (silica gel)	5–25 μ	5–25 μ	50–500 μ
Quality of solvent required	10–100 ml	2–51	5–1001
Purity of substances	high purity	high purity only small inter-mixed zones	less purity larger inter-mixed zones
Duration of development	less than 1 hr	less than 1 day	1 to several days

Thin-layer Chromatography on 250 μ Thick Layers

Preparation of solvent extract of polymer

Polymer extraction procedures using organic solvents do not extract all types of organic additives from polymers; also many inorganic compounds and metallo-organic compounds (e.g. calcium stearate) are insoluble. The presence of metals will have been indicated in the preliminary examination of the polymer. Most types of organic polymer additives, however, can be readily extracted from polymers with organic solvents of various types.

The first step is to solvent-extract the total additives from the polymer in high yield and with minimum contamination by low molecular weight polymer. Extracts should be used for analysis without delay as they may contain light- or oxygen-sensitive compounds. When delay is unavoidable, storage in acitinic glassware under nitrogen in a refrigerator minimises the risk of decomposition.

Additives coated on to the surface of a polymer (e.g., antistatic additives) can be removed easily by washing with a low-boiling solvent which does not attack the polymer. These extracts can be concentrated *in vacuo* at 35–40° prior to further analysis. Total internal plus external additives can be extracted from low and high density polyethylene and polystyrene by procedures involving solution or dispersion of the polymer powder or granules (3 g) in cold redistilled sulphur-free toluene (50–100 ml), followed in the case of polyethylene by refluxing for several hours. Rubber-modified polystyrene does not completely dissolve in toluene if it contains gel. Methyl ethyl ketone or propylene oxide are alternative suitable solvents for polystyrene. Dissolved polymer is then reprecipitated by the addition of methyl alcohol or absolute ethanol (up to 300 ml), and polymer removed by filtration or centrifuging. The additive-containing extract can then be concentrated to dryness as described previously.

Alternative procedures for the extraction of polyethylene and polypropylene involve refluxing with chloroform for 6 hr or contacting with cold diethyl ether for 24 hr or Soxhlet extraction with diethyl ether, methylene dichloride, chloroform or carbon tetrachloride for 6–24 hr followed by concentration of the extract.

Methylene dichloride is a particularly good solvent for polypropylene extractions because of its high volatility and also its small extraction of atatic material from the polymer, compared with other solvents. In addition to additives, most solvents also extract some low molecular weight polymer with subsequent contamination of the extract. To overcome this, polymer-free additive extracts from polyethylene based on low temperature extraction with *n*-hexane at 0° can be obtained. This procedure is also applicable to polypropylene and polystyrene.

The possible complications in additive identification that can arise in thin-layer chromatography due to additive degradation by light, heat or oxygen during solvent extraction operations have been well illustrated in the case of the phenolic antioxidant Ionox 330 (1,3,5-trimethyl-2,4,6-tri-(3,5-di-t-butyl-4-hydroxy benzyl) benzene).

When 10 µl of a 1% methanol solution of this compound is applied immediately after the solution has been prepared at the base of a silica gel coated plate and the chromatogram then developed by elution with 4:1 cyclohexane:benzene and then a suitable spray reagent applied it is found that only one spot, due to undegraded Ionox 330, appeared part way up thc plate. When, however, a 1% methanolic solution of Ionox 330 which has been allowed to stand in air for 10 days is chromatographed under the same conditions it is found that two compounds are present, Ionox 330 and also a degradation product of Ionox 330 which remains unmigrated at the baseline on the chromatogram. Approximately 50% of the Ionox 330 degrades during the 10-days standing period, presumably due to oxidation by air possibly assisted by the ultraviolet component of daylight.

Preparation of thin-layer plates for analysis

Plates (20 cm × 20 cm) coated with 250 µ thickness of Merck silica gel G 254 or GF 254 (ultraviolet fluorescent) are suitable for the separation of polymer additives. These plates can either be prepared in the laboratory using an adjustable spreader or, prepared plates can be obtained commercially.

The glass used to prepare plates should be cleaned with chromic acid, then with deionised water, avoiding detergents, and dried in an air oven prior to coating with silica gel in a clean laboratory atmosphere; conditioning in an air oven for 30 min at 120°C in a vertical position should be followed by storage in a desiccant box until required for use. The plates should preferably be used within 1–2 hr of preparation so that their activity does not change appreciably and to reduce the possibility of contamination of the silica gel layer by volatile impurities in the laboratory atmosphere which might interfere in the subsequent evaluation of the plates.

Thin-layer chromatographic grades of silica gel usually contain traces of organic impurities. If during development of a chromatogram, these impurities migrate to regions of the plate which coincide with the R_f values of separated additives, then the impurities interfere in the interpretation of the plate following spraying with aggressive detection reagents, such as concentrated sulphuric acid and antimony pentachloride, as the organic impurities in the adsorbent also react with these reagents and show up on the sprayed chromatograms. Also, the impurities absorb strongly in the ultraviolet region, especially below 250 mμ. The adsorbent impurities do not have an appreciable adsorption in the infrared region. Thus concentrated chloroform or carbon disulphide extracts of silica gel G have a negligible adsorption over the whole infrared region.

As these impurities are extracted from the silica gel with organic solvents such as diethyl ether, ethanol, acetone, benzene, chloroform and many others, they could occur as contaminants in some of the fractions of separated additives isolated from the plate by solvent extraction of the gel and consequently interfere in the interpretation of the spectra of the additives, particularly in the case of additives which absorb below 250 mμ in the ultraviolet. For this reason an identical blank chromatogram (only sample absent) should always be run in parallel with the sample chromatogram in order to check whether such interference effects exist.

The ultraviolet absorbing impurities in adsorbents are influenced by the nature of the migration solvent. Depending on the polarity of the migration solvent used, the impurities migrate to a greater or lesser extent up the plate towards the solvent front with the result that the lower part of the chromatogram nearer the baseline is cleared of impurities, and the impurities become concentrated in the upper section of the plate nearer the solvent front.

Subsequently, when a section of the absorbent is removed and eluted with a further solvent to recover a separated compound the amount of impurities contaminating the compound will depend on the location of the compound on the chromatogram (R_f value) and contamination may range from negligible to substantial.

The extent to which this redistribution impurity occurs will be influenced by a number of factors, including the type of absorbent, the particular impurities present, and the migration solvent used. In circumstances where slow moving compounds are being separated, the impurities may move away from the polymer additives towards the solvent front and thus not interfere in the subsequent examination of the separated compounds.

The following procedure has been found to be particularly suitable for the premigration of adsorbent impurities from thin-layer plates prior to the use of the plates for separations of polymer additives. In the case of silica gel, at least, methyl alcohol is the recommended solvent and is much superior to *n*-hexane. Methanol is poured into a glass tank with a ground glass lid to a depth of 1–2 cm and the walls of the tank are lined with sheets of filter paper dipping into the solvent. The tank is left for 30–60 min in a draught-free area until the interior has become saturated with solvent vapour. The coated 20 cm × 20 cm plate is then supported vertically in the methanol layer and left until solvent has ascended to the top of the plate (presaturation of the tank facilitates the development of migratory matter in a straight line and speeds up the chromatographic process).

The plate is then removed from the tank, conditioned at 120° for 30 min and the top 5–10 cm of adsorbent containing the impurities scraped off with a sharp instrument and thrown away leaving the remainder of the adsorbent coating free from impurities. A second solvent treatment of the plate can be carried out, but is rarely needed. Plates should be used as soon as possible after activation; if a delay is inevitable, then they should be stored in a desiccant box containing silica gel.

Descending solvent migration from the top edge to the bottom edge of the plates, for removal of adsorbent impurities has also been in use. This provided effective adsorbent cleaning with little attention. The solvent and migrating impurities are continually removed from the lower edge of the chromatogram and the process continued indefinitely. This method is recommended when substantial numbers of chromatoplates are to be used because it is effective and needs little attention.

A similar technique is suggested in which the direction of washing is across the chromatogram. Their description is sketchy, but apparently solvent flow is maintained across the chromatogram and off the edge as their photographs of developed chromatograms show no concentrated impurity zone along the edge.

Application of polymer extract to plate

Regular application of the sample is particularly important for successful separations. For this purpose, 'spotting' as a band is not recommended, and manual application using a pipette requires considerable skill. The sample is sprayed on to the layer as a band by means of an automated applicator. With this instrument a band upto 35 cm long can be applied in a short time.

The reversing points of the applicator are practically free from hysteresis so that undesired thickening of the band at the ends does not occur.

In order to minimise washing-out of the sprayed solution at high sample volumes the most volatile solvent possible is chosen, e.g. diethyl ether or methylene dichloride. Where appropriate the plate may be warmed before hand so as to achieve rapid evaporation of the solvent.

A convenient trial sample size of solvent solution of a polymer extract for application as a band on a 250 μ thick 20 cm × 20 cm plate is 1 ml of a 1% solution, i.e. 10 mg. Depending on the type of separation obtained, larger or smaller sample volumes can then be applied as seems necessary. A further approximate guide for polymer extract sample size for a 250 μ thick 20 cm × 20 cm plate is 0.1–1 mg of polymer extract dissolved in 1 ml of solvent per gram of gel loading on the plate. These sample sizes can be increased by a factor of approximately 100 (i.e. 10–100 mg) dissolved in a few millilitres of solvent when carrying out preparative layer chromatography on 20 cm × 40 cm plates with gel thicknesses in the range 750–1500 μ.

Selection of chromatographic solvent

Finding a chromatographic development or mixture of solvents for the separation of unknown mixture of additives is not always easy. In some cases a complete separation is not obtained with a single solvent or solvent combination but necessitates the preparation of several chromatograms using different solvents.

An unknown mixture should be first chromatographed on 20 cm × 5 cm plates with solvents of different polarities to obtain an idea of the types of compounds present in the sample and to reduce the possibility of missing any of the sample components.

Solvents of low polarity, such as *n*-hexane, tetrachloroethylene and carbon tetrachloride, cause polar sample constituents on silica gel to migrate to only a small extent but will cause the less polar sample constituents to migrate more readily. Solvents of intermediate polarity such as toluene, benzene, chloroform and methyl cellosolve have a greater elutive effect on polar sample components, whereas highly polar solvents such as dioxan, methylene dichloride, ethyl acetate, nitromethane, acetone, lower alcohols and water elute polar sample constituents towards the solvent front, i.e. R_f values near unity. Mixtures of 40/60 petroleum spirit and up to 10% (v/v) ethyl acetate are very useful general solvents for the separation of unknown mixtures.

Various solvent systems for the separation of polymer additives on thin-layer plates are referred to in Table 14.11.

Table 14.11. Separation of antioxidants—thin-layer chromatographic methods.

Substances separated	*Stationary phase*	*Mobile phase*	*Detection*
Phenolic antioxidants	Silica Gel G	Methanol-cyclohexane (1:24)	30% Molybdophosphoric acid + ammonia vapour
Organo-tin stabilisers	Not stated	Acetic acid-isopropyl ether (1.5:98.5)	20% Molybdophosphoric vapour acid + ammonia
Antioxidants	Not stated	Light petroleum-ethyl acetate (9 : 1)	(a) Ethanolic 2,6-dichloro-*p*-benzoquinone-4-chloramine + 2% aq. $Na_2B_4O_7$ (b) Diatzotised *p*-nitroaniline
Organic stabilisers	Kieselgel G	Ethanol-free chloroform	
Phenolic antioxidants	Polyamide powder	Methanol water (3 : 2) or Methanol–carbon tetrachloride (1 : 9)	Diazotized sulphanilic acid
Phenyl salicylate Resorcinol benzoate	Kieselgel G	Dichloromethane or isopropyl ether–light petroleum (40–60°) (7:3)	Ultraviolet light
BHA, 2,6-di-t-butyl-*p*-cresol	Silica gel	Chloroform	20% Molybdophosphoric acid + ammonia vapour
Antioxidants	Polyamide powder	Methanol-acetone-water (6:1:3)	Diazotized sulphanic acid or molybdophosphoric acid
Antioxidants	Kieselgel G		α,α'-diphenyl-β-picryl hydrazyl (free radical)
Antioxidants	Alumina +5% Plaster of Paris on microscope slide	Petrol-dioxane (10:1)	5% Ethanol, phosphomolybdic acid
Antioxidants	Silica gel	Acetone, chloroform, benzene, carbon tetrachloride or binary mixtures	
Antioxidants	(a) 10% starch in polyamide powder (b) 10% PVC in polyamide	Methanol-acetone-water (3:1:1) light petroleum-benzene-acetic acid-DMF (40:40:20:1)	
Antioxidants	Silica Gel G	Benzene	0.5% $Fe_2(SO_4)_3$ in sulphuric acid + 0.2 % $K_4Fe(CN)_6$ (1:1)

Detection of separated compounds on the plate

Detection techniques should be carried out immediately after the chromatogram has been developed, in order to reduce to an absolute minimum any opportunity for volatile sample constituents to be lost by evaporation from the plate. Detection of the separated compounds on the plate is achieved by examination under ultraviolet light which locates some, but not all, types of compounds, and by spraying with a range of general or specific spray reagents. Merck GF 254 silica gel contains an inorganic fluorescing additive which is not extracted in the methanol pretreatment discussed previously. Silica gel is suitable for most types of separations. Alumina and cellulose powders are both available in fluorescent forms, respectively, alumina Fluka type D5F, MN cellulose powder 300 F254. Exposure of plates to 254 mμ radiation permits visualisation of substances on the plate which absorb above 230 mμ as a dark area on a blue fluorescent background. Fluorescent indicators which are activated by long-wavelength ultraviolet light may also be incorporated in the adsorbent.

The sodium salts of hydroxypyrene-sulphonic acids are particularly suited for this purpose. The separated polymer constituents show on the chromatogram, partly as dark zones, partly as zones which fluoresce brightly. The Merck silica gel HF 254 + 366 contains two fluorescence indicators, long and short ultraviolet light sensitive. To incite fluorescence, apparatus is needed which will allow a choice of short or long wave ultraviolet radiation. The Desaga-Uvis lamp is well suited to this purpose. The non-fluorescent silica gel, Merck G254, reveals the presence of polymer constituents which have an intrinsic fluorescence under short wave (254 mμ) or long wave (366 mμ) ultraviolet light.

To locate separate compounds on the plate with a minimum risk of missing any sample constituents, apply 1 ml of a 1% solution of the polymer extract in a low-boiling solvent such as diethyl ether or methylene dichloride as a continuous band along the base line of each of two 20 cm × 20 cm plates coated with GF 254 and G 254 silica gel, allow the solvent to evaporate, develop the chromatograms and examine the resulting plate under 254 and 366 mμ ultraviolet light sources. After marking off the position of any compounds seen with stylus, spray the plates with the general spray reagents shown in Table 14.12 to reveal the presence of any compounds which were not visible under ultraviolet light.

Table 14.12. General spray reagents for location of compounds on 20 cm × 20 cm silica gel coated thin-layer chromatography plates.

*A. Reagents applied to GF 254 plate without subsequent heating**	
(i) Potassium permanganate	(0.1 N) in aqueous sodium carbonate (5% w/v)
(ii) Potassium permanganate	(2% w/v) in aqueous sulphuric acid (6% v/v)
(iii) Potassium permanganate	(0.1% w/v) in sulphuric acid (96%)
(iv) Antimony pentachloride	(2% w/v) in carbon tetrachloride
(v) Phosphomolybdic acid	(3% w/v) in ethanol, then expose plate to ammonia vapour
*B. Reagents applied to G 254 plate with subsequent heating**	
	Heat treatment of plate
(i) Sulphuric acid aqueous (20% w/v)	5–15 min at 120°C, then 5 min at 150°C
(ii) As (ii) under A	5–15 min at 120°C, then 5 min at 150°C
(iii) Phosphoric acid (10%) methanolic	5–15 min at 120°C, then 5 min at 150°C
(iv) Perchloric acid (2%) methanolic	5–15 min at 120°C, then 5 min at 150°C
(v) As (iv) under A	5–15 min at 120°C, then 5 min at 150°C
(vi) Phosphomolybdic acid (20% w/v) in methanol or methyl cellosolve, then expose plate to ammonia vapour	5–15 min at 120°C

* Spray 20 cm × 2 cm wide sections of plate with each reagent using an aluminium or glass mask with suitable aperture.

A further general test for organic compounds on the plate involves holding an electrically heated 25 cm long copper wire, set at red heat, at a few millimetres above the plate along the length in which the chromatogram has been developed. After a few seconds exposure, many types of organic compounds reveal themselves by charring or otherwise discolouring.

Further information on the nature of some particular classes of additives can then be obtained by spraying fresh plates with more specific chromogenic reagents (Table 14.13).

A typical thin-layer chromatogram is obtained for three phenolic antioxidants following spraying with a 1% ethanolic solution of 2,6-dichloro-benzoquinone chlorimine.

The reproducibility of R_f values obtained on chromatoplates is generally poorer than in paper chromatography.

Table 14.13. Specific spray reagents for location of compounds by thin-layer chromatography.

Additive type	*Spray reagent*
Phenolic antioxidants	(i) 2,6-dichloro-benzoquinone chlorimine (1–2% in ethanol followed 15 min later by 2% borax in 40% aqueous ethanol). (ii) $\alpha\alpha$'-diphenyl picryl hydrazyl (0.1% in 95% aqueous ethanol). (iii) Palladium chloride. (Mix 150 ml palladium chloride with 100 ml of 2 N hydrochloric acid). (iv) Diazotized *p*-nitroaniline. (Mix 5 ml 0.5% *p*-nitroaniline in 2 N hydrochloric acid with 0.5 ml 5% sodium nitrite until colourless and 15 min later add 15 ml 20% sodium acetate).
Amine antioxidants	Diazotized *p*-nitroaniline. (Mix 5 ml 0.5% *p*-nitroaniline in 2 N hydrochloric acid with 0.5 ml 5% sodium nitrite, until colourless and 15 min later add 15 ml 20% sodium acetate).
Dialkyl thiodipropionates	Potassium platinoiodide. (Mix 5 ml 5% platinum tetrachloride in 1 N hydrochloric acid with 45 ml 10% potassium iodide and 100 ml water).
Phthalate ester plasticisers	Resorcinol. (Spray with 20% aqueous resorcinol in 2% aqueous zinc chloride and heat to 150°. Then spray with 4 N sulphuric acid and heat for 20 min at 120°. Spray with 40% potassium hydroxide to produce orange spots). Phthalic acid and phthalates also react.
Acids and bases	Bromocresol green, Bromocresol purple, Bromophenol blue (0.5% in 50% aqueous ethanol). Methyl red
Carboxylic acids	Sodium dichlorophenolindophenol (1% ethanolic).
Aliphatic (primary, secondary and tertiary) amines, long chain quaternary salts and amine oxides	Cobalt thiocyanate 10 g $Co(NO_3)_2 6H_2O$ and 10 g ammonium thiocyanate made up to 100 ml. Produces blue colour.
Alkanolamines	Ninhydrin. (Heat plate for 5 min at 110°, spray with 0.2% ninhydrin in acetone and heat 5 min at 110° to produce colours. Further colours then produced upon spraying plate with 0.2% alizarin in acetone).
Alkyl phenols	Phenols coupled as *p*-nitrophenol azo dyes applied to plate of silica gel impregnated with alkali. Separated azo dyes located as yellow/red colours upon exposure of plate to ammonia vapour.

(*Contd...*)

Additive type	*Spray reagent*
Carbonyl compounds	Carbonyl compounds in sample converted to 2,4-dinitrophenylhydrazones, applied to thin-layer plate and plate developed. Separated 24 DNPH compounds located as yellow or brown colours upon spraying plate with 2% sodium hydroxide in 90% ethanol.
Organic peroxides	Hydriodic acid. (Spray plate with a reagent comprising 40 ml glacial acetic acid and 0.2 g zinc dust added to 10 ml of 4% aq. potassium iodide, then spray with fresh 1% starch solution.) Peroxides (and certain other types of oxidising agents) revealed by liberation of free iodine.
	(a) Alternatively use 2,6-dibromo-benzoquinone chlorimine.

* Spray 20 cm × 2 cm wide sections of plate with each reagent using an aluminium or glass mask with suitable aperture.

Removal of separated compounds from plate

Examination of the plates under ultraviolet light and by the application of general and specific spray reagents usually provides full information regarding the R_f values in different development solvents of the various components of the original additive mixture applied to the plate and, possibly some information regarding the types of compounds present.

Having full information on the R_f values of the different sample components using the preferred development solvents, the next stage is to run a chromatogram on fresh 20 cm × 20 cm plates using a suitably sized sample (say 1 ml of a 1% solution) and mark off with a sharp stylus the bands corresponding to the known positions of the separated compounds. No detection reagents are applied to these plates although they may be examined under the ultraviolet light to precisely locate any components which show up under these conditions.

The simplest method of removing the zones containing the separated compounds (after allowing solvent to evaporate from the plate) is to hold the plate vertically its side resting on a sheet of paper and to scrape off the desired zone with a spatula. For substances which are not sensitive to oxidation, the zones may be sucked from the layer directly into an extraction thimble by using a small 'vacuum cleaner'. Each separated adsorbent band can then be bottled off in 5-ml polyethene stoppered tubes and retained for further examination.

Extraction of pure polymer additives from separated adsorbent bands

The separated portions of adsorbent, each, hopefully, containing a pure constituent of the original polymer extract, are now extracted with suitable solvents to isolate the additive preparatory to identification by physical and chemical methods. Each portion of adsorbent is transferred from the storage bottle to a separate small sintered glass extraction thimble (Fig. 14.3) and the organic compounds leached out with a suitable solvent such as anhydrous absolute ethanol, diethyl ether or methylene dichloride. This solvent must :

1. Be a good solvent for the additive.
2. Be sufficiently polar to desorb the additive from the adsorbent (successive desorption with different solvents may be necessary at this stage).
3. Have a low boiling point to facilitate subsequent removal of solvent and reduce to a minimum evaporation losses of any volatile sample constituents; and/or
4. Not interfere in the subsequent spectroscopic examination of extracts.

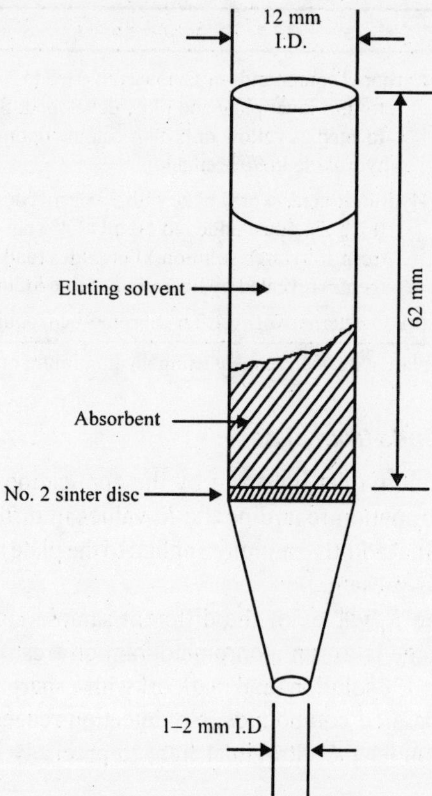

Fig. 14.3. Filtration apparatus for extracting separated additives from adsorbent. Isolated from thin-layer chromatography plates.

Provided the desorption solvent is sufficiently powerful and polar, it should recover between 50 and 100% of the additive present in the silica gel fraction and provide sufficient material for examination by ultraviolet or infrared spectroscopy, or mass spectroscopy.

During the solvent elution of compounds from the isolated bands of adsorbent, the elution solvent effectively displaces the components from the adsorbent so that all of the component is contained in the first liquid emerging from the sintered glass thimble. Loss of the component by its failure to elute is not usually serious, provided that the eluant selected is one which would elute the particular compound to the solvent front (i.e. R_f near unity) on a thin-layer plate. In fact, the behaviour of the component on the plate with plate elution solvents is a good guide to the selection of an appropriate solvent for desorbing the same component from the isolated gel in the extraction thimble. If the elution solvent used is less effective than this, then losses can be expected of the compound at the adsorbent extraction stage.

In extractions from silica gel, care should be taken to ensure that, according to the solvent used, components of the adsorbent are not co-extracted with polymer additives. Silica gel and cellulose powder adsorbents containing gypsum as a binder cannot be extracted with water or with aqueous organic solvent mixtures because calcium sulphate is quantitatively extracted at the same time. A significant extraction residue is obtained even using chloroform benzene or acetone as extractants. A significant inorganic extraction residue is also obtained from gypsum-free silica gel to which silicon dioxide of particle size less than 40 mμ is added as a binder; this extract decreases with decreasing polarity of the extraction solvent. The extract solution should, therefore, be filtered through a hardened filter at the base of the extraction thimble (Fig. 14.3), and/or through short columns of Celite. Alternatively, the solvent extract can be evaporated to dryness and the silica contaminated residue contacted with a good solvent for the polymer additive and a bad solvent for the silica. Polar elution solvents, especially if they contain water, will extract inorganic salts such as calcium sulphate from Silica Gel G; thus 10 ml of 35:65 v/v acetonitrile:water extracted 0.4% of salts from Merck AG Silica Gel G. Ultraviolet fluorescent type silica gel absorbents contain inorganic additives to induce fluorescence. After plates of these adsorbents had been premigrated with methyl alcohol, as described earlier, it was found that solvent extracts of the isolated gel did not absorb in the ultraviolet region between 200 and 300 mμ or in the infrared. Hence, this type of fluorescent activator would not be expected to interfere in the examination of polymer additive extracts obtained from this type of adsorbent.

Following irrigation of the plate with development solvent, the substrate and separated components are picked up in vacuum collector tubes fashioned from sections in 5 mm borosilicate tubing (Fig. 14.4). By varying the collector tube length and diameter, sample sizes ranging from micro to preparative can be handled. Chromatographic elution of sample from the resulting powder columns isolated in the collector tubes is accomplished through transfer of a solvent from a reservoir by means of a thread wick (Fig. 14.5). Elution volumes are minimal, 100 μl providing quantitative recovery of a separated component, thus favouring subsequent examination by micro-ultraviolet, infrared and gas chromatographic and other techniques. Using mixtures of furoic and hydroxymethyl furoic acids this technique was applied over a range of plate loading between 1 and 100 mg and excellent precision and accuracy in recoveries in the procedure was obtained.

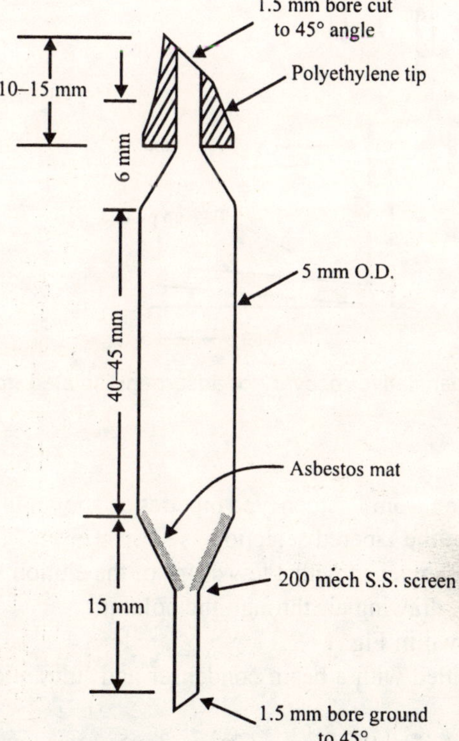

Fig. 14.4. Collector tube for quantitative removal of sample zone from thin-layer plate.

Preparation of infrared spectra of separated polymer additives

A method is described below for preparing infrared spectra of the various portions, each containing a single polymer additive, isolated from the thin-layer plate. The technique, described below, utilises an infrared cavity micro-cell.

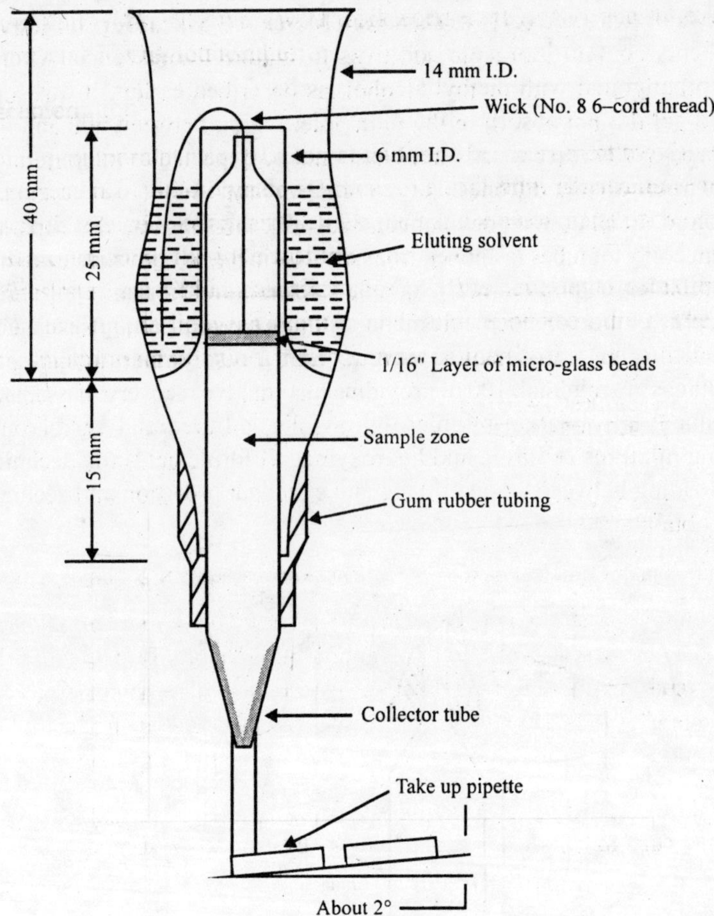

Fig. 14.5. Elution assembly for quantitative recovery of adsorbent isolated from thin-layer chromatography plates.

Apparatus

Elution columns, disposable, made from Pastuer-type pipettes as shown in Fig. 14.6. Pack a glass wool plug firmly into the larger half of the tapered selection as shown in Fig. 14.6. A more retentive pad is obtained if the glass wool pad is dampened with a few drops of the elution solvent to be used just before use. Excess liquid is removed by drawing air through the column.

Evaporation assembly as shown in Fig. 14.7.

Infrared spectrophotometer fitted with a beam condenser and 'ultramicro' cavity cells.

Centrifuge and accessories.

Micromanipulator.

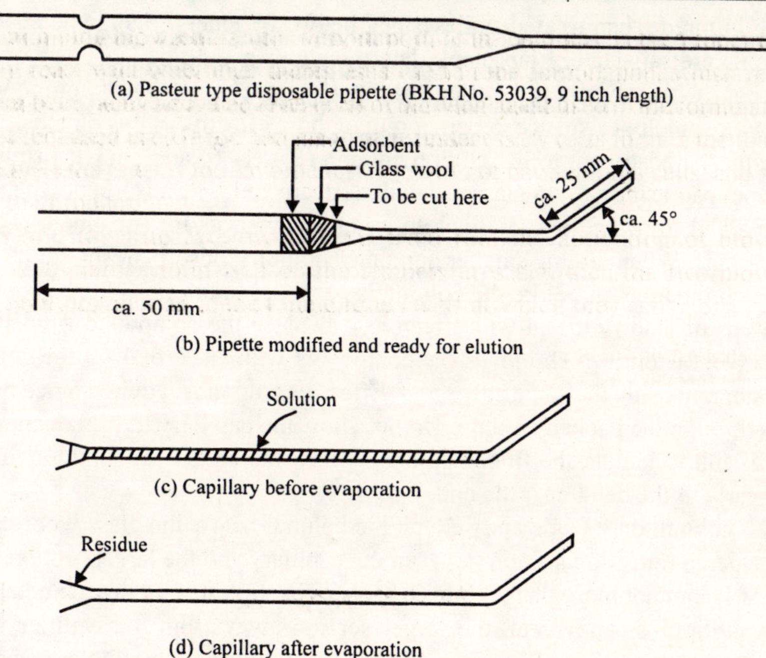

Fig. 14.6. Steps in preparation and use of elution column used for GLC fraction collecting. (a) Pasteur-type disposable pipette (BKH No. 53039, 9 in. length); (b) Pipette modified and ready for elution; (c) Capillary before evaporation; (d) Capillary after evaporation.

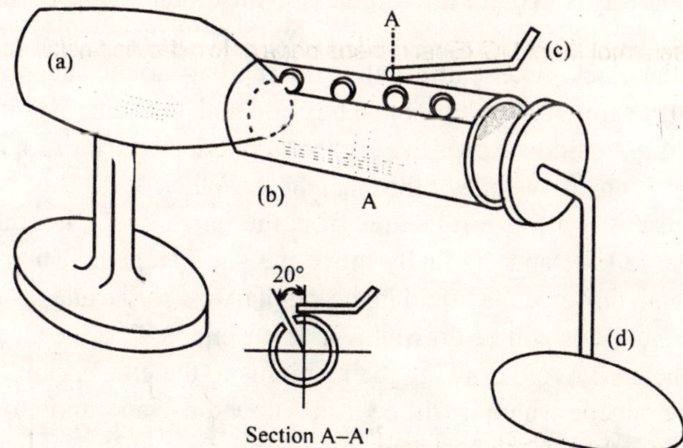

Fig. 14.7. Apparatus for evaporation of solvent from capillaries. Prior to infrared spectroscopy. (a) Hair dryer. (b) Metal or glass tube a little larger in diameter than the air outlet of the hair dryer and 6 in. long. Drill four 0.25 in. holes 1 in. apart in a straight line as shown. Mount with holes rotated 20° from the vertical as shown in section A–A'. (c) Capillary containing liquid to be evaporated. Hold horizontal in a simple micromanipulator; (d) Moveable stand holding a solid disc of same diameter as tube which serves as a damper to adjust air flow out the side holes.

Procedure

The solvents used for the development of the plate must be volatile enough to be removed from the adsorbent by evaporation and the adsorbent itself must not contain impurities which will show a significant absorbance in the infrared when treated as described below. In general, inorganic adsorbents, such as silica gel or alumina are satisfactory but blank determinations should be made to verify behaviour of each lot of adsorbent with solvents to be used. About 100 μg of a separated component is optimum; smaller amounts will sometimes suffice depending upon how effectively it is eluted and transferred.

Elution. Transfer the adsorbent to the elution column (Fig. 14.6) and pack it firmly and evenly on top of the glass wool pad using a 5-cm glass rod with a square end. The adsorbent when packed into the tube should form a bed from 1.5 to 2 cm thick to minimise chances for "channelling" and so that the flow rate is not too fast. When eluting smaller amounts additional adsorbent should be mixed with that removed from the chromatogram before the column is packed. Mount the column in a vertical position and add a few drops of a solvent that will effectively displace the compound from the adsorbent. A suitable eluting solvent is one which moves the component with an R_f of 1 (at the solvent front) on a thin-layer chromatogram using the same adsorbent. Effective elution is required since only about 0.1 ml of solvent passes through the packed column. Do not allow the capillary to fill beyond the bend as this makes it more difficult to handle the filled capillary. Allow the elution to continue until the capillary section is filled nearly to the bend near the end.

Lay the elution column on its side and scratch the column below the glass wool pad such that the scratch is on the tapered transition section between the capillary and the larger section at a point where the diameter is nearly, but not more than, 3 mm (this end must fit into a 1/8 in. diameter hole). Hold the capillary section motionless and break the larger section away from the capillary. The scratch is conveniently made with an ampoule file. If the capillary section is held motionless when the column is broken, no liquid will be spilled except a little which is close to the break. This portion is not important as most of the solute will be in the lower section of the capillary. The capillary containing the solution can be handled without spilling the contents by holding it at a slight angle from the horizontal with the bent end low. A convenient way is to place the capillary on the mounting block of the micro-manipulator so the point to be cut is at the end of the block. Clip it down, make the scratch, and break the column by bending over the end of the block. After cutting, slide the capillary so the cut end protrudes about 1.5 in.

Evaporation. Mount the capillary in a horizontal position with the bent tip pointing upwards using the micromanipulator to hold it. Position it over the evaporation assembly (Fig. 14.7) so that the flared end is directly over and about 2 mm above one of the small holes. When doing this, position the liquid in the capillary so that the meniscus is a few millimetres from the flared end. Direct the warm air flow from the hair dryer into the large tube and gradually move the glass stopper which serves as a damper to partially close the other end. Observe the liquid in the capillary carefully, and, as the air flow through the side holes increases, the meniscus will be drawn towards the end. Adjust the air flow until the meniscus is about 2–3 mm from the end. Avoid drawing the meniscus to the end as capillary forces will transfer part of the solution to the outside where it will be lost. Allow the evaporation to proceed. Normally the solvent will evaporate smoothly without further attention except in the last stages; if necessary, the capillary may be tilted a little to maintain the meniscus in the desired position.

Resolution and transfer. Remove the capillary from the evaporation assembly, hold it in the air stream for a few seconds so any vapour present is flushed out, and inject 5 μl of solvent, usually carbon tetrachloride or carbon disulphide, into the bent end of the capillary. Manipulate the capillary which is held nearly horizontal by tilting and rotating it so that the droplet of solvent moves and contacts the entire inside surface to dissolve the residue. Be sure to contact the zone near the flared end where the evaporation occurred as a large part of the residue is deposited at this point. The presence of adsorbent in the capillary at this stage may prevent transfer of the residue to the infrared cell. This can happen when the adsorbent adsorbs the residue so strongly that the solvent used for the transfer does not elute it.

Place an 'ultramicro' cavity cell in a plastic centrifuge tube, hold the tube in a horizontal position and insert the flared end of the capillary into the filling hole of the cell. Place the tube in the centrifuge and centrifuge the solution into the cell. Remove the capillary and examine the cell to see if more sample solution is present than is needed to fill the cell. If so, evaporate the excess volume of solvent by drawing a stream of air across the top of the cell. Seal the cell with a drop of mercury.

The infrared spectrum of the solvent solution can then be obtained by normal infrared spectroscopic techniques.

Alternatively, especially if the compound is insoluble in the usual spectroscopic solvents, the solid can be dispersed in well-ground solid dry potassium bromide using a dental mixing machine and the mixture pressed into a 1 mm thick 5 mm diameter disc. This disc can then be mounted in a cardboard or plastic holder and used to prepare a spectrum.

Preparation of ultraviolet spectra of separated polymer additives

Apparatus

Elution columns, disposable, made from Pasteur-type pipette (230 mm size) by cutting the larger section about 2 cm below the constriction and shortening the capillary end to about 5 cm in length. Mount in a vertical position (Fig. 14.6).

Volumetric flasks, 0.5- and 1-ml sizes.

Microspatula. Hayman type, or other similar tool having a sharp, square end suitable for scraping the adsorbent layer from the glass backing.

Ultraviolet spectrophotometric equipment (Cary) Models Nos. 14 or 15, the Beckman Model DU or others which can be used with the micro cells described below.

Micro cells and accessory equipment. 'Rectangular Micro Cells' (Code S-18-120). A simple mask as shown in Fig. 14.8 is needed to use these cells. After chromatography, the chromatogram is air-dried as necessary to remove residual migration solvent. If ultraviolet transparent solvents have been used that will not interfere with subsequent measurements, only a few minutes are needed to dry the adsorbent enough for the next step. When the solvent itself would interfere subsequently in ultraviolet spectroscopy, thorough removal by evaporation is necessary; several hours may be required. Illuminate the chromatogram with an ultraviolet light (2537 mμ) and mark the locations of the separated compounds which are visible as dark or fluorescing spots on the fluorescent background. Mark areas a little larger than those on which sample can be seen, if possible, to be more certain of including all of the compound. At this stage, compounds that absorb in the visible can be seen directly and need not be exposed to the ultraviolet light.

The amount of sample required for the preparation of a spectrum depends on the absorptivity of the particular compound. 10 μg of a compound having an absorptivity of 10 l/g-cm will give an absorbance reading of 0.1 when dissolved in 1 ml of solvent and measured according to this procedure.

Elution. Prepare the necessary number of elution columns (Fig. 14.6) as follows: Hold a tuft of glass wool in a pair of tweezers and rinse it with absolute ethanol. While still dripping wet, insert it into the top of a column and pack it into the tapered section with a 2-mm glass rod. Pack firmly. At this stage the capillary in the lower section fills with ethanol displaced from the glass wool; leave the capillary filled at this stage and prepare the rest of the columns needed.

Just before transferring the adsorbent, supported on a piece of creased glazed paper, to the elution column, shake the column to remove the ethanol in the capillary but do not dry the glass wool. Transfer the adsorbent to the column by folding the paper slightly and, while holding the crease at the top of the column, tapping the paper gently. Pack the adsorbent evenly but not too firmly with a 2-mm glass rod.

If the glass wool is damp when the dry adsorbent is added it effectively stops any of the dry powder from sifting through the wool. In addition, the adsorbent becomes a little damp and packs more easily. Packing should be firm enough so that from 15 to 30 min are required for 1 ml of elution solvent to pass through. Too firm packing causes the flow rate to be too slow; if this happens, application of moderate air pressure can be used to speed the elution. Some practice in packing and eluting is recommended to learn how to achieve proper packing. The more probable error is in packing too tightly; in handling the rod the maximum force that should be exerted is just about the minimum that can be felt in the fingers holding the rod.

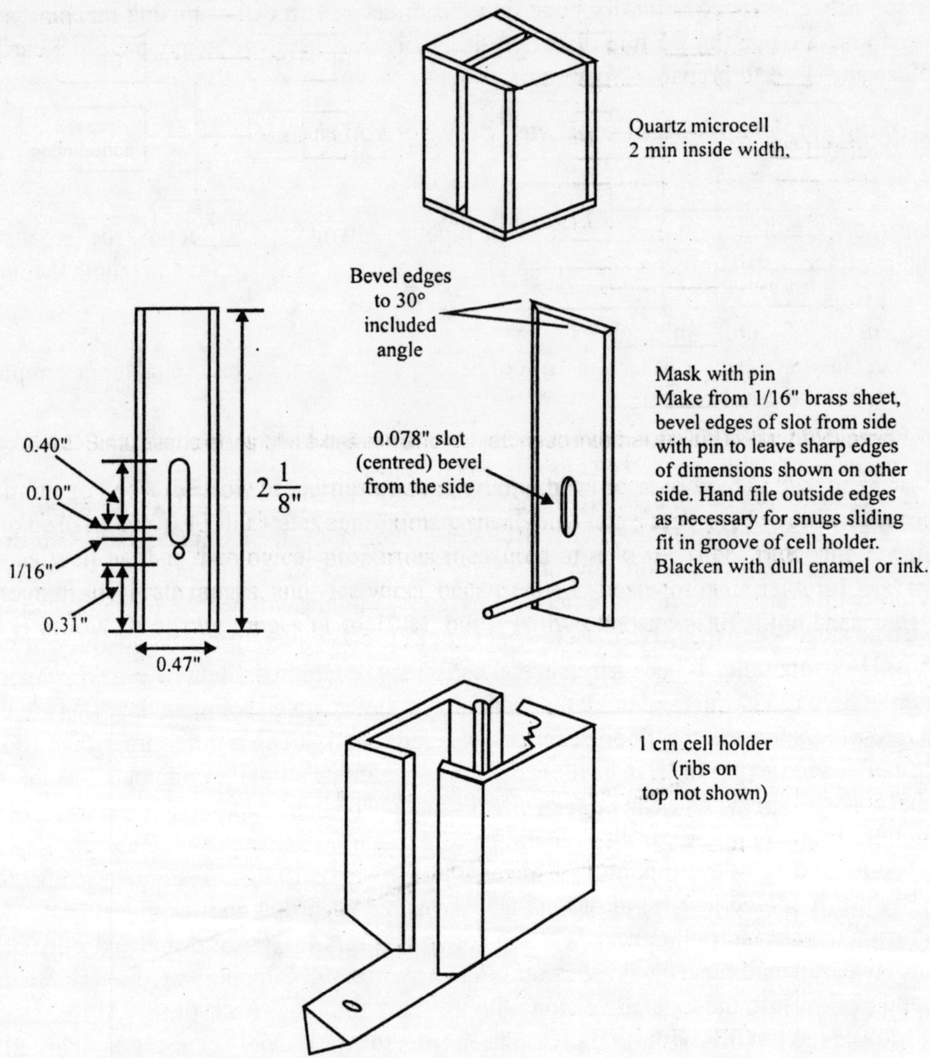

Fig. 14.8. Microcell, mask, and cell holder used in ultraviolet spectroscopy.

Place a 0.5- or 1-ml volumetric flask beneath the column and add a little elution solvent to the column so that the walls and glass rod are rinsed. Allow all of the liquid to pass into the adsorbent and add a little more. The total volume added should be a little less than is needed to fill the flask and it should be added in about three increments when quantitative recovery is desired. After the liquid has ceased emerging from the column, apply gentle air pressure to empty the capillary, remove the column, and dilute the volumetric flask to the mark. Stopper and mix.

Measurement. Measure the absorbance of the solution at the desired wavelength or record its spectrum using small volume micro cells. Either air or the elution solvent may be used as the reference. Blank determinations which were carried through the entire procedure, including chromatography, should always be made and measured or recorded.

It is important that the sections of adsorbent eluted for the blanks come from chromatograms prepared and handled in the same manner as the sample chromatogram and that they represent the same areas and locations of adsorbent. This is necessary because impurities, if present, are not distributed uniformly on the chromatogram.

Removal of impurities from the region of the adsorbent layer to be used for the sample separation by washing the layer prior to migration of the sample using the methanol premigration technique described earlier is a convenient and effective method of reducing blank absorption.

Standards. When quantitative determinations are desired using this technique, it is necessary to know or to determine the absorptivities of the particular compounds. This is done by preparing and measuring solutions of known concentration of the pure compounds. It is also advisable to chromatograph known amounts of the pure compounds to verify the applicability of the technique to the particular compounds. This is recommended because unexpected errors can occur if compounds have enough volatility to escape from the adsorbent or are unstable and change during the chromatography and drying.

Mass Spectrometry

Examples of identification of additives by combined thin-layer chromatography/infrared spectroscopy

The first example concerns three polyolefins suspected to contain one or more of the following additives: Topanol CA (tris(2-methyl-4-hydroxy-5-tert-butyl phenyl), butane), DLTDP (dilauyl thiodipropionate) and Ionox 330 (1,3,5-tri-methyl-2,4,6-tri (3,5-di-tert-butyl-4-hydroxy benzyl) benzene).

Additives are removed from 30 g of each of the polymers by a 6-hr Soxhlet extraction with diethyl ether. The extract residues are made up to 10 ml with chloroform, and 10 μl of these solutions and solutions of authentic specimens of the three additives mentioned above are spotted on to Merck GF 254 silica gel plate and the chromatograms are developed using 9:1 (v/v) 40/60 petroleum spirit:ethyl acetate solvent. Viewing the plates under 254 mμ ultraviolet light located Ionox 330 and Topanol CA but not DLTDP.

All three additives are, however, located with aggressive spray reagents and with a reagent for phenols (2:6-dibromo-benzoquinone-4-chlorimine, which gives strong colours with hindered phenols and also a yellow colour with DLTDP). The chromatograms in Fig. 14.9 show that polymers 1 and 2 both contain DLTDP and Ionox 330, polymer 2 also contains Topanol CA, whilst polymer 3 contains DLTDP and Topanol CA. Both the aggressive spray reagents reveal the presence of low molecular weight polymer at the solvent front (spot vi). These reagents also reveal the presence in the authentic specimen of Topanol CA of an impurity which appears at the solvent front. Obviously, this impurity is of low polarity and is not phenolic as evidenced by the fact that it does not produce a colour with the phenol reagent. It is probably a hydrocarbon.

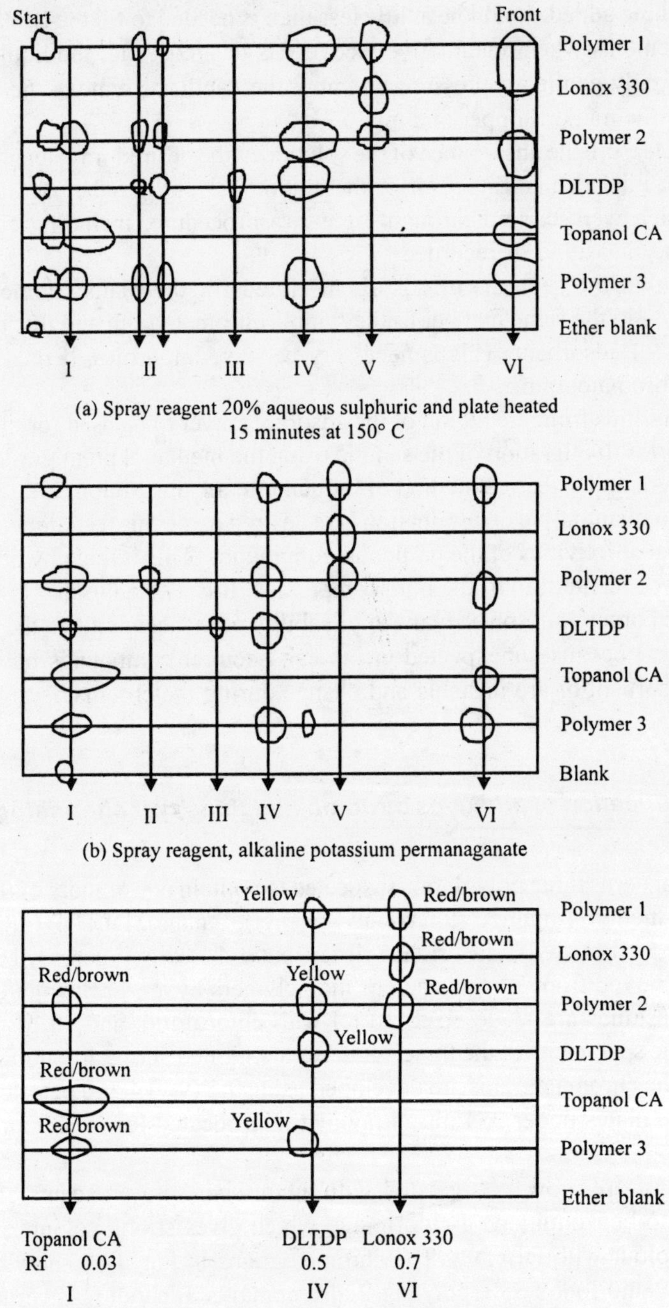

Fig. 14.9. Thin-layer chromatograms of solvent extracts of three-polyolefins. Plate: Merck silica gel GF 254; development solvent: 40/60 petroleum spirit:ethyl acetate (9:1 v/v).

A further example concerns the identification of an ultraviolet light stabiliser and an antioxidant in a sample of a polyolefin. Again the polymer is extracted with diethyl ether to isolate total additives and a portion of a chloroform solution of the extract and of various known light stabilisers and antioxidants run in parallel on a silica gel coated plate.

The chromatograms in Fig. 14.10 show that the polymer contained two additives appearing at R_f 0.6 and 0.85 which coincided in R_f value and in the colour obtained with 2,6-dibromobenzoquinone-4-chlorimine with known specimens of UV 531 (2-hydroxy-4-*n*-octoxy benzophenone) and Ionol CP (2,6-di-tert-butyl-*p*-cresol). Spraying the plate with 2,6-dibromo-benzoquinone-4-chlorimine also reveal an additional orange coloured spot at R_f 0.8 which did not coincide with any of the known additives examined.

Next, an attempt is made to identify unequivocally these three polymer components by comparing their infrared spectra with those of authentic specimens of the suspected compounds. Chloroform solutions (1 ml) containing 15–30 mg of the polymer extract and of authentic UV 531 and Ionol CP are applied along the edge of three 20 cm × 20 cm plates and the chromatograms developed using 40/60 petroleum spirit:ethyl acetate (9:1 v/v).

The three ultraviolet adsorbing bands on each plate are then marked off and the silica gel corresponding to these zones removed from the plate and the additive extracted from each portion of gel with anhydrous diethyl ether. After removing ether, the residues are intimately mixed with dry potassium bromide and small discs prepared for infrared spectroscopy. As a control a further blank chromatogram is developed, omitting the addition of chloroform solution of sample. The gel from this plate corresponding in R_f value and area to the R_f 0.6 and 0.8/0.85 bands observed in the polymer extract, are isolated and ether extracted.

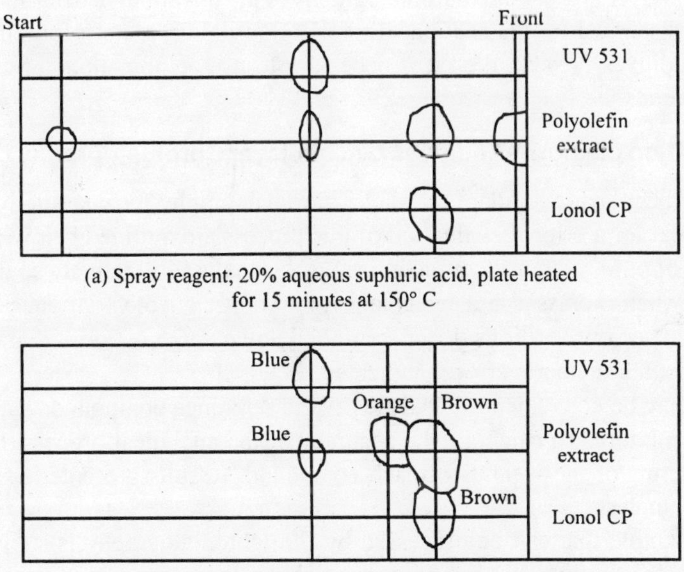

Fig. 14.10. Thin-layer chromatograms of solvent extracts of polyolefin nibs. Plate: Merck silica gel GF 254, development solvent: 40:60 petroleum spirit:ethyl acetate (9:1 v/v).

A further example concerns the identification of additives in a sample of polystyrene. Preliminary thin-layer chromatography on silica gel of an extract of the polymer revealed the presence of four additives with R_f values of 0.0, 0.45, 0.8 and 0.9. Obviously, the substance that had not migrated has a high affinity for silica gel and, in fact, a very polar solvent combination (1:1 v/v chloroform:ethanol) was needed to desorb this compound. The other three less polar compounds were easily desorbed from the adsorbent bands with chloroform. The separated extracts were weighed and infrared spectra is prepared.

APPLICATION OF CHROMATOGRAPHY FOR THE ANALYSIS OF KNOWN ADDITIVES

Direct spectroscopy, although useful on occasions, has severe limitations concerning lack of sensitivity, possible degradation of additives or of polymer during the film forming operation at elevated temperatures and interference effects by the polymer itself or other additives present in the polymer when seeking the spectral features associated with a particular additive.

Examination of solvent extracts by various chemical or chemical methods of analysis also has its limitations. If the additive to be determined in the polymer extract is of known identity and providing that any other additives present in the polymer formulation and hence in the extract are known, then it is often possible to choose a suitable method of analysis which is both sufficiently sensitive and specific for the quantitative determination of the additive in question to be achieved. This approach has been adopted to be carried out on ultra-violet or infra-red spectroscopy. If, on the other hand, as is often the case, a suitably specific chemical or physical method is not available for determining a known component in a mixture without interference from other additives present, then it is necessary to resort to the inclusion of a chromatographic step between the preparation of a solvent extract of the polymer and the final analytical determination.

The application of thin-layer and column chromatography to the quantitative determination of known additives in mixtures is discussed in this section. If, as is often the case, no information is available regarding the types of additives present, then it is necessary to use combinations of chromatographic and spectroscopic techniques.

Quantitative Determination of known Additives by Thin-layer Chromatography

Numerous attempts have been made to adapt thin-layer chromatography for quantitative analysis. The features which are essential for a critical evaluation of this process are outlined below :

1. Known amounts of pure reference compounds must be analysed iteratively so that the standard deviation, which when expressed as a percentage is the coefficient of variation, can be calculated. If the mean of a set of observations differ from the known value by more than three times the standard deviation, a systematic error is indicated.
2. To save time, it may be convenient to analyse several reference compounds simultaneously by using a standard mixture. To calculate the accuracy of the analytical process, the results must be presented in terms of the weights of each compound. Results presented as percentages of the mixture are valueless.
3. The quantity of sample that can be processed by film chromatography is subject to an upper limit. The reason for the limit is that, as the weight of the sample is increased, so the area occupied by each fraction on the chromatogram increases. Ultimately, the resolution becomes unacceptable because contiguous fractions overlap. There are several methods of quantifying thin-layer chromatography and these are discussed below under separate headings.

Spectroscopic methods

In this method a known weight of sample mixture is chromatographed on a thin-layer plate, usually coated with silica gel, by normal procedures. The test compound (or compounds) is extracted from the plate with a suitable solvent and the extract made up to a standard volume with solvent prior to spectrophotometric examination. If the compound contains a chromophore it may be determined directly; if not, before application to the plate, it must first be submitted to a chromogenic reaction if it is intended to use visible spectroscopy or, alternatively, ultra-violet spectroscopy can be used if the compound absorbs in this region of the spectrum. For spectrophotometric analysis the fractions from a single chromatoplate will provide sufficient material.

Several difficulties which are specific to the spectrophotometric method might be encountered. One of them arises from the fact that substances which absorb electromagnetic energy are invariably extracted from the absorbent material as well as the sample. The absorption spectrum of the impurity is non-specific; the intensity falls steadily as the wavelength increases, and at 400 mμ it becomes small enough to be ignored. If the extent of interference by adsorbent impurities is small, then its effect may be cancelled out by suitable calibration procedures.

By chromatographing known weights of pure reference compounds and determining the optical densities of the final solutions obtained from the plate, specific extinction coefficients may be calculated in terms of the weight of sample applied to the chromatogram. To minimise the errors introduced by extractable impurities, three refinements can be considered.

A portion of the adsorbent, equal in area to that of the sample, is removed from an unoccupied part of the chromatogram. It is processed in the standard manner and the extract is used as the blank solution in the spectrophotometer. The contribution of the blank in the ultra-violet region of the spectrum is by no means a negligible proportion of the actual measurement and it is also rather variable.

To diminish further both the magnitude and the variability of the blank, the film may be pre-washed by allowing suitable solvents to ascend the film as in the development of a chromatogram. Alternatively, the absorbent may be treated with a solvent in a *Soxhlet* extractor before the films are prepared.

Alternatively for eliminating the optical contribution from the adsorbent, pure solvent is used as the spectroscopic blank instead of an extract from the adsorbent, and the absorption spectrum is recorded over a wide waveband. Subsequently a line is drawn across the base of the absorption peak (Fig. 14.11) and it is assumed that the gap between this line and the corresponding line for the unchromatographed compound represents the optical density curve of the impurity from the adsorbent. The contribution of the sample is measured from the maximum to this baseline.

Another way of diminishing the effect of the impurity in the absorbent is to ignore the specific chromophore of the test compound and to subject it to a reaction which will enable the optical density measurements to be made at a longer wavelength. The longer the wavelength, the less will be the optical contribution of the adsorbent impurity. It is, of course, necessary to verify that the systems obeys Beer's Law.

The application of thin-layer chromatography combined with ultra-violet spectroscopy to the analysis of a known mixture of diphenyl, mixed hydrocarbons and a group of mixed oxygenated compounds which appear as three separate spots on the chromatogram. An excellent correlation is obtained between weights of diphenyl in synthetic samples spotted on to the plate and the absorbance in the ultra-violet at 248 mμ obtained in solvent extracts of the diphenyl containing area of the plate (Fig. 14.12).

In experimental details for the determination of individual components of mixtures of known esters, the esters are separated on a silica gel plate; the zones corresponding to different types of esters are scraped from the plate and the esters extracted with diethyl ether. The iron hydroxamate complexes of the individual pure esters are then formed and estimated spectrophotometrically at 520 mμ.

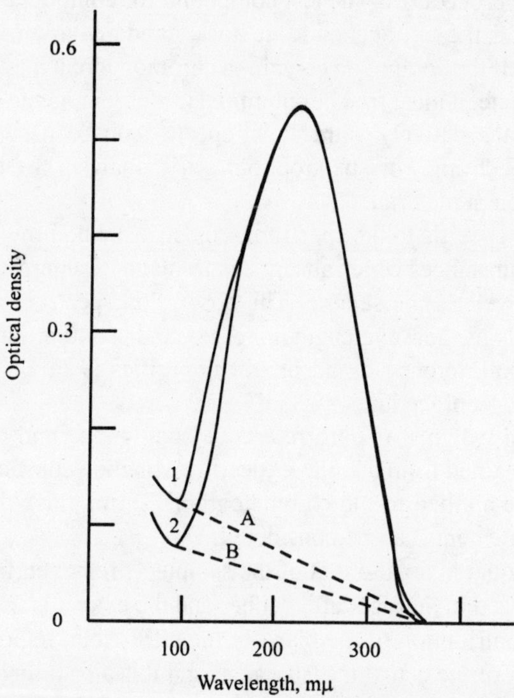

Fig. 14.11. Elimination of adsorbent impurity contribution in spectroscopy of thin-layer chromatography isolates. Optical density of propyl *p*-hydro xybenzoate: 1, sample recovered from chromatogram; and 2, original sample. The gap between the base-lines A and B represents the contribution of the impurities originating in the absorbent.

Optical densiometric analysis

A linear relationship between the 'extinction area' of a spot on the densiometric trace of a chromatogram and the weight of material in the spot was discovered in early sixties. To make the chromoplate translucent the adsorbent was sprayed with an ethereal solution of paraffin and then was measured the optical density of the chromatogram as a function of the distance from the origin (Fig. 14.13). The area under the peak, measured to the dashed baseline, is a linear function of the weight of material in the spot. Scanning photodensitometers for thin-layer chromatography plates are now commercially available. The extinction area varies from one chromatogram to another so a calibration standard is required for each chromatogram prepared.

Determination of Esters by Thin Layer Chromatography

Optical density and spectrophotometric methods

In this method glass plates coated with silica gel G chromatoplates are developed by the ascending technique, using benzene or trifluoro-trichloroethane: methylene chloride (60 : 40) as developing solvents. Detection sprays are mixtures of potassium permanganate and sulphuric acid and also vanillin in 96% sulphuric acid.

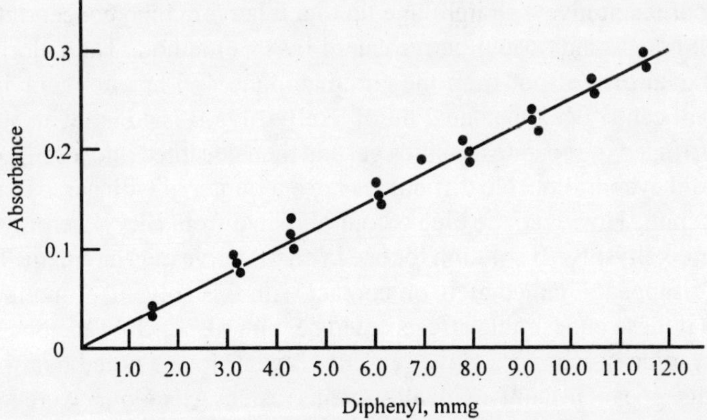

Fig. 14.12. Ultra-violet spectroscopy of thin-layer chromatography extracts. Calibration curve for mg diphenyl from ultra-violet absorbance readings taken at 248 mμ. Extraction from the silicic acid was with 5 ml 95% alcohol throughout.

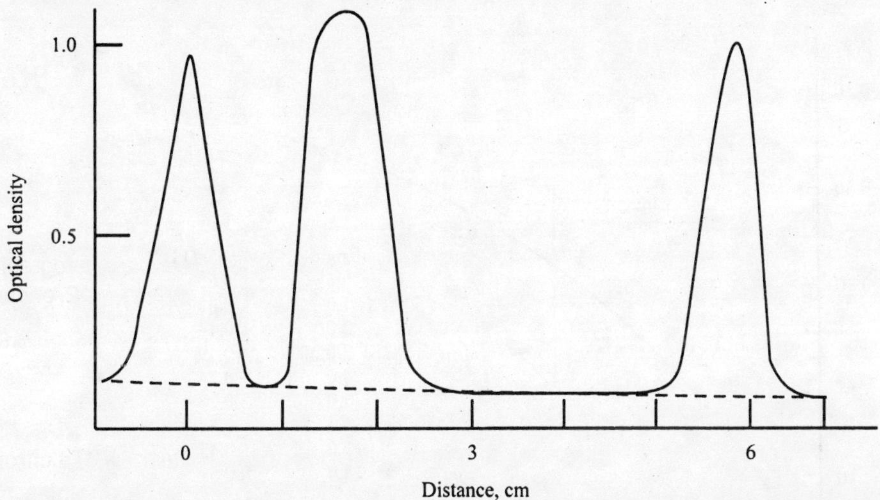

Fig. 14.13. Optical densiometric analysis of thin-layer chromatograms. Optical density of a chromatogram as a function of the distance from the origin.

To minimise differences between individual chromatoplates, standard R_f values of the esters are computed using citronellal as an internal standard.

Quantitative estimations are based on optical density measurement using a densitometer for spots of lowest concentration, 0.5–5 nl (nanolitre) per spot and on spectrophotometric measurements for spots of higher concentration. Only those compounds reacting with vanillin/H_2SO_4 could be studied accurately with the transmission densitometer as the spots formed with $KMnO_4$/H_2SO_4 can not be measured by the light transmission method. The spectrophotometric procedure is not only limited to materials sensitive to vanillin, but could only be used satisfactorily with a portion of the vanillin-sensitive substances. Fig. 14.14 shows a typical cuve obtained by reading a set of varying concentration spots using the densitometer. The graph is seen to be linear to a concentration of about 5 nl/spot. Plots of data from spectrophotometric measurement give a straight line up to a relatively high concentration. However, as mentioned above, not all compounds could be measured by this method. The colorimetric procedure involved the careful removal of the spot from the chromatoplate 4–6 hr after spraying. The coloured silica gel placed in a 12-ml centrifuge tube and 2 ml of Skellysolve B is added with stirring to elute the colour. The colour is centrifuged to separate the silica gel and then decanted into a cell for the colorimetric reading taking care to avoid evaporation. No difficulties are encountered with geranyl, neryl, linalyl, and other esters giving green spots. However, the blue colour obtained from carvyl, terpinyl, and some other esters faded too rapidly in Skellysolve B solution for use in quantitative measurement. The yellow colour from anthranilate esters disappears immediately on contact with this solvent. A number of solvents are investigated to determine if these undesirable effects could be minimised. Only acetone seems to improve results, the blue colours fading much more slowly in this solvent. It was noted that the yellow colours from anthranilate esters disappears instantly with all solvents tested. All colours were totally insoluble in both benzene and carbon tetrachloride. Cholesterol can be determined by thin-layer chromatography on silica gel G, silicic acid, alumina and Florisil coated plates by using 25% perchloric acid as detection reagent and heated the plates to develop colour. Transmission photodensitometry was used for quantitative estimation.

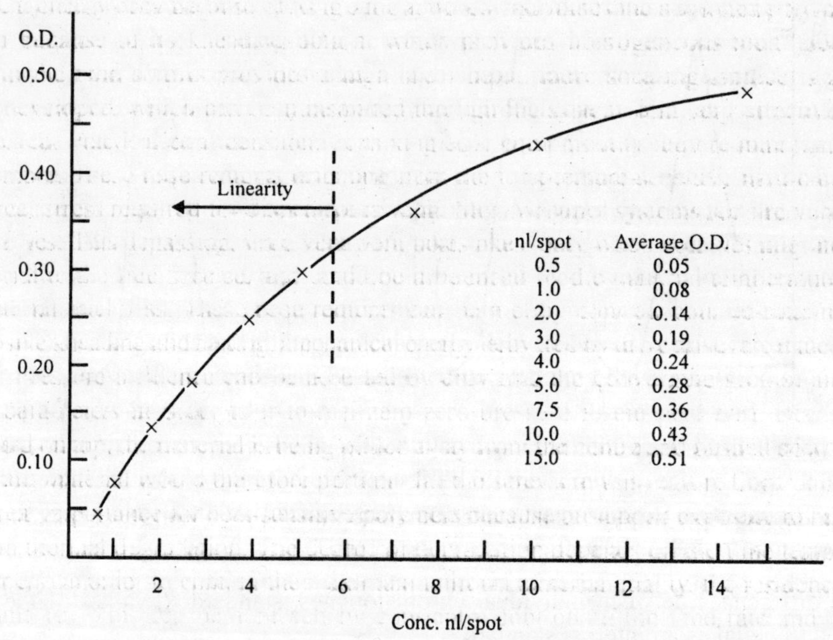

Fig. 14.14. Quantitative densitometric estimation of carvylacetate spots on thin-layer plate. Spots visualised with concentrated sulphuric acid-vanillin spray.

Determination of Santonox R

Polyethylene extraction procedure

Accurately weigh 5 g of polymer into a 500-ml round-bottomed flask. Add 65 ml of toluene and heat on a boiling water-bath for 90 min using a reflux condenser. Remove the flask from the water-bath and immediately pour 85 ml of absolute ethanol down the condenser to precipitate the dissolved polyethylene from the hot toluene solution. Allow the flask to cool to room temperature, remove the condenser, stopper the flask and shake well. Filter the solution through filter paper into a 250-ml beaker. Wash the flask and residue with a further 100 ml of ethanol.

Evaporate the toluene/ethanol solution almost to dryness on a boiling water-bath with the aid of a stream of nitrogen and finally to dryness using only the nitrogen stream. If more polymer is precipitated during the evaporation, re-filter the solution into smaller beaker. Wash the residue into a 2-ml volumetric flask with chloroform and make up to the mark.

Santonox R is extracted from polyethylene by precipitation of the polymer with ethanol from a hot toluene solution as described above. The extract is evaporated to dryness, dissolved in chloroform and an aliquot applied to a thin-layer plate and the chromatogram developed. The concentration of Santonox R in the polyethylene is estimated by visually comparing the intensity of the spot obtained with corresponding spots from known quantities of Santonox R.

Apparatus

Thin-layer chromatography tank. Thin-layer plate 20 cm × 20 cm, pre-coated with a 250 μ layer of Merck GF254 silica gel. Hypodermic syringe —25 μl capacity. Ultra-violet lamp—wavelength 254 mμ.

Reagents

Chloroform, Petroleum ether, 40/60, Santonox R (4,4′-thio-bis-3-methyl 6-t-butyl phenol). Ethyl acetate, 2, 6-dibromo-*p*-benzoquinone-4-chlorimine, Sodium tetraborate (borax).

Place 20 μl of the chloroform of the polyethylene sample as a spot on a thin-layer plate. Also apply 20 μl aliquots of standard solutions of Santonox R in chloroform, 0.05%, 0.04%, 0.03% w/v. Develop the chromatogram to a distance of 10 cm in a chromatography tank containing petroleum ether (40:60)/ ethyl acetate (5:1 v/v mixture) as the eluent.

Inspect the plate under ultra-violet light (254 mμ) and compare the intensity of the Santonox R spot from the polyethylene extract with the intensities of the standard spots. If the spot is of lower intensity than that of the 0.03% w/v standard then a new chromatogram should be developed using standards 0.025%, 0.015% and 0.005% w/v of Santonox R in chloroform.

Visually compare under ultraviolet lamp the intensity of the spot from the polyethylene extract with the standard spots and thus estimate the percentage level of Santonox R in the chloroform solution. Spray the thin-layer plate with a 2% w/v ethanol solution of 2,6-dibromo-*p*-benzoquinone-4-chlorimine. Allow the thin-layer plate to dry and re-spray using a 2% w/v aqueous solution of borax. Re-estimate the amount of Santonox R present in the chloroform solution by visually comparing the purple spots produced.

Calculation

Santonox R from 5 g of polyethylene is concentrated into a 2 ml chloroform solution. Thus the wt. % of Santonox R in high density polyethylene :

= 2/3 × the level of Snatonox R present in the 2-ml chloroform solution (as determined by thin-layer chromatography in % w/v).

Determination of UV 531

This method describes a thin-layer chromatographic spectroscopic procedure for the determination of ultra-violet Cyasorb UV 531 in high density polyethylene extracts. No interference is encountered from a number of additives, viz., dilauryl thiodipropionate (DLTP), 330, Ionol CP, Topanol CA, Santonox R, Polygard (TNPP).

UV 531 is extracted from the polymer by precipitation of the polymer with ethanol from a hot toluene solution as described above. An aliquot of the extract is applied to a thin layer plate and chromatographed. The zone corresponding to UV 531 is removed from the plate and extracted with ethanol. The concentration of UV 531 is determined by measuring the ultra-violet absorption peak near 295 mμ in ethanol solution and referring to a prepared calibration graph.

Apparatus

Thin-layer chromatography tank. Thin-layer plate 20 cm × 20 cm pre-coated with a 250 μ layer of GF silica gel. Hamilton syringe (to measure 25 μl). Ultra-violet lamp (wavelength 254 mμ). Sintered glass filter sticks, 1/2 diameter. 10-ml standard flasks. Ultra-violet spectrophotometer. 3-cm silica spectrophotometer cells.

Reagents

Methylene chloride. Chloroform A.R. Standard UV 531 solution, 0.2% w/v in chloroform, for use as a 'marker' to locate correct position of UV 531 on plate.

Standard UV 531 polymer blends—prepare by milling a series of blends of 0.05, 0.1, 0.2, 0.3, 0.4 and 0.5% w/w UV 531 in additive-free polyethylene powder. To calibrate the procedure, weigh out 0.5 g of each of the standard polymer-UV 531 blends and 0.5 to 1.0 g of the unknown polyethylene sample, and put each through the polyethylene solvent extraction procedure.

Apply 25-μl portions of each of the standard polymer extracts and the sample extract to the base line of a thin-layer plate in the form of a spot, keeping the spot size at a minimum by means of a stream of nitrogen. Develop the chromatogram to a distance of 10 cm in a chromatography tank using methylene chloride as eluant. Remove the plate from the tank and allow the solvent to evaporate.

Inspect the plate under ultra-violet light (254 mμ) and locate UV 531 as a dark blue spot. Mark with a pointed marker the zones corresponding to the UV 531 in the standards and the sample and also an equivalent blank area of silica gel.

Carefully scrape the silica gel zones from the plate and transfer to separate sintered glass filter columns. Extract the UV 531 from each with absolute ethanol into a 10-ml standard flask and finally make to volume.

Measure the optical density at 295 mμ of the standard and sample extracts against the gel blank extract in the reference cell in 3-cm silica cells. Plot a calibration curve for nett optical density at 295 mμ against micrograms UV 531 in final solution (10 ml). Hence, deduce the concentration of UV 531 in the unknown polyethylene sample.

Column Chromatography

This method is very important and is applied with some success for the determination of known phenolic antioxidants in the polyethylene and other industrial polymers, applicable to the analysis of mixtures of Santowhite powder (4,4′-butylidene-bis-(6-tert-butyl-*m*-cresol)) with B.H.T. (butylated hydroxy toluene) and mixtures of Santonox R (4,4′-thio-bis-(6-tert-butyl-*m*-cresol) with BHT. (butylated hydroxy toluene) in the concentration range 0.01–0.3%. This method should be equally applicable to other types of polyolefins. Total additives are first removed from the polyethylene sample by extraction with warm chloroform. The chloroform extract is then transferred to a column comprising a slurry of aluminium oxide in the same solvent.

Elution with chloroform removes B.H.T. only as the first fraction. Continued elution of the column with solutions of 10% water in methanol removes Santowhite powder or Santonox R as a separate pure fraction, free from B.H.T. Additives can then be determined in the respective extracts after they have been made up to a standard volume with solvent by ultra-violet spectroscopy. The specific absorptivities can be calculated from the slope of linear absorbance versus concentration plots for B.H.T. in chloroform, Santowhite powder, and Santonox R in 10% water in methanol. The ultra-violet spectra of the antioxidants are depicted in Fig. 14.15.

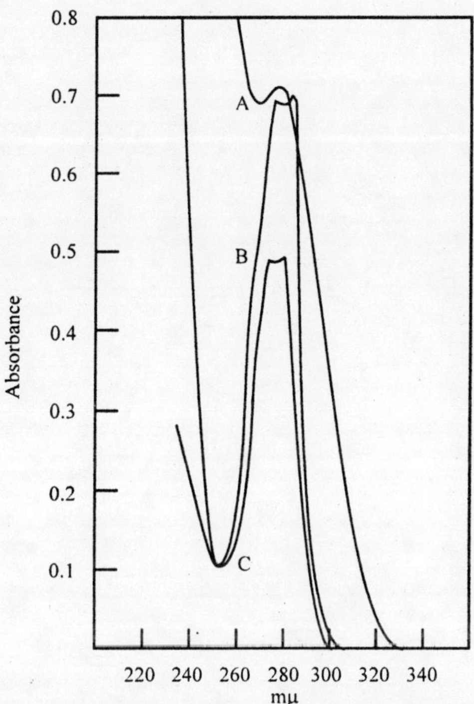

Fig. 14.15. Ultraviolet spectra of some antioxidants. A, Santonox R (0.0368 g/l. in methanol). B, Santowhite powder (0.0585 g/l. in methanol), C, B.H.T. (0.052 g/l, in chloroform).

Typical elution chromatograms show the separation of mixtures of B.H.T. -Santonox R and B.H.T. -Santowhite powder. It is noted that the water-methanol effluent front elutes the bulk of the Santonox R or Santowhite powder. A large error would result in some of the first part of the aqueous methanol fraction were discarded. The procedure can be verified by analysing polyethylene samples containing known amounts of antioxidants. Some standard samples were prepared by milling weighed amounts of B.H.T. Santowhite powder, and Santonox R in polyethylene at 130°. The analyses of these samples agreed satisfactorily for Santowhite powder and Santonox R, but the B.H.T. concentrations were approximately 30% low. A feasible explanation would be that B.H.T. has a greater volatility than the others, and some B.H.T. was lost due to evaporation during the hot milling process.

SECTION III

Applications of Polymers

CHAPTER 15

Applications of Polymers In Automobile

INTRODUCTION

The automobile combines many different materials that must be integrated with the ever-changing design and manufacturing processes. Without rubber components, the automobile in its current state of technology could not exist. Very few people are aware of the role that rubber plays in the total transportation industry. To put everything in proper perspective, consider that elastomers account for approximately 5 per cent of the material content of an automobile. These percentages vary from year to year, especially with increasing government regulations reflecting fuel economy, safety, and environment concerns, but the following breakdown is typical:

Material	%
Iron	60
Glass	3
Rubber	5
Plastics	7
Fluids and lubricants	7
Non-ferrous metals	7
Miscellaneous	1

APPLICATIONS OF RUBBER COMPONENTS

Approximately 500–600 applications in the automobile involve the use of elastomers. These uses vary from model to model and from year to year. Some critical uses having safety implications are as follows:

1. Tyres
2. Fuel hoses
3. Brake hoses and cups
4. Various fluid gaskets
5. Engine and TBI gaskets.

Additional uses include weatherstrips, windshield wiper blades, radiator hoses, and electrical wiring insulation. From the application examples, one can see how unique and important elastomers are to the automobile.

Some important automotive areas that need and use elastomers are as follows:

1. Traction for power transmission
2. Reduction or isolation of noise, vibration, and harshness
3. Seals against air, fluids, or other contaminants
4. Electrical insulation
5. Binders for friction materials, adhesives, sealers, and gaskets
6. Energy absorption (safety padding, air bags)
7. Conduits for liquids and gases
8. Flexible body panels
9. Movable impellers or diaphragms for fluid or gas transfer.

No single type of elastomers can fill all the areas or needs by itself. This is why there are so many commercial elastomers, each having a unique property profile. Rubber components need to survive and function under a wide variety of service and environmental conditions. These can include combinations of the following:

1. Extreme hot or cold temperature
2. Exposure to various chemical substances
3. Fuels and lubricants of various types
4. Other fluids, such as hydraulic, brake, and transmission
5. Exposure to ozone
6. Intermittent or continuous mechanical stress
7. Other miscellaneous conditions and environments.

Automotive people are continually faced with material, design, process, and cost decisions in choosing elastomers to meet established performance goals. These goals can include durability, economy, and meeting government regulations, depending on the applications. Table 15.1, which was assembled to assist in the elastomer selection process, lists the general types of currently available commercial elastomer in order of generally increasing temperature resistance, together with the ASTM D2000/SAE J200 type/class designations the particular elastomer meets. Also listed are qualitative resistances to various conditions and environments important to the automotive industry.

Ethylene Propylene Diene Monomer (EPDM)

The advantages of EPDM include the following :

1. Resistance to weathering, temperature, and ozone.
2. Resistance to break and engine coolant fluids.
3. Cost: compounds can accept high loadings of fillers and plasticisers.

The main disadvantages are fuel and oil resistance and iridescence.

EPDM is used for weatherstrip (primary, secondary, glass run channels, etc.), exterior rubber parts (bumpers), coolant hoses, brake system parts, and ignition cable insulation, and some exhaust hanger isolators.

Butadiene Acrylonitrile (NBR)

The advantages of NBR include the following :

1. Resistance to solvents, fuel, and oil.
2. Mechanical and physical properties: heat ageing, abrasion resistance, compression set, and flexibility.

Table 15.1. Elastomer selection chart.

Polymer type	Material designation: ASTM D2000 SAE J200	Gum polymer density range (Mg/m^3)	Hardness range (Shore A)	Original tensile strength (psi)(max.)	Original elongation maximum (%)	Compression set resistance	Service temperature (°F): Continuous low temperature 1000 H/Air	Dynamic	Static	Ozone and weather resistance
		Physical properties								
Polybutadiene (blends)	AA	0.090–0.091	45–80	2500	450	Good	158	–60	–80	Poor
Polynorborene (Norsorex)	AA	–	15–90	2500	600	Fair	158	–50	–60	Poor
Natural and synthetic polyisoprene	AA	0.91–0.94	30–95	4000	700	Very good	158	–50	–70	Poor
Polysulphide (Thiokol)	AK BK	1.34	20–80	1500	400	Poor	158	–40	–60	Very good
Styrene butadiene (SBR)	AA BA	0.93–0.94	40–95	2500	600	Good	158	–40	–70	Poor
Polychloroprene (Neoprene)	BC BE	1.23–1.25	20–95	3500	700	Fair-good	225	–40	–65	Fair-good
Polyurethane (ester, ether) (caprolactone)	BG	0.80–1.21	50–100	5500	600	Poor	235	–50	–70	Good-excellent
Polypropylene oxide (Parel)	BC	1.01	40–90	2500	600	Fair	235	–40	–70	Good
Nitrile (butadiene acrylonitrile)	BF BG BK CH	0.97–1.00	40–95	300	600	Good	235	–30	–65	Poor
Carboxylated nitrile	BF BG BK CH	0.98–1.00	50–95	3500	500	Good	235	–30	–70	Poor-fair
Isobutylene isoprene copolymer (butyl)	AA BA	0.92	30–95	2500	700	Good	250	–50	–70	Very good
Epichlorohydrin ethylene oxide copolymer (ECO)	CH	1.25–1.36	40–90	2500	400	Fair-good	275	–40	–60	Good
Chlorosulphonated polyethylene (Hypalon)	CE	1.10–1.27	50–95	2500	500	Poor-fair	275	–40	–60	Outstanding
Chlorosulphonated polyethylene (Tyrin)	CH	1.16–1.25	40–90	2500	500	Poor-fair	275	–40	–60	Outstanding
Halogenated isobutylene isoprene copolymer	AA BA CA	0.92–0.93	30–90	2500	700	Good	275	–50	–70	Very good

(Contd ...)

Polymer type	Material designation: ASTM D2000 SAE J200	Physical properties: Gum polymer density range (Mg/m³)	Hardness range (Shore A)	Original tensile strength (psi)(max.)	Original elongation maximum (%)	Compre-sion set resistance	Service temperature (°F): Continuous 1000 H/Air	Low temperature: Dynamic	Low temperature: Static	Ozone and weather resistance
Hydrogenated nitrile	CH	0.97–1.00	40–95	3000	600	Good	275	–35	–60	Good
Ethylene propylene diene monomer (EPDM)	AA BA CA	0.85–0.87	30–90	2500	500	Good	300	–50	–70	Very good
Epichlorohydrin homo-polymer (ECH)	CH	1.25–1.36	40–90	2500	400	Fair-good	300	–35	–55	Very good
Ethylene propylene copolymer (EPR)	BA CA DA	0.86–0.89	30–90	2500	600	Good	325	–50	–70	Very good
Vinyl acetate ethylene copolymer (VAE)	DH	0.98–1.02	50–90	2500	350	Fair	325	–30	–40	Very good
Polyacrylate ethyl, butyl	DF DH	1.10–1.13	40–90	2000	400	Good	350	–20	–40	Very good
Ethylene acrylic (Vamac)	DF DH	1.03	40–90	2000	400	Good	350	–40	–60	Very good
Phosphonitrile fluoro-elastomer (PNF, Eypel)	EF	170–180	35–90	2000	250	Good	350	–65	–80	Very good
Fluorosilicone (fluoro-substituted polysiloxane)	FK	1.3–1.8	30–80	1250	400	Good	450	–65	–80	Very good
Silicone dimethyl, diphenyl, phenylmethyl, and divinyl polysiloxanes	FC FE GE	0.97–0.99	25–90	1500	800	Good	525	–80	–180	Very good
Fluorocarbon fluorinated hydrocarbons, (Aflas, Fluorel, Kel-F, Viton)	HK	1.40–1.95	60–90	2500	350	Very good	525	–20	–50	Very good
Perfluoroelastomer (Kelrez)	HK	2.01–2.02	70–95	2500	200	Good	600	–10	–30	Very good

Its disadvantages are low flammability resistance, low resistance to ozone and weathering, and presence of oxygenated solvents (alcohols and alcohol-containing fuels).

NBR is used for O-rings, diaphragms, gaskets (binders in various gasket materials), oil seals, and hose liners.

AUTOMOTIVE STANDARDS FOR ELASTOMERS

To have uniform quality rubber components that meet the material, design, process, and other goals, specifications are required. A complete rubber component standard is necessary for the following reasons:

1. To ensure selection of materials that meet requirements.
2. To make certain that the materials are commercially available.
3. To prevent changes in production material(s) without the users' knowledge.
4. To maintain uniform quality in production shipments.
5. To correlate design requirements with material properties.

The requirements of a rubber component specification are as follows :

1. *Material requirements,* such as colour, type of material, finished part quality, and identification markings.
2. *Performance requirements,* such as density, tensile, elongation, and tear.
3. *Description of test methods,* including SAE, ASTM, and other laboratory procedures.
4. *Construction requirements,* for composite materials such as hoses, head gaskets, and weatherstrips.
5. *Construction requirements* (normally shown on the drawing), any information on tolerances or thickness of materials referred to in the standard is required, along with a statement acknowledging the dimensions on the drawings.
6. *Sampling instructions,* including sample size and number of samples to be submitted.
7. *Packaging instructions,* to ensure clean, distortion-free parts after shipment and storage.

The automobile company's own standards for rubber and elastomer are used, along with the classification systems of the Society of Automotive Engineers (SAE) and the American Society for Testing and Materials (ASTM).

THERMOPLASTIC ELASTOMERS (TPE)

Faced with ever-increasing demands for safety and fuel efficiency as a result of government regulation and consumer demands for quality and value, automotive engineers consider the total elastomer spectrum: thermosets and thermoplastics.

Thermoplastic elastomers were developed to bridge the gap between rubber and plastics. Thermoset materials are rubber-based compounds that are capable of becoming permanently rigid when heated or cured. They cannot be re-reprocessed.

Thermoplastics on the other hand, are plastics that are capable of softening or fusing when heated and of hardening again when cooled. Thermoplastics can be re-processed. The physical structure of a TPE has two components:

1. Hard blocks—plastic material
2. Soft blocks—elastomer material

Thermoplastics elastomers are formed in two ways: by chemical reaction and by blending. In the first case chemical bonds are formed between the soft and hard blocks during manufacturing process. In blending, fully cured elastomers are dispersed within the hard thermoplastic matrix.

Thermoplastic Elastomer Processing

TPEs are processed on thermoplastic moulding equipment by injection moulding, extrusion, blow moulding, calendering, or thermoforming. The hard segments melt and flow during the injection process. When the material contacts the cool platens, the melt re-solidifies to hard segments. There is no physical change in the soft segment.

Advantages and Disadvantages of Thermoplastic Elastomers

TPEs offer the following advantages:

1. Fast, easy processing on thermoplastic equipment: 45 seconds TPE; 5–10 minutes for the same part in a thermosetting elastomer.
2. No scrap: all parts can be re-cycled if not correct.
3. Greater design freedom.
4. Generally a lower part cost.

The disadvantages are as follows:

1. Higher compression set than some conventional elastomers.
2. Lower chemical resistance than some conventional elastomers.

THERMOPLASTIC ELASTOMERS FOR AUTOMOTIVE APPLICATIONS

The 1990's have witnessed the emergence of thermoplastic elastomers (TPEs) as a highly significant segment of the non-tyre rubber product industry, a result of three decades of concerted research and development. In 1996, TPEs enjoyed approximately 12 per cent of the worldwide non-tyre rubber usage, which should grow to 20 to 25 per cent by 2010. Ironically, these materials have found no significant uses in the pneumatic tyre area. However, over the past two decades, non-tyre automotive applications have provided one-fourth to one-third of the usage of TPEs.

The extensive adoption of thermoplastic elastomers (TPEs) in automotive uses is derived from its advantages and disadvantages relative to thermoset rubber vulcanisates.

Advantages

1. TPEs are fully formulated and require no compounding or mixing. This leads to greater consistency of composition, with batch-to-batch variation eliminated.
2. Processing has fewer steps and is much simpler. Fabrication is more reproducible, with the rejection of fewer rubber parts.
3. Thermoplastic processing enables faster production rates. TPE cycle times are measured in seconds versus minutes for thermoset rubber.
4. Process scrap can be re-cycled. Thus start-up material, off-spec parts, sprues, runners, etc., can be re-used. With thermoset rubber this scrap is discarded.
5. It has been shown that TPE parts which have given a normal service lifetime can be cleaned, re-ground and re-fabricated to give quality use in a new rubber part.
6. Less energy is consumed in fabrication, due to shorter cycle times and no vulcanisation step.
7. Thermoplastic fabrication processes—such as blow moulding, thermo-forming and heat welding–are practical with TPEs, but not with thermoset rubber.
8. TPE parts are better suited for high-speed, automated assembly operations since their dimensional tolerances are tighter than those of thermoset rubber.

9. Quality control of fabricated parts is better, and at reduced cost.
10. Most TPEs have a lower density than that of a competing thermoset rubber. These materials are purchased on a weight basis but used on a volume basis, giving more parts per kilogram.

Offsetting these advantages are certain disadvantages which must be considered:

Disadvantages

1. *TPEs embrace new technology.* Innovating with them requires fabrication skills foreign to rubber people and marketing expertise unfamiliar to plastics people.
2. TPEs generally should be dried before processing. Thermoplastics processors are familiar with this step, whereas thermoset rubber processes are not.
3. They do not function as a rubber above their melting point. This precludes their use near or above that temperature for even a short period of time.
4. Favourable economics generally require volume production, to justify the necessary engineering, tooling and capital costs.

The suitability of a thermoplastic elastomer (TPE) for a given end-use will be determined by the relative importance of these advantages and disadvantages.

Thermoplastic Vulcanisate and Thermoplastic Elastomers

All commercial TPEs may be classified as either a block copolymer or a rubbery polymer/thermoplastic combination. In the latter category, the rubbery polymer (such as EPDM) will tend to be dispersed in a continuous matrix of thermoplastic (such as polypropylene). Cross-linking of the rubbery phase, usually by a process of dynamic vulcanisation, will give rise to a thermoplastic vulcanisate (TPV) with properties more closely approaching those of thermoset rubber than any other type of TPE. The improved oil resistance, compression and tension set, and properties retention at elevated temperature qualify the TPVs for numerous automotive uses for which the same uncross-linked combination fails to perform.

Specific Automotive Application of TPVs

Rack and pinion steering gear boots

A highly successful automotive use of these TPEs is in rack and pinion boots to protect the moving parts of the steering system in lubricating grease within the boot. The convoluted design of the blow moulded boot enables the use of the harder (40 Shore D) TPV with sufficient flexibility to adapt to the movements of the steering system. The TPV offers the advantages of the speed and economy of thermoplastics blow moulding, resistance to the environment and lubricating grease, excellent flex fatigue life and puncture resistance. Replacement of a thermoset rubber by a TPV reduced the finished boot cost by a factor of approximately 3, with massive savings resulting.

Air intake ducts

A rapidly growing automotive use of TPVs is in the rubber duct which serves as a conduit for the air (purified or unpurified) to the engine. Unfiltered air is directed to the filter from which clean air is conveyed to the engine. The rubber in the duct must perform the following basic functions :

1. Without collapsing at negative pressure, the clean side duct must convey filtered air at 100 to 125°C. A convoluted design with a harder TPV aids this function.
2. Seals must be air tight, thus requiring low compression set and stress relaxation.

3. At temperatures as high as 125°C, the duct must resist fuels, oil, grease, radiator fluid, air oxidation and ozone, and be flexible down to –40°C.
4. The use of convolutes must allow ready installation and provide sufficient fatigue resistance and flexibility to allow for engine roll.
5. The modulus and coefficient of friction must enable a sufficiently high boot pull-off strength.
6. The duct must lessen engine noise.
7. Duct appearance must be acceptable, with embossed lettering feasible. New car buyers expect a clean, nicely-appearing air duct.

A blow moulded EPDM/PP TPV has been found to meet or surpass all of the above requirements. The lower density of a TPV, relative to a carbon black-loaded thermoset rubber, gives a 10 to 20 per cent weight reduction. Further, the blow moulding of the TPV allows design improvement in the convolute system and offers opportunity for generating a composite duct consisting of segments of flexible TPV heat welded or sequentially co-extruded to segments of rigid PP.

Air bag cover

TPVs have been found to be a material of choice for air bag cover in automobile safety systems, providing significant cost savings over reinforced polyurethane. The performance requirements of these covers are highly stringent. When called upon to perform, they must never fail since a human life may be at stake. The air bag components must function predictably when the vehicle impacts an object.

The deployment of an air bag is shown schematically in Fig. 15.1, with the sequence of events :

1. Impact activates sensors, sending an electrical impulse to the actuator, igniting a flammable solid in the housing to generate nitrogen gas for inflating air bag.
2. Air bag inflation causes the cover to rupture at the designated area and tear along a specific path through which bag erupts and becomes fully inflated.
3. The person in the vehicle impacts the air bag, which then deflates leaving the empty bag in his/her lap.

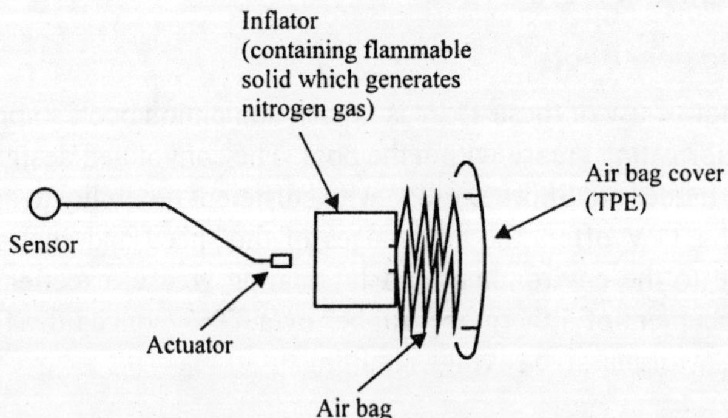

Fig. 15.1. Driver side air bag system.

This entire sequence must take place within one second, in the predicted manner.

The material selection for the air bag is both critical and difficult. At low temperatures the rupture mode of the cover is brittle fracture, whereas at high temperature it is tear. The only reliable way to successfully screen materials for this demanding use is to mould actual parts and deployment-test them at both low and high temperature. Part design is of critical importance.

Other automotive applications

Of the approximately 600 non-tyre rubber articles in a new automobile, several of the more demanding ones have been captured by TPVs. Control cables—such as the one connecting the parking brake handle to the rear mechanical brake—give better, longer service if they are protected from corrosion and mechanical abuse by a rubber cover. Successful crosshead extrusion of a TPV over these brake cables have allowed replacement of EPDM thermoset rubber with major improvements in cost, abrasion resistance, durability and service lifetime.

Engine compartment timing belt cover—with a flexible segment of rubber and a rigid segment of Polypropylene (PP) have successfully employed a TPV as the rubber. PP is insert injection moulded over a metal insert (threaded to allow fastening to the engine body), followed by overmoulding of the TPV onto the edges of the cover. The 'adhesive-free' application is based on the ability of the TPE to heat weld to the PP and thus perform the sealing function of the cover.

Fuel line covers have also been found to exploit the properties and processing of a TPV. Here, a flame-retardant TPV is co-extruded over the nylon fuel line, alleviating a series of fabrication problems previously encountered with thermoset rubber, as well as giving a cost saving.

USE OF SPECIALITY ELASTOMERS FOR AUTOMOTIVES

Various elastomers are used especially for automotive components like belts, seals, gaskets, o-rings, hoses and diaphragms. The materials are selected for their resistance to fuels, oils, and heat; as well as their cold flexibility and sealing ability. However, different materials are sometimes selected for the same device depending upon such factors as applied temperature class of fuels and oils, and the engineering design of the car. Nowadays automotive industry is facing many environmental and safety problems. Elastomers for automotive applications have been changing to solve those many kinds of problems. In this section, the trends of rubber materials being used for automotive parts and character and typical applications of high performance rubbers like HNBR, ACM and ECO are described.

Automotive Elastomer Parts and Applied Elastomers

The classification system of SAE J200 is intended for use with rubber for automotive. Rubber materials are designated on the basis of 'Type' indicating their heat resistance and 'Class' designating their oil resistance. Fig. 15.2 shows classified elastomers that will be classified in the near future.

The many problems confronting the automotive industry these days are shown in Table 15.2.

Table 15.2. Automotive programmes for environmental problems.

Environmental problems		*Automotive programmes*
Earth warning by CO_2		Fuel economy regulation
Ozone layer erosion by CFC's		Exhaust gas regulation
Forest attack by acidic rain		Alternative fuels
Nitrogen oxide	⟹	Electric car
Diesel smoke		Low emission car
Scrap cars		Taxation for diesel fuel
		Recycle
		CFC regulation
		Reduced sulphur in diesel fuel

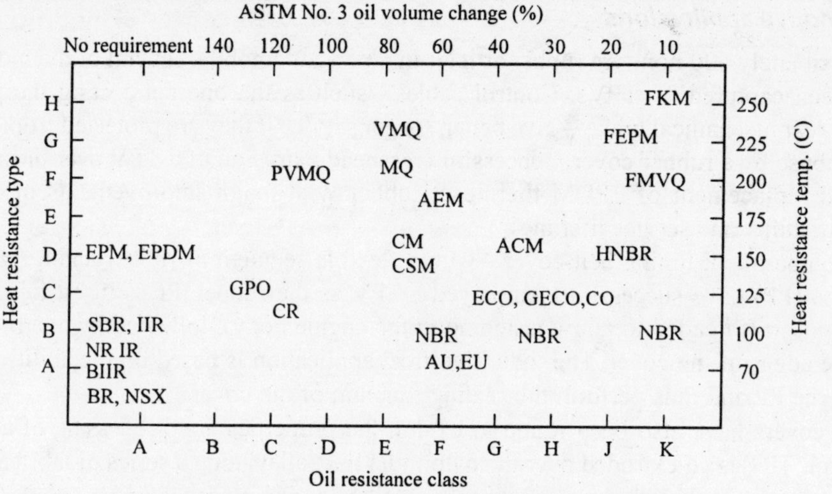

Fig. 15.2. Classification of rubbers on SAE J200.

There is earth warning by CO_2, ozone layer erosion, acidic rain, nitrogen oxide, scrap cars. Table 15.3 shows trend of rubber materials being used for automotive parts. NR is used for engine mount. NBR is used for oil hoses, seals and gaskets. EPDM is used for water seals and hoses, brake seals and hoses. The trend shows shifting to the higher performance elastomers, which have heat resistance, long life, low permeability, good abrasion resistance, etc. For some of non-oil applications, NR and SBR had changed to EPDM. Oil resistant and special like fuel hose, ATF hose and timing belt are changing from NBR and CR to AR, ECO, FKM and HNBR. Additionally, low cost is very important.

Table 15.3. Trend of rubber materials being used for automotive parts.

Seals parts name		*Current materials*	*Needs*	*New materials*	*Trend*
Engine	Crank shaft-rear	VMQ	Long-life	FKM	–
	Crank shaft-front	ACM	Long-life	FKM	–
	Diaphragms	ECO, NBR, FVMQ	–	–	–
	Valve stem seal	FKM	–	–	–
	Valve cover gasket	NBR	Heat resistance	ACM	–
Body	Weather strip	EPDM	–	–	Long-life
	Glass run	EPDM	–	–	Low abrasion
	Air conditioning seal	NBR	New freon gas	HNBR	–
	Filler cap O-ring	NBR/PVC	–	–	–
Chassis	Transmission oil seal	NBR	Heat resistance	ACM	–
	Power steering oil seal	NBR	–	(HNBR)	Heat resistance
	Ball joint dust cover	CR, U	–	–	–

(Contd...)

Seals parts name		*Current materials*	*Needs*	*New materials*	*Trend*
	Constant velocity joint boot	CR	Heat, ozone	TPEE	–
	Rack and pinion boot	CR	Heat ozone	TPO, TPEE	
	Brake master cup	SBR	Heat resist., long life	EPDM	
	Caliper piston seal	SBR	Heat resist, long life	EPDM	
Hoses parts name		*Current materials*	*Needs*	*New materials*	*Trend*
Engine	Fuel hose	FKM/ECO/ GECO	–	–	Low cost
	Emission control hose	NBR/CR, ECO, ACM	–	–	–
	Air duct	NBR + PVC	Heat resistance	TPO, ACM, ECO	–
	Water hose	EPDM	Heat resistance	EPDM (aramid)	Heat resistance
	(Radiator, etc.)	(Polyester)	Heat resistance	–	–
Body	Fuel hose	NBR/CR	Permeability	FKM/ECO/ ECO	–
	Filler hose	NBR + PVC	–	–	Low permeability
	Air conditioning hoses	NBR/CR	Permeability	PA6/CIIR/ EPDM	–
Chassis	A/T oil cooler hose	ACM	–	–	Heat resistance
	Power steering hose	NBR/CR	Heat resistance	ACM, (HNBR)	Heat resistance
	Brake hose	SBR/NR/CR + EPDM	Water permeability	EPDM	–
	Clutch hose	SBR/NR/CR + EPDM	Water permeability	EPDM	–
	Master vacuum hose	NBR/CR	Heat resistance	ECO	–
Others parts name		*Current materials*	*Needs*	*New materials*	*Trend*
Engine	Synchronous belt	CR	Heat resistance, life	HNBR	Long-life
	Accessory drive belt	CR	–	–	Heat resistance
Body	In-tank pump cushion	NBR + PVC	–	–	–
	Muffler hanger rubber	EPDM	–	–	–
	Wiper blade	CR, NR	–	–	Long-life
Chassis	Engine mount	NR (SBR, BR)	–	–	Heat resistance

Comparison among typical speciality elastomers being used for automotive parts along with main required properties is shown in Table 15.4.

Table 15.4. Comparison table of main speciality rubbers.

	NBR	*NBR/PVC*	*ECO*	*ACM*	*HNBR*	*EPDM*	*Q*	*FKM*
Tensile strength	O	O	Δ~O	Δ	**O**	O	Δ	Δ~O
Service temperature	120	120	140	175	150	150	200	250
Engine oil resistance	**O**	**O**	Δ	**O**	**O**	× ×	Δ	**O**
Gasoline resistance	O~**O**	**O**	**O**	×	O~**O**	× ×	×	**O**
Water resistance	O	O	Δ	Δ	**O**	**O**	O	**O**
Ozone resistance	×	O~**O**	**O**	**O**	O	**O**	**O**	**O**
Cold resistance	O	Δ	**O**	Δ	O	**O**	**O**	Δ
Cost	O	O	Δ	Δ	× ×	O	Δ	× ×

O: Excellent; O: Good; Δ: Fair; ×: Poor

Automotive Applications of HNBR

Structure and properties of HNBR

Hydrogenated Acrylonitrile Butadiene Rubber (HNBR) was developed as a high performance materials by a selective hydrogenation reaction of the butadiene units in base NBR. HNBR provides improved resistances to heat, ozone, oxidised fuel, and lubricating oil containing aggressive additive packages. High tensile strength and excellent abrasion resistance are also provided. Since HNBR appeared on the market, it has been utilised in many applications because of its highly balanced properties. These include fuel hoses, oil hoses, oil seals, grommets, diaphragms, printing rolls, blankets, brake shoes, adhesives, and belts, etc. The structure and characteristics of HNBR are shown in Fig. 15.3.

[NBR] —(CH=CH) — $(CH_2$–CH$)$—
C≡N

H_2 ⇩ Catalyst

[HNBR] —$(CH_2$=$CH_2)$— $(CH_2$–CH$)$— (CH=CH)—
C≡N

Ethylene	Acrylonitrile	Butadiene
Elasticity Heat resistance Ozone resistance Cold flexibility	Oil resistance Fuel resistance High strength	Cross-linking

Fig. 15.3. Structural unit and each function of zetpol (HNBR).

Synchronous belt

These belts synchronize crankshaft and camshaft. In Japan, it comes into general use in the 1970s. Synchronous belts are superior to metal chains in several respects: reduce noise when in contact with the sprocket, no lubrication needed, lighter weight because of low density, multi-axle drives are an option because of flexibility. A synchronous belt and its structure are shown in Fig. 15.4 along with its composed fibreglass reinforcing cord, synthetic rubber matrix and polyamide fabrics cover.

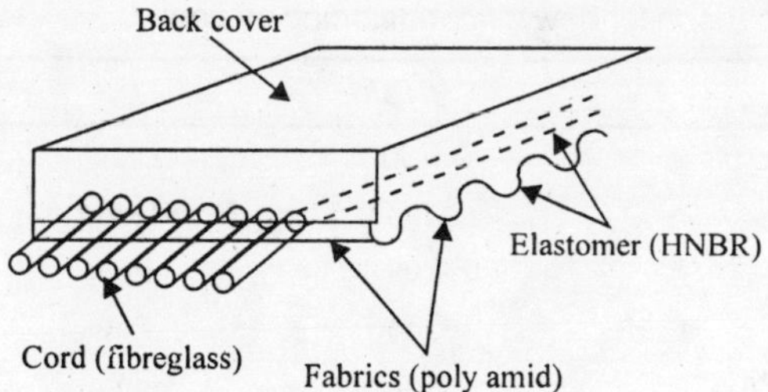

Fig. 15.4. Structure of synchronous belts.

In order to replace the chain, special consideration should be given to mechanical load and degradation in ambient temperature. Twelve properties of an elastomer regarded as significant for synchronous belts are shown in Fig. 15.5. It can be seen that HNBR has much better balance of properties than CR because of its stable complex modulus over wide temperature range, low-temperature properties, better heat and ozone resistance, much better resistance to flex crack initiation after ageing, and good oil resistance: HNBR can provide double the life of CR. Due to its superior characteristics, today, more than 90 per cent of all synchronous belts in Japan are made of HNBR.

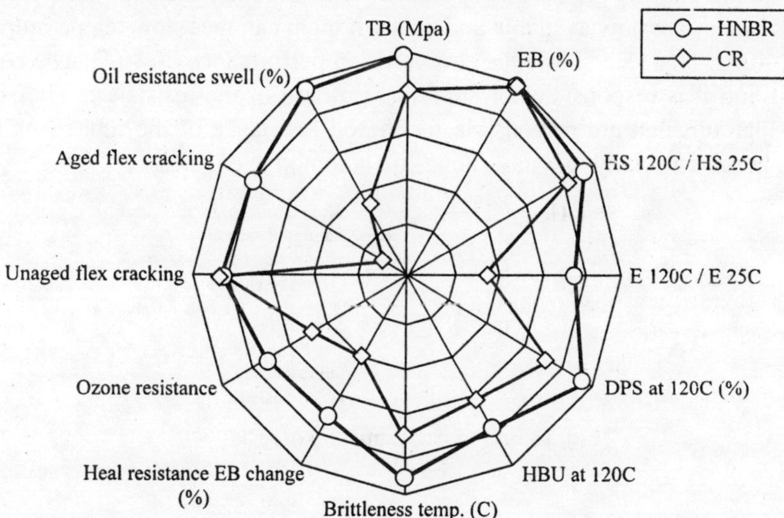

Fig. 15.5. Quality requirements of synchronous belt.

Air con seals

Reactive chlorine, containing chlorofluorocarbon for use as automotive refrigerants had been considered to contribute to the break down of ozone in the upper atmosphere. The automotive companies therefore adapted newly designed air conditioning system based on R-134a, new non-ozone depleting refrigerant. For this new refrigerant, new lubricant, polyalkylene glycol (PAG) was developed for use in the system. The elastomer, which was used if old refrigerant like NBR could not be used for those new refrigerant nor additional new severe requirements like high temperature resistance up to 150°C. The best elastomer for this new system was HNBR, as shown in Table 15.5.

Table 15.5. Comparison for new air conditioning system.

Properties	*HNBR*	*CR*	*Nylon*
Service temperature	150	120	120
Strength	Good	Good	Excellent
Compression set	Excellent	Poor	Poor
PAG resistance	Good	Poor	Poor
Volume change in PAG	Good	Good	Poor
Brittleness after heat ageing	Excellent	Poor	Poor
R-134a resistance	Good	Good	Excellent
Permeation of R-134a	Good	Good	Excellent

Automotive Applications of ACM

Structure and properties of ACM

Polyacrylate elastomers (ACM) are widely used in the automotive industry because of their excellent resistance to lubricating oils and greases. They can be formulated to resist high temperatures of 150 to 190°C. Numerous grades are now available and some of them can meet low-temperature performance specifications to –40°C. Fig. 15.6 shows the structure and performance of ACM. The backbone of the ACM is saturated and it is responsible for superior heat and ozone resistance. The side chains are composed of ester groups that are responsible for the oil resistance of the rubber. ACM is used for gaskets, seals and hoses especially in today's long-life automotive engines.

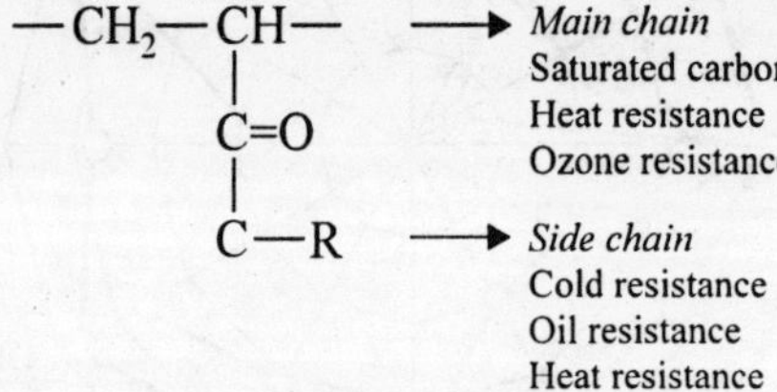

Fig. 15.6. Chemical structure of ACM.

Engine gaskets

The largest area for use of ACM is in engine gasket. ACM is the major polymer for this applications both in Japan and Europe. The gasket must have the ability to perform in the temperature range from –35°C to 150°C. And the material should show a slight positive volume swell in the engine oil and it must have good compression set. It should also show an ability to dampen the noise that may be generated. Fig. 15.7 shows heat resistance of speciality elastomers. At 150°C, ACM can stand for 1000hrs.

Automotive Applications of CHR

Structure and properties of Epichlorohydrin elastomers

Epichlorohydrin elastomers (CHR) have a good balance of low temperature flexibility, oil and fuel resistance, and heat and ozone resistance. These polymers are obtained by ring-opening polymerisation of cyclic ether monomers; they are epichlorohydrin (ECH), ethylene oxide (EO), and allylglycidyl ether (AGE).

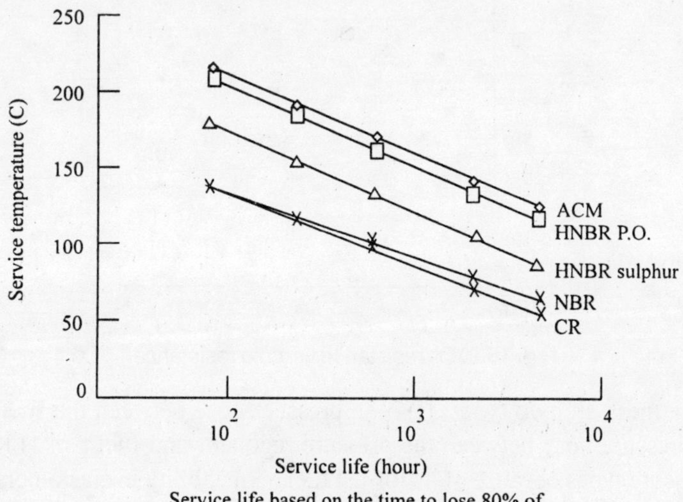

Fig. 15.7. Service temperature and service life for various kinds of speciality elastomers.

The structures of those monomers are shown in Fig. 15.8. From these monomers, various of polymers are available, like epichlorohydrin homopolymer (CO), epichlorohydrin/ethylene oxide (ECO), epichlorohydrin/allylglycidyl ether (GCO), epichlorohydrin/ethylene oxide/allylglycidyl ether (GECO).

Polymer type	Polymer structure
CO	$-(CH_2-CH-O)-$ CH_2Cl
ECO	$-(CH_2-CH-O)-(CH_2-CH_2-O)-$ CH_2Cl
GCO	$-(CH_2-CH-O)-(CH_2-CH-O)-$ CH_2Cl $\quad$ $CH_2-O-CH_2CH=CH_2$
GECO	$-(CH_2-CH-O)-(CH_2-CH_2-O)-(CH_2-CH-O)$ CH_2Cl $\quad$ $CH_2-O-CH_2CH=CH_2$

Fig. 15.8. Polymer types and structures of CHR.

Fig. 15.9 shows brittleness point vs. volume swell of ASTM # 3 oil. ECO shows very good balance of cold flexibility and oil resistance.

Fuel hose

There are two fuel systems, carburetor and fuel injection. Fuel hoses for carburetor system were made from NBR or NBR/PVC for inner and CR for outer. But the outer layers are changed from CR to CSM (chlorosulphonated polyethylene) or GECO of the elevated temperature in the engine compartment.

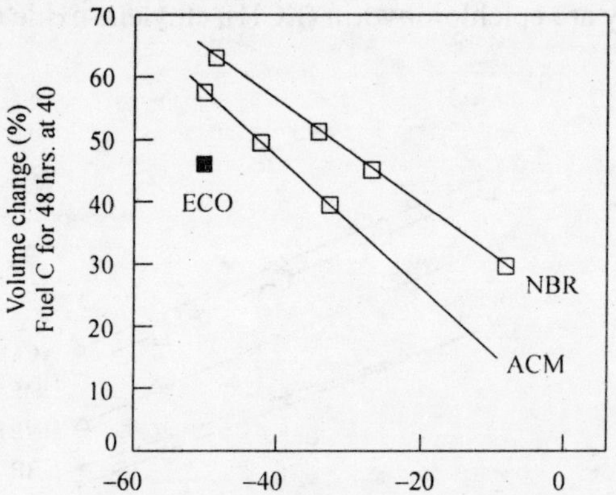

Fig. 15.9. Oil resistance vs. cold resistance.

For fuel injection, there are two types: (i) high pressure hose between the fuel pump and injection valve; and (ii) low pressure hose between the pressure regulator and fuel tank. FKM is used for inner layer of the high-pressure types because of its lowest fuel permeability in elastomers. ECO or GECO are used for cover layer of the hose. The materials used to cover the hose must have a good heat and ozone resistance. Fuel resistance is also desirable property. It acts as a back up to the veneer FKM, in case of a leak though a pinhole. Fig. 15.10 show fuel permeation resistance for some kinds of hoses. CO (ECO) hose shows second lowest permeation of the fuel.

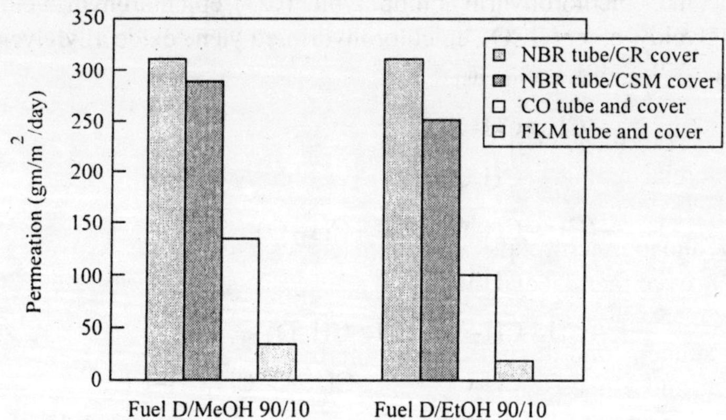

Fig. 15.10. Fuel permeation performance of hoses constructed with several kinds of elastomers.

Thus various elastomers are used for automotive components because of their physical properties, chemical properties, durability, etc. and elastomers for automotive have been changing to solve many kinds of problems such as quality, safety and environment. Situation of rubber manufacturing is changing in the near future, speciality elastomers such as HNBR, AR, ECO should be more necessary for rubber manufactures who producing products for automotive fields.

CHAPTER 16

Some Novel Applications of Polymers In Aerospace and Railway Engineering

INTRODUCTION

The use of non-metallic materials in aerospace applications shows a healthy and continuing growth, with much of the increase in recent years resulting from the very exciting developments in the fields of fibre-reinforced plastics and engineering thermoplastics. Any discussion of recent developments in the aerospace use of rubbers and composites is, therefore, inevitably dominated by a consideration of structural composites, their diversity of forms, and their enormous potential for engineering application.

By comparison with composites, the usage of rubbers is well established in both fixed wing aircraft and helicopters, and consumption can be regarded as being relatively stable in terms of volume per aircraft. Indeed significantly increased application for this class of material is unlikely without the introduction of a rubber with substantially better high temperature properties than can be offered by the best of existing commercial products.

Aerospace requirements certainly exist for such rubbers, and what is needed is an improvement in resistance to both short and long-term exposure to high temperatures, that is, improved strength properties at temperature and improved thermal and thermo-oxidative stability. Notwithstanding this lack of novel and new high temperature elastomers, there remains a considerable impetus for research and development on existing speciality rubbers, and this arises from the persistent pressures from aircraft operators for improved performance and longer service life for existing rubber components on airframes and in aircraft systems. Without doubt the single most striking influence on the usage of non-metallic materials in aerospace applications over the last fifteen years has been the discovery in the mid-60s of lightweight, strong, and extremely stiff carbon fibres. These materials, which were a product of research at the Royal Aircraft Establishment, Farnborough, quickly became available in continuous filament form, and when used in combination with epoxide resins, they led to products which made it possible, for the first time, to envisage a fibre-reinforced plastic giving serious competition to the light alloys traditionally used as materials of construction in airframes.

In the wake of the carbon-fibre development, there has come a flurry of developments in strong stiff lightweight synthetic fibres (notably the aramids), in speciality resins suitable for use in high performance composites, and in engineering thermoplastics. These developments taken as a whole, represent a substantial challenge to metals in a range of applications going far beyond the aerospace field.

The effective utilisation of carbon fibres in structural composites called for, amongst other things, the ability to produce sheets of accurately aligned fibres pre-impregnated with resin, i.e. uni-directional pre-preg, and the consolidation of stacks of this material into a composite in which the alignment of fibres was retained and the volume fraction of fibre closely controlled.

This 'pre-preg' technology, developed for carbon fibres, has since been transferred to the glass-reinforced-plastics field with excellent results, the upshot being that high quality GRP is also beginning to realise its very considerable potential as a structural material. This is particularly evident in the helicopter field where many of its characteristics are especially attractive.

Composites also feature in another important area of aerospace engineering application, i.e., dry bearings. Something of a quiet revolution has been taking place over the years with the introduction of ever-more-effective plastics composites as liners for flight control bearings in fixed wing aircraft and helicopters.

The obvious advantage of dry bearings is that oil and grease lubrication can be dispensed with, and this in turn should lead to lower costs of maintenance. The life of such dry bearings is governed by the rate of wear of the composite liner in service and it is in this area that substantial research and development has been and still is being applied with notable success.

RUBBER COMPONENTS

Rubbers find wide application in aircraft of all kinds, and although the volume of material used is not particularly impressive, the number and variety of components is. A range of such applications, listed below whilst it is probably not an exhaustive one it does give a feel for the diversity of use.

Tyres, cabin and window seals, seals for fuel, oil, and hydraulic systems, flexible hose, flexible fuel tanks, integral fuel tank sealants, sealants for air-ducting, diaphragms in fuel pumps and pneumatic devices, electrical insulation, and erosion protection.

Apart from the usage in tyres, the requirements are largely for 'speciality' rubbers, and in many of the applications, such as seals, sealants, and hose, crucial properties are resistance to a wide range of aircraft fluids and retention of highly elastic properties over a wide temperature range.

Speciality rubbers finding the most significant usage are the medium and low nitriles, polysulphides, neoprenes, methyl silicones, fluorosilicones, ethylene propylenes, and fluorohydrocarbons, with the nitriles still being the principal work-horse because of their inherent fluid resistance and the ability to formulate to meet an appreciable part of the wide temperature range requirement.

Of the newer speciality rubbers the fluorohydrocarbons, as exemplified by the commercially available Vitons and Fluorels, have probably attracted the most attention, their particular virtues from the standpoint of aerospace application being an exceptional resistance to numerous chemicals and to ageing, and their outstanding high temperature stability.

These properties singled them out for use in aircraft fuel and hydraulic systems (excepting those operating with Skydrol fluids) where they would find application as sealants and static and dynamic seals.

In the latter applications a very influential property in determining in-service life is the resistance of the seal to compression set, and initially fluorohydrocarbon vulcanisates were somewhat disappointing in this respect.

The contrast between the resistance to thermal ageing of the fluorohydrocarbon rubbers as shown by conventional oven ageing and compression set testing was striking (Table 16.1).

Table 16.1. Ageing of fluorohydrocarbon vulcanisates in air at 200°C.

	Conventional oven ageing test		*Compression set test*
Time (h)	*% loss of UTS*	*% loss of EB*	*% set*
24	0	0	47
120	0	0	73
336	0	0	92
5,000	27	7	—
10,000	42	30	—
20,000	58	47	—

Over the time-scale for almost complete set to occur, and by implication the time for a complete loss of sealing force in a static or dynamic seal, there is no apparent loss of mechanical properties in the conventional oven ageing test. Clearly substantial changes are occurring in the vulcanisate structure which are having virtually no effect on the condition of the rubber as judged by the results of the latter tests. It follows that whatever changes are taking place in the rubber network structure are not influencing the overall elastic properties of the material but are leading to a permanent change in dimensions of the compressed rubber with a consequent degradation in sealing force.

Research on the mechanism of thermal ageing in these early fluorohydrocarbon vulcanisates show that the site of instability was the cross-link structure introduced during the vulcanisation process, and that the reaction is thermo-oxidative in nature. Development work subsequently concentrated on introducing modified cross-link structures and stabilising the network by appropriate heat treatments, and the outcome has been a major improvement in resistance to compression set in these types of materials.

Composite Materials

Composite dry bearings

An increasingly important, but relatively unsung area of aerospace application for composites is in dry bearings, and the scale on which such bearings are now being used is indicated in Fig. 16.1. The practical advantage of dry bearings over those lubricated with oil or grease can only be fully realised when composite liner materials with suitably low wear rates become available, and this has proved an extremely fruitful area for research and development on speciality composites.

The composite is used in the form of a thin liner for the bearing and life in service is governed by its rate of wear. As wear proceeds, a gradually increasing backlash develops in the bearing, and in practice this reaches an unacceptable level long before the liner has become worn through. An important area for research has, therefore, been that of studying the effects of variations in materials and forms of construction of the composite liner on the rate of bearing wear. For these thin composite liners the number of possible variations in materials and modes of construction is large, and they include, for example, the weave structure of the reinforcing fabric, the type of fibre and the size applied, the method of incorporating solid lubricant, the type of solid lubricant, and the type of resin binder.

For the last few years, the most widely used aerospace dry bearing material was composed of a thin liner of interwoven PTFE fibre and glass-fibre impregnated with a phenolic resin and adhesively bonded to a metal substrate. A typical interwoven structure is shown in Fig. 16.2. However, a second generation of liner materials began to emerge with a significantly greater degree of wear resistance, particularly in the early stages of sliding. Some of these materials are variants of the original type incorporating different weave structures, different resins, or with the glass fibre replaced by synthetic fibres. Other types are basically synthetic fibre fabrics reinforcing thin layers containing dispersed PTFE particles or flock.

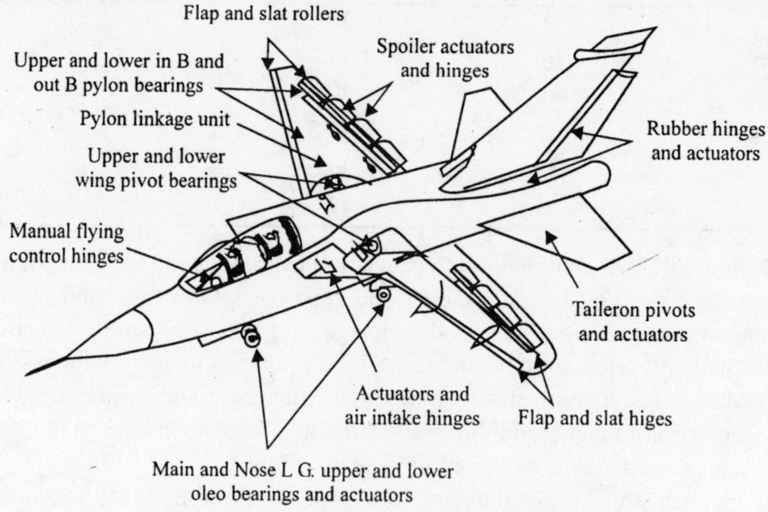

Fig. 16.1. Use of dry bearings in fixed wing aircraft.

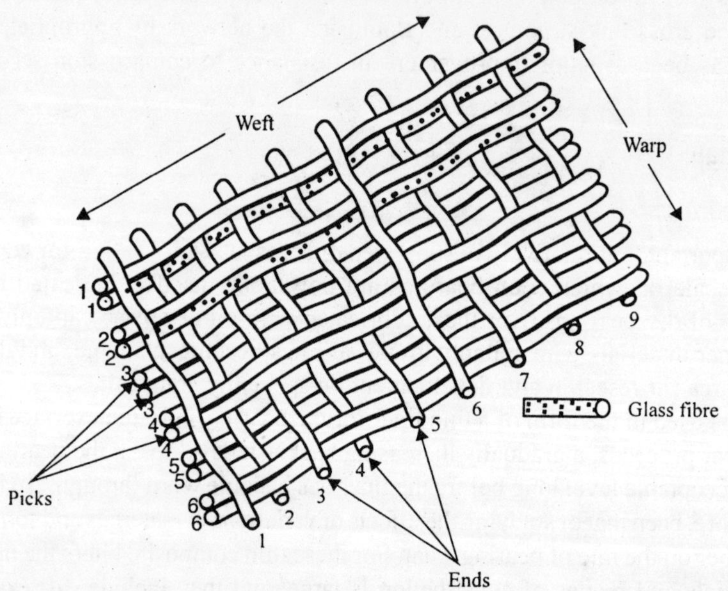

Fig. 16.2. Weave structure for standard dry bearing liner material; PTFE interwoven glass fabric; and PTFE/glass interwoven fabric.

Research has shown that perhaps the most important single factor influencing wear in fabric-based composite bearing liners is the weave structure of the fabric. Some results obtained using different weave structures incorporating both PTFE and glass fibre, are shown in Fig. 16.3, all the liners having been fabricated using the same resin binder. Adopting the commercially available double plain weave material as a yardstick for comparison (Fig. 16.3 curve 1), it can be seen that when the outermost PTFE fibres are replaced by glass it is no longer possible to establish a wear plateau (Fig. 16.3 curve 2). This suggests that too little PTFE is now available to form the desirable 'third bodies' on the rubbing counterface or on the linear surface or on both. This trend is further accentuated when the PTFE content is reduced even further by halving the denier of the PTFE warp fibres (Fig. 16.3 curve 3). By contrast a very slight reduction in the original PTFE content, by replacing 10 per cent of the warp fibres with glass, reduces wear to a small extent (Fig. 16.3 curve 4). With a four-end satin weave (PTFE warp and glass weft) a large amount of wear occurs before the wear plateau is established (Fig. 16.3 curve 5), but with a plain weave the converse is true (Fig. 16.3 curve 6). However, none of the composites containing interwoven PTFE fibres performs as successfully as one containing an all-glass fabric with PTFE particles dispersed throughout the resin matrix (Fig. 16.3 curve 7). This particular configuration has also been found to give low wear when the glass is replaced by other such fabrics, as polyamide, aromatic polyamide, or carbon.

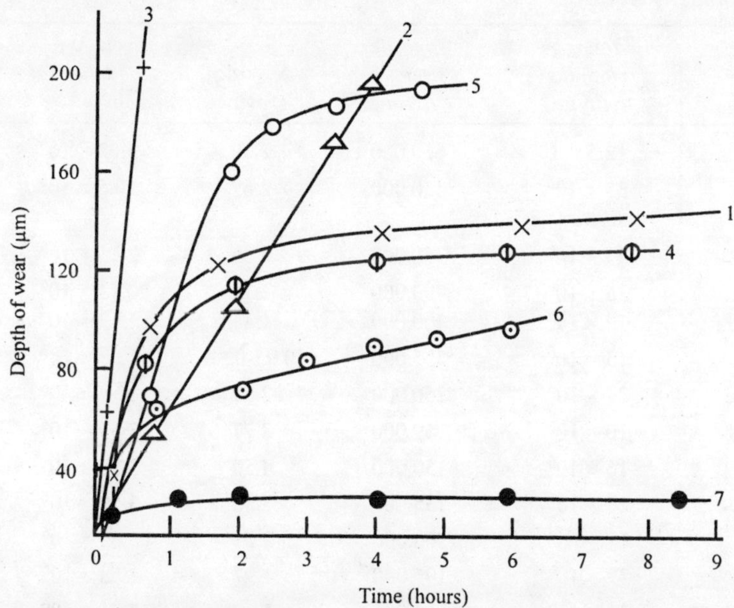

Fig. 16.3. Effect of weave structure on wear rates in PTFE/glass fabric 1. Double plain weave – PTFE warp and top weft; 2. double plain weave – PTFE warp, both wefts glass; 3. as (2) but PTFE warp denier halved; 4. as (1) but 10 per cent of PTFE warp replaced by glass; 5. four-end satin weave – PTFE warp, glass weft; 6. plain weave – PTFE warp, glass weft; 7. all glass plain weave + PTFE flock in resin.

Structural Composites

One of the many attractions of organic polymers and polymer-based composites for aerospace applications is their generally low specific gravity. What has inhibited their wider use has been the inability to match the strength and stiffness of traditionally used metallic materials. However, with the advent of carbon fibres and aramid fibres, and the further extremely exciting prospects for the production of other strong, stiff, and even lighter organic polymer fibres, the whole position is changing. Table 16.2 shows mechanical property levels for a selection of commercially available synthetic fibres and composites based on them, and for some metallic materials favoured for aerospace applications.

The materials which commend themselves most strongly for aerospace structural applications are unquestionably the composites based on carbon fibres, the preferred variant at present being high strength fibre in combination with an epoxy resin matrix. The high levels of strength and stiffness possessed by these materials, coupled with their low specific gravity leads to remarkably high 'specific property' levels. High strength CF/epoxy composites, for example, show a specific modulus 5.6 times greater than that of aluminium alloy (RR58) or titanium alloy (Ti-6Al-4V) and a specific tensile strength 10 times greater than light alloy and 7 times greater than titanium alloy. Apart from carbon fibre, the most commercially advanced of the strong, stiff synthetic organic fibres are the aramids, and in composite form they are also making inroads in the aerospace field, albeit in secondary structure. The other synthetic organic fibres included in Table 16.2 that is ordered polyethylene and ordered polybenzothiazole, are still at the research stage but are nonetheless displaying extraordinary mechanical property levels at low specific gravity. These types of material undoubtedly have potential for further development and could well make an impact in the structural composite field in the not-too-distant future.

Table 16.2. Mechanical property levels.

Materials	*Young's modulus E (lb/in²)*	*Tensile strength (lb/in²)*	*Specific gravity*	*Specific modulus, lb/in²*	*Specific strength, lb/in²*
S-glass fibre	12.5×10^6	650,000	2.49	5.02×10^6	761,000
Boron fibre (tungsten substrate)	35×10^6	450,000	2.69	20.5×10^6	167,000
Carbon fibre, high modulus	55×10^6	250,000	2.00	27.5×10^6	125,000
Carbon fibre, high strength	35×10^6	400,000	1.74	20.11×10^6	230,000
Aramid fibre	19×10^6	400,000	1.44	13.2×10^6	277,777
Ordered polyethylene	13×10^6	435,000	0.95-0.98	13.5×10^6	453,125
Ordered polybenzothiazole	27×10^6	250,000	2.54	10.6×10^6	98,425
Aluminium alloy (RR58)	10×10^6	59,000	2.77	3.59×10^6	21,300
Titanium alloy (Ti-6Al-4V)	16×10^6	150,000	4.50	3.55×10^6	33,333
Steel (FV 520)	28×10^6	155,000	7.80	3.59×10^6	19,870
S-glass/epoxy composite	7.6×10^6	260,000	2.08	3.65×10^6	125,000
Boron/epoxy composite	39×10^6	195,000	1.97	19.8×10^6	99,000
C fibre/epoxy, high modulus	30×10^6	150,000	1.60	18.75×10^6	93,750
C fibre/epoxy, high strength	19×10^6	220,000	1.55	12.26×10^6	141,900
Aramid/epoxy	11×10^6	200,000	1.38	7.97×10^6	144,927

Early exploitation of continuous carbon fibres was principally in the form of highly aligned pre-impregnated tape or sheet material, which could be cut and stacked as required. Pre-impregnation was provided by a partially cured epoxide resin which gave the pre-preg adequate coherence and tackiness for subsequent handling and laminating. Laminates were prepared by the compression moulding of stacks of pre-preg sheet with individual layers arranged to give the fibre orientation desired in the finished component. With regard to this technology one could either be an optimist and see it as providing the designer with a new and considerable freedom to incorporate reinforcement only in those directions where it was required, or be a pessimist and see the requirement to use a highly anisotropic material as being disadvantageous and something giving rise to particular and undesirable difficulties in design. However, in some areas of application, as for example, in the fabrication of helicopter rotor blades, all were agreed that the use of unidirectional pre-preg technology, offered positive advantages in allowing the mechanical properties of the blade to the optimised in particular and desirable directions by appropriate orientation of the fibre layers.

Carbon fibre is now available commercially in a wide range of tow sizes and a wide range of forms. In high strength fibre, tows of 1000–160,000 filaments are available, and in high modulus fibre, the tows can be obtained in counts 400–40,000. Forms available are unidirectional sheet or tape, woven cloth (plain, square, twill, satin, and bias weaves), random mat (short-fibre sheet or felt), aligned mat, knitted and braided, and the way is, therefore, open for widespread application. Interest is also growing in the use of 'hybrid' composites wherein more than one reinforcing fibre is used, the incentive to usage being either one of cost reduction or of enhancement of specific properties. Commercially available hybrid fabrics, in which the fabric is woven from a mixture of fibres, feature carbon with aramid, E-glass, S-glass, or polyester fibre. The use of carbon fibre in woven form is technically acceptable in a wide range of aerospace applications, and in practice it can ease difficulties in fabrication and reduce component costs appreciably. A move towards the more widespread use of carbon cloth is already becoming evident in the aerospace industry.

A reduction in weight of an airborne vehicle is welcome from whatever source. The incentive to use lightweight composites in primary and secondary structures is, therefore, particularly strong, and nowhere stronger than in the fields of satellites and other space vehicles and in vertical lift aircraft. Not surprisingly, therefore, the largest pieces of CFRP structure in service to date are to be found on a space vehicle, i.e., the NASA space shuttle, and the most widespread use of composite on a high performance aircraft is at present envisaged for the advanced Harrier Av-8B jump jet. The former vehicle has its payload bay doors, which are 60 ft. long and 1500 sq ft. in area, fabricated in CFRP, with an estimated weight saving of about 23 per cent over the same structure in aluminium alloy, whilst in the latter aircraft, advanced composites are expected to account for about 26 per cent of the airframe structure.

Composite Applications in Helicopters

Lightweight structural composites are naturally attractive for use in helicopters. The most significant area for application so far has been in the main and tail rotor blades where the prospects for improved fatigue resistance over all-metal blades, and the greater freedom in design and fabrication have been quickly recognised and exploited. The use of composites based on unidirectional pre-preg has allowed for a better optimisation of rotor blade mechanical and dynamic characteristics, and compared with metal blades they have demonstrated better static and fatigue strengths. In state-of-the-art composite blades, the principal material of construction is GRP, with discrete areas of usage for CFRP. Substantial numbers of blades have been manufactured using these materials and are fitted to production aircraft. European manufacturers have been in the forefront of composite rotor blade development and the most impressive in-service experience has been gained on helicopters of French and German origin which have seen service world-wide. Many millions of flying hours have been accumulated with high life blades now approaching 7000 hours, and in-service performance is coming fully up to expectations. The blades are demonstrating increased performance and greater reliability than their all-metal counterparts, with the result that significant reductions in direct operating costs are being realised in practice. An attractive feature of GRP for this type of application is the inherently slow rate of crack propagation which is found to occur. This gives the blades a 'fail-safe' quality in that a period of gradual deterioration precedes any major failure.

For the future, consideration is being given by constructors to a much wider use of composites in the airframe structure, and in some quarters it is expected that composites will comprise upto 50 per cent of the empty weight of future helicopters.

Composite Applications in Fixed Wing Aircraft

Following closely on the commercial development of continuous carbon fibre, steps are taken in the UK to design, fabricate, and flight test a limited number of CFRP components of a non-critical nature (i.e., not critical to flight safety). These components are fitted to operational aircraft in order to begin the long process of building confidence in the service usage of such novel materials. The earliest such components were the rudder trim tabs for the Jet Provost trainer aircraft in use with the RAF; six in number, and of thin-skinned honeycomb construction, they are fitted in an unpainted condition. The aircraft fitted with these components spent most of their service life with the RAF Red Pelicans' aerobatic flight, and are subjected to the full range of weather and flying conditions without being given any special treatment. Regular visual and non-destructive inspection confirmed that in-service performance had been entirely satisfactory.

It gives considerable satisfaction that each one of the oldest of the UK flight test components has performed very well under the rigours of in-service use. Of particular note is the absence of corrosion problems on the Vulcan air-brake where CFRP is attached to light alloy, the absence of foreign object impact damage on all the components, and the apparent resistance of the CF/epoxy to aircraft fluids in service.

In parallel with the UK programme, flight tests have also been carried out in the USA on a similar range of non-critical components, but as might be expected, their tests were on a much larger scale. The ability to fit composite components to civil aircraft proved a great boon to the American programmes and has resulted in the accumulation of flying hours at a very high rate, the two most outstanding examples being the Boeing 737 spoilers and the DC-10 upper aft rudder segment. In the former trial 108 spoilers were manufactured and 3 were fitted to each of 27 in-service aircraft. Having this number of components allowed a programme of planned retrievals from flight service and also a limited amount of destructive testing. Six spoilers are selected annually for removal and taken back to the Boeing factory for inspection, using ultrasonic techniques. A further 3 spoilers are selected for destructive testing each year and are tested in the same static test jigs as were used for the initial certification tests on the as-manufactured components. Since flying started in 1973, well over a million flight hours have been accumulated, no significant changes have been observed during non-destructive testing, and the results of destructive tests have been found to fall within the scatter band of strengths of the as-manufactured components. The other component which has been accumulating flight hours at a great rate is the DC-10 upper aft rudder segment, also made in CF/epoxy.

Ten of these components were fitted and flying commenced in 1976, the utilisation rate on some of these has been extremely high, sometimes amounting to 400 hours per month. Overall the USA civil aircraft trials involving CFRP components have accumulated an enormous number of hours, and as yet no significant incidents or damage have been reported. The UK and USA experience with these non-flight-critical composite components has been good, and the in-service durability has come fully up to the expectations of those protagonists of composite materials who expected them to show lower maintenance costs than similar standard production metal items. The range of composite items either in use or projected for early usage on production aircraft is indicated in Table 16.3.

Table 16.3. Aerospace applications for composite.

Wing to body fairings	Vertical and horizontal stabilisers
Doors and access panels	Wing skins
Flooring	Engine nacelles and cowlings
Radomes	Helicopter rotor blades
Air brakes	Dry bearings
Air ducting	Drive shafts
Control panels	Leading and trailing wing surfaces

The major step of introducing advanced composites into primary flight critical structure is understandably proceeding more slowly, but significant advances are nevertheless being made. There have been, and continue to be, a number of major structural demonstrator programmes in both the USA and UK wherein design and fabrication processes are validated. Major ground tests of static and fatigue strength and environmental resistance have been carried out successfully and this has led, more particularly in the USA, to the manufacture of some large aircraft components. The Av-8B wing structure has already been mentioned, and another heavy user of composite structure in production is the McDonnell Douglas F-18 Hornet.

Thus, it is fitting to mention what is by far the most ambitious use of structural composites (carbon fibre/epoxy resin) in aircraft construction to date, that is the Lear Fan 2100 seven-seater light aircraft. This aircraft has an all carbon fibre/epoxy airframe which gives it an empty weight of only 3850 lbs, and it made its maiden test flight successfully earlier 1997.

APPLICATIONS OF RUBBER IN RAILWAY ENGINEERING

The long range elasticity and resiliency of rubber and rubber-like materials, offer advantages in eliminating undesirable shocks, vibrations and noise associated with vehicles moving at high speed. Shocks and vibrations in rail vehicles are much more severe due to rail-wheel inter-sections, track irregularities, crossings and rail joints, firing impulses of engines, etc. Railway engineers too have taken advantages of rubbery state of matters in designing and construction of all types of rolling stock and track. The Indian railways utilises a large number of functionally different rubber components of which one of the oldest use of rubber which immensely interested the railway men is buffer spring against the longitudinal shocks.

Indian railways recently adopted ballastless track structure especially for Metropolitan Transport System. Function of the rubber pad is critical, providing entire elasticity to the track and minimising undesirable vertical and lateral vibrations. Different design features and physical characteristics which have been optimised for the purpose have also been discussed here.

The ability of rubber to change shape under load, to resist this change in shape and subsequently to recover from the change in shape left no competitors for the purpose of elimination of shocks. In Indian railways, helical metal springs inside buffer of wagons have been replaced by metal bonded rubber springs with improved serviceability and durability.

The design has undergone some modifications and high capacity buffer spring of more than 1280 kg-m has come to stay. These rubber spring elements are assembled in series with varying numbers depending on the type of rolling stock, to form a pack as shown in Fig. 16.4. The nominal thickness of each spring is 32 mm and the free height of the assembly of 16 springs is 512 mm. The stroke of assembly is limited to 127 mm and pre-compression 45 mm nominal.

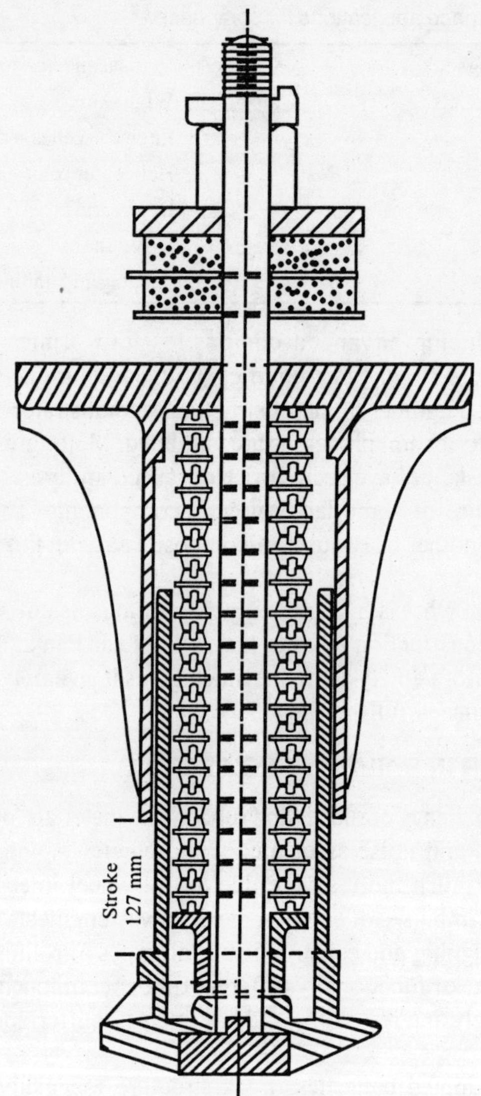

Fig. 16.4. Rubber springs in buffer assembly

Rubber Buffer Spring

This design is considered to be unique having 'U' section circular channel and optimum 'shape factor' for improved performance and durability. Load-deflection characteristics as shown in Fig. 16.5 nearly follow the pattern of a metal spring upto a load of 12-15 tonne which is within the normal operating zone.

This type of load deflection characteristics in rubber is attributable to the combination of air cushioning of the entrapped air and the 'shape factor'. These combinations of design features also enhanced the capacity more than 1280 kg-m. The impact endurance test on the assembly indicates retention in capacity more than 85 per cent. The various parameters stipulated for the assembly of 16 springs are given as under:

1. Capacity for a pack of 16 springs, minimum 1030 kg-m
2. Load to compress the free height to 467 mm, maximum 2 T
3. Load to compress the free height to 340 mm, maximum 40 T

The typical physical properties of the rubber are as follows:

1. Hardness (IRHD) 70 – 80
2. Tensile strength (kg/sq cm) 150
3. Elongation at break (per cent) 250
4. Compression set (per cent) at 70 ± 1°C for 24 hrs. 30

The introduction of the rubber buffer spring in place of steel spring has resulted in substantial economic gain to the railways in terms of improvement in service ability, operational efficiency and reduction in maintenance and on material cost. The service ability of these springs under Indian climatic conditions is observed to be around 8 years and in some cases beyond 12 years.

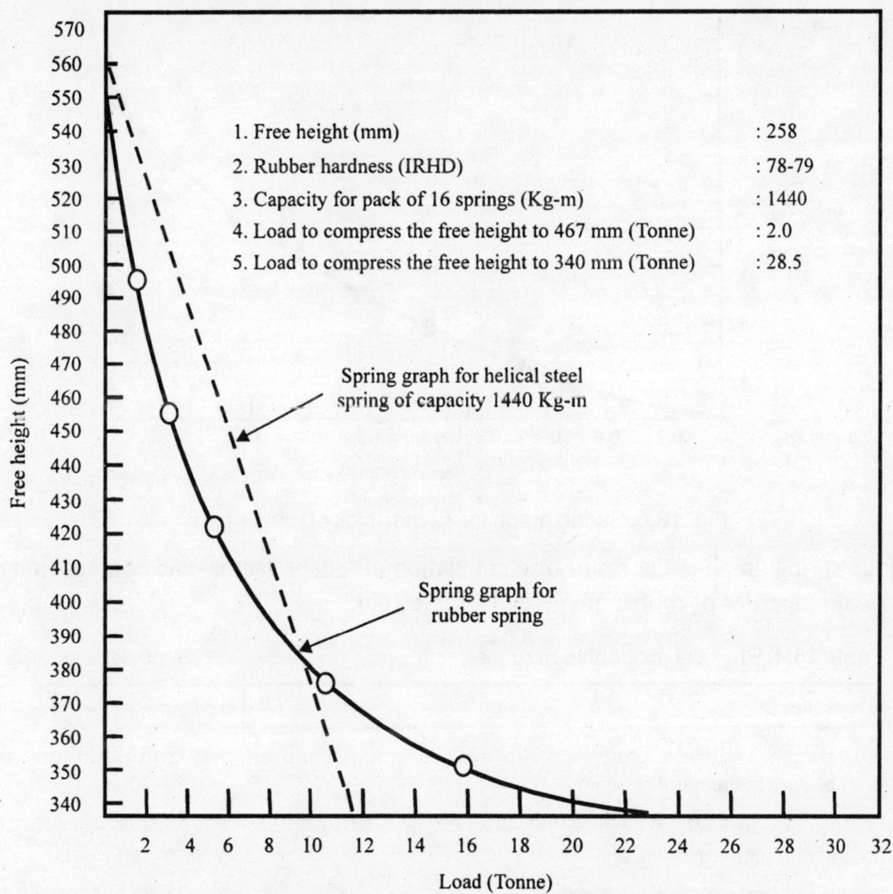

Fig. 16.5. Load and deflection characteristics.

Rubber Pads for Ballastless Track

In ballastless track assembly, rubber pad provides entire elasticity to the track. The rubber pad is placed between the bearing plate and the concrete bed. From assembly design consideration, thickness of the pad has been kept 12 mm nominal. Semi-circular grooves in the transverse direction have been provided on the pad to obtain 'shape factor' and serviceability. With this typical design, the fatigue is expected to be minimum.

The static spring graph is shown in Fig. 16.6. The spring characteristic is nearly linear except in initial range of loading. This non-linear portion of the curve is compensated by the pre-compression of the assembly and therefore, spring characteristic of the pad is linear under actual service conditions.

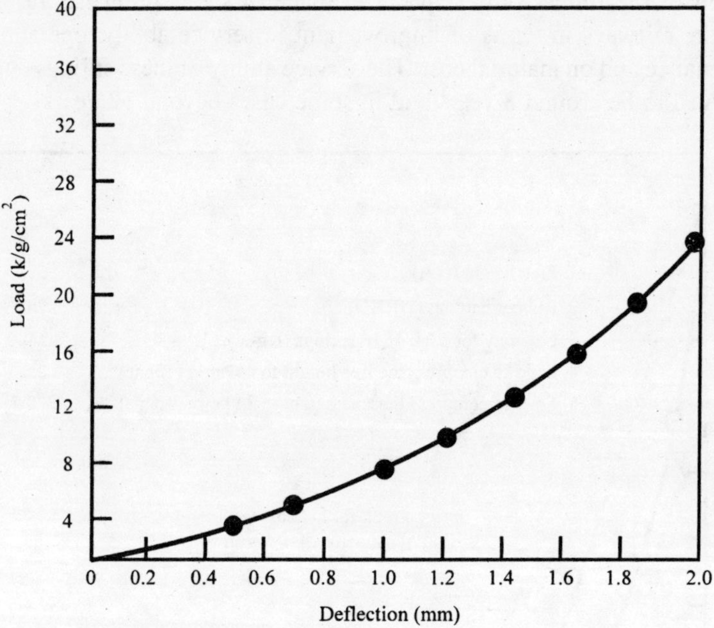

Fig. 16.6. Spring graph for 12 mm thick of rubber pad.

This unique spring graph is the result of combination of 'shape factor' and composition of rubber. Various physical properties of rubber are given in Table 16.4.

Table 16.4. Physical properties of rubber.

	Properties	*Observed values (range)*
i.	Hardness (IRHD)	55–65
ii.	a. Tensile strength (kg/sq.cm)	150–200
	b. Per cent retention after ageing*	80–90
iii.	a. Elongation at break (%)	450–500
	b. Per cent retention after ageing	75–85
iv.	a. Stress at 100% elongation (kg/sq.cm)	20–35
	b. Change after ageing (%)	± 25

(Contd ...)

Properties		Observed values (range)
v.	Set (%)	
	a. Subjected to 50% strengthening	15–20
	b. Subjected to 50% compression	15–25

* Air ageing at 100 ± 1°C for 96 hrs.

In consideration of durability in service, the stipulations of the ageing characteristics have been kept stringent and achieved after long strive.

With preliminary assessment of the spring characteristics and the physical properties of rubber, the pads are subjected to various simulated service tests in Vibrogir Pulsator and limited service trial on experimental track. Test parameters for Pulsator and Vibrogir are given in Table 16.5.

Table 16.5. Test parameters for Pulsators and Vibrogir.

Pulsator		
a.	Vertical load	Varying between 7.5 to 2 T
b.	Lateral load	Varying between 4.6 to 1.2 T
c.	Frequency	300 cycles/minute
d.	No. of cycles	2.5 million
Vibrogir		
a.	Pulsating load	± 4.3 T
b.	Induced cyclic load	+ 8.3 to –0.3 T
c.	Frequency	50 cycles/sec.
d.	Acceleration	100 'g'
e.	Test duration	50 hours

Evaluation of the Pads

For evaluation of the pads during procurement besides physical properties and spring characteristics it is intended to incorporate certain dynamic properties based on Yerzley oscillograph studies. The parameters obtained in this test are as under:

1.	Resilience (per cent) at 20 per cent deformation	80–85
2.	Static modulus (kg/sq.cm) at 20 per cent deformation	55–65
3.	Effective dynamic modulus (kg/sq.cm) at 20 per cent deformation	70–90
4.	Oscillating decay	0.35–0.40

The test results are compared with the rubber pads developed by the advanced railway systems. Pads developed indigenously are found to be most suited one as compared to others.

In test track, pads were laid for extended field trials in tunnels. After 8 years of service it was observed that the retention in physical properties is more than 80 per cent even with pads laid outside the tunnel having exposure to direct sunlight and rain fall—most adverse conditions. The above application is also considered as a major indigenous invention.

CHAPTER 17

Special Applications of Polymers and Rubber

INTRODUCTION

In some areas of applications, polymer/plastic materials have been long established. A well-known example is in the electrical industries where the excellent combination of properties such as insulation characteristics, toughness, durability, and, where desired , flame-retardation capacity have led to increasing acceptance of plastics for plugs, sockets, wire and cable insulation, and for housing electrical and electronic equipment. This has lead to the increased use of the more general purpose plastics.

Plastics have been widely accepted in industrial equipment development as materials for many critical components and for accessories where such features as toughness, abrasion resistance, corrosion resistance, electrical insulation capability, nonstick properties, and transparency are of importance. For example, PVC, polyacetal, and PTFE are used for pipes, pumps, and valves because of their excellent corrosion resistance. Nylons are used for such diverse applications as mine conveyor belts and main drive gears for knitting machines and paper making equipment because of their excellent abrasion resistance, toughness, and low coefficient of friction.

In agriculture and horticulture, the most widely known use of plastics is in film form. But plastics also find use for automotive watering equipment, water piping containers, and potato chitting trays, to name but a few uses.

In the medical sector, plastics uses range from medical packaging and disposables (e.g., bandages and injection syringes) to nontoxic sterilisable items (such as tubing and catheters) to spare-part surgery (such as hip joints and heart valves). A wide-range of polymeric materials are found in these applications though medical grades of PVC and polystyrene are the most commonly used. Engineering resins, however, are finding an increasing number of uses in this area.

While natural and synthetic fibres have remained the major materials for clothing, plastics have also found increasing use in this area. Footwear has provided a major outlet for plastics as they are used not only in soles and uppers, but increasingly in an all-plastics moulded shoe that is in high demand in underdeveloped regions of the world. In rainwear, plastics and rubbers continue to be used for waterproof lining and in the manufacture of all-plastics packable raincoat. In cold-weather apparel, polyurethane foam has found increasing use as an insulation layer and for giving 'body' to clothing. While the above paragraphs indicate some of the major uses of polymers in a number of areas, the more recent trends in polymers applications in these areas form the subject of subsequent discussion in the chapter.

POLYMERS IN ELECTRICAL AND ELECTRONIC APPLICATIONS

Polymers have played critical roles in every aspect of the ongoing revolution in electronics. They are now found in everything from computer housings and floppy discs to encapsulants and connectors. The wide range of electrical and electronic applications of polymers includes wire and cable coverings, housing adhesives, printed circuit boards and insulators.

One of the key trends in electronic components today is miniaturisation,. This development is bringing new opportunities for plastics in the thin-walled and downsized products now reaching consumers. In consumer and office electronics, portability is a key design and marketing criterion. Portable computers, printers, copiers, cellular phones, and measuring and monitoring devices are highly plastics intensive. Today's portable (laptop and notebook) computer, for example, has 50–75 per cent of its weight (and even more of the volume) represented by plastics.

Wire and Cable Insulation

Neoprene synthetic rubber has been used for many years in the cable industry as it provides good low voltage insulation and mechanical toughness as well as good resistance to oil, high temperature, and oxidation. Alcryn, (from Du Pont), is a melt-processable rubber cable sheathing for signal and power transmission. It is claimed to have excellent oil resistance even at high temperatures, and good resistance to severe mechanical treatment. Another key product (from Du Pont) is Hypalon, a chlorosulphonated polyethylene that exhibits high resistance to oils and a host of chemicals. One application of this product is the jacketing of cables in chemical plants.

The largest single outlet for plastics in electrical and electronic applications—roughly half of the volume consumed—is the wire and cable, where PVC dominates on the basis of its formulating flexibility, ease of processing, and low cost. However, the halogen-containing polymer systems tend to be more corrosive to copper than the non-halogen-based systems. Chlorinated compounds can also result in unacceptable emissions from conventional incinerators. Because of industry and regulatory concern over combustion, corrosively, toxicity and environmental impact at the points of manufacture and disposal, extensive use of PVC in future thus appears uncertain.

Fluoro-polymers such as Teflon and Tefzel that have high mechanical strength and dielectric properties with heat resistance enable thin wall insulation for high current densities. Du Pont's fluoro-elastomer Viton is used for insulation and heat shrinkable sleeving. In addition to excellent heat and chemical resistance it has high resistance to fuels, oils, and lubricants. Fluon, a polytetrafluorethylene (PTFE) product from Fothergill Cables, which specialises in electronics cables for aerospace and defence industries, is applied to thin wall insulation for equipment wires, multicore cables, thermocouple wires, and radio-frequency coaxial cables. Fluon has a service temperature range of –75 to +260°C, non-flammability, high chemical resistance, good dielectric properties, and resistance of damage by soldering.

Polyesters in wire and cable applications include Hytrel (Du Pont), a thermoplastic polyester elastomer for the offshore industry; Kapton (Du Pont), a polyimide film offering weight saving and chemical resistance; and Mylar (Du Pont), used in primary insulation, shield isolation, chemical barrier, and mechanical protection.

Large number of products to suit different applications in the wire and cable industry have been developed. While for general purpose insulation and sheathing, poly (vinyl chloride) (PVC) is used with chlorinated plasticisers such as polyether ether ketone (PEEK) wire and cable covering is designed to operate in high-demand environments. PEEK has high-performance electrical properties that permit the use of smaller diameter cable covering without loss of electrical power or reduction in insulation.

Because of the product's good inherent flame retardancy, low smoke emission, and low toxic and corrosive gas emissions it is used in underground mass transit wiring and cable insulation. PEEK has also been used for insulation a coaxial control cable operating in a high radiation environment at a nuclear installation in the United Kingdom.

A new technology has been introduced for silane cross-linking in low-voltage cable insulations to produce more uniform cross-linking and greater shelf-life. The new silane-linked polyethylene compounds can be run on conventional extrusion equipment. A strippable double layer cable shield in cross-linkable polyethylene for medium voltage cables has also been produced. The shield has a strippable outer layer formed by a coextrusion of 20 per cent strip material and 50 per cent standard semi-conductive material.

Polymer Insulators

Transmission insulators provide mechanical support and electrical isolation for overhead, open wire, power transmission lines. The primary insulation on these lines is air. All insulators use an innovative and unique silicone compound to provide an unbonded interface between insulating rubber housing and load-bearing fiberglass core. This interface has proven to be more reliable than the chemical bonded type of interface.

Printed Circuit Boards

Developments in the electronics industry are focused on two main areas—miniaturisation and improved cost/performance ratios—creating the need for polymers that can tolerate higher temperatures for longer periods with good dimensional stability and mechanical strength.

Though the majority of printed circuit boards (PCBs) at present are two-dimensional and made of glass-filled epoxy resins, increasing miniaturisation is leading to more PCBs being multi-layered with several copper circuits sandwiched between insulating layers. In these applications, because of the added stress due to thermal expansion, speciality materials such as polyamides and fluoro-polymers are being increasingly used.

The present trend is towards injection-moulded thermoplastic boards in place of thermoset laminate PCBs. The two main reasons for this shift are the difficulty of etching on thermoset laminates, and the greater design flexibility and potential cost saving in high-volume production by injection moulding. A further advantage of injection moulding is the ability to produce circuit board with three-dimensional features such as ribs, bosses, heat sinks, and holding devices for snap-parts. Engineering polymers that can be used for moulded circuit boards include polyethylene terephthalate (PET) and glass-reinforced grades of polyethersulphone (PES).

In the electronic industry, one of the most successful uses of advanced composites has been in PCBs. Polyetherimide is being used in transport PCBs, especially aerospace and missile guidance systems. Polysulphones and other high-performance thermoplastics are being used in multilayer PCBs for mainframes and supercomputers operating at gigahertz frequencies.

Process developed for the injection moulding of the circuit boards include PCK Technology's mould-n-plate process and an in-mould metallisation process. The mould-n-plate process uses two shots, the first being one of catalytic resins, to form paths on plastics parts that are then processed in a copper bath to form circuitry.

Special grades of polysulphone, polyethersulphone, and polyarylsulphone are used for the process. In the in-moulding metallisation process, various conductive, resistive, capacitive, or inductive components are embedded directly into the plastic part in the tool.

Connectors

The miniaturisation of PCBs has created the need for connectors to be made from special materials suitable for shaping into thinner parts that can be packed into higher-density spaces and that can withstand higher soldering temperatures. Such materials available on the market include tailored grades of polyphenylene sulphide (PPS), PEEK, PET, polyetherimide, sulphone-based alloys, and liquid crystal polymers.

Enclosures

Plastics are increasingly being used for electrical/electronic enclosures. Those used for enclosures of circuit chips should, however, be of high purity and high flow. Since miniaturisation reduces the surface area of heat dissipation, the materials used are required to have high heat resistance in addition to high purity. While epoxy resins dominate the market, other materials used include polyurethanes, polyimides, and silicones.

Plastics used in electrical housings are now required to provide protection against electromagnetic interference (EMI), radio frequencies, and electrostatic charges that could damage sensitive equipment circuits. To impart EMI shielding properties to a plastic component, a coating of conductive layer may be provided by post-moulding or in-moulding processes or polymers may be used that have been metal or carbon black filled to make them conductive.

The methods mostly used to produce EMI shielding on plastic moulding are conductive paints, zinc arc spraying, vacuum metallising, electroplating, and foil application. There are also a range of filled plastic compound on the market that are based on metal fillers such as aluminium flake, brass fibres, stainless steel fibres, graphite-coated fibres, and metal-coated graphite fibres. The most cost effective conductive compounds are, however, based on carbon black.

One of the key determinants of future material selection for enclosures will be the growing global demand for recycling of these units. The recycling of electronics equipment enclosures and other moulded electronic parts, however, presents formidable obstacles, since there are many different resins in use, as well as a wide range of performance-enhancing additives and functional coatings for EMI shielding and ESD protection. To overcome these problems, industry is likely to narrow the enclosures resin menu to the usable plastics stream and focus on a well-defined groups of additives, coatings, adhesives, and other ingredients.

Optical Fibres

A major advance in communications has been the introduction of optical fibre transmission. In optical fibre transmission, voice and data signals are converted into beams of light that then travel along extremely thin glass fibres. Major advantages of this transmission over the conventional current carrying metal wire are higher message speed and density, lighter weight, and immunity to electromagnetic interference. The main growth in the use of optical fibres is expected to be in telephone networks.

Plastics find extensive use in several areas of fibre optic cable. Buffer tubes, usually extruded from high-performance plastics such as fluoropolymers, nylon, acetal resins, or polybutylene terephthalate (PET) are used for sheathing optical fibres. A blend of PVC and ethylene vinyl acetate (EVA) polymer, does not require a plasticiser, which helps the material maintain stability when in contact with water-proofing materials. PVC and elastomers blends are also used for fibre optic applications. (Stiffening rods for fibre optics are either pultruded epoxy and glass or steel. Around these is the outer jacketing, which is similar to conventional cable).

Japan's Mitsubishi Rayon Co. Ltd. developed a plastic optical fibre cable 'Eska Premium' with a wide range of freedom, excellent durability and which can be used for wiring in narrow spaces. It is claimed that the cable can be bent or wound without deteriorating the characteristics of the fibre cable. The cable is lighter and easier to work with than glass-based cables, and is ideal for short communications and transmission wiring between the controllers of machine tools, personal computers, and peripheral equipments.

Information Storage Discs

In recent years, information has begun to be stored magnetically, on tape and on disc. Today their position is being challenged by optical discs, because an optical disc can store 1,000 times more information than its magnetic equivalent, has a longer guaranteed lifetime (usually 10 years), and more easily accessible data.

Although magnetic storage discs are beginning to be regarded as archival as compared to optical discs, the advantage they have over most optical systems is that information can be added and erased.

The compact audio disc (CD) based on digital recording and playback technology is 4.75 inches in diameter and can store 16 million bits of information in the form of minute pits in the substrate. It is the presence or absence of the pits that is read by the laser. The pits are 0.1 microns deep, 0.5 microns wide, and between 0.833 and 3.56 microns long. Each 'track' comprises a spiral of these pits. The track is laid in polycarbonate that is backed with reflective aluminium and coated with a protective acrylic lacquers. The CD can thus be played only on one side.

The CDs are manufactured either by injection compression moulding or injection moulding. Injection compression moulding minimises moulded-in stress but requires longer cycle times than injection moulding and produces flash. For both processes very clean room conditions, up to 1000 times cleaner than medical moulding, are essential.

The processability of the polycarbonate, or any other material used as the substrate, is crucial in the manufacture of all optical discs. The material must have low melt viscosity and high melt flow to minimise internal stresses caused by molecular orientation of the polymer during filling. Also important are high long-term dimensional stability, resistance to moisture and gas permeability, low birefringence, and superior optical properties. The molecular weight of the polycarbonate has been reduced to give an exceptionally high melt flow rate compared to a standard polycarbonate. In addition to polycarbonate, a number of other materials including acrylic resins, polyacrylate, thermoset epoxies, and blends on co-polymers have been tested for their suitability in optical discs.

Polymeric Field Effect Transistor (FET) and Light Emitted Diodes (LED)

The fact that many conducting polymers are semiconductors in the undoped or slightly doped state has prompted attempts by both industrial and academic groups to use them in microelectronics such as semiconductor devices and especially field effect transistors (FETs). For the semiconducting layer between the source and drain electrodes of a FET, micrometer-thick films are required. Film as thin as this has been prepared by spin coating a solution of precursor polymer on a substrate having the required electrode pattern and subsequent heat treatment in a stream of gaseous dopant to convert it into a conducting polymer.

Improvement of technique has led to carrier mobilities as high as 10^{-1} cm^2/V-s in polymeric FETs. This has opened the way to uses such as flat-panel colour displays for computers and possibly flat colour televisions when tied to a liquid crystal matrix.

Light emitting diodes (LEDs) have also been fabricated from the conjugated polymers such as poly(p-phenylene-vinylene) as the emissive layer. The advances in the chemistry of processible conjugated polymers and focused work on the physics of electroluminescence in these materials have led to the development of flexible, almost entirely metal-free LEDs,. These polymers-based LEDs could be competitive in display applications because of the potential ease, low cost of fabrication, and large surface area of devices based upon processible polymers.

ANTI-STATIC AND CONDUCTIVE USES OF POLYMERS

We all are aware of the insulating properties of rubber—most of our domestic and industrial electrical cables are made using rubber as insulant. Rubbers can be modified to make them conduct electric charges or currents, with the usage of certain additives and modifiers. The major use of conductive or anti-static rubber is the dissipation of electrostatic charges, usually known as static charges. This is the reason why rubbers for this type of application are termed 'anti-static'.

Rubbers can be rendered conductive by incorporating adequate loadings of carbon black, i.e., 20 phr to 100 phr depending upon carbon grade. One other method, though not common, is by usage of fine metal powders like that of nickel and the resistivities by this method may be brought down substantially. With metal powders, especially when used in a liquid polymer, the conductivity may be increased in one direction by applying a magnetic field during hardening.

The terms 'anti-static' and 'conductive' must be defined with respect to the resistance between certain points on a product. When static electricity is merely a nuisance, resistivity values of 10^{10} to 10^{12} ohms may be tolerable, but in case it poses a hazard, the value must be below 10^8 ohms. The conductive products which are also used to prevent static electrification would not have any lower limit of resistance. However, these should not exceed a resistivity limit of maximum 10^5 ohms, though this value is somewhat arbitrary. British standard 2050 defines the dividing line between conductive and anti-static products at a resistance of 5×10^4 ohms.

These conductive rubbers, in some cases, exhibit a thermo-emf and also Hall effect. Hall effect is the generation of an emf in a body when this is subjected both to a magnetic field and to a perpendicular current flow.

Generation of Anti-static Charges

Whenever, two dissimilar solids or a solid and a liquid are placed in contact, there may be re-distribution of charges between them owing to transfer of electrons or ions, so that when they are separated, one surface may carry an excessive positive charge, while the other an excess of negative charge. It is often thought that rubbing of surfaces is necessary and hence the term tribo electricity has come, but this is not the case.

Charges may be separated when the surfaces are separated, even if no sliding has taken place. Owing to the sliding, contact area of two surfaces increases, thereby, enhancing charge generation. Besides the above, due to sliding a slight transfer of material may take place as well as localised heat generation which also plays a role in charge re-distribution.

When an article on which a charge has developed, is discharged rapidly, voltages less than the original voltage and of the opposite signs are developed by electrostatic induction on neighbouring objects. This may result in an incendiary spark discharge to the earth from one of these neighbouring objects even if the resistance to earth of the latter is sufficiently low to prevent the energies of directly generated static charges from reaching incendiary levels.

Applications of Anti-static

Electrostatic charges of the sparks resulting, as mentioned above, are normally quite harmless and unimportant. However, these become considerably important under any of the following conditions.

1. Where inflammable liquids or gases are being handled.
2. Where explosive dust, e.g., coal dust or flour are being handled.
3. Where detonators are being handled.
4. Where the continuity of production of a light material depends on relatively accurate registration between the material and machine, e.g., textile fibres, which are required to pass evenly through spinning machinery and which bow apart when they are charged.
5. Where an insulating material becomes charged and attracts dust.
6. Where highly insulating covers or parts are used for mechanical or electrical equipments which must respond to very slight forces and these forces may be effected (countered or simulated) by electrostatic forces e.g., in very sensitive meter movements.
7. Where sensitive electronic equipment, such as communication equipment, is disturbed by sparking.
8. Where a slight shock or appearance of sparks is disturbing to a person.

Usage of anti-static paints, ionising the ambient air, usage of surface treatments which reduce transfer of charges, are some of the other methods of reduction of static electrification.

Products with Anti-static Usage

Anti-static vehicle tyres

In earlier years, it was common for motor vehicles to become charged due to separation of the tyre tread and the road by rotation of the tyres. Though the problem is normally mildly unpleasant, it is aggravated if the climate is dry. These complaints led to the introduction of anti-static tyres, based on special carbon blacks, but the use of furnace black in place of channel blacks led to all tyres having sufficient conductivity to prevent shock from vehicles.

Static charge interferes with radio communications in vehicles. A further problem was the occurrence of ozone punctures or spark-bite of the inner tube, where the voltage gradient along the internal surface of the tyre was locally sufficient to cause ionisation and the production of ozone, which caused cracking of the strained rubber of the tube.

When an aircraft lands, it may carry a considerable charge (acquired during flight). This could be dangerous to the first person bridging a gap between the aircraft and an earthed object and, therefore, an anti-static tyre is used normally on the tail or nose-wheel of the craft. Vehicles handling explosives should have tyres with resistance below 5×10^4 ohms.

Belts

Voltage builds up rapidly on conveyer and transmission belts and may be vary high in case of V-Belts due to their high rate of separation of surfaces. It could be dangerous in usages like coal mines and places where explosive vapours and dusts are handled, if anti-static compounds are not used.

Hoses

Hoses carrying inflammable powders or liquids or which pass through an inflammable atmosphere are big static hazards. For such cases both hose and container are required to be anti-static. It is required to use single or multi-static conductive wires and fibres in the hoses and these should protrude out at the ends so as to make contact with the hose nozzles. In such applications it is required that the resistivity should not be more than 10^4 ohms at a time. The tyres of the vehicles for this type of applications should also be anti-static.

Material processing machinery

Anti-static rubbers have been used for roll coverings and other drafting items in textile and other light material processing machinery as static charges would tend to bow the materials apart.

Hospital equipment

In earlier years, a number of cases of explosions were reported in hospital operating theatres, resulting from electrostatic discharges. This is critical in the case of some anaesthetics which are highly explosive and have very low ignition energy, especially when mixed with oxygen. With anaesthetic equipment it is necessary to use considerable lengths of rubber tubings in intimate contact with unanaesthetic gases. Some items such as warm blankets, when pulled on an insulated trolley, leave considerable static charges on the trolley, giving rise to the need for the trolley wheels to be conductive.

The floor of an operating theatre should also be anti-static. Other articles of hospital operating theatre equipment which should be anti-static include footwears, articles made from sheeting including aprons, cellular rubber pads and mattress covers, tyres of all trollies and beds, furniture feet, corrugated tubings, anaesthetic and other tubings, anaesthetic masks, bellows, buffers and non-rubber articles also.

An important factor is the increase in resistivity during service and care should be taken for this. For instance, anti-static rubber footwear may be within specified limits when new, but they may leave this range during service. It is recommended for anti-static floorings, that resistance should be so low that use or ageing does not result in resistance rising above 10^6 ohms. However, any change in resistance of flooring are likely to be smaller than those items which are deformed considerably in usage.

Cables

Conductive rubbers are used in cables for various purposes. Where a cable is constructed of a central conductor, an insulant and an outer conductive sheath, minute voids occur at the metal/insulant interfaces. Under the influence of high voltages, discharges take place, causing deterioration of the insulation. If, however, a layer of conductive rubber or plastic is interposed between each of the metallic members and the insulant and is firmly bonded to the insulant, the conductive material takes up substantially the same voltage as the adjacent conductor and hence the air in the voids against metal is not stressed electrically and the voids are innocuous. At the new interface between the conductive and insulating polymers no voids need occur to avoid corona effects in the voids. This makes possible the satisfactory use for power cables up to at least 30 KV and for X-ray cables up to 200 KV. In case of high voltage rubber insulated submarine cables, dry cotton tapes frictioned with an unvulcanised neoprene compound containing acetylene black, are commonly used around metal strand conductors.

Conductive rubber usage has been proposed in microphone cables also. Owing to the local separation of the dielectric from the core and sheath which takes place when an instrument cable is flexed, spurious signals known as 'noise' appear at the end of the cable. By using conductive rubber at each interface, this noise is eliminated. Usage of conductive rubber in a highly flexible conductor and high voltage cables is common in other applications also. Such materials may be used as sheathings, which can carry sufficient current to operate earth leakage trips.

Conductive rubbers have been used as heating elements (especially silicone rubbers), and have been proposed as solders, conductive adhesives, radio frequency screening, battery components and optically transparent conductive layers, but the extent of such uses is not known.

Compounding

Carbon blacks with small particle-size, a high specific surface, permanent chain structures and low volatile content, givé most conductive rubber compounds. Graphite, though giving good conductivity, is rarely used because of the poor reinforcement provided.

For many years acetylene black held the field as the only carbon black conferring a high degree of conductivity at moderate loadings and was, therefore, termed a conductive black. With the introduction of furnace blacks in place of channel blacks, it is now rather difficult to produce insulating black mixes, except when carbon blacks of large particle-size or very low concentrations of other carbon blacks used. Fig. 17.1 illustrates resistivities vs. concentration values for channel blacks while those in Fig. 17.2 are for furnace blacks.

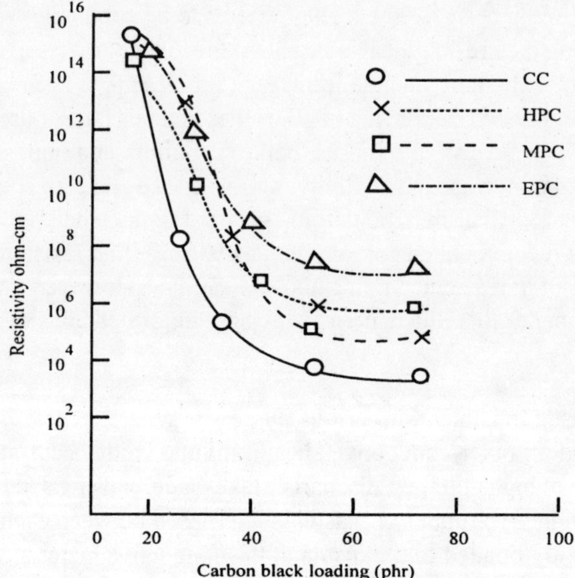

Fig. 17.1. Effect of channel blacks on resistivity.

Resistivity change with progressive replacement of a HAF by a thermal black is given in Fig. 17.3. It may be noted that anti-static protection is not sufficient with HAF or ISAF type blacks for aircraft tyres, footwears, floorings, beltings and hoses and it is required to use conductive furnace (CF) and a super-conductive Furnace (SCF) blacks to get resistivity values 20-200 ohm-cm in tread type stocks, as compared with 1000-5000 ohm-cm given by HAF. Other physical properties of compounds are similar. Carbon blacks which exhibit considerable structure tend to form harder and more conductive rubber. To get resistivity below 10^6 ohm-cm, 50 or more phr loadings of an ISAF, CF, SCF, and extra-conductive furnace (ECF) is required.

In case of certain carbon blacks, resistivity increases with the initial few phr loadings and starts decreasing after reaching a minimum value. In certain formulations it has been reported that around 25 phr of a FEF or HAF is required to reduce the resistivity to that of the unloaded mix. This increase in resistivity at low loadings is attributed to the adsorption of moisture and/or electrolytic impurities in the carbon blacks.

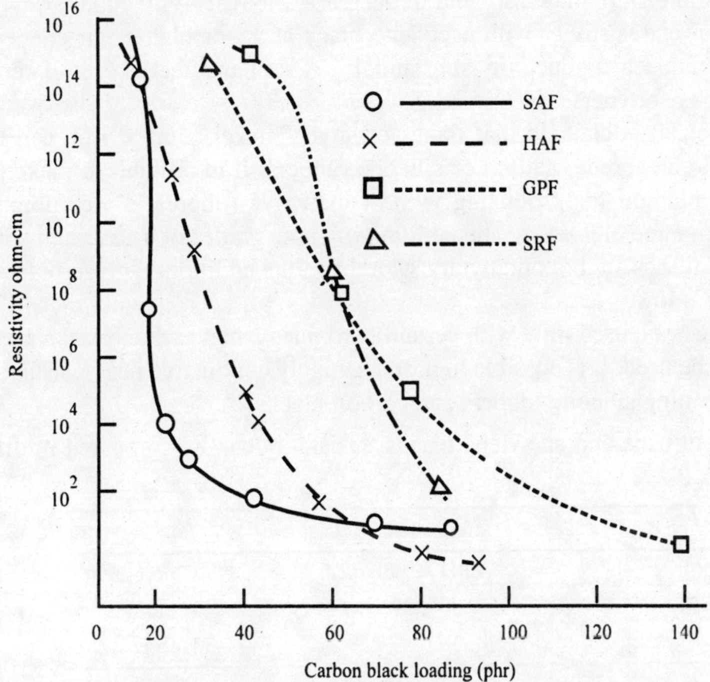

Fig. 17.2. Effect of furnace blacks on resistivity.

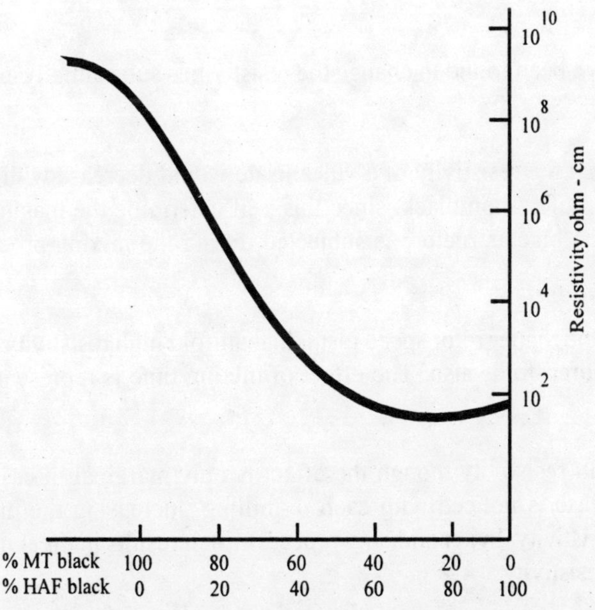

Fig. 17.3. Effect of progressive replacement of HAF by MT black.

The addition of a softener to a mixing generally increases the resistivity. However, if the softener is added before the carbon black during the mixing, the latter may suffer less structure breakdown and give high conductivity.

Certain softeners like DOP may also tend to decrease, the resistivity. Pine-tar, though increasing in other cases, gives lower resistivity with acetylene black at 15 or higher phr concentrations. Further, mineral rubber also reduces conductivity substantially. A similar effect is seen if certain other coal-tar derivatives are used as softeners.

Apart from carbon, it is claimed that flakes of silver, nickel, zinc or iron can be used to impart conductivity to nitrile, neoprene, natural or silicone rubbers. It is difficult to make highly conductive white rubbers. One method for producing white conductive rubbers is incorporating glass powder treated with conductive metallic salts. The addition of some grades of aluminium silicate may make it anti-static, if the conditions are not unduly dry. Moreover, the addition of glycerol lowers the resistivity even when humidity is low.

Silicone rubbers can be used only with certain peroxides when carbon blacks are incorporated and acidic blacks cannot be used. It is possible to produce highly conductive heat resistant compounds from the vinyl group containing silicone rubbers and carbon blacks.

With the uses of furnace and acetylene blacks, resistivity can be compared in different rubbers as follows:

High	*Medium*	*Low*
Nitrile	EPD terpolymer	Chloroprene
SBR		Natural
		Butyl

Chloroprene gives higher resistivity than natural rubber but at higher loadings of carbon blacks, the situation reverses.

Processing

Processing conditions have been found to change the resistivities substantially and sometimes manifold.

Mastication

It would be anticipated that the resistivity of a vulcanisate would decrease with increasing plasticity of the rubber before addition of carbon black, since this will determine the magnitude of the mechanical forces to which the carbon black structure is subjected during the mixing process.

Mixing time and rotor speed

Both higher mixing time and higher rotor speed result in manifold high resistivity values and the effect is, in some cases, a few hundred-folds also. The effect of mixing time is represented in Fig. 17.4.

Re-milling

This results in increases in resistivity though the effect is only marginal in case of high black loaded compounds and the increase is noticed with each re-milling. Increasing the mixing temperature also results in an increased resistivity. In general, any process which results in breaking of carbon structures, results in an increase in resistivity.

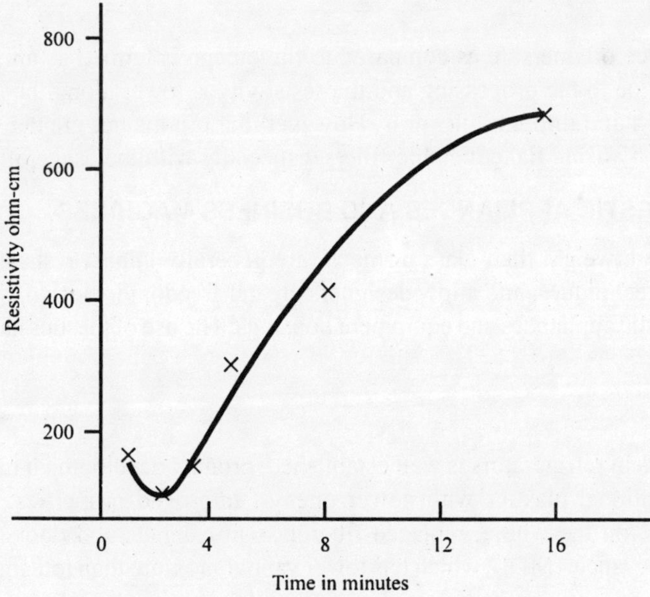

Fig. 17.4. Effect of mixing time on resistivity.

Heat-treatment

Heat treatment is desired for technical reasons in many mixes, especially butyl mixes. It has been found that the resistivity values increase invariably with this. In the Table 17.1 below are shown resistivity of different black field normal and heat-treated stocks.

Table 17.1. Resistivity (ohm-cm) of vulcanisates from normal and heat-treated N.R. stocks.

Carbon blacks	*Normal*	*Heat treated*
HMF	1×10^7	2×10^3
HAF	4×10^4	7×10^7
Acetylene	1×10^3	5×10^3
C.C.	9×10^2	1×10^4

Extrusion

Drawing down the material after it leaves the die reduces the longitudinal resistivity, thus, partly off-sets the increase in resistivity due to the extrusion.

Moulding and vulcanising conditions

The resistivity increases with the increasing period of standing before vulcanisation. However, the resistivity of press-cured articles generally increases with the increasing moulding pressures. The effects quoted are relatively slight over the range of pressures generally employed. Mould flow also increases the resistivity greatly. It is preferred that the vulcanisates used for anti-static purposes should be press-moulded as compared to open steam/air cure systems as the resistivities obtained are lower in the former case.

Anisotropy

To show more resistance on one side as compared to the others, is termed as anisotropy in a product. This generally results due to the processing and the resistivity is lower along the grain as compared to the side across the grain and through thickness. However, the ratio is not greater than 1:6. Anisotropy can be made very high by using flake metallic fillers, especially with the usage of magnetic fields.

POLYMERS IN DOMESTIC APPLIANCES AND BUSINESS MACHINES

As plastics are more lightweight than glass or metal, are generally impact resistance, lend themselves well to manufacturing techniques, and allow designers greater freedom in design, they are continuing to replace metals in domestic appliances and equipment housings. The use of plastics has been well exploited in both these areas.

Large Appliances

While the use of plastics in refrigerators is well established, product development has led to replacement of some of the conventional plastics with newer ones of improved properties. For insulation rigid polyurethane foam has, on the whole, replaced fiberglass in cabinets and doors. In this application, diphenyl-methane diisocyanate (MDI), which has lower vapour pressure than toluene diisocyanate (TDI), is now preferred.

For refrigerators lines low-coast polystyrene (PS) was the first choice, but this was soon replaced by ABS, which has higher stress crack resistance to trichloromethane, a foaming agent that is used to blow polyurethane insulation behind the liners. However, because of the higher cost of ABS there has been a swing back to PS with special grades of PS now being available.

Plastics are used to make a number of dishwasher components, such as impeller and other pump parts, inner door, tubs, and consoles. Polyphenylene sulphide is used to make dishwasher impellers. The material shows processing ease, good durability, and resistance to high heat and detergents. For dishwasher pump parts, tubs, and door liners, glass-filled polypropylene has all but replaced diecast zinc. Noryl-modified polyphenylene oxide (PPO) is replacing stamped metal for the dryer for blower wheels. Both rigid PVC and PC are used for consoles. Rigid PVC is used because of its inherent flame retardancy, rigidity, and high heat distortion temperature. Polycarbonate (PC) is used for its high heat resistance, dimensional stability, and good electrical insulating properties. Plastics are replacing metals in washing machines.

Small Appliances

The market for plastic cookware that can be used in the freezer and microwave is set for a big expansion. Most cookware is made up of fiberglass reinforced thermoset resin. High-barrier multilayer containers that have gained popularity in the market are heat retortable, microwaveable, and provide a shelf life of up to two years without refrigeration for a variety of convenience foods. Liquid crystal polymers have also been used to make cookware as they are transparent to microwave, offer high heat resistance, and give an excellent finish. The most notable development in small appliances has been that of the electric kettle. Nearly all new kettle bodies, lids, and spouts are moulded from plastics.

Business Equipment

Business machine and equipment housings represent a fast growing area for ABS polymers that are commonly described as tough, hard, and rigid—a combination that is unusual for thermoplastics. It's light weight and the ability to economically achieve a one-step finished-appearance part have contributed to large volume applications of ABS. A variety of special ABS grades are used for equipment and housings. These include high-temperature resistant grades (for instrument panels, power tool housings), fire-retardant grades (for appliance housing, business machines, television cabinets), electroplating grades (for exterior decorative trim), high-gloss, low-gloss, and matte-finish grades (for moulding and extrusion applications), and structural foam grades (for moulded parts with high strength-to-weight ratio).

Air filters

High-efficiency particulate air (HEPA) filters almost entirely use extrafine glass fibres and generally feature excellent functions, but if the fibre diameters are made finer to improve particle collection, the power consumption rate is increased considerably, clogging occurs, the filter becomes vulnerable to chemicals, and combustion treatment becomes difficult. Japan's Toyobo Company Ltd. has developed for air conditioning units a super-high-performance filter, the Eliton Super, made of a polypropylene polymer fibre. It is to be used as HEPA filter in clean rooms for removing over 99.97 per cent of dust particles of 0.3 μm and as an air-cleaning system. The fibres in this new filter are charged with static electricity to improve the particle-collection efficiency. Compared with glass fibre, it consumes less power, has greater resistance to chemicals, and has a substantially extended service life. The filter can be combusted with ease and reduced into a small volume. The company also supplies filters for industrial clean rooms, primarily to semiconductor and air-conditioning systems manufacturers.

Solar systems

Polymers have a share in the development of each major solar technology because they offer not only potentially lower costs, easier processing, and light weight, but also a greater design flexibility. Polymers are used in most of the solar systems and equipment such as covers, thin-film honeycombs and housing for flat-plate (non-concentrating) collectors, reflecting surfaces and optical lenses for concentrating collectors, reflector shells, structural and support members, insulation, piping, moisture barriers, adhesives, and sealants.

POLYMERS IN MEDICAL AND BIOMEDICAL APPLICATIONS

A wide variety of polymeric materials are used in various medical and biomedical applications, which range from medical packaging and products used in medical treatment and diagnosis equipment to items that are actually implanted in the body, e.g., pacemakers, hip joints, and artificial heart valves. These products have unique requirements from absolute non-toxicity and compatibility with human tissue to mechanical strength, resistance to sterilisation techniques, and, in the case of disposable end products, mass production on a large scale. Due to differing requirements a wide variety of materials are used. While medical grades of poly (vinyl chloride) (PVC) and polystyrene (PS) are the most commonly used polymers in the medical sector, high-performance engineering resins are now finding an increasing number of uses in this field. The areas in which plastics find use can be divided broadly into the following: (i) medical packaging; (ii) nontoxic sterilisable items; (iii) appliances; and (iv) disposables.

Medical Packaging

Significant advances have been made in recent years in the packaging of medical products requiring medium and high barrier protection. The demand for sterilizable packing for hospitals and industry is being met by special paper, plastic films, and other non-woven materials.

The chlorinated tetrafluoroethylene (CTFE) fluoro-polymer is one of the highest barrier materials against moisture transmission. It is used in an antibiotic test kit where even trace amounts of moisture would violate the accuracy of the test.

A large number of small medical containers are now moulded in poly (ethylene terephthalate) (PET). Its advantage over glass include no breakage of filling liners and higher output. Tamper-evident closures, droppers, and dispensers are usually moulded in polypropylene and high-density polyethylene.

Nontoxic Sterlisable Items

Polymers used in nontoxic sterilisable items, such as tubing, artificial organs, and wound coverings, must be able to withstand sterilisation by ethylene oxide, steam autoclave, or gamma radiation. For medical products sterilised by gamma irradiation, only plastics that do not degrade or discolour on exposure to radiation, such as polyester and polycarbonates, can be used.

The biological compatibility of materials used in the nontoxic category is vital and is the subject of continuing research. Since no material is completely inert, it is the level of interaction between an implant and the surrounding tissue that determines the acceptability of the material. Several other facts that are important in determining acceptability include mechanical properties of the polymer, e.g., wear resistance and fatigue, and bulk chemical properties such as resistance to degradation by hydrolysis, sensitivity to enzymes, and the way it reacts to the deposition of protein.

PVC is used for nearly all surgical tubing. Recent developments include flexible PVC compounds with greater resistance to body fluids, and X-ray-opaque PVC compounds that enable PVC catheters and tubing to be traced after insertion into the body.

The development of artificial wound coverings has received much attention. The major requirements for such coverings are protection from microbial attack from the environment, optimal water permeability, capability of adhering well to the wound, and ready removability without causing tissue damage. Other important factors are prevention of excessive formation of granulation tissue and optimal elasticity to facilitate an intimate cover of the wound.

Sustained-release wound dressing capable of delivering antibiotics or other biomedical agents in small concentrations have been obtained by micro-capsulating a variety of drugs into a UV curable urethane elastomer. Special UV-curable polyurethane elastomers has been developed for such applications.

Development related to medical implants, artificial organs, and prosthesis have improved the quality of life and increased the life expectancy of many individuals.

Silicone rubber is a highly biocompatible thermosetting elastomer that has found applications in prosthesis for opthalmology, neurology, facial reconstruction, replacement of finger, toe and wrist joints, cardiovascular applications, such as pacemaker coatings and lead wires, and tendon replacements. It is also used in drug delivery systems and tubes for carrying blood, drugs, and nutrients.

The artificial heart, which has been successful in keeping a recipient alive for more than a year, has valves made from modified polypropylene and the two ventricles made of polyurethane supported on an aluminium base. For arterial replacements or by-pass of clogged blood vessels, polyester fibre remains the preferred material.

Ultrahigh molecular weight polyethylene (UHMW PE) is most commonly used for articulation surfaces in joints. It is also used for prosthesis components in total hip replacement. In the latter application, it may be reinforced with carbon fibres to increase wear properties. Expanded polytetrafluoroethylene (PTFE) grafts have gained increasing popularity as synthetic or non-textile grafts for reconstructive procedures, such as, above or below the knee bypasses for limb salvage.

Appliances

Plastics components are now used in a number of appliances such as crutches, walking sticks, and wheelchairs. The comfort of the product is an important consideration in all of these applications.

Thermo-formed high density polyethylene (PE) and low density PE foams that can be moulded into a variety of complex shapes are being increasingly used in supporting devices in the treatment of orthopaedic and disabling conditions. Some examples are wheelchair seats, spinal jackets, medical shoe inserts, and splint supports. Specialised high-temperature plastics such as polyether sulphone and polyetherimide are used for components for medical analysis and diagnosis equipment.

Disposables

Disposable medical products such as syringes, test tubes, petridishes, dialysis products, and drainage devices are proving to be high-volume markets for transparent plastics such as rigid PVC, polystyrene, styrene acrylonitrile (SAN) co-polymer acrylic resins, and polycarbonate. The possibility of mass production is an important consideration in these applications.

POLYMERS IN MARINE AND OFFSHORE APPLICATIONS

Polymers find a wide range of applications in offshore and marine industry. The properties of materials used in these applications invariably include resistance to marine fouling, sea water, and crude oil. The main applications are in marine cables, coatings, and adhesives.

Cables

Due to the damage caused by smoke-filled or halogen-based fumes from cables in the event of a fire, thc offshore industry is showing increased interest in low smoke, zero toxic gas emission cable sheath. Research has centred on obtaining halogen-free polymers having fire retardance as well as oil, chemical, and weathering resistance. One example is a series of formulations called NONHAL-E (elastomeric) and NONHAL-T (thermo-plastic) developed for the offshore industry. The elastomeric formulations are based on nitrile rubber and are resistant to oil and other commonly encountered fluids. Cables insulated with ethylene propylene diene terpolymer (EPDM) and sheathed with NONHAL-E are suitable for very tough applications. NONHAL-T was developed to replace PVC in cable sheathing. Halogen-free cables produced with a cross-linked polyethylene and bedded and sheathed with NONHAL-T gives excellent performance in fire conditions.

POLYMERS IN PACKAGING

The packaging industry is the most important user of plastics both in Europe and America. The five most common polymers in use today as packaging are polyethylene, polypropylene, polystyrene, poly(ethylene terephthalate), and poly(vinyl chloride).

LDPE, the oldest of the polyethylenes, has bcen used extensively in packaging, but due to its higher price and lower mechanical performance compared with LLDPE, it is being used increasingly in speciality and niche areas where its unique properties add value. In blown and slot-cast film, LDPE offers excellent clarity and easy processing. It is often used in blends with LLDPE for these reasons. The rheological characteristics of LDPE and its ease of extrusion suit is for extrusion coating of various substrates. And because of its lower melting point at a given density, it is used where ease of heat sealing is important.

LLDPE has major outlets in film and sheet markets. In film, LLDPE offers greatly improved mechanical performance compared with LDPE, and it has steadily displaced the latter in many uses. However,

LLDPE has significantly different processing characteristics than LDPE due to the absence of long-chain branching; this requires different extrusion equipment in order to produce film at high rates. The trend in LLDPE film resins for packaging has been consistently toward higher performance. These improved products range form butene co-polymers to the higher-alpha-olefin (HAO) hexene and octene grades, and most recently the 'superstrength' HAO grades.

The penetration of LLDPE in the market of *stretch-cling wrapping* is increasing rapidly. Stretch-cling wrapping, used for consumer as well as industrial packaging, is a process in which a pre-extruded tacky film is wrapped around any product, such as bundles, rolls, and pallet loads, of any shape and size to impart a firm grip to the package. The tacky characteristics of the film helps in self-adhesion of the film without any external aids.

LLDPE has all the required properties for manufacturing stretch-cling film. With excellent drawdown characteristics, the film offers high elongation in machine direction, good tear strength at cross direction, and excellent puncture resistance. A required type of tackifier is used to impart the cling effect. Depending on the end-use requirements, both HAO-grade LLDPE and butene LLDPE are commercially used, the former offering superior strength but at a higher cost. To avail cost benefit and achieve the property, processors commonly offer coextruded stretch films with one layer of HAO grade LLDPE and the other layer (or layers) of butene LLDPE. This type of films are tougher than 100 per cent butene LLDPE films.

Very-low-density polyethylenes (VLDPEs), ranging in density form 0.880 to 0.912 g/cm^3, are ideal products for a variety of low-temperature film applications, such as ice and frozen-food packaging. The low crystallinity of the VLDPE results in energy-absorbing characteristics that yield outstanding low-temperature impact properties. The lower-crystallinity products of the VLDPE family has also shown increased permeability to oxygen and other gases, while their water vapour transmission rate is moderate. Thus, VLDPE offers a novel balance of barrier properties necessary for fresh produce packaging film.

Polypropylene (PP) homopolymers in oriented and cast film, provide clarity and gloss, high tensile and tear strength, stiffness, and a wide heat-sealing range. Other features of homopolymers include good moisture-barrier properties and printability. Cast film is used in food packaging, in bags for textiles, and (in combination with other materials) for composite packaging film laminates. Oriented polypropylene film achieves its high strength, clarity and impact resistance due to mechanical processing (orientation) that aligns the polymer molecules. It is often used in food packaging, particularly for snacks, cigarette packaging, and pressure-sensitive tapes.

Random co-polymers of ethylene and propylene produced by random addition of ethylene to a propylene chain as it grows are noted for high clarity, and lower and broader melting range than homopolymer grades, and reduced flexural modulus. At refrigerator temperatures, these materials also have impact resistance greater than that of homopolymers. Orienting the material further improves its clarity and impact resistance. Applications include blow-moulded bottles, cast film, and injection-moulded products, such as food storage containers, which take advantage of the excellent clarity and good balance of impact and stiffness of this material. The high gloss and very broad heating-sealing range of this resin is useful in such cast-film applications as trading cards and document protectors.

The melt-flow rate of ethylene-propylene random co-polymers ranges from 1g/10 min for a blow-moulding grade to 30g/10 min for an injection-moulding grades. The density is about 0.90 g/cm^3 and the notched Izod impact strength of the materials rages from under 1 to more than 5 ft-lb/ in. Special grades of co-polymers with a melt-flow rate of 35g/10 min or higher find applications in thinwall parts, usually for injection moulded food packaging such as delicatessen containers or yogurt cups. Such containers have walls with a length-to-thickness ratio as high as 400:1, yet they retain the properties of top-load strength, impact resistance, and recyclability that are typical of polypropylene. Injection moulders have these specialised co-polymers to reduce containers weights by 30 per cent over containers produced with other materials. Meanwhile extrusion coaters can now make use of high-melt-strength PP resins with melt-flow rates as high as 40 g/10 min. These materials resist pinholes and curling.

Polypropylene is making inroads in the packaging market at the expense of other materials, a process aided by new technologies that extend polypropylene properties. High-crystallinity polypropylene, for example, is targeting such materials as polystyrene and PVC in packaging. High-modulus grades of PP, as produced via Himont's Catalloy polymerisation technology are expected to challenge clear plastics (such as PET) and glass in container applications.

Major markets of rigid PVC include bottles and packaging sheet while flexible PVC finds major packaging uses in film, sheet, blood and other bags.

New technologies are broadening the range of available thermoplastic elastomers (TPEs). These developments include low-stiffness reactor grades of thermoplastic polyolefin (TPO) co-polymers, which are likely to make inroads in packaging markets at the expense of PVC. Potential end products of reactor-based TPOs include packaging films and sheets, shipping sacks, medical-waste-disposal bags, and blood bags.

Major innovations in the packaging field are retortable plastics as a replacement for the can, hot-fillable rivals to glass, high-barrier-coated or laminated bottles for carbonated drinks, and biodegradable or ecofriendly packaging. These development are highlighted below.

Retorting

Retorting is the most widely used method for processing food. In this method, the raw or partially cooked food material is placed in a container with sufficient liquid to fill it. Prior to sealing the container, it is heated under pressure to at least 121°C to kill all bacteria.

American Can's Omni retortable plastic can marks a major development in food packaging. The five layer can, used by Hormel, Del Monte, and Campbell Soup Company, consists of polyolefin structural layers, adhesive layers incorporating desiccants, and ethylene vinyl alcohol (EVOH) polymer as the barrier layer.

A newer development in this area is the production of retortable, multilayer, blow-moulded containers that incorporate polycarbonate as the tough outer layer, with EVOH or poly(vinylidene chloride) (PVDC) as the barrier, and polypropylene (PP) as the food contact layer. The advantage of polycarbonate are it light weight, improved heat stability, and food optics compared to the polyolefins that are normally used.

Asceptic Packaging

For asceptic packaging, both the food produce and the container are sterilised separately and the brought together in a sterile environment. Both low-acid foods, like yogurt and soup, and high-acid foods, such as juices and drinks, are packaged by this method.

Hot-Filling

In the hot-filling technique, which is being used as a cheap alternative to asceptic packaging, food is cooked or pasteurised by heating at a high enough temperature to kill the bacterial and then placed into the container at a temperature above 80°C to kill all the bacteria in the container. With this method flexible materials can be used to replace metal cans at a reduced cost.

In the Q-box system introduced by Mead packaging for hot-filling high-acid juices, a web made of LDPE, paperboard, ionomer, foil, polyesters, and LDPE is used. Products are pasteurised at 90°C before filling. Also used for high-acid products, is an oriented polyester bottle from Monsanto Company.

Hot-fill pouches are used for both low- and higher-acid applications. Heinz Company's form-fill seal Pouch Pack, which is claimed to give a shelf life of one year, has a web of 3-mil LLDPE, ethylene-acrylic acid co-polymer, 35-gauge foil, ethylene-acrylic acid co-polymer, and biaxially oriented nylon, from inside outward. The Pouch Pack system from Du Pont offers pouches of sizes 4 oz to 1.5 L to fill a range of high- and low-acid food applications at various temperatures, including fruit toppings at 80°C.

Controlled Atmosphere Packaging

By the turn of the century, controlled-atmosphere packing (CAP) is expected to consume more fill and sheet than all other types of plastic-based barrier structures combined. Most CAP used until recently PVDC and EVOH polymer-based oxygen barrier films to package red meat and baked goods. These films still allow the food to continue to respire after being packed, thus altering the balance of CAP gases introduced at the time to packing. However, selectively permeable films can now be used to control the amount of carbon dioxide, oxygen, and moisture in contact with the produce after it has been packed. In this way, CAP can more than double the shelf life of certain fruits and vegetables.

One such film that has been successfully used to package tomatoes is DRG's Ventaflex selectively permeable film. It is a polyester and HDPE strip-laminated film that has a laminating adhesive applied in diagonal strips. The film allows the carbon dioxide produced by the fruit to pass through the HDPE layer into the atmosphere while air penetration is limited by the outer polyester film. Besides tomatoes, other products commercially packaged in Ventaflex film include cheese, coffee beans, and bean-sprouts.

High-Barrier Films

Resins with high-barrier properties are being developed by a number of companies in response to a big increase in the demand for high-barrier films. Amorphous nylon resins offer a good oxygen and flavour/odour barrier, high clarity, good mechanical properties at elevated temperatures, and good processability. The films are used for meat wrap, snack food bags, and cereal box liners.

Amorphous poly(ethylene terephthalate) (PET) blends combined the inherent barrier properties of PET with the ability to be processed on conventional equipment without orientation or crystallisation steps. Besides food packaging grades, other industrial grades of amorphous PET are available that rival PP or HDPE for paint pails and other containers that requires a hydrocarbon or solvent barrier.

Dow's Saran wrap, which is a co-polymer of vinylidene chloride and vinyl chloride, is a useful packaging material that possesses exceptional clarity, toughness, and impermeability to water and gases. Also available from Dow is a range of Saran resin and films with improved barrier properties. For example, Saran HB monolayer lamination films are claimed to exhibit an oxygen barrier 10 times greater than the conventional Saran wrap films.

In the snack food area, metallised plastics are being increasingly used. Metallised oriented PP is particularly popular for this application.

Plastic Bottles

Plastic bottles for carbonated beverages are mostly made of PET as it offers high tensile strength and good barrier and properties. Blow-moulded PET bottles account for 33 per cent of the total carbonated soft drink market. For packaging beer, which is oxygen sensitive, PET bottles are coated with PVDC, while the base cups that support the bottle are usually made of HDPE.

However, since PET barrier properties decrease with a decrease in size of the bottle, smaller size poses a problem. Two possible ways to solve this problem are to employ coated PET or to use coextrusion technology. As an example of the latter case, PET has been successfully injected with nylon to form a PET-nylon-PET construction, with nylon improving the barrier properties.

In addition to carbonate drinks, PET bottles are also being used in other markets, which include toiletries, cosmetic ranges, spirit miniatures, and PET aerosols for cosmetic and pharmaceutical applications.

Biaxially stretched oriented PVC (OPVC) pressure bottles have been used as an alternative to PET for packaging some carbonated drinks. In this application, the level of carbon dioxide inside the bottle and the temperature at which the bottles are kept are important. Producing OPVC by extrusion stretch blow moulding results in greater clarity, increases impact resistance, and reduces cost by up to 14 per cent compared to ordinary PVC. Fruit, squashes, and edible oils are commonly packed in PVC bottles.

American Can pioneered the concept of multilayer squeezable plastic bottles with the introduction of its Gamma bottle. Another such produce is Continental Plastic Containers' Lamicon bottle. The bottle are being used for food products such as ketchup, jelly, mayonnaise, and barbecue sauce. In a new development, polycarbonate has been used by Holmicon Plast as the outer layer, rather than polyolefin, which is normally used. Polycarbonate gives the bottle a more shiny appearance.

Chemical Containers

Household and industrial chemical containers are increasingly being made of plastics. HDPE has almost completely replaced glass for packing household liquids. However, for solvents and flammable liquids, a plastic is required with higher barrier properties than conventional HDPE.

Three methods can be employed to obtain higher barrier properties. These are fluorination, coextrusion, and laminar blends. Fluorination of HDPE bottles is used most commonly for packaging solvent-based products and agricultural chemicals in particular. One process consists of treating the surface of HDPE bottles with a mixture of fluorine and nitrogen gas to produce a fluorocarbon barrier layer. The process developed by Air Products Limited, creates and internal barrier layer during the blow-moulding process, while Bettix Limited's fluorination process, Fluoro-seal, is post-moulded and particularly suited to short runs. The fluorination process suffers form the disadvantage that it is potentially hazardous.

Coextrusion to form an effective barrier layer has been used by Plysu Containers Limited and Reed Plastic Containers, both producing tough solvent-resistant containers. The five-layer construction used by Plysu has HDPE both on the inside and outside, and a central barrier layer with adhesive layers on either side to allow lamination to HDPE. The Reedpac range of blow-moulded containers developed by Reed Plastic Containers have polyethylene, polypropylene, polycarbonate, or PET combined with barrier resins such as polyamide, PVDC, and EVOH polymers.

The laminated resin blend technology using Du Pont's barrier resins, has been employed to produce barrier bottles. Blowmocan Limited has used the Selar RB polyamide barier resin in a one step process that blends with polyethylene to form overlapping discontinuous plates within the wall system. The container can be used for many cleaning chemicals and agricultural products.

Metal Box has developed a co-polymer polypropylene container, Polycan, which has an antistatic additive and is suitable for water-based paints and coating products. Another development is the first bag-in-box paints from Ashwell Paints. A smooth flow is obtained in this system by the use of an open bung and screw top rather than a tap.

Dual Ovenables

Dual ovenable containers must be able to perform under freezable conditions as well as in both microwave and conventional ovens. Materials commonly used for dual ovenables are temperatures-resistant plastics such as crystallised PET (CPET), thermoset polyesters, and polycarbonate. CPET, the dominant materials, has a temperature stability up to 232°C.

Attention has been paid to the possible use of multilayer coextrusion of engineering resins for dual ovenable containers, as multilayer packaging could increase shelf stability and thus eliminate the need for freezing or refrigeration. One multilayer combination in commercial use is a sheet of Lexan polycarbonate sandwiched between layers of Ultem polyetherimide. Thermoformed trays of this material are reported to have better break resistance at lower temperatures and higher stiffness at temperatures up to 232°C, as compared to CPET. They are also inert, compatible with polyester lidding materials, and have lower oxygen permeability than PET materials.

Most plastics used in dual ovenables are two to six times more expensive than the aluminium containers they replace in frozen food applications. Less expensive plastic containers are made of moulded fibre and polyester and paper laminates. However, due to the popularity of the microwave oven, in many cases microwaveable-only containers, instead of dual-ovenables, are used.

With the increase in the market of dual-ovenable and microwaveable trays there has been an increase in the demand for special lidding materials, as the lid has to provide an oxygen barrier and a 100 per cent hermetic steal to prevent food spoilage, besides being easy to remove. Grades of Du Pont's Mylar OL polyester lidding can be used to seal CPET, thermoset polyesters, and paper-board trays.

Closures

Tamper-evident closures are used to prevent customers from opening containers before they buy them. Several tamper-evident features were originally introduced including neck bands, tamper-rings, and shrink wrapping. More recently, a number of new tamper-evident and child-resistant closures (CRC) have been introduced.

Clic-loc is CRC available form United Closures and Plastics. A specialist version is provided with a thermoset inner linear for chemical applications.

A double tamper-evident cap sealer has been introduced by Paklink. It has a ring style enclosure incorporating an inner foil seal which, once heat sealed, provides a hermetic barrier suitable for products in the food, drug, and dairy industries.

A number of CRCs including the Childgard and Safecap, are available, the Childgard is a one-piece jigger cap style CRC used for pharmaceuticals, while Safecap is a two-piece palm and turn style CRC used for a variety of pharmaceutical and household products.

Biodegradable Packaging

Earlier, products were designed to balance cost and performance—during their useful life. Customers, end-users, legislators, and environmentalists now demand a new type of products, one that is also designed for responsible disposal. With the number of approved landfills diminishing each year, the coasts of disposal at remaining sites will escalate annually. That is why recycling and composting have a useful place in an overall solid-waste strategy. This has provided impetus for the development of biodegradable, and therefore compostable, polymeric products.

Incorporation of starch, a biodegradable natural polymer, into synthetic plastic imparts biodegradability. Shopping bags containing up to 15 per cent starch in a matrix of polyethylene are said to be biodegradable.

Microbes digest the starch and leave a flimsy plastic lace that disintegrates mechanically. Ferruzzi, an Italian firm that dominates the European cornstarch industry, has launched a polymer that is more than 50 per cent starch, and it hopes eventually to raise the content to 90 per cent.

Novon, a biodegradable plastic made by Werner Lambert, based in New Jersey, is reported to be composed entirely form starch. Through the use of different starches and nontoxic biodegradable additives, Novon speciality polymers can be formulated to provide a wide range of properties and still biodegradable completely in biological active environments, ultimately yielding carbon dioxide, water, and minerals, and leaving no toxic, hazardous, or synthetic residues, Any consumer packaging products with a high potential of being dropped on the ground may thus be considered in Novon polymers. For example, biodegradable loosefill packaging 'peanuts' are made of Novon 2020. They are compostable and compatible with existing sewage treatment facilities so that they can be safely washed down the drain or even flushed down the toilet.

Poly(vinyl alcohol) is one polymer that poses no litter problem, because it dissolves in water and is readily biodegradable. It makes ideal packaging for such things as swimming pool chemicals, detergents, descalers, seed strips, sanitary items, and even oxygen tenting.

The polymer polyhydroxybutyrate (PHB), a polyester of 3-hydroxy-butanoic acid, is completely biodegradable as it is an ideal food for microbes. By co-polymerisation with another hydroxyacid, such as 3-hydroxypentanoic acid, the polymer can be tailored to make it suitable either for moulded articles such as shampoo bottles or thin films for plastic envelopes or carrier bags.

Rubber in Packaging

The control of shock impulses on packaged goods has traditionally been a matter of incorporating into the pack shock absorbing or cushioning material, usually chosen by type and quantity and arranged according to the packer's knowledge of existing practice and of the handling and storage hazards expected. Particularly, in the industrial and the military field, packs are becoming increasingly complicated and costly products, to which purchasers such as government departments wish to apply strict tests of performance.

An important problem is that of discovering what material and what thickness of pad, or what type of solid mounting, will give protection against damage to an article when its container is dropped. The complete problem is complex because the result depends, among other things, on the shape and rigidity of the container, how it drops, whether flat, edgewise or cornerwise, and so on.

Typical shock and vibration environment

(i) Transportation by rail; (ii) transportation by road; (iii) transportation by sea; (iv) transportation by air; and (v) climatic conditions.

Bulk cushioning materials

(i) Latex foam; (ii) cellular rubber (open cell); (iii) cellular neoprene; (iv) foamed polyurethane (polyester); (v) closed cell rubbers; and (vi) rubberised hair.

Solid rubber mats

(i) Low cost; (ii) high shock absorption efficiency; (iii) low crepe; (iv) characteristics which do not change very much over a wide temperature range; (v) availability in a mat form; (vi) non-susceptibility to varying humidities.

Additional air damping

Additional air damping can be introduced with bulk cushioning and mats by enclosing the cushion in an envelope of PVC sheet, or similar material, with suitably sized holes in the side surfaces of the sheet. The ratio of the area of these holes to the volume of cushioning is critical and depends on the type and nature of the cushion used.

Buckled level systems

Buckled lever systems, which have been developed for packaging and vibration absorption applications, are systems in which a suitable material or spring is compressed (or tensioned) between jointed levers that buckle under load.

Buckled columns

The elastic instability of columns under compressive loads is also used for packaging purposes. A column becomes unstable and buckles if the applied compression load exceeds a certain critical value.

Solid rubber mountings

The biggest difficulty encountered in designing isolators for portable equipment is allowing for effective isolation in a wide range of possible situations. For instance a shock isolator employing rubber in compression and having a natural frequency of approximately 25 cycles per second affords satisfactory protection against shocks of the type met in naval service, but if this same isolator is subjected to the vibration experienced in aircraft, which includes the frequency range of 10–50 cycles per second, any mounted equipment may be subjected to excessive vibration unless the isolator contains sufficient damping. Unfortunately to obtain sufficient damping usually requires added complex equipment, and this is often not possible because of space and cost.

Shear mountings

The simplest type of unit used in packaging applications is the straight-forward shear mounting, consisting of metal plates with rubber sandwiched in between. Suitable lugs or holding bolts are provided for fixing to the supported equipment, and to the outside container.

Shear mountings are used for a variety of shipping container applications protecting missile and aircraft components, delicately balanced jet and reciprocating aircraft engines, and extremely fragile electronic equipment and parts. (i) tube form mounting; (ii) metalastic buckling type shock mounting; (iii) silentbloc delta mountings; (iv) lord 'dynashock' mounting; (v) stud ring mountings.

POLYMERS IN BUILDING AND CONSTRUCTION

The building and construction industry is being transformed by economic and demographic changes that are creating increased opportunities for plastics products. It is next only to the packaging industry in order of importance as a user of plastics, and has proved to be the fastest growing market for plastics use with increased sales in roofing, flooring, pipe and conduit, windows and doors, plumbing fixtures, and insulators. With its high potential for replacement of nonplastic materials, this market is expected to grow more rapidly. One of the relatively new areas of building applications has been in sports surfaces, which are discussed in a later section at the end of the chapter.

The use of nonwood materials in building and construction has been stimulated by the escalating prices of wood construction products. However, a key factor in the growth of plastics in residential construction is consumer acceptance. The advantages of plastics over conventional material, including improved quality, reduced installation cost, better appearance, durability, and low maintenance requirements, have hastened their acceptance.

Poly(vinyl chloride) (PVC) is the predominant plastic materials consumed in building and construction; it accounts for almost half of this industry's total consumption of resin. The largest PVC application is in rigid pipe and tubing, where it accounts for some 37 per cent of total industry consumption. Specific PVC piping uses include sanitary sewer lines, storm water lines, and potable water mains. These uses reflect a classic example of new-product acceptance, where a superior quality, moderately priced, and high perceived-value plastic product can supplant existing building materials and achieve a dominant market position. Other major PVC uses include siding, accessories, and profile extrusions for such end product as windows and doors, accounting for some 11 per cent of total consumption of resins in building and construction.

The commodity thermosets of urea, melamine, and phenolics account for about one-third of building and construction industries' resin consumption. Most of this is used for resin-bonded woods such as plywood, particle board, and oriented strand board. High-density polyethylene (HDPE), used primarily for rigid piping, accounts for about 5 per cent of consumption. Other significant resins, accounting for 3 per cent to 5 per cent of total consumption, include other polyethylenes, unsaturated polyester, and polystyrene.

Roofing

The basic parts of a roof are the deck, the thermal barrier (insulation), and the impervious *roofing membrane* that seals the roof complex structure. The built-up roofing membrane is made of (i) bitumen (asphalt) or other more modern materials; (ii) the roofing felts (reinforcement); and (iii) the aggregates for protection of bitumen against ultraviolent light and oxidation.

The molten asphalts used for waterproofing are a mixture of mineral fillers and bitumens. Because of its limited characteristics, bitumen was the subject of many studies aimed at improving its physical and mechanical properties. Several approaches were tried, such as chemical treatment and blending with natural or synthetic rubbers or latices.

So far, in practice, there are various polymer-bitumen mixtures, such as PE-bitumen or elastomer-bitumen, including styrene-butadiene-styrene block co-polymer (SBS)-bitumen or styrene-butadiene-styrene block co-polymer (SBS)-bitumen or styrene -butadiene co-polymer (SBR)-bitumen. The use of waterproof roofing membranes based on the thermoplastic elastomer-bitumen mixtures is increasing, especially in northern climate countries.

The most dramatic change in commercial roofing has been the recent shift from bitumen-based built-up roofing (BUR) to single-ply membrane roofing, particularly in the United States and other European countries. The ease of installation and durability of single-ply roofing together with its resistance to weathering, chemicals, and ozone make it a cost-effective roofing system. A distinctive feature of single-ply roofing is that it cures with weathering and increases its elasticity as it ages. Fire-retardant grades of EPDM are also being developed.

Although the material most used for single-ply membranes is EPDM, in competition with it are several other materials—Du Pont's Hypalon (a chlorosulphonated polyethylene), its Neoprene products, and modified PVC. It is believed that future growth in the roofing market will depend on products, such as Hypalon, that are easy to handle and have good ozone resistance, and on PVCs modified with polymeric plasticisers.

Flooring

The range of flooring material increases in direct proportion to the number of new polymers daily added to the market. In addition to the common floor surfacing materials (e.g., ceramic tile, wood block, wood strip and board, marble, granolithic, terrazzo, linoleum, and cork tiles), new materials are used, such as PVC tiles, vinyl chloride-vinyl acetate co-polymers, vinyl-asbestos tiles, PVC welded sheet, synthetic fibre-epoxy polymers, PP, and polyurethane.

PVC and vinyl-asbestos tiles are found to be the most economic materials for use in flooring applications. The most widely used synthetic flooring material is produced by calendering compounds based on vinyl chloride-vinyl acetate co-polymers filled with asbestos fibbers. In general, the co-polymers for flooring applications contain about 13–15 per cent vinyl acetate.

The so-called *homogeneous vinyl* is processed by laminating precalendered webs produced in a common PVC calendering operations. The laminated product is made of several sheets of varying thickness: the top layer is a clear, relatively stiff formulation; the middle one contains more filler and is usually opaque; and the bottom sheet contains higher filler loadings. After laminating, the product is die-cut into tiles and packaged.

PVC flooring materials are widely used in hospitals and schools, and domestic kitchens and bathrooms, where they offer a wide choice of colours and patterns, ease of cleaning, good cushioning, insulation, and reasonable price, the last three features representing advantages over linoleum.

Heavy-duty, lightweight PP duck boarding is claimed to provide a versatile, easily cleaned work platform, increasing operator comfort and safety. PP flooring is noncorroding and resistant to bacteriological attack. The upper surface is ribbed in order to impact nonslip characteristics, with slotted, self-drawing squares allowing free passage for wash solution and contaminants.

Due to its low level of sound insulation and lack of pleasing appearance, epoxy flooring is not used for domestic purposes. But for industrial flooring, epoxy floors are clearly competitive on a performance basis with most of the other floor types. There are three types of epoxy flooring systems: pourable self-leveling seamless floors, troweled floors, and terrazzo floors. These epoxy flooring systems can be used as new floor coverings over a subsfloor of concrete, wood, or steel, and can also be used for remedial work and applied over existing floors. Epoxy flooring has a proven durability of over 25 years.

The growth of seamless floors has had an exciting and profound effect on polyurethane flooring industries. This unique flooring concept can produce durable, attractive, and imaginative effects at the job site, by embedding a variety of different coloured particles or fillers into a liquid polyurethane resin just poured from the can. When cured, the lay-up is sanded to remove the sharp edges and smooth the surface, and then vacuumed to remove the sanding dust. Depending on the application and the amount and type of traffic, two or more urethane glaze coats are applied as the wear surface. However, for industrial floors where appearance is not critical, sanding and glazing are not necessary.

The features of seamless floors are well established. Not only are these floors easily installed, they are durable, lightweight, flexible, slip and dent resistant, scratch and scuff resistant, stain and dirt resistant fungus resistant, heel-mark resistant, and have superior chemical resistance compared to many floor materials. There are a number of coloured materials that the imaginative mind can use to provide decorative effects for seamless floors.

Windows

The use of PVC window frames is increasing at a steady pace. In Germany nearly 43 per cent of all windows and 70 per cent of all replacement windows are in PVC.

PVC formulations are generally ethylene vinyl acetate (EVA) polymer-based, chlorinated polyethylene (CPE)-based, or acrylic-based. The majority of new grades are acrylic-based and are claimed to exhibit greater impact strength than the other two types.

Grades of PVC are being tailored to meet different requirements of impact strength and weathering resistance for different climates. Solvay, for example, has developed an unmodified materials in granular form that is suitable for use in European countries, an acrylic-modified powder for moderate and Mediterranean climates, lead-stabilised grades for extremely hot climates, and an unmodified grade with very higher weathering resistance.

The majority of window profiles, until fairly recently, were white. Colouring techniques were developed to produced coloured window profiles, thus widening the market. Hoechst, for example, uses in-line bonding of acrylic foil to PVC profiles, allowing a wide range of colours and designs including wood-grain finishes. Both production and precision in window extrusion has been improved with the development of new screw designs, better dies, and microprocessor control of production parameters.

Pipes

Plastics pipe demand has been growing and the factors contributing to the accelerating use of plastic pipes include their low cost relative to conventional materials, excellent corrosion resistance, ease of installation, and wider building acceptance. The plastic pipe market is dominated by PVC pipes, which hold 80 per cent of the market. Polyethylene, principally HDPE, is used for about 12 per cent of pipe production. Following polyethylene are other higher priced, speciality pipes made from ABS terpolymer, chlorinated PVC, polypropylene, polybutylene, and fibre-reinforced epoxy and polyester.

Insulation

The major uses of insulation in the construction industry are in roofing, residential sheathing, and walls. In these applications, polymeric foams offer advantages over traditional insulation such as glass fibre, and these include higher insulation value per inch of thickness and lower costs. The use of polymeric foam for insulation increased markedly due to increased awareness of the need for energy conservation.

Foams are available as rigid sheets or slabs (which are used in the majority of roofing systems), as beads and granules (used in cavity wall insulation), and also as spray and pour-in applications. The market is dominated by polyurethane (PU) foams, in particular polyisocyanurate products, expanded polystyrene (PS), and extruded polystyrene.

Polystyrene foam (PS) holds much of the sheathing market. In masonry and brick walls, PS foams are mainly used because of their better moisture resistance. In cavity walls, loose fill PS is used, while exterior wall applications, which have no limitation on space, use low-cost expanded PS.

Polyurethane/polyisocyanurate products have higher insulation value and good flammability ratings and are expected to continue to be the leading products in plastics foam market as sheets and slabs.

A shift toward single-ply roofing as compared to built-up roof systems has a important influence on the type of foam being utilised. Thus the lower cost of expanded PS has promoted its use in preference to PU foam and extruded PS in single-ply applications. This facilitated by the fact that the problems of damage to expanded PS foam from hot pitch when used in built-up roof systems are not encountered in single-ply systems.

Polymers-Concrete Composites

The mechanical properties, the corrosion stability, and some useful properties are the reasons for the continuous interest shown in polymer-concrete composites by various design, research, and production organisations. The most important types of polymer-concrete composites are polymer-impregnated concrete (PIC), polymer-cement concrete (PCC), and polymer concrete (PC).

PIC is a precast and cured hydrated cement concrete that has been impregnated with a monomer that is subsequently polymerised *in situ*. This type of cement composite is the most developed of polymer-concrete products. PCC, on the other hand, is premixture of cement paste and aggregate to which a monomer or polymer latex is added prior to setting. PC is an aggregate bound with a polymer binder. This product may be obtained on site.

The largest improvement in structural and durability properties has been obtained with the PIC system. With the impregnation of a macro-molecular compound, the compressive strength can be increased four times or more, the modulus of elasticity increased at least two times, the water and salt permeability reduced by 99 per cent, the freeze-thaw resistance improved enormously, and —in contrast to conventional concrete—PIC exhibits essentially zero creep properties. Both strength and durability are strongly dependent on the fraction of the porosity of the cement phase that is filled with polymer.

PIC is generally prepared by impregnating dry precast concrete with a liquid monomer and polymerising the monomer *in situ* by thermal, catalytic, or radiation methods. Some of the most widely used monomer for PIC systems include methyl methacrylate, styrene, butyl acrylate, vinyl acetate, acrylonitrile, methyl acrylate, and trimethylpropane trimethacrylate (TMPTMA). These monomers may be used alone or in mixtures. TMPTMA is a cross-linking agent used to decrease the time of polymerisation. The polymerisation of the monomer used for impregnation is realised usually by thermal polymerisation (commonly with water or steam) or radio polymerisation, especially with Co^{60} gamma radiation. The initiators most commonly used in polymerisation processes for producing PIC composites are 2,2′-azobisisobutyronitrile, benzoyl peroxide, lauroyl peroxide, and methyl ethylene ketone peroxide.

With impregnation by an appropriate monomer, the main effect after polymerisation is the sealing of the main effect after polymerisation is the sealing of the continuous capillary pore system, which reduces the effect of stress concentrations from pores and micro-cracks, thereby increasing the strength. A reduction of porosity by 10 per cent doubles the strength. PIC must be considered a new complex materials with specific characteristics, which place it in a position, from the viewpoint of quality and cost, between traditional concrete and other groups of engineering materials such as metals and ceramics.

The most important applications of PIC are bridge decking, tunnel support-lining systems, pipes (hydrostatic tests shown that PIC supported about twice as much hydrostatic pressure as the unimpregnated pipe), desalting plants, beams (ordinary reinforced and post-tensioned), underwater habitats, dam outlets, off-shore structures, underwater oil storage vessels, and ocean thermal energy plants, among others.

Unlike PIC, which requires a precast structure, PCC has the advantage that it can be cast in place for field applications. Most of the PCC composites are based on different kinds of lattices obtained especially by emulsion polymerisation. A latex is a stable dispersion of fine polymer particles in water, also containing some nonpolymeric constituents used in emulsion polymerisation recipe. The lattices obtained through emulsion polymerisation contain small polymer particles of 0.05–5 μm. Styrene-butadiene rubber latex (SBR, GRS) and acrylonitrile-butadiene rubber latex (NBR) are two of the earliest to arrive on the market. Since then, many other types have appeared, with poly(vinyl acetate) and co-polymers, acrylics (generally polymers and co-polymers of the esters of acrylic acid and methacrylic acids), and carboxylic-SBR types being the major products. Since latices are aqueous emulsions, less water is usually needed in the production of PIC. The addition rates of SBR, poly(vinyl acetate), and acrylic latices are comparable, ranging from 0.10–0.20 part latex solids to 1 part Portland cement.

The particular characteristics of compatibility of poly(vinyl acetate) with cement have led to wide use of this latex as a main component in polymer-mortar and polymer-concrete composites.

In all cases, the operations for preparing PCC are similar to those used for common concrete and mortar. Curing, however, is different. After only 1 day of moist cure, the surface of PCC work can be uncovered. The film already formed on the surface retains the necessary amount of moisture for the full hydration of the cement. In principle, after only a few days of air cure at around the normal ambient temperature (20°C), the PCC can be put in service. The polymer latex used for making a PCC must be able to form a film under ambient conditions, coat cement grains and aggregate particles, and form a strong bond between the cement particles and aggregates.

Investigations have demonstrated that PCC has a higher corrosion resistance as compared with ordinary concretes and can effectively be used for floor coating in a moderately aggressive atmosphere, at milk processing factories and breweries. However, the presence of cement in PCC is a source of corrosion destruction under the action of more aggressive and concentrated chemical media at sugar refineries and meat processing enterprises, among others. In such cases, PCC may be recommended only for underfloors. The problem may be radically solved by producing floor coatings with a purely polymeric binder. For example, investigations and practice have revealed that the use of epoxy alkyl resorcinol–based polymer concretes (PC) for floor coatings in production shops at food industry enterprises increases the corrosion resistance of the floors to a great extent.

Polymer concretes (PC) may be considered as an aggregated filled with a polymeric matrix. The main technique in producing PCs is to minimise void volume in the aggregate mass so as to reduce the quantity of the relatively expensive polymer necessary for binding the aggregate. By carefully grading the aggregate, it is possible to wet the aggregate and fill the voids by the use of as little as 7–8 per cent polymer. With high degree of packing, high compressive strength can be obtained.

A wide variety of monomers, prepolymers, and aggregates have been used to realise PC. The list includes epoxy polymer unsaturated polyester styrene system, methyl methacrylate, and furane derivatives, usually in conjunction with cross-linking agents. To obtain the best chemical resistance, complete curing of the polymer is necessary. This usually is achieved by careful heating, or by using an appropriate cross-linking agent. In order to improve the bond strength between the macro-molecular matrix and the aggregate, a silane coupling agent is added to the monomer before the polymerisation process. The nature of the aggregate influences the hydrothermal stability of PC composites. Aggregates commonly used include quartz, silica, fly-ash, cement, and their combinations.

The PC systems based on unsaturated polyester and wet aggregates of cement and silica result in significant strength improvements. The mechanism of chemical bonding is believed to involve cross-linking reaction between Ca^{2+} ions of cement particles and carboxylate anions of unsaturated polyester brought about by a hydrolytic reaction. The addition of MMA to a polyester-styrene system provides a hard, clear mirror finish, improves the workability without reducing the strength, and enhances durability.

Besides fast curing, PCs offer several other advantages. Thus, because the polymer is virtually impermeable to moisture, there is very little cracking of the concrete caused by freezing and expansion of moisture within the cured mix. Polymer concretes also resist salts and other agents that eat away at the reinforcing steel within the concrete. In Germany, these properties have led to the use of PCs in water-treatment and sewage plants.

Although PCs do cost more initially, they can be put down in thinner layers than Portland cement concrete to give the same strength at lower volume, thus allowing a weight reduction of up to 50 per cent. PC overlays are normally 3/8–1/2 in. thick, while Portland concretes average 2 1/2 2–3 in. thick.

for bridge overlays, large stretches of epoxy and polyester resins are combined with Portland cement. These mixtures have a stress failure point of 10,000 psi (69 MPa) compared to 2,000–5,000 psi (13.84–34.5 MPa) for Portland cement. Polymer concretes can be mass produced, taking advantage of short curing times. With special processing techniques and moulds it is also possible to cast equipment such as wash basins and bath tubs, plain or with a marble or onyx effect, slabs with many different surface effects, and even marbles with two or more colours. Finished PC products are much lighter than the cement-bonded material because they have thinner walls. The excellent resistance to chemicals allows many applications in the construction of sewer systems, sewer treatment plants, animal stables, oil separations, chlorine electrolysis, machine foundations, high-resistance floors, and so on.

NITRILE RUBBER AND ITS APPLICATION IN CONSTRUCTION INDUSTRY

Construction activities in the country are seeing its boom time at present. The stress by the government is on road infrastructure development. There are a number of projects such as Golden Quadrangle (Joining India across its length and breadth by expressways) or flyover projects. These projects are using state of art technologies involving more sophisticated machineries, improved concrete mix or moulding and forming patterns. The essence is also on aesthetics, fine finish and strength of concrete design. With these objectives, newer products and technologies are being used in individual construction projects. In one of the flyover project site, appropriate shuttering fluid along with rubber as liner (in particular nitrile rubber liner) in the concrete moulds was used.

It was established that whenever quality shuttering fluids along with nitrile rubber liners were used, one could achieve superior surface finish, coupled with its antistatic properties, low compression set etc., as other benefits for concrete moulding operations. It was also found that there stands a wide gap in technology upgradation in the normal construction patterns presently followed and the present day knowledge available. It was established an ease in mould release, superior surface finish ; but it was also established that the nitrile rubber liners had more than doubled the life and provided a clean, shining surface as compared to the natural rubber liners in use resulting in overall significant savings.

Construction methods and construction activities are undergoing sea change, with new, modern, faster and more efficient techniques of construction being adopted and applied every day. This has led not only to the development of more efficient techniques of construction but has also led to equally high performance product requirement as shuttering liners, fluids etc.

Shuttering liners are desired in the manufacture of pre-cast concrete products and are used in building and construction applications for various reasons. Among these, the most prominent is the easy release of concrete block from the mould and also the surface finish of such mouldings. In the past, various types of shutter liners and release agents were used, with major use being wood and had dominated this industry for quite some time until the polymer liners made an entry. Today, the vast majority of liners are made out of polymers such as rubbers, epoxy, polyolefins, nylons etc. In India, the first rubber lined concrete moulding concept was introduced by the construction major Larsen and Toubro Limited in the construction of flyover at Bangalore in the late 90's.

With this background, detailed and exhaustive studies were undertaken at a site provided by a Construction major using both Power shuttering fluids and nitrile rubber based liners.

Oils and Lubricants are in use in large quantities in the construction industry. These are used either by way of lubricants for plant and equipment or as mould release agent. Except few major players and considering the unstructured and un-organised that this industry is, oils in particular for use in mould have received little attention or had any significance.

People use diesel, kerosene, spent engine oils etc for the same purpose. However, these products have their own drawbacks and deficiencies, which over the years have been given little importance; however, presently, there is a growing realisation, that with upgradation of construction techniques, the need for use of quality construction materials including mould liners, shutter release oils and process techniques are imperative.

Rubber in Concrete

Rubber in Concrete can be classified into following major groups.

1. As a quality improver to concrete such as impact modifier.
2. As an 'aid' in the processing and application of concrete etc.

The use of rubber lattices, both synthetic and natural and many block polymers find their way in improving the impact properties besides other qualities such as adhesion, water proofing etc. There are special products available for each of the above specific requirement. Presently, in-order to improve the aesthetics of the concrete surface, use of polymer lined moulds have become increasingly popular compared to the conventional methods of using wood or metal. Though, the use of natural rubber liners predominated the market, it became mandatory to improve the quality and cost performance of these liners. Today, use of synthetic rubbers such as nitrile rubbers and their blends, polychloroprene etc., were tried and have became popular.

Nitrile rubbers can be successfully used as a shuttering-aid in the construction of high quality finish bridges, where the concrete moulds are provided with rubber. The rubber liners are stuck to the mould surface using special adhesives , the gaps and joints are filled with good caulking compounds Before the concrete is filled inside the mould, the liners inside the mould are cleaned from past residues and a fresh coat of release agent is applied. These release agents are normally based on synthetic oils. Several types of packages based on synthetic/mineral, oils or greases are available today. Typical detail is available in Table 17.2.

Table 17.2. Properties of various release agents.

Release Agent	*Conventional Shuttering Fluid – 1*	*Conventional Shuttering Fluid – 2*	*Shuttering Fluid (Spl.)*
Base material	Kerosene-silicone combination	Mineral oil Base	Power oil
Viscosity @ 40°C, c.St	Irregular	8–9	24–26
Stability	Non uniform showing separation	Homogenous solvent Based product	Homogenous uniform product

Selection criteria for a good liner material

The selection criteria for a good liner material are based on following characteristics:

1. It should provide excellent abrasion resistance during moulding cycles.
2. Should possess good weather and age resistance properties.
3. Should have good oil and solvent swell resistance properties.
4. Excellent tensile and flex properties.
5. Excellent compression properties.
6. Good and easy adhesion to metal surfaces of mould.

7. Should be economical and easily Available.

As nitrile rubbers meet all the above requirements, they were ideally selected for this work. Some of these properties are evident from the results obtained and are discussed later in this section.

Experimental

Release Agents: The following options were used in this work.

1. Conventional Shuttering Fluid 1 -Kerosene and Silicone combination.
2. Conventional Shuttering Fluid 2 -Shuttering Fluid in use.
3. Power Shuttering Fluid (Spl).

All the release agents were tested for following and the results are tabulated below in Table 17.2.

Rubber liners: The following options can be used:

1. Natural rubber liners.
2. Nitrile rubber liners

Both the rubber liners were compression moulded in standard contoured surface type and are approximately 6–8 mm thick. The typical geometry of rubber liner is given in Fig. 17.5.

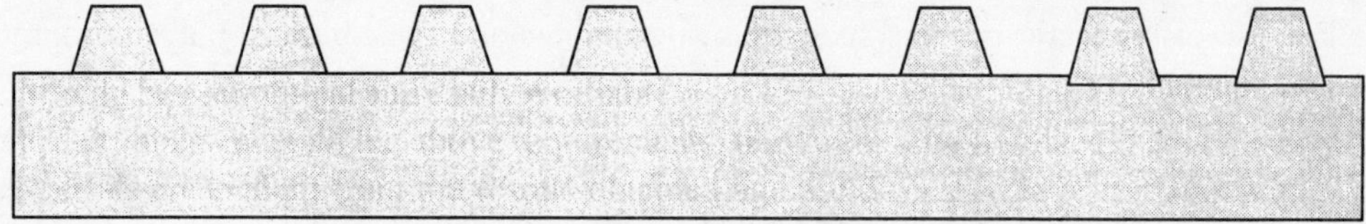

Fig. 17.5. Typical cross section view of a rubber liner.

It was observed that natural rubber, which was used as lining material for concrete moulds was found to absorb and swell with petroleum products. Hence, the best answer was to use a more viable option – Oil Resistant Synthetic Rubber -Aparene-NBR.

As discussed earlier, the rubber compound used in the manufacture of liners shall not only provide good ageing and oil resistance properties but shall also provide excellent durability. Normally, good compression set properties give better ageing resistance properties to the compound; however, this is not reliable. Hence, besides, compression set, stress relaxation can be used as a tool in predicting the life of rubber compounds in various applications and in particular for seals.

Extensive studies were carried out on stress relaxation properties on many rubber compounds. From these studies, it can be revealed that many synthetic rubbers such as EPDM, CR, NBR perform much better in service than their counterparts. Nitrile rubbers are also found to be far superior than natural rubbers. As a rule of thumb, the relaxation properties of nitrile rubbers can be two to four folds better than those of natural rubber based compounds. This should improve the life cycle of the product in use. It also provides better dimensional stability to the product which will enable to provide consistent concrete surfaces in repeated cycles. A typical recipe for the manufacture of rubber liner is given in Table 17.3.

Table 17.3. Recipe for rubber liner for concrete moulds.

Aparene NBR	100.00	Hardness, Shore A	75
Zinc oxide	3.00	100% Modulus, kg/sq.cm	28
Stearic acid	1.00	300% Modulus, kg/sq.cm	53

(Contd...)

Carbon black (HAF)	20.00	Tensile strength, kg/sq.cm	64
China clay	100.00	Elongation at break, %	401
Calcium carbonate	30.00	Specific gravity	1.52
Di-Octyl phthalate	10.00	Ash content, %	45.02
SP	1.00	*Oil swell properties @ 100°C, 70 h*	
Sulphur	1.00	Volume swell in P-50* (%)	+ 2
CBS	2.00	* Paraffinic oil of 50 sus viscosity.	
TMTD	0.35		
Total	268.35		

Operations

The operations involved casting of parapets and splines for a major flyover are given below:

1. Mould preparation : wherein the moulds were cleaned and liners were stuck on it. These liners were approximately 120 cm in length and 30 cm in width.
2. Application of release oil : Mould release oil was applied on the rubber lined surface to achieve effective and easy release during the de-moulding of concrete. It was found that approximately 3 litre of oil was required for a spline and 1.5 litre for a parapet mould assembly.
3. Mould assembly: Here various parts of the mould were assembled to provide it the proper shape as per the requirement.
4. Moulding: In this step, the concrete was poured into the mould assembly. Thereafter it was left for 48–168 hours depending on the product.
5. Demoulding and lifting: It was subsequently demoulded, lifted, kept in the stock yard where remaining curing was effected or sent for rework etc.

The mould to concrete bond/grip after curing is very strong , resulting in difficulties in demoulding; besides, the liner channels were very narrow which aggravated the problem further, So much so that in case of parapet, if an incorrect shuttering fluid was used, the mould release was difficult and it shook the complete mould from its base apart from pulling out the stuck rubber thereby spoiling the rubber liner assembly.

This not only added to delay in operations but further re-work in mould preparation for subsequent cycles. Thus, this operation can be an extreme test for a shuttering oil to perform and deliver.

Experimental Trials

Exhaustive trials were undertaken involving shuttering fluids and nitrile rubber liners.

The following steps were involved.

1. Application of shuttering fluid and analyses of the result.
2. Analyses of the properties of rubber liner as well as shuttering fluid being used along with their compatibility.
3. Developing appropriate nitrile rubber liner and compatible shuttering fluid for use in this operation.
4. Undertaking trials in actual condition to study and establish the product both technically and commercially for its use in the concrete moulding operation.

To improve performance properties and ease in shutter release, quality of the rubber liner were examined and subsequently use of aparene nitrile rubber liner along with power shuttering fluid were recommended. The detailed analyses of results are discussed in the following section.

Results and Discussions

Prior to initiating the trials exhaustive studies on the type of rubber being used, the shuttering oil in use, their demoulding characteristics etc were conducted. studies are also undertaken to check the compatibility of the rubber in use with the shuttering fluid.

These studies helped us to develop an appropriate and cost effective nitrile rubber formulation along with shuttering fluid. The detailed studies of various physical and ageing properties on these liners in the laboratory is provided in Table 17.4.

Table 17.4. Physical and ageing properties of different liners.

Aparene NBR based shuttering liners		
Hardness (Shore° A)	75	76
Tensile strength (kg / cm^2)	64	134
Elongation at Break (%)	401	392
Ash content (%)	37	20.8
Change in hardness (Pts)	–29	–44
Change in volume (%)	+25	+111
Observation: conventional shuttering oil in use effected natural rubber much more than nitrile rubber		
Change in hardness (Pts)	–7	–29
Change in volume (%)	+2	+102
Observation: NBR liner had better oil compatability than natural rubber		

These results indicate that the solvent used in the conventional oil was swelling the rubber more and bringing down the hardness significantly, thereby affecting it's overall consistency and the life. It was decided to overcome this problem by providing nitrile rubber liner and shuttering fluid that would provide quicker release, better surface finish ease of operation with higher consistency levels etc. These were evident from our observation and are listed below.

Observations from the experiments conducted on NRL and NBR

Some of the observations from the experiments conducted on Natural rubber liners and NBR using the facilities of construction major are given below. The construction company were moulding concrete cantilevers, spline, and parapets at their construction yard site.

1. It was observed that natural rubber has more surface adsorption and gripping tendency thereby creating difficulties in shutter release. This effect was much less pronounced in nitrile rubber liners.
2. The currently used shuttering fluid mixture as well as shuttering oil provided by other oil company had tendency to swell the rubber liner and affect the rubber consistency, thereby affecting its overall performance in shutter release.
3. Both nitrile rubber liner and shuttering fluid, performed effectively and satisfactorily by providing better surface finish and quicker mould release.

Because, the size of the rubber liners are much smaller (120cm × 30cm) and constrained as per the mouldings and maufacturing capabilities, number of liners are required in the specific type of mould depending on its size.

Assumptions

1. Life of NBR Liners are at least twice that of conventional system, though in practice it can be much higher.
2. The cost of NBR compound is approx. 20 per cent more than the natural rubber.

The above results indicate that in shuttering operations, it is essential that we use rubber liner technology with effective and efficient shutter fluid as mould release. This will help to obtain good surface finish and effective shutter release. The following conclusions were arrived at:

1. Nitrile rubber liner alongwith appropriate shuttering fluid are economical both in short and long runs.
2. Nitrile rubber can provide better quality concrete surface in repeated cycles of moulding in conjuction with compatible shuttering fluid.
3. Nitrile rubber liners and shuttering fluids can be successfully employed in the manufacture of concrete shuttering operation.

USES OF POLYMERS IN BEARINGS PIERS AND FENDERS

There are a number of polymeric materials that distinguish themselves from others by virtue of their limited use, high prices, or very specific applications or properties. In this section the applications of speciality rubbers which are high pressure, high temperature and weather resistant are discussed. These rubbers find wide applications in: bearing and bridges; piers and fenders; structures; aircrafts, ships and machines; buildings and packaging etc.

Rubber in Bearings for Bridges

Bearings for bridges must be so designed that they permit expansion and contraction of the bridge with changing temperature in summer and winter, they must be able to rotate slightly from deflection under load, and they must transfer thrusts from heavy traffic smoothly to the sub-structure of the bridge.

Modern vehicles travel at high speeds and have improved braking and acceleration. These factors, together with increasingly heavy loads, are increasing the problems to be overcome. In the past bridge bearings have been mainly in the form of mechanical devices designed to roll or slide, in order to permit movements of the ends of bridge.

Such bearings are difficult and expensive to maintain, and need frequent lubrication and attention to keep them free from corrosion. They suffer from the very small movements which they are required to perform, which result in a high rate of wear over small areas. Concrete bearings have been used also, but these take up a large amount of space and are expensive. Bearing pads consisting of layers of canvas and red lead have been used as well.

These are able to compensate to some extent for minor irregularities in alignment, and to prevent moisture, grit, and dirt from getting between the bearing surfaces and causing abrasion and rusting. They have, however, little or no cushioning effect, and as they are built up on the spot, considerable time and labour is involved.

They are also susceptible to weather conditions during installation, and should only be built up during good weather to obtain the best results.

Advantages of rubber bearings

The comparatively recent introduction of rubber bridge bearings has enabled engineers to obtain the necessary flexibility in bearings without the attendant disadvantages of frequent attention, and/or high cost, cited above. There are four main types of bridge bearings which make use of rubber.

1. Fabric-laminated pads designed primarily for load distribution.
2. Moulded pads of solid rubber which are designed to support moderate loads on large bearing areas and to provide small expansion movements.
3. Pads to improve bearing surfaces and to prevent electrolytic action (e.g., between a rail and a bridge structure).
4. Steel-reinforced pads designed for heavy loads on limited bearing areas.

Types of Bridge Bearing

(i) Steel-reinforced neoprene; (ii) steel-reinforced natural rubber; (iii) butyl rubber bearing pads; (iv) fabric-laminated pads; (v) butyl bearings; and (vi) neoprene bearings.

Rubber in Piers and Fenders

For energy absorption of either impact or vibration rubber is pre-eminent. It provides a variable spring rate which, whilst yielding to a light blow, stiffens up as the shock pressure increases. It is largely self damping, its reaction to a blow dying out quickly, and it needs no lubrication, painting or other attention. Rubber correctly compounded for the job has stood up for years, even decades, to tropical sunshine or arctic cold, to salt water, oil scum, Toredo worm and other agents to be found in enclosed waters, with the result that there is now a diversity of fender designs.

Simple Fenders

Simple fenders of the type consisting of baulks of timber, often further protected by rows of old tyres or other scrap rubber, or by rope fenders are very common. Other non-rubber fenders which have been used for a very long time are floating wooden rafts, which are moored to the jetty and keep the ship away from the main structure, and wooden piles which are driven into the harbour bed in front of the jetty, and slightly away from it. A more modern development is the gravity fender, consisting of a heavy block of material such as concrete, suspended from the jetty by chains, with its outer face projecting beyond the face of the jetty. The weight of the block exerts an outward force on the side of the ship if the ship moves towards the jetty and pushes against the block. Simple fenders using rubber are usually heavy rubber tubes, rubber balls, fender blocks, or devices using old tyres.

More Complicated Systems

While the use of rubber in the simple fenders noted above leads to considerably increased protection of both ships and jetties, more complicated systems have been specially devised to give maximum absorption of impact.

Tweddell fender

A more sophisticated fender based on the use of tyres, in combination with rubber blocks and baulks of timber is a floating fender of the type known as dummies or catamarans, and constructed of baulks of timber.

Compression cylinder type

In this fender the fender pile is recessed into the jetty and the shock of a vessel coming into dock is taken by the main rubber cylinder, while secondary cylinders control and restrain sideways movement. The rubber acts purely in compression or rolling compression and there is no mechanical connection with either the jetty or the fender pile.

Rubber buffer systems

Notable features of this design of fender are that the fender face can move in any plane to suit the varying shape of the ship's sides and its angle of approach, and that it provides smooth operation and a complete absence of moving parts requiring maintenance.

Buckling column fenders

Buckling column type rubber fenders have been developed in the United States, and possess certain advantageous characteristics. When used in fenders, these units will :

1. Build up relatively high load for small initial deflection
2. Collapse at relatively small initial deflection
3. Maintain a constant force over the range of buckling defection
4. Buckle in a pre-selected direction.

Raykin fender

This fender unit, which is manufactured in the United Kingdom, consists of a series of sandwiches of rubber bonded between steel plates and assembled together to form a V-type assembly. The apex of the V-faces seaward and is thrust inwards by berthing ships. The rubber components are stressed in both shear and compression and it can be readily seen that the flexibility of the unit is such that it will yield to either a direct or glancing impact.

Andre-neidhart system

The unit consists of square inner and outer metal parts with cylinders of rubber forced into position between them but not bonded. The compression and torsional rates are dependent on the ratios between the dimensions of the metal parts and the diameter of the rubber parts.

Shear type fenders

These units consist of a rubber sandwich bonded between two metal plates. One metal plate of each unit is fixed to the wharf or jetty and the other to either a pile or, when using two units, to a suspended timber or metal fender.

POLYMERS IN DAMPING PANELS AND STRUCTURAL MEMBERS

The capacity of a homogeneous engineering structure to dissipate damping energy is often seriously limited unless high stress amplitudes are involved. Material composites, particularly those having a high damping visco-elastic adhesive in the composite, provide new possibilities and are becoming increasingly important.

Types of Visco-elastic Damping Treatment

There are two main types of damping treatment, the type in which extensional deformation of the damping layer is dominant, and the type in which mainly shear movements occur. Extensional deformation accounts for the losses when a single layer of damping material is applied to a plate (e.g. mastic on a car door) while shear damping is likely to dominate whenever a viscoelastic material is used between the plate to be damped and a relatively stiff cover plate or septum.

POLYMERS AS ISOLATORS IN AIRCRAFT, SHIPS AND MACHINES

Manned Aircraft and Guided Vehicles

Some general factors are common to the majority of requirements, which can be extremely wide and varied.

1. Any type of power plant will produce excitations extending to relatively high frequencies, although reciprocating engines and propellers produce a more pronounced periodic vibration at significantly lower frequencies than the more random characteristics of rocket type power plants. The disturbances found in guided missiles may be mechanically or aerodynamically generated and are generally anharmonic and aperiodic. The frequency spectra involved usually cover a range from 5 to 2000 cycles per second. Other vibratory excitation comes from aerodynamic conditions such as buffeting and boundary layer noise.
2. The shock occurring in airborne vehicles is not usually severe, although in manned aircraft landing or the firing of armaments may produce shocks which have to be taken care of in design. In guided missiles shock may result from action of the guidance system, separation of booster sections inflight, and handling hazards prior to launching.
3. Guided missiles and several types of manned aircraft can operate at any altitude, and the forces on the equipment may become several times that of gravity.
4. Of necessity aircraft structures are light, so in consequence there is probably a relatively low stiffness in the members to which the isolators are attached. The lower stiffness must be considered in isolator design.
5. In many situations the isolators may be required to withstand temperatures down to –54°C (–65°F) especially for components outside the pressure cabins, etc. The effect of aerodynamic heating of guided weapons, space vehicles or high speed aircraft may require the rubber to work at high temperatures. High temperature performance is often required for isolators situated near the propulsion units.

The installation and servicing of equipment in aircraft and missiles raises several problems of isolator design, both because space is at a premium and because the vehicle is meanwhile unavailable for use. Therefore, equipment must be mounted with the minimum number of thumb-screws or other quick release devices.

Often it is simpler to regard the isolators as a permanent part of the structure and to release the equipment itself rather than try to release the whole assembly.

Ships

Typical shipboard vibration factors to be considered in isolator design are as follows :

Firstly, the predominant vibration of the ship is that induced by the propeller shaft and depends on the frequency and the number of blades involved. The transmitted frequency will vary from a minimum upto approximately 20 cycles per second.

Secondly, the predominant shocks will probably arise from underwater explosions or from the ship's own armament, and the severity of these shocks will obviously depend on the type of ship involved.

Thirdly, all the machinery operating on a ship creates an ambient vibration which is not usually objectionable. However, in some types of naval shipping, such as submarines, this ambient vibration may be capable of radiating acoustic power into the water, and this may constitute a hazard to the ship as it can be detected by underwater listening devices.

Grinding and Shaping Machines

The grinding machine operates in a steel works, where it can maintain the necessary standard of accuracy and finish only if it is insulated from vibration due to a rolling mill. This installation could be mounted on a small number of self-contained isolators. Shaping machines are sometimes mounted as being treated as systems with two degrees of freedom.

Reciprocating Machines

Machines involving reciprocating motion include internal combustion engines, air compressors, vacuum pumps, etc. The vibrations in a reciprocating machine result from two unrelated sources, namely torque impulses and mass unabalance.

Lathes

When lathes are mounted upon isolators there is the problem of a possibly high out of balance force at the highest part of the machine (i.e., on the face plate). Even though it is bad practice to have an out of balance force on the face plate, it is seldom that complete balance is obtained. This force tends to excite the rocking modes of the machine and mounts must be selected that are stiff enough to limit the rocking and flexibility of the installation whilst ensuring adequate isolation from outside sources.

Electric Motors

The problem of isolation of vibration of electric motors results from two sources,

1. Fluctuation in the torque characterised as an angular reaction upon the stator and thus upon the supports for the motor.
2. Unbalance of the rotating parts, characterised as a rotating radially directed force vector at the rotational frequency of the shaft.

RUBBER IN BUILDINGS

Acoustics

The term 'acoustics' in building practice refers to the study of listening conditions in auditoria. It is concerned with the perfection of desired sounds and the elimination of echo. Rubber possesses properties which command its use for such purposes in wall and ceiling linings, upholstery and floors.

Where interiors of large buildings are finished with hard surfacing materials such as glazed tiles, hard plast r, stone and glass, they have a long reverberation period. This is due to the sound waves impinging on the non-absorbent surfaces and being reflected from surface to surface for some seconds before gradually fading away to inaudibility. In such auditoria normal rate speech and rapid passages in music are not clearly defined due to the confusion of initial and reflected sound waves. To correct this condition the reverberation period must be reduced, and this is achieved by the careful use of sound absorbing surfaces. Porous or cellular materials are ideal for this purpose; when sound waves impinge on them, they penetrate the surface, causing vibration of the air in the pores of the material. The sound is weakened by frictional losses, sound energy becoming heat energy.

Sound Insulation

Sound insulation deals with the insulation of parts of a building either from internal or external noises. It is concerned with the elimination and reduction of unwanted sound. Sound can be transmitted through a structure by wave transmission, by transmission due to resonance of the structure or structural élement, or by a combination of these two.

In the case of airborne sounds generated in a room and having frequency components remote from the resonant frequency of the structure, most of the transmission is wave motion. If an airborne sound has a frequency identical to, or very close to, the resonant frequency of the structure, then most of the transmission is due to the movement of the structure as a whole. Unwanted sounds from sources either inside or outside a structure to be insulated are dealt with as follows.

1. By isolating the source of the unwanted sound.
2. By constructing all floors, ceilings, partitions and walls to prevent the transmission of noise from one part of the building to another. Vibration will travel long distances through homogeneous structures with very little attenuation, consequently the only way to reduce structure borne or airborne transmission is to provide discontinuous construction as far as possible.
3. By making all openings airtight and, therefore, soundproof. In sound insulation it is obvious that the final result is only as good as its weakest link. Noise is conveyed about buildings by the air and it is important that free air paths should not be overlooked. Doors and windows should be made to fit closely and shut tightly, and holes for the passage of services through the structure should be adequately stopped.

Antivibration pipe fastenings

Vibration and noise carried along piping for water supplies, radiators and air conditioning systems are an everyday nuisance in buildings. To prevent the transmission of such vibrations to the pipe supporting structure, anti-vibration pipe fastenings have been devised.

Mounting of a complete building on rubber

One of the most difficult problems in building technology has been the design of bearing support pads which could withstand the normal expansion and contraction of metal structures, as well as being able to stand up to the stresses and strains arising from vibrations, shocks and tremors. Recently, however, some bearings have been produced which are claimed as suitable for general building construction.

POLYMERS IN CORROSION PREVENTION AND CONTROL

World consumption of plastics has significantly increased in the past 20 years because of their numerous advantages, including excellent corrosion resistance, competitive cost, low weight, easiness of installation, and higher specific strength than many other materials. Plastics have extended the choice of corrosion-resistant materials by an order of magnitude. (Note: The mechanism of corrosion of plastics is quite different from that of metals. Unlike metals, plastics deteriorate by loss in mechanical properties, softening, swelling, hardening, spalling, or discoloration). One special advantage of polymers over metals is the maintenance cost after their installation.

This cost is absolutely lower with polymers because they basically do not require any of those maintenance and corrosion-control systems normally used with metals, such as cathodic protection, external coating, or corrosion inhibitors. All these reasons have increased the use of polymers in corrosion prevention and control.

The thermoplastics used for corrosion resistance are fluorocarbons, acrylics, nylon, chlorinated polyether, polyethylenes, polypropylenes, polystyrene, and poly(vinyl chloride); important thermosetters used in this field are epoxies, phenolics, polyether, polyethylenes, polypropylenes, polystyrene, and poly(vinyl chloride); important thermosetters used in this field are epoxies, phenolics, polyesters, silicones, and ureas. All plastics shown useful tolerance to temperature. Fluorocarbons are the 'noble metals' of plastics; they are corrosion resistant to practically all environments up to 290°C.

Fluoroelastomers, which include elastomeric fluorocarbons, are a family of carbon-or silicone-based polymers containing significant amount of fluorine. Fluoroelastomers products are a family of specialty materials valued for their elastic behaviour under comparatively harsh conditions involving a broad thermal range and/or aggressive environments. Unique features and benefits afforded by specific fluoroelastomers have lead to their diverse industrial applications as in seals, linings, tubing, hose, belting, fabrics, caulks, adhesives, laminates, and vibration absorbers, for service in harsh chemical and thermal conditions. Sound knowledge of all types of candidate materials is fundamental to making the 'right' choice.

Flue Gas Desulphurisation

Flue gas desulphurisation (FGD) is an extremely important technology enabling power plants to produce electricity in an environmentally acceptable manner. A number of polymers that can provided protection from corrosion are used with varying degrees of success in FGD systems. They include chloroprene, epoxies, fluorocarbons, polyester, and vinyl ester systems. However, each system has its advantages and disadvantages and should be carefully scrutinised before selection. For example, although a coating formulated utilising the epoxy technology can provided resistance to corrosion in FGD system, the condensate formed in a stack or outlet duct generally provides a high concentration of acid, which results in blistering of most epoxy systems.

Similarly, a fluorocarbon system can provide sufficient acid and temperature resistance, but has a major disadvantage in its ability to adhere to a substrate, which can result in premature blistering or disbondment of the system.

Vinyl ester systems have proven to provide a protective barrier to the various components in FGD systems. These linings can be cost effectively applied with standard spray equipment. They exhibit excellent resistance to the reaction of chlorides, fluorides, and sulphur under high temperature and abrasive conditions during the burning of fossil fuels.

Generally, a vinyl ester could be classified as methyl acrylic acid extended bisphenol A epoxy, acrylic acid bisphenol E epoxy, a bisphenol F novolac epoxy, or a urethane modified ester. Each will provide a different degree of resistance. Vinyl ester lining systems have been successfully applied to provide protection to stack liners, ducting, scrubbers, gas-gas heaters, thickener tanks, and other vessels in FGD plants, stacks and ducts, electrostatic precipitators, and bag house environments.

Chemical Resistant Masonry

Chemical resistant masonry (CRM) is used to provide thermal and mechanical protection of the interior of tanks which are normally in a severe corrosive service at elevated temperatures. CRM is made up to three components: acid brickwork (block), acid proof cement (mortar), and a lining or coating system (membrane). The membrane can be either a sheet lining, a coating, a true pinhole-free membrane, or a semimembrane, depending on the actual service conditions. The membrane system is selected on the basis of its ability to prevent any corrosive materials from contacting the substrate. Because of their excellent resistance to a broad range of corrosive conditions at elevated temperature, the fluoropolymer plastic membranes have become an excellent choice for use as a membrane beneath chemical-resistant masonry. Most fluoropolymers currently being used as membrane system are poly(vinyl fluoride) (PVF), ethylene chlorotrifluoroethylene (ECTFE), ethylene trifluoroethylene (ETFE), fluorinated ethylene propylene (FEP), and perfluoroalkoxy (PFA). Lack of melt-flow processability restricts the selection of PTFE as a viable membrane system.

There are basically three types of mortars in use in CRM. They are thermoset resin based mortars (such as furan, phenolic, epoxy, polyester), silicate mortars, and sulphur mortars).

Furan mortar is the most commonly used mortar in the chemical process industry. It is a two-component system containing a liquid binder and a powder that is mixed in a mortar box to produce a material of trowelable consistency. Furan mortar is suitable for a wide range of corrosive service conditions including various concentrations of acids and alkalies at temperatures up to 170°C. It is not recommended for strongly oxidising services.

Epoxy mortars are less frequently used in CRM than other resin bonded systems due to higher cost, lower maximum service temperature, and narrower range of chemical resistance. However, because of their excellent bond strength to concrete, epoxides are frequently used for acid brick floors over concrete. They also have applications in alkali exposure.

Polyester mortars are used in oxidising services such as nitric and chromic acid. They are, however, more expensive than furan and also exhibit a narrower range of chemical resistance. They are usually a three component system consisting of a resin, a powder, and a curing agent.

Piping Systems

The cooling or heating systems in commercial or institutional buildings are frequently fouled by the accumulation of deposits containing organic material, silt and/or iron oxide. A conventional off-line chemical cleaning in these facilities with an acid or other strong solvent is a very complex and expensive process since the piping is usually not designed for such cleaning procedures and it may be enclosed within walls. In many cases, high-molecular-weight water-soluble polymers can be used as an on-line method for removing substantial quantities of deposits from piping systems. While the method will not work in all case, it is so inexpensive compared to off-line chemical cleaning that it must be considered.

A slow on-line cleaning over a long period of time has been found to be effective in several applications. Such a cleaning use high-molecular-weight anionic polymers to slowly soften and disperse the deposits over time. The process may take from several months to up to two years. A high-molecular weight anionic polyacrylamide with molecular weights in the range of 15–18 million and charge densities between 30 and 100 per cent anionicity have been used at doses as low as 20 parts per million. The on-line cleaning is a physical process that 'pries' small pieces small pieces of the deposit from the pipe surface. The charged portion of the polymer attaches to surface charges on the deposit. The polymer chains acts like a 'sail' and allows the velocity of the water to exert a physical force on the deposit.

Boiler and Cooling Water Treatment

Programmes such as phosphate and zinc treatments of boiler and cooling water require the use of appropriate polymer to demonstrate satisfactory corrosion control. These polymers also prevent bulk precipitation of the corrosion additives in the circulating water. Sulphonated co-polymers have thus proven to be effective for both boiler and cooling water treatments.

Possibly the most widely used sulphonated co-polymers, used both as dispersants in boiler treatment and as cooling water additives, are composed of acrylic acid and 2-acrylamido-2-methylpropane sulphonic acid (AA-PSA).

A novel sulphonated co-polymer, claimed to be superior to existing sulphonated polymers, has recently been introduced for boiler and cooling water treatment. The polymer is said to have a unique composition based on multiple sulphonic acid containing monomers which provide hydrophobic character and high charge density. This composition has been shown to provide a range of exceptional properties. The new polymer demonstrates not only equivalent or better performance for calcium phosphate and zinc stabilisation, but also superior thermal stability when compared to existing sulphonated polymers. A notable property of the new polymer is its ability to disperse and resolubilise pre-existing scale and corrosion product.

Biodegradable Scale Inhibitor

In the area of water treatment, polyacrylate scale inhibitors and dispersants, based on homopolymers of polyacrylic acid, are commonly used. Although generally of low toxicity, these polymers are not biodegradable.

In some sensitive areas, such as the North Sea Oil fields, this has led to the search for biodegradable alternative to polyacrylates.

Polyaspartates, as exemplified by the parent polymer, polyaspartic acid, offer an attractive biodegradable alternative to polyacrylates. In addition to their activity as scale inhibitors and dispersants, polyaspartates have also been shown to have corrosion inhibition activity, particularly for the corrosion environments frequently found in oil-filed applications. Polyaspartates have very high biodegradability as indicated by OECD guidelines for aquatic biodegradability.

Sodium polyaspartates are readily obtained by thermal polymerisation of aspartic acid followed by hydrolysis with sodium hydroxide:

H_2N O OH OH O → Δ With or without catalyst → O N O

(I) (II)

Hydrolysis → H O NH O OH n   H NH O O OH m

α β

(III)

Heating aspartic acid (I) with or without catalyst leads to a linear thermal polycondensation polymer known as polysuccinimide (II), conversions of monomer to polymer of greater than 95 per cent being easily accomplished. Hydrolysis of polysuccinimide with base such as sodium hydroxide, leads to a random co-polymer of α and β aspartate units (III), with the β aspartate comprising about 70–75 per cent of the repeating units.

The performance of polymers as mineral inhibitors is dependent on the ability of the polymer to interact with growth sites on the surface of growing mineral crystals. Generally, the molecular weight of the polymer plays a significant role in this interaction. Polyaspartates have been shown to be very good inhibitors for both calcium and barium scales, the optimum inhibition activity being in the range of 1000–4000 molecular weights (Mw). This corresponds to 7–30 mer units in the polymer. For calcium carbonate and barium sulphate scale inhibition, the optimum is in the range of 2000–4000 Mw (15–30 mer units). In comparison with commercial scale inhibitors, the polyaspartates shown comparable activity, besides being biodegradable, and in many cases better activity than polyacrylates of similar molecular weight.

Reinforcing Steel in Concrete

Corrosion of steel in concrete has been growing at an alarming rate and has become a very costly problem over the past three decades. In attempts to solve this corrosion problem, the successful application of epoxy coatings on underground transmission pipes drew a lot of attention, and it was pronounced by researchers back in 1974 that fusion-bonded epoxy-coated reinforcement (ECR) could significantly extend the time before deterioration of reinforced concrete bridge decks beyond the 5–10 years experienced with uncoated steel bar in area where deicing salts were used. Since then, ECR has been extensively used in the United States, especially in bridges and viaducts. For the most part, the performance of ECR in highway bridge decks has been successful as relatively little corrosion-induced concrete deterioration has been found on the great majority of ECR bridge deck structures.

Recent studies have established the importance of coatings defects to the performance of ECR. If there are no defects, the corrosion protection barrier is effective. In the real world, however, there will be defects in placed ECR and the coating's resistance to disbondment from these defects is where the performance can be improved. Several ways have been suggested to improved the long-term adhesion of epoxy coating to steel reinforcing bars. These include: alternative application methods, chemical treatment of blasted steel surface prior to coating application, and evolving a strong quality assurance programme for coating application plants.

POLYMERS IN AGRICULTURE AND HORTICULTURE

The application of polymeric materials in agriculture and horticulture has increased considerably in recent years, not only as replacement for traditional materials but also as a means of effecting improvement. For example, the application of plastics has led to a significant improvement in technological processes in the growing and storing of agricultural crops, construction of storages for fertilisers, vegetables and animals, and in agricultural equipment and drainage technology.

Plastic Film

The main uses of plastic films in agricultural and horticultural applications are for covering fodder and cultivation units such as seed-beds, for mulching and warming soil, for the building of canals and reservoirs, and as packaging materials for storing and transporting food.

The most widely used plastic film is polyethylene (PE). PE film tunnels and perforated flat PE film allow better use of natural resources such as solar energy, water, and soil. Crops may be made to ripen early making cultivation possible all year round. In addition to covering produce, PE film has a number of other uses. Livestock feeding stuff such as green foliage grass and maize can be preserved by covering the forage in a silo with large gastight black or white PE films. Shrinkable PE films are used for sheet steaming in horticulture.

Polyethylene films for agricultural applications needs to have high strength and elasticity, resistance to wind forces, and a long service life. They are mostly produced from low density PE combined with linear, low density PE and ethylene-vinyl acetate co-polymer.

As PE films, like all synthetic polymers, suffer from decomposition by environmental influences such as light and atmospheric oxygen, special photoprotective systems are added to delay these effects. Systems that are commonly used are ultraviolet (UV) light absorbers, quenchers, radical scavengers, and hydroperoxide decomposing agents.

Photodegradable PE film is used for mulching the soil in vegetable growing. It improves soil drainage and discourages weed growth. The film breaks down within two weeks to three months as a result of solar radiation, combines with the soil, and is broken down by micro-organisms.

Pipes and Hoses

Plastic pipes are being increasingly used in the agricultural industry for drainage and irrigation. The improved strength and greater crack resistance of new grades of polyethylene have allowed the wall thickness of such pipes to be reduced and their diameters increased.

Polyethylene has been irradiated to produce cross-links thus giving it improved strength. Drip irrigation for raw crops uses such irradiated LDPE tubing. For vine and tree crops the drop emitter is made of PE moulded in large multicavity moulds.

Plastics are now being used in a variety of ways to replace metal components in different irrigation systems. Interestingly, one of the first uses of PVC pipe in the 1960s was in sprinkler irrigation system as a replacement of brass, gun-metal, or zinc alloys.

Hoses used in irrigation are invariably made of PE. Crimped PE pipes are being more frequently used for drainage and irrigation of soils. Flexible hoses that can be rolled up are used in spraying and sprinkling machines. Hoses reinforced with a rigid spiral are being used with machines to introduce chemicals into the soil for the protection of plants.

Greenhouses

The use of plastics in horticultural industry has grown substantially in recent years. Several plastics, and PVC in particular, are used in applications such as hosepipes, glazing strips, paint, and wiring. However, it has come to light that as greenhouses become airtight, some chemical vapours released by plastics may build up to levels toxic to plants. The plasticiser dibutyl phthalate (DBP) used in PVC has thus been found to the phytotoxic in a greenhouse environment. Research has shown that a fire retardant used in polystyrene trays and granules reduced the growth of some seedlings. Similarly, a silicone joining compound used in green houses and fumes from a paint used on an electric boiler were also found to be harmful to test plants.

Responding to the toxicity problem of plastics in greenhouse applications, the Dutch horticultural industry developed a test for screening. Plastics that pass the test qualify as safe horticultural plastics can be used in a greenhouse environment.

POLYMERS IN SPORT

The properties of polymers—resilience, toughness, and good strength-to-weight ratio—ensured the quick entry of polymers into the sports industry. In most cases, the polymer has been used as a replacement for another material and, being a competitive business, the sports industry has also been very receptive to new developments in polymer formulations and advances in design that contribute to improved sporting performance. Uses of polymers in the sports industry fall into three main categories: synthetic surfaces, footwear, and equipment.

Synthetic Surfaces

In sports such as athletics, traditional materials have been almost completely replaced by synthetic materials. Sports surfaces in modern indoor stadiums are usually made from a porous structure of rubber crumb and a binder such as polyurethane. Other polymeric materials used include flexible PVC or rubber sheet, polyolefins, polypropylene (PP) grasses, and foam laminates.

Footwear

There has been significant development in sports footwear in recent years, greatly aided by polymeric materials that allow for better shape, manufacturing techniques, and material performance.

In the design of running shoes, much emphasis has been placed on anatomical comfort leading to the development of toe bars and heel canters, wedges, and midsoles, usually moulded in polyethylene foams. Greater attention has been focussed on the outsole of sport shoes in order to cope with the increased use of indoor facilities and artificial surfaces. Ski boots use a wide variety of plastics adapted for low temperatures. Thermoplastic polyurethane/ABS blends and polyamides are used for the outer shell of ski boots and modified polyethylene terephthalate for the binding. Polyurethane (PU) foam is often used to line the boots. Micro-sphere fillers may be incorporated into foam-lined boots to add further thermal insulation to the wax binder. The antivibrational characteristics of PU foams have also led to their use in ski fittings. A sandwich construction of polyurethane elastomer and aluminium alloy has been fitted between the ski and the binding to reduce shock and vibration.

Equipment

Developments in plastics sports equipment are mainly aimed at improving player performance and increasing player safety. Items that fall into the first category include balls, bicycles, tennis rackets, and windsurfing boards. Crash helmets belong in the second category.

Since the maximum performance of sports balls is governed by strict specifications, the scope of using plastics to improve them is rather limited. Improvements can however still be made. For example, micro-sphere fillers have been used in golf balls and bowling balls. The travel distance of golf balls is decreased by up to 30 per cent by the incorporation of micro-spheres. In bowling balls, incorporating of micro-spheres reduces the density of the ball's core enabling lightweight balls to be made for children and women players. Plastics are being used to make a number of bicycle parts replacing metals. Examples include polyamides for structural components such as the bike frames and wheels, and polyacetals for tyre valves, brake levers, and gear change and power switches. Sailboard accessories use thermoplastic blends; for example, boom fittings and mast steps that have to withstand very high dynamic loads are made of glass-fibre reinforced polyamide blends. Polycarbonate/ABS blends are commonly used for daggerboard cases. The most popular type of sails are film sails, which are laminates of a plastic film, usually a polyester such as bonded to woven fabric. They are claimed to offer superior tear resistance.

The tennis racket is another example of sports equipment that has been improved significantly due to the availability of high-performance plastics. The carbon fibre-reinforced nylon tennis racket frame first introduced by Dunlop has become the industry's standard for this equipment. An improved design incorporates Atochem's Pebax nylon ether block co-polymer into the racket frame where maximum vibrations are concentrated, i.e., between the two profiles of the frame in the handle and across the throat of the frame.

An item of sports equipment that has been improved greatly with regard to player safety is the crash helmet. American football helmets are polycarbonate or ABS shelled, while impact nylons are often used on motorbike crash helmets. Another item of sports equipment in the United States where safety has been improved is a Reduced Injury Factor baseball that has a novel polyurethane core material.

SPECIALITY POLYMERS IN ROAD APPLICATIONS—SOME NOVAL TECHNOLOGICAL ADVANCES

Laboratory and field studies conducted at Central Road Research Institute confirm that ethylene vinyl acetate co-polymer with vinyl acetate content ranging 18 to 28 per cent, when added to bituminous binders to the extent of 2.5 per cent to 5.0 per cent improve flexibility at low temperature, prevent permanent deformation of road surfacing at high temperature and improve fatigue life leading to safer, durable, longer lasting pavement with low maintenance cost. Performance of road surfacings modified with ethylene vinyl acetate co-polymer as an ingredient have been recorded superior under heavy loading.

Increased number of traffic loads with overloading of axles in excess of permissible limits and higher tyre pressure have caused widespread problems with bituminous roads on National Highways and available 60/70 grade bitumen is not suitable for heavy duty roads in industrial complex, ports and airfield pavement.

The accelerated deterioration of bituminous surfacing prematurely burdens the maintenance budget and pose funds constraints on new development. It is a global view that the quality and longevity of pavement must therefore be restored in order to reduce maintenance budget and to save road user's cost. Polymer modified binders are premium binders, which allows an engineer to design and build a stronger and durable pavement. Scientists have conducted extensive laboratory and field studies in recent years using a variety of polymeric additives for their use in heavy duty road and airfield surfacings. Ethylene vinyl acetate co-polymer having wide range of properties, which depending upon the vinyl acetate content, extends from plastic to rubber-like products. At a very low vinyl acetate content, the co-polymer behave like LDPE and with increasing vinyl acetate content amorphous character become predominant resulting in flexible rubbery material. Hence, such wide range properties of ethyl vinyl acetate (EVA) make it, a polymer with diverse applications. Flexibility, polarity, low temperature toughness, shock-resistance and good weathering has made it a material suitable for some potential applications in highway and transportation engineering viz., bitumen modification, bitumen packing, traffic cones and traffic lane dividers.

The physical and engineering properties of modified bitumen are better than conventional 80/100 or 60/70 bitumen. Higher value of penetration ratio indicates improved temperature susceptibility of modified bitumen. Improved value of low temperature ductility further indicate resistance of modified binders to abrasion and thermal cracking. Higher values of softening point for modified binders indicate reduced chance of bleeding and fatting up at higher temperature besides resistance to deformation.

SECTION IV

Environmental Aspects Related to Polymers

CHAPTER 18

Green Chemistry and Its Role in Designing Safer Polymers

INTRODUCTION

Green chemistry is the design of chemical products and processes that reduce or eliminate the use or generation of hazardous substances. Through careful planning, chemists are able to manipulate molecules and develop products and processes that minimise harmful effects to human health and the environment. Green chemistry employs the fundamental principles of chemistry in reducing pollution at its source. It embraces chemistry as the solution to many environmental problems.

Green chemistry represents a paradigm shift from pollution remediation, exemplified by the 'command and control' approach, to pollution prevention. Historically, environmental problems have been addressed after they have occurred; waste treatment, for example, occurs at the end of the pipe. By preventing pollution, green chemistry eliminates the need for end-of-the-pipe treatment.

BACKGROUND

Chemistry is often perceived as being both beneficial and detrimental to human health and the environment. Chemistry has provided products that have improved quality and duration of life, from polymers to textiles to pharmaceuticals. Antibiotics destroy bacteria that once were fatal; vaccines have eradicated diseases that previously caused epidemics; air conditioning prevents deaths in the elderly during prolonged heat waves. Accompanying these advances, however, have been unwanted and unanticipated side effects––chemical accidents, contamination of air and water, destruction of ecosystems, and depletion of the ozone layer by chlorofluorocarbons (CFCs). Designing safer chemicals and chemical reactions can minimise the unintended consequences of technological advances.

The effort to minimise the impact of pollution on human health and the environment has resulted in an explosion in the number of environmental regulations in recent years. Public awareness of environmental problems has been heightened by well-known incidents, such as the birth of thalidomide babies in Europe in the 1960s; the dumping of chemicals in Love Canal, New York, and Times Beach, Missouri; and the decline in the bird population due to widespread use of DDT. The deadly methyl isocyanate gas leak in Bhopal (India), in 1984 promoted the passage of the Emergency Planning and Community Right-to-Know Act (EPCRA) by the United States Congress in 1986. EPCRA provides citizens and communities with information related to chemical hazards in their area.

The increase in environmental laws has been accompanied by the added cost of compliance with these regulations. For some companies, the cost of compliance equals its research and development budget, which is not the best use of a company's resources. Pollution prevention through green chemistry is an economically sound strategy for allocating a company's resources.

Environmental regulations aim to prevent disasters and ensure a safer workplace. However, regulations can simply minimise risk, but they cannot eliminate it. Risk is a function of both hazard and exposure. The traditional approach to risk management has been to limit exposure: given the hazard inherent in a particular substance, risk can be minimised by restricting concentration or exposure time. Green chemistry advocates a different approach to risk management. Exposure becomes immaterial if the substance poses no hazard. The use and generation of innocuous substances remove hazard from the equation.

PRINCIPLES OF GREEN CHEMISTRY

Twelve principles of green chemistry have been identified by *Anastas* and *Warner* which are given below:

1. It is better to prevent waste than to treat or clean up waste after it is formed.
2. Synthetic methods should be designed to maximise the incorporation of all materials used in the process into the final product.
3. Wherever practicable, synthetic methodologies should be designed to use and generate substances that possess little or no toxicity to human health and the environment.
4. Chemical products should be designed to preserve efficacy of function while reducing toxicity.
5. The use of auxiliary substances (e.g., solvents, separation agents, etc.) should be made unnecessary wherever possible and innocuous when used.
6. Energy requirements should be recognised for their environmental and economic impacts and should be minimised. Synthetic methods should be conducted at ambient temperature and pressure.
7. A raw material of feedstock should be renewable rather than depleting wherever technically and economically practicable.
8. Unnecessary derivatisation (block group, protection/deprotection, temporary modification of physical/chemical processes) should be avoided whenever possible.
9. Catalytic reagents (as selective a possible) are superior to stoichiometric reagents.
10. Chemical products should be designed so that at the end of their function they do not persist in the environment and break down into innocuous degradation products.
11. Analytical methodologies need to be further developed to allow for real-time, in-process monitoring and control prior to the formation of hazardous substances.
12. Substances and the form of a substance used in a chemical process should be chosen so as to minimise the potential for chemical accidents, including releases, explosions, and fires.

These principles serve as a blueprint for designing chemical products and processes that are inherently safer. In many instances, a green chemistry initiative may address several of the principles, resulting in added environmental benefits. It is important to note that trade-offs may be necessary in achieving the goals of green chemistry. For example, a new catalyst may yield a product with higher atom economy and percent yield than a traditional catalyst but may be a more toxic catalyst. The synthetic chemist must balance the advantages and disadvantages when designing a reaction.

The twelve principles of green chemistry are briefly discussed below and include examples that illustrate the application of each principle. Although the examples are not perfect, they represent an improvement over existing processes.

Principel 1: It is Better to Prevent Waste than to Treat or Clean up Waste after it is Formed

Waste prevention makes sense from both an environmental and economic standpoint. Treatment and disposal of chemical waste pose potential hazards to human health and the environment from accidental spills, leaks, and releases. The cost of separating, treating, and disposing of waste products is enormous. Eliminating waste, such as unreacted starting material, unwanted by-products, and contaminated solvents, from a chemical reaction removes the treatment and disposal steps at the end of the process.

Eastman Kodak has decreased the amount of waste generated in the synthesis of polycarbonate/polydimethylsiloxane copolymers by 600,000 lb/year. The new process (Fig. 18.1) avoids the use of bisphenol A bischloroformate, a reactant that is synthesised from phosgene. The aqueous waste stream produced is less hazardous, as triethylamine replaces pyridine as the acid acceptor. The co-polymer produced retains the same properties as the original.

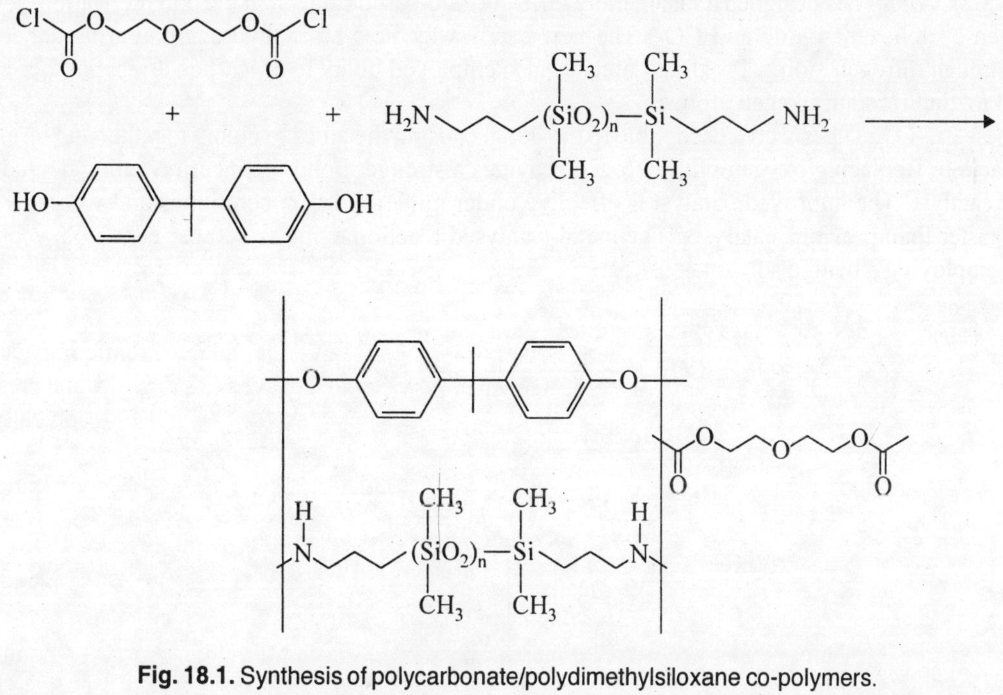

Fig. 18.1. Synthesis of polycarbonate/polydimethylsiloxane co-polymers.

The Nalco Chemical Company has developed a new water-based process for manufacturing liquid polymers. High molecular weight, water-soluble polymers are commonly used to remove contaminants from industrial waste-water. Traditionally, these polymers have been produced as a dry powder or as a water-in-oil emulsion. The powder form presents exposure hazards and is energy intensive to produce and use, whereas the emulsion employs large quantities of hydrocarbon solvents and surfactants. Nalco's new process manufactures water-based acrylamide polymers in an aqueous environment, eliminating the use of oils and surfactants.

In addition, ammonium sulphate, a waste by-product in the synthesis of caprolactam, is used in the manufacturing process. Less waste is generated by avoiding the use of oils and surfactants in the manufacture of water-soluble polymers and by recycling a waste product (ammonium sulphate) from another industrial process).

Principle 2: Synthetic Methods Should be Designed to Maximise the Incorporation of all Materials Used in the Process into the Final Product

Traditionally, the measure of the success of a chemical reaction has been the per cent yield of the desired product. Although yield is important, it does not take into account the fate of all the atoms participating in the reaction. The concept of atom economy or atom utilisation, as proposed by *Trost* and *Sheldon*, emphasises the importance of incorporating the maximum number of atoms into the final product as a way of minimising waste. Atoms that are not contained within the final product become part of the waste stream. For example, the Diels–Alder reaction is 100 per cent atom efficient; all of the atoms in the starting materials are assimilated into the final product. Elimination reactions, however, are inherently wasteful, as an unwanted by-product is always formed.

Carbon dioxide has been used as both comonomer and solvent in the synthesis of polycarbonates (Fig. 18.2). *Coates* has designed a faster, more efficient zinc-based catalyst in the polymerisation reaction between cyclohexene oxide and CO_2. The new catalyst features bulky hydrocarbon arms tethered to zinc through nitrogen atoms. Previous catalysts also employed bulky hydrocarbon arms, but these were linked to zinc through oxygen atoms.

These Zn–O linkages were susceptible to breaking, causing the zincs to clump together and terminate the reaction. Replacing oxygen with nitrogen provides a stronger bond to zinc, preventing degradation of the catalyst. The improved catalyst is effective under milder reaction conditions and works over 50 times faster than previous catalysts. The metal-catalysed reaction exhibits 100 per cent atom economy while employing a benign solvent.

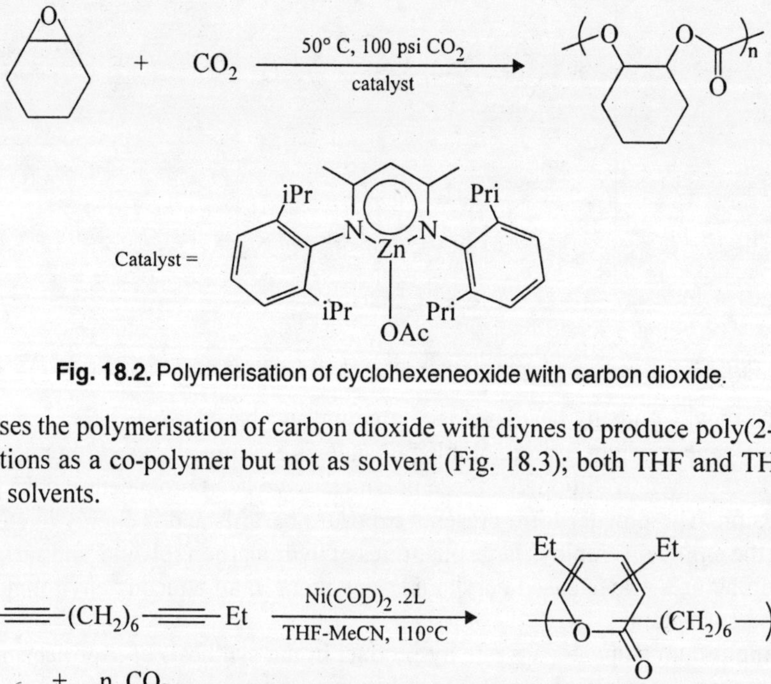

+ CO_2 —(50° C, 100 psi CO_2; catalyst)→

Catalyst = iPr, Pri, N, Zn, N, iPr, Pri, OAc

Fig. 18.2. Polymerisation of cyclohexeneoxide with carbon dioxide.

Ni(0) catalyses the polymerisation of carbon dioxide with diynes to produce poly(2-pyrones). In this case, CO_2 functions as a co-polymer but not as solvent (Fig. 18.3); both THF and THF–CH_3CN have been utilised as solvents.

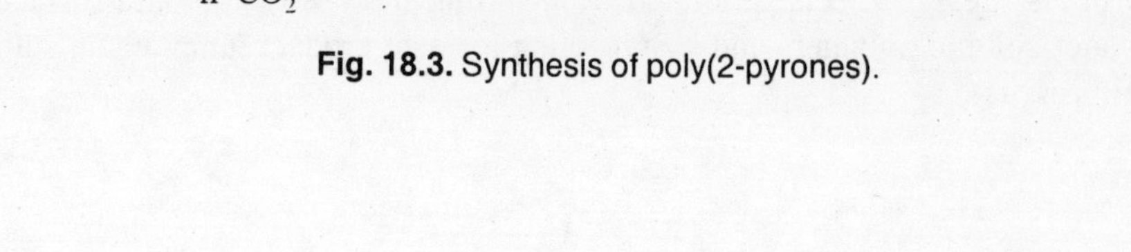

Et—≡—$(CH_2)_6$—≡—Et + n CO_2 —($Ni(COD)_2$ .2L; THF-MeCN, 110°C)→ Et, Et, O, O, $(CH_2)_6$, n

Fig. 18.3. Synthesis of poly(2-pyrones).

Principle 3: Wherever Practicable, Synthetic Methodologies should be Designed to Use and Generate Substances that Possess Little or no Toxicity to Human Health and the Environment

Because risk is a function of both hazard and exposure, it can be managed by minimising the hazardous nature of a substance or by limiting the exposure time or concentration of a substance. Limiting exposure may require the use of special equipment or protective clothing, safeguards that increase the cost of a given process. Exposure controls also have the potential for failure, thereby increasing risk to human health and the environment. Risk management is better addressed by focusing on the hazard of a particular substance. A benign reagent or product presents no risk to human health or the environment regardless of the length of exposure. Obviously, it is desirable to use nontoxic substances wherever possible.

Polyurethane foam has typically been manufactured using CFCs and HCFCs as blowing agents, substances that contribute to ozone depletion and global warming. The Stepan Company has designed a new process that utilise water as the blowing agent in the manufacture of rigid polyurethane foam. During the blowing process, isocyanate and water combine to form polymer and carbon dioxide gas, which becomes trapped in the cellular structure of the foam product. The improved process eliminates the use of ozone-depleting substances while yielding a product with properties that are virtually identical to those of the polyurethane foam manufactured following the traditional procedure.

Sequa Chemicals uses starch graft polymer as phenolic resins extenders to decrease formaldehyde emissions. Aqueous phenol-formaldehyde (PF) resins contain approximately 2 per cent free formaldehyde, which is released into the atmosphere and workplace each year. Replacing 50 per cent of the phenol-formaldehyde resin with starch grafts cuts formaldehyde emission in half. The extended resins exhibits improved properties over the original PF resins. The starch graft polymers are water-based, nontoxic, and non-irritating and utilise a renewable resource, starch, in their synthesis.

Principle 4: Chemical Products should be Designed to Preserve Efficacy of Function while Reducing Toxicity

A better understanding of the mechanism of toxicity allows structural changes to be made that do not affect the function of a product. For example, toxicity may be avoided by eliminating or modifying a functional group. Physical properties may be altered to minimise bio-availability. A parallel may be found in the pharmaceutical industry, where the effectiveness of a drug is balanced against side effects. In some cases, simply changing a functional group on an aromatic ring may decrease an undesirable side effect yet not diminish the efficacy of the pharmaceutical agent. Structural modifications that decrease toxicity while maintaining function are highly desirable. The chemical industry needs to examine products currently in use for less toxic alternatives.

Coating metals with paint helps to prevent corrosion, a significant problems that costs industry and consumers millions of dollars each year. To remain effective, coatings must be continually stripped and reapplied. Organic solvents, notably methylene chloride, are commonly employed in the stripping process. Because methylene chloride is a known carcinogen, dry-blasting has gained acceptance as suitable alternative. Laevo-polylactic acid (1-PLA) has been identified as an appropriate dry-blast medium. The polymer can be chemically depolymerised after depainting. PLA can be converted to ethyl lactate, a solvent with potential applications in the electronics industry. This technology reduces the amount of solid waste produced by the dry-blast process by 90–97 per cent.

Asahi Chemical Industry Company has developed a solid-state process for the synthesis of polycarbonate (Fig. 18.4). Traditional polycarbonate synthesis employs both phosgene and large quantities of methylene chloride as solvent. Asahi's new process combines bisphenol A with diphenyl carbonate to yield a prepolymer. Diphenyl carbonate itself is also synthesised without the use of phosgene. The crystallised prepolymer is heated under a stream of inert gas in the polymerisation step. Higher quality polycarbonates are realised as a results of the solid-state process.

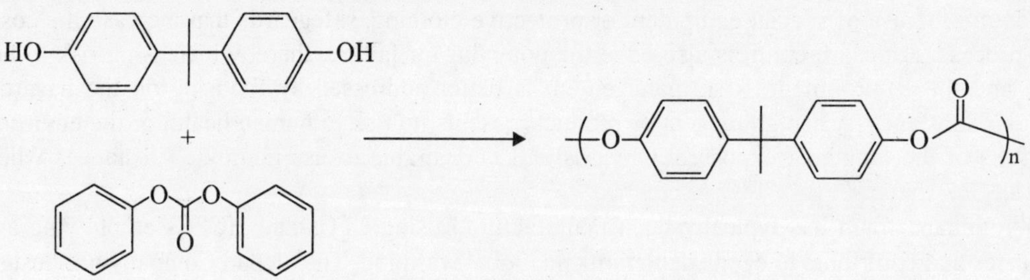

Fig. 18.4. Nonphosgene synthesis of polycarbonates.

Principle 5: The use of Auxiliary Substances (e.g., Solvents, Separation Agents, etc.) should be Made Unnecessary wherever Possible and Innocuous when Used

A major source of waste in the chemical industry is the use of solvents in running a reaction and separating products. Many organic solvents, such as benzene and methylene chloride, are known or suspected carcinogens. Volatile organic compounds (VOCs) are associated with the generation of smog; chlorofluorocarbons (CFCs) have been cited for their role in stratospheric ozone depletion. Although organic solvents may be needed to facilitate a reaction, less hazardous solvents are now being utilised and are proving effective in mediating organic reactions. Benign alternatives to organic solvents include supercritical CO_2 and water; some reactions are even run in the absence of a solvent. The need for separation agents can be minimised by increasing selectivity, often through the use of catalysis.

Buelow has polymerised 1,3-butadiene in supercritical CO_2 using a nickel catalyst (Fig. 18.5). The *cis*-1,4 product was highly favoured, exhibiting a slight dependence on pressure. Supercritical CO_2 replaces toluene or heptane as the solvent in the polymerisation reaction.

cat = $[(\pi - \text{allyl})\text{Ni}(\text{CF}_3\text{CO}_2)]$

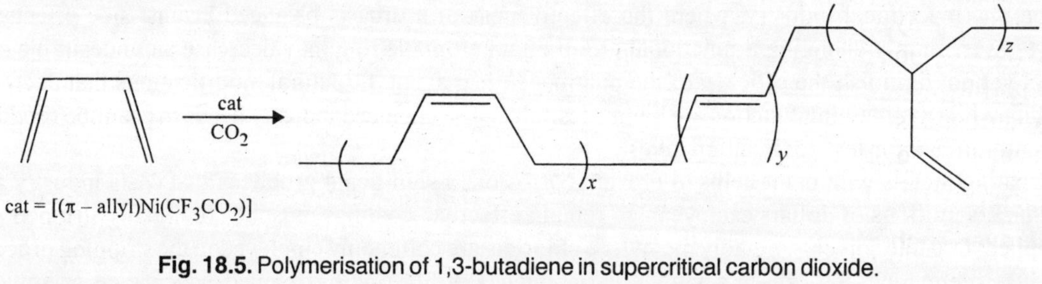

Fig. 18.5. Polymerisation of 1,3-butadiene in supercritical carbon dioxide.

Mistele and *DeSimone* have conducted numerous polymerisation reactions in supercritical CO_2. For example, supercritical CO_2 is an effective solvent in the synthesis of fluoropolymers, materials that are traditionally synthesised in ozone-depleting CFCs. 1,1-Di-hydroperfluorooctyl acrylate (FOA) has been co-polymerised with a variety of hydrocarbon-based monomers, such as styrene and methyl methacrylate, in supercritical CO_2 (Fig. 18.6). The homogeneous reaction proceeds via a free-radical process using azobisisobutyronitrile (AIBN) as initiator. When supercritical fluids are used as solvents, the same high-quality products may be synthesised without the use of environmentally damaging solvents.

$$H_2C{=}CH{-}C({=}O){-}O{-}CH_2(CF_2)_6CF_3 + H_2C{=}C(CH_3){-}C({=}O){-}OMe \xrightarrow[CO_2]{AIBN/\Delta} {+}CH_2{-}CH(C({=}O){-}O{-}CH_2(CF_2)_6CF_3){+}_m{+}CH_2{-}C(CH_3)(C({=}O){-}OMe){+}_n$$

Fig. 18.6. Polymerisation of 1,1-dihydroperfluorooctyl acrylate with methyl methacrylate in supercritical carbon dioxide.

Principle 6: Energy Requirements should be Recognised for their Environmental and Economic Impacts and should be Minimised. Synthetic Methods should be Conducted at Ambient Temperature and Pressure

The energy demands of a reaction represent hidden economic and environmental costs. Heating, cooling, separation, and purification of products require energy in one form or another. Typically, the energy used is supplied by fossil fuel combustion, a process that contributes to numerous environmental problems while depleting a limited natural resources. One approach to minimising energy requirements is through catalysis, which lowers the activation energy of the reaction, thereby decreasing the amount of thermal energy required to complete the transformation.

The General Electric Corporation has developed an improved catalytic process for the synthesis of its ULTEM polyetherimide resin. The new process requires 25 per cent less energy to produce each pound of resin when compared with the old procedure. Additional environmental benefits have also been realised: the volume of the organic waste streams for off-site disposal has been decreased by 90 per cent, the volume of the water-based organic waste for on-site thermal oxidation has been reduced by 60 per cent, the amount of catalyst used has been decreased by 50 per cent, and the amount of waste generated in the synthesis of the catalyst itself has been reduced by 75 per cent. The new synthetic protocol yields both environmental and economic advantages.

Thermal polyaspartic acid (TPA) is a biodegradable polymer that is a suitable substitute for polyacrylic acid (PAC), a non-biodegradable polymer. The Donlar Corporation has designed a catalytic synthesis of this polymer. Use of a catalyst allows the polymerisation to occur at a lower temperature and produces a product with superior performance characteristics. *Donlar* has developed a second synthesis of TPA that employs no organic solvents and produces water as the only by-product.

Principle 7: A Raw Material of Feedstock should be Renewable Rather than Depleting wherever Technically and Economically Practicable

The finite supply of fossil fuels, the primary feedstock for the chemical industry, has spurred interest in developing alternative, renewable feedstocks. Attention has centred on plant-based feedstocks as viable options to petroleum products. Currently, it is difficult for biomass to compete economically with petroleum. As oil becomes more scarce, however, the price of petroleum-based products is expected to rise, making alternative feedstocks economically competitive. Natural polyhydroxylkanoates (PHAs) are microbial polyesters that are synthesised by bacteria. Two approaches to PHA biosynthesis have been employed: production by recombinant plant systems and fermentation in aqueous media. Renewable resources, such as fructose and glucose, can serve as the carbon source in the fermentation process. *Gross* has found poly(ethylene glycol) to be effective in controlling both molecular weight and composition of PHAs. These polymers have the added advantage of being biodegradable.

Some products may be converted back to feedstock at the end of their useful lifetime. Dupont has designed the Petretec process to recycle polyethylene terephthalate (PET). As shown in Fig. 18.7, PET can be broken down into dimethyl terephthalate and ethylene glycol, which can then be used to synthesise PET intermediates. This process decreases the amount of petrochemical feedstocks needed to synthesise PET and decreases the amount of scrap PET that is land filled.

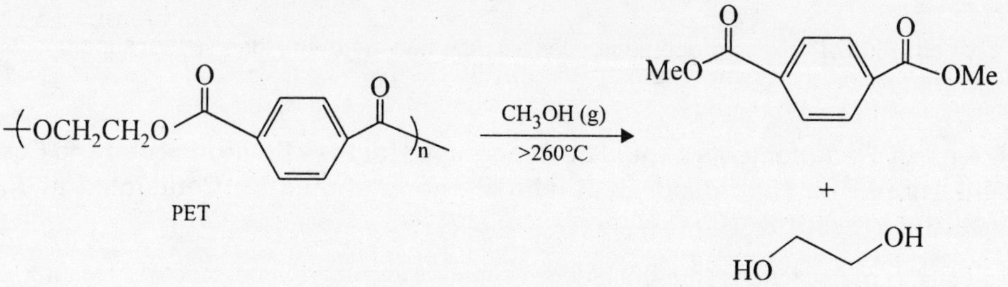

$$\left[OCH_2CH_2O-C(=O)-C_6H_4-C(=O) \right]_n \xrightarrow[>260°C]{CH_3OH\ (g)} MeO-C(=O)-C_6H_4-C(=O)-OMe + HO-CH_2CH_2-OH$$

PET

Fig. 18.7. Depolymerisation of polyethylene terephthalate (PET).

Principle 8: Unnecessary derivatisation (blocking group, protection/deprotection, temporary modification of physical/chemical processes) should be avoided whenever possible

It is common practice, particularly in multistep syntheses, to mask sensitive functional groups with protecting groups. Although this prevents unwanted side reactions from occurring, it also adds additional steps that carry with them added use of reagents, solvents, and energy. Converting a product to its salt for ease of separation is a process that also requires additional materials. Minimising the number of unnecessary steps will decrease the amount of waste that is generated.

β-Substituted β-lactones serve as monomers in the synthesis of poly(β-hydroxyalkanoates), PHAs, a class of natural polyesters cited in the previous section. Synthesis of enantiopure monomers requires multistep syntheses that employ protection/deprotection steps. *Gross* has reported the lipase-catalysed resolution of β-lactones, providing enantioenriched monomers for the ring-opening polymerisation reaction. The chemoenzymatic resolution of the β-lactones offers a more efficient route to the starting monomers.

Principle 9: Catalytic reagents (as selective as possible) are superior to stoichiometric reagents

Catalysis offers distinct advantages over stoichiometric reactions. A catalyst may provide enhanced selectivity, eliminating the necessity of separating reaction products. Catalysis may allow a reaction to be performed in a benign solvent, rather than an organic one. Less energy may be required to carry out a catalysed reaction.

Many advances in polymers chemistry have been achieved through the development of new catalysts. Metallocene, or single-site, catalysts permit the synthesis of polyolefins with improved physical properties. These catalysts typically feature a metal in a restricted environment, allowing single access by monomers to this catalytically active site. As a result, polymers grow by a single mechanism, forming a more uniform and reproducible polymer structure. Metallocene-catalysed polyolefin synthesis yields a product with increased strength and toughness, better clarity and gloss, and improved handling characteristics. Linear low-density polyethylene (LLDPE), high-density polyethylene (HDPE), and polypropylene have all been manufactured with metallocene catalysts.

Swift has designed poly(aspartic acid) to degrade in a sewage treatment plant. The polymer has been synthesised both with and without a catalyst. The acid-catalysed process yields a more linear polymer that exhibits 90–100 per cent biodegradation in sewage treatment plants. Without a catalyst, the polymer is more highly branched and undergoes less degradation. Poly(aspartic acid) is a replacement for poly(acrylic acid), which is neither biodegradable nor water soluble.

Principle 10: Chemical products should be designed so that at the end of their function they do not persist in the environment and break down into innocuous degradation products

An ideal chemical product would break down into benign substances at the end of its useful lifetime. The persistence of plastics in the environment is well known, and much of the research on biodegradable materials has centred on the polymer industry. Identification of the decomposition products is essential, as the possibility exists that the products could be more harmful than the original substance.

Eastman Chemical Company has developed a co-polyester that degrades to water, carbon dioxide, and biomass. The polymer consists of adipic acid, terephthalic acid, and 1,4-butanediol, exhibits tensile properties similar to low-density polyethylene (LDPE), and may be blended with natural polymers, such as starch. The compost that results from biodegradation is a soil amendment that adds water retention to soil. Biodegradable polymers such as this one also decrease the amount of waste sent to landfills and incinerators. Another example of biodegradable polymers is provided by Mater-Bi, a resin developed by Biocorp, Incorporated. This polymer is designed for use in a variety of disposable products, such as trash bags, food service ware, and packaging. When used in films and sheets, Mater-BNi biodegrades under composting conditions in 20–45 days. Biodegradation for injection-moulded items typically takes 75–120 days. The compost acts as a soil amendment for agricultural and horicultural use.

Principle 11: Analytical methodologies need to be further developed to allow for real-time, in-process monitoring and control prior to the formation of hazardous substances

Pollution prevention can be facilitated through on-line monitoring. Process analytical chemistry can be used to detect the generation of hazardous by-products, permitting reaction conditions to be adjusted to minimise or eliminate the formation of these unwanted substances. In-process monitoring may also be employed to follow the progress of a reaction, reducing the need for excess reagents.

Georgakis has used on-line monitoring to study semibatch emulsion polymerisation of styrene. Conversion is monitored via an on-line digital densitometer, providing near-continuous feedback. A variation of capillary hydrodynamic fractionation (CHDF) is used for on-line particle size measurements. The data obtained from the on-line sensors were used to support a dynamic growth model but could also be used to direct the process evolution.

Principle 12: Substances and the form of a substance used in a chemical process should be chosen so as to minimise the potential for chemical accidents, including releases, explosions, and fires

Toxicity and pollution are not the only concern with regard to chemicals: flammability, volatility and shock sensitivity present potential problems as well. Because accidents do happen, it is important to use the safest form of a chemical possible to minimise harm in the event of an accident. At times, pollution prevention measures may increase accident potential, as in the case of recycling a highly flammable solvent. Compromise may be necessary to realise the most environmentally sound process available.

Nowhere is the safety of materials used more important than in the interior of an airplane. Polymers are ubiquitous in airplances. When a fire breaks out in an airplane, lives may be lost as the result of inhalation of toxic gases released by combustion of polymers. An ideal material would resist decomposition and avoid the release of toxic and combustible gases upon burning. Westmoreland has identified polyhydroxyamide (PHA) as a fire-resistant thermoplastic that can be moulded into seats, overhead bins, and other airplane parts. When heated, PHA degrades into water and polybenzoxazole (PBO), a hard, inert polymer (Fig. 18.8). The use of safer materials in airplanes may permit a higher survival rate in the event of a plane crash and fire.

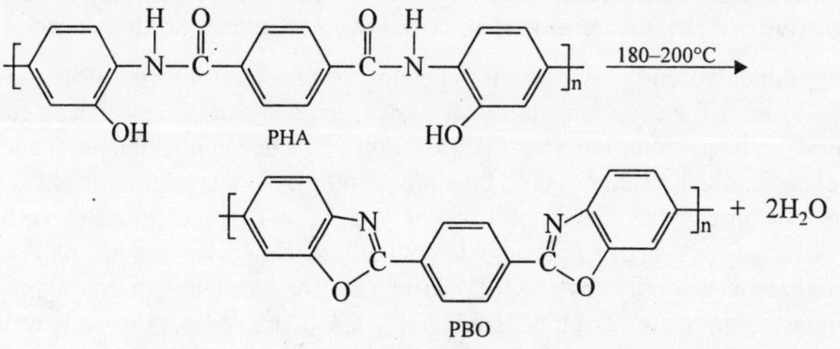

H O O H

OH PHA HO

180–200°C

N N + $2H_2O$

C C

O O

PBO

Fig. 18.11. Thermal degradation of PHA to PBO and water.

Although widely used in a variety of applications, PVC (polyvinyl chloride) presents environmental concerns. PVC contains phthalates and chlorine, which are possible endocrine distruptors, and generates dioxin by-products during manufacture and combustion. An alternative to PVC is provided by metallocene polyolefins. These polymers demonstrate superior physical properties to PVC and are manufactured without toxic plasticisers. The metallocene catalysts allow the polymer properties to be tailored during the polymerisation process. Potential applications for this new technology include, flooring, siding, and automotive finishes.

DESIGNING SAFER CHEMICALS

When designing a reaction, chemists need to evaluate all aspects of the process: Are the reagents used or products isolated toxic? Is an organic solvent necessary? Does the reaction require heating or cooling? How will products be separated? Is a protecting group needed? Although it is tempting to stick with familiar reactions, rather than search for benign alternatives, gains in pollution prevention will remain elusive if chemist are satisfied with the status quo.

Safer chemicals are those that are less toxic than those in current use or those that are safer to handle (lower accident potential). The substitution of dimethylcarbonate for phosgene in carbonylation reactions provides an excellent example of using a less toxic substance to synthesise the same product. Recently, the conversion of a ketone to a lactone has been accomplished using genetically engineered baker's yeast, a safer alternative to the shock-sensitive and explosive *m*-chloroperoxybenzoic acid commonly in use. Developing safer chemicals draws on our current understanding of functional groups and structural features that may pose a threat to the environment or human health. A safer chemical is one that does not include functional groups that are known to be toxic to living organisms. A chemical that breaks down into harmless degradation products at the end of its useful lifetime is better than one that is persistent in nature. A safer chemical presents a reduced risk of fire or explosion, thereby decreasing the potential for accidents. By evaluating the hazard potential for reactants and products, chemists can make informed decisions regarding safer alternatives. A safer chemical will perform the same function as a chemical currently in use but will do so in a more benign manner.

POLYMER SUPPORTED REAGENTS: A WINDOW TO GREEN CHEMICAL SYNTHESIS

Now a days the whole world is facing an acute environmental problem. With the increase in population, the technocrats are putting their brains together for maximum productivity with cheaper technologies. But the problem is that every process has some environmental problem regarding waste disposal. In chemical processes, the toxic un-reacted reagents always pose problem to our human physiology.

If we think in a newer way, reagents can be recycled very easily if supported, minimising the handling and disposal problems. The use of polymers supported catalysts have been used for many years in industrial chemical synthesis. However considerable research is being undertaken to extent the use of supported reagents in the field of commercial chemical synthesis. This section is devoted to brief discussion on polymer supported reagents in chemical reaction engineering.

Advantages of Supported Reagents over Commercial Reagents

1. Less operation hazard.
2. Less waste disposal problem.
3. Regenerateable.
4. May give rise to regiospecfic reactions.
5. Easy separation.

Polymer Supported Reagents and Organic Reactions

Chiral reagent

Polymer supported chiral reagents should find increasing application in asymmetric synthesis owing to the advantages typically associated with solid phase organic chemistry. These benefits include: easy separation and recovery of the expensive chiral species; possible application to a continuous flow system; and simple isolation of the desired chiral product.

Two chiral allylating reagents have been prepared by treating their parent *N*-sulphonylamino alcohols with functionalised boranes. *N*-(Trimethylsilyl) benzaldehyde imine was treated with either reagent to furnish primary homo allylamines as shown below:

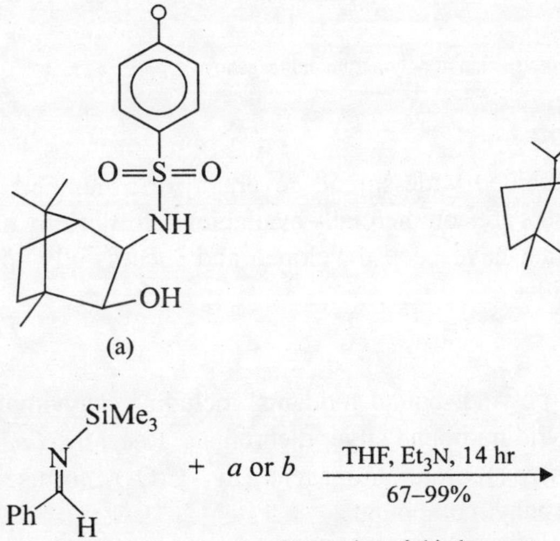

O=S=O NH OH

(a)

O S O N H OH

(b)

$SiMe_3$ N Ph H + *a* or *b* $\xrightarrow[67–99\%]{THF,\ Et_3N,\ 14\ hr}$ NH_2 Ph

Preparation of chiral reagent

Chiral polymers

Molberg has investigated the preparation of resin-bound, entatiopure amino alcohols. These were synthesised by treating protected phenols with optically pure amino alcohol, followed by anchoring onto merrifield resin as shown below.

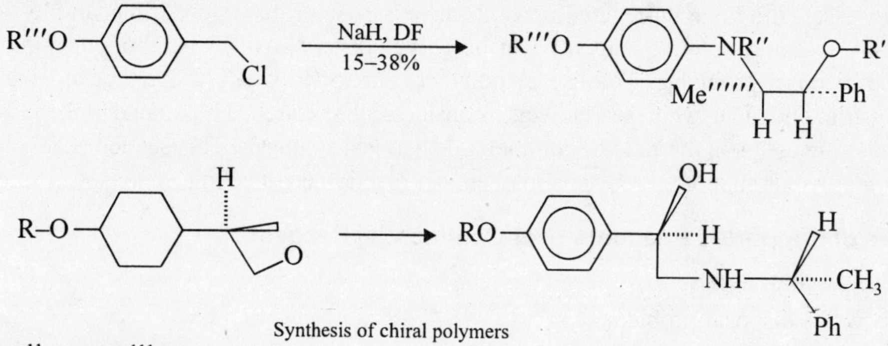

Synthesis of chiral polymers

R^{l} = H, R^{ll} = Me, R^{lll} = Ts
R^{l} = H, R^{ll} = H, R^{lll} = Ts
R^{l} = Me, R^{ll} = Me, R^{lll} = Ts
R^{l} = H, R^{ll} = Me, R^{lll} = H
R^{l} = H, R^{ll} = H, R^{lll} = H
R^{l} = Me, R^{ll} = Me, R^{lll} = H

A polymer bearing a chiral diol pendant, has been prepared by asymmetric dihydroxylation of an immobilised parahydroxycinnamic acid derivative. Entatioselctivity of this reaction was determined prior to the immobilisation steps.

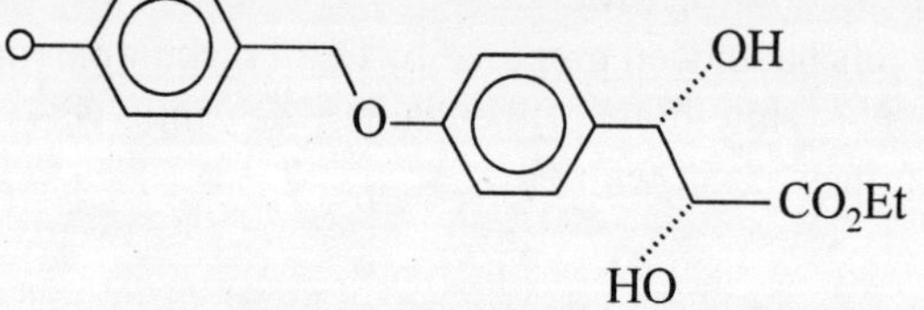

Cross section of polymer bound reagent

General polymer-supported reagents

Polymer supported reagents have received wide-spread coverage for a number of year and many developed through the 1970's and 1980's are commercially available. Below briefly highlighted a cross-section of polymer-bound reagents that have been developed and successfully applied to synthetic organic chemistry since the early 1980's.

Oxidants

Tamami has developed a number of polymer-bound oxidants, including polyvinylpyridine *N*-oxide-supported dichromate and polyvinylpyridine-bound silver dichromate. Recently, *Tamami* has disclosed the preparation of a polymeric oxodiperoxochromium reagent. $(Py_2–CrO_5)_n$, for use in the conversion of alcohols to their corresponding carbonyl compounds.

Reducing agents

Resin-bound borohydride has developed many years ago for use in a number of reduction processes. Recently the highly successful utilisation of this reagent for reduction of arylazides and aryl sulphonyl azides has been reported. The same complex has also been used to effect the rapid reduction of conjugated olefinic bonds in high yield, and for taking down nitroalkenes to nitro alkanes. Thus a co-ordination polymer of zinc borohydride with pyrazine for use as a stable and reliable reducing agent has been prepared which is shown below:

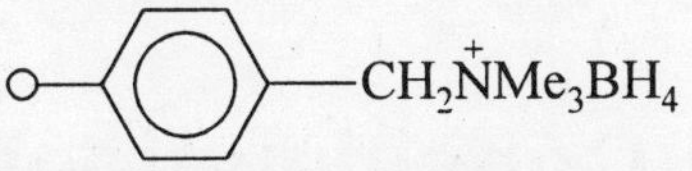

$$\text{—}CH_2\overset{+}{N}Me_3BH_4$$

Soluble polymeric reagent

There are three recent examples of soluble polymer reagents. Firstly, a sialicacid glucosyl donor supported on MeO-PEG, has been developed for use in glycosylation of galactose analogues to give a-linked disaccharadies. Secondly a PEG supported Burgess reagent has been developed for the use in the cyclodehydration of *b*-hydroxy amides and thioamides. Finally a PEG triaryl phospine complex has been developed and applied to Staudinger and Mitsunobn reactions as shown below:

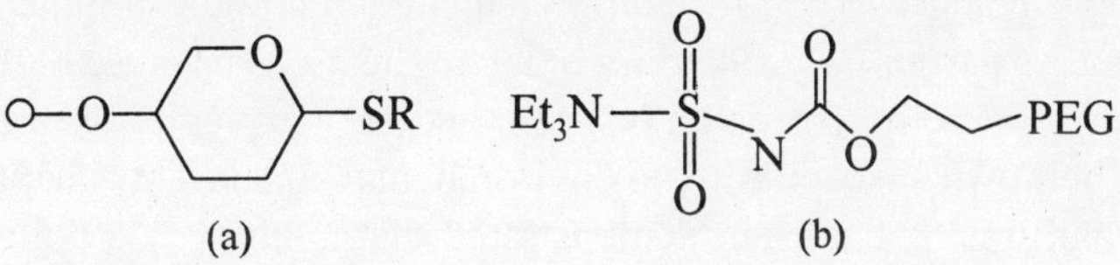

O—O—(ring O)—SR Et_3N—S(O)(O)—N—C(O)—O—PEG

(a) (b)

Miscellaneous

Polymer-bound organoalkyl compounds have been prepared using radical anion induced reductive cleavage of polymeric phenyl thio-ethers for the use in the broad range of organometallic chemistry. Polymer supported alkyl metal naphthalenides have been prepared and used for the preparation of lithium and sodium reagents. Highly reactive polymers-bound copper reagents have been created and their NMR properties studied in detail.

Parlow has devised a three step synthetic sequence, involving three polymeric species with each step being mediated by one polymer-bound reagent. This is a unique example where three different polymeric species may be used in one pot process and isolated product yield of 48 per cent was obtained.

Thus the use of polymer supported reagent is a new technology which will be taken to industrial production scale in near future. The elimination of effluent, by-products and unwanted isomer has long been the target of chemical industry and we believe this approach has taken us nearer to that goal.

CHAPTER 19

Environmental Considerations in Designing with Polymers

INTRODUCTION

Polymers are increasingly being used as structural elements. While these materials are often used as metal replacements, they are particularly suited for other applications because of their unique properties. Polymers are often viewed as being corrosion resistant and are frequently used as protective coatings. However, they exhibit their own susceptibility to environmental attack. This chapter focuses on some of these degradation mechanisms. It has been demonstrated that atmospheric pollution such as nitrogen dioxide (NO_x), sulphur dioxide (SO_x), and ozone (O_3) as well as UV radiation can have deleterious effects on some polymers. For oriented polymers such as Nylon and Kevlar as well as rubbers, a synergistic effect between applied stress and the environmental agents has been observed. Ageing is recognised to be accelerated by increases in normal operating conditions such as temperature, environmental chemical concentrations, radiation intensity, or loads. To conduct real time tests on materials under combinations of these conditions is excessively time consuming and likely to produce useful results long after they are needed. There is, therefore, a critical need for reliable accelerated test methods and means of analysis. The challenge is to relate the accelerated rate of degradation to the increased conditions, while ensuring altered mechanisms do not occur which would invalidate the approach. This study focused on the combined effects of particularly deleterious atmospheric pollutants and stress on the failure of selected polymers. This chapter includes four parts. A brief review of past research is presented first. Ongoing research in three areas is then presented: (i) stress and NO_x or O_3 degradation of natural rubber; (ii) stress and NO_x degradation of non-oriented nylon-6 film; and (iii) stress and SO_x degradation of highly oriented polyethylene fibres (Spectra-1000).

EXPERIMENTAL

Natural rubber, nylon-6, and Spectra-1000 specimens were exposed to combinations of stress and/or aggressive environments in closed, glass test chambers. Stress was applied to the specimens using aluminium load frames and lead weights. NO_x or SO_x was measured from a cylinder and injected into the test chamber using a piston cylinder device. Ozone was introduced into the chamber at a constant flow rate. Some specimens fractured in place while others were broken in an Instron testing machine subsequent to exposure.

RESULTS AND DISCUSSION

Review of Past Research on Rubber

Ozone, in combination with stress, has long been known to cause rapid cracking of natural rubber and many other elastomers. This lab's contribution to this important area of inquiry was to apply chemical analytical methods, Electron spin resonance (ESR) techniques that facilitated direct experimental observation of atomic occurrences associated with the cracking process. Major findings in the study can be summarised as follows. For the rubbers studied, there was a threshold stretch ratio below which cracking was very slow. Above this elongation, cracking increased rapidly and roughly in proportion to the increase in stretch ratio.

Measurements of rubber samples in an ESR microwave cavity indicated the production of free radicals (Fig 19.1), presumably resulting from rupture of molecular chains. The number of radicals was concomitant with the amount of cracking.

A 'Griffith' type fracture model, using a fracture energy, g (J/bond) [analogous to Griffith's fracture energy, G (J/m^2)] determined from a single relaxation test, could be used to reliably predict behaviour for other relaxation test as well as crepe and cyclic loading.

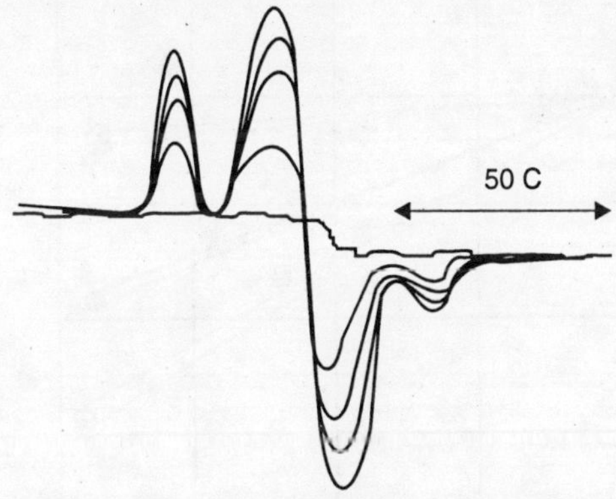

Fig. 19.1. Growth of the ESR spectra of acrylonitrile-butadiene during ozone attack with the lower most curve being the residual signal. Each spectra was drawn 5 minutes after the preceeding one.

Stress and NO_x/O_3 Degradation of Natural Rubber

In current studies the affects of NO_x, SO_x, and O_3 on natural rubber are being explored. Natural rubber exhibited the following: (i) after exposure to 1 per cent NO_x /99 per cent N_2 for 24 hours under no strain it retained 98 per cent of its ultimate tensile strength (UTS); (ii) after 24 hours in lab air at 250 per cent strain it retained 94 per cent of its UTS; (iii) after 24 hours in 1 per cent NO_x/99 per cent N_2 at 250 per cent strain its strength had dropped to 78 per cent of its UTS; (iv) specimens that were exposed to 1 per cent NO_x /99 per cent lab air at 250 per cent strain typically broke within 2 to 4 hours; (v) after 2 hours exposure to 0.5 per cent NO_x /99.5 per cent lab air and 250 per cent strain the strength was reduced to 39 per cent of its UTS: (vi) after 1 hour exposure to 0.25 per cent NO_x /99.75 per cent O_2 and 350 per cent strain the strength was reduced to 19 per cent of its UTS.

Scientists also reported the affects of O_2 on natural rubber which they attributed to free radical scavenging. They reported that in low O_2 concentrations (0.05 per cent in argon) mastication generally produced branching and crosslinking while at higher concentrations chain scission occurred.

Stress and NO_x Degradation of Non-oriented Nylon-6 Film

This study is an examination of how molecular orientation effects the extent and nature of degradation by comparing the degradation of highly oriented nylon-6 fibres to non-oriented nylon-6 film due to the combination of mechanical stress and exposure to NO_x. Smith and Perry, have studied the combined effects of NO_x and stress on nylon-6 and Kevlar fibres. The strength of these fibres diminished rapidly in NO_x. Even more impressive was the synergistic effects noted.

The nylon-6 film used in this study was tubular blown extruded film with an average thickness of 0.0025 mm. Specimens were prepared following ASTM D882. Fig. 19.2 shows the fraction of the original yield strength, ultimate strength and strain at break of stressed and unstressed nylon-6 film versus exposure time for a concentration of 0.25 per cent NO_x by volume. The stressed specimens were loaded to 50 per cent of their original yield strength during environmental exposure. This shows that reduction in the material's yield and ultimate strength increases with exposure time. In contrast, the material's strain at break is much less influenced by NO_x exposure. The material's tensile properties are also observed to decrease with increasing NO_x concentration. This demonstrates that non-oriented films do not exhibit a synergism between an applied stress and NO_x exposure.

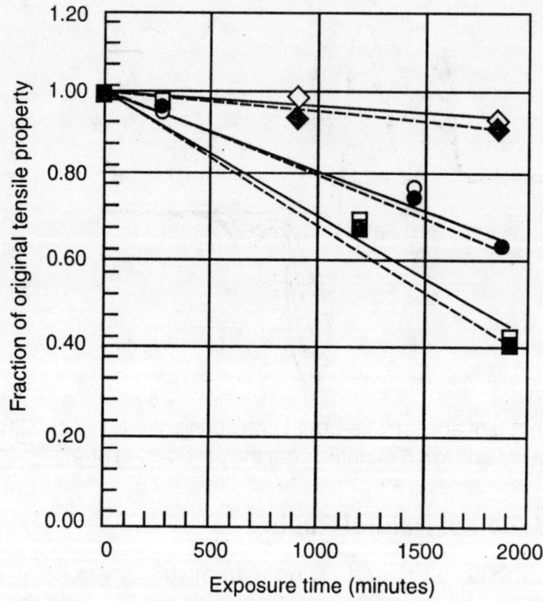

Fig. 19.2. Degradation of stressed (solid symbols) and unstressed (open symbols) nylon-6 film versus exposure time for a concentration of 0.25% NO_x by volume: (◆) strain at break, (●) ultimate strength; (■) yield strength.

This is in contrast to the previously studied highly oriented nylon-6 fibres which have been shown to exhibit a strong synergism between an applied stress and NO_x exposure as previously discussed. The lack of synergism in the non-oriented nylon-6 film is attributed to the fact that non-oriented nylon gets most of it's structural, load carrying strength, from secondary hydrogen bonding. Therefore, in a non-oriented nylon, a given applied load is primarily carried by the secondary bonds leaving the primary covalent bonds much less susceptible to NO_x attack and chain scission than in the highly oriented fibres.

The change in short term tensile properties of non-oriented nylon-6 film is, therefore, highly dependent on NO_x concentration, the ambient exposure environment and exposure time and is largely independent of the state of stress during exposure.

Stress and SO_x Degradation of Spectra–1000

The material used in this study was Allied Signals Spectra-1000 fibres, which are highly oriented, ultrahigh strength/modulus polyethylene fibres. The Spectra-1000 used in this study was a 68 filament, 42 tex yarn. The specimens were prepared as fibre bundle 'dogbones' with a gage section consisting of a single yarn that was 5 cm long. A baseline ultimate tensile strength (UTS) of 3.1 GPa was determined for the Spectra-1000 fibres. Fig. 19.3 is a plot of the residual strength versus the exposure time.

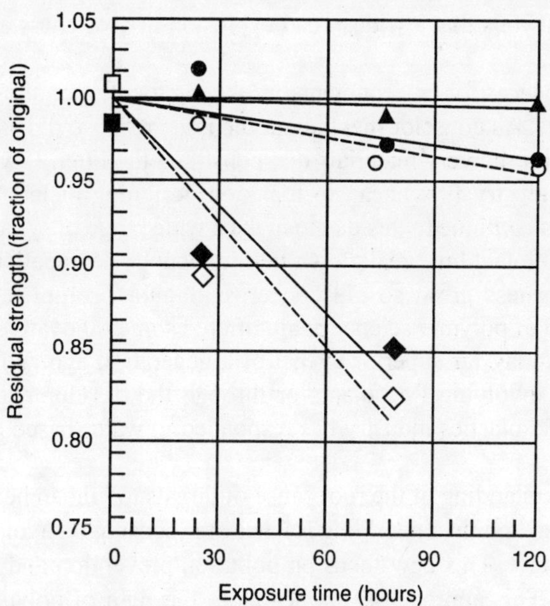

Fig. 19.3. Degradation of stressed and unstressed Spectra-1000 fibres versus exposure time in environments of laboratory air (open symbols) and laboratory air with 5% SO_2 by volume (solid symbols): (▲) 0% stress; (●) 25% stress; (◆) 50% stress; (■) 75% stress.

To examine the effect of stress on the Spectra-1000 fibres, specimens were placed under crepe loading conditions. Relatively short crepe rupture lifetimes of Spectra-1000 were observed: approximately 2 hours for a 75 per cent UTS crepe load; approximately 80 hours for a 50 per cent UTS crepe load; while the crepe rupture lifetime for a 25 per cent UTS crepe load was not reached in the 120 hour time frame. For the specimens that were placed under a crepe load and then broken, the residual strength exhibited a linear decrease. For 72 hours of crepe loading at 50 per cent UTS the strength dropped by 18 per cent.

To examine the effect of SO_x on Spectra-1000 fibres, specimens were placed in a lab air and 5 per cent SO_x environment for several time durations. The residual strength of all these specimens after exposure was 3.1 GPa. This is identical to the baseline UTS and it was concluded that SO_x by itself does not degrade Spectra-1000. To examine the effects of stress and SO_x on Spectra-1000, specimens were simultaneously placed under crepe loading with a lab air and 5 per cent SO_x environment. These specimens followed the same trends as the specimens subjected to crepe loading.

The residual strengths were similar with a maximum difference of 5 per cent. Fig. 19.3 indicates that mechanical degradation was responsible for the decrease in the strength and that chemical degradation was not significant.

Significant mechanical degradation was observed for Spectra-1000 fibres as evidenced in short crepe rupture lifetimes and a decrease in the ultimate strength after crepe loading. No chemical degradation or mechanical/chemical synergism was observed.

THE ROLE OF POLYMERS IN SOCIETY AND THE ENVIRONMENT

The growth in the use of polymers and plastics has had remarkable parallels to the growth of the environmental movement. Both of these movements have their origins in the early part of this century; however, it was during the 1960s that a widespread explosion in awareness and adoption of these two new frontiers occurred.

Many of the earliest concerns for environmental problems focused on those that were most visible, such as litter and solid waste. As consciousness about the issue rose, so did a desire to change people's littering behaviour along with the very materials that constitute the litter. It was during this period that the polymer and plastics industry first began to focus on designing materials that biodegrade in the natural environment. Efforts continue to this day to make a wide range of synthetic materials that do not persist in the environmental intact but break down into innocuous degradation products.

As environmental awareness grew, so did concerns for more complex problems such as water quality. As a result, the field of polymer science again offered some innovative technologies that helped make the water quality of today far superior to that of a generation ago. Whether through the use of water purification systems involving flocculants or through the development of complex filters and membranes, the polymer and plastics industry has responded to water-based environmental issues in a visionary way.

Today, with greater understanding of the root cause of threats to human health and the environment, the environmental movement and the underlying science are focusing on avoiding problems rather then cleaning them up after the fact. This new focus on pollution prevention and a method to achieve it is known as *green chemistry*. This approach is applicable to the field of polymers and plastics, and the scientific and industrial communities involved with polymers are responding with vision and innovation.

The polymers should be designed in such a way that reduce and eliminate the use and generation of hazardous substances throughout all stages of the polymer life cycle. Thus, one can generate general guidelines that provide a blueprint for designing substances that not only reduce the impact on human health and the environment but also reduce the regulatory burden on manufacturers.

It is difficult to imagine our society without polymers. From plastic soda bottles to automobile parts, from medical implants to bulletproof vests, polymers are an integral part of our daily lives. New polymers and new uses for existing polymers are constantly being developed. Fully half of the new chemicals entering commerce in the United States each year are polymers. In addition ethylene, primarily used as a monomer for plastics, is second only to sulphuric acid in production volume. The plastics industry plays a major role in the economy of India and the world.

The versatility of polymers has been central to their extensive use in a wide variety of applications. For example, consider the use of plastics as beverage containers. The strength of plastics allows less packaging to be used, decreasing production, shipping, and energy costs. Improvements in plastics themselves have resulted in soda bottles and milk jugs that use less material today than used 20 years ago. In addition to consumer products, plastics perform a critical function in life-saving devices used in hospitals and emergency rooms. Syringes are used to deliver essential drugs, IV bags dispense vital fluids, and plastic tubing transfers oxygen to assist with breathing. The inertness of plastics makes them ideal candidates for use in prosthetic limbs and pacemakers. Biodegradable stitches are commonly used to eliminate the need for subsequent removal. Improved medical devices and procedures have benefited from better polymer products.

Polymers share several common features that distinguish them from other molecules, such as high molecular weights and the presence of repeating units. A high molecular weight limits transport across biological membranes; therefore, a polymer with a high molecular weight may not pose a threat to human health or the environment through absorption. A typical polymer sample, however, contains polymer molecules that exhibit a range of molecular weights. This lack of uniformity necessitates reporting the molecular weight of the sample as either the number-average molecular weight (M_n) or the weight-average molecular weight (M_w). The lower molecular weight species present in a polymer sample may be transportable across biological membranes, presenting a potential concern to living organisms.

The polymerisation process itself decreases the number of reactive functional groups present in the final polymer. The bond-forming reactions incorporate the functional groups of the monomers into the polymer backbone, often leaving reactive functionality only as end groups. These polymers are more inert than the starting monomers, decreasing potential health and environmental problems. However, some polymers are synthesised with reactive functional groups imbedded in or pendant to their backbone. The presence of such groups facilitates cross-linking between polymer chains, reactions with other molecules, or complexation with added reagents. Increased toxicity may be a concern when certain reactive functional groups are present.

A potential hazard to human health and the environment is posed by macroscopic properties of polymers as well as microscopic properties. Consider, for example, a polymer that is inhaled and leads to respiratory distress because it fails to clear the lungs. The biological effects may have less to do with the molecular structure than with the gross physical features. Problems associated with macroscopic properties are evident in the well-publicised case of the environmental persistence of plastics six-pack rings. The failure of this product to degrade created a hazard to marine life that became entangled in the ring. Consumers were encouraged to cut up six-pack ring that degrades in the environment.

A key positive characteristic of polymers in a variety of applications, however, is their durability. This feature, which makes polymers so versatile, also creates a problem due to persistence in the environment. Many polymer products do not degrade in landfills or sewage treatment plants. Recycling of plastics helps to minimise the disposal problem; however, not all plastics are currently capable of being recycled. The design of biodegradable polymers demonstrates industry's practical approach to addressing the issue of persistence in the environment.

Another important feature is the use of various additives. In many cases, product performance may be optimised by combining additives with the polymer during the processing stage. Fire retardants, for example, may be added to clothing. Antioxidants may be used to increase the durability of a product, and dyes can be added to give the polymer colour. Although additives enhance polymer properties, they can also leach from the polymer product into the surrounding environment. The polymer and plastics industry continues to identify new additives that give the same desirable properties without posing a threat to human health and the environment.

Environmental concerns extend beyond the polymer products themselves, to synthesis and manufacturing. These processes traditionally have utilised non-renewable feedstocks, volatile organic solvents, or hazardous reagents. In some cases, the reaction may require a large input of energy or may generate a significant quantity of waste.

Most polymer syntheses, like 98 per cent of all organic chemical products, being with petroleum-based feedstocks. Concerns for the dwindling supply of oil necessitates the development of alternative feedstocks. Many of the starting materials currently in use are also of concern because of carcinogenicity. Consequently, utilisation of biomass presents an alternative to current feedstocks that is both renewable and relatively innocuous to human health and the environment.

Polymer synthesis often requires the use of large amounts of organic solvents. Numerous problems are associated with the use of these solvents: toxicity, volatile organic compound (VOC) emissions, disposal, and contamination of the aqueous waste stream. These solvents are often highly regulated under the Clean Air Act, the Clean Water Act, and the Toxics Release Inventory. Compliance with these regulations can be very costly to a company. Polymerisation reactions that use an aqueous environment or no solvent at all may be designed, as in a solid-state polymerisation. Environmentally friendly solvents, such as supercritical CO_2, may also appropriate.

Polymers properties are fine-tuned through the addition of auxiliary substances in a process known as compounding. These additives are used to improve a polymer's stability, processability, or physical properties. Compatibilisers, for example, allow two or more materials, such as virgin and post-consumer plastic resins, to coexist indefinitely. Heat stabilisers reduce thermal degradation during processing and product lifetime. Fillers may be incorporated into the polymer to improve strength and thermal stability. Flame retardants are essential when processed into polymers that are used in confined spaces, such as airplanes. Antioxidants prevent breakdown of the polymer through oxidation; phenol derivatives are commonly used in this capacity. Plasticisers improve the flexibility of the polymer by lowering the glass transition temperature below room temperature. Unfortunately, these additives can leach out of the polymer product and into the surrounding environment. Recently, concern has been revised about the potential leaching of phthalates, used as plasticisers in children's toys. The additive that makes a teething ring soft and pliable may present a health risk to the infant chewing on it.

Modifiers, which improve impact strength or flexibility, are the largest category of additives. Plasticisers, impact modifiers, and organic peroxides provide examples of modifiers. Property extenders, such as antioxidants, light stabilisers, and flame retardants, impart stability during processing and use. Processing aids improve the mouldability of a plastic and include mould-release agents and lubricants.

A polymer that, by itself, may be considered safe, may contain comonomers and additives with potentially harmful effects. For example, polyvinyl alcohol is approved by the Food and Drug Administration (FDA) for use in cosmetics and food wraps yet is manufactured with substances that present a health risk to plant employees. Comonomers used in the synthesis of polyvinyl alcohol include acrylamide, a neurotoxin; formaldehyde, a suspected carcinogen and powerful mutagen and teratogen; and vinyl chloride, a known human carcinogen. The cross-linking agent, divinyl sulphone, is highly toxic, causing skin and eye burns similar to those caused by mustard gas. Materials used in manufacturing and processing also need to be considered when designing safer polymers.

Numerous polymerisation reactions utilise hazardous reagents, such as phosgene. The transportation and handling of such reagents may require special equipment, clothing, and safety precautions. These additional safeguards increase the costs associated with the manufacturing process and have the potential to fail. New and existing technologies should be examined to identify safer reagents to those currently in use. The polymer manufacturing process often requires a large energy input. A prolonged heating period may be necessary to effect polymerisation; cooling may be required to maximise product yield; energy-intensive separations may be needed to isolate the desired product. High energy requirements further deplete existing fossil fuels. Ideally, polymer processing and manufacture should be conducted at ambient temperature and pressure. Catalysts have proven especially useful in minimising the energy needed to carry out specific transformations.

It is not uncommon for batches of polymers to be 'off-spec,' resulting in a product that may be commercially unusable. Unless another use is found for this off-spec material, this substandard product will become part of the waste stream. The problem can be minimised by increasing the use of in-line monitoring. Continuous feedback on the reaction progress and polymerisation conditions permits real-time process adjustments, decreasing the probability of producing substandard polymers.

A final factor to consider in polymer manufacturing is atom economy, the efficiency with which atoms in the starting materials are incorporated into the final product. Polymerisation reactions frequently demonstrate high atom economy. Addition polymerisation adds monomers and reactions without loss of mass, whereas condensation polymerisation yields products with loss of a small molecule, such as water. Addition reactions inherently demonstrate high atom economy than elimination reactions, thereby decreasing the number of atoms contributing to the waste stream.

The polymer industry has revolutionised the quality of life through the generation of new materials that have transformed everything from transportation to communication to recreation. However, it is also recognised that there are still challenges to meet in the development of new polymers and polymer-manufacturing technologies. Through the discovery, development, and implementation of innovations to reduce hazardous substances in all stages of the polymer life cycle, the polymer industry is achieving its dual goals of both protecting human health and the environment and being more profitable by reducing a variety of costs, for example, regulatory costs. These techniques are finding widespread use and will continue the traditional parallels between the fields of environmental protection and polymer science.

WEATHERING OF POLYMERS

Polymers often deteriorate rapidly in outdoor applications. The major agent of degradation is solar ultraviolet radiation which promotes photo-oxidation involving radical chain reactions. Elevated temperatures accelerate the chemical reactions involved in degradation and plastics are most vulnerable in hot sunny climates. The chemical reactions that are involved in degradation are complex and different polymers display different weathering sensitivities. Studies of the chemistry of degradation have led to the development of stabilising systems that produce spectacular improvements in the service lifetimes of many polymers. Chemical degradation mechanisms and the action of stabilising systems have been reviewed extensively.

Experimental

The discussion below is based on results obtained with thermoplastics in which artificial weathering conditions were applied. Samples were usually in the form of injection moulded bars but some studies were conducted on polypropylene bars cut from compression moulded plaques and the rubber samples were cut from sheets provided by SP Tyres (Fort Dunlop, Birmingham). The polymers studied included polycarbonate, polyethylene, polystyrene, polypropylene and polypropylene compounds containing talc or short glass fibres. The thermoplastics were supplied as European grades, with no ultraviolet (UV) stabiliser unless specified otherwise. The rubber was a sidewall composition containing a blend of natural rubber and butadiene rubber. Rubber compounds were also provided with carbon black and with carbon black plus an antioxidant system.

Samples were exposed to UV illumination provided by fluorescent tubes with spectral UV output matching that of terrestrial solar radiation for various periods of time at 30°C. Samples were exposed on open racks or loaded in various ways including uniaxial tension, uniaxial compression and bending. In most cases the exposure was continued for a pre-determined time then the sample was withdrawn for testing. Tensile tests were performed on the weathered bars and analyses of the material at various locations (different depths from the surface) were made by gel permeation chromatography.

Results and Discussion

Fracture mechanisms

A common feature was that the side exposed to UV became embrittled and in many of the materials investigated a pattern of fine cracks formed when a tensile deformation was applied. This replicated observations made with samples exposed outdoors. Cracks formed in the embrittled surface often propagated into the interior of the bar and could be recognised as the origin of failure. This occurred within 4 weeks exposure with polypropylene. After longer exposures it was discovered that the crack was arrested when it reached the relatively undegraded material inthe interior, and a ductile zone developed before the crack reinitiated and propagated to failure.

This is believed to happen either because the surface layer becomes so degraded that it is too weak to transmit stresses on into the relatively undamaged material in the interior or because the multiple cracks that form in the surface mutually unload to such an extent that the crack cannot be propagated effectively. A recovery in the breaking strength and in the elongation before failure occurred when this condition developed. It was sometimes possible to identify crack initiation in the surface facing away from the UV source. The UV level at this surface (transmitted through the polymer and/or reflected by surfaces near to the sample) was sufficient to cause slow degradation and embrittlement occurred after the front surface lost its potency. Some examples were found in which bars made from polypropylene containing light stabiliser failed from internal sites when tested in tension following a prolonged period of outdoor exposure.

Diffusion-controlled degradation

The rate of photo-oxidation in the interior of plastics articles is often limited by a shortage of oxygen. This happens when the reactions near the surface proceed so rapidly that oxygen is consumed before it can diffuse far into he interior. High UV levels favour this and lead to steep oxidation profiles near to the surface. This probably contributed to the failures described above in which fracture nucleated at an internal site after prolonged outdoor exposure.

Effect of stress on degradation rate

Stress can alter the rate of chemical reaction and may directly increase the rate of any thermally activated process, including those involved in photo-degradation. It may influence the diffusion rate of oxygen (or other reactants) or it may influence the rate of recombination of radicals formed by photo-chemical scission reactions, effectively pulling them apart before the molecules have a chance to repair themselves. Over an extended period of time it is possible that stress may promote an increase in crystallinity or of molecular orientation in non-crystalline regions, and this is expected to influence the rate of degradation. The application of tensile stress during UV exposure was found to accelerate molecular scission in polystyrene, polypropylene and glass fibre reinforced polypropylene. It has been shown that compressive stresses reduce the rate of photo-degradation. Residual stresses that form during cooling of injection moulded articles are compressive near to the surface, where photo-oxidation is concentrated, and it has been demonstrated that a useful reduction in the rate of degradation can be obtained by using conditions that maximise the magnitude of residual stresses. Recent experiments with rubber compounds have demonstrated that the application of tensile stress during UV exposure causes accelerated degradation and the formation of surface cracks that may lead to the eventual failure of the part.

Thus, unmodified polymers are very vulnerable to photo-oxidation through a complex set of reactions. The reaction rate can be very fast near the surface and is often limited by lack of oxygen in the interior of thick sections. Extremely effective UV stabilisers have been developed but when they are included as UV the reactions are even more complex. As a result the degradation is very difficult to model. Changes in the chemical degradation can lead to changes in the (engineering) failure mechanism. For these reasons it is difficult to devise satisfactory accelerated weathering tests: it is impossible to find a procedure in which acceleration leaves the reaction pathway unchanged. Laboratory results can rarely be extrapolated to provide a reliable guide to lifetime under service conditions.

GLOBAL APPROACH TO CATASTROPHIC FAILURE OF POLYMERIC MATERIALS

E.H. Andrews initiated what can be referred to as a 'global approach' of stress waves and fracture surfaces of brittle rubber. Although not identifying his studies by that name, he was intrigued by the 'similarity of fracture markings in brittle materials of very different compositions'.

This section seeks to extend the knowledge of failure surface topographies of polymers to include the surfaces generated under real life conditions. In so doing, it has been found that the fracture of rubbery polymers under real life conditions contain features that invite favourable comparison with the fracture behavior of other materials even though there are differences in elastic properties. By considering this a global (or universal) aspect of all failures, the failure analysis can benefit from a stimulating cross fertilisation of ideas and can achieve a greater understanding of failure processes.

Tyres, being complicated composites, are good sources of studying failure surface topographies of real life rubber failures. It has been found that these failures generated in this manner reveal complex interactions between components. This results in torn surface topographies that have received little attention to date. Since the goal of the failure analyst is to develop a statement that adequately describes failure histories, there is a need for awareness of all topographies and the circumstances giving rise to them. Those who devise dynamic physical tests need to know whether their tests produce meaningful surfaces. Reports on polymer fracture are laden with terms borrowed from metallography, such as mirror, mist, hackle, conics, etc., with others added. This makes the terminology of failure quite confusing and the situation should be reviewed by an authoritative organisation with the goal of recommending standard descriptive terms for describing fracture surface topographies.

Modes of Separation in Polymer Composites

In the simplest case, separations are based upon two components, either one of which can be categorised based upon their relative stiffness, i.e., be they 'rigid' or 'flexible'.

Case 1 components: rigid-rigid global aspect: fatigue striations

In this case, composite components are fixed in place restricted to only the most minute (micron sized) deformations so they both effectively behave as rigid components even though they may be rubbery by nature. Examples of tyre failure in this category are: initial separations at ends of steel cords in wire belted tyres, interply shear failure between wire belts, crack growth from non-lethal tyre punctures, and stone damage in treads. Examinations of failure surfaces from these conditions reveal the formation of features resulting from a stepwise propagation emanating from an initiating flaw. These are fatigue striations analogous to fatigue striations formed in metals and plastics. Fatigue striations that develop in wire coat compounds that initiate at the cut ends of wire cords in belted, radial ply tyres present a sawtooth pattern. The smallest identifiable sawtooth markings have a 'pitch' on the order of 25 microns with some evidence that even smaller ones may have been present but subsequently obliterated as the failure opened up the structure. The striations increase in size as the distance from the initiating point increases with the largest striations being on the order of one millimeter. From that point on massive tearing is the propagating mode. The rubber surfaces of the striations are smooth and uneventful–characteristic of advancing tear that has a sharp profile.

Fatigue striations can also be seen adjacent to paths of a non-lethal punctures and to failure surfaces known as 'stone damage'—damage caused in tread rubber due to embedded stones. Aside from the characteristic stepwise features the surfaces of these fatigue striations are uneventful.

In all materials the initial fatigue striations are extremely sensitive to surface-surface abrasion (polishing) when the failed structure increases to a critical size that allows for deformations higher than initial. The rigidity of metals and hard plastics increases the longevity of fatigue striations.

Case II components: rigid-flexible global aspects: tear lines

In this case, one component is capable of extreme elongation whereas the other component is fixed in place, resulting in a peeling mode. Examples of this situation is the peeling away of a flexible strip of tread or wire belt from an otherwise functional tyre that remains rigidly inflated and represents the rigid component. The surface features are tear lines and the metallurgical analogs are conics, and hackle. A salient feature undocumented in failure literature is that of *directional anisotropy*. Such a tear texture derives from the mechanism of steps and welt formation. The essential feature of directional anisotropy is revealed by low angle illumination and observance of presence or absence of shadows.

Conics (parabolas) observed on these fractures are formed in the same manner as conics found on metal and polymer fractures for glassy materials and for rubber embrittled by low temperatures. They are caused by deviation of the advancing crack by an imperfection or by the interaction of the main crack front with a secondary crack front induced by stresses around an embedded flaw ahead of the main crack. Indeed, conics on rubber surfaces exist even at a level that requires the scanning electron microscope for their resolution and they are a valuable clue as to the direction of propagation. The focal point of the parabola faces the direction of propagation.

Fine hackle markings are sometimes related to rough areas of brittle fracture faces and also called many other names when referring to certain markings characteristic of glass reinforced resins. The formation of fine hackle markings in rubbery polymer fractures is not as easily seen due to the presence of reinforcing fillers that can disguise fine hackle. Knotty tear can be considered as a gross form of hackle. Additional features on tyre surfaces arise from ghosting of wire cord textures on the three paths possible through a laminate. These three paths are: (i) close to the outermost wire belt ply; (ii) midway between plies; and (iii) close to the wire belt adjacent to the carcass. Each path displays a unique texture.

Case III components: flexible-flexible global aspects: ductile tearing

All bulk rubber tear can be considered as form of tearing for comparison with ductile tearing in rigid plastics and metals. Examples of this are found in bulk rubber tearing in tyre sidewalls where the tear progresses between cords. Metallurgical and plastic analogs would be the processes that form herring bone patterns. The path of rubber tear down a sidewall involves directly at least three or four individual compounds along with the influence of the textile cord comprising the sidewall. In addition there is considerable influence from the surface of the sidewall. If the sidewall is severely cracked due to ageing, the cracks become initiators of secondary tears and herring bone textures develop due to a coalescence of primary and secondary tear fronts. These interactions coexist with the development of conic tear lines unless the thickness of the rubber component is small compared to the secondary crack initiators. Linear compounds are relatively thin and compounded with larger particle size fillers so conics do not develop.

Failure surfaces of large components, such as found in the gross peeling away of one wire belt from another, will often display markings analogous to beach marks. These can be conical in nature, often shaped as arcs that extend from one shoulder to the other and profile growth and arrest of the propagating separation. These growth-arrest markings are also characteristic in directional anisotropy as observed with low angle illumination.

Comparisons of conjugate failure surfaces presented on torn rubber surfaces present an enigmatic deviation from other materials. The shear lip, so often found in metal and plastic fractures is absent on torn rubber due to the complete retraction of necked rubber. That is to say, conjugate surfaces fit together well despite the fact that when the tearing took place the cross-section of the failed area would have been considerably less than the final relaxed cross-sections. Computerised images of conjugate fracture faces produced from ductile plastic fractures show a residual pixel count when the images are subtracted. This is ascribed to the amount of ductile tearing in the plastic. With the exception of the very small step and welt tear line formations, conjugate images of rubber tears are highly superposable.

Catastrophically fractured composites can be extremely complicated and difficult to portray using two dimensional photography. In those cases when the fracture components interfere with even illumination of the subject, three dimensional, or stereo photomacrography, is relied upon. For producing these views the camera is held in a stationary position and the specimen is rotated and photographed at each extreme of the stereographic angle.

Thus, features found on rubber fracture faces can relate directly of the features generated on fractures of other materials, be they plastic, glass, or metal. All that remains is a consensus upon the names to be used to describe the features. Since the main physical difference of rubber is a high elongation, a method of categorising torn rubber is based upon the relative flexibility or rigidity of the parting members.

There are two aspects unique to rubber fracture: (i) the predominance of directional anisotropy of rubber tear; and (ii) the fact that strained rubber is not permanently necked. The study of conjugate fracture faces is often beneficial in analysing rubber tear.

Closeup stereoscopic photography (stereomacrophotography) is often beneficial in presenting information. The resulting stereo pairs can be mounted and viewed in a Holmes type viewer or projected onto a screen as anaglyphs using red and blue or polarised filters and appropriate glasses.

CHAPTER 20

Control of Biological Degradation of Polymers by means of Biocides

INTRODUCTION

This chapter deals with the control of biological degradation of polymers by means of biocides.

Many natural and synthetic polymers are subject to attack by biological agents. At some time during the life of a polymer, it may be a possible source of food for living organisms, a barrier to be penetrated in the search for sustenance, a substrate subject to enzymatic or chemical by-product attack, or simply a material for shelter or attachment during the life cycle of certain organisms. This may or may not manifest itself in an attack on the object, depending on the inherent properties, including the susceptibility of the polymer or its component parts to penetration or degradation. The degree of susceptibility, the anticipated end use of the material, cost, and other factors determine whether or not a chemical or other protective treatment should be applied. Active chemical compounds used as preservatives control degradation in several ways; those which produce death are termed biocides. Many chemical agents, however, may inhibit the reproduction and growth of organisms without causing death. These may be exemplified in the two common classes of control agents known as bacteriostats and fungistats. With higher organisms, some of the mammals, for example, the repellent qualities of a chemical agent may minimise damage adequately. These protective agents make the polymer less susceptible to biological attack and prolong the useful life of the polymer material by preservation of its physical properties and esthetic appearance.

The more rigorous definition of a biocide as an agent which kills living organisms precludes the somewhat more common usage of the term, which implies that a biocide is an agent which kills, repels, or inhibits the growth of the biological environment in which the polymer exists.

The subject of natural and synthetic polymers is a broad one covering a multitude of substances and products. Natural polymers include cellulosic materials such as wood and natural fibres. Also included are polysaccharides and their derivatives (industrial gums), waxes, natural rubber, resins, leather, and other proteinaceous materials.

Synthetic polymers, such as polyolefins, polyesters, or vinyl polymers, are considered, for the most part, particularly in relation to the polymeric constituents of the compounded polymers, to be resistant to biological attack. However, this is a maxim which has not been scientifically investigated in depth. As more general use is made of these polymers, they may be subjected to different and exotic biological environments which may affect their resistance to biological attack.

Synthetic polymers are seldom used alone. Other materials are mixed with polymers in amounts from minor to major proportions. These include plasticisers, lubricants, stabilisers, fillers, and other materials which may change the biological resistance of both the polymer and the finished composition. Present indications are that these materials particularly should be reviewed for their vulnerability to biological attack. Considerable work with plasticisers has been reported and the indications are that the improper choice of plasticiser can directly affect the durability of the finished product, specifically by consumption of the plasticiser by bacteria and/or fungi.

The history of biocide use for polymer preservation dates to antiquity. Application of cedar and other essential oils was known and used by embalmers in Egypt, ca 3000 BC, to protect mummies and their cloth wrapping from degradation by micro-organisms. In 1000 BC, *Homer* mentioned the use of sulphur to protect against 'pests'. *Moses* was reported to have instructed Joshua to preserve the books of the Pentateuch with cedar oil. Other examples can be found in which chemical treatment was used against the ravages of insects, rodents, and micro-organisms.

After the fall of the Roman Empire, references to the use or development of biocides in the literature are difficult to uncover. This condition lasted until the start of the Renaissance. The biocides used then were inorganic or natural 'extractions' including sulphur, copper, mercury compounds, and arsenical compounds. Bordeaux mixture (slaked lime in a copper sulphate solution), elemental sulphur for mildew control, and arsenical compounds for insect control emerged as important chemical agents.

The investigations of many men including *Pasteur*, *Prevost*, *De Bary*, and *Semmelweis* were instrumental in establishing the concept of micro-organisms as the causative agents of disease and microbial degradation. Foremost among these distinguished scientists was *Pasteur*. His work with the problems of wine souring, silkworm disease, cholera, anthrax, and rabies are milestones in the annals of science.

The importance of the inorganic chemicals for biological control was successfully challenged by the development of the organic biocides. The discovery of the fungicidal activity of dithiocarbamates and of the insecticidal properties of phenothiazine (9-thia-10-azaanthracene, thiodiphenylamine) initiated investigations into the value of organic compounds as biocides. Since that time, many chemicals have been developed and marketed, including DDT (1,1,1-trichloro-2,2-bis(*p*-chlorophenyl)ethane), DDD (1,1-dichloro-2,2-bis(*p*-chlorophenyl)ethane), aldrin, chlordane, dithiocarbamate salts, phenylmercury compounds, organocopper compounds, organophosphorus compounds, substituted urea compounds, and a multitude of others.

World War II stimulated the study of deterioration and the use of biocides for preservation. A combination of material shortages and microbial degradation of military equipment and clothing triggered a comprehensive search for causative agents and effective treatments. One result of this study was the establishment of the Prevention of Deterioration Centre in Washington, D.C., a unit of the National Research Council. The literature of biological degradation, including that of polymeric materials, continues to grow as new problems are encountered, old problems are solved or continue to elude solution, chemicals are developed, and the biological environment flourishes. A valuable source of information is the abstracts issued by the Prevention of Deterioration Centre.

BIOLOGICAL ENVIRONMENT

The biological environment, that is, the biological surroundings in which polymers are present, includes the biological agents responsible for the deterioration of polymeric substances; these agents are listed in the following sections.

The Plant Kingdom

Eumycetes, Schizomycetes, Actinomycetes, and algae are considered here.

Eumycetes

Eumycetes or true fungi, are micro-organisms of particular importance in causing the degradation of materials. Fungi are nucleated, spore-forming, non-chlorophyllous organisms, which reproduce both sexually and asexually; most of them possess filamentous, somatic structures, and cell walls of chitin and/or cellulose. More than 80,000 species are known.

The true fungi are present everywhere. Their importance as deteriorative agents stems mainly as a result of the production of enzymes which break down non-living substrates in order to supply nutrient materials for support of fungal growth. These substrates can be polymers or other materials present in polymer compositions. Certain environmental conditions are essential for optimum growth and degradative activity. These include an optimal ambient temperature, presence of nutrient materials, and high humidity. Fungal growth is slow in arctic areas and more restricted or absent in low-humidity areas. The growth of fungi is probably greatest in the tropics, although in relation to economic losses, the problem is significant in temperate zones particularly where high humidities prevail. It is interesting to note that some environmental conditions which appear highly unlikely to support fungal growth do allow fungi to exist. Few would suspect that hides in brine solution would be infected by fungi, but there are halophilic and facultative species living and thriving in this hostile environment.

The eumycetes vary widely in their morphology, ranging from microscopic single-cell yeasts to the more highly evolved macroscopic structures of the mushroom type. Representatives from most fungal genera have been isolated from polymeric materials. Many fungi play an important role in degradation, although others grow on polymer surfaces with no known effect on the substrate. However, adventitious growth, that is, growth on the dust particles, inorganic salts, hair, dead insects, epidermal cells, and similar materials deposited on the surface of polymers, can greatly detract from the esthetic properties and other characteristics of end-use items.

Schizomycetes

Schizomycetes the bacteria, and *Actinomycetes*, a type of organism intermediate between fungi and bacteria, have played an undetermined role in relation to fungi in polymer deterioration. The importance of bacteria in some materials such as water-based paints is usually recognised. Bacteria can be single-cell rods, cocci, or spirilla; others are chainlike or filamentous. Bacteria can be either aerobic, or anaerobic; in contrast, the actinomycetes and fungi are necessarily aerobic. Some bacteria are motile; bacteria are predominantly non-chlorophyllous. Their degradative action also is chiefly a result of enzyme production and resultant breakdown of non-living substrates in order to obtain nutrient materials.

Bacteria

Bacteria present in soil are important agents in material degradation. Particularly affected are cellulosic plant life, wood products, and textiles subject to cellulytic degradation. Some actinomycetes cause discoloration of vinyl films. There are excellent references on the subject of bacteria and related organisms.

Algae

Algae differ from fungi and bacteria in that they are chlorophyllous and capable of photosynthesis of nutrient materials. Species types range from single cells to large multicellular seaweeds and kelps.

The algae are not associated with deterioration of polymers to any large extent. Although they were found on deteriorated tentage material in the tropics, this breakdown may have been caused by other organisms. Cellulolytic activity of algae is limited or non-existent. The chief problem with algae is their unsightly and odorous nature when present on polymeric materials, and the fact that they may constitute a source of food for other organisms. They are present as fouling organisms in salt and fresh-water environments.

Plant life representative of the higher phyla in the plant kingdom are not thought of as degradative agents to polymers.

The Animal kingdom

Insects, birds, and rodents are considered here.

Insects

The most varied and prolific organisms of the biological environment are the insects, which number in the hundreds of thousands of species. Insects of particular interest are those which affect cellulosic and proteinaceous materials, such as termites, carpenter ants, boring beetles, moths, silverfish, and grasshoppers.

Birds

Material damage by bird life is of minor importance. There is interest in bird-repellent chemicals to prevent the secondary effects of fungal and bacterial attack after exposure resulting from initial bird damage.

Rodents

Rodents have played an important role in man's struggle for survival. In particular, the rat has played an important part both as a disease bearer and as an agent of material destruction. Rodent damage to polymers is of importance when polymers are used as barriers or packaging materials impeding access to foodstuffs. Rodents and other members of the animal kingdom have had considerable biocidal research directed toward their control or elimination.

Marine Fouling Organisms

Marine fouling organisms include members of both the plant and the animal kingdoms. Some 2000 species have been reported to cause fouling. Many are unique to marine environments. Representatives include algae, bacteria, fungi, protozoans, barnacles and other arthropods, molluscs, tunicates, hydroids, and annelids. Extensive damage, direct or indirect, to wood pilings, hulls, buoys, other immersed materials and devices, and organic coatings, are a result of the activity of fouling organisms.

SUSCEPTIBILITY OF POLYMERS TO BIOLOGICAL ATTACK

Natural and synthetic polymers, or materials containing them, may be attacked by a variety of species within the entire environmental complex to which they are exposed. The introductory statements briefly presented the obvious and important reasons for direct attack on these substances. The following sections elaborate on these reasons and present a summary of polymer susceptibility to the biological agents of degradation. The discussions of attack by these agents will be divided into attack as nutrients, as attachments, and as barriers.

Nutrients

Extensive and costly damage to polymeric materials results from the attack of organisms in the biological environment in search of nutrients. Damage by insects and insect larvae to the natural polymers of fruits, vegetables, ornamental plants staple crops, and stored grains, as well as damage by rodents and birds in similar areas, is well documented. The most prevalent type of degradation is caused by bacteria and fungi. Many instances of microbial deterioration are encountered in daily experiences with bread mould, cheese fungi, putrid meat, 'athlete's foot,' sour wine or cider, etc., but the extent and variety of attack by micro-organisms is seldom suspected by the layman.

Fermentation, putrefaction, and associated chemical changes are carried on by many species of micro-organisms through catalytic agents called enzymes. The action of enzymes is specific to a substrate (eg., cellulases degrade cellulose, lipases split fats, proteases attack proteins). The susceptibility of a polymer to microbial (or, actually, enzymatic) attack is a function of enzyme availability; co-enzyme presence, if required; availability of a point in the polymer susceptible to enzyme attack; and specificity of enzyme for the type of polymer. The polymer may be protected by altering or blocking the point of attack by the enzyme (eg., acetylation of cellulose, resin impregnation), or the biological agent may be killed or inhibited through the use of biocides.

Attachments

The organisms in a biological environment may adhere to polymers without causing degradation of the substrate. Certain marine fouling organisms attach themselves to ship bottoms and may be considered as doing little direct damage; lichens grow on roofing materials with little effect. However, problems of esthetics, loss of fuel, economy and speed in ships, secondary infestations, or impairment of function may arise. For example, fungal and bacterial slimes may attack papermaker's pulp and cause a physical plugging of the felt which reduces water drainage from paper pulp. Bacterial slimes on ship bottom coatings may initiate the growth of other more degradative agents. In such cases, the use of biocides for protection of the preventative measures is necessary.

Barriers

A polymer may act as a barrier to the organisms in a biological environment in the search for food. Rodents and insects will attempt a penetration through any material to get at food. The producer of packaging materials knows only too well the extensive damage caused by such pests. Crickets have damaged plastic-glass fibre screening. Damage to trees and telephone poles by woodpeckers in search of insects and larval food is also well known. Insect penetration of poly (vinyl chloride) joint seals on clay pipes carrying water has been reported. The use of biocides or repellent chemicals is required here also. Not all polymers will be used or found under circumstances where they are subject to attack by biological agents. Thought and planning are required to determine the potential forces of biological deterioration polymeric materials will encounter in actual use. Preventative treatment should be applied where required, not only to polymers that are sources of microbial nutrients, but also to those functioning as points of attachment or as physical barriers.

BIOLOGICAL DEGRADATION

The following section is a summary of the susceptibility of polymers to biological degradation. Rodent and insect susceptibility is assumed to be general for these materials. Therefore, the main topic covered relates to microbial deterioration except in cases of particular interest.

Natural Polymers

Resins

Dammar, East India, Copals, and miscellaneous resins have no history of microbial degradation. There may be a possibility of superficial fungal growth occurring on these resins without degradative effects.

Lignin

Native lignin, an insoluble polymer of phenolic composition found as a major component of plant cell walls, is somewhat resistant to microbes, although some white-rot fungi and certain soil micro-organisms may attack it. Chemically processed lignin is attacked by bacteria and fungi.

Industrial gums

Plant polysaccharides or their derivatives are referred to as industrial gums. They are usually plant exudates, seaweed extracts, seed extracts, starches, and other naturally occurring polysaccharides.

Natural rubber

Natural rubber is the latex exudate of a group of plants, the most common of which is the rubber tree (*Hevea brasiliensis*). After collection, the latex is subject to microbial attack. Slime formation, darkening, decomposition, and spontaneous coagulation occur partly, if not totally, through bacterial action. Crude rubber is discoloured by microbial pigments, attacked by some actinomycetes, and supports surface mould (*Penicillium glaucum*), developing discoloured, soft, tacky areas. Natural rubber products are completely consumed by soil micro-organisms. Some biocidal protection is given by mercapto-benzothiazole and tetramethylthiuram disulphide, which are used as vulcanisation accelerators. The latter is the more efective biocide when zinc oxide is included in the rubber compounds.

Starch

Starch, amylose, amylopectin, dextrins, and their ethers are degraded by numerous bacteria and fungi, but by few yeasts.

Proteins

Purified native proteins are not substantial nutrients for the growth of bacteria, and are not readily attacked even by strongly proteolytic organisms. A nitrogen source other than native protein must be present for active proteolysis to occur. Few bacteria attack elastin or keratin, although the native keratin of hair and wool is susceptible to attack by certain insects, fungi, streptomycetes, and actinomycetes. Collagen (a native protein not resistant to microbial attack) is subject to hydrolysis. Living bacterial cells and their proteins are not normally hydrolysed by enzymes, although heating seems to stimulate enzyme attack. Denatured proteins, casein, and gelatine are readily attacked as are the peptides (peptones and proteoses).

Leather

Leather is subject to microbial degradation both during processing (tanning) and in service. Both bacteria and fungi are involved. Biocides used include phenylmercuric acetate, coconut amine salt of tetrachlorophenol, sodium *o*-phenylphenoxide, sodium 2,3,4,6-tetrachlorophenoxide, sodium pentachlorophenoxide, *o*-phenylphenol, pentachlorophenol.

Cellulose

Cellulose is the most abundant organic material in nature. It makes up the greater portion of the plant body of wood, seed hair, bast fibre, straw, etc. Cotton contains about 91 per cent cellulose, and wood about 40–50 per cent. Cellulose is found in minute quantities in the animal kingdom. The products manufactured from cellulose and its derivatives are utilised as fibres, textiles, in home and boat construction, in paints as thickening agents, as paper and paper products, and in numerous other ways.

Cellulose and cellulose derivatives are susceptible to attack by biological agents through enzymatic hydrolysis or physical penetration.

Biocides used include copper naphthenate, copper oleate, bis(8-quinolinolato)-copper, zinc naphthenate, 2,2′-dihydroxy-5,5′-dichlorodiphenylmethane, sodium salt of 2-mercaptobenzothiazole, quaternary ammonium carboxylates, coconut amine salt of tetrachlorophenol, sodium *o*-phenylphenoxide, sodium pentachlorophenoxide, chlorinated phenols, sodium 2,4,5-trichlorophenoxide, *o*-phenylphenol. Cotton textiles have been reported to be completely resistant to mildew after impregnation with urea-formaldehyde or melamine-formaldehyde resins.

Synthetic Organic Polymers

Synthetic resins or polymers show an inherent resistance to attack by biological agents. This resistance is a result of their hardness, limited moisture absorption, surface smoothness, the absence of enzymatic systems capable of action on the chemical compositions of some polymers, or the chemical inertness of certain polymers.

Inorganic Polymers

Although many of the phosphonitrilic, siloxane, boron-phosphorus, boron-boron, titanium-nitrogen, and other inorganic polymers and their organic derivatives are not commercially important at the present time, their unusual properties have warranted increasing investigation. Data on biological attack on the inorganic polymers are scanty.

The nature of the inorganic bond eliminates the utilisation of this type of polymer for food by the micro-organisms. Microbiological degradation through conventional enzyme systems would not be expected, but a possible attack on the polymer used as a barrier could be anticipated based on the total degradative effect of environmental factors.

Polymer Additives

A product may be highly susceptible to biological attack even though the polymer upon which it is based may itself be inert or only mildly susceptible, since other compounding ingredients may be of nutritive value to pests and form a focal point of growth or attack. Although the early history of polymer degradation related the degree of fungal growth to the type of polymer used in the formulation, actually the plasticiser, lubricant, or filler used was the susceptible moiety. Other examples of susceptible additives are the thickening agents in latex emulsion paints, starch sizings on synthetic fibre textiles, and cellulosic fillers in urea-formaldehyde and other resins.

It must be stated that not all additives render the product or polymer more susceptible to attack by biological agents. Some may increase the resistance of the substance to attack or degradation through an antibiological activity inherent to the additive. For example, a compounded plastic may contain an amount of certain stabilisers sufficient to exert a fungal-resistant effect.

ADDITION OF BIOCIDES TO POLYMERS

Biocides are usually added to the polymers themselves, rather than being incorporated during polymerisation of the monomer. They may be added during processing, such as during mastication of rubber, or during mixing and fusion of poly(vinyl chloride) compounds, or be incorporated by post-addition to finished products such as impregnation of textiles, wood and cordage. Biocides may be applied as undiluted active compounds, as solutions in water or organic solvents, as emulsions or dispersions in volatile solvents, or as dispersions in organic vehicles or binders.

EVALUATION OF BIOCIDES

The ultimate effectiveness of a biocide can be determined only by actual use in full-scale field exposure under a variety of conditions. The record of its performance guides the manufacturer's recommendations for dosage, areas of application, and duration of effectiveness of the biocide. Properties of ideal biocide are shown in Table 20.1.

Table 20.1. Properties of the ideal biocide

A broad spectrum of activity	Stability in use environment
High toxicity to the deleterious biological organisms	Ease of application
	Compatibility with other chemicals (stabilisers, lubricants, antioxidants, etc.)
Low toxicity to the host or neutral biological environment	Permanency, when required
Low toxicity to humans	No negative effects on the use system
Low use cost	No dermatitic manifestations
Stability in storage	

A series of screening tests are applied to a candidate biocide to eliminate those having no activity and to establish a spectrum of effectiveness against the biological agents of interest. The test procedures used during the screening programme are varied and depend on the type of chemical, its potential use, the industry served, and the testing facilities available to the researcher. Many governmental and private agencies, such as the American Society for Testing and Materials, have established test methods to evaluate biocide efficiency. Screening tests follow this basic pattern.

Preliminary Screening (*In Vitro*) Tests

Laboratory tests using high dosage levels or range-finding procedures usually establish an approximate effective dose. These tests are followed by more sensitive or sophisticated methods to establish definitive dose levels.

Activity spectra against a variety of biological agents from one or more of the plant or animal species is then developed. A typical preliminary test is the modified paper disk bioassay method.

Secondary Screening (*In Vivo*) Tests

Based on preliminary screening data, the candidate is tested under simulated use conditions. The chemical is applied to a textile, incorporated in a plastic or paint formulation, or subjected to plant-scale tests; functionality under use conditions is established. A typical secondary test is the Soil Burial Method 5762.

Field Exposure

Actual field studies or short-term in-plant studies before or during commercial development yield further information about the condidate biocide. Table 20.2 lists some of the means of evaluating the suitability and efficacy of biocidal treatment of polymers.

Table 20.2. Some properties measured in evaluation of biocidal treatments of polymers.

Presence of organism growth or damage by organisms estimated by visual observations (microscopic and macroscopic)
Phytotoxicity, either selective or general
Destruction or inhibition of organisms through interference with essential life functions or reproduction odours indicative of spoilage or fermentation
Weight gain or loss by the polymer
Changes in mechanical properties, eg, tensile strength, bursting strength, ultimate elongation, elastic modulus, stiffness, torsional modulus, etc.
Changes in permeability to solvents, plasticisers, moisture, or gases
Changes in electrical properties
Chemical changes in the polymer
Colour changes

Whatever the results of its laboratory tests, a successful biocide must function effectively under the conditions of its actual use, and with the minimum number of negative characteristics, including its cost.

TOXICOLOGY

Biocides are, of course, toxic materials. Ideally, a biocide for use with polymers should have a high degree of toxicity to the organisms responsible for degradation, but should be harmless, or relatively so, to man. In actual practice, many biocides do present varying types and degrees of hazards towards man, and must be handled with due care. The fact that such hazards may exist should not, and does not, preclude the commercial use of these biocides where they are required, or where a less biologically susceptible polymer does not exist for use in a given application. The hazards may exist during the manufacture of the biocide itself, during the process of applying it to a polymer, or during end-use exposure of the treated polymer. They may be minimised by the use of protective clothing and equipment, observation of good habits of cleanliness, judicious choice of dosage, and knowledge of the use to which the polymeric products are to be put. The stigma of toxic hazard does not rest solely on the biocide. Examination of the industrial hygiene literature shows that many chemicals (including the polymers themselves) used in polymer manufacture and processing possess toxicity equal to or greater than that of many biocides. Man may be injured by toxic materials, by contact with skin and mucous membranes, by inhalation, and by ingestion.

It is impractical in a short article to discuss the hazards associated with each individual biocide. It is suggested that when the use of a specific biocide is contemplated, some excellent literature sources be consulted, as well as the supplier of the material in question. Some slight toxic hazard to the consumer of biocidally treated polymeric products may exist, primarily through misapplication or improper choice of a biocide. The greater hazard, however, lies in the handling of biocides during polymer processing. A possible source of exposure by ingestion could be via extraction of biocides by and into foods from packaging materials made from polymers. It is suggested that the section of the Food Additive Regulations pertaining to the intended use, component, article of manufacture, or manufacturing aid be consulted for the prescribed conditions concerning the use of the biocide.

INDUSTRIAL BIOCIDES

In 1959, the damage caused by microbial agents to natural and synthetic polymers was estimated to be in excess of 500 million dollars per year. The bulk of the industrial biocides have been developed to counter fungal and bacterial attack on wood, textiles, and agricultural products (both in the ground and in storage). Many of these materials are useful for preservation of protective coatings, rubber, and other natural and synthetic polymers. Approximately 100 different basic chemicals or combinations are offered for use as biocides. Table 20.3 contains a partial list of commercially available biocides.

Table 20.3. Some industrial biocides

Biocide	*Uses*
l-acetoxymercuric-2-hydroxyethane	Paper-mill slimicide
Acrolein	Paper-mill slimicide
Alkyl benzyldimethylammonium bromides	Textile fungicide
Bis(2-hydroxy-5-chlorophenyl)methane (dichlorophene)	Textile, rope, cordage, and paper fungicide and bactericide
Bis(8-quinolinolato)copper (copper 8-quinolinolate)	Textile, rope, cordage, wood, plastic, paper, adhesive, and paint fungicide
Bis(tri-*n*-butyltin) oxide	Paper, pulp, textile, and plastic preservative
Chloromethoxypropylmereuric chloride	Slimicide, paint emulsion preservative
Creosote (coal tar)	Wood preservative
Cupric oleate (copper oleate)	Textile, rope, and cordage fungicide
3,5-dimetyl-4-chlorophenol (*p*-chloro-*m*-xylenol)	Paint, adhesive, and industrial protein preservative
3,5-dimethyl-4-chlorophenol, derivative of (*p*-chloro-*m*-xylenol)	Fungistat and fungicide for vinyl compounds
Dimethyldithiocarbamic acid and 2-mercapto-benzothiazole, sodium and zinc salts	Slime control in pulp paper manufacture, preservativefor cotton fabrics, adhesives
3,5-dimethyl-1,3,5(2H)-tetrahydrothiadiazine 2-thione	Paper-mill slimicide
di(phenylmercuric) dodecenylsuccinate	Paint preservative
p-hydroxybenzoic acid, ester of (paraben)	Vinyl compounds
Laurylpyridinium 5-chloro-mercaptobenzothiazole	Slimicide, textile preservative
2-mercaptobenzothiazole (MBT)	Textile and leather preservative
2-mercaptobenzothiazole, sodium salt of	Adhesives, textiles, paper
Monoethanolammonium salt of 2-mercapto-benzothiazole	Slimicide, textile preservative
Naphthenic acids, copper salt of (copper naphthenate)	Wood, cordage, and textile fungicide
Naphthenic acids, zinc salt of (zinc naphthenate)	Wood and textile preservative
Pentachlorophenol	Wood, fibre, textiles, leather cordage, and paint preservative

(Contd ...)

Biocide	*Uses*
Phenylmercuric acetate	Water-based paint bactericide and fungicide, used in leather processing
Phenylmercuric dimethyldithiocarbamate	Paper-millslimicide, paper and paperboard fungicide
Phenylmercuric oleate	Paint preservative
Quaternary ammonium 2-ethylhexoate	Textiles
Quaternary ammonium naphthenate	Vinyl film, vinyl coating, fabric bactericide, fungicide
Salicylanilide	Textiles, varnishes for electrical equipment, leather
Salicylanilide, brominated	Paint and textile fungicide
Sodium *o*-phenylphenoxide (sodium *o*-phenyl plenate)	Textile, leather, adhesive, paint, paper, natural and synthetic gum preservative
Sodium 2,3,4,6-tetrachlorophenoxide	Adhesive, leather, paint, wood and cordage preservative
Tetrachloro-*p*-benzoquinone (chloranil)	Leather
2,3,4,6-tetrachlorophenol	Paint, cordage, wood, leather, textile preservative
Tetrachlorophenol, coconut amine	Adhesives, cordage, textiles, paint, leather preservatice
N-trichloromethylmercapto-4-cyclohexene 1,2-dicarboximide (captan)	Paint, lacquer, paper, plasticiser resins
N-(trichloromethylthio)phthalimide	Paint and vinyl fungicide
2,4,5-trichlorophenol	Paper-mill slimicide, paint, wood, adhesive and textile preservative

CHAPTER 21

Effect of Ionising Radiation on Polymers and Packaging Materials

INTRODUCTION

Packaging is an essential terminal process in the manufacture of any article before distribution. The term package is synonymous with packaging material, which may be the container itself or a wrapper. Pharmaceuticals, medical and biological products, employed for topical application and oral or parenteral administration, are supplied in a carefully packaged form. Some essential requirements of packaging materials are:

1. Protection of the contents from dust, moisture.
2. Prevent rusting or tarnishing of metallic surgical devices.
3. Prevention of entry of micro-organisms if the contents are for topical or oral use.
4. Be robust to withstand handling and transportation.
5. Permit ease of opening and retrieval of sterile contents without contamination.
6. Allow penetration of the sterilising agents.
7. Ensure, in the case of pre-sterilised materials, that the sterility of the product is maintained till it is opened.

The term sterility connotes a very small probability of finding a viable micro-organism in a product. The probability could be as low as one in a million.

CLASSIFICATION OF PACKAGING MATERIALS

The various packaging materials may be classified as follows:

Metals: as collapsible tubes and envelopes.

Glass: as bottles and ampoules.

Polymers: cellulosics—natural or regenerated cellulose, esters of cellulose in the form of cartons or envelopes.

Plastics: as bottles, bags, ampoules, envelopes.

Laminates: metal-plastics, cellulose-plastics.

The choice of the polymer depends upon the required physical protective and aesthetic properties of the packaging required, and the quality and toxicity requirements as determined by performance tests.

Polymers Used for Packaging

The most frequently used polymers for these purposes are the following thermoplastics:

Polystyrene, polyethylene (low, medium, high density), nylon 6 and 66, polyester (mylar or melinex), polypropylene, polyvinyl chloride (PVC) and its co-polymers, ethyl vinyl acetate co-polymers (EVA), polycarbonate, cellulose estersacetates propionate and butyrate, cellophane etc.

STERILISATION METHODS

The packaging materials should be free from substances which can be extracted by the contents in such quantities as to cause any hazards of toxicity.

The method employed for sterilisation depends on the type of packaging material employed and the product packaged in it. Metals and glass can be sterilised by dry heat, autoclaving, ethylene oxide, and radiation.

As a majority of the cellulosics and plastics are sensitive to heat, sterilisation would be possible only by gas or radiation. If the contents themselves are sensitive to heat, sterilisation would be possible either by ethylene oxide (if penetration is possible), or by radiation.

The factors which influence the various sterilisation processes applicable to packaging materials are given in Table 21.1. Gas and moisture permeabilities are very important when steam and ethylene oxide (ETO) sterilisations are employed.

Table 21.1. Factors affecting the sterilization processes.

Factor	*Dry heat*	*Moist heat*	*Ethylene oxide (ETO)*	*Radiation*
Temperature	+	+	+	–
Time	+	+	+	+
Pressure	–	+	+	–
Vacuum	–	+	+	–
Humidity	–	NA	+	–
Post-sterilisation drying/degassing	NA	+	+	NA
Interaction with product	Oxidation degradation	Hydrolysis	Hydroxy ethylation	Radiolysis
Residual toxicity	Nil	Nil	Yes : due to ETO, ETG, ETCH	Rare
Product density	+	+	+	+
Choice of packaging	Narrow	Narrow	Narrow	Wide
Sterilisation in final packaging for shipping	Not possible	Not possible	Not possible	Possible
Type of process	Batch	Batch	Batch	Continuous
Inactivation factor		10^{15-20}	10^{9}	10^{9-860}

(+) Affects the process (–) Does not affect the process

(NA) Not applicable; ETG: Ethylene glycol; ETCH: Ethylene chlorohydrin.

Note: Both ETO, and ETCH are toxic, base-pair mutagens.

Sterilisation by ETO is best with some problems such as retention and toxicity. The retention of ETO expressed in terms of mg. of ETO/g of the material is as follows: polyethylene (5–10), polystyrene (15–20), polyvinyl chloride (10–30), rubbers (15–35). Even after a lapse of 24 hours after sterilisation there can be a high percentage of retention of the absorbed gas (01): low density polyethylene (12 per cent), high density polyethylene (20 per cent), polyvinyl chloride (70 per cent), polypropylene (58 per cent), polycarbonate (57 per cent). It is therefore necessary to remove ETO by aeration for 24 to 168 hours at room temperature or for 8 to 12 hours at 5–60°C (E1). ETO reacts with moisture and chloride ions to produce ethylene glycol (ETG) and ethylene chlorohydrin (ETCH) respectively.

Toxicity of ETO Sterilised Products

The toxicity of ETO sterilised products is well documented. ETO and ETCH are base-pair type mutagens, causing damage to cytoplasm and chromosomes. As there is a high correlation between carcinogenicity and mutagenicity, the residues of ETO and ETCH in medical products may give rise to clinical complications as authenticated in literature. The permissible limits of ETO, ETCH and ETG in ETO sterilised medical devices are as in Table 21.2.

Table 21.2. Limits of ETO, ETCH and ETG in drug products and Medical devices.

Product/Device	*ETO*	*ETCH* *ppm*	*ETG*
Drug product:			
Ophthalmics (for topical use)	10	20	60
Injectables (including veterinary intramammary infusions)	10	10	20
Intrauterine device (containing a drug)	5	10	10
Surgical scrub sponges (containing a drug)	25	250	500
Hard gelatine capsule shells	35	10	35
Medical device			
Implant: Small (< 10 g)	250	250	5000
Medium (10–100 g)	100	100	2000
Large(>100 g)	25	25	500
Intrauterine device	5	10	10
Intraocular lenses	25	25	500
Devices contacting-mucosa	250	250	5000
blood (ex vivo)	25	25	500
skin	250	250	5000
Surgical scrub sponges	25	250	500

In Japan the limiting concentration of ETO in implants is about 2 ppm. In France the residual should not be more than 2 ppm. The problem of residuals, therefore, necessitates their determination in the product before release. It must be borne in mind that the limits mentioned are below the level that will usually be found in an ETO sterilised plastic material and the prescribed analytical method does not appear to have the sensitivity to reliably detect such low levels of residual gas. The unique advantage of radiation over the conventional methods of sterilisation lies in the unhindered penetrability, especially of gamma radiation. In addition, there is only a slight increase in temperature (of about 5°C above ambient temperature in organic materials), and the sterilisation is effected on finally packaged products.

SOURCES OF RADIATION

Two practical sources of radiation are electron accelerators and radioisotopes. Electron accelerators produce high-energy electrons whose degree of penetration is directly proportional to the electron energy and inversely proportional to the material density. Their use requires careful control of many variables affecting sterilisation efficiency. Among these variables are electron energy, electron current, scan width, exposure time and density of the material. The accelerators provide very high dose rates of several megagrays per second. Substantially monochromatic and relatively low energy pulsed electron beams are considered to be very efficient for surface sterilisation.

Radioisotope sources, among which cobalt-60 and caesium-137 are the most commonly employed for sterilisation purposes, emit gamma rays, whose energies are in the region of million electron volts. For example the gamma rays of cobalt-60 have 1.17 and 1.33 MeV and those of caesium-137 have 0.66 MeV. An electron volt is the energy gained by an electron when it falls through a potential difference of 1 volt. Its equivalence to other more familiar energy units are:

$$1\text{ev} = 1.602 \times 10^{-12}\ \text{erg} = 1.602 \times 10^{-19}\ \text{J}$$
$$= 3.83 \times 10^{-20}\ \text{cal.}$$

Gamma radiations are more penetrating than high energy electron (Fig. 21.1). They produce a more reliable radiation output than electron accelerators. These attributes make it possible to sterilise bulky products with fewer control operations for a reliable performance.

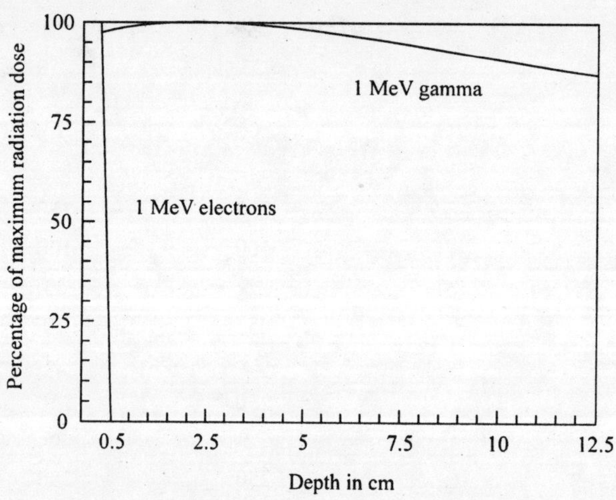

Fig. 21.1. Attenuation in water of 1MeV electrons and gamma rays.

The average dose rate in industrial type gamma ray sterilisation facilities may be so low as to necessitate duration of radiation sterilisation of the order of 7 to 70 hours. In the radiation sterilisation facility, ISOMED, installed at Trombay, cobalt-60 is employed as the radiation source. The duration of sterilisation is about 38 hours at a source strength of KCi (kilocurie). The higher the source strength the shorter is the duration of sterilisation.

The plant consists essentially of:

1. The radiation source housed in a concerete cell.
2. An automatic system for conveying the product boxes into the irradiation cell, exposing the boxes to the radiation field for the specified period and taking them out of the cell.
3. Laboratories for microbiology, dosimetry, etc. The products to be sterilised are packed in standard corrugated cardboard cartons of outer dimensions 59 cm length × 34 cm width × 43 cm height. These boxes are loaded in carriers, each having five shelves. Each carrier travels at a controlled speed, on an overhead mono-rail, enters the irradiation cell through a labyrinth, passes each side of the source four times, and returns through the exit labyrinth (Fig. 21.2). Five such passes through the cell ensure the exposure of the products to a minimum dose of 25 kGy (or 2.5 Mrad). A complete system of interlocks protects personnel from radiation exposure, and also prevents the products from receiving an overdose or underdose in the event of a mechanical failure.

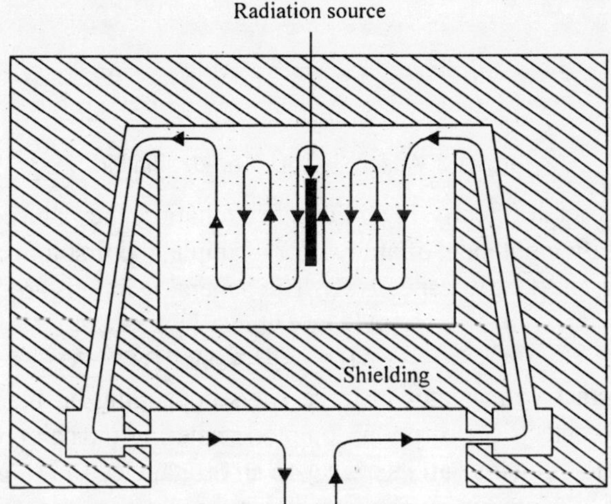

Fig. 21.2. Movement of carriers around the radiation source.

INTERACTION OF RADIATION WITH MATTER

Absorption of infra-red radiation results in transitions in a molecule between rotational and vibrational energy levels of the ground state (Fig. 21.3). Visible and UV radiations are absorbed selectively and cause transitions between vibrationals and rotational energy levels of different electronic states as in Fig. 21.4. M_g, M_s, M_t and M_r represent the potential energy curves of molecules in the following states respectively: ground, singlet excited, triplet and repulsive states. Excitation occurs from A to B. The excited state may revert to the ground state by internal conversion i.e. without the emission of radiation (Fig. 21.4a); here curve M_s, crosses M_g at a vibrational level lower than that to which the molecule was first excited. In Fig. 21.4b excitation has occurred with excess vibrational energy, which is soon lost by collision, and the excited state will soon achieve its lowest vibrational level; de-excitation occurs by emission of fluorescent radiation and a ground state is produced with excess vibrational energy, which once again is lost by collision. In Fig. 21.4c excitation is from M_g, to M_s; M_s, crosses M_t; this is analogous to internal conversion, but here the conversion is between two states of different multiplicity, and is therefore called intersystem crossing; although the transition from the triplet excited state to the singlet ground state is optically forbidden, the transition occurs but with a low probability; the triplet state returns to the ground state with the emission of phosphorescent radiation. In Fig. 21.4d the singlet excited state crosses the potential energy curve of a repulsive state; there is then the possibility of transfer of the molecule from the stable state to the repulsive state and dissociation will ensue; the two electrons forming the bond which is broken are usually divided one to each of the two fragments, and thus two free radicals are formed.

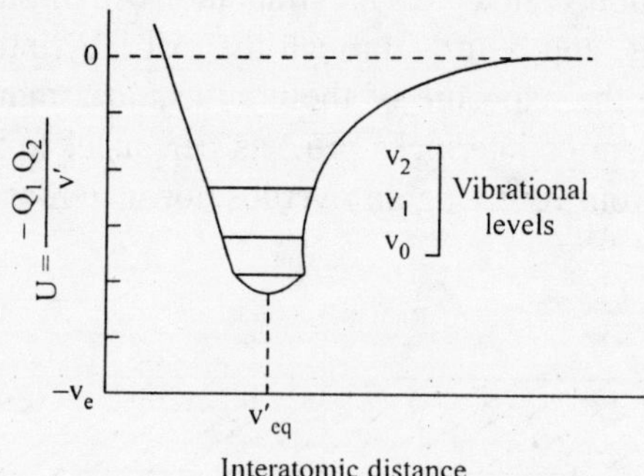

Fig. 21.3. Absorption of infrared radiation.

X-rays and gamma rays act differently. In addition to excitation, they cause ionisation as their energy is far in excess of the ionisation potential of atoms. For example, the first ionisation potentials of carbon and hydrogen are about 11 and 14 eV respectively. The general reactions are :

$$AB \longrightarrow AB + e$$
$$AB \longrightarrow AB^* \longrightarrow A^o + B^o$$

Gamma radiations interact with matter in three ways depending on their energy and the atomic number of the material: Photoelectric absorption, compton scattering and pair production. In photoelectric absorption the gamma photon transfers all its energy to an orbital electron (Fig. 36.5c). This interaction is greatest for materials of high atomic number and gamma energies of less than 0.1 MeV. In compton scattering the photon transfers part of its energy to the electron. The electron and the de-energised gamma ray move in different directions (Fig. 21.5b). This interaction is greatest for material of low atomic number and energetic gamma radiations. Pair production occurs when a very high energy gamma photon (>1.02 MeV) interacts with the field of a nucles and produces a positron electron pair (Fig. 21.5c). Excitation can sometimes lead to spilitting of the molecules.

In packaging materials which contain mostly light elements, C, H, N, O, Cl the gamma photons of cobalt-60, 1.17 and 1.33 MeV, undergo predominantly compton scattering. The absorbed and scattered photons produce high energy or primary electrons. These, in turn, ionise the molecules in their path and produce secondary electrons. The primary and secondary electrons soon lose their energy to a level below which they can no longer ionise or excite and become sub-excitation electrons. These may be captured by positive ions to give neutral molecules, or captured by neutral molecules to give negative ions, which may later decompose. While high energy electrons produce only singlet excited states, slow electrons produce both singlet and triplet states (Fig. 21.4).

Gamma radiation from cobalt-60 does not induce radioactivity in naturally occuring elements, as its energy is well below the threshold energy level for photon activation. For example the minimum energy for making oxygen and carbon radioactive is about 19 MeV.

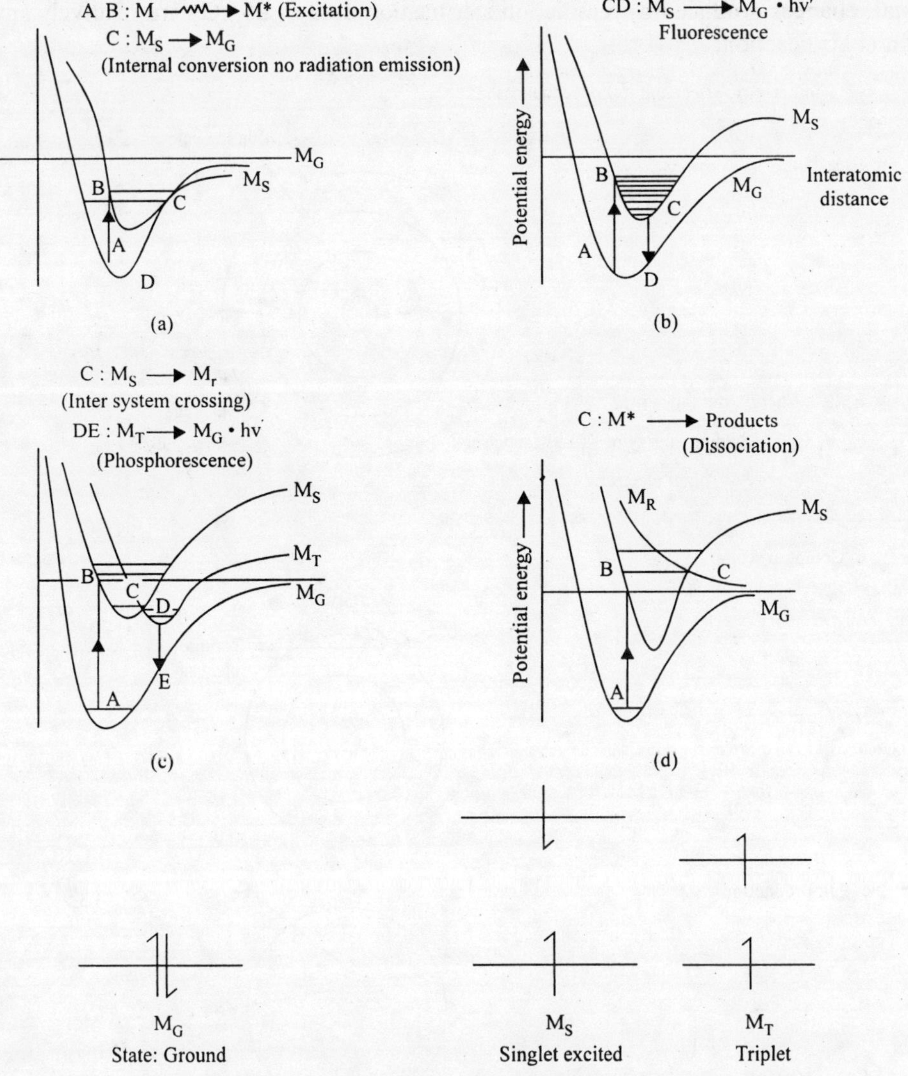

Fig. 21.4. Excitation and de excitation phenomena.

The unit of absorbed energy is Gray. The abbreviation for Gray is Gy. One unit of Gray is equivalent to the absorption of 100 rads, where 1 rad is equal to 100 ergs/g. Thus

$$1 \text{ Gy} = 100 \text{ rads} = 10^4 \text{ ergs/g} = 1 \text{ J/kg}.$$
$$= 6.2415 \times 10^{15} \text{ eV/g} = 2.389 \times 10^{-4} \text{ cal/g}.$$

If 25 kGy energy is absorbed in water for example, its temperature would rise by 6°C.

Since bond energies in molecules are of the order of a few eV, chemical bond rupture is one of the likely consequences. The number of such radiation induced chemical changes occuring per 100 eV of absorbed energy is usually expressed as a 'G' value. This also means that $0.1035 \times G$ micromoles of a substance are produced per gram of a material, which has absorbed 1 kGy or 6.24×10^{15} eV radiation.

The chemical changes produced by a radiation sterilisation dose of 25 kGy are relatively small in the absence of a chain reaction.

a. Photoelectric effect and its consequences

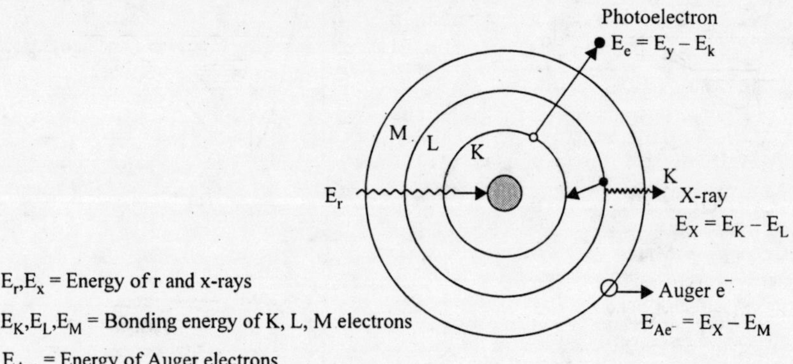

b. Compton scattering

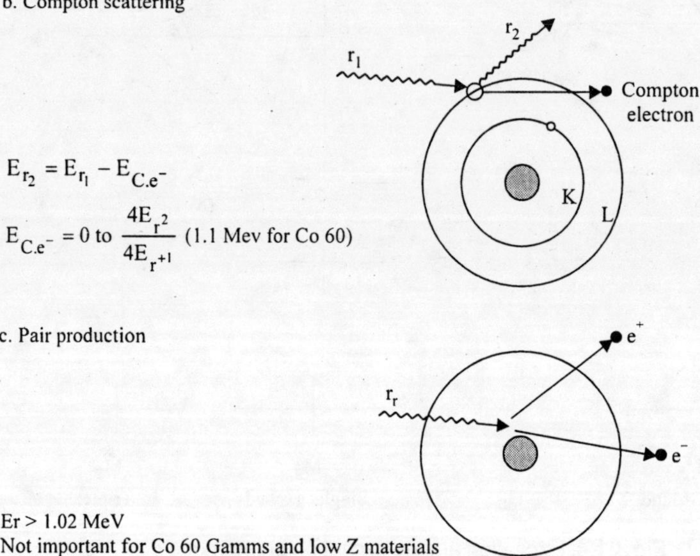

Fig. 21.5. Interaction of gamma radiation with matter.

Production of Free Radicals

Another unique phenomenon is the production of free radicals, which may be ions, atoms or molecules with a free electron. Normal oxygen, which is in a triplet state, behaves, to some extent, as a free medical, i.e. diradical. Hydrogen atom is a free radical by virtue of its posscessing a single electron. Removal of a hydrogen atom from a water molecule produces (H°) and hydroxyl radicals (°OH). Electron irradiators deliver radiation at dose-rates 10^5 to 10^{12} times faster than the gamma irradiators. Consequently the density of free radicals produced per unit time would be very much higher during high dose-rate electron irradiation than during low dose-rate gamma irradiation. These radicals may be short-lived or long-lived. Short-lived radicals disappear by reactions such as depolymerisation, cross-linking, recombination etc. The long-lived radicals which are detectable even after a few days after termination of irradiation, are responsible for post-irradiation ageing effects noticed in some plastics. The free radicals are mobile in 'rubbery' polymers such as natural and artificial rubbers, silicone rubbers and plasticised polymers.

They are relatively immobile in 'glassy' polymers such as polystyrene, perspex, PVC, and in 'crystalline' polymers such as polyethylene, cellulose, teflon etc. In the latter two cases, a high concentration of trapped long-lived free radicals may pose post-irradiation problems such as slow degradation processes. Such effects can be obviated by post-irradiation heat treatment which will eliminate the free radicals and thus increase the ultimate utility and efficiency of sterilisation by radiation.

Chemical Changes by Irradiation

The chemical changes brought about by irradiation in packaging materials can be ascribed to a few basic processes such as cross-linking, degradation, oxidation, unsaturation and gas evolution. The polymers, which are of two types, predominantly cross-linking or degrading.

Polymers whose molecules contain repetitive units, $–CH_2\,CHR–$, cross-link upon irradiation. Those containing the repetitive units, $–CH_2C\,(CH_3)\,R–$, degrade. Location of at least one H on the alternate carbon in the polymer chain favours cross-linking. If the alternate carbon atoms are fully substituted, degradation predominates. Cross-linking polymers have heat of polymerisation in excess of 16 kcal/ mole.

Cross-linking is a process involving the joining of the two polymer chains and may involve the following reactions.

1. A polymer molecule may lose a hydrogen radical and yield a polymer radical (Fig. 21.6a)
2. The hydrogen radical can abstract hydrogen from a neighbouring polymer molecule (Fig. 21.6b).
3. The polymer radicals may recombine (Fig. 21.6c).
4. This process may lead to formation of a three dimensional network (Fig. 21.6d).

Degradation

Degradation may result from chain scission (Fig. 21.6e) or from interaction between oxygen and polymeric radicals (Fig. 21.6f). In products with regions of low crystallinity or ordering and with irregularly arranged molecules, degradation reactions can predominate.

Unsaturation

Unsaturation is induced by elimination of two substituent atoms on adjacent carbon atoms resulting in a double bond formation (Fig. 21.6g).

Gas evolution

A majority of the polymers yield hydrogen and low molecular weight hydrocarbons, the branching chains being usually ruptured.

The gaseous products of some polymers are shown in Table 21.3. Chlorine containing polymers yield HCl and polymers containing esters yield CO, CO_2 in addition to hydrogen and low molecular weight hydrocarbons. In thin films and sheets this may not pose any problem as the gas transmission rate can be sufficiently high (Table 21.4).

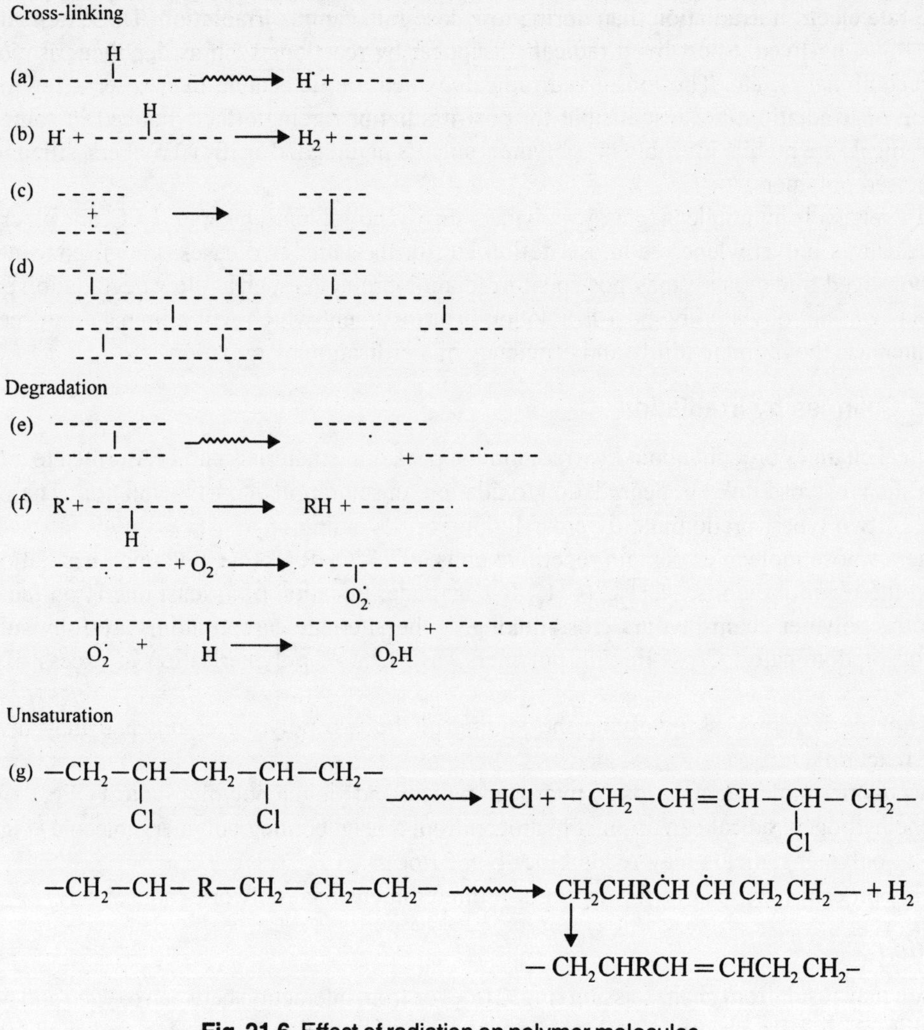

Cross-linking

(a) $-----\overset{H}{|}-----\rightsquigarrow H^{\cdot} + ------$

(b) $H^{\cdot} + ----\overset{H}{|}---- \longrightarrow H_2 + ------$

(c) $\begin{matrix}-----\\ + \\ -----\end{matrix} \longrightarrow \begin{matrix}-----\\ | \\ -----\end{matrix}$

(d) Cross-linked network

Degradation

(e) $\begin{matrix}--------\\ | \\ --------\end{matrix} \rightsquigarrow \begin{matrix}-------\cdot \\ + \\ \cdot-------\end{matrix}$

(f) $R^{\cdot} + \underset{H}{---|---} \longrightarrow RH + ------\cdot$

$---\cdot--- + O_2 \longrightarrow \underset{O_2^{\cdot}}{--|-----}$

$\underset{O_2^{\cdot}}{--|--} + \underset{H}{--|--} \longrightarrow \underset{O_2H}{---|---} + ---\cdot---$

Unsaturation

(g) $-CH_2-\underset{Cl}{CH}-CH_2-\underset{Cl}{CH}-CH_2- \rightsquigarrow HCl + -CH_2-CH=CH-\underset{Cl}{CH}-CH_2-$

$-CH_2-CH-R-CH_2-CH_2-CH_2- \rightsquigarrow CH_2CHR\dot{C}H\,\dot{C}H\,CH_2CH_2- + H_2$

$\downarrow$

$-CH_2CHRCH=CHCH_2CH_2-$

Fig. 21.6. Effect of radiation on polymer molecules.

Table 21.3. Gaseous products from some polymers.

Polymer	*H*	*CH_4*	*C_2H_4*	*C_2H_6*	*C_3H_6*	*C_4H_9*	*iso-C_4H_{10}*	*C_4H_{10}*
Polyethylene (a)	99.3	0.37	0.32	–	–	–	–	–
(b)	85.8	1.30	7.00	–	–	–	–	4.80
Polyisobutylene	69.6	9.40	1.10	2.60	1.40	11.40	3.80	–
Polypropylene	85.5	12.00	9.64	0.87	–	–	–	–
Polystyrene	97.3	2.50	0.14	–	–	–	–	–

(a) Low pressure, high density, linear chain.

(b) High pressure, low density, branched chain.

Table 21.4. Permeability rates of selective plastic packaging materials.

Plastic	*g/100 in² 24 hr./mil @37.8°C* *Water vapour*	*Oxygen*	*cc/100 in²/mil/ 24hr./atm @ 25°C* *Carbon-dioxide*
Nylon – Type 6	16–22	2.6	10–12
– Type 12	4	52–92	156–336
Polyethylene			
low density	1–1.5	500	2700
medium density	0.7	250–535	1000–2500
high density	0.3	185	580
Polypropylene	0.7	150–240	500–800
Polystyrene	7–10	250–350	900
Vinyl			
non-plasticised	4-5	8–30	20–30
plasticised	6–10	30–160	30–1900
Vinyl chloride acetate co-polymer			
non-plasticised	4	15–20	40–70
plasticised	5–8	20–150	70–800
Polyvinylidene chloride	0.2–0.6	0.8–6.9	3.8–44
Polycarbonate	11	300	1075
Cellulose acetate	90	100	560
Cellophane	–	0.2–5.0	0.4–6.0

The radiation stability of some thermoplastics, thermosets and elastomers, some of which are of importance in packaging, are given in Table 21.5.

The physical properties studied before and after irradiation depend on the desired property of the polymer.

Some of the properties usually of interest for plastics are elastic modulus, shear strength, tensile strength, per cent elongation at a particular load, and notch impact strength.

For some of the packaging materials and for some end uses, not all of these properties would be important, as for instance for low density polyethylene films the Indian Standards Institution (BIS) lays down specifications for tensile strength and elongation at break, and optional tests for clarity, and impact resistance and slip. For see-through packages optical transmission would be a significant property. Each plastic must, therefore, be studied individually using these references as a guide to determine its suitability or unsuitability.

This is necessary as the plastic with the same name may have a unique make-up of additives such as plasticiser, etc., which might behave differently. It may be mentioned that thin films of materials may behave differently from thick pieces for the same dose, as for example the greater degree of diffusion of oxygen in thin films of polypropylene can accentuate the interaction between oxygen and short-lived radicals.

Table 21.5. Radiation stability of polymers.

Thermoplastics			
Acrylonitrile butadiene styrene (ABS)	Good	Polycarbonate	Good
		Polyesters	
Acrylics		aromatic	Good
polymethyl methacrylate	Fair	Polyolefins	
Amides		polyethylene	Good
aliphatic	Fair	polypropylene	Fair
aromatic	Excellent	polymethylpentene	Good
Cellulosics	Fair	copolymers	Good
Fluoroplastics		Polystyrenes	Excellent
polytetra fluoroethylene	Poor	Polystyrene acrylonitrile	Good
polychlorotrifluoroethylene	Fair	Polysulphones	Excellent
polyvinyl fluoride	Good	Polyvinyls	
polyvinylidene fluoride	Good	PVB	Good
Co-polymer of ethylene and terafluoroethylene	Good	co-polymers	Fair
Thermosets		*Elastomers*	
Epoxies	Excellent	Polyacrylic	Good
Phenol (or urea) formaldehyde	Good	Butyl	Good
		Fluoro	Good
Polyesters (unsaturated)	Good	Natural	Excellent
Polyimides	Excellent	Nitrile	Excellent
Polyurethanes	Excellent	Polychloroprene (neoprene)	Good
		Silicones	Excellent
		Styrene-butadiene	Good
		Urethane	Excellent

Poor: Non-sterilisable; Good: More than 25 kGy; Fair: About 25 kGy; Excellent: Very much more than 25 kGy.

POLYETHYLENE

Though it is principally represented as a primarily linear polymer with $-CH_2-CH_2-$ units repeating, it contains some branch chains, a few methyl groups, double bonds and carbonyl groups. The polymer contains crystalline as well as amorphous zones.

Low density polyethylene (0.910–0.935 g/cc) is obtained from ethylene monomer and small amounts of oxygen or peroxides which generate free radicals for the initiation of the polymerisation at high pressure (15,000 to 45,000 psi).

High density polyethylene (0.940–0.965 g/cc) is obtained at low pressures (1,500 psi) with the aid of special catalysts.

Both low and high density polyethylenes withstand several megarads without affecting the packaging quality of films, bags, sachets, moulded containers, squeeze tubes, closures as well as their laminates.

The G (cross-linking) has been cited as 2 to 3 and the ratio of degradation to cross-linking probabilities (*beta/alpha*) is about 0.2 to 0.3. The heat of polymerisation is 22 kcal/mole. Irradiation of polyethyl in air can lead to the following reactions:

1. Radiolysis of polyethylene (Fig. 21.7a).
2. Addition of oxygen to the polymeric radical (Fig. 21.7b).
3. Splitting of the polymeric peroxide radical (Fig. 21.7c).
4. The polymeric radical abstracting hydrogen from a neighbouring molecule (Fig. 21.7d).

The degree of crystallinity (per cent) decreases as irradiation dose increases and the values at various doses (MGy) are: /L 5(68 per cent), 10(46 per cent), 15(36 per cent). Change in IR spectral pattern or fringes in the region 700–200 cm^{-1} indicates the change in refractive index of polyethylene film.

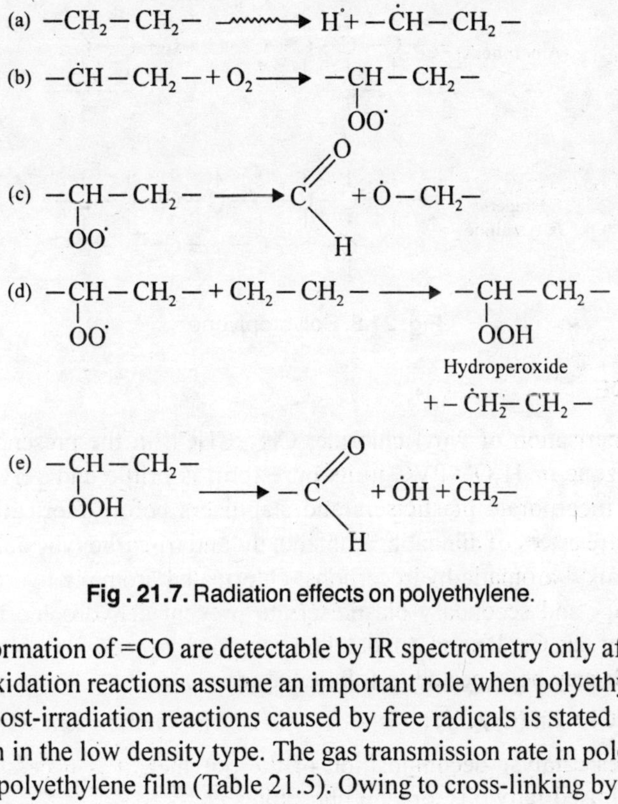

(a) $-CH_2-CH_2- \rightsquigarrow H^{\cdot} + -\dot{C}H-CH_2-$

(b) $-\dot{C}H-CH_2- + O_2 \longrightarrow -CH(OO^{\cdot})-CH_2-$

(c) $-CH(OO^{\cdot})-CH_2- \longrightarrow C(=O)H + \dot{O}-CH_2-$

(d) $-CH(OO^{\cdot})-CH_2- + CH_2-CH_2- \longrightarrow -CH(OOH)-CH_2-$ + $-\dot{C}H_2-CH_2-$

Hydroperoxide

(e) $-CH(OOH)-CH_2- \longrightarrow -C(=O)H + \dot{O}H + CH_2-$

Fig. 21.7. Radiation effects on polyethylene.

Unsaturation and formation of =CO are detectable by IR spectrometry only after irradiation to a dose of several MGy. The oxidation reactions assume an important role when polyethylene is irradiated in air at low dose rate. The post-irradiation reactions caused by free radicals is stated to be more pronounced in the high density than in the low density type. The gas transmission rate in polethylene film decreases with the density of the polyethylene film (Table 21.5). Owing to cross-linking by irradiation, permeation is further reduced.

POLYPROPYLENE

Catalytic polymerisation of propylene yields polypropylene (PP). Depending on the polymerisation conditions, the polypropylene molecules may be atactic with all the methyl groups lying at random or isotactic with all of them lying on one side of a theoretical plane containing the main carbon-carbon chain (Fig. 21.8). The isotactic form is crystalline and the atactic form is amorphous. The permeability in polypropylene to gases or moisture is very much lower than that of low density polyethylene (Table 21.5), but it has a high resistance to dry or moist heat. In pure form it is brittle below room temperature and it is therefore plasticised for commercial use. The presence of hydrogen attached to tertiary carbon makes it more susceptible to oxidation than polyethylene thereby making necessary the inclusion of antioxidants. However, antioxidants are said to induce formation of yellow colour with time in irradiated polypropylene. Radiation sterilisation in a non-oxidising atmosphere such as in an evacuated glass container or in nitrogen in a laminated package substantially reduces brittleness and polymer breakdown. The G (cross-linking) is about 0.6 and *beta/alpha* is about 0.8 to 1.0. The heat of polymerisation is greater than 16.5 kcal/mole. Cross-linking is favoured by absence of air and degradation by presence of air. While it is just about satisfactory to a sterilising dose of 25 kGy, it becomes brittle at 50 kGy and turns into flakes at 100 kGy. Biaxially oriented PP appears to be more radiation stable than ordinary PP.

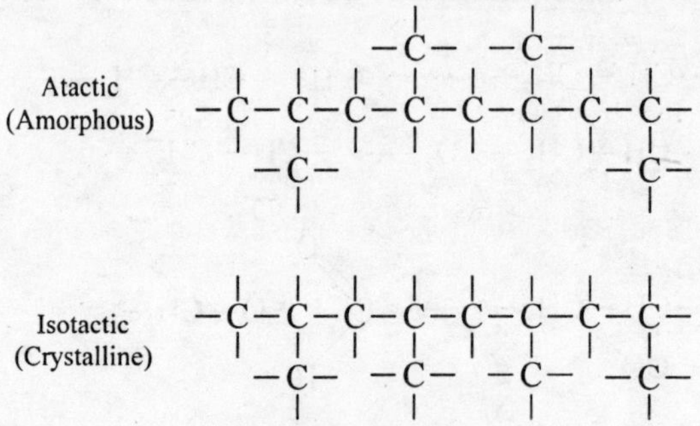

```
                                |           |
                              —C—         —C—
                    |     |     |     |     |     |     |     |
Atactic            —C— —C— —C— —C— —C— —C— —C— —C—
(Amorphous)         |     |     |     |     |     |     |     |
                        —C—                                 —C—
                          |                                   |

                    |     |     |     |     |     |     |     |
Isotactic          —C— —C— —C— —C— —C— —C— —C— —C—
(Crystalline)       |     |     |     |     |     |     |     |
                        —C—         —C—         —C—         —C—
                          |           |           |           |
```

Fig. 21.8. Polyproplyene.

POLYVINYL CHLORIDE

It is prepared by polymerisation of vinyl chloride, CH_2=CHCl, in the presence of catalysts such as organic persulphates, ozone or H_2O_2. PVC in its pure form is brittle and very unstable to heat. It is therefore necessary to incorporate plasticisers and stabilisers before fabrication. Some plasticisers mentioned in literature are esters of aliphatic, aromatic, di- and tri-carboxylic acids, organic phthalates, high molecular weight alkyl aromatic hydrocarbons, chlorinated aromatic hydrocarbons etc. It is also necessary to add primary and secondary plasticisers to prevent dehydrochlorination. Some primary stabilisers are stearates of Na, Ca, Ba and Zn, octyl tin compounds, cadmium salts etc. The stabilisation, for example, by zinc salt may be expected to follow the reaction:

$$—CH_2CHCl + Zn(OOCR)_2 \longrightarrow —CH_2CH(OOCR) + ZnCl(OOCR)$$

Since metal chlorides catalyse decomposition of the polymer, it is necessary to have secondary stabilisers such as epoxidised fatty oils, organo phosphites etc.

A proper choice of plasticisers and stabilisers has to be made in order to satisfy the ultimate requirement of non-toxicity to physiological systems and non-corrosiveness to metallic objects for which it is often used as a packaging material. European pharmacopoeia recommends a suitable composition of plasticised PVC for containers.

PVC is moderately resistant to radiation sterilisation doses with G (cross-linking) of 0.2 to 0.5 irradiation causes a yellow to brown colouration which is somewhat deeper than that caused by heating PVC.

The colouration is ascribed to polyene formation which occurs via intramolecular dehydrochlorination (Fig. 21.6g). Radiation resistant formulations are, however, available in some countries including India.

POLYSTYRENE

When styrene is heated to 200°C it is converted into the polystyrene. The common packaging uses of polystyrene are thermo-formed containers and injection moulded containers. It is the most radiation stable polymer known, capable of withstanding several hundred MGy with no change in colouration. G (cross-linking) is only 0 to 0.06 and beta/alpha is about 0 to 0.2. The heat of polymerisation is 17 kcal/mole. Biaxially oriented polystyrene is reported to be suitable for blister packs.

CELLULOSICS

Cellulose is a polysaccharide with the glucose linked in 1,4 positions as in cellobios. It has three chemically reactive hydroxyl groups in each anhydro glucose unit. Irradiation causes depolymerisation and an increase in water solubles and carboxyl content. The moisture content in the cellulose has a profound influence on the irradiation behaviour of cellulose. Dry cellulose turns yellow-brown in hue on irradiation. Post-irradiation humidification partially reverses this change. Humidified cellulose undergoes very little colour change on irridiation. Also the number of free radicals is less in pre-humidified irradiated cellulose and in cellulose humidified after irradiation, than in irradiated dry cellulose. The most commonly referred free radicals are in Fig. 21.9. Humidification removes radicals IV and V, leaving I, II and III.

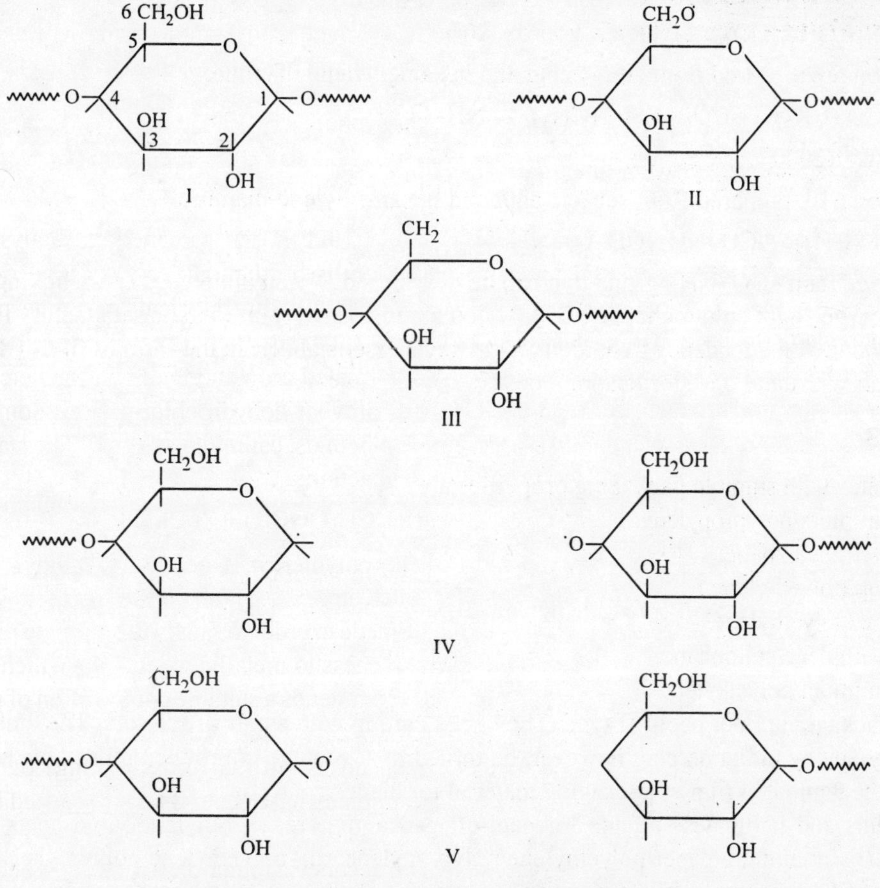

Fig. 21.9. Free radicals in irradiated cellulose.

Cellulose products such as paper and cardboard are stable up to 100 kGy and cellophane or regenerated cellulose becomes brittle at 50 kGy.

Since paper packages are liable to a considerable risk of microbial contamination in humid atmosphere or indirect contact with water, waterproof packaging materials such as laminates of cellulose with plastics, and cellulose esters are to be preferred.

Organic cellulose derivatives such as the acetate, propionate and butyrate are prepared by esterification of the three reactive hydroxyl groups. The triester is prepared at first and by controlled hydrolysis the mono and di-esters are obtained. For instance, in the case of cellulose acetate there would be more acetyl groups at C-2 than at C-6 if the degree of substitution is greater than 1.25. The cellulose esters are tough, transparent and resistant to water. Cellulose butyrate films are reported as suitable for blister packs. Cellulose acetate can withstand radiation doses of 50–100 kGy. Acetylation has been shown to offer protection against hydrolytic breakdown.

NYLONS

The word nylon covers a number of polyamides which are protein-like in structure. The most generally used are:

1. Nylon 6, a linear polymer obtained by polymerisation of α-caprolactam.

   $H\,[NH(CH_2)_5CO]_n$ —OH; n = approx. 200

2. Nylon 66, prepared from adipic acid and hexamethylene-diamine.

   $[—CO\,(CH_2)_4 —CO—NH—(CH_2)_6 —NH—]_n$

   m. wt. 12000

3. Nylon 610, prepared from sebacic acid and hexamethylene-diamine.

   $[–CO(CH_2)_8—CO–NH–(CH_2)_6—NH—]_n$

All these contain –CO–NH– group that cab be hydrolysed. Nylon undergoes cross-linking (G = 0.3) and may undergo slight colour change (C 3). Nylon 6 and 66 are reported to be satisfactory upto a dose of 1000 kGy but with a tendency to harden and to have lesser stability in the form of films (R 3). Nylon in general can withstand one sterilisation does (P 2).

LAMINATES

Some laminates with suitable packaging properties are:

1. Aluminium/polypropylene.
2. Polyethylene/paper.
3. Nylon/polyethylene.
4. Mylar/polyethylene.
5. Polyethylene/cellophane.
6. Aluminium/polyethylelne.

These packages are not peel-off type. They need cutting with a pair of scissors. The embrittlement of polypropylene by radiation can, however, be turned to advantage in producing peelable aluminium-polypropylene laminated film as packaging material for blades.

Some films and laminates suitable for peel-off packaging are: (i) ethylene vinyl acetate/ethylene isobutylene; (ii) aluminium/paper/polyethylene/polypropylene; (iii) polyethylene/polyvinylacetate/mylar; (iv) high density polyethylene/low density polyethylene; and (v) polypropylene/biaxially oriented polypropylene.

The laminates have, in general, better barrier properties against microbes, and are stronger than either of the constituent components alone. For improved peeling and heat-sealing, lacquer coatings are recommended. Excepting polypropylene all of the other ingerdients of the laminates withstand the radiation sterilisation dose.

Thus irradiation by gamma rays produces some physico-chemical changes in many polymers due to cross-linking, degradation and oxidation depending on the type of polymer and the environment in which irradiation is carried out. Irradiation is an inert atmosphere is not practicable for packaging materials and therefore a proper choice is to be made after taking into account the various changes due to irradiation. It must, however, be noted that the extent of any change in a variety of packaging materials at a sterilisation dose 25 kGy is low enough to permit their successful use. Changes in colour or odour, which are often seen to occur in some packaging materials even by exposure to sunlight, should not be mistaken as a deleterious effect of gamma radiation. The colour change does not affect in anyway the quality of the contents.

CHAPTER 22

Carbon Black Recovery from Polymer Waste by Vacuum Pyrolysis

INTRODUCTION

The enormous number of used tyres has been threatening the environment. Especially in USA and Canada one used tyre is dumped per head per year. Most of these used tyres are not recycled but are simply dumped in open or landfill sites, where they breed mosquitoes and rats and can be the cause of uncontrolled fires. To solve this environmental problem, waste tyre is to be reused in such a manner that it will be environmentally friendly as well as economically viable.

There are several disposal methods for waste tyres like (i) landfill, dumping, stockpiling; (ii) incineration; (iii) retreading; (iv) reclamation; and (v) others.

In the above described disposal methods, reclamation is the most environmentally friendly method. However, its market is limited and economic is debatable.

Pyrolysis of scrap rubber to remove and recycle raw material meets the ideal objective and can be operated economically. Since 90 per cent of carbon black is used in tyre industries, so, if we can regenerate carbon black from tyres than 90 per cent of the used carbon black would be regenerated. This work is mainly based on regeneration of carbon black from waste automobile tyres.

Tyres contain 25–30 per cent carbon black and 3–7 per cent oil, which are valuable and can be regenerated easily be simple pyrolysis under controlled conditions. Since raw materials required for regeneration of carbon black are mainly waste tyres (only transportation and labour costs are required for its collection, i.e., cheap materials) and electrical energy (some what costly), the process should also be economically viable.

However, regenerated carbon black is different from the virgin commercial carbon black. Different blacks may be mixed up along with the inorganic components and additives of the tyre may change the morphology of the regenerated black. Regenerated black may contain adsorbed different oxygenated or acidic groups, hydrocarbon or carbonaceous deposits on the surface.

So, the work mainly based on regeneration of carbon black from waste automobile tyres and its characterisation to find out its properties and morphological change from that of the virgin commercial black. For this study, some required formulations with virgin, heat treated and regenerated blacks are made to study the processability, curing characteristics and physical properties.

The flow chart of the process for regeneration of carbon black is given in Fig. 22.1

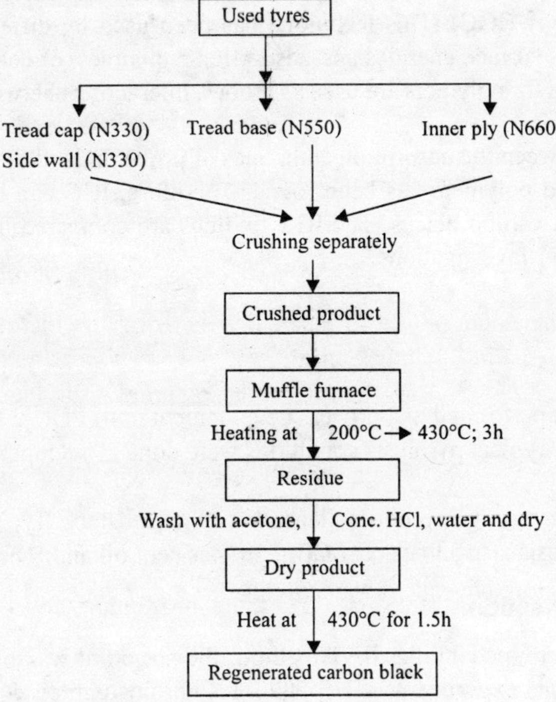

Fig. 22.1. Regeneration of carbon black from waste tyres.

HEATING OF THE VIRGIN BLACK

During pyrolysis, regenerated black experiences heat. So, this heat has a effect on the properties and morphology of the regenerated black. To find out the effect of heat on properties and morphology, virgin black was also heated under the same conditions (at 430°C for 1.5 h). To find out the change in properties and morphology due to pyrolysis or heat treatment all the properties for the virgin blacks (VB), heat treated blacks (HTB), and regenerated blacks (RB) were measured and compared. The probable reasons for the change in properties were explained accordingly.

During the pyrolysis the rubber polymer is cracked to shorter fragments which are recovered in condensers as oils. The solid residue consists of the recovered carbon black filler, inorganic tyre components (such as ZnO and ZnS) and the steel belt. After mechanical separation of the steel the pyrolytic carbon black (CBp) still contains inorganic compounds which can be removed by a demineralisation treatment. The vacuum pyrolysis process is protected by international patents.

By vacuum pyrolysis scrap tyres and other waste rubbers can be transformed into useful products. The commercial value of the recovered carbon black (CBp), i.e. its ability to replace commercial grades, is very important for the economics of the vacuum pyrolysis process. The most important carbon black application is rubber reinforcement. For this application the carbon black surface area, structure, surface activity and surface chemistry are very important. In this work the properties of CBp originating from different rubber materials (e.g. whole passenger and lorry tyres, tyre sidewalls and cable insulation) and of commercial carbon blacks are compared. The carbon black surface area and structure was determined by nitrogen and dibutyl phthalate (DPA) adsorption, respectively. The surface activity was studied by inverse gas chromatography (IGC). This technique has been used by different groups to investigate surface properties, such as surface energies and adsorption enthalpies, of commercial carbon blacks. If suitable model compounds for polymers are used as probes, interactions between the filler and polymers can be estimated using IGC.

A good correlation between the adsorption enthalpies of probe molecules and rheological properties of carbon black reinforced polymers has been found. Therefore, IGC can be used for predicting the reinforcement potential of carbon blacks. The IGC findings are compared in this chapter with results from surface spectroscopic investigations.

EXPERIMENTALS

Materials

The vacuum pyrolysis was performed in a process development unit (PDU). The PDU has a capacity of 100 kg of rubber shreds/h. Typical pyrolysis conditions were done at a temperature of 500°C and a total pressure of 10 kPa.

The yield of the vacuum pyrolysis products depends on the feedstock. Typical yields for tyre vacuum pyrolysis on a steel-free basis are: 33 per cent CBp, 59 per cent oil and 8 per cent gas.

Carbon Black Characterisation

The surface area was determined by the BET method, the structure was obtained following ASTM method D2414. For the IGC experiments a HP 5890 gas chromatograph equipped with an HP 3390 integrator and a flame ionisation detector was used. The carbon black particles were pressed, then gently crushed and a fraction with particle diameters between 600 and 850 μm was separated by sieving. These particles were filled into a stainless steel column with an outer diameter of approximately 3.2 mm (1/8 in.).

The column length was selected so that the total BET surface area of the carbon black particles in the column was approximately 10 m^2. The pressure drop in the column was smaller than 5 kPa. Prior to the IGC experiment the carbon black filled column was heated to 200°C under a helium flow rate of 30 ml/min for 10 h. The same helium flow rate was used during the experiment. Approximately 1 μl of the vapour phase above the liquid probes, stored in vials, was injected.

RESULTS AND DISCUSSION

Surface Area and Structure

Some standard properties of a demineralised carbon black sample from lorry tyre pyrolysis (CBp lorry) are presented in Table 22.1, in comparison with two commercial grades. The CBp (run H22) has a higher surface area than that of commercial grades N330 and N539. Assuming that the lorry tyres contained a mixture of high and medium carbon blacks, (e.g. N115 and N330) this indicates that the surface area of the carbon black was 'conserved' during the vacuum pyrolysis process. The recovered CBp had a similar structure to that of the two high-structure commercial blacks N330 and N539.

This suggests that during the vacuum pyrolysis no significant break-up of the carbon black aggregates took place. The CBp ash content was somewhat higher than in the commercial blacks, whereas the sulphur concentration was similar.

Table 22.1. Properties of pyrolytic and commercial carbon blacks.

Property	*CBp (Lorry)*	*N539*	*N330*
Surface area (m^2/g)	95	43	85
Structure, DBP No. ($cm^3/100g$)	102	111	102
Ash	0.7	0.2	0.4
Sulphur	0.5	0.6	0.6

Surface Chemistry

The surface chemistry of carbon black is very important for its use as reinforcing filler because the interaction between the carbon black filler and the polymer takes place at the carbon black surface. During rubber pyrolysis, the carbon black surface can be modified by the formation of carbonaceous deposits. The atmosphere in the pyrolysis reactor consists of hydrocarbons originating from the decomposing elastomers. These hydrocarbons may adsorb on the carbon black surface where they polymerise and finally form carbonaceous deposits. Carbonaceous deposits are defined here as chemically bound (mostly) polyaromatic hydrocarbons on the carbon black surface. The presence of carbonaceous deposits was indeed confirmed by ESCA. The ESCA spectra of CBp show a peak (C_1) which is not present in the spectra of commercial carbon blacks (Fig. 22.2). The peak can be used to monitor the amount of carbonaceous deposits as a function of the pyrolysis conditions. The amount of carbonaceous deposits on the CBp decreases with decreasing pyrolysis pressure and increasing pyrolysis temperature. Under reduced pressure conditions, the hydrocarbon concentration in the gas phase decreases, less hydrocarbons are adsorbed on the surface and, therefore, less carbonaceous deposits are formed.

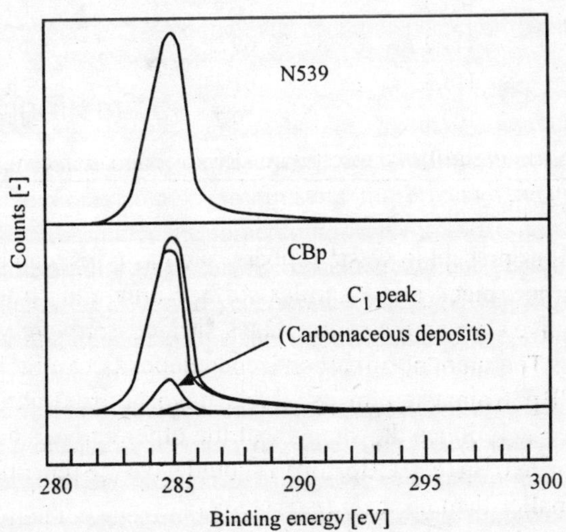

Fig. 22.2. ESCA spectra of commercial carbon black and CBp.

Surface Energy and Adsorption Enthalpy

The carbonaceous deposits influence the CBp surface properties. Surface energies of pyrolytic and commercial carbon blacks and adsorption enthalpies for different probe molecules on these samples are presented in Table 22.2. The surface energy of the CBp was smaller than those of commercial carbon blacks. The CBp (polyisoprene) sample was recovered from a rubber feedstock which contained N330 as filler. Comparison of the two samples showed that the recovered CBp lost approximately one-third of its surface energy during vacuum pyrolysis (103.7 vs. 150.4 mJ/m^2). The surface energy is the excess of the total energy that a solid has over the value it would have if the surface were in the same thermodynamic state as the interior. To minimise the surface energy, surface atoms may rearrange or substances with a lower surface energy may be adsorbed on the surface. Exactly this happens during pyrolysis when carbonaceous deposits are formed. These deposits cover a portion of the active sites on the carbon black surface which are then 'lost' for rubber reinforcement. A correlation between the amount of carbonaceous deposits, measured by ESCA, and the surface energy was observed.

An important question is: what happens to active sites which are not covered by carbonaceous deposits? The adsorption enthalpy (ΔH) can be regarded as a measure for the strength of the active sites. The adsorption enthalpies of different probe molecules on N330 and on recovered N330 (CBp polyisoprene) were very similar (Table 22.2). Also for the other CBP samples ΔH was at least as high as for commercial grades.

Table 22.2. Carbon black surface energies (non-specific component) at 150°C and adsorption enthalpies of probe molecules.

Sample	*Surface energy* γs^d, *(mJ/m^2)*		*ΔH (kJ/mol)*	
		n-Hexane	*1-Hexane*	*Benzene*
N 330	150.4	55.6	61.8	65.2
N 774	118.1	49.5	49.6	49.6
CBp (cable)	125.4	50.1	57.6	48.1
CBp (polyisoprene)*	103.7	56.8	63.9	66.6
CBp (sidewall)	111.2	69.2	62.2	58.2
CBp (passenger)	113.3	61.3	58.3	57.7
CBp (lorry)	109.4	79.2	74.5	67.9
CBp (passenger, 1 atm)	28.0	51.5	52.5	53.3

* The feedstock contained N 330 as filler.

Thus, the carbon black filler present in waste rubbers, such as tyres, can be recovered by pyrolysis. The recovered pyrolytic carbon black has a similar surface area and structure as the virgin black. However, a portion of the active sites on the carbon black surface is covered by carbonaceous deposits formed during the pyrolysis. The amount of carbonaceous deposits can be limited by performing the pyrolysis under vacuum.CBp from vacuum pyrolysis have approximately one-third less active sites per m^2 than the virgin black. Since the strength of the other active sites is not influenced by the carbonaceous deposits a 95 m^2/g CBp (e.g. CBp, lorry) should have the potential to replace a commercial grade with a surface area smaller than 63 m^2/g (e.g. N539).

Thus from the above it can be concluded that: (i) regeneration of carbon black is environmentally friendly; (ii) regenerated carbon blacks have higher surface area due to increase in surface roughness, compared to respective virgin blacks; (iii) structure of regenerated black remains almost same; (iv) agglomerate size of regenerated black remains almost same; (v) some polar acidic groups like carbonyl, hydroxyl, carboxylic and chlorine are incorporated in regenerated carbon blacks; (vi) compounds containing regenerated black gives comparable state of cure (but low cure rate), dispersion, hardness, tensile strength, EB per cent, tear strength with respective virgin black; and (vii) compound containing regenerated black gives lower modulus compare to respective virgin black.

CHAPTER 23

Environmental Implications of Polymer Degradation

INTRODUCTION

Biodegradable polymers offer an attractive alternative to traditional petroleum based non-biodegradable polymers from an environmental perspective. The principal driving force behind this technology is the solid waste problems, particularly with regard to decreasing availability of landfills and litter problems. Additional impetus driving this technology is the problem with non-biodegradable practice in marine environment which results in entrapment or ingenious hazards to marine life.

Biodegradable polymers are a particularly attractive option for addressing the solid waste and marine pollution concerns. Since they do not need to be land filled, they will re-enter normal geochemical cycles over time and many of the polymers are derived from renewable resources. Aside from being biodegradable, the polymers must also perform their intended functions as they must undergo biodegradation once their intended use is complete (controlled lifetime) performance and safety issues must be met for the material to be acceptable.

BASIC METHODS OF DEGRADATION

The two basic means of degradation that can be achieved through nature are: (i) biodegradation; and (ii) photo self-destruction.

Biodegradation

Biodegradation is defined as a process carried out by biological systems (usually bacteria or fungi) wherein a polymer chain is cleaved via enzymatic activity. In some cases the hydrolysis of the chains can be catalysed by chemical processes due to pH shifts as a result of secondary effects due to the organisms. The two principal mechanisms by which enzymes degrade polymers are hydrolysis and oxidation. The two steps occurring in this process are:

1. Depolymerisation or chain cleavage step.
2. Mineralisation.

The first step often occurs outside of the organism (usually bacteria or fungi) due to the size of the polymer chain and the insoluble nature of many of the polymers. Extra cellular enzymes are responsible for this step acting either: (i) endo (random, cleavage on the internal linkages of the polymer main chains); or (ii) exo (sequential cleavage of the terminal monomer units in the main chain).

Once sufficiently small size, oligomeric fragments are formed, they are transported into the cell where they are mineralised. At this stage the cell usually derives metabolic energy from the mineralisation process and the products, apart from ATP (adinosine tri-phosphate) are gases (e.g. CO_2, CH_4, N_2), water, salts and minerals and biomass. The complex products are biodegradable although their turnover rates are significantly reduced in comparison to the original polymer.

Mineralisation is defined as the conversion of the polymer into biomass minerals and salts, water and gases such as CO_2, CH_4 and N_2. Terms or data including fragmentation, disappearance, disintegration, deterioration, solubilisation, consumption, depolymerisation, weight loss, loss of mechanical properties and mineralisation are often used interchangeably to describe biodegradation.

Photo-Self Destruction

Photo oxidation and subsequent loss of durability of plastic formulation has been postulated as a mechanism for destruction of commercial plastics. Depending on temperature, normal plastic films have been estimated to take 3 months to 3 years to photo degrade. However, plastics disposed off in landfills are rarely exposed to sunlight for any appreciable period before additional garbage or soil is added for cover. Therefore, photo-oxidation, often does not play an important role in plastic disposal in garbage. Plastics that photo-self destruct are important especially in the agricultural field which plastic films serve a variety of specialised purposes, many intended to create controlled environments for growing crops.

These thin plastic films incorporate a photo-stabilising agent with a predetermined active period. After the predetermined period the photo-stabiliser is inactive and further exposure to sunlight causes photo-oxidation of the plastic polymers and loss in structural integrity. Photo-oxidation results in carbonyl end formation in the polymer chain which increases the further geological breakdown of the plastic film.

Advantage of biodegradable polymers over photo-degradable polymers

Although photo-degradation is a very viable process in many situations the materials are not exposed to light when discarded or re-used as coatings. In case of biodegradable polymers light is not required hence under proper conditions of temperature, moisture and with proper organisms the polymer is easily degraded. In case of biodegradable mechanism, the micro-organisms are ubiquitous in nature and are active with or without oxygen or light over a wide range of temperatures provided that other requirements such as moisture are met.

Key Elements to Biodegradation

There are essentially three key elements to the biodegradation process. If any one of the elements is lacking, then the entire process will be hampered. An example of the importance of this interplay is illustrated by the data collected during landfill investigations. Newspapers and food items are often cited as not being biodegradable because they are found virtually unchanged after burial for 40 years or more in a landfill. This is a misinterpretation of the findings, since these materials are clearly biodegradable, however, one of the three key elements described below is lacking.

Organisms

The basis for any biodegradation process is the existence of micro-organisms with the appropriate biochemical machinery to synthesise enzymes specific for the target polymer to initiate the depolymerisation process. These two steps in the central dogma of biodegradation may be carried out by the same organism or different organisms. In some instances, a polymer may be depolymerised or metabolised as a secondary effect. In a scenario such as this (co-metabolism), the organism is not gaining any metabolic benefit from the process and the process simply occurs due to non-specificity of a given enzyme formed by the organism for other metabolic processes.

Some of the additional issues to consider here are the types of organisms (e.g. aerobic, anaerobic, facultative), the nature of the enzymes in these organisms (e.g. hydrolytic, oxidative), the level of enzyme product (e.g. constitutive, inducible), the stability and location of the enzyme (e.g. intracellular, extra cellular, peri plasmic space), the kinetics of the process and the influence of inhibitors or inducers in the process. New opportunities for genetic level tailorability of the biodegradation process is a viable option just beginning to be explored. Once the metabolic processes involved in the biodegradations of a target polymer are elucidated, the genetic material responsible can be selectively modified (e.g. to enhance encoded protein, stability or substrate specificity) or cloned into alternative host organisms perhaps better suited for a given treatment scenario.

Environment

There are many factors that must be tuned in a given environment for the biodegradation process to function. Some of these factors would include temperature, moisture (liquid and vapour), salts (conc. and type), oxygen (aerobic to anaerobic), availability of alternative electron acceptors, trace metals, trace nutrients, pH, redox potential, environmental stability of flux, pressure, predators (e.g. protozoa), inhibitors, alternate carbon and light. All of these elements have to be present within the window of acceptability for the organisms producing the appropriate enzymes to degrade the target polymer. If any of these elements falls outside the required window, the process can stop, albeit temporarily, until proper conditions are re-established. Fortunately most organisms can tolerate relatively wide range of environmental conditions for many of the factors listed.

Substrate

We can consider the substrate from a series of levels of interactions or hierarchies involved in the biodegradation process. At the lowest level of hierarchy the polymer chain must contain chemical bonds susceptible to hydrolysis or oxidation by enzymes for the biodegradation process to initiate. At this level the degree and type of branching can automatically impact rates of chain, scission, particularly because additional enzymes may be required that are specific to the side chain chemistries or specific to the branch linkage.

In addition, the degree of hydrophobicity or hydrophilicity of the chains will impact directly on the rates of depolymerisation. Usually the rates of hydrolysis or oxidation will correlate positively with the degree of hydrophilicity. The stereochemistry of the monomer units in the polymer chain will also impact rates. Since an inherent property of many enzymes in their stereochemical selectivity, this property can be used to control or tailor the kinetics of depolymerisation. The molecular weight distribution of the polymer can have a dramatic effect on rates of depolymerisation as in the case of polyethylene. Studies with polyethylene samples ranging from molecular weight 4,800 to 41,000 supported bacterial growth, with the greatest growth occurring on the low molecular weight samples.

At the next level of complexity or hierarchy, the degree of crystallinity and other aspects of polymer morphology, such as domain size and distribution, need to be considered. Usually the higher the degree of crystallinity the lower the rate of hydrolysis since the crystalline domain, must be accessed by enzymes for the degradation process to start. At the next higher level of complexity, the interactions with other polymers (blends) and coatings applied to the polymer must be considered. These additional materials may act as barriers to prevent the migration of moisture into the polymer domains of interest, or may act as barriers to the permeation of gases such as oxygen, indirectly inhibiting the degradation process if the organisms responsible for the degradation require oxygen.

Potential factors impacting polymer degradation

Biological factors	Bacteria, fungi, predators, higher organisms
Chemical factors	Hydrolysis and oxidation
Physical phenomena	Leaching, sunlight, weathering, mechanical abrasion

There are two types of biodegradable polymers: (i) natural and; (ii) synthetic. Some natural biodegradable polymers are lignin, lipids, shellac, elsinon, collagen, poly (gamma-glutamic acid), natural rubber. Synthetic biodegradable polymers are polyamides, polyanhydrides, polyvinyl alcohol, poly-(3hydroxybutyrate-CO-3-hydroxyvalerate) and PHBV.

Processing of PHBV

Production technology

PHB is produced by the fermentation of glucose by *Acaligenes eutrophus.* In the first stage of the process *Acaligenes eutrophus* is inoculated into a balanced glucose salts medium in which the quantity of one nutrient is such that only a certain amount of cell growth occurs. Subsequent addition of further glucose leads to accumulation of upto 80 per cent by dry weight of polymer in the cell. Copolymers with PHV are produced by feeding a mixture of glucose and propionic acid during the polymer accumulation phase. By controlling the ratio of glucose to propionic acid it is possible to control the hydroxyvalerate content and hence to produce a range of copolymers from the same micro-organisms. Recently an aqueous extraction process has been developed which removes the cell components and leaves behind purified polymer in the form of white powder.

Physical properties

The physical properties of 'BIOPOL' copolymers vary with the 3-hydroxyvalerate content of the polymer. As the HV content increases in the range of 0-20 per cent, polymer flexibility and toughness increases and the melting point decreases. In fact the melting point reaches a minimum at about 30 mole per cent HV. The incorporation of HV into PHB allows flexibility, tensile strength and melting point to cover a range from, values typical of polypropylene (low HV) to those typical of polyethylene (high HV).

To show how these properties can be exploited in practice, consider the example of a bottle to be made from 'BIOPOL'. The bottle body must be flexible and resistant to cracking when dropped, whereas the cap needs to be more rigid to accommodate the thread and to form a good closure. This can be achieved by using a higher 3HV content copolymer as the principle component of the bottle body formulation whilst a lower 3HV copolymer is used to join the cap. The chemical resistance of PHBV to oils is very good and the polymer is moisture resistant and impermeable. However, it is susceptible to hydrolysis above pH 9 and below pH 3 and can be dissolved in number of chlorinated solvents such as chloroform and methylene chloride.

Biodegradation

PHBV has been shown to be completely biodegradable in both aerobic and anaerobic microbial active environments (aerobic and anaerobic sewage and compost, estuarine sediment, soil, river water and sea water). Apart from small amounts of biological material, the final products of biodegradation are carbon dioxide and water for aerobic condition and methane and some carbon dioxide for anaerobic conditions. The rate of degradation depends on such factors as moisture level, nutrient supply, temperature and pH. PHBV samples stored in humid air alone do not biodegrade.

Key Biodegradation Test Methodologies

It has become necessary to develop accelerated laboratory test procedures to evaluate potential biodegradable matérials which can simulate and be correlated to degradation in various natural environments. Before choosing an assay to simulate environmental effects in an accelerated manner, it is critical to consider the closeness of fit that the assay will provide between substrate, microbial agent (organism of enzyme) and the environment.

Enzyme assays

Enzyme assays are very useful in examining the kinetics of depolymerisation or oligomer or monomer release from a polymer chain, under different assay conditions. These assays do not determine mineralisation rates. Caution must be used in extrapolating this test as a screening tool for different polymers since the enzyme used in the assay is isolated based or exposure to only one polymer. If modified polymers or different polymers are to be screened, then new enzymes should be sought or isolated for the process since actually only the original polymer substrate enzyme pairing is optimised. Quantitative aspects of the assay will depend on the polymer being screened.

Often changes in pH, or changes in chemical constituents can be monitored to assess kinetics. Caution must also be used if the enzyme preparations are not purified or appropriately stabilised or stored, since inhibitors and loss of enzymatic activity can occur.

Plate Tests

Plate tests are used to screen micro-organisms against a given substrate to see if the polymeric material will support growth. A positive result means that an organism can grow on the substrate. A positive result does not mean that the polymer is biodegradable since growth may be on contaminants, a plasticiser present, portions of the polymer that perhaps were not representative of the substitution pattern on the rest of the polymer chain, growth only on oligomeric factions present in the polymer and other reasons. If the plate test is coupled with further analytical assays (e.g. gel permeation chromatography, mechanical performance) the additional conclusion can be drawn.

Method	*Advantages*	*Disadvantages*
Enzyme assays	Specific enzyme-polymer pairing	Limited use for modified polymers
	Rapid (minutes to hours)	Inhibitors
	Quantitative	Enzyme stability, co-factor requirement
Plate tests	Surface growth	Growth or additives
	Relatively rapid (days to weeks)	Qualitative
	Films, powders, objects can be tested	Inhibitors
BOD.	Rapid (hours to days)	Non-specific, indirect measure
Radiolabelled polymers	Sensitive	Needs specialised equipment
	Quantitative	Relatively small
	Axenic or mixed cultures	Difficulty in making labelled polymers

(Contd ...)

Method	Advantages	Disadvantages
	Relatively rapid (days to weeks) Water soluble or insoluble polymers can be tested	
Automated respirometry	Sensitive Relatively rapid (days to weeks) Water soluble Natural microbial populations can be used	Specialised equipment Water-soluble polymers can be tested Powders cannot be tested
Field exposures	Natural conditions	Slow, variable conditions

Biological Oxygen Demand (BOD)

Biological oxygen demand provides an indirect measure of the relative oxygen requirements of a sample measuring the oxygen requirements for the biochemical degradation of the organic material. Samples are incubated under specific conditions for specific time frames with dissolved oxygen, measurements taken before and after incubation.

Radioactively labelled polymers

Assays with radioactively labelled polymers are the most definitive approach to the determination of mineralisation rates of polymers. Collection of radioactively labelled gases or low molecular weight volatile products can provide extremely sensitive and reproducible methods to assess the mineralisation process.

Respirometry

It provides a practical, sensitive, reproducible approach to the assessment of mineralisation. The system can be tailored to a specific environmental scenario (e.g., soil, fresh water, marine water, sewage, landfill, compost) and the system can be run for extended periods of time (days, weeks, months). There are chambers in which the specific environment is created. The atmosphere can also be exchanged which avoids build-up of inhibitory gases in the head space. Water soluble or insoluble films, fibres, powders, gels or other shaped objects can be evaluated in these systems. These techniques, when combined with organic carbon analysis, can be used to determine carbon mass balances so that precise biodegradability measurements can be made.

Problems. (i) the technique is a batch run process and build-up of metabolic products in the media can result in inhibition of activity; and (ii) the assessment of methane release is cumbersome since the system is not automated for this gas and samples must be run independently by gas chromatography.

Laboratory scale simulated accelerated environments

Simulated laboratory-scale accelerated systems provide useful series of environmental scenarios (e.g. sewage, marine water, landfills, compost, soil) to screen polymer films and materials. These would be similar to environmental microsomes but with accelerated conditions.

Advantages of these systems are their similarity to natural environments, the control of environmental variables, and the ability to accelerate biodegradation process. A drawback to the system is that only water insoluble polymeric films or, objects can be studied. These systems do not provide direct measures of biodegradation, but provide the standardised environmental exposure conditions for the polymer.

After exposure, the samples can be removed and characterised for the impact of the biodegradation process which follow-on characterisation such as weight changes, morphological changes, changes in molecular weight (e.g., gel permeation chromatography, viscometry, light scattering), changes in mechanical properties (e.g., tensile strength, elongation per cent and visual appearance).

Natural environments

Exposures in natural environments provide the best measure of the environmental fate of a polymer because these tests include a diversity of organisms and achieve a desirable natural closeness of fit between the substrate, microbial agent and the environment. However, the results of that exposure are only relevant to the specific environment studied and it is known that every environment differs in some way from every other environment. Therefore, the lack of control of variables in the natural systems lead to difficulty in extrapolating the results from the specific environmental extremes are studied (e.g., tropical, arctic, desert) to look at the rates for given material. Another problem, is the time scale for this method since the degradation process, depending on the environment may be very slow. This time frame may extend to months, years and decades, thus making assessments of the biodegradability of new materials unrealistic. With these length of exposure it becomes virtually impossible to distinguish biological vs. non-biological effects since sterile controls cannot be run.

Biodegradation Testing Summary

A combination of the above methods is needed to adequately assess the environmental rate of a polymer according to the data required. For example for many of our interests, we need to validate that a target polymer can be: (i) mineralised; and (ii) will lose mechanical integrity (e.g., tensile strength) within a certain period of time. As this data is collected for a given polymer under a variety of different conditions, then extrapolations can be made to additional environments. With this data set available, informed decisions can be made concerning the acceptability of a specific polymer for a specific application and disposal route.

Biodegradable polymers are characterised by their ability to be broken down directly or indirectly by the enzymatic action of micro-organisms, eventually to gaseous carbon-containing products such as carbon dioxide or methane. The lack of standardised technical test procedures to measure biodegradability constitutes to ambiguities and disagreements within the technical literature regarding the degradability of synthetic and natural plastics. For example, criteria such as loss of tensile strength or flexibility may be good indicators of disintegration but not necessarily of biodegradation of mineralisation.

Toxicity testing

"PVC items contain toxic substances, warns green peace campaigner". To protect environment, it has become necessary to know not only how materials degrade but what becomes of them (environmental fate). 'The softeners and plasticisers like phthalates added to PVC were loosely bound to the actual PVC and thus are harmful to humans'. Therefore biodegradable polymers also have to go through these tests so that it can be proved that the by-products or additives released from biodegradables are also non-toxic.

Food contact

For biopolymers that are intended for consumption or food contact, FDA toxicity tests must be performed. Materials that may be designed for environmental degradation upon disposal, such as some shipboard plastics, agricultural mulch films, or compostable items are regulated by the environmental protection agency. In addition to food contact biopolymers, the testing of toxicity is also carried out on marine and soil systems and their effects on the environment are studied.

Biodegradation Kinetics in Marine and Soil Systems

Recently developed biodegradability tests examine the weight loss per surface area of biopolymer films in simulated marine and soil environments. Weight loss per surface area data provide better kinetics for hydrophobic bipolymer films because it normalises differences in film thickness which can change relative rates of biodegradation.

Advantage

Ability to evaluate soluble polymers or powders and the elimination of variation due to loss of sample or accumulation of biomass on samples when weight loss is used the biodegradability measurement criterion. During the course of performing these tests, the criteria including weight loss, loss of tensile properties, changes in morphology depolymerisation and mineralisation will enable users to choose the criteria that best fit their application and disposal requirements in a particular environment.

Materials and methods

The marine biodegradation apparatus consists of 76 litre aquaria, aquaria lights, pumps and is identical for both.

1. Accelerated simulated natural.
2. Accelerated synthetic marine laboratory scale testing system.

The testing is done at 30°C, with 12 hours light and dark cycles, and with standardised salt concentrations in the aquatic water. The water is exchanged at the rate of 15 per cent volume per week to minimise potential build-up of toxic biodegradation by-products in the systems. The marine systems are aerated at a rate of 1 litreper minute using standard aquarium air pumps.

The major differences between the natural and synthetic systems are the sediment materials used and the inoculation method. In natural marine systems marine sediments are collected from a local beach whereas in synthetic marine systems the sediment is marine agar enriched with commercial sand. The synthetic sediment and water are inoculated with lyophilised cultures of 11 marine organisms isolated from past experiments using sea water of sediment collected from temperature and tropical region of Atlantic and Pacific oceans. After exposure in different systems control the test samples are evaluated for changes in tensile strength, ultimate elongation, molecular weight and gravimetric weight loss.

Sample preparation

Sample films are cut from commercial or prepared film prior to exposure in accelerated marine and soil testing systems. The polymer films studied include cellophane uncoated, bacterial polyhydroxy butyrate/ polyhydroxy valerate (PHBV), polycaprolactone and polypropylene as a control. Some organisms identified are *Altermonas haloplankis, Vibrio proteolytics* and *Pseudomona creostinsis*.

Results

The data collected from the different test methods provide a good indication of preliminary biodegradation (defined as weight loss per surface area) of polymer (Fig. 23.1 and 23.2) samples evaluated. In the marine and soil accelerated systems the weight loss per surface area data provided a reasonable indication of biodegradation rate kinetics where surface hydrolysis is the mechanism of degradation. Representative data for a marine simulate natural system, water and sediment phases are found in the figures. PHBV appears to lose weight at a faster rate than the other polymers. Polypropylene serves as the negative control and did not lose weight in any test systems. The data also shows that some of the polymers do not last the entire test cycle before they fragment and cannot be recovered.

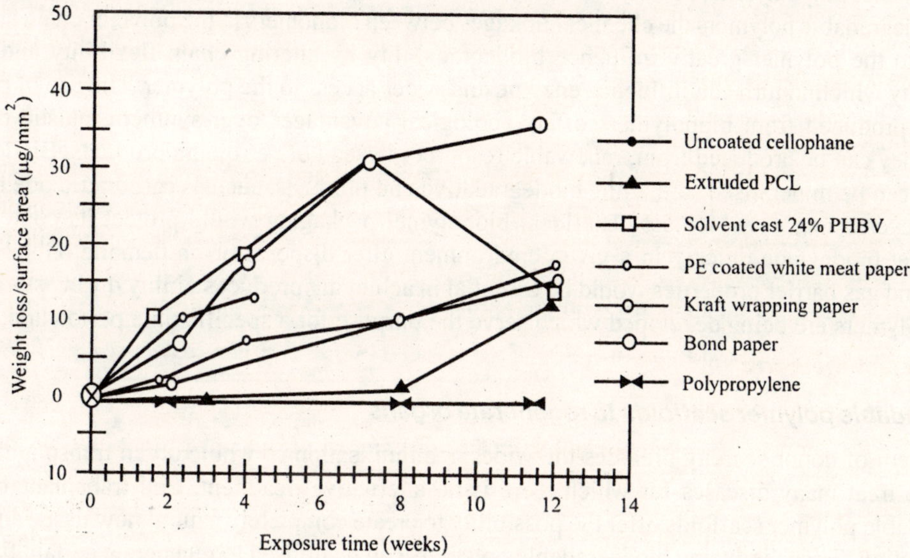

Fig. 23.1. Weight loss per surface area data for a variety of polymeric materials in the sediment phase of the stimulated natural marine system.

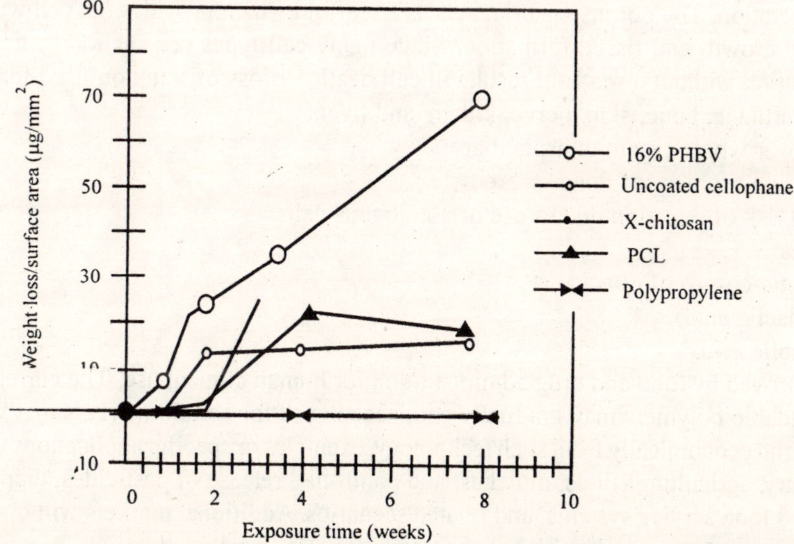

Fig. 23.2. Weight loss per surface area data for a variety of polymeric materials in the simulated soil system.

Applications of Biodegradable Polymers

(i) in packaging; and (ii) in the human body as scaffolds to regenerate organs.

Application as packaging materials

Most current packaging materials are composed of non-biodegradable synthetic polymers. These include polypropylene, polyethylene, polystyrene, polyvinyl chloride, polyamides and aromatic polyesters.

In biodegradable polymers the chemical linkages between monomers in the polymer chain and groups attached to the polymer greatly influence biodegradability by altering chain flexibility and polymer crystallinity which in turn can influence enzyme and water access to the polymer. The use of packaging materials produced from biopolymers offers ecological advantages over synthetic plastic packaging because they can be produced from, renewable resources instead of petroleum products. Biodegradable polymers can be mineralised so that the biodegradative end products, such as carbon dioxide and water can be recycled into new biomass. The ideal biopolymer packaging would protect its content during storage, yet biodegrade quickly in a given environment after disposal or on demand. Good moisture stability and gas barrier properties would be essential in achieving product stability during storage. Thus special polymers are being developed which serve the purpose for a specific time period and then start degrading.

Biodegradable polymer scaffolds to regenerate organs

The problem of donor scarcity preludes the widespread utilisation of whole organ transplantation as a therapy to treat many diseases for which there is no alternative treatment. Cell transplantation using biodegradable polymer scaffolds offer the possibility to create completely natural new tissue and replace organ function. Tissue inducing biodegradable polymers can be utilised to regenerate certain tissues and without the need for *in vitro* cell culture. Biocompatible biodegradable polymers play an important role in organ, regeneration as temporary substrates to transplanted cells which allow all attachments, growth and retention of differentiated function. The scaffold releases chemotactic agents which attract cells to the area of regeneration. The porous material acts as a scaffold which provides the cell with a substrate and organises all growth and tissue formation. Since many cell types are anchorage-dependent their direct transplantation without a scaffold results in cell death, or loss of function used for regeneration tissues such as cartilage, bone, skin, nerve, kidney and liver.

Advantages over whole organ, transplant are:

1. Overcoming problem of donar scarcity.
2. Reduced risk of rejection due to use of autologous cells.

Polymers uses:

1. Poly (lactic-co-glycolic) acid.
2. Poly (L-lactic acid).
3. Polyglycolic acid.

These are approved by food and drug administration for human clinical use. The current projections are that biodegradable polymers may gradually gain acceptance for certain markets that would benefit environmentally and economically from such technology, examples or specific applications would include the fishing industry, agricultural films, mulches, and controlled release of pesticides, diapers, fast-food service, shipboard food service systems, and related scenarios. Additional markets will open even if the cost of these polymers is somewhat higher than traditional polymeric materials because the public demand is higher. The biomedical applications for these polymers are extensive, including controlled release, implant, and wound healing. A number of key issues must be resolved for biodegradable polymers to gain more widespread acceptance. The water stability issue for many of the polymers continues to pose problems and has to be adequately addressed. The costs for some of the polymer remain too high, for general use. The disposal systems needed to handle these types of materials, (e.g., compost, sewage, new types of accelerated depolymerisation systems) must be brought into the loop to close the cycle from raw material to mineralisation.

INDIGENOUS DEVELOPMENTS IN BIODEGRADABLE POLYMERS

As the debate grows over post-consumer plastics in municipal solid waste, litter, and marine pollution, there has been an inevitable focus of attention on making polymers degradable. Since the early eighties a number of developments have taken place in the area on the presumption that these plastics could replace their synthetic counterparts in a significant number of applications. While many of the forecasts have been overly optimistic—with some saying they will capture half the plastics market and others suggesting even 5–10 per cent more—there is no denying that biodegradable plastics have a niche market, which in many of the advanced countries is seeing above average growth.

Much of the developments in this area have taken place in the developed countries, in particular the triad of North America, W. Europe and Japan, where most of the markets for these products are to be found. Hence it is heartening to note that an Indian laboratory, the Central Tuber Crops Research Institute has now made a significant contribution in the field. The laboratory, a part of the Council for Scientific and Industrial Research (CSIR), has developed the technology for a starch-based biodegradable plastic, which it claims will degrade within 2–6 months in the soil, depending on the conditions.

The formulation contains up to 40 per cent of starch—which contributes to the biodegradability—with a host of other additives such as plasticisers, coupling agents etc. The concept is hardly new, and it has been known for some time that synthetic polymers, filled, grafted or blended with starch, either in the native or modified form, have the potential to degrade rapidly. Indeed the first products in this category were introduced as far back as the late eighties in the US.

Applications envisaged for the starch-filled plastics include disposal carry bags, agricultural mulches, and controlled release formulations of drugs and agrochemicals, besides medical implants. The polymer can be blown into films, with adequate strength, and could offer an alternate packaging option.

An agreement has been signed to transfer the technology to the National Research and Development Corporation (NRDC), which has announced plans to license it on a non-exclusive basis to private entrepreneurs, marking a first for the indigenous availability of the material.

Cost is a factor that has so far restricted the use of these materials, and it is expected that the biodegradable plastic available locally will be at least 20–30 per cent more expensive than the commodity thermoplastics. This will restrict use to small niche markets where price is not a deciding factor.

The promotion of biodegradability of thermoplastics by the addition of starch is but one approach that has been taken in order to enhance the biodegradability of plastics. Truly natural materials—biopolymers–based on natural raw materials have been synthesised. These polymers of starch, glucose, lactic acid, and hydroxyalkanoic acids are not only increasing the range of opportunities, but as the feedstock used to make them are derived from biomass and other agricultural products, they represent an opportunity to create new markets for agri-business companies.

Since its discovery in France 60 years ago, polyhydroxybutyrate (PHB) was little more than a laboratory curiosity, till Marlborough Biopolymers, a subsidiary of UK's, ICI, began efforts at commercialisation. Curiously the company signed a letter of intent with a major chemical company in India to produce the polymer (a linear polyester of 3-hydroxybutyric acid and hydrovaleric acid), trademarked Biopol, in India for general use in the country and elsewhere. While this was not to be, ICI took a big step in May 1990 with the announcement of a first commercial application for this polymer–a shampoo bottle that went on the market in Germany–and set up the first plant at Billingham (UK).

Biopol polymers gained approval for sale in Japan in July 1991, and in 1992 entered the US market once again as packaging container for shampoos. Since then a number of other applications have been developed, but so far its has not been possible to improve the productivity of bacteria sufficiently to make Biopol economic for bulk usage in competition with polyethylene and polypropylene. In early 1996, Monsanto took over the business from Zeneca, and is probably the only chemical company to have one of its products—a credit card made from Biopol—endorsed by Greenpeace.

Recently, Dow Chemical and Cargill formed a joint venture for the development of polymers of polylactic acid (PLA) (branded EcoPLA), which can be used to produce film, fibres, non-wovens, rigid containers and coatings for paper and cardboard. The aim is to develop PLA grades for new applications, and polymers modified with other co-monomers that can have characteristics similar to or better than those of other polymers like polyolefins, PET and PS. High-molecular weight polymers can now be produced via low molecular weight polymers. A semi-commercial plant will be doubled in size to 16 million lbs/year, and a 125,000 MT/year plant is planned by 2001. Both companies forecast a very large market for PLA plastics (over 1 billion lbs/year).

Other companies involved in the business include Mitsui Toatsu producing PLAs at a semi-commercial plant in Japan, Chronopol which produces 2 million lbs/year in the US, Shimadzu and Neste. The number of companies with research interest in this area has grown significantly in the last few years, and the first international trade organisation for this sector, the International Biodegradable Product Manufacturers Association, was established in 1996.

An increase in oil prices often gives biodegradable plastic development a boost, and with oil prices expected to stay low in the foreseeable future, this might not be the best of times for these materials. At about ten times the cost of existing plastics, biodegradable plastics cannot compete on cost, and this is expected to remain the case for a long time to come.

DEGRADABLE POLYMERS

Increasing concern has been expressed by Environmentalists and Public Health Authorities about the escalation of plastics litter on land, in sea and other water bodies. With it has come the need to find more effective ways of dealing with plastics waste and litter. Degradable plastics are of utmost importance in light of this common concern.

Essential requirements are that the degradation process should be controllable and that it should lead to lower molar mass oxidation products which can be assimilated by micro-organisms to give biomass and ultimately carbon dioxide and water.

Since last decade, there has been considerable debate and controversy on the definition of 'biodegradable' plastics. It is pertinent to recall the International Seminars and Workshops on Biodegradable Plastics held at Toronto 1989, Tokyo 1990, Montpellier 1991, Osaka 1993 and elsewhere. While the concept of biodegradability is broadly accepted, most of the debate has focused on reducing the broad concept with all its ramifications to a narrow statement, which tends to be accepted by very few.

The issues pertaining to degradability in aquatic and terrestrial environments are sufficiently complex that a structured approach to define degradability has been found useful; by first defining the concept of degradation, followed by degradable plastic, the type of degradable plastic by the process which causes the degradation, as biodegradable, photodegradable, chemically degradable, and hydrolytically degradable plastic may then be qualified.

From their own debate and wisdom from international forums on biodegradable plastic ASTM has adopted the following definitions:

1. Degradation is defined as a deleterious change in chemical structure, physical properties, or appearance of a plastic.
2. Degradable plastic is defined as a plastic which is designed to undergo a significant change in its chemical structure under specific environmental conditions resulting in a loss of some properties that may vary as measured by standard test methods appropriate to the plastic.
3. Biodegradable plastic is defined as a degradable plastic in which the degradation results from the action of naturally occurring micro-organisms such as bacteria, fungi and algae.
4. Photodegradable plastic is a degradable plastic in which the degradation results from the action of natural sunlight.

Other definitions are also used by standardisation bodies and regulatory agencies of different countries. In another commonly used definition, 'degradable polymers are materials which undergo changes as a result of chemical, physical and biological interactions between active living environment and the material'. However, while variation may exist in definitional phraseology, little disagreement exists on the concept of biodegradable plastic.

It has also been recognised that degradable polymers may be required to degrade in very different environments. For example, polymers that end up in a sewage system are not exposed to sunlight, but are expected to degrade rapidly in an aqueous environment. For this application, hydrolytically unstable polymers offer the best solutions, provided this is consistent with the ability to fulfil their required function without deterioration during service.

At the other extreme are polymers used in agriculture or discarded as packaging litter which end up in the outdoor environment in contact with the soil. Here, hydrolytic biodegradability is a positive disadvantage since it is likely to occur at too early a stage during its useful life. Initial controlled photo-oxidation can provide the required useful life followed by microbial bioassimilation after disintegration.

Between the above extremes are polymers which are required to biodegrade in a controlled aerobic compost. This is quite a different environment from sanitary landfill. It has been observed that neither paper biodegradable plastics will biodegrade in a properly prepared landfill. Furthermore, it is now considered important that waste products should not biodegrade in anaerobic landfill as the subsequent methane evolution is associated with explosion hazard.

All organic polymers are in principle degradable, differing only in the degradation mode. In practice, nearly all bulk polymers are inert materials either by themselves or by addition of compounds rendering them insensitive towards ageing. Growing environmental issues have created a demand for new materials with low environmental burden.

The creation of polymeric materials having strength and functionality while in use, but will degrade after use is a novel concept in material science. The concept of degradability has initiated the design of new plastic products. At the same time, an additional demand is widely proposed that the polymeric material should come from renewable resources instead of from petrochemical sources. A third requirement is to save materials by improved product design and by recycling. Recent trends in polymer research point to an increasing demand for the development of a range of biodegradable polymer products in mainly four fields: packaging materials; mulching films in agriculture; disposable items totally degradable in litter and waste; and in medical application; *in vivo* in human or animal body.

Several interesting solutions have been proposed to make filled polyethylene to degrade faster than previously observed. Incorporation of a chromophoric group, which increases the susceptibility of polyethylene towards photo-oxidation, is one choice.

Metal salts are also useful and potent additives, and they give room for radical formation within the polymer chain and again increase the photo-oxidative behaviour of these materials. It is now widely accepted that the conservation of resources is important, and this has led to the use of renewable materials. The knowledge that the biopolymers usually degrade quickly by microbiological action has led to the use of agricultural products as additives for synthetic polymer with the intention of producing biodegradable polyethylene.

Starch is an abundant, inexpensive, natural raw material which permits the development of products recyclable to atmospheric CO_2 when biodegraded or incinerated. These characteristics have excited a renewed interest in starch-based plastics in recent years. Products are available based on thermoplastic starch in combination with polymers or copolymers of vinyl alcohol and with aliphatic polyesters and copolyesters. Their biodegradation kinetics, mechanical properties and reduced sensitivity to humidity make these materials suitable for industrial development beginning with film and foam applications. Thermoplastic starch alone can be moulded as traditional thermoplastic materials; its sensitivity to humidity, however, makes it unsuitable for most applications. Starch-filled degradable polyolefins show good disintegration properties. There is no evidence, however, of a significant biodegradation of the fragments produced by disintegration. An overall 'ideal' degradable plastic, biologically or otherwise, is not within reach in near future, it seems.

ENVIRONMENTAL IMPLICATIONS OF POLYMER DEGRADATION

The three major trends of twenty first century are competition, collaboration and conservation. The conservation of natural resources, the environment and human resources will become increasingly important ingredient of successful strategies for growth of companies and nations. Due to proliferation of polymers and composites in all spheres of industrial and human activity domains in end of twentieth century and beginning of twenty first century, the environmental issues also concern the reclaim of polymers and composites. All organic polymers degrade differing only in degradation rate and mode. All bulk polymers are inert materials either by themselves or by the addition of compounds rendering them insensitive towards ageing. The environmental issues demand for new materials which continue to have strength and functionality while in service but which degrade after use. In the first category of polymers, post-use disposition is also under environmental scrutiny. The degradation can be performed by physical, chemical, mechanical and biological agents. The environmental implications of recycling and degradation need to be discussed and debated. Here an attempt has been made to examine various environmental concerns regarding the recycling and degradation of polymers and also on biodegradation of polysaccharides and degradation of polymers in turbulent flows.

Polymers are endowed with low weight, corrosion resistance, cost effectiveness, capability of fabrication in integrated complex shapes, electrical and thermal insulation (exception being in case of conducting and conductive polymers), energy efficiency one step fast processing, moulds and fungus resistance impermeability to gases and aesthetic appeal etc. Their use has therefore proliferated in every walk of our life, clothing, automotive parts, boat hulls, tennis rackets, aircraft frames, telephones, computers, human body prosthesis and numerous other materials goods. The polymers are manufactured from petroleum feedstock by process that are environmentally benign. The metal refining and glass making require large amounts of energy and pollute environment. Energy saving favour plastics when energy requirements for production, transport and recycling are considered in comparison with glass and metals. Furthermore when polymers are substituted in aircraft and automobiles, weight reduction results that translate to fuel efficiencies over the life of vehicle. Polymeric based synthetic fibres are better than natural cotton and wool fibres which make heavy demand on land use, fertiliser (also from petroleum) and fuel for transportation. Ecofriendliness of polymers can be demonstrated by many other examples. However a challenge to the use of all materials is posed by their disposal after use. Two approaches have been adopted. One concerns with waste disposal after use in four fundamental ways. It can be burned, buried or reused or less of material is used in the first place. They are also categorised as primary, secondary, tertiary and quaternary recycling. The second approach is to develop material which degrade after use. The degradable polymer so developed are of following category :

1. Pho-oxidisable degradable polymers.
2. Directly biodegradable polymers.
3. Autooxidisable polymers which then biodegrade.
4. Water soluble polymers.
5. New form of natural polymers.

In both approaches, environmental issues are involved. In various laboratories, materials science and engineering approach has been adopted in developing natural polymers and grafted natural polymers for their applications in turbulent drag reduction and flocculation. The approach has been entirely environmental friendly. This section describes first polymer degradation then their environmental implications in details along with environmental friendly approach adopted in development of drag reducers and flocculating agents for industrial effluent treatment.

Polymer Degradation

The degradation of polymers is possible by physical, chemical, mechanical and biological means. The chemical and physical degradation of polymers implies that micro organism, macro organisms or enzyme are not present and ageing is dependent on physical, chemical and or mechanical influences. The degradation refers to any chemical change in polymer structure. The biodegradation involves living organism or their secretions and can be defined as a molecular phenomenon in which biological systems mediate at least one step of the overall transformation of the molecule into gases products and biomass. Polymers may be degraded as a result of exposure to various agents such as heat, light, atmosphere, hydrolysis and biodegradation.

The degradation may take place by exposure to one agent or combination of them such as photo-degradation can be defined as breakdown of polymeric molecules initiated by sunlight. The photo-oxidative degradation involves sunlight and oxygen. Photooxidation decreases the life time of most materials used in outdoor applications. For biomedical materials, the body fluids and compound dissolved in these fluids decrease the lifetime of the polymers. For clinical applications, it is often to distinguish clearly between 'biodegradable polymers' and 'bioabsorbable polymers'. Biodegradable polymers are polymers which are decomposed in the living body but whose degradation products remain in tissues long-term. On the other hand bioabsorbable can be defined as polymers which degrade after implantation into non-toxic product which are then eliminated from body or metabolised therein.

The bioabsorption of polymeric materials can be divided into two distinct stages : decomposition and absorption. The former stage involves bond breaking in the main chain, a decrease in molecular weight and production of oligomer and monomers. The main reaction for this process is hydrolysis.

Polymers that are decomposed by enzyme-specific reactions are called enzymatically degradable polymers while polymers which are decomposed by contact with water or serum are non-enzymaticaly degradable polymers. Hydrolysis affects a number of synthetic polymers as well as such as polyesters, polyanhydrides, polyamides (proteins) and polyurethanes.

The degradation can be artificial using phosphate buffers simulating body fluids or medicated by different enzyme (hydrolases). The biodegradation attack by living organism can take place on biopolymers as well as synthetic polymers having hydrolysable backbone chains like aliphatic polyesters.

Factors affecting degradability

The degradation times of polymers vary considerably from a couple of days upto several years depending upon the type of bond in the repeat unit and the environment. It also depends on several other factors.

1. Amorphous regions are more susceptible to hydrolysis reactions and microbial attacks than crystalline.
2. Hydrophobic polymer is often less degradable than a hydrophilic one. Hydrophobicity substantially decreases the hydrolyse rate and microbial attack.
3. The relationship of surface/volume dictates the degradation rate-small polymer granule degrade faster than the corresponding more compact film.
4. Linear polymers are generally more subject to degradation than are the branched polymers in biodegradation test.
5. The most important factor that influences controlled degradation is degradation in presence of oxygen in some form.
6. Hydroxyl, hydroxyperoxide and carbonyl groups formed during photodegradation may lead to increased hydrophilicity which leads to biodegradation. Ester and anhydride groups containing polymers that represent a substantial portion of macromolecular unit result in highly oxygen containing units and are easily hydrolysed.
7. Polymers that are hydrolytically unstable and water permeable give structures that we subject to rapid degradation e.g. aliphatic polyesters. Some polyamides and polyurethanes might also be hydrolytically unstable but they usually have a lower water penetration and permeability and subsequent degradation is slow.
8. Some polymers are highly resistant to hydrolytic degradation. They include polyacrylates and silicones. Both are essentially inert to hydrolysis but at the same time water or liquid can penetrate them and cause some swelling or they might be subject to cracking in cycling experiment.

BIODEGRADABLE POLYMERS

The use of biodegradable polymers in single use disposable packaging in concert with composting infrastructure is an ecologically sound approach to management of the resultant waste and its diversion from land fill. Composting biodegradable materials alongwith other organic waste represent biologically recycling of the polymeric carbon by way of carbon cycle (Fig. 23.3).

Different approaches have been followed to make classical polymers such as PE and PP (bio) degradable by physical incorporation of starch as biodegradable element. Three approaches have been followed (Table 23.1). The granular starch has been used as filler in polyolefins. Gelatinised starch plastics composites and grafted starch itself have been proposed plastics materials. In the first case, the incorporation of starch is limited to 6–12 per cent while in the second case it may be upto 40 per cent. The phase transfer agents along with other ingredients are needed to produce the films. The graft copolymers of starch with PMMA and PS have shown potential for production of extrusion shaped products and films. However, the mechanical properties of these materials are not upto the mark. The constituent plastics biodegradation is also slow. The current trend is to use 100 per cent biodegradable materials. Polymers such as poly(caprolactone), poly(lactic acid) and poly(hydroxy butyrate-CO-hydroxy valerate) have been introduced into the market place.

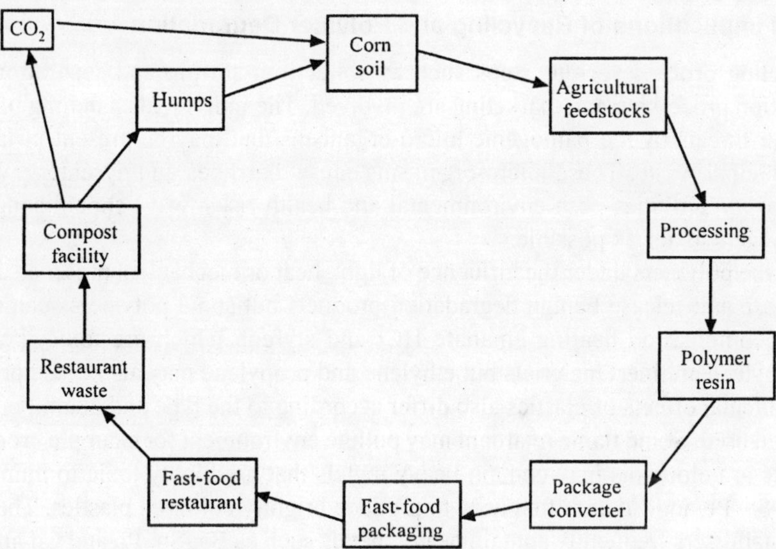

Fig. 23.3. Cradle-to-grave concept for biodegradable materials in fast-food restaurant packaging applications.

Table 23.1. Examples of degradable polymers.

Polymer	*Degradation*	*Type of product*
Starch based	Biodegradation diaperbacking	Loose-fill, plates, consumer and medical products,
Starch-filled	Biodegradation (auto-oxidation)	Mulch films, diaper backing bottles, bags and stretch films
Natural polyesters	Biodegradation, hydrolysis	Shampoo and motor oil bottles, disposable razors
(PHB/PHB/HV)	Hydrolysis, biodegradation	Medical products
Polyactic acid		
Polycaprolactone (TONE polymers)	Biodegradation	Biodegradable films
Polyvinyl alcohol	Hydrolysis	Laundry bags, packaging for agricultural chemicals and dyes and pigments

The last one has mechanical properties comparable to those of polypropylene. However, these materials are costly. Hence these materials can be blended with starch to reduce their cost and enhance their flexibility. It has been shown that although the addition of 0–50 per cent starch granules to P(HB-CO-19 per cent HV) results in an increased biodegradation rate, it also results in reduced tensile strength and flexibility because of formation of composite rather than blend. This lacuna can be partly removed by using thermoplastic starch in place of granular starch. Mixing of thermoplastic starch with poly(hydroxy butyrate-CO-12 per cent hydroxy valerate) resulted in blends. Nevertheless, the blends are still brittle despite the presence of flexible starch. Improved properties of the incompatible blend can also be obtained by optimising the blend morphology during processing. There is considerable scope to enhance the compatibility of these blends by incorporating grafted starch in the blends. Such blends will have less brittleness and higher mechanical strength.

Environmental Implications of Recycling and Polymer Degradation

1. In recycling process various steps such as collection, sorting and separation, washing and purification processing and marketing are involved. The individuals handling urban solid waste are under threats by the pathogenic micro-organisms that may be present in faceal materialin raw municipal waste. These micro-organisms cause diarrhoea and dysentery. Waste collection and recovery activities pose environmental and health risks, ways should sought to minimise these dangers as far as possible.
2. Most of the polymers under the influence of light, heat or mechanical presence during recycling decompose and release benign degradation products but some polymers such as PVC and PS when decompose on heating emanate HCl and styrene which are toxic. Polyethylene and polypropylene are inert materials but ethylene and propylene may also cause problems.
3. Environmental effects of plastics also differ according to the type and quantity of additives that have been used. Some flame retardant may pollute environment for example bromine emissions. Pigments or colourants may contain heavy metals that are highly toxic to humans such as Cr, Cu, Co, Se, Pb and Cd are often used to produce brightly coloured plastics. The additives used as heat stabilisers frequently contain heavy metals such as Ba, Sn, Pb and Cd and sometimes in combination. These metals are serious environmental pollutants and have deleterious effect on human health. They also interfere with processes in plant, animal tissues and in the soil when present at or above some specific concentrations.
4. Taking into consideration the process of plastic recycling, the most important problem caused by the polluting substances is air pollution either within the reprocessing units or in the open air. During extrusion process several substances such as additives may be released. In polymers like PVC along with additives, vinyl chloride and HCl are also emanated.
5. During incineration unless the combustion is complete burning plastics within the waste may release considerable quantities of polluting substances. With complete combustion 90 per cent of plastic materials is converted to CO_2 and H_2O. Incomplete combustion of PE, PP, PS and PVC can cause further problems as CO and smoke may be produced. PVC when burnt emanates HCl. During waste incineration chlorodioxins are formed which are highly toxic.
6. During the recycling, combustion of plastics should be avoided at all cost. In reprocessing machines, PVC may decompose and give off fumes of HCl which may damage the machine and discolour and spoil finished product. The processing temperatures of PVC range between 150-210°C. Heat decomposition may occur at temperatures as low as 180°C. These difficulties can be overcome by keeping the moulding period short and adding a heat stabiliser.

Precautions during recycling

It is, therefore, advisable that following precautions should be undertaken during recycling.

1. All waste should be thoroughly cleaned and sorted before it is reprocessed. This will help to minimise odours and will improve the quality of the end product.

2. Do not allow plastics to burn; the gases released may be toxic and pose a health risk to the workers.
3. Avoid the use of PVC for recycling.
4. Some recycled plastics should not be used in products that will come in contact with food. Waste plastics from unknown sources should not be used for toys, kitchen utensils and drinking water pipes.
5. Plastic waste from hospitals must be approached with caution as they may contain dangerous or unhygienic materials.

Biodegradation of polysaccharides

A large number of polysaccharides such as guar gum, xanthan gum, carboxymethyl cellulose, starch, amylose, and amylopectin have been investigated for their effectiveness in turbulent drag reduction, flocculation and shear stability. The solutions of these polysaccharides biodegrade. However, when grafted with polyacrylamide branches, these graft copolymers become highly efficient drag reducers and flocculants. They have robustness of rigid polysaccharide chains and efficiency of flexible polyacrylamide in causing drag reduction. In case of flocculant, the dangling polyacrylamide chains on rigid polysaccharide chains, increase the approachability to contaminants hence they are very effective flocculants.

When polysaccharides are grafted with polyacrylamide chain, their regularity is diminished.This decreases their biodegradability which is observed in all cases. The branched polysaccharides are less susceptible to biodegradation. Grafted and ungrafted amylopectin are clear examples of this. In case of drag reducers and flocculants, the grafting reduces biodegradation and thus increases their self-life which is a desirable attribute for their applications in these phenomena.

BIODEGRADABLE POLYMERS IN POLYMER WASTE MANAGEMENT

Polymeric waste management requires sound complementary practices of conservation, recycling, incineration and biodegradation-bioconversion. Since biodegradation is potentially the most environmentally friendly of all these practices, there is increasing activity in the area of development of biodegradable polymers as packaging materials.

Among the most studied polymers are bio-polymers, modified bio-polymers and blends containing bio-polymers. Many microbial systems, especially bacteria, use poly-R-3- hydroxy alkanoates as energy storage materials. Data to date shows that –3- hydroxyalkanoates are hydrolysed slowly and biodegradation proceeds rapidly in the presence of suitable micro-organisms. The knowledge that bio-polymers usually degrade quickly by microbiological action has led to use of agricultural products as additives in synthetic polymers with the intention to produce biodegradable polymers. Development of polymers and blends that exhibit high susceptibility to microbial attack is a good approach towards relieving environmental concerns caused by the current use of non-biodegradable polymers.

Current Approaches towards Biodegradability and Biodegradable Polymers

Biodegradability corresponds to the ability of polymer being chemically transformed by the action of the biological enzymes or microorganisms into products which themselves are capable for further degradation. Two possible approaches currently being adopted to reduce the 'vices of plastics' are to (i) develop biodegradable commodity materials; and (ii) identify potential microorganisms and develop protocols to effectively biodegrade polymeric materials.

Systematic approaches have been made in research laboratories to induce biodegradability in some polymeric materials and blends by employing micro-organisms having hydrocarbon utilising as well as polymer degrading ability with a view to developing biodegradable polymeric materials. During studies on the behaviour of carbon fibre composites, as influenced by microbiologically induced degradation due to fuel residing micro-organisms, possible utilisation of chemical components as energy and carbon sources by micro-organisms was noted. It has also been observed that when glass fibre reinforcements in epoxy and vinyl ester resins are exposed to sulphate reducing bacteria, physical and chemical changes in the composite materials are induced. The vinylester resin based composites were found more biosusceptible then epoxy based composites.

The biodegradability of polyurethane is also known for quite some time. Polyurethanebased coatings are increasingly being used in military transport systems and war equipments due to its low porosity and good adhesion properties. Increased biogrowth in terms of ATP, carbohydrates and protein contents alongwith loss in adhesion and porosity was observed due to fungal growth on the coatings suggesting biodegradable nature of these coatings in presence of microbial cultures. Such studies have helped in establishing parameters for assessing the extent of biodegradation of polymeric materials. Further, studies with a view to acclimatising the growth of selected fungal isolates on polyethylene-glucose blends reveal that certain fungal cultures do possesses potential to biodegrade polymeric blends.

The current approach in this area of research is that the performance of degradable polymers should be assessed in the context of their intended purpose and environment in which they are required to degrade. Biodegradable polymers should be developed and carefully designed to degrade in a desired controlled manner. This will help in alleviating problems during their use and or during execution of any waste management procedures.

SECTION V

Recycling of Polymers

CHAPTER 24

Recycling of Polymers: An Overview

INTRODUCTION

It is certainly true that plastics left lying around after use do not disappear from view and such post-consumer waste as foam cups, detergent bottles, and discarded film is a visual annoyance. This is because plastics are not naturally biodegradable. However, to consider this a detriment is a questionable argument. Rather, it may well be considered an advantage. This is borne out by the fact that recycling of plastics materials is now an important field in the plastics industry, not just an activity born under environmental pressure.

Although the plastics industry practiced recycling for many years, attention was mainly focused on the recycling of industrial scraps and homogeneous post-consumer plastics, which are easy to collect and reprocess. However, more recently the plastics industry accepted the challenge of recycling of heterogeneous plastics waste based on new technologies of separation and reprocessing. Scientific research, scarcely visible only a few years ago, is now a very active, fast-growing discipline, contributing to the development of newer processes.

According to the type of product obtained from the recycling process and the percentage of the economic value recovered, the following broad classification of recycling technologies can be made: (i) *primary recycling,* the reprocessing of plastics waste into the same or similar types of product from which it has been generated; (ii) *secondary recycling,* the processing of plastics wastes into plastics products with less demanding properties; (iii) *tertiary recycling,* recovery of chemicals from waste plastics; and (iv) *quaternary recycling,* recovery of energy from waste plastics.

The processes mainly used to these ends are: direct reuse after separation and/or modification, chemical treatment or pyrolysis for recovery of monomers and/or other products, and burning or incineration. Primary recycling is used when the plastic waste is uniform and uncontaminated and can be processed as such. Only thermoplastic waste can be directly reprocessed; it can be used alone or, more often, added to virgin resin at various ratios.

The main problems encountered in primary recycling are degradation of the material resulting in a loss of such properties as appearance, mechanical strength, chemical resistance, and processibility. Contamination of plastic scrap and handling of low-bulk density scrap such as film or foam are additional problems in primary recycling. Primary recycling is widely performed by plastics processors; it is often considered an 'avoidance of waste' rather than recycling.

For post-consumer, mixed plastic wastes (MPW), which are unsuitable for direct use, the industry resorts to secondary recycling methods. There are various technical approaches to secondary recycling of MPW. These include reprocessing based on melt homogenisation using specialised equipment; use of ground plastics waste as filler; and separation into single homogeneous fractions for further processing, such as partial substitution of virgin resins and blending with other thermoplastics using suitable compatibilisers.

In tertiary or chemical recycling of plastic wastes, polymers are chemically unzipped or thermally cracked in order to recover monomers or petrochemicals indistinguishable from virgin materials. Thermal cracking procedures offer viable alternatives by utilising comingled plastics without decontamination. In quarternary recycling, energy content of plastics waste is recovered. In most cases, plastics are burned, mixed with other waste. Incineration of plastics alone creates a number of problems and requires the use of specially designed incinerators.

OUTLINE OF RECYCLING METHODS

Postconsumer plastic wastes can be divided into two different groups depending on their source: (i) mixed plastics from the household waste; (ii) plastics from the industrial sectors. The first category involves the medium/short-life articles that are used in food, pharmaceutical, and detergent packaging, shopping, and others. The majority of these articles are composed of thin protective films: a variety of bottles for soft drinks, food, and cosmetics, sheeting for blisters, strapping and thermo-formed trays.

There are basically five different polymers that contribute to the total amount of domestic plastic waste, namely, PE, PP, PS, PVC, and PET. The composition of this MPW can change depending on the regional habits and seasons of a year, and also on the mode of waste collection. A typical composition may be PE 39%, PVC 22%, PET 19%, PS8%, and PP 12% (by wt.).

The collection of plastics wastes always yields a polluted product, and this fact poses the need for the first operation of the recycling process, namely the cleaning of foreign bodies. The machinery required at this stage may be either manual or automatic type, the former being simpler from the standpoint of installation. The operations following the first step of cleaning are determined by the type of recycling process to which the material is to be subjected. There are basically two main recycling processes; recycle of heterogeneous MPW and recycle of selected polymers separated from MPW.

A direct solution to disposal of domestic MPW can be the reuse of the heterogeneous mixture by processing through extrusion or injection moulding technologies using traditional machineries. However, when MPW is processed, one of the main problems is to find the best compromise between homogenisation and degradation. The optimal processing condition must ensure a good dispersion of the materials with high melting point (such as PET) in a continuous phase of molten polymers (such as PVC), avoiding gas bubbles, low-molecular-weight compounds, and cross-linked residues that are formed by thermal degradation. Some possible applications of such moulded mixed plastics are injected tiles for paving, and extruded profiles for making structural articles such as benches, garden tables, bicycle racks, fences, and playing facilities. However, because of the incompatibility of the various components in mixed plastics, the mechanical properties of the moulded or extruded products are rather poor.

The market of park benches, playgrounds, fences, and so on, cannot absorb, in the long run, the massive amounts of MPW that are produced every year. Hence the possible route to recycling of MPW to obtain secondary materials with acceptable mechanical properties could be to blend them with virgin polymers, or, at least, with recycled homopolymers. For example, experimental results of processing and properties of blends of virgin LDPE and MPW have shown that all mechanical properties, with the exception of elongation at break, are very similar to those of the virgin material if the MPW content does not exceed 50%.

The possibility of using MPW as 'filler' for both LDPE and HDPE has been considered as such an approach, and may offer two important advantages: (i) improvement of the use of huge amounts of MPW that are generated by municipalities and industries; and (ii) savings in non-renewable raw materials and energy, both associated with the manufacturing of the virgin materials that can be replaced by plastics waste. Even if the percentage of plastics waste used as a filler cannot be high, its common use may absorb sizable amounts of waste.

A widespread solution, in terms of application and market volume, can be the recycling of single materials or homogeneous fractions obtained from a 'differentiated' collection system and/or a separation process of the mixture. Moulded products from single or homogeneous fractions usually show a general performance far greater than that of products from mixed plastics. To obtain single or homogeneous fractions, it is useful to separate the mixed domestic plastics into four fractions, namely, polyolefins, PS, PVC, and PET.

An important preliminary to separation of mixed domestic plastics is the cleaning and selection operation. A simple method to perform this operation consists of a selection platform where a number of trained sorters separate the different types of plastics by visual assessment. Because manual selection is liable to human error, selection platforms may be equipped with detectors such as electronic devices to check the quality of the selected materials.

The drawbacks of the manual platforms—which range from high labour cost to the complexity of labour management—may be avoided by resorting to automatic platforms. The machines required for such automation are manifold and the necessity to employ them is related to the quality of the collected material. Essential machines are rotary screen, light-parts separation equipment, heavy-parts separation equipment, and aluminium rejection equipment. All such machines are preliminary to the stage of separation into homogeneous plastic fractions.

Bottles constitute the largest high volume component of postconsumer plastics and need special attention in reclaim operation. The developments in bottle reclaim systems have made recycling postconsumer plastics more efficient and less costly. Municipalities, private organisations, universities, and entrepreneurs have worked closely to develop new collection, cleaning, and sorting technologies that are diverting larger portions of plastics from landfills to recycled resins and value-added end products.

To collect the high volume to weight ratio postconsumer plastics economically, truck-mounted compactors have been developed that seem to have the most promising future for mobile collection. They are self-contained and offer, on average, a reduction ratio of 10:1. Simple to operate, compactors accept all types of plastics, including film, and perform equally well with mill jugs and PET bottles as with mixed plastics.

Using compactors for on-board truck densification can thus be a cost-effective part of multimaterial collection programmes. Another noteworthy development is that of flatteners and balers, which have also proven cost-effective under certain conditions. An integrated baler developed by Frontier Recycling Systems (USA) is fully automatic and capable of handling the plastic throughput of larger and costlier systems without the corresponding expenditures of space and labour. By producing smaller, high density bales, it allows for lower transportation costs of recyclables shipped to market.

In keeping with the progress in densification options, efficient sortation systems have also been developed. The Poly-Sort integrated sorting line developed by Automation Industrial Control (AIC), Baltimore, is capable of sorting mixed stream of plastic bottles at a baseline rate of three bottles per second, or 700 kg/hr. With expected advances in scanning and detection, the sorting rate of the system could double to 1400 kg/hr.

Designed to sort compacted bottles, the Poly-Sort system employs conveyors for singulation, and two devices for colour and chemical composition identification. A vibratory conveyor singulates bottles; a 'read' conveyor transports bottles to an ultrasonic sensor that detects their position; a near-infrared system detects the resin type; a camera detects the colour of the container; a computer integrates data and makes an identification; air jets divert bottles to the appropriate segregation conveyor or hopper.

The above type of separation is a macroseparation. It may be noted that the methods of separation into homogeneous fractions fall into three groups: molecular separation, microseparation, and macroseparation. Molecular separation is based on the dissolution of the various plastics in selective solvents, a method that is promising but still in the stage of study. Microseparation is a method by which a suspension medium is used to separate plastics with density higher or lower than the suspension medium. Macroseparation, which is the separation of plastics when waste materials are still in initial form, appears to be the most conveniently applicable system, considering the increasing possibilities of automation it offers. The key to this separation process is the development of an efficient detector system that can distinguish between type and quality of different plastics in waste materials.

Different types of detectors have been developed and many are under development. These are based on distinctive physico-chemical properties of plastics and employ different techniques such as X-ray, near-infrared spectrophotometry, fluorescence, and optical measurement of transparency and colour. Automatic systems consisting of a platform for selection according to plastics topology, a number of identification and detection steps, and adequate checks on the efficiency of separation following detection have been developed. The Poly-Sort system described above is one such example.

Recycle installations take up the separated plastics flakes for further processing. Various elements that normally compose the item to be recycled are caps made of PE, PE with PVC gaskets, aluminium, labels of tacky paper with different types of glue, and residues and dirt that have been added during the waste-collection phase. Various operations that are carried out in a specific sequence because of the problems posed by the type of material are: grinding to ensure homogeneity of the product, air flotation for separation of flakes with different specific weight and removal of parts of labels freed by grounding (such as separation of PVC labels from PET bottle flakes), and finally washing to remove residues. The washing system consisting of a range of equipment that includes centrifugal cleaners, washing tank, settling tank, and scraping machines is part of a know-how of various manufacturers.

The majority of municipal solid waste consists of plastics waste, which is often contaminated with significant amounts of paper. This is not only the case with plastics fraction of municipal solid waste (PFMW), but also with such industrial waste as used packaging materials, laminates and trimmings. The reprocessing of plastics waste contaminated with more than 5% paper is difficult using conventional plastics processing machinery, and becomes almost impossible at paper levels exceeding 15%. The sorting operation at a municipal plant normally aims at removing the paper component from the light plastics fraction to a level well below 1%. However, this operation has not been quite successful because the material handling side has been difficult and the costs have far exceeded the price of virgin polyolefins.

A simpler solution to the problem contamination may be to allow for a paper component in the plastics fraction and to use a processing method that can disintegrate the cellulose fibres into small fragments such that they act as particulate fillers in the plastics. Such a method has been developed at Chalmers University of Technology, Gothenburg, Sweden. Known as the CUT-method, the process makes it possible to reprocess both the PFMW and a number of different industrial plastic waste materials contaminated with paper. The CUT-method, consisting of a prehydrolytic treatment of the paper component, is an industrially applicable method of reprocessing paper-contaminated plastics waste of various origins.

The main advantage of the CUT-method is that the plastics fraction and the paper component do not need to be separated and the hydrolysis does not degrade the plastics component but reduces the chain length of the cellulose component to a level at which the cellulose fibre becomes extremely brittle and the shear forces generated in normal plastics processing machinery (compounding extruders and moulding machines) can easily disintegrate the paper parts into small fibrous fragments. It is the disintegration of the embrittled paper component into an almost pulverised substance that is the key to the success of the method, since this results in greatly enhanced melt flow properties, better homogeneity, and thus in improvement in the mechanical properties of the material.

The method of hydrolysis used in the CUT-method offers an efficient and economical way of processing plastic waste, both postconsumer municipal waste and industrial waste, contaminated with a cellulose component. The presence of cellulose gives a desired stiffness to the final product, as studies have shown. Such plastics product can be used in several applications, such as artificial wood.

Plastic wastes from industrial sectors concern mostly the medium-/long-life articles, as plastics have played a fundamental role in the exceptional growth of production technology seen during recent years in these sectors, and in particular the automotive industry. Because of the advantage in design and functionality, plastics are now an indispensable part of any kind of car; the amount of polymers employed to build a car has risen to about 20% from a mere 5%, with a corresponding increase in the quantity of non-metallic waste during scrapping. The main problem of plastic wastes from all industrial sectors, and in particular the car industry, is the large variety of materials employed to build a single component or system, for example, a dashboard. This takes place because of the sophisticated and complex mission that the system must perform. The large number of plastics used and the disproportionately high costs in the dismounting of the different plastic pieces of a car represent an intractable waste-recovery problem and thus have a negative impact on the recycling process. As a result of this, only the metallic fraction is recovered, while the plastic materials continue to be eliminated by deposition in refuse dumps.

An alternative approach to the recovery of automation plastics is therefore to use them as large, easily removable components that offer potential for reclamation as well-characterised individual polymers. Some particularly complex components such as vehicle front and rear end systems, exhibit special suitability for manufacture in plastics instead of metals because of their ease of production and assembly. It is generally recognised that improvements in automotive scrapyard economics may be best achieved by the prior removal from vehicles of such large polymeric components and their recycling as well characterised plastic fractions. For example, plastic fuel tanks of HDPE are now in common use and represent the most common recyclable plastic component. Trials with material re-covered from used plastic fuel tanks have shown promising results for the manufacture of new tanks.

A concept that is being developed to solve the recycling problems of plastics from industrial sectors, and in particular the car industry, is based on the use of materials of the same family for all components of the plastic systems to be recycled at the life end. This allows an easy and direct recycling of the scraps and the recovery of the whole system. Greater recycling efficiency can be obtained when the following two basic requirements are satisfied: (i) materials compatibility through materials homogeneity; and (ii) easier disassembly through planned design. This concept has been first applied to the automotive sector, where the environmental problems have become of primary importance; however, it could be also applied to other products, i.e., appliances and building materials. There are two tasks in developing this concept: to develop new advanced materials in individual categories of polymers and to promote new technologies.

Consider, for example the automotive industry. Although many polymers are used in cars today, the industry tends to favour more and more polypropylene use due to a large range of properties available. New developments in polyolefin-based materials have thus created a family of polypropylene products with a wide range of physical properties, including the ability to be easily recycled. When utilised by automotive and product designers as a part of a 'design for disassembly' strategy, these compatible materials will yield large sub-assemblies that can be reclaimed with a minimum of handling. In each project, the design incorporates readily identifiable hard point connections between the polypropylene components and the metal automobile subframe. This allows personnel in recycling centres to remove these parts quickly and in large pieces that can be completely reground and recycled. This concept has been applied to car dashboards and interior vehicle components like floor covering, trim, and door panels, as well as bumpers.

Blends of EPDM rubbers with polypropylene in suitable ratios have been marketed as *thermoplastic elastomers* (TPE), also commercially known as *thermoplastic polyolefin elastomers* (TPO). These heterophasic polymers, characterised by thermo-reversible interactions among the polymeric chains, belong to a broad family of olefinic alloys that can now be produced directly during the polymerisation phase, unlike blended TPE and TPO, and various compositions (with various compounding additives) can be formulated which are primarily tailored to meet different requirements of most of car applications. The TPE-based synthetic leather and foam sheets are typical examples.

In order to obtain all-TPE recyclable applications, different assembly techniques have been specifically studied to obtain the basic composite structures. The most interesting technique is one that allows simultaneously thermoforming, embossing, and coupling to be obtained in one stage of operation, yielding a foamed synthetic leather bilayer on a rigid support (all TPE based) without adhesives. With new designs for recycling, dashboard, floor covering, and other interior components such as door panels, pillar trim, and rear shelf have been made of the same chemical material (polypropylene) in different forms, thus providing an important aid to the recycling of plastic.

Lead-acid batteries from automotive applications normally have a shorter service life than the car itself. The logistics system for used car batteries is geared to lead recycling. However, the first battery reprocessing step yields not only lead but also polypropylene in a form of the casing fragments. Accordingly, the polymer is available without additional cost. As the casing makes up a substantial part (7%) of the total battery, the quantities of polypropylene obtained are sufficient to warrant the operation of a plastics recycling plant. For example, BSB Recycling GmbH in Braubach, Germany, a subsidiary of Metall-gasellschaft AG, operates secondary lead smelter for lead recovery from postuse lead-acid batteries. They process some 60,000 tonnes of batteries per annum, which accounts for half of the used battery volume to be disposed in the western states of Germany. BSB started to segregate the polypropylene from the battery casings and route it to a separate recycling process as far back as 1984. For the recycling process, a quality assurance system geared to the specific requirements of the application has been developed and implemented.

Polyolefins and poly(ethylene terephthalate) (PET) are the most frequently recycled polymers obtained from both the domestic and industrial plastics wastes, and as such they have received most attention in the recycling research and technology. PET is one of the largest recycled polymers by volume, because it is suitable for practically all recycling methods. Over 50% of the PET film produced in the world is used as a photographic film base. The manufacturers of these materials have long been interested in

PET film recovery. An important motivation for this has been the fact that photographic films are usually coated with one or more layers containing some amount of rather expensive silver derivatives.

Silver recovery makes PET-base recovery more economical. In a typical way of operation, PET film recycling is thus coupled with the simultaneous recovery of silver, for example, by washing with NaOH and follow-up treatment. PET-recycling by direct reuse, if the washed PET-film scrap is clean enough to be recovered by direct re-extrusion, is by far the most economical process. However, this process is most suited for the recovery of in production wastes. For customer recollected PET film, which may have a higher degree of contamination, other technologies are to be applied.

There exists a hierarchy in PET-film and plastics recycling technologies depending, first of all, on the degree of 'purity' of PET scrap to be handled, and secondly, the economics of the process. For the cleanest PET grade, the most economical process, i.e., direct use in extrusion, is self-explanatory. For less clean PET waste, it is possible to reuse them after a modification step (partial degradation, e.g., by glycolysis) at a reasonably low price. More contaminated PET waste must be degraded into the starting monomers, which can be separated and repolymerised afterwards, of course at a higher cost.

Polyethylene films from greenhouses, although highly degraded by UV radiation, are recycled by various means leading to manufacture of films and moulded products with low mechanical properties. Problems in the recycling of greenhouse films arise from the presence of products of photo-oxidation, which significantly affect the properties of a recycled material. An interesting possibility of use of photo-oxidised PE in blends with nylon-6 to improve blend compatibility has been demonstrated. These follow from the earlier efforts to compatibilise blends of polyamides and polyolefins (which are potentially very interesting, but, because of the strong incompatibility of both polymers, yield products having poor properties) with the use of functionalised polyolefins that can react with the amino groups of polyamides, giving rise to copolymers and thus stabilising the blend.

Such functionalisation, in general a long and extensive step, is mostly performed by chemical modification of the polyolefin structure. However, studies have demonstrated that photo-oxidised PE offer similar results. Thus the use of recycled (photo-oxidised) greenhouse PE in blends with nylon give rise to PE/nylon graft copolymers during processing, which improve the mechanical properties of the resultant material. The graft copolymers act as compatibilising agents; the properties of nylon-rich blends (80% by wt. nylon-6) thus are found to be very similar to those of blends compatibilised by PE, and which is initially functionalised by chemical means. Moreover, in co-extrusion, a good adhesion between the two layers (nylon and recycled PE) of coextruded films helps to avoid a need for the addition of a third layer binding two incompatible phases.

Chemical means such as glycolysis, methanolysis, and hydrolysis are good at unzipping only the condensation polymers—such as polyester, nylon, and polyurethanes—to facilitate chemical recycling. Addition polymers, such as vinyls, acrylics, fluoroplastics, and polyolefins, can hardly be reprocesed except that, if they are sorted, they may be converted into powder by grinding operation and mixed with respective virgin resins for remoulding into finished good or, in some cases, blended with other resins using suitable compatibilisers to make useful end-products of commercial value.

Tertiary recycling of addition polymers require pyrolysis, which is a more aggressive approach. For mixed or unsorted plastics, in particular, it is a practicable way of recycling. Pyrolysis is the thermal degradation of macro-molecules in the presence of air. The process simultaneously generates oils and gases that are suited for chemical utilisation.

The advantage of pyrolysis over combustion (quaternary recycling) is a reduction in the volume of product gases by a factor of 5–20, which leads to considerable savings in the gas conditioning equipment.

Furthermore, the pollutants are concentrated in a coke-like residue matrix. It is possible to obtain hydrocarbon compounds as gas or oil.

The pyrolysis is complicated by the fact that plastics show poor thermal conductivity, while the degradation of macromolecules requires considerable amounts of energy. The pyrolysis of mixed plastic wastes and used tyres has been studied in melting vessels, blast furnaces, autoclaves, tube reactors, rotary kilns, cooking chambers, and fluidised bed reactors.

Rotary kiln processes are particularly numerous. They require relatively long residence times (20 min or more) of the solid wastes in the reactor. Moreover, due to the large temperature gradient inside the rotary kiln, the product spectrum is very diverse. For this reason, the gases and oils generated by the pyrolysis are normally used for the direct generation of energy and the process may well be considered as a quaternary recycling process.

For chemical recycling of mixed plastics, the fluidised bed pyrolysis has turned out to be particularly advantageous. The fluidised bed is characterised by an excellent heat and mass transfer as well as constant temperature throughout the reactor. This results in small dwell times (a few seconds to a 1.5 min maximum) and largely uniform product spectra. The fluidised bed is generated by a flow of air or an inert gas (nitrogen) from below through a layer of fine-grained material, e.g., sand or carbon black. The flow rate is sufficient to create turbulent motion of particles within the bed. Using a fluidised bed pyrolysis, 25–45% of product gas with a high heating value and 30–50% of an oil rich in aromatics could be recovered. The oil is comparable to that of a mixture of light benzene and bituminous coal tar. Up to 60% ethylene and propylene are produced by using mixed polyolefins as feedstock. Moreover, depending on the temperature and the kind of fluidising gas (nitrogen, pyrolysis gas, steam) different variants of the fluidised bed pyrolysis process can be carried out, yielding only monomers, BTX-aromatics, high boiling oil, or gas.

A promising concept that is receiving increasing attention is recycling plastics to refinery cokes, where pyrolysis units and a well developed infrastructure are already in place. The main hindrance to the execution of this concept is the presence of contaminants (including chlorine and nitrogen) in the plastics stream, as well as the need to turn plastics into a liquid form that the refinery can handle. Projects are in place to address these issues. Initial small-scale pyrolysis, dissolving plastics into other refinery feedstocks, or turning solid wastes into a slurry, are some of the options that have received attention. Efforts have also been made in some refineries to convert mixed plastics into a petrochemical feedstock by catalytic hydrogenation. In the refinery, the aim of tertiary recycling is not to displace regular refinery capacity, but to use plastic waste as a very minor stream. However, even if all refineries with cokers took only 2% of their capacity as plastic waste, it would be extremely significant.

Recently a new process has been developed which can treat polyolefinic plastics (polyethylene, polypropylene, and polystyrene) or their mixtures by a combination of thermal and catalytic cracking to produce gasoline, kerosene, and gas oil fractions of about 85%. Recovered liquid and gas are separated by cooling and the gas is used as in-house fuel.

POLYMERS FROM RENEWABLE RESOURCES

A large number of thermosetting resins have been synthesised by condensing diazotised cardanol and prepared by diazotising cardanol with an amino compound, formaldehyde and substituted aromatic compounds in the presence of acid as catalyst. The resins have been characterised by FTIR spectra. The thermal behaviour of the resins have also been studied. The kinetic parameters for the degradation of resins have been evaluated by using computer analysis, the *Lotus* package method.

Phenolic resins were the first synthetic polymeric materials to be synthesised and they have multifold applications in different fields. They have excellent ablative properties and structural integrity, also they can be used as high temperature resistant polymers. The research work with the thermosetting polymers using renewable resources has been accelerated in the current years. Over the years, a large number of papers have been published in this field relating to the work with the biomonomers derived from the renewable resources. One of such renewable resources is cashwnut shell liquid, the major component of which is cardanol, a phenolic compound. A programme has been under taken to synthesis and characterisation of some high temperature resistant resins from the cashwnut shell liquid (CNSL). The major components of the CNSL have been characterised using UV, IR, H^1NMR etc. and kinetics of formaldehyde condensation of cardanol has been reported. The oligomerisation of cardanol using acid as catalyst has been studied.

The structure and properties of CNSL-novalac resins prepared by using succinic acid as catalyst, has been studied and it has been found that the synthesis of a vinyl monomer, cardanyl acrylate and its suspension polymerisation to get cross-linking beads. Cardanol acrylate and its homopolymer has been synthesised and Lenk has reported the synthesis and characterisation of a large number of resins using formaldehyde/furfural and substituted aromatic compounds.

Experimental

Preparation of resins

Cardanol was diazotised with *o*-anisidine. The co-polymer resins were prepared by refluxing a mixture of the diazotised cardanol (4m.mole), a substituted organic compound (4m.mole) and formaldehyde/ furfural (40m.mole) with 1 per cent dydroquinone in the presence of 3ml. of 6N dydrochloric acid at 90°C for 4–6 hours.

During the course of reaction, periodic shaking was needed to ensure thorough mixing. After completion of the reaction the containts were collected with ice cold water. The product was filtered and washed repeatedly with hot water to drive out unreacted material and then dried in vacuum. The physiochemical properties of the resins are furnished in Table 24.1.

Table 24.1. Physiochemical properties of resins.

Sample resins	*% of yeild*	*Colour*	*Structure*	*Solubility*
DC+FD+PCBA(I)	61	Black	Crystalline	Toluene, DMF
DC+FF+ α-Naphthol(II)	64	Black	Crystalline	Toluene, DMF

DC-Diazotised Cardanol, PCBA-*para* chloro benzoic acid, FD-formaldehyde, FF-furfural

Solvent absorptivity of resins

Different amounts of the prepared resins were stored in 3ml. of different solvents for 24 hrs. Excess of solvent present on the surface of resins were removed with the help of filter papers. Each of them was then weighed and the solvent absorptivity (SA per cent) was calculated using the equation:

$$SA\ \% = \frac{W_2 - W_1}{W_1} \times 100$$

where, W_1 is the weight of dry sample and W_2 is the weight after absorption of the solvent. The absorptivity data are furnished in Table 24.2.

Table 24.2. Absorptivity data of solvents.

Names of resins	*Names of solvents*					
	DMF			*Toluene*		
	W_1*(mg)*	W_2*(mg)*	*SA%*	W_1*(mg)*	W_2*(mg)*	*SA%*
DC+FD+PCBA (I)	420	550	30.9%	450	530	17.7%
DC+FF+ α α -Naphthol (II)	470	620	35.7%	350	420	20%

Results and Discussion

From the FTIR data it is concluded that, the compound (I) is an azo dye based resin containing the phenolic group and an aliphatic side chain. This copolymeric compound contains a CH_2 group of formaldehyde at 2923.95 cm^{-1} and C–O–C stretching of OCH_3 group of anisidine at 1287.49 cm^{-1} and 2561.11 cm^{-1} . The resin (II) also an azo dye based copolymeric resin containing a peak at 2922.34 cm^{-1} for CH stretching of furfural ring and 1392.94 cm^{-1} and 1342.40 cm^{-1} for in plane OH bending of α-naphthol.

From the thermograms and the data showing the per cent of decomposition at various temperature ranges it is observed that around 10 per cent weight loss of two resins upto 200°C is due to removal of moisture contained. Resin (II) has higher thermal stability compared to resin (I), because the resin (I) has simple formaldehyde (CH_2) linkage and a benzene ring of benzoic acid but in case of the resin (II), it has a furfural ring and a naphthol group. It is also observed that the activation energy of the first phase is greater than the second phase, which indicates that the depolymerisation occurs at a faster rate in the second phase. This is in good agreement with the thermogram patterns.

CHAPTER 25

Equipment and Machinery for Recycling of Polymers

INTRODUCTION

While the polymers recycling activity, driven by consumer and legislative pressures, is all but certain to increase, the key variables in the rate of growth are the polymers industry's ability to develop an economical material-collection infrastructure and to improve the methods for handling and processing of contaminated scrap. Techniques for selection and recycling of postconsumer plastics are, however, closely related to the characteristics of plastics containers consumption, which vary greatly according to the geographical areas and the relevant law regulations governing activities in this sector.

Consumption features play a major role in the choice of the materials to be recycled. In the United States and Canada, the materials chiefly recycled are PET bottles and PE containers; in France, on the contrary, recycling of PVC got priority on account of the large quantity of such material used in the packaging of drinks. In Australia, recycling includes primarily PET and PE, whereas in Japan it is mainly PET. In Italy, the plastics recycled are mostly PET, PVC, and PE. In short, the material to be recycled and the enforced legislation determine the choice of collection system. In many countries such as United States, Canada, Australia, France, Austria, and Switzerland, some fractions of postconsumer plastics are collected in the most homogeneous way possible. In other countries, plastics are collected more heterogeneously, that is, different types of plastics out of different types of manufactured articles, such as foil, containers, bottles, are collected together.

The outcome of the collection system constitutes the raw material for the recycling process. The degree of purity of this raw material evidently depends on how selective the collection is.

PLASTOCOMPACTOR

For voluminous scraps such as light film, textiles, fleece, and foam, it is advantageous to increase the density, which may be in the range of 20–40 kg/m^3, to about 400 kg/m^3 for transport reasons and for further processing. With thermoplastics and thermoplastic mixtures, a *plastocompactor* can be used for this purpose. It is, however, more useful for homogeneous fractions of thermoplastics. The process agglomerates the material without plastifying it. By heating the material locally for a short period to a temperature above the softening point, the soft components begin to adhere. The material is then compacted into a condition very similar to the virgin material.

In a typical agglomeration plant, the loose material is usually fed to a granulator, which is also fitted with a nip roll device for feeding continuous material such as fleece or film from the roll. A blower transports the flakes from the granulator to a holding silo. The discharge screw in the silo transports the flakes to the feed hopper from where they are carried by a blower to the plastocompactor. A dosing and pressing screw feeds the actual agglomerator part of the machine. This comprises essentially two discs—one rotating and the other stationary—between which the flakes are compacted by using heat from friction and pressure. The agglomerate leaves the discs through the outer gap in the form of warm soft sausages and is cooled immediately by an air stream. It is then fed, in a semisoft state, to a hot melt granulator where it is reduced to a free-flowing granulate.

DEBALING AND INITIAL SIZE REDUCTION

The first operation of the recycling process is the cleaning of foreign bodies. It requires a number of operating steps, the first being normally a debaling operation as the collected material, for transport reasons, is reduced into bales. Debaling is still often carried out manually. The reason often given for using a manual method is that the workman can also check the baled scrap for large pieces of foreign matter at the same time. However, there exist very efficient debalers for making the task automatic, and the best brands are equipped with specific devices designed in accordance to the composition of the bales to be loosened. Such factors as the forms of plastics items; the proportion of PE, PET, or PVC in the bales; the collection features; the container typology; and the share of foil plastics determine the type of debaler construction technique.

Two simple debaling services are a *grab truck* and a *screw shredder*. A grab truck can normally be used for bales of film. Sitting in his machine, the workman can break open the bales using three hydraulically operated grabs fitted to an extension arm. Checks for large pieces of foreign material can also be conducted. Using the same grabs, the loosened material can then be placed onto the feed conveyor fitted with a metal detector.

For bales containing individual items of scrap such as bottles and other hollow items, the use of a screw shredder offers many advantages. This machine is fitted with a very large feed hopper and can be fed directly with a large bucket loader or similar device. The feed material is reduced by a tearing process in the shredder between independently driven screw shafts fitted with shredding teeth. Screw shredders are manufactured with up to six adjacent shafts and are thus suitable for the feed of very large bales or a large number of bales at one time.

After debaling, the material passes through initial size reduction units to separation and selection operations. Most size reduction tasks can be performed by the following machines: shredder, cutter or guillotine, screw shredder, and granulator.

Shredder

Shredders have been in use for a long time in many sectors for the recovery of scrap. To a large extent they draw in the material automatically, and are suited for film, sheets, solid pieces, hollow items, cables, etc. The stresses caused by the tough and partly high strength material are enormous. Extremely sturdy units designed for this technology are thus required for the treatment of plastics. It is important that the cutting shafts run at a suitable speed so that cutting and tearing processes occur.

Models of shredders are offered with from one to six cutting shafts. Machines with a capacity of many tonnes per hour are available. They are able to reduce complete bales of film fed by forklift without any difficulty. Also, hollow items such as rubbish bins and barrels can be reduced when a ram is fitted to the feed. Another field of application is the shredding of cables to allow the separation of plastics and metal. Generally, a trough fed by a forklift or conveyor is located above the shredder shafts. After the material is reduced to a practical size in the shredder, it is transported to subsequent process stages by means of a conveyor or other mechanical device.

Cutter or Guillotine

Some plastics scraps are not suited for initial reduction in a shredder described above. These include fibres, long pieces of material, rolled strips, and lumps of rubber. A guillotine is better for these applications.

The material is fed manually in a trough or on an open conveyor to the guillotine. The latter operates as opposed to the shredder, on a stroke principle. A cutter, usually hydraulically operated, is lowered from above to cut the material in slices of the desired thickness. The complete cutting process including the material feed is best operated with a programmed control.

Screw Shredder

The screw shredder mentioned earlier as a machine for debaling is also used for the initial size reduction of plastic items, and in particular when these are very voluminous and not too tough. It is suited for very large items or bundles of material and has the advantage of being fitted with a very large feed hopper that can be filled by bucket loader or similar device.

The machine is fitted with two shafts, rotating independently. It can be constructed, however, with up to six adjacent screw shafts, each shaft having its own drive via gears and electric motor. The shafts are equipped with shredder teeth for reducing the feed material in a crushing and tearing process. Not being a cutter, the machine is best suited for materials that can be broken or torn. When overloaded, a special control stops the respective shaft and switches it into reverse gear for a set time before returning to normal operation.

Granulators

Granulators can be seen as the most versatile size-reduction machine for the complete sector of plastics size reduction, and are used for the dry reduction of plastics. The machines used for this application are therefore designed to meet the special demands of job conditions, which are sturdy mechanical design, quick knife replacement, easy cleaning, and high capacity.

Since the reduction process is subject to the generation of a considerable amount of heat, it is necessary to water-cool parts of the machine or remove the heat with it. It is advantageous to fit the granulator with an open or semiopen rotor and a strong suction device to ensure that the grinding chamber is cooled intensively so that water-cooling is not required and the air is used at the same time for discharging the size-reduced material. All granulators should be equipped with a screen that can be easily removed. The screen opening determines the top size limitation of the size-reduced product. Material is fed to a feed hopper manually, or on a feed conveyor, screw, or similar device.

Mostly, all granulators have the following characteristics: (i) welded steel construction; (ii) externally mounted bearings; (iii) hinged two-piece housing with split point around the shaft; (iv) easily replaceable screen; (v) double cross-cutting action; and (vi) preadjustable knives. These characteristics offer a number of advantages, as explained below.

Due to the welded construction, the machine is resistant to extreme stresses caused by any foreign matter that may enter the granulator despite all precautionary measures taken, and fractures are avoided although the housing may be deformed. As the bearings are mounted outside the granulator housing, it is not possible for the feed material to enter the bearings or for grease to contaminate the material. The hinged housing allows easy and quick cleaning necessary when feed is changed and simplifies servicing and knife replacement. This is a significant advantage particularly in scrap recycling, where increased wear and more frequent knife replacement are to be reckoned with. For double cross-cutting action, all knives on the rotor are mounted at an inclined angle in a straight line to the rotor axis while all bed knives are set at the same angle but in an opposed inclined direction, also in a straight line. Complete sets of resharpened rotor and bed-knives can be readjusted to the exact gap required between them to achieve the desired reducing action. This advantage, in conjunction with the good accessibility due to hinged housing, allows knife replacement to be carried out very quickly.

Fine Grinding

Fine grinding also offers solutions for the recycling of plastic scrap. Different types of machines are used, two common types being *universal blast mills* and *disç pulverisers*. In a universal blast mill, plastic scrap is reduced between the beater wings of a blast disc and a screen or the grooved grinding track of a grinding chamber. The hole size of the screen determines the fineness of the plastic powder. This usually has an upper size limitation of 500–800 μm.

In an impact disc pulveriser, which has much lower power consumption than the blast mill, the material is size reduced to an upper limit of about 800 μm between a fixed and a rotating disc. If a lower top size limitation is required, a screener may be used to return the coarse material continuously to the grind process. The finely ground powder is separated in a cyclone.

CLEANING AND SELECTION

Cleaning, separation, and selection operations that usually follow the initial size reduction are determined by the type of recycling process to which the material is to be subjected. Basically there are two main recycling processes: recycle of heterogeneous plastics and recycle of selected polymers. The former process leads to the manufacture of extruded or injected products by direct reuse or in mixture with other components. The latter process consists of separation of the mix of collected plastics into homogeneous fractions, subjected to further processing that brings their characteristics and purity as near as possible to those of the original polymers.

The simplest method to perform the cleaning and selection operation consists of a selection platform where a number of trained sorters separate the different types of plastics on the basis of visual assessment. Though this is a hard and unpleasant job, the advantage of manual selection is that sorters operate to a degree of intelligence that the automatic equipment cannot reach. On the other hand, manual selection is, understandably, always liable to human error. To counter this problem, selection platforms are often equipped with detectors to check the quality of the selected material. These may be electronic appliances capable of recognising, for example, PET in flux of PVC and vice versa, and detectors able to identify traces of metal overlooked during manual sorting, such as aluminium from caps and rings. The material manually selected and then electronically checked is therefore of best quality and can be sold at the maximum market price.

The most serious drawback of manual platforms lies in the high cost of labour and the need to manage a large number of workers when considerable quantities of material are to be sorted. Such drawbacks may be avoided by resorting to automatic platforms.

Automation is introduced at the stage of debaling. In order to obtain a product suitable for the recycling process, operations to remove undesired impurities must be carried out. The machines required are manifold and the necessity to employ them is related to the quality of the collected material. A few essential machines are: (i) rotary screen, by which parts of the desired dimension are sorted out, separating them from smaller and larger ones; (ii) light-parts separation equipment, in which lighter parts such as films are separated by air blowing from the plastic material to be recycled; (iii) heavy-parts separation equipment, in which heavy particles are separated and the operation is carried out by means of air that shifts the material selectivity; and (iv) aluminium rejection equipment, which normally consists of an electromagnetic drum placed in a suitable location on the train of operations. All such machines are preliminary to the stage of selection into homogeneous plastics fractions.

Dry Separation

The cleaning of plastics is often combined with the separation of other types of plastics and is performed by either dry or wet process depending on the quality of the collected material. A significant advantage of dry cleaning with air, as compared to wet cleaning, is that it has a lower power consumption. Loose adhering dirt—and this is the only type of sorting that can be removed in a dry process—is loosened and pulverised by the impact and rubbing caused during size reduction. The dirt, then as dust, can be separated by using an appropriate equipment. Examples of this are screen units and air stream separator.

The screen unit is the most economical means of removing the dirt. By selecting a screen with suitably sized openings, it is possible to minimise the amount of plastic discharged with the dust. Electrostatic charge can however cause too many dust particles to adhere to the plastic. The air separator (Fig. 25.1) is then the more suitable device to use. Lighter-weight dust is carried out of the unit by the opposed directional air stream.

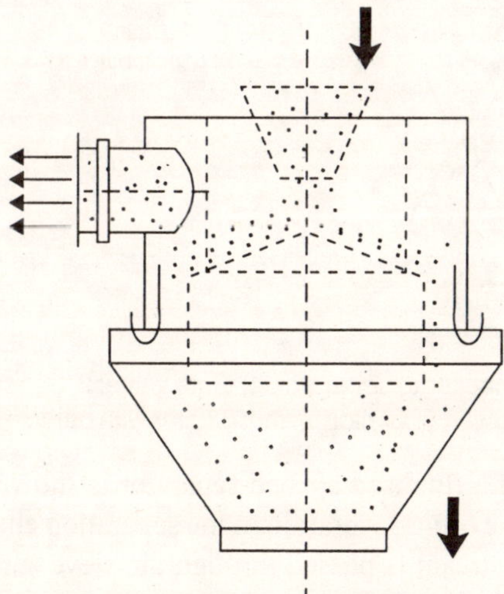

Fig. 25.1. Schematic of air-stream separator.

Mixed plastics, including various types of composite materials, that are to be dry-separated are first size-reduced (e.g., in a granulator) before the different constituents are separated. The separation is done in a process based on differences in material densities or shape and size of particles.

Screening may follow size reduction, depending on the material and the particle size distribution. In the screen unit, all of the mixed material is divided into two or three size fractions, e.g., 0 to 2 mm, 2 to 4 mm, and 4 to 6 mm. This is necessary since small heavy particles and large light pieces behave in the same manner, as do heavy flakes and light conical pieces. The separation based on density difference is easier or only then possible when all particles are of similar size.

Dry separation using air can be repeated several times, and the process is then classified as a cascade separator. A cascade separator, also known as a zigzag separator, uses an air stream passing through a rectangular zigzag channel from below (Fig. 25.2). The material to be separated is fed to the top end of the channel into the air stream. The separation point is set by adjusting the air flow rate. Material turbulence occurs in each section of the air stream channel, with lighter material being carried upward by air stream to discharge and heavy material moving downward from step to step. Fine particles are loosened from the larger ones and also from each other each time impact with side wall occurs. Cascade separators produce more efficient separation than single-stage units due to dispersion taking place at each kink in the channel. Typical applications for the cascade separator are the separation and fibres and insulation film or foam and soft film.

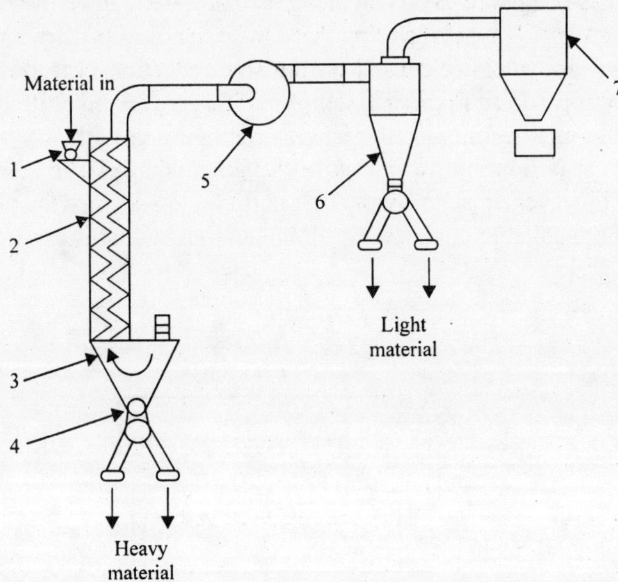

Fig. 25.2. Schematic of air-stream (cascade) separator. 1, Gate valve; 2, Cascading (zigzag) channel; 3, Container with air suction; 4, Gate valve; 5, Blower; 6, Cyclone; and 7. Filter with particle container.

The principle of operation of a fluidised air bed separator is shown in Fig. 25.3. The material to be separated is carried 'uphill' by the orbital vibration in the separation channel designed with a rectangular cross-section. An adjustable air stream is passed through the sieve surface in the channel where it lifts the material. The particles that are lifted higher in the air stream (that is, jump higher due to elasticity) flow downhill and are discharged from below. This type of fluidised air bed separation process can also be enhanced by using a multistage plant. A typical application for this unit is the separation of rubber from rigid thermoplastics or aluminium from plastics.

Wet Separation

Wet separation of plastics is a *microseparation method* in which a suspension medium is used to separate plastics with density higher or lower than the suspension medium. For example, water can be used as medium to separate PE from PVC or PET. In this case, special tanks are used in which various types of plastic flakes are mixed with water and then given a sufficient time to position themselves in the most suitable way according to their density. Materials are subsequently extracted separately from the top or bottom. This method is, however, not suitable for separating PVC from PET, because they have similar density.

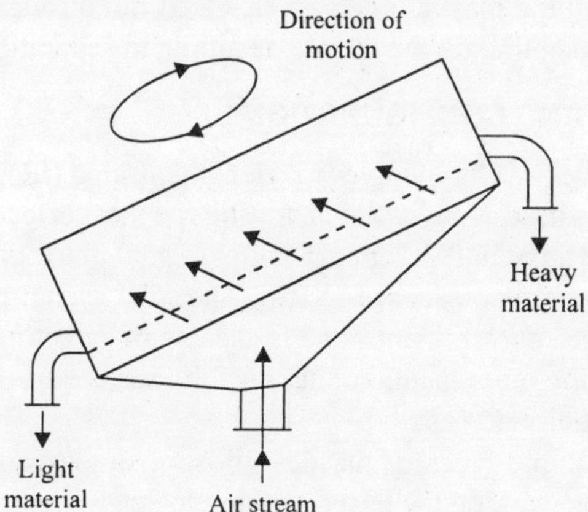

Fig. 25.3. Schematic of vibrating air separator.

Researchers at Rutgers University studied a method of PVC microseparation from PET, by which PVC is subjected to a process of selective bulking that causes it to float. Such a method may be applied for separating small quantities of PVC from large quantities of PET, as normally is the case in the United States, but is not suitable when PVC is a major component of the containers mix. Another wet method applicable to PVC and PET containers, previously reduced to flakes, is flotation with surface active agents.

Other Methods

Need for using microseparation techniques arises in critical situations where minute concentrations of identified contaminants adversely affect the postconsumer resin's usefulness in high-value end-uses. In many cases, microseparation, or the ability to sort resins by type, can be accomplished by air elutriation techniques (e.g., dry separation techniques, described above), wet separation techniques (such as sink/float tank technologies and hydrocyclones), magnets, electrostatic and electrodynamic methodologies, and optical scanners.

Air aspiration and elutriation systems work well for separating light-density foams and films from denser reground plastics, while density-based methods are better for separating polyethylene from PET and denser resins.

Challenges to separate materials having similar densities, for example, PVC from PET regrind, or polypropylene from polyethylene, remain. Examples of microsorting techniques commercialised so far include electrostatic separating devices designed to sort by way of resin's conductivity, supercritical fluids which alter the separation fluid's density, froth flotation using the alteration of a liquid's surface tension to separate various solids, and chemical dissolution based on the difference in solubility of various plastics in selective solvents. Another development is a novel method to separate diverse resins by taking advantage of their differences in stick temperature.

BASF's Kali and Salz AG company, which has extensive experience in electrostatic separation of salts, employs its own electrostatic separation process (ESTA) to the separation of plastics. Using density separation, paper and plastic residues from labels and crowns are separated first. Then, following pretreatment with surface active substances designed to enhance the electrostatic properties, the homogeneously milled particles are charged electrically as they rub against each other. The extent of electrostatic charge depends on the plastic. The particles then fall through a high tension field and are diverted at different angles depending on the charge, resulting in separation.

RESIN DETECTORS: TYPE AND CONFIGURATION

Detectors fall into four categories—X-ray, single-wavelength infrared (IR), full-spectrum IR, and colour. The earliest automated systems used X-rays, which are still the most effective means of determining the presence of PVC. The chlorine atom in PVC emits a unique signal in the presence of X-rays by either X-ray transmission (XRT) or X-ray fluorescence (XRF). The XRT signal passes through the container, ignoring labels and other surface contaminants, and is capable of detecting a second container that may be stuck to the first. XRF, on the other hand, bounces off the surface of the container and is useful for finding any PVC, including labels and caps.

Systems for separating multiple types of plastics utilise a single wavelength of the near infrared (NIR) spectrum. These systems work on the basis of a simple determination of opacity and separate the stream of mixed containers into clear (PET and PVC), translucent (HDPE and PP), and opaque (all pigmented and coloured materials) streams.

The most sophisticated detectors, however, employ full spectrum NIR. Since all plastics absorb IR to different degrees and each resin has a unique 'fingerprint', these detectors can accurately separate each of the resins. In later developments, filters for individual wavelengths are used for rapid identification and there is promise for even faster, lower-cost systems.

The colour detectors are, in fact, very small cameras capable of identifying a number of colours. When combined with a resin-specific detector, they allow a variety of sorts based on both resin type and colour.

Containers must be separated and presented to a detector in order to collect data on each unit. That information is then integrated via computer, and the container is tracked down the conveyor until it reaches the appropriate ejection point, where it is removed by a timed blast of air. The two primary techniques for presenting containers to a detector are full-conveyor or single-file systems. Containers delivered in a full-conveyor system can achieve higher throughput rates and are normally used to remove a single material from the mixture. A singulated stream, on the other hand, has lower throughputs, but allows sorting into a number of streams on a single pass.

Detectors are usually arranged in one of three configurations. The first is *single detector/single container* (Fig. 25.4a): this is the simplest setup for singulated containers. As each container passes the detector, several readings are taken instantaneously and a decision is made by the computer. While usually accurate, this process is subject to error if the container has a large label blocking the signal and thus restricting data input to the computer. The second configuration is *multiple detector/single container* (Fig. 25.4b): as each container passes the detector assembly, it is 'read' by a number of detectors resulting in a more accurate reading. The third is *multiple detector/multiple container* (Fig. 25.4c): This is the standard configuration for a mass-flow system and has detectors spaced to cover the width of the conveyor. When the target material is spotted, its position on the belt is noted and accordingly an ejector removes it before falling off at the end of the belt.

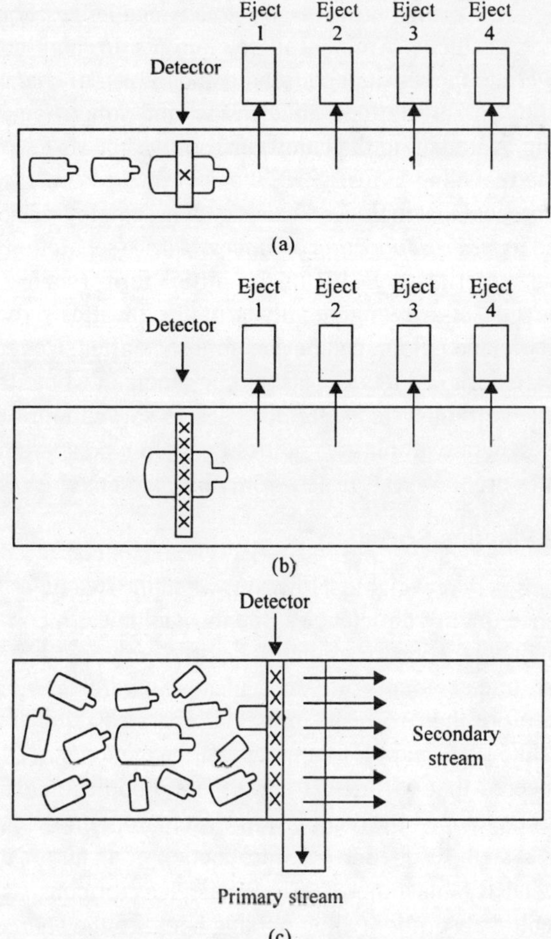

Fig. 25.4. Typical separation and sorting setups using three main detector systems: (a) single detector/single sample; (b) multiple detector/single sample; and (c) multiple detector/multiple sample.

Further development in autosort technology is represented by particulate-sorting units capable of sorting by colour. Applied in combination with the aforesaid resin detectors this facilitates autosorting according to both resin-type and colour.

AUTOMATIC SORTATION

Most end-use markets of recycled plastics require that they be separated by resin type and colour. For postconsumer bottles, all lids, caps, and closures should also be removed because they are often of different colours, and made of different resins than the bottle. The goal of any sorting process is to obtain the highest purity, consistency, and quantity of a particular consumer resin type. This ensures the highest end-use value for new products incorporating substantial amounts of the recycled resin.

In the widely used manual methods of sorting, employees known as material handlers are stationed at predetermined locations alongwide the sort conveyor to remove the desired bottles for recycling. All caps and closures still affixed to the bottles are manually removed by the handlers prior to or during sorting. These manual methods of sorting, however, face many challenges, some of which are economic and others environmental and aesthetic. Among the key hurdles are high cost of the labour-intensive process, exposure of employees to the residual household and industrial chemicals contained in some of the collected bottles, difficulty in sorting look-alike resins, and subjective material quality standards resulting from manual sorting. Automating the container sorting process to overcome these hurdles has thus been a major goal of the recycling industry.

An automatic separation process that includes various systems employing detectors currently available in the market is ideally suited to *macroseparation* or *macrosorting* (separation of plastic fractions before size reduction) when waste materials are still in the initial form (such as postconsumer bottles). Macrosorting is one of the fastest-growing segments of the plastics recycling industry. Although automated sorting, by resin type and colour, can be accomplished after size reduction, most commercial automated-sorting machine are designed to sort plastics containers in whole form before size reduction. Some of the automated bottle sorting systems are designed to sort certain bottles such as look-alikes, while others are designed to separate all plastics bottles by resin type and colour. The most commonly encountered look-alike bottles are those fabricated from non-pigmented PVC and clear PET.

PVC/PET and Comingled Plastics Sortation

Both PVC and PET bottles are very recyclable. However, when the two types are received commingled, the reprocessor can experience quality deficiencies due to rheological incompatibilities between these two resins. This has been of special concern to the PET bottle reclaimer because, as the PET is heated to its processing temperature, trace amounts of PVC can cause severe deterioration in the quality of the reprocessed PET resin. Therefore the two resins must be separated prior to recycling.

Manual sorting of look-alike PVC and PET cannot meet the market's needed quality. Therefore new techniques have been engineered that will detect and separate bottles made from either of these two resins. The first detector was developed by Tecoplast in Casumaro, Ferrara-Italy, to separate PVC from PET. The application of this system resulted in the introduction of an automatic plant, processing drink plastic bottles using the AZZURRA machine. The Tecoplast detector consists of an X-ray source and a receiver that measures the bottle absorption while passing between the source and receiver. PVC has a higher absorption compared to the other plastic due to the presence of a chlorine atom. The value of PVC absorption, electronically processed through algorithms, makes it possible to detect its presence and consequently to eject bottle.

Another detector is employed to line up bottles using a suction robot. The aligning process enables bottles to be arranged in a suitable way for measuring transparency and colour. Using this technology, Tecoplast developed the first optical detector capable of establishing the quality of PET, thus allowing the separation between clear PET and coloured PET, besides the aforesaid separation from PVC. In a later development, Govoni's technology employs detectors performing these separations without alignment.

Other detectors have been developed in the United States. A leading manufacturer of plastics sortation equipment, ASOMA Instrument, Inc. (Austin, Texas), developed a simple operator-friendly device for sensing the presence of chlorine atoms contained within the PVC resin. This sensor identifies PVC by X-ray fluorescence as bottles fall through a chute or move on a belt conveyor, at the end of which either the PVC or PET bottles are extracted—in most cases by a burst of compressed air-jet actuated by the device, although mechanical devices have also been employed. The X-ray fluorescence sensitivity is so reliable that a 10-ms analysis is all that is needed to make proper selection. Single-unit systems have been developed that both detect and separate commingled bottles at production rates of 800–1200 lb/hr, which corresponds to between two and three bottles per second. As is the case with most macro-separation devices, singulation (that is, alignment of bottles in a single file) is critical for this system to function at high production rates while maintaining rigorous quality standards. With X-ray detection, sorted PET streams having less than 50 ppm of PVC have been consistently produced, while, in contrast, manual separation generally results in PET with 2000 ppm PVC.

Another example is a system made by National Recovery Technologies (NRT), Inc. (Nashville, Tenn.). This apparatus incorporates a proprietary electromagnetic screening process that also detects the presence of chlorine as found in PVC bottles. Once detected, PVC bottles are pneumatically jettisoned from the comingled bottle feed stream by a micro-processor-based air blast system. The NRT technology permits bottles to be delivered to the unit in a mass sort concept in either crushed or whole form. This system does not require any special positioning or orientation of the bottles in order to achieve high efficiency rates.

An optical detector developed by Magnetic Separation Systems (MSS), Inc. (Nashville, Tenn) incorporates an optical sensing device with a transmission output range of 200–1500 nm to detect both the resin composition and the shape of the inspected container. Additionally, a video camera is employed to identify coloured containers via computerised spectrographic matching. This information is also processed through a high-speed micro-processor that has the ability to perform algorithmic analyses and alarm the programmable logic controller (PLC) to actuate an ejection apparatus to sort the desired bottle. Finally, an X-ray fluorescence sensor is used to sort PVC bottles from the PET bottle fraction. The MSS detector system is thus designed to obtain separation of comingled plastics into homogeneous material fractions, including PVC, clear PET, coloured PET, multicoloured HDPE, and translucent HDPE.

A modular sorting system of MSS, BottleSort, incorporates a sensory apparatus designed to detect and mechanically separate commingled plastic bottles in a process that includes several functions: debaling, screening, sensing, separation, and electronic control. Sensing is performed both optically and with X-ray fluorescence.

Each BottleSort modular unit can process 1250 lb/hr. Systems have been commercially installed incorporating four units having a combined capacity to sort 5000 lb of commingled bottles each hour. A near-infrared spectrophotometry detector, developed by Automation Industrial Control (AIC) of Baltimore, Maryland, allows identification of resin type, such as PET, HDPE, PVC, PP, LDPE, and PS. The equipment is connected with another detector for colour determination and the resulting data are processed by computer with highly sophisticated software.

The equipment thus enables separation of a container mix into various component with a high degree of selectivity in regard to typology and colour. This type of detector, however, requires material singularisation and lining up. The PolySort automatic plastic bottle sorting system introduced by AIC is designed to receive comingled plastic bottles in either baled or crushed form. At the heart of this sorting system is a sophisticated video camera and colour monitor incorporating a strobe to detect and distinguish colours in the inspected plastic bottle. This optical scanning device interfaces with a computer to match the colour of the bottle against a master. The detector is reputed to detect and match up to 16 million shades of colours.

In addition, the system can be programmed to disregard labels on the bottles. Following colour detection, a near-infrared detection system scans the single-bottle stream at a rate of approximately 3000 times a minute, to determine in less than 19 ms the primary resin found in each one. This is achieved by matching the interferogram produced by the bottle to a known 'master' for each base resin as stored in the system.

Computations are coordinated through the use of a rotary pulse generator and sensing light curtain to impel the qualified bottle to a discharge chute located on the sort conveyor. Although the standard PolySort system is designed to detect and sort about 1500 lb of compacted bottles each hour, higher production rates can be obtained by feeding multiple lines from one debaler.

A point of consideration is the efficiency of separation that could nullify the high efficiency of a detection equipment. Detection must be unfailingly followed by rejection of the detected bottle, and in reality this does not always occur. Delivery systems in use today are effective but not 100 per cent accurate. Most errors that occur with an autosort process are due to mechanical delivery errors rather than error related to detection. Therefore, a check on the operations of selection of the detected product is of primary importance. Detectors with 99 per cent or greater efficiency, if installed in series and in number of at least two units, can bring the level of impurities within the limits required. For example, with a mix in the proportion of 90 per cent PET and 10 per cent PVC, two detectors with 99.5 per cent efficiency placed in series enable one to obtain a level of impurities of 2.5 ppm, whereas with efficiency of 99 per cent the value of residue is 10 ppm.

RECYCLE INSTALLATIONS

Special importance is generally attached to the techniques of electronic selection of homogeneous fractions, as discussed above, while disregarding the phase of regeneration of selected plastics. This would be justified if plastics articles were manufactured following criteria of perfect recyclability. However, such criteria are not yet universally adopted or followed. Therefore, an accurate response must be given, in any recycling plant design, to the problems posed by various elements that normally compose the item to be recycled. Referring to liquid containers, in general, these consist of, in addition to the body made of plastics, other foreign bodies that are to be removed. Such elements may be caps made of PE, PE with PVC gaskets, aluminium, labels of PVC or tacky paper with different types of glue, and residues and dirt that may have been added during the waste collection phase.

Various operations, to be carried out in a specific sequence because of the problems posed by the type of material, are grinding, dry separation, and wet separation. Machines to be installed for these operations are dictated by the typology and quality of recycle items. Grinding is the first step following selection and requires attention and accuracy in design to ensure optimum homogeneity of the ground product. Several types of grinding equipment in common use have been described earlier.

The purpose of dry separation is the removal of a part in the dry phase. The process allows avoiding problems of dissolution in water and relevant contamination. An air flotation method is used for dry separation. Specially designed machines combining the effects of vibration and air flotation ensure separation of flakes with different specific weight. Such machines are very useful for removal of parts of labels that were freed by grounding. An extremely, interesting application of this method is the separation of PVC labels from PET bottle flakes.

Residues are normally washed out of material flakes using a class of equipment that includes centrifugal cleaners, washing tanks, autoclaves, settling tanks, combined-action machines, scraping machines (mechanical friction), and centrifugal machines (for water separation). The construction details of the machinery and their installation according to a specific sequence are in the know-how of various manufacturers and very little is revealed. In view of complex problems posed by postconsumer plastics installations, it may, however, be said that it is not possible to expect miraculous results from key processes carried out in a single passage and therefore the efficiency is maximised by repetition of the same operation in more than one phase.

CHAPTER 26

Recycling of Poly(Ethylene Terephthalate) and Polyurethane

INTRODUCTION

Plastic materials are increasingly replacing conventional materials such as paper, wood, metal and glass in a variety of volume applications. They are essentially used in the form of fibres, films, moulded and extruded goods. Desirable combination of their properties such as good mechanical strength, chemical inertness, water resistance and some times, their limited permeability to water and oxygen have led to the extensive use of polymers in a variety of end consumer articles. However, these very properties make polymers non-biodegradable materials resulting in a solid waste disposal problem. Discarded plastic waste is accumulating at a rapid rate and has adverse effects on the environment because they are not degraded biologically in sanitary landfills and garbage dumps, especially, under anaerobic condition and in the absence of light.

Plastic waste management has become a major strategic issue for many developed and developing countries. Plastic producers and end users are currently experiencing intense pressure of environmental issues that are directed primarily at reducing post - consumer plastic waste as well as clean scrap generated at production units as part of the polymer processing and conversion steps. Thus the investigations concerning the utilisation on disposal of polymeric waste by way of recycle are of considerable technological significance and social relevance. Here we wish to focus our attention exclusively on recycling of (PET) and Polyurethane (PU).

POLY(ETHYLENE TEREPHTHALATE)

The largest use of poly(ethylene terephthalate) (PET) is in the fibre sector, with PET film and PET bottles representing only about 10 per cent each of the total PET volume produced annually. A large percentage of the total PET output comprising films, plastics, and fibres is recycled by various methods and for several applications, which makes PET one of the largest in volume of recycled polymers in the world. Contributing to this is the suitability of PET for practically all recycling methods, which include direct reuse, reuse after modification, recovery of monomers and other low-molecular-weight intermediates, and incineration.

Any particular method is selected on the basis of the quality of waste and scrap, the economy of the process, and the convenience of the operation.

Direct Reuse

This method, also called recycling by re-extrusion or melt recovery, is used for relatively pure PET waste such as cleaned consumer bottles or in-house waste. The method is based on the same principles as the original equilibrium polycondensation reaction:

$$H{-}[{-}ET{-}]_m{-}OCH_2CH_2OH \; + H{-}[{-}ET{-}]_n{-}OCH_2CH_2CH \rightleftharpoons$$
$$H{-}[{-}ET{-}]_{m+n}{-}OCH_2CH_2OH \quad + EG$$

$$[ET] \equiv [{-}OCH_2CH_2OOC{-}C_6H_4{-}CO{-}] \quad \text{(ethylene terephthalate)}$$

$$EG \equiv HOCH_2CH_2OH \qquad \text{(ethylene glycol)} \qquad ...(26.1)$$

As polymer buildup and polymer degradation are taking place in the melt simultaneously, the reaction conditions have to be controlled very carefully in order to obtain the desired molecular weight and molecular weight distribution for the end use. In theory, this seems rather simple; in practice, however a large amount of determining parameters (temperature, environmental atmosphere, holding time in a melt state, amount of impurities, type of used catalysts, stabilisers, etc.) have to be kept under control.

A practicable re-extrusion process with some modifications—is still being used. The greatest recycler of fibre waste in the United States in Wellman; they recover PET fibre and bottle waste for home furnishing and non-woven materials by a similar method.

Customer-recollected waste from fibre and textiles consists mainly of continuous filaments or staple fibres, which may be contaminated with dyestuffs, finishes and knitting oils, and other fibres such as cotton, wool, rayon, nylons, and acrylics; they are the most difficult to recover products.

A different picture can be presented for the PET bottles. In the environmentally active states in the U.S., 80–95 per cent of the PET bottles sold are recollected and recycled. In Europe and in Japan where recycling has started earlier than in the United States, various reclamation and reprocessing methods have been worked out and applied in practice. Because these processes are usually proprietary, the details of their operation are not known.

The largest use of PET film is as a photographic film base, which accounts for over 50 per cent of the PET film produced in the world. The manufacturers of these materials, mainly Agfa-Gaevert, Eastman Kodak, du Pont de Nemours, Fuji, 3M, and Konishiroku, have long been interested in the recovery of PET film because of its content of rather expensive silver derivatives. Recycling of PET-film waste in production, which may amount to 25–30 per cent of the total output, is almost complete by these manufacturers.

In a typical way of operation, PET film recycling is coupled with the simultaneous recovery of silver, as represented schematically in Fig. 26.1. In the first step of the process, photographic emulsion layers containing silver are washed with, for example, NaOH, and after separation, silver is recovered on one side and cleaned PET waste on the other side. Careful analysis is necessary to ensure that the washed PET-film scrap is clean enough to be recovered by direct extrusion.

The most obvious way of adding the recycled PET flakes is after the usually continuous polymerisation and before the PET melt enters the extruder screw. Such a procedure, however, has two main drawbacks: first, the highly viscous melt is difficult to filter (to eliminate possible gels or microgels); and second, other impurities (e.g., volatiles, oligomers, coloured parts) cannot be eliminated any more. In order to remove these disadvantages, several alternative modes have been worked out. A method to add recycled PET to the polymerisation batch reactor during the esterification step was described by du Pont. Such a method shows the following advantages over the method described above: filtration can take place in the low-viscosity phase, and volatiles can be eliminated during the pre-polymerisation phase.

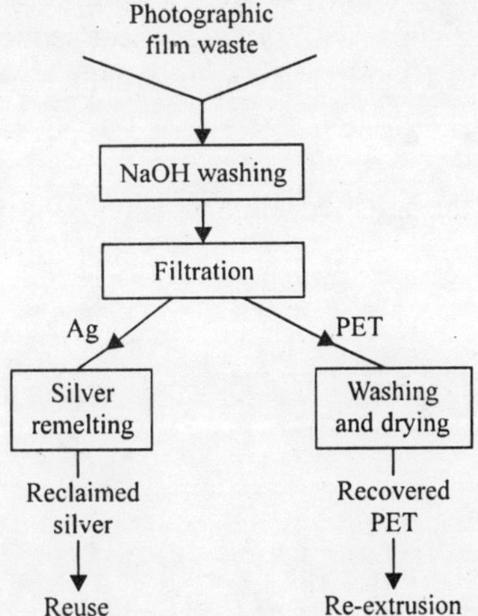

Fig. 26.1. Combined recovery of silver and PET.

PET recycling by direct reuse, as described above, is by far the most economical process. However, it is useful in practice only for well-characterised PET wastes that have exactly known chemical composition (catalysts, stabilisers, impurities). Therefore, the method is ideally suited for the recovery of in-production wastes, but it may not be suitable for postconsumer PET film.

Reuse After Modification

For postconsumer PET wastes having a higher degree of contamination, technological processes based on degradation by either glycolysis, methanolysis, or hydrolysis can be used. These yield products that can be isolated. The principles of chemical processes involved in these methods are schematically represented in Fig. 26.2.

$$HO\!\left[CO-C_6H_4-COO(CH_2)_2O\right]_n\!H$$

PET

Glycolysis

$PET + HO-CH_2-CH_2-OH \rightarrow$

Hydrolysis

$PET + NaOH\ (H_2O) \rightarrow$

or

$PET + H_2SO_4\ (H_2O) \rightarrow$

Methanolysis

$PET + CH_3OH \rightarrow$

Mixture of a score of intermediates of varying molecular weights

$\rightarrow HO-CH_2-CH_2-O-OC-C_6H_4-CO-O-CH_2-CH_2-OH$

$\rightarrow HO-CH_2-CH_2-OH + HOOC-C_6H_4-COOH$

$\rightarrow HO-CH_2-CH_2-OH + H_3C-OOC-C_6H_4-COOCH_3$

Fig. 26.2. PET degradation by glycolysis, hydrolysis, and methanolysis.

Hydrolysis and methanolysis of PET regenerates the starting monomers. Thus, terephthalic acid (TPA) along with ethylene glycol (EG) are obtained by hydrolysis, while methanolysis yields EG and dimethyl terephthalate (DMT) among other products. Stopping short of complete depolymerisation, glycolysis degrades long polymer chains (with typical repeat sequences of 150 units) into short-chain oligomers (repeat sequences of 2 to 10 units) having hydroethyl end groups.

Glycolysis

The addition of EG to PET reverses the polymerisation reaction. This can be stoichiometrically represented by

$$\left(\frac{x}{y}-1\right) \underset{\text{(EG)}}{HOCH_2CH_2OH} + \underset{\text{(PET)}}{HO\left[CH_2CH_2O-\overset{O}{\overset{\|}{C}}-C_6H_4-\overset{O}{\overset{\|}{C}}O\right]_x CH_2CH_2OH}$$

$$\xrightarrow[\Delta]{\text{Cat.}} \left(\frac{x}{y}\right) \underset{\text{(Monomer)}}{HO\left[CH_2CH_2O-\overset{O}{\overset{\|}{C}}-C_6H_4-\overset{O}{\overset{\|}{C}}O\right]_y CH_2CH_2OH} \quad \text{... (26.2)}$$

where x = average number of repeat units in polymer and y = average number of repeat units in monomer. When $y = 1$, 'monomer' = dihydroxyethyl terephthalate (DHET).

Glycolysis thus represents a compromise between regeneration of starting ingredients by methanolysis or hydrolysis and direct melt recovery. It is less costly than the former and more versatile than the latter. The resultant, easily filtered, low viscosity 'monomer' can be repolymerised to a useful higher molecular weight product. A typical flow sheet of the process is shown in Fig. 26.3.

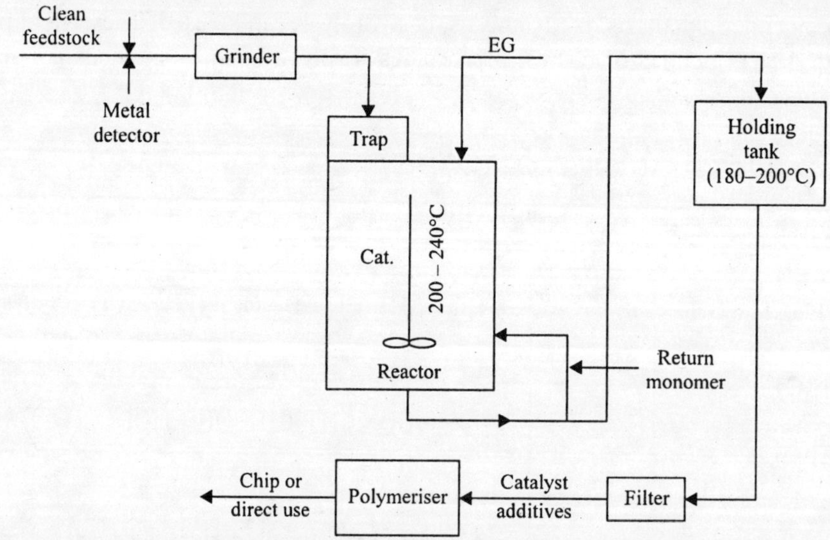

Fig. 26.3. Flow diagram of a typical system for glycolytic recycling of PET waste.

PET scrap suitable for glycolytic recycle includes production waste, fibres, film, flake, and bottles. In a practical system, major contaminants are separated from feedstocks, e.g., bottle waste is cleaned and separated from a polyethylene base, paper labels, metallic caps, and liners. For many end uses, coloured PET must also be segregated. (Highly modified co-polymers, glass-reinforced resin, fibre, or fabric blends are not suitable for glycolysis. These can only be recovered by methanolysis/hydrolysis). Since reaction time depends on surface area, PET feedstocks must be reduced to relatively small particles by grinding, cutting, etc.

The du Pont company published many details covering the glycolytic recycling of PET. Goodyear has also developed a PET recycling process based on glycolysis that is called REPETE. In a batch process, a molten 'monomer' heel is left in the reactor to allow the feedstock/glycol mixture to reach optimum reaction temperatures. In a continuous process (Fig. 26.3) some of the molten 'monomer' is recycled to a stirred reactor to accomplish the same function. High glycol/terephthalate (G/T) ratios lead to more complete glycolysis but lower the maximum temperature, increasing the reaction time. A ratio of 1.7 to 2.0 G/T is a practical compromise. An ester exchange catalyst as zinc or lithium acetate is usually added to increase rate of glycolysis. Reaction temperatures of 220–240°C and times of 60–90 min are typical. The reactor is operated under a positive pressure to prevent forming an explosive mixture of air and glycol vapours.

The major side reaction is the production of ethers:

$$2HOCH_2CH_2OH \xrightarrow{Acid} HOCH_2CH_2OCH_2CH_2OH \longrightarrow HO\text{-}[\text{-}ET\text{-}]_n\text{-}CH_2CH_2O\text{-}[\text{-}ET\text{-}]_m\text{-}OCH_2CH_2OH$$

$$ET \equiv \text{-}[\text{-}OCH_2CH_2OOC\text{-}C_6H_4\text{-}CO\text{-}]\text{-} \quad \text{... (26.3)}$$

Since this reaction is acid catalysed, it can be minimised by adding a buffer such as sodium acetate or by adding water. Lithium acetate catalyst also produces less ethers than zinc acetate. Some other side reactions are the formation of aldehyde, cyclic trimer of ET, and dioxane. Oxidation of glycol ends produce aldehydes that lead to coloured compounds. Traces of dioxane can form from the cyclisation of glycol.

If other glycols, such as diethylene glycol, are substituted for ethylene glycol, the corresponding oligomers are formed. These can subsequently be polymerised with aliphatic diacids as adipic or 4,4'-diphenylmethane diisocyanate to give rigid elastomers. Additives to control luster, colour, and so on can be added in the usual manner before and after polymerisation.

Primary uses for PET from glycolytic recycle are geotextiles, fibres for filling products, non-wovens, and moulding resins where colour, strength, and control of dyeability is not important. Recovered polymer can be added to virgin polymer for films, fibres, and moulding resins.

Methanolysis

PET waste obtained in the form of film, bottles, and fibres can be very conveniently converted into its raw materials dimethyl terephthalate (DMT) and ethylene glycol (EG) by methanolysis. The process involves heating the PET waste with methanol at 240–250°C and 20–25 kg/cm^2 pressure in the presence of catalysts such as metal oxalates and tartrates. Once the reaction is completed, DMT is recrystallised from the EG-methanol molten liquor, and distilled to obtain polymerisation-grade DMT. Also EG and methanol are purified by distillation. Eastman Kodak has been using such a process for recycling of X-ray films for nearly 40 years and it is still improving the process, e.g., by using superheated methanol vapour to allow the use of even more impure PET waste. Important factors that have to be dealt with in this process are avoiding colouration due to aldehyde formation and minimising the formation of ether glycols.

Ammonolysis

PET wastes can be converted via ammonolysis to paraphenylenediamine, which is a basic raw material for the high-modulus-fibre Kevlar or for high-value hair dyes. The chemical basis for this process is a modified Hoffman rearrangement. The synthesis may be done via the following three stages:

In the first step, granulated PET is suspended in ethylene glycol and treated with gaseous ammonia at 100–140°C. In this reaction, the ethylene glycol also acts as a catalyst. The product terephthalimide (I) is insoluble in the medium and thus may be isolated. In the second step, terephthalimide (I) is suspended in water and chlorinated vigorously with chlorine gas. The resulting terephthalic bis-chloramide (II) is treated with NaOH solution to obtain paraphenylene diamine (III). An important aspect of this process is that paraphenylenediamine so obtained is completely free from its *ortho* and *meta* isomers and its production cost is much less than the market price. ICI has reported an alternative single-step process for conversion of PET to paraphenylenediamine by ammonolysis in the presence of hydrogen gas.

Step 1

$$\left[-OCH_2CH_2OC(=O)-C_6H_4-C(=O)-\right]_n \xrightarrow{NH_3} \underset{(I)}{H_2NC(=O)-C_6H_4-C(=O)NH_2} + HOCH_2CH_2OH$$

Step 2

$$\underset{(I)}{H_2NC(=O)-C_6H_4-C(=O)NH_2} \xrightarrow{Cl_2} \underset{(II)}{ClHNC(=O)-C_6H_4-C(=O)NHCl} + HCl$$

Step 3

$$ClHNC(=O)-C_6H_4-C(=O)NHCl \xrightarrow{NaOH} \underset{(III)}{H_2N-C_6H_4-NH_2} + NaCl + Na_2CO_3$$

Hydrolysis

PET can be completely hydrolysed by water at higher temperatures and pressure in the presence of catalysts (acidic as well as alkaline) to regenerate the monomers, terephthalic acid, and ethylene glycol. While both acid and base catalysed systems are completely realistic, their usefulness under practical production conditions remain controversial. As far as acid hydrolysis is concerned, the large acid consumption and the rigorous requirements of corrosion resistance of the equipment make profitability questionable. Moreover, the simultaneous recovery of TPA and EG, requiring the use of ecologically undesirable halogenated solvents, is difficult and not economical. For the alkaline hydrolysis process, also, the profitability is strongly determined by the necessity of expensive filtration and precipitation steps. In spite of the fact that the majority of newer industrial PET-synthesis plants are based on the TPA process rather than on the DMT process, the hydrolytic method of PET recycling has not gained favour.

Incineration

For PET wastes containing a large amount of impurities and other combustible solids it is more profitable to resort to 'quaternary recycling', that is, energy recovery by burning. Research along this line has been performed, particularly in Europe and Japan. Strong emphasis has been laid on the optimisation of incinerators with regard to higher temperature of their operation and reduction of the level of air pollution.

Having a calorific value of *ca* 30.2 MJ/kg, which is about equivalent to that of coal, PET is readily suited for the incineration process. However, like other plastics its combustion requires 3 to 5 times more oxygen than for conventional incineration, produces more soot, and develops excessive heat that thus calls for special incineration equipment to cope with these problems.

Several processes have been developed to overcome the technological drawbacks of plastics incineration cited above. These include continuous rotary-kiln processes; a process for glass-reinforced PET; a combined system for wood fibre and PET to provide steam to power equipment; and a fluidised system for pyrolysis, in combination with silver recovery from photographic film. Incineration of photographic film raises the additional problem of the formation of toxic halogenated compounds due to the presence of silver halides.

Incineration of PET is usually carried out at temperatures around 700°C, since at lower temperatures waxy side products are formed, leading to clogging, while at higher temperatures the amount of the desirable fraction of mononuclear aromatics in the condensate decreases. A representative sample pyrolysed under optimum conditions yields, in addition to carbon and water, aromatics like benzene and toluene, and a variety of carbon-hydrogen and carbon-oxygen gases. Studies have been made relating to the formation of dioxines and residual ashes containing heavy metals and other stabilisers. While most problems arising during incineration of PET can be resolved, it is evident that quite a few hurdles remain to be overcome before an economically feasible and ecologically acceptable industrial technical process becomes available.

In conclusion, it may be said that there exists a clear hierarchy in PET-film recycling technologies. Two most important criteria of classification are the degree of purity of PET scrap to be handled and the economics of the process. While for the cleanest PET grade the most economical process is direct reuse in extrusion, for less-clean PET samples it is still possible to reuse them after the modification step (partial degradation, e.g., by glycolysis) at a reasonably low price. More-contaminated PET waste must be degraded into the starting monomers, which can be separated and repolymerised afterwards, of course, at a higher cost. For this operation, mostly the methanolysis process has been exploited industrially. Finally, the most heavily contaminated PET wastes have to be incinerated or brought to a landfill.

PET films

PET film does not require the use of solid state polymerised resin. The intrinsic viscosity of PET resin used for film application varies from 0.6 – 0.65 dL/g. PET films are available in a variety of grades suitable for: (i) packaging; (ii) photographic and x-ray film, and as (iii) magnetic recording film used in video and computers as recording tapes.

Recycling of photographic and x-ray film had received importance initially for the recovery of silver content. Normally this type of film has a composite structure containing gelatin, silver and a coating of polyvinylidine chloride (PVDC). The film is first passed through a bath preferably comprising a hot caustic solution for precipitating silver layered on the film. The film then passes through a second bath of hot caustic until an adhesive sheet disposed on the film has been dissolved. This adhesive sheet is made of PVDC that help bind the silver to the film. After a second caustic bath, the film is dried and available for recycling process. Besides using alkaline solution, another common desilverising process uses enzymes to destroy the gelatin on the film and thus liberate the silver. The use of cyanides to complex the silver is also cited as an alternative process.

Films that are silver free and also free of other contaminants can be used to produce value added products such as polyols by glycolysis, or to regenerate the monomers by methanolysis process.

PET fibres

The PET waste generated during the production of PET fibres and textiles has been investigated for its chemical conversion. In a typical polyester staple fibre plant, homogeneous waste is generated at the polymerisation (1.0 wt per cent), spinning and take up (3.5 wt per cent). This waste is a clean scrap which is well suited as feed stock for chemical processes as well as recovery of spinnable resin.

Fibre waste can be reextruded followed by granulation, or by dissolution in a solvent from which the polymer can be precipitated, dried and ground to a powder. The properties of fibre obtained by repeated extrusion and pinning are satisfactory upto three recycling steps. However, during the extrusion melting, PET is degraded to some extent. By heating the fibrous waste below its melting point in an inner and dry medium, the reduction in molecular weight can be compensated.

Spinnable polymer can be obtained from polyester waste by heating the waste in a solvent, e.g. naphthalene which is preferred. The polymer precipitates as an amouphous gel or paste when the polymer solution is gradually cooled to about 100°C.

RECYCLING OF POLYURETHANES

Polyurethanes are by far the most versatile group of polymers, because the products range from soft thermoplastic elastomers to hard thermoset rigid forms. Although polyurethane rubbers are speciality products, polyurethane foams are well known and widely used materials. While the use of plastics in automobile has increased steadily over the years, a major part of these plastics is polyurethane (PU), which is used for car upholstery; front, rear, and side coverings; as also for spoiler. In fact, about half of the weight of plastics in modern cars is accounted for by PU foams. Accordingly, in addition to production scrap, large quantities of used PU articles are now generated from automotive sources. Though most PU plastics are cross-linked polymers, they cannot be regarded as ordinary thermosetting plastics, owing to their chemical structure and physical domain structure. Thus in contrast to typical thermosetting plastics, various methods are available today for recycling PU scrap and used products.

There are basically two methods for recycling polyurethane scrap and used parts, namely, *material recycling* (primary, secondary, and tertiary recycling) and *energy recycling* (quaternary recycling). The former methods are preferred since in this way material resources are replenished. After multiple uses the material can finally be used for energy recovery by high-temperature combustion or gasification.

Processes of Recycling Polyurethanes

Among several processes described for PU material recycling, *thermopressing* and *kneader recycling* have attracted much attention. By the thermopressing process, granulated PU wastes can be converted into new moulded parts, while in the kneader recycling process a thermomechanical operation causes partial chemical breakdown of PU polymer chains that can be subsequently cross-linked by reacting with polyisocyanates. Hydrolysis and glycolysis are important tertiary recycling processes for PU wastes.

Thermopressing process

Thermopressing, or moulding by heat and compression, is a direct method of material recycling that is designed such that elastomeric, cross-linked polyurethanes can be recycled in much the same way as thermoplastic materials. The principle of thermopressing is based on the realisation that polyurethane and polyurea granules are capable of flowing into each other and building up new bonding forces under the influence of high temperature (185–195°C), high pressure (300–800 bar), and strong shearing forces. The granules generally used for this purpose have a diameter of 0.5 to 3 mm. They completely fill the cavities of a mould meaning that mouldings with new geometries can also be manufactured. Unlike injection moulding of thermoplastics for which a cold mould is used, in the thermopressing process, the mould is kept constantly hot at a temperature of 190 ± 5°C and no release agent is used for demoulding. This relatively simple technique will permit 100 per cent recycling of polyurethane RIM and RRIM mouldings, particularly when the formulations of RIM systems to be used in future have been optimised for recycling. The steps in the thermopressing process are shown in Fig. 26.4.

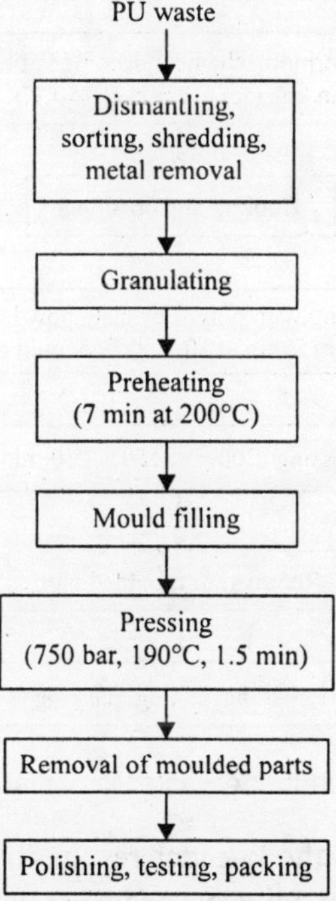

Fig. 26.4. Reprocessing of polyurethane waste by thermopressing.

The moulded parts obtained by thermopressing of granulated PU waste exhibit only slight reduction in hardness and impact strength but significant reduction in elongation at break. The last named property, for example, drops to about 10 per cent of the original value if painted PU wastes are used. Moreover, because of the use of granulated feed, the resulting moulded parts lack surface smoothness and thus should be used preferably in those areas where they are not visible. In a passenger car, there are many such parts that are not subjected to tensile stress but require dimensional and heat stability—properties fulfilled by PU recycled products. Examples of application are wheelboxes, reserve wheel covers and similar other covers, mudguard linings, glove boxes, and casings.

Kneader process

The basis of the kneader recycling process is a thermomechanical degradation of polymer chains to smaller-size segments. The hard elastic PU is thereby converted into a soft, plastic (unmolten) state, which is achieved with a kneader temperature of 150°C and additional frictional heating. This leads to temperatures above 200°C and causes thermal decomposition into a product that is soft at 150–200°C but becomes brittle at room temperature, enabling it to be crushed to powder in a cold kneader or roller press. The resulting powder can be easily mixed with a powder form polyisocyanate (e.g., Desmodur TT or 44 of Bayer) and moulded into desired shapes by compression moulding at 150°C and 200 bar pressure. The scheme of the recycling process is shown in Fig. 26.5.

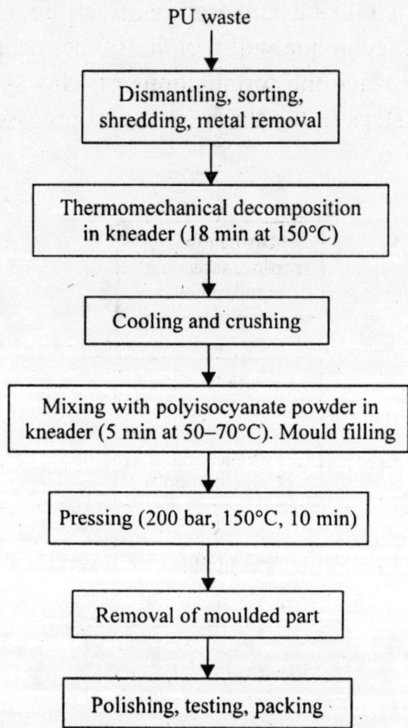

Fig. 26.5. Recycling of polyurethane waste via partial decomposition in kneader.

Partial breakdown of PU network in the kneader results in highly branched molecules with many functional groups necessitating addition of polyisocyanate in relatively high concentration for subsequent cross-linking to produce moulded articles. The process thus yields products of high hardness (with Shore hardness up to 80) and high tensile strength (30 MPa), but small elongation at break (6 to 8 per cent).

Hydrolysis

Hydrolysis of PU waste results in the formation of polyethers and polyamines that can be used as starting materials for producing foam. In this process, powdered PU waste is reacted with superheated steam at 160–190°C and the polymer gets converted in about 15 min to a liquid heavier than water. The liquid is a mixture of toluene diamine and propylene oxide (polyether diol), the former accounting for 65–85 per cent of the theoretical yield:

$$\sim\sim\text{R–NH–}\overset{\overset{\text{O}}{\|}}{\text{C}}\text{–OR}'\sim\sim \xrightarrow{+\,H_2O} \sim\sim \text{RNH}_2 + \text{CO}_2 + \sim\sim\text{R}'\text{OH} \qquad ...(26.4)$$

The recovered polyether can be used in formulations for making PU foam, preferably in admixture with virgin polyether.

A continuous hydrolysis reactor utilising a twin-screw extruder has been designed that can be heated to a temperature of 300°C and has a provision for injection of water into the extruder at a point where the scrap is almost in the pulp state. Polyurethane scrap in powder form is fed into the extruder and residence time is adjusted to 5–30 min. Separation of the two components, polyether and diamine, in the product may be effected by fractional distillation, by extraction with a suitable solvent, or by chemical means. The PU foams made from these recycled products can be used in several applications, one example being protection boards for construction sites. Hydrolytic recycling has not, however, found much application, since virgin raw materials are cheaper than the regenerated products.

Glycolysis

Extensive studies have been made on glycolytic degradation of PU wastes. In a glycolytic process, powdered PU waste is suspended in a short-chain glycol and heated to a temperature of 185–210°C in nitrogen atmosphere. The glycolysis reaction takes place by way of transesterification of carbonate groups in PU (Fig. 26.6).

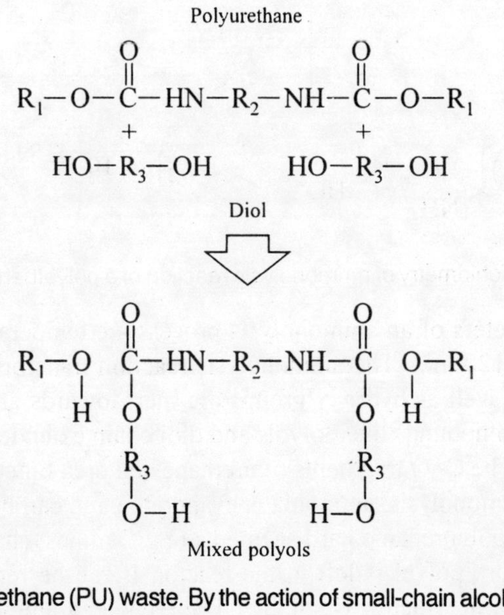

Polyurethane

$$\text{R}_1\text{–O–}\overset{\overset{\text{O}}{\|}}{\text{C}}\text{–HN–R}_2\text{–NH–}\overset{\overset{\text{O}}{\|}}{\text{C}}\text{–O–R}_1$$

$$+ \qquad\qquad +$$

$$\text{HO–R}_3\text{–OH} \qquad \text{HO–R}_3\text{–OH}$$

Diol

$$\Downarrow$$

$$\underset{\text{H}}{\underset{|}{\text{R}_1\text{–O}}}\;\;\underset{\underset{\underset{\text{O–H}}{|}}{\underset{|}{\text{R}_3}}}{\underset{|}{\underset{\text{O}}{|}}}\!\!\overset{\overset{\text{O}}{\|}}{\text{C}}\text{–HN–R}_2\text{–NH–}\overset{\overset{\text{O}}{\|}}{\text{C}}\!\!\underset{\underset{\underset{\text{H–O}}{|}}{\underset{|}{\text{R}_3}}}{\underset{|}{\underset{\text{O}}{|}}}\;\;\underset{\text{H}}{\underset{|}{\text{O–R}_1}}$$

Mixed polyols

Fig. 26.6. Alcoholysis of polyurethane (PU) waste. By the action of small-chain alcohols (e.g., diol), PU is decomposed yielding homogeneous, liquid, mixed polyols.

The reaction product is predominantly a mixture of glycols and does not need any further separation of the components, unlike in the hydrolytic process. The cost of producing such recycled polyol is reported to be low enough to make the process economically viable.

The mixed polyols resulting from glycolytic degradation of PU waste is suitable mainly for the production of hard foam, such as insulating foam for houses.

Ammonolysis

Chemical recycling of polyurethanes by ammonolytic cleavage of urethane and urea bonds under supercritical conditions has been described. It is well known that a number of low-boiling materials give enhanced solubility and reactivity under supercritical conditions. Ammonia has a critical point at 132.45°C and 112.8 bar (11.28 MPa) with a density of 0.235 g/cm^3. Being able to act as hydrogen-bond donor and acceptor, it provides good solubility for polyurethanes and dissolves their hard segment domains thus enabling a homogeneous reaction.

Ammonia is also a reagent having greater nucleophilicity than, for example, water or glycol is; since it is added in a huge molar excess compared to the urethane or urea groups of the materials to be cleaved, the equilibrium is shifted towards the ammonolysis products. The stoichiometry of ammonolysis reaction of a polyetherurethane is shown in Fig. 26.7.

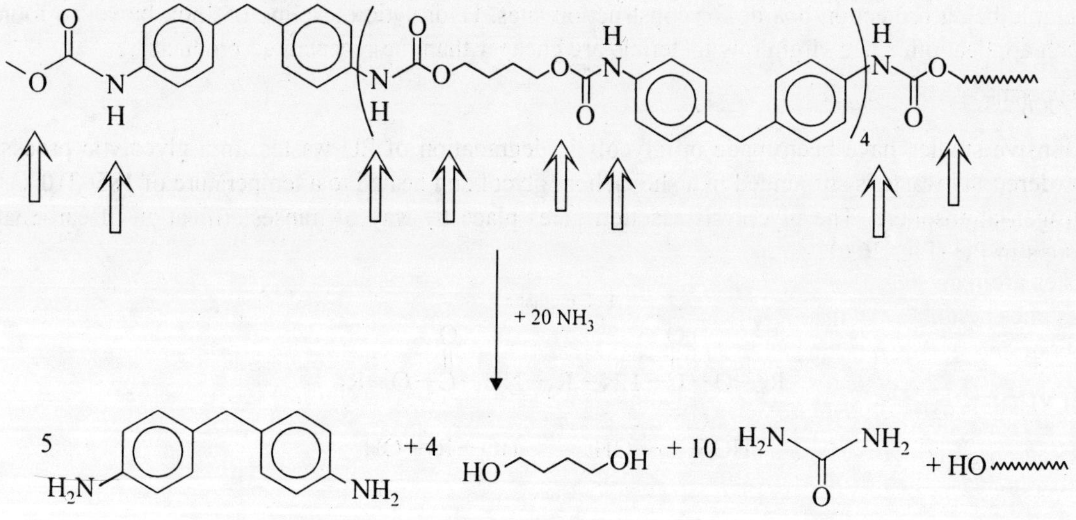

+ 20 NH_3

5 H_2N … NH_2 + 4 HO…OH + 10 H_2N–C(=O)–NH_2 + HO～～～

Fig. 26.7. Stoichiometry of ammonolysis reaction of a polyetherurethane.

The typical reaction parameters of an ammonolysis process are temperature of 139°C, pressure of 140 bar, and reaction time of 120 min. The ammonolysis reaction transforms derivatives of carbonic acid into urea. Ether bonds as well as hydroxy groups are inert towards ammonia under the reaction conditions applied. Hydroxy compounds like polyols and diol chain extenders that do not contain ester groups are recovered as such. The C=O fragments of urethane and urea functional groups are converted to unsubstituted urea. After ammonolysis, ammonia is evaporated and can be reused after liquefaction, while degradation products of polyurethane hard segments (e.g., amines, chain extenders) and urea are removed by extraction. The pure polyol is left in the reactor. It can be removed mechanically or by extraction with liquid ammonia in which it is soluble. The recovered amines can be converted to the corresponding isocyanates and can be reused, along with polyols, in the same applications as before. A flow scheme of the recycling process is shown in Fig. 26.8.

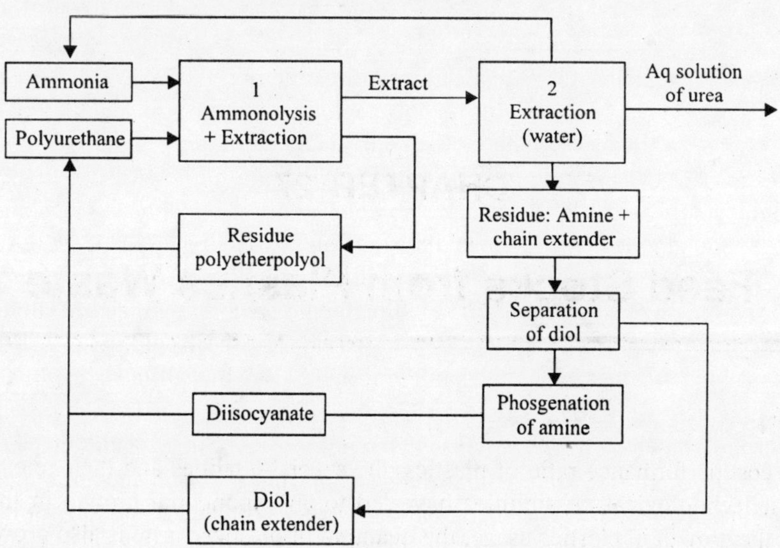

Fig. 26.8. Flow scheme of a chemical recycling process based on ammonolytic cleavage and separation of polyol by supercritical ammonia.

Among the various material recycling methods for PU scrap and wastes described above, the thermopressing and kneading processes are especially significant, because these simple processes render the recycling of cross-linked PU products equivalent to that of thermoplastic products. Lack of surface smoothness and some reduction in mechanical properties are to be tolerated, especially when painted PU wastes are recycled. However, good values of E-modulus, structural rigidity, and hot and cold impact resistance permit use of the moulded components of recycled PU in many applications, e.g., in unsighted parts of automobiles, instruments, and machineries.

POLYURETHANE FOAM MADE FROM WASTE VEGETABLE OIL

A research group led by Assistant Professor Katsuji Matsunaga of the Faculty of Engineering at Toyo University has succeeded in synthesising polyurethane foam (PUF) and polyurethane elastomer (PUE) from waste vegetable oils such as salad oil.

The research group modified the waste vegetable oil, and employed it as a polyol in the synthesis of polyurethane. Their success brings hope as an effective means of recycling waste cooking oil.

Using a polyol blend comprised of 100 per cent modified waste soy oil-based polyol, polyxypropylene triol, and castor oil, the research group synthesised elastomers and polyurethanes and investigated their properties. They found that while the elastomers were brittle in terms of properties and in need of further application development, the properties achieved for the urethane foams were in no way inferior to products based on virgin raw materials. In addition, the research group reports that rigid urethane can be synthesised through addition of hydrogen peroxide and sulphuric acid, as well as changing the formulation ratios. On this account, waste cooking oil can be reused and recycled in various urethane applications, including cushioning materials, coatings and adhesives. The research groups plans to push ahead further with application research and development.

CHAPTER 27

Feed Stocks from Plastics Waste

INTRODUCTION

The favourable cost/performance ratio of plastics, the special qualities and their relevance in modern society and their technological possibilities have led to an exponential growth in their usage. As a consequence of the growth in plastics usage, the quantum of plastics waste is also growing rapidly and this growth is inevitable. Plastics waste comes from various sources, thus segregation of the waste is an important step in plastics waste re-processing/disposal. There are four widely accepted plastics waste processing techniques, brief details of which are given in Table 27.1.

Recycling of plastics waste in conventional ways may not make sense:

1. When it uses more resources than it saves.
2. If there is insufficient demand for the recycled material.
3. When the plastics waste has gone beyond a particular grade, after which such wastes are resorted to for energy recovery by incineration.

There is a need for recycling schemes to be market driven. Recycling should be recognised as just one of the tools to help recover and re-utilise resources. Harmonisation alone can encourage industry and public authorities to work together to determine the best mix of waste management options, including ways to develop markets for the recycled materials. Further, as life cycle analysis (LCA) develops, the benefits of plastics will become more apparent.

Table 27.1. Plastics waste reprocessing techniques.

Primary recycling	Plastics waste is uniform, uncontaminated and can be processed as such.
Secondary recycling	Plastics waste is a mixture of plastics items and cannot be re-processed as such.
Tertiary recycling	Thermal and chemical decomposition methods are used to convert plastics waste into raw materials for plastics and other items.
Quarternary recycling	Incineration of plastics waste to recover energy.

Two obstacles remain in the path of developing a large scale, commercially economic business, based on recycled plastics from municipal solid waste (MSW).

1. The cost of recovery and purification.
2. The deficiencies in the properties impairment by contaminants and by reprocessing.

Considerable work is being done around the world to reduce these obstacles. An alternative approach to simultaneous solution of the two problems is also being developed.

TECHNICAL OPTIONS

All the approaches to recycling involve only physical manipulation of thermoplastics. Another approach is to use chemical means to separate various resins in mixed plastics waste, still another process involves using chemical and/or thermal means to break the polymers into useful by-products. However, thermoset plastics do not flow when heated and cannot be recycled by the physical methods. These can, however, be subjected to pyrolysis. Very little commercialisation exists because of the high production costs that would be incurred for scaling up. These processes may be more useful in the future, when the supply of petrochemical raw materials becomes scarce or when the cost or regulation of petroleum feed stock becomes prohibitive. Production of acrylic monomer feedstock from acrylic scrap is one of the most important and widely accepted processes of this type.

FEED STOCK RECOVERY

Besides conventional recycling methods, polyurethanes, polyesters and polyamides can be degraded by chemical processes to low molecular weight products. Hydrolysis and alcoholysis are well known reactions for pure polymer waste, but the commercial application has only been realised on a small scale.

The trend to multiphase polymers makes it more difficult to get materials with high quality by usual recycling methods. A typical example is wastes of automotive seats, which consists of flexible polyurethanes in combination with polyamide and polyester fabrics.

The advantage of alcoholysis is that by variation of reaction parameters the properties of polyol can be flexibly adopted to the requirements of polyurethanes. Also, the above mentioned wastes do not find a purchaser when they are mixed, soiled or have network structures. A physical separation of these materials is impossible. Only the chemical pecularity of the above mentioned polymers can be used to regain original substances by depolymerisation with special chemicals. In the case of polyesters, the end products of this so called saponifying reaction are monomeric alcohols and carboxylic acids.

To illustrate, a great part of the quantitatively dominating flexible polyurethane waste can be refined profitably to composite foams, elastomers and resilient materials. Recycling by melting is not possible, because of their chemical network structure. The hydrolysis reaction can be applied with different technical methods at high temperature and under pressure. Similarly polyamides and polyesters can be alcoholised to low molecular weight products. When these polymers are alcoholised with diols or polyols, the resulting product is a mixture of hydroxyl-terminated polyether and polyester polyols; in the presence of polyamides, the products are hydroxyl and amino terminated oligomers.

CHEMICAL/TERTIARY RECYCLING

Chemical recycling may involve either pyrolysis or chemical decomposition of plastics waste. Chemical decomposition of plastics can be carried out by one of the following ways: (i) hydrolysis; (ii) hydrocracking; and other processes.

Pyrolysis is the physical and chemical decomposition of organic materials caused by heating in an oxygen-free or oxygen-deficient atmosphere. This process is also called destructive distillation whereby volatiles are driven off leaving a char consisting of carbon and quite often a fairly large ash content.

Chemical decompositon of plastics waste by hydrolysis, hydrocracking and other processes has definite advantages over pyrolysis. The products by decomposition are more uniform, control is easier, there is lower capital investment and even smaller plants can work economically. It involves basically three major chemical processes:

1. Hydrolysisthe— thermocracking of polymers into hydrocabons (HC).
2. Hydrocracking— the thermal processing under extreme pressures in the presence of hydrogen to yield different and more functional HCs.
3. Depolymerisation—the processing of polycondensation polymers into starting or similar materials for re-use in polymerisation.

Chemical recycling of plastics covers recovery of polymer raw materials or other useful chemicals from plastics waste. Decomposition of waste plastics by chemical means is possible for a variety of plastics (Table 27.2).

Table 27. 2. Tertiary processes of re-processing of plastics

Condensation polymers		
Polymer type	*Process*	*Product (s) recovered*
Polyamide	Acidolysis hydrolysis	Repolymerisable
Polycarbonate	Hydrolysis	Bisphenol A
Poly(ethylene terephthalate)	Hydrolysis	Terephthalic acid ethylene glycol
Poly (ethylene terephthalate)	Alcoholysis	Dialkyl terephthalates
Poly (ethylene terephthalate)	'Alcoholysis' with ethylene glycol	Bis (hydroxyethyl) terephthalate
Polyurethane	Hydrolysis	Polyols, amines
Polyurethane	'Alcoholysis' with glycol	Polyols, amines
Vinyl and cyclic monomers		
Polymer type		*Process*
Poly (alkyl methacrylates)		Pyrolysis
Polystyrene		Pyrolysis
Poly (tetra fluoroethylene)		Pyrolysis
Poly (vinylidene cyanide)		Pyrolysis
Polycaprolactam		Catalytic depolymerisation
Polycaprolactone		Pyrolysis

PYROLYSIS OF PLASTICS WASTE

The basic concepts of plastics pyrolysis are similar to those involved in the pyrolysis of municipal refuse. The difference lies in the handling characteristics of plastics compared to those of municipal solid waste, and the types of products obtained in the process. The pyrolysis products of plastics can be used either as fuel or as feedstock for chemical industry.

PYROLYSIS REACTION

The depolymerisation of polymers is somewhat similar to the process of polymerisation. Polymers decompose into smaller molecules or monomers, depending on their structure and the conditions of reaction. Polymers that decompose to monomers are shown in Table 27.2. Although the mechanism of polymerisation and depolymerisation are similar, the energy required for depolymerisation is greater than that for polymerisation. If the rate constants of both processes are plotted as a function of inverse absolute temperature, they will intersect at a certain temperature. Above this ceiling temperature a net depolymerisation will take place.

During the pyrolysis process, the following reactions takes place: (i) depolymerisation, producing monomers; (ii) chain fragmentation, producing low-molecular weight materials; and (iii) production of unsaturated compounds, crosslinking of the polymer and char formation. The kinetic behaviour of polymers during thermal degradation is intermediate between the following extreme possibilities.

1. A large initial production of gaseous products with a very slow decrease in the molecular weight of the polymer (the monomer being removed from the ends of the chains—a typical example is polymethylmethacrylate.
2. A very slow initial production of gaseous products with a rapid decrease in the molecular weight of the polymer (the polymer chain is broken in random locations: a typical example is polyethylene). Thus, depending on the design of the process, the resulting product can be in the form of a monomer, a low-molecular weight polymer, or a mixture of various hydrocarbons.

TECHNICAL OPTIONS AVAILABLE FOR SELECTED PLASTICS

Polyolefins (PO)

Polyolefins such as polypropylene (PP), polyethylenes (PE) can be pyrolysed to hard waxes which are superior in hardness and colour to carnauba wax. These waxes find potential applications in polishes, printing inks, release agents and lubricants. Decompolsition of LDPE results in the production of greases and waxes which can be made emulsifiable in water and can be used in polish formulations, textiles, finishing agents, adhesives and lubricants. During the pyrolysis of polymers some degree of unsaturation is produced, which provides reactive sites for further chemical modification of the product. Depending on the temperature, tar, grease, liquids or gases can be produced. Low molecular weight pyrolysate is similar to petroleum and can serve as a petrochemical raw material source.

Reaction products resulting from the degradation of polyethylenes contain mainly paraffins (molecular weight upto 50) rather than monomer. Once the chain is initiated, the mechanism probably consists of breaking off large molecular fragments in rapid succession. Fig. 27.1 gives probable species in thermal decomposition of polyethylenes, which equilibrates at 773K and l atm. The reaction intermediates include toluene, ethane, propane, ethylbenzene and *o*-, *m*-, *p*-dimethyl benzene.

Polystyrene (PS)

Initially the decomposition of polystyrene occurs randomly: weak links present in the polymer matrix break first. During this period inhibitors give rise to an induction period. At the end of this induction period, depolymerisation takes place. The results indicate that benzene is in equilibrium with styrene at 1073k and l atm pressure. Since the formation of that mixture represents the minimum free energy of the system, it is the most probable distribution of the species at equilibrium. Fig. 27.2 gives the probable species formed during thermal decomposition of polystyrene.

Polyvinyl Chloride (PVC)

The thermal degradation of polyvinyl chloride is mainly by way of the dehydrochlorination reaction. After a brief initial period, HCl is evolved and after HCl has been removed from PVC, a secondary decomposition of PVC takes place which yields a variety of organic products. The major equilibrium products of thermal decomposition of PVC at 773K and 1 atm pressure are benzene and hydrogen chloride. Fig. 27.3 shows the probable species formed during thermal decomposition of PVC. Fig. 27.4 shows the major products formed during thermal decomposition of PE, PS, and PVC.

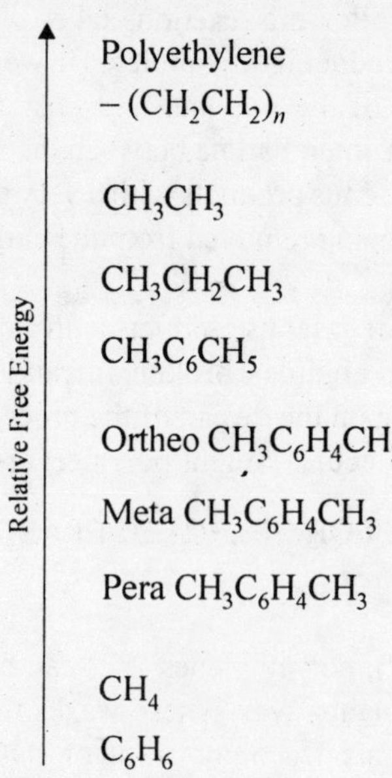

Fig. 27.1. Probable species in thermal decomposition of polyethylene.

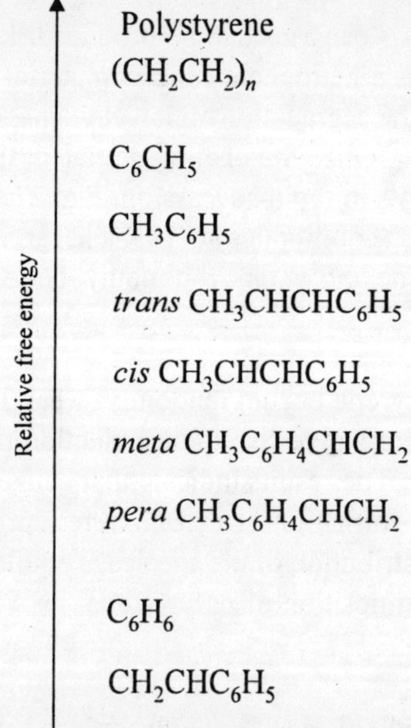

Fig. 27. 2. Probable species in thermal decomposition of polystyrene.

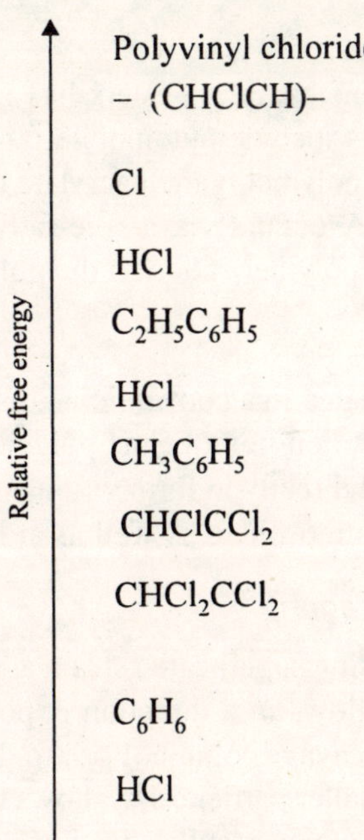

Fig. 27.3. Probable species in thermal decomposition of polyvinyl chloride.

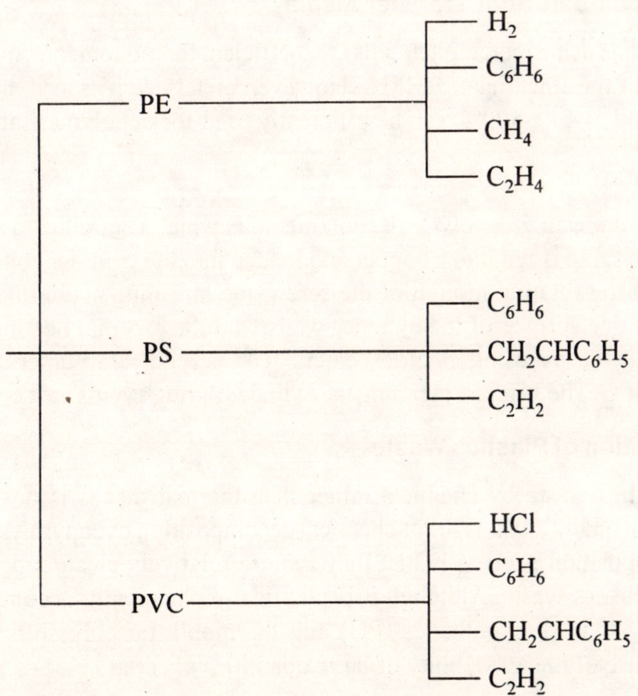

Fig. 27 .4. Probable products of thermal decomposition of PE, PS, and PVC.

Acrylics

The use of polymers of acrylate and methacrylate esters has increased over the last decade. Some special applications of acrylic polymers include automobile tail lights, sanitaryware, signboards etc. Most often, the term 'acrylic' refers to polymethylmethacrylate (PMMA).

The thermal degradation of PMMA occurs via a free-radical chain reaction at approximately 400–450°C. The free radicals initiate depolymerisation of the polymer.

Dry distillation

In dry distillation, the acrylic scrap is heated in a conventional distillation flask at atmospheric pressure to a temperature above the depolymerisation point of the polymer. The monomeric vapours are either condensed and collected or polymerised directly to form virgin acrylic polymers. Further purification if required can be carried out in reduced atmosphere as well as at high pressures.

Super-heated steam as heat transfer agent

The depolymerisation of acrylic scrap using superheated steam as heat transfer agent involves the use of an inert gas as a carrier. The inert gas flows in a direction opposite to the steam, resulting in a wind sifting action within the steam-heated cleavage column. Prior to depolymerisation, PMMA is ground to a particle size of less than 6 mm. The smaller particle size allows the scrap to be carried into the column by the inert gas. The monomer vapours may be stabilised using an inhibitor to limit the potential for re-polymerisation.

Molten Metal and Metal Salt Heat Transfer Media

Depolymerisation methods using inert molten salts are inefficient for monomer recovery due to insufficient heat transfer medium for the cracking of PMMA. However, metals such as lead, bismuth, cadmium, tin, thallium, zinc etc. heated at 400–500°C can be efficiently used for depolymerisation of PMMA.

Extrusion Methods

Extrusion is one of the interesting depolymerisation methods which uses either a single or a twin screw vented extruder. Acrylic scrap is fed into a hopper and loaded into the cylinder containing a single screw. The scrap is heated via the rotating motion of the screw, the kneading action that occurs between the surface of screw and inside surface of the cylinder walls, and the external heating of the cylinder. The external heating is achieved by using electric heaters. The acrylic scrap enters various zones and is decomposed at 550–600°C. The vapours escaping the cylinder through vents are condensed and collected.

Chemical Decomposition of Plastics Waste

Decomposition of plastics waste by chemical rather than thermal means is possible for a variety of plastics. Though there are many advantages of chemical decomposition over pyrolysis, the main drawback of the chemical decomposition process is that they require relatively clean and uniform feed and are unsuitable for mixed plastics waste. Although it is possible to chemically decompose many polymers, the main interest today is in polyurethanes (PU) and thermoplastic polyesters (PET). These are of particular interest because of the availability of large quantities of scrap.

Hydrolysis

Hydrolysis is one of the ways of chemical decomposition of plastics containing chemical groups susceptible to such reactions. Hydrolysis is a reverse reaction to condensation; it is only those polymers obtained by polycondensation that can be hydrolysed. Examples of such polymers are polyurethanes, polyesters, polycarbonates and polyamides. Of the hydrolysable plastics, the recycling of PUs is of the greatest industrial interest.

Glycolysis

The necessity of separating amines and glycols produced by the hydrolysis of polyurethanes is a serious drawback to the hydrolysis process. This drawback can be eliminated by a polyurethane recovery process based on glycolysis. Glycolysis is a relatively simple process in which polyurethane foam is decomposed at 185–200°C in the presence of an appropriate glycol. The chemistry of the glycolysis process involves transesterification of the carbonate groups in the polyurethane foam with glycol solvent. Here no separation is necessary, and the recovered mixed polyols can be re-used in combination with virgin material. Glycol recovered by glycolysis produces foams practically indistinguishable from those made with virgin glycol. Glycolysis can be used for decomposing rigid and flexible polyurethane as well as isocyanurate foams. The cost of producing recycled polyol is low enough to make the process economically attractive. Further glycolysis of polyurethane waste may be catalysed by certain organometallic compounds and/or tertiary amine catalysts.

Other Processes

Broeck and *Peabody* describe a process for reclaiming polyurethane foam by dissolving it in liquid pre-polymer and using the solution to produce a new foam. The dissolving of the foam is carried out at temperatures of 250°C. It is possible to produce mixtures containing upto 50 per cent dissolved scrap, but concentrations below 25 per cent are recommended.

Heissal developed a process for the liquification of scrap polyurethanes and polyureas by dissolving them in carboxylic acids. Examples of reagents which may be used in the process are fatty acids, polymerised unsaturated acids, polycarboxylic acids, and the mixture containing suitable acids such as tall oil, stearine pitch, fish oil pitch.

Matsudaira described a process wherein polyethers and diamino-toluenes can be recovered from waste PU foam or elastomers by thermal decomposition at temperatures below 250°C in the presence of an aliphatic amine, a cycloaliphatic amine, a heterocyclic amine, or derivative of these amines and one of following types of compounds: an alkali metal oxide, an alkali metal hydroxide, or an aqueous solution of these compounds.

Macdowel developed a two-stage process for the reclamation of waste terephthalate polyester films or fibre, by heating the waste with glycol and converting it to diglycol terephthalate. This monomer can then be added to the virgin monomer and polymerised to new polyester resin.

Gruschke developed a two-stage process for depolymerisation of polyethylene terephthalate. The first stage is carried out in an autoclave in which polyester waste is melted and mixed with 2–8 times the amount of methanol. The residence time in the autoclave is about 10 min. The second stage is carried out under the same pressure conditions but at slightly lower temperatures in a reaction tube without stirring. After the reaction is completed, the solution of diethyl terephthalate in methanol is transferred to a stirred vessel at 100°C, and the pressure is reduced to atmospheric. In most cases the decomposition product is so pure that polyethylene terephthalate or degradation products of low molecular weight cannot be detected by the usual analytical methods.

According to *Stevenson*, a process for the recovery of high purity dicarboxylic acid and glycol from scrap polyesters containing dyes and other impurities is based on the reaction of waste polyester in the form of fibres, films, pellets, moulded pieces, and so on with an alcohol in an autoclave at an elevated temperature, followed by catalytic hydrogenation. Depolymerisation products, while in solution, may be purified by filtering or by the absorption of impurities in activated charcoal. After depolymerisation and purification, the material is subjected to a selective hydrogenation conducted under such conditions that unsaturated colour forming impurities are hydrogented while the main product is not. The unsaturated colour forming impurities thus become colourless.

Specific Examples

Conversion of polyester waste to phenylene diamine

Polyester waste obtained in the form of film, bottles and fibres can be very conveniently converted into raw materials like dimethyl terephthalate (DMT) by the process of methanolysis involving heating of polyester waste with methyl alcohol at a temperature of 240–250°C in the presence of catalysts such as metal oxalates and tartrates at a pressure of about 20–25 kg/cm^2. The reaction may be represented by following flow diagram (Fig. 27.5).

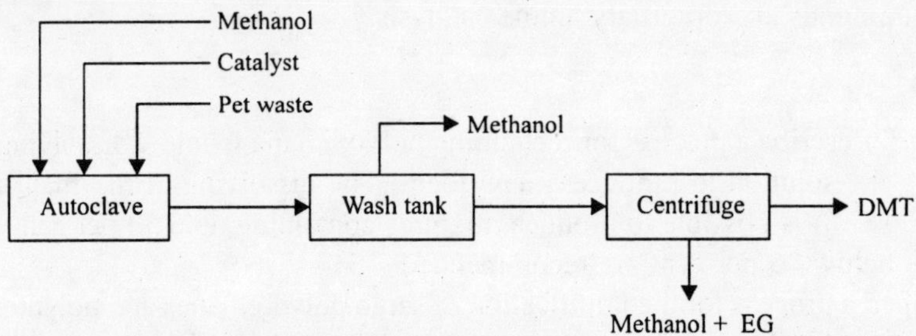

Fig. 27.5. Conversion of polyester waste to phenylene diamine.

However, in comparison to methanolysis, when the polyester waste is subjected to ammonolysis one can selectively obtain paraphenylene diamine which is the basic raw material for high modulus fibre, Kevlar, or for high-value hair dyes. The chemical basis for this process is a modified *Hoffman* rearrangement of acidamine to amine. This synthesis is reported to be progressing via the following three stages.

Ammonolysis of polyester waste to terephthalamide

Granulated PET waste is suspended in ethylene glycol and treated with gaseous ammonia at 100–140°C.

$$\left[-CH_2CH_2OOC-C_6H_4-COO \right]_n \xrightarrow{+NH_3}$$

$$HOH_2CH_2COO-C_6H_4-CONH_2 \xrightarrow[-HOCH_2CH_2OH]{+NH_3} H_2NOC-C_6H_4-CONH_2$$

In the above reaction, ethylene glycol acts as a catalyst. The product terephthalamide is insoluble in the medium and may thus be isolated.

Conversion of terephthalamide to terephthalic bis (chloramide) (TBC)

It is a simple reaction and is carried out by suspending terephthalamide in water by HCl and chlorinating the same vigorously with chlorine gas. The TBC formed is isolated by way of filtration.

$$H_2NOC-C_6H_4-CONH_2 \xrightarrow[-\,HCl]{+\,Cl_2} H_2NOC-C_6H_4-CONHCl$$

$$\xrightarrow[-\,HCl]{+\,Cl_2} ClHNOC-C_6H_4-CONHCl$$

(TBC)

Conversion of TBC to paraphenylene diamine (PPD)

TBC is reacted with caustic soda to yield PPD in the following manner.

$$ClHNOC-C_6H_4-CONHCl \xrightarrow{6NaOH} H_2N-C_6H_4-NH_2 + 2NaCl + 2Na_2CO_3 + 2H_2O$$

A rough calculation of cost of the raw materials indicates that PPD produced from waste shall be available almost at 50 per cent of the market cost. The most important aspect of this process is that the PPD so obtained is completely free from its *ortho* and *meta* isomers.

An alternative single-step process for conversion of PET to PPD by ammonolysis in presence of hydrogen gas has been developed.

Fluoro lubricating materials from teflon waste

Teflon finds extensive applications as Teflon sheet, tape and rope and is a difficult-to-process plastic. Due to extremely high melt flow viscosity, Teflon is processed by a special technique called sintering, involving conversion of Teflon powder into various items under temperature and pressure.

Thus, Teflon collected in the waste can hardly be re-processed except that it may be converted into powder by grinding operation and sent back to Teflon processing unit, which may utilise this powder for mixing in fresh polymer. Other than that, Teflon waste has probably limited scope for being utilised by plastic waste processing industry. However, if we utilise the unique behaviour of Teflon towards high energy radiation we may end up with highly valuable lubricating materials like high performance greases.

It is well known that polymers like polyolefins, when exposed to high energy radiation like gamma rays and electron beams get cross-linked and turn hard and brittle. Contrary to this, when Teflon is exposed to gamma rays (10^7 rads) it loses its molecular weight due to chain scission. This is reflected in loss of tensile strength and the material turning into soft powder having much higher impact strength. It is thus possible to convert lubricants like soft powders into greases by dispersing them into suitable media. Such greases are likely to replace similar materials being used in space and atomic energy areas where reliable high-performance materials are required.

Further, the powders produced from waste Teflon may also be blended with asbestos or graphite to produce gasketing and gland-filling materials for reactors operating at high temperatures.

Polyurethanes to polyols and diamines

Polyurethane foam has attracted the attention of chemists and engineers engaged in devising ways to reutilise foam waste. Conversion of foams into processable polyurethanes has a number of practical limitations, three of which are more prominent: (i) chemically cross-linked nature of foam; and (ii) very low density; large volume of entrapped vapours of blowing agents.

Difficulty is faced by workers when foams are sent for quarternary recycling process, because incineration produces highly toxic gaseous products. Keeping these limitations in view, attention has been focussed on devising chemical techniques such as hydrolysis or glycolysis. These processes are based on the fact that urethane groups present in PU foams are susceptible to attack by hydrolysing agents and hence foams may be converted into basic chemical raw materials such as polyols and diamines.

In India, though the plastic waste re-processing industry is quite developed, waste upgradation efforts are either quite limited or totally non-existent and therefore, re-processors must appreciate the need for upgradation of wastes and devise suitable measures in this direction. Tertiary plastics re-processing methods can yield valuable products but require a judicious selection of feed stock and technology.

CHAPTER 28

Recycling of Miscellaneous Polymers

INTRODUCTION

Aside from the polyolefins, poly(vinyl chloride) (PVC) and some other chlorine-containing polymers belong to the most widely applied thermoplastic materials. There are many applications of rigid and plasticised PVC. In the building sector, for example, very large amounts are used for pipes, profiles for windows, floor coverings, roofing sheets and so on. By the end of the lifetime of these articles, large amounts of scrap have been produced. It is of economic and environmental interest to recycle this PVC waste as much as possible. Disposal of PVC waste by incineration has its special problems. Due to high chlorine content of PVC, its incineration yields large amounts of HCl gas in addition to the possibility of formation of toxic dioxines and furans. On the other hand, it is a great advantage that many sources produce large amounts of PVC scrap of the same origin and with similar composition, which simplifies the reuse possibilities from a logistic point of view.

Dealing with postconsumer mixed PVC waste involves special considerations. Reprocessing PVC-containing plastics waste without separation will normally entail dealing with mixtures in which large proportions of polyolefins (mainly polyethylene) are present. In view of the poor compatibility of polyolefins with PVC, this is not a particularly attractive practical proposition, with respect to processing and the resulting product. Selective reclamation, i.e., separation from waste mixtures with other plastics, and subsequent reprocessing are complicated by the wide variety of PVC formulations, and the increased susceptibility to heat degradation in reprocessing. The main factors in the latter are the 'heat history' already acquired; the possible presence of polymer already partly degraded in the course of past heat treatments and/or service; and the remaining stability of PVC articles before their recycling, which often necessitates an additional stabilisation by addition of heat stabilisers. Moreover, about 1/3 of the used PVC is plasticised by various types of plasticisers. Therefore, for the recycling of such PVC types the concentration of plasticisers should be known. Due to these considerations, it is important to have rather detailed information about a PVC scrap before use.

RECYCLING OF PVC

Since several chemical reactions occur during processing and use of PVC, which can change the properties of the polymer, it is necessary to characterise PVC scrap before deciding about the reusability. Under the influence of heat and light (and also oxygen), PVC chains can be degraded or even cross-linked, which results in changes in the molecular weight and distribution and thus in the mechanical properties of PVC. For determining the molecular weight distribution, gel permeation chromatography is the most applied method, but in many cases the measurement of solution viscosity after separation of all insoluble components, including cross-linked PVC, will suffice. For practical purposes, the determination of the K-value can give sufficient information in most cases.

Because practically no PVC is processed and used without the addition of stabilisers, one should know the residual stability of a used PVC product. For this the best way may be the determination of the hydrogen chloride elimination at 180°C under air or nitrogen. The conversion-time curves so obtained provide indication of the residual stability from the induction period and also enable calculation of the rate of HCl split-off after consumption of the stabilisers. In some cases, however, it may be sufficient to use a simple Congo Red test (e.g., according to DIN 53 418) instead of the apparatus for measuring the HCl elimination.

The dehydrochlorination of PVC results in the formation of polyene sequences that can be responsible for discolouration and also act as starting sites for further degradation and cross-linking reactions. For some applications it may thus be useful to have some knowledge about the unsaturated structures that have been formed in PVC during the use. For this purpose, the investigation of the UV-VIS spectra that give at least semiquantitative information about the dehydrochlorination and the application of ozonolysis, which results in cleavage of the unsaturated sequences in PVC, may be useful.

In the case of reuse of plasticised PVC, it is important to determine the residual plasticiser content. This can be obtained by extraction with ether or similar nonsolvents for PVC and determination of the chemical nature of the plasticisers by thin-layer or gas chromatographic methods. The determination of the glass transition temperature by differential thermal analysis also gives information on the efficiency of the residual plasticisers.

In-Line PVC Scrap

Normal recirculation, in the same process, of the clean PVC scrap generated (e.g., edge trim in calendering) is widely practiced, in particular with PVC for noncritical applications. General PVC scrap, both from internal and external sources, is also converted by some processors into such products as cheap garden hose or core composition for cables.

Two possibilities have been investigated for the recycling of die-cutting scrap produced in the processing of PVC sheet: production of secondary sheet and production of extruded profiles or pipes. First the scrap is ground up in a grinding mill. The stabiliser used in the original process is added again in a premixer, and pigments are often added to achieve a uniform, desired colour. This premix is fed into the compounding unit.

If the regrind includes rigid, semirigid, and plasticised PVC, the rigid and semirigid fractions can be charged into the compounding unit through a first-inlet opening and plasticated in an initial kneading zone. The plasticised PVC scrap is then fed into this fluxed stock, which ensures the gentlest and most homogeneous processing (Fig. 28.1). Any additional plasticiser required is injected directly into the kneading zone of the compounding unit by a pump. This is done because the plasticiser cannot diffuse in the PVC regrind within a reasonable time, which it can in the case of virgin PVC. It is advisable to use a strainer in order to remove any contamination from the stock. Afterwards, the calendering process is carried out as usual.

For the production of profiles and pipes, the homogeneous stock is pelletised following compounding. The pellets are fed to an extrusion line.

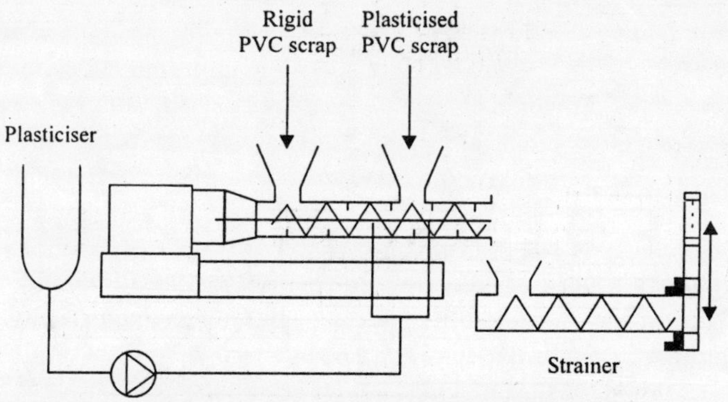

Fig. 28.1. Recycling of PVC film scrap.

In the cable sector, compounders are often confronted with the problem of recycling copperless insulation and sheathing scrap. An approach that may be taken in this case is to use this scrap for producing filling core mixtures. The purpose of the filling cores is to fill out the cavities between a cable's conductors (Fig. 28.2). Since their composition is not subject to any special electrical or mechanical specifications, it is normally made as inexpensive as possible, usually receiving a high level of chalk filler. The PVC in this case acts mainly as binder for the filler. For compounding such cable filler cores, the reground PVC scrap, with a particle size of 5–10 mm, is fed into the first inlet of a compounding unit designed specifically for this application (Fig. 28.3). The reground scrap is plasticated homogeneously in the first kneader zone, enabling it to absorb the high filler loading fed into the second inlet opening without any difficulty. For increased flexibility of the filler cores, plasticiser may be injected into the kneading chamber by a pump (Fig. 28.2). The homogenised stock is pelletised following compounding.

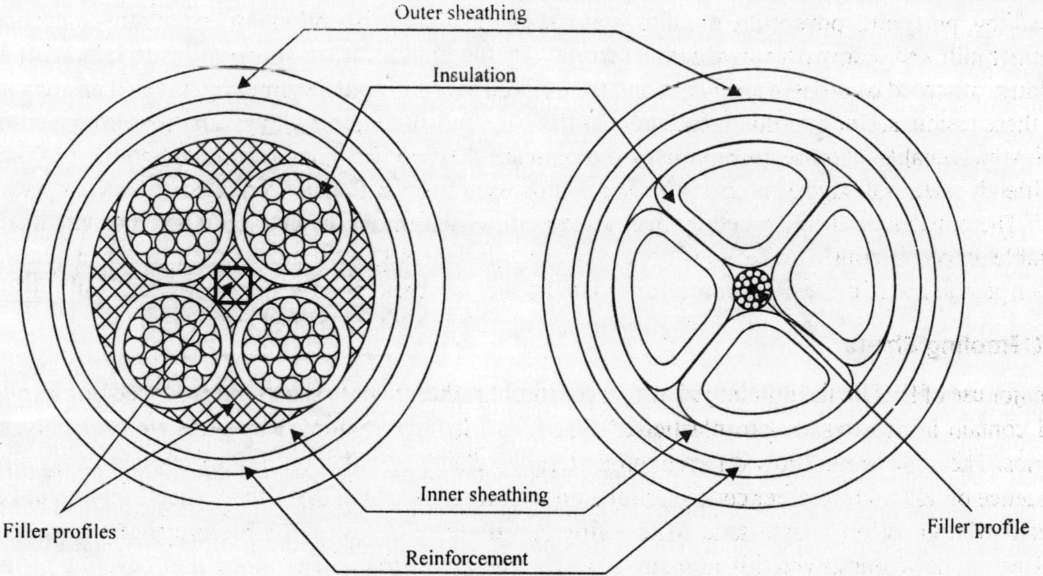

Fig. 28.2. Schematic of cable design.

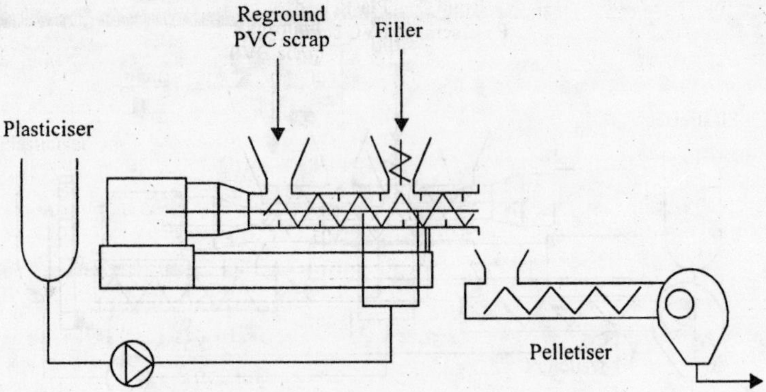

Fig. 28.3. Compounding of cable filler cores.

PVC Floor Coverings

PVC floor coverings are a combination of a number of constituents that together comprise the recipe for a floor covering. Typical floor covering recipes are for the most part made up as follows: PVC 28–50 per cent, plasticiser 10–20 per cent, stabilisers 0.5–1 per cent, slip agents < 1 per cent, filler 25–60 per cent, pigments 1–5 per cent. Further, many floor coverings are provided with additional textile or glass-fibre–containing carrier layers. Given an average service life of 10–17 years, old PVC coverings represent a large reserve of raw materials.

To exploit these large reserves of raw materials on an industrial scale, one needs a network for systematic collection of old coverings, a system of transport logistics, and the ability to build technically feasible recycling plants.

The main operations carried out in a recycling plant for old PVC floor coverings are sorting, cleaning, shredding, purifying, powdering, mixing, and packaging. The purification unit essentially comprises a hammer mill and a downstream vibrating screen. The function of the hammer mill is to knock off any residues of screed or adhesive still adhering to the floor covering, and the vibrating screen then separates off these residues. The shredding material that has been purified in this way is first passed to a cutting mill, which enables it to be precomminuted to granules that are first homogenised in a mixing silo and purified by means of zigzag sifters before being processed to PVC floor covering powder in the powder mill. The powder can be upgraded by the addition of plasticiser, PVC, or filler to give powder recipes suitable for calendering to make new floor coverings; this depends on the quality of the batch in question. Other possible uses of the powder are for products, such as mats for cars, mud flaps, and soft profiles.

PVC Roofing Sheets

A major use of PVC in the building industry is for roofing sheets. These sheets are produced on calenders and contain in most cases two plasticised PVC foils that are reinforced by glass fibre or polyester fabrics. The used sheets show different properties, depending on whether they are applied under direct influence of light and weather conditions and on the fact that in some cases these sheets on the roof are loaded with gravel or coarse sand. After a lifetime of 10 to 20 years, the roofing sheets have to be replaced by new ones, which is normally done by the same firm(s). Therefore, the recycling of used roofing sheets is rather easy from the logistic point of view and has been common practice for many years. New PVC roofing sheets may contain up to 10 per cent of the recycled material.

Postconsumer PVC

Some postconsumer PVC sources are water, food, pharmaceutical, and cosmetic bottles, and film. Another significant source of postconsumer PVC is used electric cable, coming principally from plant demolition and, to a lesser degree, from manufacturing scrap and offcuts.

Most end-use markets for recycled plastic bottles require that they be separated by resin type and colour. This ensures high end-use value for new products incorporating substantial amounts of the recycled resin. PVC bottles, like PET bottles, are very recyclable. Manual sorting of nonpigmented PVC and clear PET bottles is difficult because they look alike. When the two types are received comingled, the reprocessor can experience quality deficiencies due to rheological incompatibilities between these two resins. Therefore, all attempts to separate and remove these two resins must be made prior to recycling. Manual sorting techniques are inadequate to meet the market's needed quality standards, so new techniques have been engineered that will detect and separate bottles made from either of these two resins. A simple device senses the presence of chlorine atoms as a means to detect PVC bottles. Once detected, PVC bottles are pneumatically jettisoned from the comingled bottle feed-stream by a microprocessor based air blast system. The step after sorting is baling or granulation. Granulation is the preferred method of intermediate processing since the material so processed commands the highest market value. For upgradation of the resin from the recycled PVC bottles, several steps are explored depending on the results of characterisation tests as discussed earlier. These include incorporation of virgin resin (10–90 per cent), restabilisation against UV and heat, and incorporation of processing aids, impact modifiers, lubricants, plasticisers, and antioxidants. The recovery of electric cable is long established because of its valuable copper content. After this conductor material has been extracted, the residue consists of sheathing and insulation that may contain rubber and polyethylene as well as PVC. These other materials can be largely removed from grinding, by flotation, vibration, and filtration, but rubber is especially difficult to remove entirely, so that applications for material recycled from cables containing it are limited to areas such as car mats and carpet underlay.

RECYCLING OF HIGH DENSITY POLYETHYLENE CONTAINERS—HDPE

High Density Polyethylene Containers find usage in packaging of variety of products like organic/inorganic chemicals, petroleum products, vegetable/vanaspati oil, soaps/detergent, cosmetics, drugs and pharmaceutical products. The size of HDPE containers varies with products as well with the kind of end user industry.

HDPE containers are fast replacing the conventional glass, tin, steel containers and today it is perhaps the fastest growing application segment of HDPE. This increasing popularity of HDPE containers over conventional containers can be attributed to the following basic properties and its resulting advantages.

Base properties	*Functional Advantages*
Superior impact strength	Drop strength, reusability
Greater rigidity	Stackability
Enhanced downgaugability	Reduced wall thickness/weight
Excellent chemical resistance	Safe packaging - wide range of chemicals
Improved ESCR	Safer storage/transportation-hazardous chemicals
Non-toxic	Suitable for food packaging
Light weight	Easy to handle - Energy savings in transportation
Corrosion resistance	Non-rusting
Good plastic memory	Non-denting, shape retention

The name of commodities which can be packed in HDPE drums/barrels are given in Table 28.1.

Table 28.1. Commodities that can be packed in HDPE drums/barrels.

Chemicals	*Processed food*
Petrochemicals	Edible oil
Organic chemicals	Vanaspati
Petroleum products	Fruit products
Inorganic chemicals and acids	Vegetable products
Dyes and dye intermediates	Dairy products
Surface coatings	Additives
Pesticides	Pharma
Auxiliaries	
Oil field/Construction chemicals	
Adhesives	
Surfactants	
Solvents	

Although the processing industry consumes by and large the entire quantum of waste generated during the processing, the industry needs to device and follow an integrated model which will ensure the reuse and recycle of the containers in a structured manner to prevent any pressure on the waste management system and damage to environment.

The 3-dimensional approach to eco-friendly product management system:

1. Source reduction.
2. Structured recycling.
3. Comingling.

This system is aimed at extending the product life cycle and to minimise the load on the waste management system. Following flow chart illustrates the proposed product management system :

Drums/Barrels for chemicals

↓

Multi trip/Product usage

↓

100 per cent Structured recyclate

↓

3 Layer Containers

↓

Comingled plastic

Source Reduction

Source reduction means maximising the usage of material or getting more out of less. In the proposed model the source reduction can be achieved by following ways.

More bulk packaging

This can be practised while supplying to a large customer so that it would reduce the number of containers required and thus the amount of material consumed.

Multiple usage

The containers used to supply a customer can be reused to pack the similar material to avoid any contamination. With kind of impact properties and ESCR properties HDPE containers exhibits, it is possible to reuse them at least for second loading.

Product modification

In this the product supplied in the containers can be modified to reduce the overall number of containers consumed. e.g., supplying concentrates of surface coatings, two-in- one products/packs like adhesives.

Structured Recycling - Industrial Container Park

This will involve a methodical collection of used containers from the industry which can be stored in an open industrial container park (Fig. 28.4) which can be used as the storage yard. These used containers can then be sent to washing and laundering yard for cleaning. As this operation will require lot of water, a wash water treatment plant is recommended for recycling of the water. The cleaned containers can then be supplied to the recycling unit for converting it into granules.

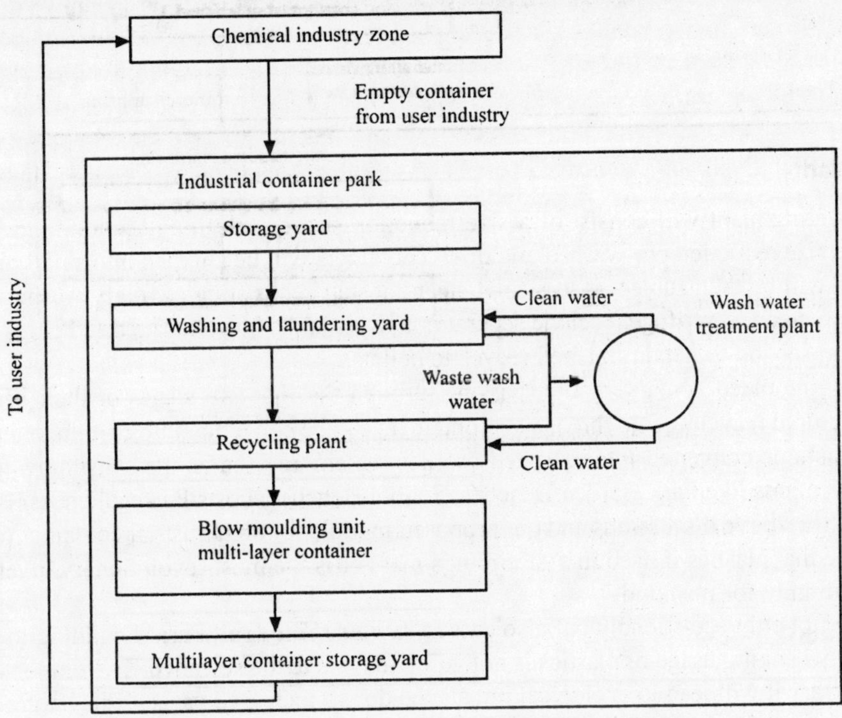

Fig. 28.4. Close loop recycling HDPE-industrial container.

A multilayer blow moulding container with a middle layer (50 per cent) from the recycled material and inner and outer layer from virgin material can then be produced without sacrificing the functional requirement. The industrial container park should be located in the proximities to chemical zone and downstream users to minimise the transportation cost.

Pre-Washing and Laundering

This step is essential as containers coming from various end uses will be containing residuals of chemicals, oils, etc. These residuals need to be cleaned with cold and hot water jets using appropriate detergents and 5 per cent NaOH solution. Table 28.2 shows the typical steps involved in the industrial drums recycling pre-washing and laundering.

Table 28.2. HDPE industrial drums recycling pre-washing and laundering.

Typical steps process	*Times (Sec)*
Cold water	30
Detergent wash	30
Cold water jet	30
NaOH (5% solution) and water at 80°C	30
Live stream	30
Hot water stream	60
Drip dry	300
Wipe	30
Air jet drying	30
Total	10 minutes approx.

Recycling Plant

A typical recycling plant will consist of a size reduction section, cleaning section, and an extrusion section. In the size reduction pre-washed and dried containers will be chipped into small pieces using a grinder which will be then subjected to a final cleaning stage consisting of rotary washers, separation tank, vibratory screen, centrifuge, cyclone separator, a dryer and a blower. The dried chips are then fed to the extrusion section to obtain finished recyclate pellets.

HDPE blow moulded containers after passing through the first two stages of their life cycle viz., reuse and structured recycling into three layer containers can be converted into co-mingled plastic. This co-mingled plastic is converted into extruded profiles which are used as the substitutes to wood and timber in applications like Park Garden Benches, School Benches, Posts/Pickets/Fences, Pallets etc.

Thus, from the above discussions and the proposed module for an industrial container park one can easily conclude that plastics if used in a disciplined manner is the most revolutionary invention by the scientific community for mankind.

Plastics are not only eco-friendly but also an eco-saviour. You don't burn a building to kill the rats inside. Similarly banning usage of plastics is not going to resolve the so called environmental problems but on the contrary it will lead to ecological imbalance due to increased usage of its natural substitutes like wood and paper. What one need to do is to create social awareness to change the littering habit and the NIMBY (Not in My Back Yard) approach. Structured collection, sorting and recycling and social code of conduct are the true answers towards efficient usage of a wonderful material like plastics.

POLYSTYRENE DISPOSABLE SMALL CONTAINERS

Recycling of polystyrene food service disposables are now-a-days a common practice. In our country a large quantity of such disposable is generated which largely comes from the food packaging sector.

The waste collection system are same as what we have discussed in case of PVC. Once separation is accomplished, the segregated polystyrene is collected by recycles. After collection the material can be intermediately reprocessed by further source separation.

The feedstock is then metered into the recycling system whereby the polystyrene is wet granulated in a specially modified granulator. The wet ground washed polystyrene is dewatered in a vibratory screener and rinsed with clean hot water. This is necessary to remove trace element of organic oils and food grease from the polymer. Following washing the regrind is dried to a moisture content of less than 0.5 per cent. Drying of the washed polystyrene is complex job due to the exposed cell structure of the granulated foam. Once dried, the regrind is once again comminuted in an effort to increase the apparent bulk density of the regrind. The clean dry regrind is now reclaimed. Finally the molten devolatised polystyrene is filtered in a screen changer and pelletised.

The quality of finished Polystyrene pellets are acceptable for the injection moulding of numerous durable goods. Examples are office supplies, non-disposable food serving trays, rulers, pen barrels flower pots and insulation board. The application requires certain compounds containing butadiene rubber in order to enhance the ductility of the finished product. Colour concentrates are also added to meet customer requirements.

RECYCLING OF NYLON 6

Recycling of plastics is emerging as the best way to get rid of plastics pollution and to conserve the precious resources. Plastics waste originates from a number of sources, and the technology is being constantly upgraded for the efficient disposal of the waste by means of recycling and reusing. Technological upgradation of the recycling processes has become the focal point to meet the stringent quality requirements needing more research in this area. In this context, the recently inaugurated Centre for Research into the Recycling of Plastics Waste (CEREMAP) in France, has enlivened the hopes of plastics processors. It claims to be the only research organisation of its kind capable of developing methods to different groups of waste polymers to arrive at recycling solutions. For any recycling scheme to be commercially successful, the following criteria must be satisfied :

1. Consistent and abundant supply of feedstock. The feedstock need not be monopolymeric and uncontaminated.
2. There must be significant economic advantage in using recycled rather than virgin material. Otherwise, operators who manufacture high value added products (e.g., Engineering Plastics) may be reluctant to use recycled feedstock, even though there may be no technical reason preventing them from doing so.
3. Reprocessing technology must be developed and available.
4. There must be wide markets for the products.

Types of Recycling

Recycling is of four types:

1. *Primary recycling:* This is the processing of scrap plastics into the same or similar types of products from which the scrap was generated, using standard plastics processing methods. It involves uniform, uncontaminated plastics waste.
2. *Secondary recycling:* This is the reprocessing of scrap plastics into plastics products with less demanding properties. It involves a mixture of various plastics and non-plastics materials.
3. *Tertiary recycling:* This is the recovery of chemicals from the waste plastics. e.g.: pyrolysis and depolymerisation.

4. *Quaternary recycling:* This is the recovery of energy from the waste plastics.

Chemical recycling (tertiary recycling) involving the breakdown of polymeric waste into reusable fractions for reincarnation as polymers, monomers, fuels, or chemicals is emerging in a viable way to reconcile plastics and the environment, thereby bringing real versatility to solid-wastes management.

The fundamentals of chemical recycling are well established. The chemical processes are used to depolymerise or 'unzip' nylon, polyurethane and polyester back into their chemical building blocks. The bonding of molecules in these materials is such that, if appropriate heat and pressure are applied in the presence of a reactive chemical agent, they break into relatively simple and shorter chains. Technologies for the breakdown of such polymers (hydrolysis, glycolysis, methanolysis) are already proved, and are viewed as relatively cost effective.

Nylon 6 Polymerisation - Industrial Process

Nylon 6 is a homopolymer of Caprolactam obtained by the ring opening polymerisation of the latter whose melting point is 68-69°C. The general industrial route involves water as the initiator of the polymerisation reaction. However, cationic or anionic initiators can also be used. The processes most widely used on an industrial scale are batch polymerisation in an autoclave or continuous process in a VK tube.

Batch polymerisation

The conditions of polymerisation are closely related with the chemistry of ring opening polymerisation of Caprolactam. In a typical process, the raw materials are charged into the autoclave and the heating is started. The temperature is raised to 230–240°C over several hours, during which time, the pressure shoots up, depending upon the amount of water. After maintaining the pressure for sometime, it is slowly reduced until a residual pressure of 0.1–0.5 atm. is reached and the temperature is increased simultaneously to 250–270°C.

This operation lasts for 5–10 hours. The pressure is then raised to ambient by supplying purified nitrogen and polymerisation is continued under these final conditions for 2–3 hours in order to reach equilibrium. 5–10 per cent (wt.) of water is generally added as initiator.

Continuous polymerisation

This is carried out in a VK tube provided with a heating jacket divided into several sections. The temperature of the polymerising mass is maintained at 250–270°C throughout the process. Usually a pre-polymer is made and then fed into the VK tube. The tube dimensions and the output per day are regulated so that retention time is longer and speed of the moving front is low. As the degree of polymerisation at equilibrium is dependent upon the amount of water remaining in the caprolactam, which, in turn, is regulated by the temperature at the top of the tube, the degree of polymerisation can be varied by the temperature at the top of the melt. The retention time in the tube must be appreciably longer than the time needed to reach the polymerisation equilibrium.

Origination of Nylon 6 Waste in Industry

1. Waste generated during the polymerisation process (reactor scrappings, unsuccessful runs, etc.).
2. Waste generated during processing operations.

Waste generated during nylon 6 polymerisation

The continuous polymerisation of Caprolactam to Nylon 6 is controlled by various factors including temperature of various zones of prepolymerising unit and VK tube, pressure of prepolymerising unit and VK tube and the rate of flow of the polymerising mass in the prepolymerising unit. The various operational parameters are fixed to obtain optimum polymer viscosity suitable for the intended application. Any disturbances in any of these parameters controlling the polymerisation process might lead to quality variations, resulting in sub-standard generation.

For example: power failures, frequency variations in power supply, deliberate process changes, pack changes in metering pumps, fuel oil burner failure or trip, etc. will affect the relative viscosity of the polymer. Too much of variations will lead to polymer of low viscosity or oxidised polymer resulting in its rejection as waste/sub-standard.

The Nylon 6 polymer melt is extruded as strands and granulated into chips for further processing. The waste generated during this operation is classified as strand waste. The cut chips are sampled for quality by opening the granulator and collecting them. Waste may also occur due to spill-over of the chips during this operation.

In any industrial polymerisation of Caprolactam, the conversion goes upto 90 per cent and roughly 10 per cent of monomer/oligomers are left out in the polymer. The left out monomer and oligomers need to be removed to produce polymer of higher relative viscosity and quality. The monomer also must be recovered and recycled for cost control.

Also, in the continuous polymerisation, the water generated during the condensation process is constantly let out from the top of prepolymerising unit and VK tube to push the equilibrium in the forward direction. This eliminated water contains dissolved monomer and other impurities, from which the monomer needs to be recovered. Lactam lost during the sampling of molten lactam must also be recovered.

Waste generated during processing operations

Waste is generated at various stages in the conversion of Nylon 6 chips to fabric. This includes the following:

1. *Spinning waste:* Waste generated during the melt spinning process arises due to yarn break, low mechanical properties, stains etc. This is classified into two categories :
   (a) Yarn waste without oil.
   (b) Yarn waste with oil.
2. *Textile waste:* This includes waste generated during the conversion of yarn to fabric. The operation passes through three stages, viz., plying, cabling, and conversion. Waste is generated in all the three operations either due to quality problem or due to the process limitations.
3. *Post process waste:* Waste generated during the dipping process.

Waste Generated in Other Operations

1. *Injection moulding:* Waste comes from the sprues and runners (routinely reground and can be added to the virgin resin in various amounts).
2. *Extrusion:* Waste originates during equipment start-up, purging and trimming operations. In the continuous extrusion process, power failures lead to waste generation. In film extrusion, film scrap is generated during roll change-overs.

Methods for the Recycling of Different Wastes

Waste during polymerisation

1. The condensate collected from the top of the prepolymerising unit and VK tube contains Caprolactam and minor impurities dissolved in water. It is subjected to a monomer recovery operation which includes concentration of the condensate and distillation. The monomer thus recovered will be inferior in quality as compared to virgin monomer because of the presence of oxidised impurities. The recovered monomer can be polymerised as such or mixed with virgin monomer in different ratios depending upon the quality requirements of the final polymer product.
2. The nylon 6 invariably contains roughly 10–12 per cent of unreacted monomer and oligomers (linear and cyclic), extractable in hot water with about 85-90 per cent of monomer and the rest oligomers. This is done by treatment with a stream of hot water for several hours. This wash water containing dissolved monomer and oligomers is sent along with the condensate from prepolymerising unit and VK tube for the monomer recovery.
3. Lactam spill-over during molten lactam sampling is also mixed with the above for monomer recovery through distillation.
4. The strand wastes go to the conduit process where they are cut into granules and used for engineering plastics production. Even waste in the form of lumps is granulated by this process.
5. The chip waste from sampling for QCD as well as Chips rejected as waste due to the disturbances in the operational parameters are also sent for depolymerisation for the recovery of monomer.

Waste during processing operations

1. All the spinning waste (except yarn waste without oil) and the textile waste are subjected to a plast-compaction process, in which the waste is shredded first and then subjected to the 'plast-compaction' where they are softened, compressed and cut into granules. The granules are then used for engineering plastics production. The waste is subjected to a magnetic scanning to remove 'non-plastic and magnetic' contaminants before being subjected to conduit or plast-compaction process.
2. The yarn waste without oil is reprocessed by first 'compression-moulding' followed by the grinding/chopping operation to granules meant for engineering plastics generation.

Use of Reground Material in Processing Industry

1. *Injection moulding:* Control of feed stock involves control: (i) of the ratio of virgin to reground resin, (ii) of the size of chips used and (iii) the level of material in the hopper. Changes in processing characteristics (melt viscosity due to a change in molecular weight, melting rate because of variation in chip size) can be caused by random introduction of virgin and reground materials. These changes can result in quality fluctuation in the final product. In general, 50 per cent reground levels will not impair the performance of the engineering parts made out of Nylon-6. Only in very specific situations like, moulding of glass fibre reinforced Nylons, it may be desirable to keep the reground ratio below this figure. Grinding equipment is selected and maintained to produce reground with an average chip size close to that of the virgin nylon to ensure uniformity of the mix. It is necessary to remove fines and shavings with cyclone separators to avoid choking of the automatic loaders and the filters.

2. *Extrusion:* Nylon extrudates can be recycled in many different ways depending upon the resin, the process involved and the tolerances which have to be held. Air or chill-roll quenched nylon extrusions may be immediately reground and re-extruded since the moisture level should be low. Thinner reground, such as that made from film will generally have to be re-pelletised before it can be conveniently extruded. Water quenched extrusions can be re-used, if the extrudate which exits the quench, is hot enough to evaporate the surface water provided the material will be reground immediately. The reground is best blanketed with an inert gas such as dry nitrogen.

Most general purpose fly-knife cutters designed for regrinding plastics will be suitable for nylon. For the best protection of nylon reground from contamination, the extrudate should be reground immediately and stored in drums lined with polyethylene film. In many extrusion processes, 100 per cent reground may be extruded directly or it may be blended with virgin resin in any proportion as long as the ratio of the reground is constant.

Depolymerisation of Nylon 6

Apart from recycling Nylon 6 waste as outlined above, i.e., by means of mechanical reworking. Nylon 6 waste of any kind can be 'unzipped' to recover the monomer, Caprolactam. From a practical point of view, this process becomes very important for the 'endless' waste recycling. Nylon 6 can be depolymerised hydrolytically in acid solutions. The depolymerisation is basically a cracking process and is done at a temperature of 280–300°C in the presence of a catalyst like phosphoric acid over super-heated steam. It has been experimentally shown that this hydrolysis under various conditions follows the behaviour predicted for polycondensation polymers, i.e., the probability of scission for every amide bond is independent of the bond position in the macro-molecule and therefore the degree of polymerisation decreases with time, following a first order reaction.

The process is done under vacuum in order to remove the caprolactam as and when it is formed, with a recovery of 90–95 per cent crude caprolactam containing many oxidised impurities. The impurities are removed by oxidation with potassium permanganate, dichromate, hydrogen peroxide, etc., followed by passing the 40 per cent solution of the lactam in water through ion-exchange resin and finally vacuum distillation. The resulting monomer still contains volatile impurities like methyl pentyl ketone and cyclohexyl amine. Its quality is inferior to that of the monomer recovered from chips and VK-tube condensate.

Other methods of recycling nylon 6

1. *Process for manufacture of Nylon 6 from its wastes:* Nylon 6 wastes, its oligomers and oligomer-lactam mix have all been used for manufacturing EP grade N-6 by passing the molten waste through a shear filter, and degrading the melt to a low degree of polymerisation (solution viscosity > 1.5) under pressure steps using water at 230–290°C. A metal fibre fleece, fine-pore sinter filter or fine sand filter are used for the filtration. Thus, in a typical case, the polyamide fibre waste was melted in an extruder (melt had solution viscosity 2.70), passed through a metallic fibre shear filter and fed through a pressure reactor to give a polymer with solution viscosity 1.55 after 5.5 hours at 280°C. This polymer was dried to remove excess moisture, heated at 250–280°C for 8 hours under dry nitrogen to give a polymer with solution viscosity 2.40 and monomer content 9.5 per cent with average degree of polymerisation 140.
2. *Method for casting Nylon 6 with monomers from its wastes:* The modified version of the previous method involves the following operations: Nylon 6 waste is depolymerised with 3 per cent KOH

at 300–320°C at 750 mm Hg. Pressure. 0.1–1 per cent of hexamethylene diamine is added to the melt and stirred for one hour. Then the melt is sprayed and washed for 2 hours in a tower at temperature of 110–120°C heated by steam at 150–200°C. Then 0.1–0.5 per cent of sodium hydroxide is added and the mass is distilled at 140°C at 5 mm Hg.pressure to obtain Caprolactam. This can be polymerised in presence of sodium hydroxide and a co-catalyst to prepare 75–80 per cent polymer having molecular weight 35,000–70,000.

3. *Nylon 6 sorbents from wastes of knitting industry:* Good quality sorbents could be obtained by the partial depolymerisation of waste nylon 6 fibres from the knitting industry, in 45 per cent sulphuric acid at 90–110°C followed by neutralisation with ammonia, ammonium carbonate or sodium carbonate, precipitation and drying. The sorbents thus obtained can be used in a variety of chemical and agriculture industries.

Thus, recovery of clean segregated production scrap is an important cost saving project of plastics processing industries. It may be used as reground in the producers own processes or through other outlet for conversion and compounding operations. The majority of post-consumer scrap is disposed of in either domestic or mixed industrial/commercial waste. Once diluted with large quantities of other waste/compounding materials, recovery of the base polymer is very difficult and expensive. Heat treatment processes such as pyrolysis are being extensively studied and many of the technical problems such as coking at heat transfer surfaces have significantly been overcome. These processes are capable of accepting a wide range of feedstocks. If high grade feedstocks are processed, it might be possible to produce chemical intermediates useful for high technology end use.

Technology for separation and cleaning of mixed and contaminated polymeric waste is gaining importance. Separation of polymeric mass from extraneous material is also a vital step in the whole recycling process. The recent development of 'High-tech Filter System' improves separation capacity of post-consumer plastics processing equipment. The laser filter consists of two steel discs with upto 1.6 million laser-drilled holes of 110–120 microns with rotating scrapers. These scrapers clean the hardened surfaces of the filter discs and direct impurities to the middle of the disc. Separated impurities (paper, aluminium, copper, wood, etc.) are extracted via two coaxial screws.

Effort should be concentrated in those industries where significant quantities of wastes, containing high proportions of plastics are generated in localised areas. Specific examples are refuse processing plant, fragmentiser operations, and the textile and related industries. Precise quantities and compositions of the wastes need to be determined following which technical feasibility of differing treatments/routes can be demonstrated. Economic assessments of processing options might then be made for the use of plastics as fuels, chemical feedstocks, low grade materials or as supplements to virgin feedstocks.

Effective means for the recycling of nylon 6 is already established and the processes need further improvement with respect to the quality. To-date monomer recovery by extraction and depolymerisation has not provided an answer to get high quality lactam. Efforts must be directed to this particular area to improve the quality of the recovered monomer, which will ultimately serve as the best way of 'endless recycling' of nylon 6.

Caprolactam is a very corrosive waste material. It has the potential to eat through concrete slabs and corrode the reinforcing iron bar. It is therefore imperative that the Nylon 6 producers take adequate steps to contain the lactam spillage and optimise the processes to get the maximum Nylon 6 output.

Five commodity polymers (HDPE, LDPE and LLDPE, PP, PVC and PS) are the largest in terms of world production. Fig. 28.5 illustrates the production steps that apply to plastics manufacture and conversion to products.

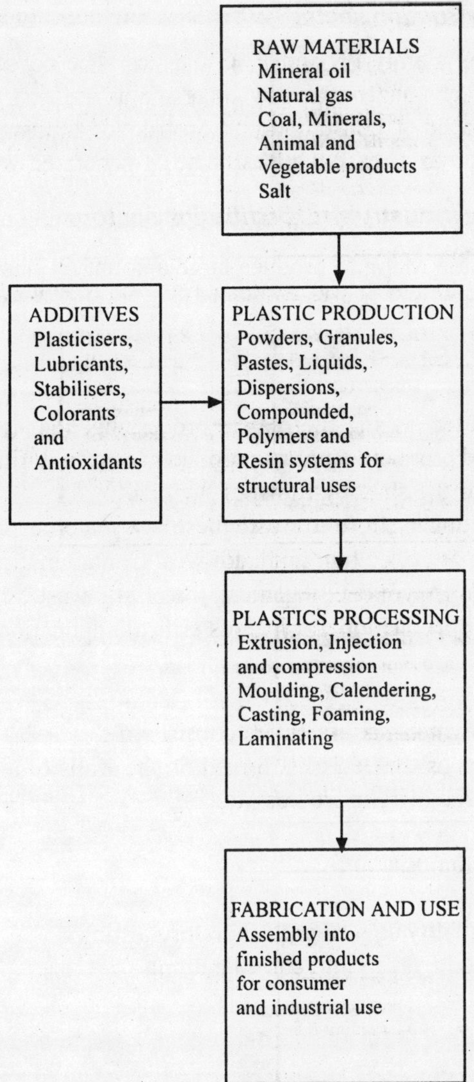

Fig. 28.5. Production steps in manufacturing plastics and conversion products.

Sources of Plastics Waste

The sources of plastic wastes are discussed below.

Plastics waste from the agriculture sector

Agriculture primarily uses PE, PP and PVC plastics. The plastic materials used in agriculture generally have a short to medium life-span.

Plastics waste from the automotive sector

Currently, 15 per cent to 20 per cent of an automobile is made of non-metallic materials (glass, rubber, plastics, etc.). It was estimated that about 30 per cent of the consumption are discarded as waste.

Plastics waste from the construction sector

The construction industry uses mainly PE and PVC plastics. The construction sector typically uses plastics in applications for much longer than any other major industry sector, making it difficult to estimate the waste generated based on consumption. Construction industry sources have established 10 per cent of current consumption as a reasonable estimate of generated waste.

Plastics waste from the large industry and distribution sectors

This sector is the second after the MSW to produce large amount of plastics wastes. It includes bags, drums, containers, packaging film etc. It was estimated that 90 per cent of the consumed plastics are discarded.

Transformation industry

This plastic wastes are generated during the plastic processing. Plastic wastes consist of, runners, distorted and incomplete formed products, surpluses products rejected during quality control and material arising during the starting-up or the shutting-down of the plant.

Even smaller firms can no longer afford to waste these raw materials. Waste is carefully segregated at the source, according to its nature, colour, and additives. Contamination by dust, oil or burned (i.e. thermally decomposed) material is avoided. In general, the plastic wastes is ground and recycled, either directly in the same or a similar production, or in a less critical application.

Manufacturers

Plastics waste generated by polymer manufacturers consists of :

1. Production wastes, such as the deposits formed on the walls of a polymerisation vessel or the driers, or sludges separated from process waters.
2. Off grade products.
3. Extrusion purging and lumps.
4. Floor sweepings.
5. Wastes arising during quality control and laboratory testing.

The following factors should be taken into consideration:

1. Reduction at source.
2. Re-use.
3. Recycling
   (a) Mechanical recycling (collection and sorting).
   (b) Chemical recycling.
   (c) Energy recycling.
4. Disposal in landfill.

But putting this rule into practice can be difficult. For developing countries, effective technology may be too expensive; for developed countries it is often cheaper to bury the waste landfill, so only environmental legislation insures the best waste management.

Reduction at source

Reducing the waste generated during the plastic-product manufacturing process and/or reducing the amount of the eventual waste of disposed product by using less plastic to produce it. Today's plastics packaging is up to 80 per cent lighter than 20 years ago through a combination of better raw materials, improved converting techniques and more efficient design.

Methods

1. Recycling in line.
2. Reduction of product weight by redesign.
3. Use a different or better plastic for the same application.
4. Improvement of machinery, process controls, and quality control.

The above mentioned methods involve a financial benefit for the manufacturers in reducing their raw material consumption and operating or waste disposal costs.

Re-use

Ensures an extended life; reducing the volume of waste. Re-use is appealing concept but the net environmental gain of re-use systems need careful consideration before making the decision that re-use is always best.

Examples

1. The carrier bag is probably one of the most popular re-use items in Europe.
2. The plastics industry is currently exploring methods of re-use in the drinking market (soft drinks).

Recovery of plastic wastes

Pre-user plastic wastes

Generated by plastics manufacturers and conversion industries. The pre-user wastes do not pose large problems in terms of recovery.

Post-users plastics waste

Collection and sorting of plastic wastes. The collection and sorting of plastic wastes are typically the first two steps involved in post-user plastic waste recycling. The plastics may be sorted (from non-plastics) before being collected, or they may be separated from a mixed waste stream (sorted after collection).

Sorting before collection appears to be the more successful method to date. The plastics are kept separate from non-plastic wastes by the end user (who generates the plastics waste). The user can then bring the plastics to a central collection point where the plastics are collected and kept separate. The advantages of these methods over sorting after collection are that the plastics are usually cleaner and the labour-intensive process of separating plastic wastes from a mixed stream of MSW is avoided since the required manpower to separate the wastes is supplied by the waste generators themselves. These methods, therefore are dependent upon household participation.

Throughout all developing countries, scavenging of solid waste is widely practiced as a means of making a living.

Recycling of plastic wastes

Recycling refers to a process in which plastic wastes are used for mechanical, chemical and energy recycling. Fig. 28.6 is a flowchart outlining different methods of recovery which exist today as explained hereafter.

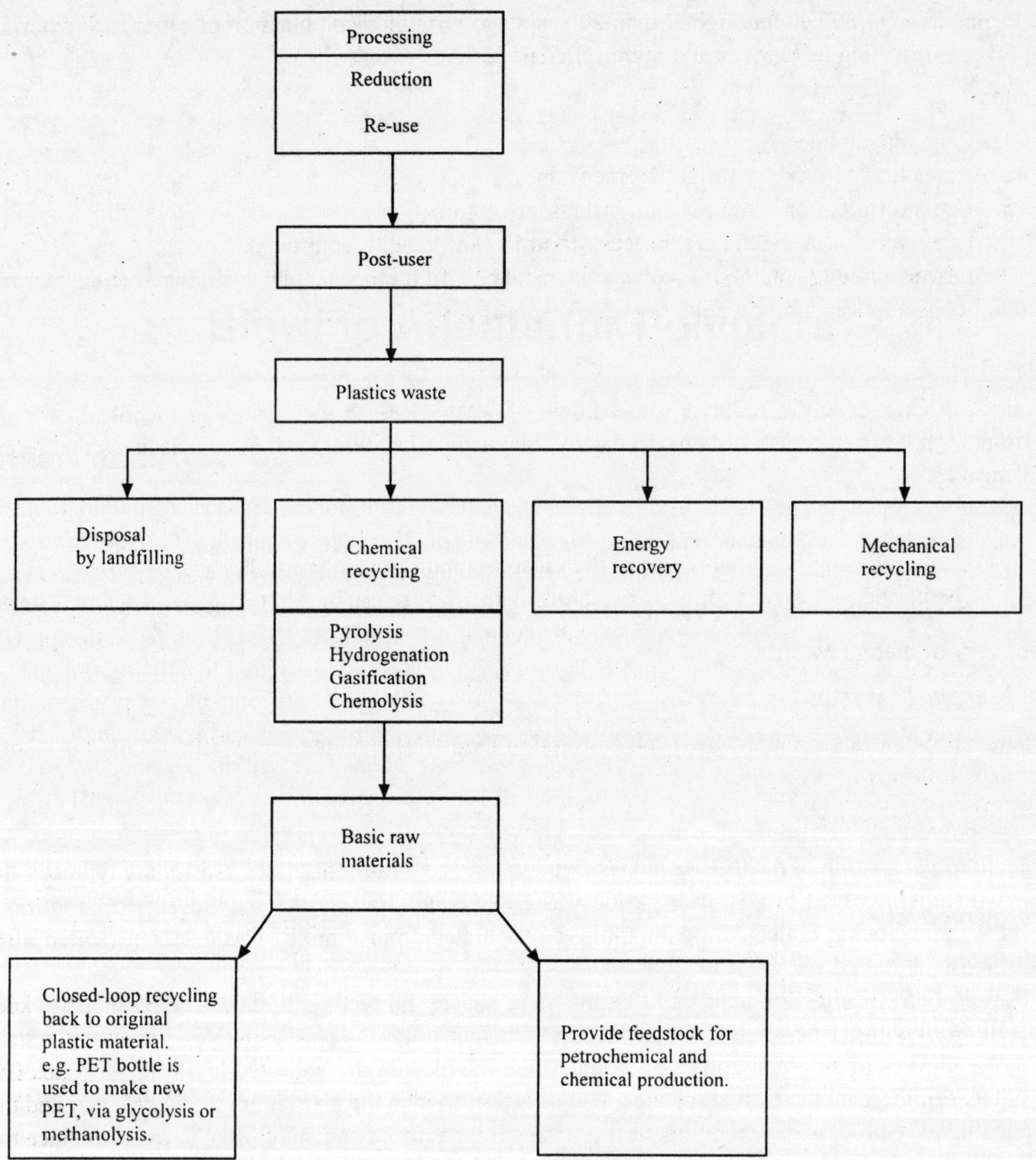

Fig. 28.6. Flow diagram showing integrated waste management.

Mechanical recycling

Mechanical recycling, or recycling into materials refers to recovery and processing of used plastics for use in new applications.

Sometimes recycling is not desirable or possible because:

1. The material is grossly polluted or burnt, or has taken up moisture from the air.

2. The material is composite, e.g. multi-layered film, insulated of electric cable, coated textile and various floorings.
3. The production method does not incorporate a step, in which the wastes can be melted. It occurs in the thermoforming of PVC-sheet, the rotation moulding of plastic powders and in the conversion of plastic semi-products to a final product by operations such as welding, gluing, cutting, drilling, etc.
4. The dimensional specifications are too strict. Milled plastic scrap has not the same bulk density and flow properties as virgin material. The addition of scrap may alter the metering of plastics by the screw, hence also the temperature and viscosity of the melt and ultimately the degree of filling and pressure in the mould. For this reason the scrap is often being regenerated or, at least, added in a fixed proportion to virgin material.
5. The raw materials cost in some cases is only of secondary importance (electronic industry).

Chemical recycling

Chemical recycling involves either the depolymerisation of long-chain plastic molecules into their original monomer molecules or the breakdown of the polymer and monomer constituents into useful industrial chemicals. There are a range of chemical recycling technologies currently being explored:

1. *Pyrolysis:* Breakdown of the molecules by heating in vacuum. These processes lead to gaseous or liquid hydrocarbons which can be further processed in refineries.
2. *Hydrogenation:* By treating plastics with hydrogen and heat the polymer chains are cracked down to a valuable hydrocarbon oil which can again be used in refineries and chemical plants.
3. *Gasification:* Plastics are heated with air or oxygen. The resulting synthesis gas, consisting of carbon monoxide and hydrogen, can be used for the production of methanol or ammonia or even as a reducing agent for steel production in blast furnaces.
4. *Chemolysis:* By applying solvolytic processes such as hydrolysis, alcoholysis or glycolysis - polyesters, polyurethane and polyamides can be recycled back to their basic monomers for repolymerisation into the original plastics.

Energy recovery

Incineration is the next most common method of disposal for municipal wastes. In Japan and Switzerland more than 75 per cent of municipal waste is incinerated. In many instances energy can be recovered from waste incineration; this practice has a number of benefits which enable to save energy resources, to reduce emissions of carbon dioxide, and to reduce demand for landfill space.

Regarded by some as a legitimate and important contributor to the recycling of plastic waste, the incineration and waste to energy solution is fiercely resisted by others who point out problems of toxic emissions and oppose the building of incineration.

Throughout Asian countries, the majority of MSW which is landfilled and some countries such as South Korea use the incineration with energy recovery.

Incineration is a unit process that uses thermal decomposition to convert a solid waste to a less bulky, toxic or noxious material.

Incineration offers the benefits of reducing waste volume, completely eliminating harmful bacterial and viral constituents, destroying many toxic organic compounds and allowing waste heat recovery.

Incineration has been a traditional disposal and volume reduction concept used to conserve sanitary landfill space in areas where suitable landfill capacity is not available within an economical hauling distance. Incineration reduces the weight and volume of solid waste while producing a residue which can be used as a fill material.

Solid waste incinerators can be design with or without energy recovery. Where revenue can be generated by selling or utilising energy, the income can be used to cover some of the operating costs of incineration.

Incineration is capable to reduce the weight of refuse by 80 per cent and the volume by over 90 per cent. The residue from the refuse is inert and may be disposed off in landfill.

Incineration can be performed and operated with safety conditions. The key is a high temperature combustion, proper controls, and trained, experienced operators. Today's modern incinerators operate with drastically reduced emission levels below those associated with risks to humans and the environment.

The incineration has turned to energy recovery.

The role of plastics content in the MSW is very important for the well-established MSW incineration programmes and facilities. The energy content of plastics renders them the most valuable of the major categories in MSW from an energy point of view. The overall energy value of the waste stream depends upon the composition of the waste. Changing lifestyles have led to an increased share of plastics and paper in the materials mix, increasing the heat value of this waste.

Indeed incinerators of the latest generation including energy saving and fumes cleaning adequately complete the sorting and recuperation operations with the burning of the elements which have a calorific power and no economical interest as for example the fragments of plastic films.

The arguments in favour of the treatment of solid waste in energy plants are not just environmental. These plants can also represent a viable economic investment offering a competitive return. Clearly a great deal depends on loçal conditions.

Disposal by landfilling

The reaction against landfill as the primary method of waste disposal is part of the general upsurge of concern about the state of the environment. More specific difficulties relating to potential shortages of capacity have resulted in a near-crisis situation in parts of the Asian Countries have given impetus to recycling and other methods of waste disposal.

Landfill will generally decline, not only because of land shortages but because of environmental concerns, which are particularly directed towards plastics. Such concerns surround the following perceptions about what happens to plastics in landfill :

1. Plastics are widely believed to comprise a major share of landfill in volume terms.
2. There has been speculation that plastics impair the structural integrity of landfill.
3. Plastics are said to slow down the degradation of other materials in the landfill.
4. The reputation of plastics for non-degradability also intensifies attacks on their landfill role. Many plastics are certainly slow to degrade, but it appears that in the increasingly dry conditions required in today's tightly regulated landfill sites, slow degradation is not only a problem for plastics. Excavations have revealed newspapers which are decades old and virtually intact.
5. The biggest concern about landfill is the possibility of toxic additives and fillers from plastic materials leaching into the ground-water. Again, as in other areas relating to landfill, knowledge of what actually occurs is limited.

Landfill is the most common method for the municipal solid waste disposal. It is also a final method, landfill being the ultimate destination of waste.

RECYCLING OF MIXED PLASTICS WASTE

Comingled plastics currently represent an estimated two-thirds of today's recycled plastics streams. The fraction can be expected to shrink somewhat with the development of more successful identification and segregation technologies in the future. However, comingled plastics streams will continue to make up a significant volume for several reasons: proliferation of grades and types of commodity; profusion of polymer blends and alloys; contamination of recycle plastic parts with metals, coatings, and laminates; and practical cost considerations. Mixed plastics wastes can be divided into two groups depending on their source: mixed plastics from household or municipal solid wastes and plastics from industrial sectors.

The first category (postconsumer mixed plastics) involves the articles that are used in food, pharmaceutical and detergent packaging, shopping, and others. The majority of these are composed of films, sheeting, strapping, thermoformed trays, as well as a variety of bottles for soft drinks, food, and cosmetics. There are mainly five different polymers—PE, PP, PS, PVC, and PET— that contribute to the total amount of plastics waste. The composition of mixed plastics can change depending on the regional habits and the seasons of the year. Also the mode of waste collection can influence its final composition.

The category of postindustrial wastes concerns articles like the products of the car, furniture, and industrial appliances. The problems of these sectors is a wide variety of engineering materials and a high number of components employed to build a final system.

Direct Reuse

A direct solution to the problem of plastics disposal can be the reuse of a heterogeneous mixture of plastics directly obtained from an urban collection. Today there are extruders specifically designed for reprocessing postconsumer and postindustrial waste materials. The waste material can have many forms and can range in bulk density from approximately 1 to 35 lb/cu ft. The form, bulk density, moisture content, contamination level, and process-temperature restrictions all affect the design of the extruder to be used. For example, due to the presence of PVC resin, the melting temperature must be kept below 210°C and the barrel residence time must not exceed 6 min. Furthermore, in the mixture, the relatively high content of semicrystalline polymers like PET, whose melting point is above the processing temperature, influences the extruder design or the final properties of the manufactured product. Larger injection gates and mould channels must be used in order to avoid undesirable occlusions in the channels.

Standard single-screw extruders are no longer adequate to recycle or reclaim this wide range of materials in a cost-effective manner. There are now special extruders designed to process the lighter-bulk-density materials. Low-bulk-density materials are the various forms of film, fibres, and foams commonly used in the packaging industry. Due to their low bulk density, such materials typically require an auxiliary device to facilitate proper feeding into the extruder throat. There are several varieties of such feeding mechanisms available. Two of them are a rotating screw-type crammer and a piston-type ram. The crammer and ram systems both act on the same principle; that is, an auxiliary feeding device is used to convey and to compact the low-bulk-density materials into the feed section of the extruder screw.

The *screw-crammer system* uses a conical hopper with a screw that is driven by a separate gear reducer and variable-speed drive motor. The output and effectiveness of the crammer are determined by the screw configuration and the available speed. The *ram-type system*, on the other hand, uses a pneumatic ram to stuff material into the screw. The ram is a piston-driven unit with the stroke timing adjustable by setting a series of timers located in the control panel. The feed section used by the ram system has an opening that is 12 to 14 times larger than that of a standard screw extruder. This allows low-bulk-density material to flow freely into the feed throat where the ram can compress it into the screw. Depending on the extruder size, the ram can compact materials with a force of 2000 to 9000 psi.

Feed materials usually need to be supplied to either the crammer or the ram system in a chopped form. The size and bulk density of the chopped particles affect the performance of the crammer and ram, and thus ultimately the output of the extruder. Both these systems can also be used to process higher-bulk-density products.

A third method of processing low-bulk-density materials is through the use of a *dual-diameter extruder*. This system has two distinct sections: a large diameter feed and a small diameter processing section. The large-diameter section acts as a cramming device—compacting, compressing, and conveying the feed material—while the smaller-diameter section is used to melt, devolatilise, and pump the extrudate into a die. In the feed section, the screw can have deep flights, allowing low-bulk-density materials to flow freely, while in the processing section the screw resembles that of a typical extruder. The screw is available in one of several configurations: single-stage, two-stage or barrier design. Depending on process needs, these designs optimise output and raise product quality.

The feed section of a dual-diameter extruder can be equipped with feed-assist components that in some cases work in conjunction with specially designed screws to allow processing of a wide variety of feedstocks that are fed to the machines in roll form. Among these possible feedstocks are loose bags, handle cutouts from bag making operations, and continuous web products such as blown and cast film scrap. This ability eliminates the cost of shredding, grinding, and densification of many materials. A dual-diameter extruder is also capable of processing materials with a high bulk density. The crammer, ram, and dual-diameter systems do not differ widely in equipment costs, installation costs, or production rates.

It is well established that a strong incompatibility is typical of polymers usually found in comingled wastes (PE, PP, PET, PVC, PS). This incompatibility gives rise to materials that have inferior mechanical properties, particularly with regard to tensile, flexural, and impact strengths. This means a strong limitation of applications, in particular in the case of thin walls and manufactured products that have to work under flexural and tensile stresses. However, by adding to the mixture specific components like other polymers from homogeneous recycling, fillers (talc), fibres, or promoter (compatibilisers) that increase the compatibility, it is possible to improve the tenacity or stiffness, product aesthetics, and processability.

Addition of glass fibres, for example, is found to yield products with very high stiffness (e.g., elastic modulus $E \simeq 2800$ MPa with 30 per cent glass fibre), higher than that with talc ($E \simeq 1250$ MPa with 20 per cent talc) and far better than that of the original mixture ($E \simeq 950$ MPa). Addition of LDPE and styrene-butadiene-styrene copolymer, on the other hand, improves the tenacity (showing, typically, a 30–90 per cent increase in elongation at break). Extruded profiles have thus been made that can be employed to build benches, garden tables, bicycle racks, fences, and playing facilities for parks. Coextrusion technology can be used very effectively to improve surface properties like puncture, impact and weather resistance, as well as appearance. One interesting application for mixed plastics, because of their large market volume, is the production of the injected tiles for paving.

A new and exciting technology has been developed in the fabrication of composite materials made with comingled plastics. It followed the discovery by the scientists and technologists at GE Plastics that useful products could be fabricated using Radlite technology (Fig. 28.7) if the powdered feed consisted of two or more resins. While compatibilisation of dissimilar resins is typically brought about by chemical means, this is not the case with Radlite technology products where compatibilisation appears to take place by physical means, i.e., the binding of dissimilar resin domains through the fibres. A conceptual model of *physical compatibilisation* is shown in Fig. 28.8.

The rolls or sheets made by Radlite technology using comingled plastics and chopped glass fibres (0.25–1.0 in.) can be converted to finished parts by conventional forming technologies such as compression moulding (typically 200–290°C, 3–5 MPa, 2–4 min). These are fully consolidated, essentially void-free products with specific gravities that would be calculated on the basis of resin type and glass fibre content. The physical properties of the consolidated structures from comingled plastics are in general similar to standard grades of SMC composites.

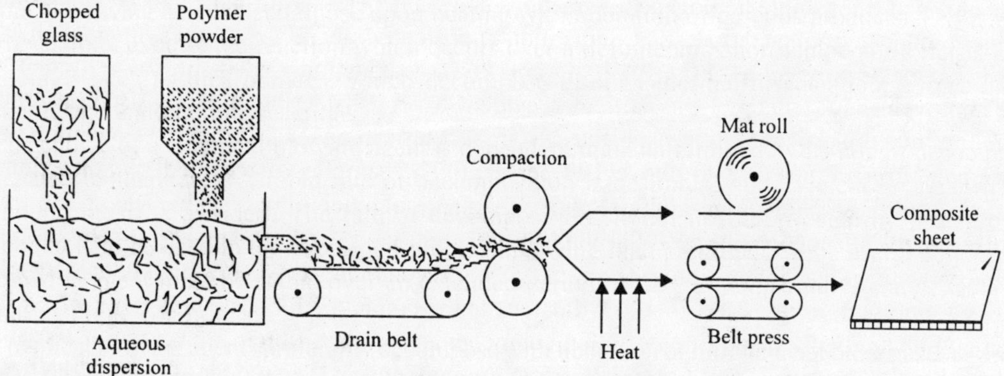

Fig. 28.7. Schematic of Radlite technology.

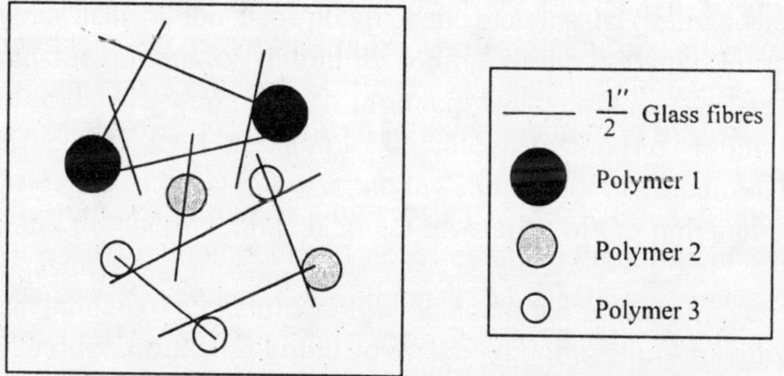

Fig. 28.8. Conceptual model of physical compatibilisation.

Foamed network structures (*loafed structures*) are also prepared in a compression press using Radlite technology. In this case, after applying full temperature and pressure for the requisite 2–4 min, the platen gap is opened 1.5 times the original setting (for primarily open-cell structure) or 1.1–1.2 times the original setting (for primarily closed-cell structure). Foamed structures have the aesthetic and physical appearance of particle board, a common material of construction. Application possibilities of products made by the Radlite technology include highway signs and sound barriers, substructures for bathtubs and shower stalls, pallets and tote boxes, and flooring products and building fascias.

A novel application of mixed plastics is to toughen road surfaces. Recently the Ragusa Laboratories of ECP Enichem Polimeri in Italy investigated the use of mixed plastics waste to reinforce bitumen. Sorted municipal waste with a polyethylene content of approximately 60 per cent was mixed with bitumen in varying proportions up to 20 per cent. The properties of the resultant bituminous concrete were improved in two important ways: better wear resistance and raised softening point. Use of bitumen modified with mixed plastics waste of high polyethylene content as an experimental road surface under heavy traffic has established its notable superiority over unmodified bitumen for road surfacing.

Homogeneous Fractions

A widespread solution, in terms of application and market volume, could be the recycling of single materials or homogeneous fractions obtained from a separation process of the mixture. In fact, the samples obtained from single homogeneous fractions show a general performance far greater than that of samples produced from mixed plastics. Separation of postconsumer mixed plastics (municipal waste) into four fractions—polyolefins (PO), PS, PVC, and PET—is commonly adequate.

The improvement of tenacity is, in particular, evident when considering impact resistance of PO as compared to mixed plastics. Samples subjected to impact tests show an increase in elongation at the breaking point from 7 per cent to above 100 per cent. The samples of recycled PVC fraction are comparable with those of a common virgin, with only marginal reduction in mechanical properties.

With regard to PET fraction, the potential applications are strongly dependent on its purity (as discussed earlier). Applications like films, fibres, or straps are not recommended when a high concentration of impurities are present. In this case, the PET fraction can be employed for structural applications as an engineering polymer with the addition of other components like glass fibres, impact modifiers, and/or nucleating systems. However, reuse of the PET fraction implies that the amount of residual PVC must be kept below 50 ppm to avoid undesirable polymer degradation that results in poor surface appearance and loss of mechanical properties of the manufactured products.

Mixed waste consisting of PET and PE can be converted into useful products using compatibilisers such as LDPE and LLDPE with acid and anhydride groups grafted on the backbone. A range of products can be made with such compatibilised blend, e.g., office partitions, roofing, slates for benches and chairs, and generally any extruded or moulded sections needing mechanical load-bearing capacity similar to aforesaid applications.

With advances in cleaning, sorting, and other recycling technologies, more products with recycled plastics content are being manufactured. Some recent developments include using recycled PP and HDPE to produce a wide range of products. For example, multimaterial PP bottle scrap (typically, 90 per cent PP, 5 per cent ethylene-vinyl alcohol barrier resin, and 5 per cent olefin adhesive) can be added in varying amounts (3–12 per cent) to recycled HDPE and processed on a single-screw extruder to form pellets for compression moulding to a range of products. The multimaterial resin can also be sandwiched between two layers of virgin HDPE using a three-layer extrusion blow-moulding process. Bottles made in this way from 75 per cent virgin resin and 25 per cent postconsumer blend are found to be suitable for normal commercial trade.

Liquefaction of Mixed Plastics

There have been many research activities on plastics liquefaction because oil is easy to store, transport, and use. Most promising among them is the liquefaction technology jointly developed by the Japanese Government Industrial Development Laboratory (Hokkaido), Mobil Oil Corporation, and Fuji Recycle. The process features a combination of thermal and catalytic cracking using a proprietary Mobil ZSM-5 catalyst. It can treat polyolefinic plastics, PE, PP, PS, or their mixtures, producing relatively low pour and highly aromatic liquid at a yield of about 85 per cent. The produced oil contains many aromatics including benzene, toluene, and xylene.

Waste plastics are crushed, washed, and separated from other plastics that cannot be liquefied (e.g., PVC) by utilising the difference of specific gravities against water. Plastics that can be liquefied mostly float in water while plastics that contain a lot of chlorine, carbon, and oxygen have high specific gravities and sink in water. However, some PVC floats and is recovered with PE or PP. Therefore, after separating, the feedstock for liquefaction may still contain 3–7 per cent PVC. Fuji Recycle has developed liquefaction technology to treat such PVC contaminated mixtures.

For liquefaction, polyolefinic plastics are warmed to about 250°C, melted and transferred to the melting vessel by a heated extruder. In the melting vessel, plastics are further heated to about 300°C by heat transfer oil and transferred to the thermal cracking vessel. In the thermal cracking vessel, melted plastics are heated to about 400°C by the cracking furnace. The thermally cracked gas phase hydrocarbon passes through the catalytic reactor containing ZSM-5, where it is cracked and converted to higher value hydrocarbon. The recovered liquid and gas are separated by cooling and the gas is used as in-house fuel.

Because of the pore structure of ZSM-5, the produced hydrocarbons are composed of low molecular species (4 carbons to about 20) which are in the gasoline, kerosene, and gas oil boiling range. In comparison, the carbon numbers of hydrocarbons produced only by thermal cracking range from 4 to 44. Polystyrene in the feedstock enhances the yield of ethylbenzene, toluene, and benzene, while producing gas that is predominantly propane/propylene.

RECYCLING OF CURED EPOXY-BASED MATERIALS

Since the oil crisis in the 1970s, the question of recycling has gained importance with the experience of increasing waste streams and with space for landfills becoming expensive and more and more scarce.

Epoxies in structural applications with their long usable life and their excellent chemical resistance offer ecological advantages and play, with less than about 0.5 per cent, only a minor role in the generation of the plastic waste streams; nevertheless they are also included in the emerging regulations and methods for recycling them have to be found.

Particle recycling, e.g., the use of milled cured epoxy-based materials as additional filler in new formulated epoxy products, is already used in the processing of epoxy moulding compounds. Up to about 15 per cent of milled runners can be added to new material. With a proper choice of particle size distribution (low fines), up to about 30 per cent of the mineral filler in quartz-filled anhydride cured epoxy-casting formulations can be replaced by recycled epoxy-based insulating materials etc., without affecting the properties of the cured materials too much. The use of milled recycled epoxy-based printed circuit boards as reinforcing filler in thermoplastic materials as well as in concrete or road building materials is under evaluation.

Feedstock recycling as way to split plastic waste back to basic chemicals is also applicable to cured epoxy-based products. Pyrolysis results in a gas fraction with an energy content similar to that of natural gas and a liquid fraction consisting mainly of aromatic compounds. Hydrolysis with alkali at elevated temperature is very easily achieved with anhydride-cured expoxies and leads to phenolic ethers with OH end-groups and acid-salts. Hydrolysis of dicyanodiamide-cured expoxies, e.g., from printed circuit boards, needs more severe conditions.

In the case of bromine-containing resins, the bromine is completely split off and aromatic compounds similar to those found after pyrolysis are formed. Catalytic hydrogenation of e.g. anhydride-cured epoxy resin at elevated temperatures under pressure results in a fraction of phenolic and other aromatic compounds, neopentylglycol-derivatives and lactones.

Energetic recycling (incineration in power plants) is no problem with epoxy-based cured materials that contain no critical by-products. Incineration of chlorine or bromine containing materials requires modern equipment with sufficient corrosion resistance and perfect cleaning of the smoke gases.

Use of waste plastics instead of coal in blast-furnaces or similar equipment is a method lying between pyrolysis and incineration. Use is already made of this technique in the case of printed circuit boards. The concept behind this to recover the copper and, if present, the noble metals contained in the electronic scrap. Surveys show that the methods used for the recycling of the commodity plastics are, with proper choice of the reaction conditions, also applicable to cured epoxies.

Toxicological Properties and Safety Precautions

Epoxy resins and curing agents are reactive chemicals and have to be handled with care. Toxicological properties of the numerous products used in epoxy resin formulations differ very much. Bisphenol-based conventional epoxy resins, for example, have a low order of acute toxicity and are mild to moderate irritants; low molecular weight resins are skin sensitisers. Skin irritation and sensitising effects are stronger, when certain reactive diluents are added.

Amongst curing agents, the aliphatic amines are strong bases and thus corrosive and volatile materials which can cause tissue damage (burns) to skin, eyes and mucous membranes. Several aromatic amines show evidence of carcinogenicity in animals and exposure should be strictly avoided. Anhydrides are eye, skin and respiratory irritants. Solvents may cause fire hazards and even mineral or reinforcing fillers may be dangerous to health. It is therefore very important in every case to consult the safety data sheet for the individual product to be used. It is also important to make a persons and workplace-oriented risk analysis followed by intensive instruction and training before work with epoxy-based products starts. To minimise exposure, the following protective equipment should be provided and used: rubber or plastic gloves and sleeves, overalls and aprons (to be washed and cleaned regularly), goggles or face shields, special protective skin cream to be used before work starts.

After work, and in the case of skin contamination, but also before eating or using the WC etc., operatives should wash with warm water and mild alkali-free soap and dry themselves with disposable paper towels. Never use solvents to cleanse contaminated skin.

The working area should be separated from other areas in the factory and be provided with general and local exhaust systems to remove vapours, mists etc., where they are generated. Rooms for eating, taking a break or changing clothes should be separated from the working area.

First aid

Contamination of the eyes by resin, hardener or product mix should be treated by immediate irrigation with running tap water for 15 minutes. A doctor should then be consulted. Contaminated skin should be treated by dabbing away the resin etc., washing with warm water and soap and applying a cleansing cream. A doctor should be consulted in the case of severe skin irritation and burns. Contaminated clothing should be removed, isolated and washed before using again.

MEDICAL PLASTICS WASTE MANAGEMENT

Increasing volumes of synthetic polymers are being used for various medical applications. Highly used products are syringes, blood bags, surgical gloves, and catheters. The used materials are becoming a serious threat to public health due to their contamination with disease causing micro-organisms and non-biodegradability. Especially, in India, this increases the risk of contamination of diseases to people associated with collection and disposal of these materials. These materials should not be used for recycling. At present, the best method of disposal is incineration under controlled condition and strict supervision. Next best way of disposal is using a waste pit.

At present, medical devices prepared from synthetic polymers are contributing significantly to the quality and effectiveness of the worlds health care system. These devices range from catheters to vascular grafts, from semi-occlusive dressings to mammary implants and from transdermal drug delivery systems to medicated patches. Besides these, polymers are also used to enhance speed and sensitivity of biosensor and dry chemistries to measure glucose, cholesterol, urea, dissolved oxygen and other important blood analytes. The American Food and Drug Administration estimated that there are now 2,700 kinds of medical devices and another 2500 diagnostic products. The majority of these devices utilise synthetic polymers. Mainly polyvinyl chloride (PVC), polypropylene (PP), polystyrene (PS), polyethylene both high density and low density (HDPE), (LDPE), nylon, silicones and polyurethane (PU) are used.

Plastics used in patient care are associated with blood, sputum, urine and many other components of the living systems. These components contain disease causing micro-organisms and sometimes they also serve as a good medium for the growth of micro-organism. Usually, at the site of collection, the number of organisms are generally small.

Later, due to cross-contamination during storage, the entire load potentially becomes harmful. Bacterial multiplication may occur in the days following the disposal due to moisture content and availability of other favourable conditions. So the clinical wastes become dangerous. The risk of infection or injury can be minimised by proper handling and careful disposal of these wastes. Table 28.3 gives major sources for clinical wastes in India.

Table 28.3. Source of clinical wastes in India.

Hospitals and associated clinics,	General practitioner (surgical),
Nursing homes,	Blood banks,
Veterinary hospitals,	V pet shops,
Medical research institutes,	Acupuncturists,
Tattoist, cosmetics piercers,	Armed force establishments,
Residential premises (from domiciliary dialysis etc.).	

Infectious waste can be disposed by: (i) chemical inactivation; (ii) autoclaving; (iii) exposing to microwave; (iv) incinerating; and (v) dumping in the waste pit. A practical disposal method should be one which is economical and involves minimum risk during the process. Besides this, the method should be sustainable for a long time without causing long term ecological hazards.

Following steps are critical for the safe disposal of infectious wastes; (i) initial storage (waste bins); (ii) segregation; (iii) transport; and (iv) final disposal treatment. Each of these steps requires proper inventory and strict supervision.

No doubt, the safest method of final disposal is to incinerate every bit of waste. But, economically and ecologically, incineration becomes unrealistic, even for rich countries. Next best way of disposal is to make a waste pit. This will be a good solution for rural health care facilities of India. WHO has suggested the following design and management for the cost affective and easy to make a waste pit.

1. The pit must be located where no ground water or surface water can come into contact with the deposit and, no leaching of liquids from the waste into ground water can occur. The waste pit must never be built close to water sources, such as wells or springs.

2. The site must not be located where ground is likely to be used for agriculture or where land development could take place.
3. The ground must be soft enough for manual digging but of low permeability.
4. The pit must be provided at least $1 m^3$ per bed to have service life of approximately 5 years.
5. The pit must be protected from scavenger (animal, birds, humans).

Finally, the most desirable long term solution is to develop biodegradable polymers, that are capable of degrading in nature after their use with out producing any toxic byproducts. Several groups around the world are in pursuit for such polymers.

Thus, it can be said that at present, infectious plastic waste disposal is grossly inadequate in India. These infectious wastes are being allowed to be mixed with house hold refuses. These find their way to rag pickers who collect and sell them to people who recycle them. In this process unknowingly these people are exposed to micro-organisms and become the victims of several diseases and they also become potential source for spreading diseases. Further, deadly disease like HIV is going to have an easy access to spread through this process. All these ultimately will become a threat to public health, especially in towns where private nursing homes are mushrooming without proper private or communal waste disposal service. This alarming situation can be avoided to a large extent by using relatively simple and cheap measures for disposal of infectious plastic waste.

SECTION VI

Special Topics

CHAPTER 29

Ferroelectric Polymers

INTRODUCTION

This chapter considers the science and application of electroactive polymers. Within this area, polymers which are intrinsically 'active' will be discussed along with polymers which themselves are inert, but have been made active by the incorporation of a second (active) phase (usually a ceramic material).

All crystalline materials may be categorised into one of 32 crystallographic point groups. Of the 21 which lack a centre of symmetry, 20 produce an electric dipole (i.e., polarisation) when mechanically stressed. These materials are termed 'piezoelectric'.

Ten of these classes possess a permanent electric dipole and (using similar nomenclature as magnetic phenomena), are called 'ferroelectric'. They also respond to changes in temperature and are called 'pyroelectric'.

The generation of an electric dipole as a result of applied stress is known as the piezoelectric effect. This was first recognised in the 1880s, but it was not until the 1940s that major breakthroughs were made with the discovery that certain ceramic materials [e.g, barium titanate and lead zirconate titanate (PZT)] were ferroelectric and could be rendered piezoelectric by application of an electric field, a process known as 'poling' (described later). PZT has been the preferred piezoceramic for many years and has been utilised in a variety of piezoelectric transducers for in-air and underwater applications.

This chapter will initially be concerned with intrinsic ferroelectric polymers and then move on to ferroelectric polymer composites. Material preparation and properties are discussed along with a comparison of the electroactive properties. After applications, a final brief section introduces so-called smart materials. These have the potential to revolutionise both the properties of devices we presently manufacture and the philosophy we use in the design process.

A full list of chemical acronyms and symbolism used in this chapter can be found at the end of the chapter.

Ceramics PZT is derived from the solid solution of two perovskite materials, tetragonally distorted lead titanate ($PbTiO_3$) and rhombohedral lead zirconate ($PbZrO_3$). The anisotropic nature of this material, especially at the morphotropic boundary between the tetragonal and rhombohedral phases, makes it a particularly strong piezoelectric.

The performance of PZT as a transducer material is none the less limited by the conflicting nature of its piezoelectric and dielectric properties:

1. In hydrostatic mode, the induced electric polarisation per unit strain (d_h) is determined by the tensor elements d_{33} and d_{31}. The piezoelectric strain coefficient is given by

$$d_h = (2d_{31} + d_{33}) \quad \text{... (29.1)}$$

and the large negative value of d_{31} (–170 pCN^{-1}) almost completely cancels d_{33} (450 pCN^{-1}).

2. The piezoelectric voltage coefficient (g_h) is given by

$$g_h = \frac{d_h}{\varepsilon_o \varepsilon_r} \quad \text{... (29.2)}$$

This gives the electric field produced in the material, per unit stress and the large dielectric constant of the material (1800) produces a very low g_h.

Further, the brittleness, high density and fabrication problems have limited the exploitation of ceramic materials. Similar electrical properties in a polymeric material would give considerable practical advantages for many applications. The search for a superior material has considered both pure polymers and polymer composites, where an active filler is used to produce the electrical properties.

Polymer investigation into piezoelectric polymers extend as far back as the 1920s when it was discovered that certain organic materials such as rubber and celluloid could be rendered piezoelectric by cooling in an electric field. Investigations have also found detectable piezoelectricity in wood, bone and tendon, DNA, polysaccharides and the polypeptide, poly-γ-methyl-L-glutamate. Small, unstable piezoelectric effects in commercially available synthetic polymers such as polystyrene, polypropylene and polymethylmethacrylate were reported. However, the major breakthrough came with the discovery by *Kawai* that stretched and poled polyvinylidene fluoride (PVDF) exhibited the largest known piezoelectric activity of all polymer investigated at that time. Since this discovery and the detection of pyroelectric activity in PVDF, extensive research and development has been conducted on this material, leading to a variety of applications detailed later in this chapter.

Investigations were extended to include co-polymer of vinylidene fluoride (VDF) with related fluorinated vinyl monomers, vinyl fluoride (VF), trifluoroethylene (TrFE) and tetrafluoroethylene (TeFE). The latter two co-polymers were found to crystallise into a polar form without the stretching stage required by PVDF. This facilitates easier processing, offering a wider range of potential device structures. Later research has therefore focused on the ferroelectric behaviour and piezoelectric activity of the vinylidene fluoride-trifluoroethylene co-polymers.

Investigations into alternative piezoelectric polymers, both semi-crystalline and amorphous have continued, and two of the most significant discoveries are worth a brief mention. Of semi-crystalline materials, the odd-membered polyamides (or nylons), particularly nylon-11, were found to exhibit reasonable piezoelectric activity especially when mechanically oriented. The highest piezoelectricity in an amorphous polymer was discovered in a co-polymer of vinylidene cyanide and vinyl acetate.

Although PVDF and the associated co-polymer have advantages over the brittle ceramics, they are also subject to limitations however. A low value of d_{33} (20 pC N^{-1}) inhibits its use as an active device. Its large negative value of d_{31} (–4.5 pCN^{-1}) lowers the hydrostatic strain coefficient, d_h, and finally, the material's low Curie temperature makes for a strong temperature dependence of the piezoelectric coefficients and problems with depoling at higher temperatures.

Composite piezoceramic/polymer materials are a means of overcoming the limitations of both ceramic and polymer materials to provide complementary properties which produce a superior piezoelectric device. The polymer phase lowers the density of the material providing a better acoustic coupling to water as well as a more easily adjusted buoyancy than that obtained for a high density homogeneous ceramic. The low dielectric constant of the polymer phase effectively increases the g_h coefficients and the figure of merit. The high elastic compliance of the polymer provides a better resistance to mechanical shock and exhibits the high damping required for a good passive device. The polymer's flexibility allows the composite to form to any reasonable profile. Finally, the piezoelectric properties of the ceramic are easily adjusted within the composite as they may be tailored via the connectivity patterns (discussed later) to impart the maximum piezoelectric response to the unit.

ELECTROACTIVE POLYMERS

Polymer Synthesis

For ferroelectric use, the only practical polymers are PVDF, formed by the linking together of simple VDF (1,1-difluoroethylene) molecules, and the VDF-TrFE co-polymer which consists of the VDF and TrFE molecules linked together in the same chain.

The vinylidene fluoride monomer ($CH_2 = CF_2$), is a relatively stable gas at normal temperatures and pressures, boiling at –85°C. Commercially it is produced by pyrolysis of more readily available precursors such as trifluoroethane.

$$CF_3CH_3 \xrightarrow[\text{heat}]{} CH_2 = CF_2 + HF$$

Laboratory preparations are normally carried out in the liquid phase, a typical route being the reaction of alcoholic potash with halogen substituted ethane.

$$CHF_2CH_2Br + KOH\ (\text{alcoholic}) \xrightarrow[\text{heat}]{} CH_2 = CF_2 + KBr + H_2O$$

Trifluoroethylene, $CHF{=}CF_2$ is also a gas at normal temperatures and pressures, boiling at –52°C. It is slightly less stable than vinylidene fluoride and a polymerisation inhibitor is normally added to bulk quantities to prevent premature reaction. Preparation is by routes analogous to those used for vinylidene fluoride.

Polymers are prepared from the monomers by free radical polymerisation in suspension or emulsion, with water as the reaction medium. These two routes are superficially similar but differ in the location of the organic peroxide free-radical initiator and in the kinetics of the reactions.

It should be noted that the correct choice of initiator, and any other additives necessary for efficient polymerisation, is crucial to the preparation of material for use as a ferroelectric polymer. This is because of the essential requirement to subject the polymer to very high electric fields to render it electrically active (a process known as 'poling'). Any conductive residues in the polymer can make poling very difficult, or in extreme cases impossible, due to the high probability of dielectric breakdown. Because of this, details of polymerisation methods used to prepare commercially available, electrically active material are not generally accessible.

Polymerisation of the fluoro monomers is carried out at temperatures in the range 20–60°C and at pressures up to 60 MPa. Even in the case of the simple self-addition of vinyl fluoride molecules, the precise reaction conditions have a significant effect on the detailed structure of the resulting polymer. Because the molecules are not symmetrical they can link to each other in the chain in three possible ways. These are the normal, and energetically favourable 'head to tail' configuration (producing electrically inactive material) or the so-called 'defect' combinations of 'head to head' and 'tail to tail'. The effect of the low fraction of these defects present in the poly(vinylidene fluoride) prepared by standard routes is small and the polymer crystallises from the melt in a non-polar form. In contrast to this, the polymerisation of a mixture of vinyl fluoride and trifluoroethylene monomers results in many 'head to head' sequences, the effect of which causes the polymer to crystallise in a potentially active form, a structure only attainable in poly(vinylidene fluoride) by vigorous mechanical re-orientation.

Polymer Structure

As stated in the previous section, there are three chain conformations of PVDF. The two most common forms are described in this section. Crystallisation from the melt or solution produces the most common polymorph of PVDF, the α-phase (also known as form II). The chain conformation, is trans gauche$^+$ trans gauche$^-$ ($tg^+ tg^-$); (Fig. 29.1). Each chain conformation possesses a net dipole moment, originating from the CF_2 bond, with a component perpendicular to the polymer chain. In the case of the $tg^+ tg^-$ conformation the dipole components normal to the chain axes are antiparallel and thus cancel each other. Therefore the α-phase is non-polar. Transformation from the α-phase to the polar β-phase (also known as form I) is achieved by mechanically stretching or rolling at elevated temperatures. This extends the polymer chains to the all-trans conformation (Fig. 29.1b). As all the dipoles are normal to the chain axes, on a microscopic scale, each crystallite has a net dipole and is piezoelectric. However, on a macroscopic scale, there is zero polarisation within the polymer film due to random orientation of the crystallites. The final stage in rendering the PVDF film piezoelectric is known as poling and involves the application of an electric field. This preferentially orients the crystallites in the direction of the poling field, thus producing a net polarisation.

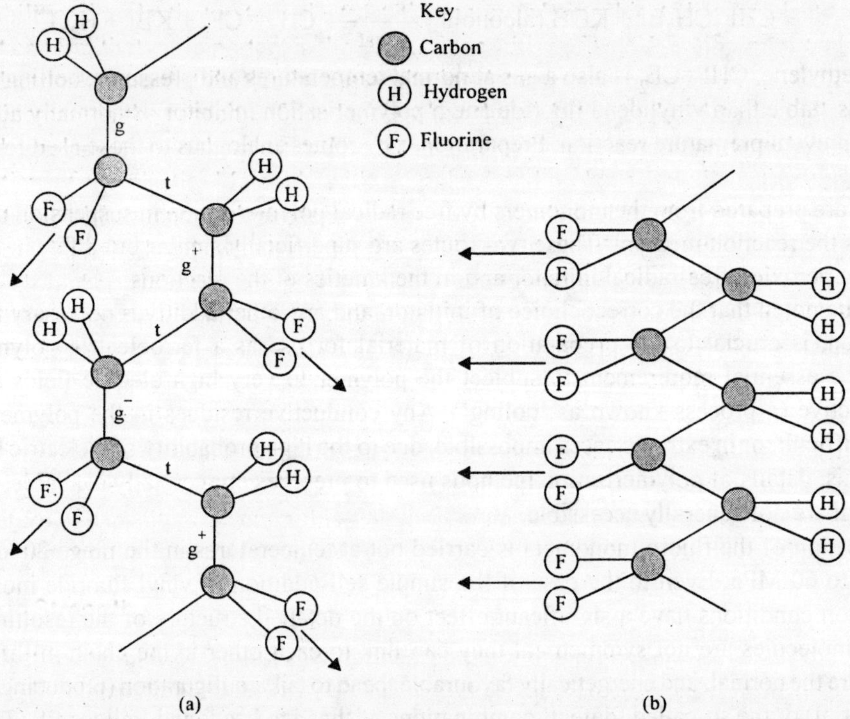

Fig. 29.1. Schematic of the two most common chain conformations of PVDF: (a) α-phase; and (b) β-phase. Arrows indicate $-CF_2$ dipole directions defined by the carbon backbone.

In the co-polymer of vinylidene fluoride and trifluoroethylene, the increased number of comparatively large fluorine atoms prevents the formation of the $tg^+ tg^-$ conformation by replacing hydrogen atoms on adjacent carbon atoms. This forces the molecular chains to crystallise directly into the polar β-form. The co-polymer film then undergoes a poling stage similar to PVDF to make it fully piezoelectric.

Mechanical Processing and Fabrication

Both PVDF and its co-polymers with TrFE are thermoplastic and as such can be easily processed in conventional thermoforming equipment. PVDF has a combination of processability, mechanical strength and chemical resistance. Because of this it has found a niche in the speciality polymers market unconnected with its unusual electrical properties. The co-polymer is a similar material to PVDF and, whilst as yet no large commercial usage has emerged, the same processing methods can be used.

The ease with which a plastics material such as PVDF can be formed into large areas or complex shapes is one of the significant advantages that polymer ferroelectric materials have over the conventional ceramic types.

Thin films

Historically, nearly all the early investigations into the ferroelectric properties of PVDF were carried out on thin (≈ 25μm) films. This was necessary to orient the material and develop an incipient ferroelectric structure. This is most conveniently done by using conventional thin film making equipment which stretches a preformed polymer strip in either one or two in-plane directions. Additionally, very high fields (typically 100 $kVmm^{-1}$) must be applied to the oriented polymer during the poling process to develop the ferroelectric properties. Practical considerations of voltage availability and safety therefore set an upper limit on the thickness of films which can be conventionally poled.

Uniaxial film

The most straightforward method of mechanically producing a thin film is to stretch it in one direction only. In the laboratory this can be done by applying a load to one edge of a preformed sheet held at elevated temperature. The load and the temperature are adjusted to produce elongation, typically about four times the original length.

Commercial production methods involve passing a continuous band of polymer through two pairs of rollers rotating at different speeds, in some variants of this process the polymer strip passes over two single rollers. If the second pair (or single) roller is rotating faster than the first than a stretching force is applied to the polymer, the amount of stretch achieved depending on the relative rotational speed of the two rollers. As with laboratory stretching, the polymer is held at a suitable elevated temperature during the operation. Material produced by stretching along one axis is termed uniaxial. The polymer is oriented by this stretching and the mechanical properties in the stretch direction are very different from those in the in-plane direction perpendicular to it. Properties, especially tensile strength, are enhanced in the stretch direction although this is accompanied by a transverse strength reduction.

Biaxial film

Film which is as strong as possible in both directions is often required. This has led to the development of techniques for stretching in two mutually perpendicular in-plane directions. Typically, a combination of longitudinal stretching using rollers followed by tentering where the edges of the film are gripped in clamps running in divergent tracks is used. Specialised machines have also been made which stretch film in two directions simultaneously. Film produced by stretching in two directions at right angles is said to be biaxially stretched. Strength is enhanced in all in-plane directions, although not to the extent achieved in a single direction with uniaxial stretching.

The salient feature of these two methods of film production is that both the uniaxial and the biaxial films are oriented in the way necessary to allow the production of electrically active ferroelectric materials by poling, so that a single process not only produces a thin film, but one which is suitable for conversion into an electrically active form.

Film casting

Thin films can also be made by flowing a solution of a polymer onto a smooth surface. After the solvent has evaporated the film is stripped from the casting surface. Film thickness can be controlled by adjusting the viscosity of the polymer solution (by varying its concentration) and by defining the depth of the deposited solution through the use of a roller or sharp movable edge ('doctor blade'). Both PVDF and its co-polymer with TrFE can be successfully cast into thin films from ketone solutions. In the case of PVDF, the resulting film is not oriented although it has been reported that oriented material can be produced under some conditions using specific solvents. The co-polymer film produced from ketone solution is oriented and needs no stretching stage before poling. This is a considerable practical advantage over PVDF.

Sheet production

For some ferroelectric applications, polymer sheets of 1–2 mm thickness are required. Commercially, this can be produced by extruding molten polymer through a die of the appropriate cross-sectional shape and size. The hot extrudate can be led over or through pairs of cooling rollers or into a cooling liquid before it is cut into individual sheets or wound into a roll.

On a small scale, compression moulding, where the requisite quantity of polymer pellets is introduced into a hot mould of the desired dimensions and pressed into a homogeneous mould-sized sheet, has been found satisfactory.

Neither of these methods produces any significant orientation; this is immaterial in the case of co-polymer which is self-orienting but a further stretching stage would be essential after sheet formation in the case of PVDF.

Simple three-dimension structures

In the case of the PVDF-TrFE co-polymer, advantage can be taken of its self-orienting property to fabricate potentially electrically active structures. Attempts to thermally form oriented PVDF inevitably lead to loss of orientation which is not easily recovered. Co-polymer is more useful in this respect as it reverts to the oriented form on cooling. Simple hemicylinders can be formed from co-polymer sheet by heating and bending round a former. Shapes such as domes can be made by using heat and vacuum to force the sheet to conform to a shaped master. Most complex shapes are possible by the use of injection moulding, although the need to pole the active area with high fields must always be considered, especially if the shape produces sharp edges or material thinning (e.g., at corners).

POLING PROCESS

The poling process is where the electric dipoles in a material are preferentially aligned in one particular direction defined by the direction of an applied electric field. This has to occur in electroactive ceramics, polymers and composites. Although this section is located in a discussion of polymers, the comments equally apply to all three materials.

The conditions with which the ferroelectric material is poled are crucial in determining its final electrical characteristics. The amount of net polarisation produced as a result of poling is directly related to the level of piezoelectricity. The higher the degree of polarisation, the greater the activity. There are two main poling techniques, conventional two-electrode poling (commonly known as 'thermal' poling), and corona poling, each requiring optimisation of poling time, temperature and field to achieve maximum polarisation.

For thermal poling, the polymer film is first electroded on both sides. A wide range of electrode materials can be employed e.g., painted-on silver solutions, pressed-on metal foils and the more commonly used evaporated or sputtered layers of chrome-gold or aluminium. A DC field is then applied across the film thickness at an elevated temperature (typically 70–100°C) for a period of 15–120 minutes and then remains applied while the sample cools to room temperature. The fields required to produce the maximum polarisation are just below the breakdown strength, typically 30–120 kV mm^{-1}. Fig. 29.2 shows the piezoelectric strain coefficient as a function of poling field for a 70:30 VDF:TrFE co-polymer, illustrating that saturation polarisation occurs around 25 kV mm^{-1}.

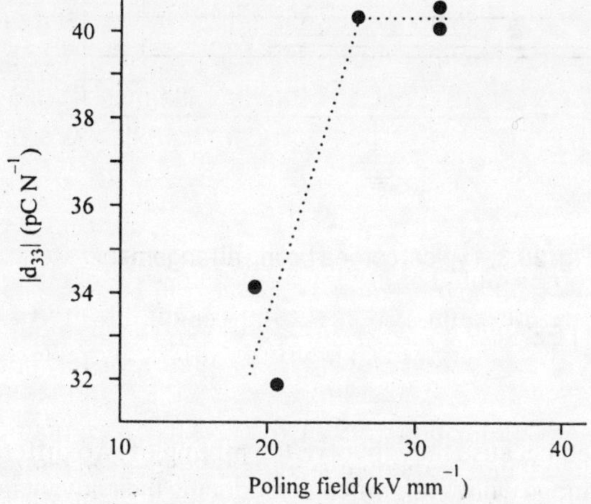

Fig. 29.2. Piezoelectric strain coefficient d_{33} as a function of poling field for 70:30 VDF:TrFE co-polymer. All film samples were poled at 100 ± 2°C.

In corona poling, a corona discharge is produced by a needle electrode (or an array of needles) at a high potential, a few centimetres above the top, unelectroded surface of the polymer film. To provide a discharge path and thus create a field across the sample, the opposite surface must be at ground potential and is therefore metallised or in contact with a metal electrode. A grid, at a lower potential, placed between the sample and the needle(s) controls the intensity of the discharge, as illustrated in Fig. 29.3. Voltages of typically 25–40 kV are applied to the needles with grid potentials of 5–15 kV. The poling process can be completed at room temperature in a few seconds, although higher and more thermally stable piezoelectricity has been obtained at elevated temperatures.

Corona poling offers advantages over conventional thermal poling. The absence of electrodes eliminates shorting problems during poling when there is an electrical breakdown or defect in the polymer film. It is also more adaptable to continuous production. However, success at achieving maximum polarisation, particularly for thicker films (>40 µm) is more dependent on experimental setup than the thermal poling technique. It should also be noted that if a material contains defects producing a relatively high conduction path, neither technique will work, the sample requiring either too much current or charge.

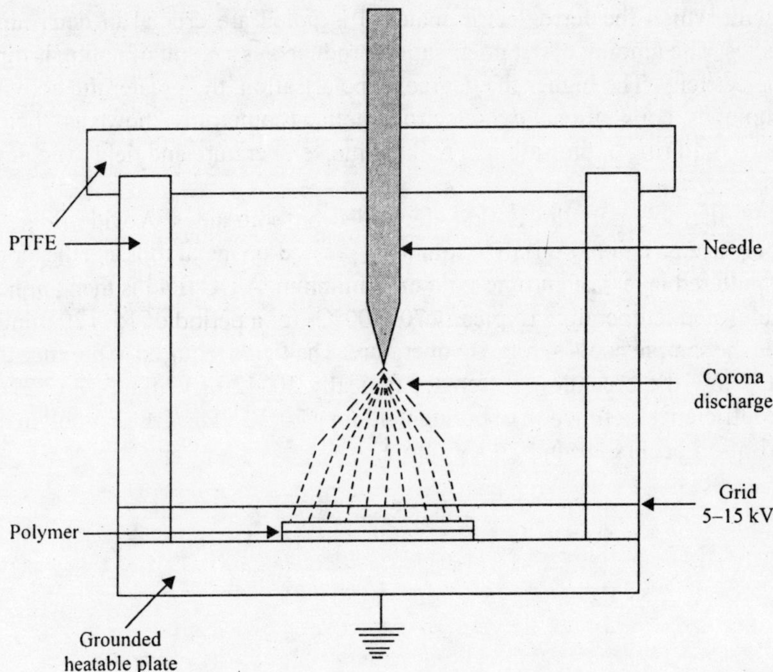

Fig. 29.3. Typical corona poling arrangement.

ELECTROACTIVE COMPOSITES

Composite Designs

Composite materials are so-called because they involve the mixing of two different materials in order to improve on their individual limitations without significantly reducing their advantages. Combining materials involves not only choosing the correct components but also coupling them in the optimum way. The type of phase coupling (or connectivity) is of utmost importance because this controls the electric flux patterns which exist within the samples as well as the mechanical properties. Symmetry is a second important consideration since symmetry and properties are interrelated through tensor coefficients. In this regard there are several levels of symmetry to be considered:

1. The crystallographic symmetry of each phase.
2. The phase symmetry after processing.
3. The combined symmetry of the composite.
4. The environmental influence on the total symmetry including electrodes and clamps.

The physical and chemical properties of composites will depend on the properties of the individual phases and may be conveniently classified as either a sum, combination or product.

Sum properties

A sum property is one where a composite coefficient depends on the corresponding coefficients of the individual phases. For example, the permittivity (ε) of a two-phase (or diphasic) composite is given by

$$\varepsilon^n = v_1\varepsilon_1^n + v_2\varepsilon_2^n \quad \text{... (29.3)}$$

where ε_1 and ε_2 are the dielectric constants of the constituent phases and v_1 and v_2 are their volume fractions. The exponent (n) is +1 for parallel mixing and −1 for series mixing. For many composites, the geometric arrangement is between these two extremes, in which case ε can often be described by a logarithmic mixing rule for which n is non-integer.

Combinational properties

A combinational property is one in which the composite coefficient depends on two or more coefficients of the individual phases (which may average in different ways). For example, the acoustic wave velocity determines the resonant frequency for a piezoelectric device. The velocity is in turn dependent on the *Young's* modulus and the density of the phases.

Product property

A product property utilises different properties in the two phases to produce a third property through the interaction of the phases. For example, a magnetoelectric composite can be made from a ferroelectric and a ferromagnetic phase. When a magnetic field is applied to the composite, the ferrite grains change shape (due to magnetostriction) and the strain is passed onto the piezoelectric grains, resulting in an electrical polarisation.

Connectivity

The physical properties of a composite can change by many orders of magnitude depending on the way in which phase connections are made. As such, connectivity is a major feature in property development for multiphase composites. Each phase may be self-connected in zero, one, two or three dimensions producing the ten diphasic connectivities (illustrated in Fig. 29.4): 0–0, 1–0, 2–0, 3–0, 1–1, 2–1, 3–1, 2–2, 3–2 and 3–3. For example, a 2–1 connectivity pattern has one phase self-connected in two-dimensional layers while the other is self-connected in one-dimensional chains (or fibres). The connectivity patterns are not geometrically unique. In the case of a 2–1 pattern the fibres of the second phase may be perpendicular to the layers of the first phase (Fig. 29.4) or they may be parallel.

3–3 Connectivity

In a 3–3 composite, both phases are continuously connected in all three dimensions to give two interlocking skeletons. This type of structure is seen in certain polymer foams and by biological substances such as wood and coral. Two basic designs are particularly noteworthy and described briefly below:

1. *Replamine composites*. This process utilises the three-dimensionally connected structure of the calcium carbonate coral skeleton as a template for a PZT replica. The structure has the characteristics of a narrow pore size distribution, a pore volume approximately equal to the solid phase volume and complete pore interconnectivity making every pore accessible from all other pores. The dimensions of the pores vary from species to species, but within one species the size range is quite narrow. Different species of coral have different degrees of anisotropy in their structure ranging from a 3–1 connectivity of nearly parallel tubes to highly isotropic 3–3 structures. The replamine process involves shaping the coral (which is easily machinable) to the desired geometry. The coral template is then vacuum impregnated with a casting wax and the wax is allowed to harden before the calcium carbonate coral skeleton is leached away in hydrochloric acid leaving a wax negative of the original coral template. The negative is reversed with a PZT slip containing PZT, water and poly(vinyl alcohol). The wax negative is burned off leaving a coral type structure of PZT which is then sintered. The PZT replica is back filled with a suitable polymer.

2. *BURPS composites*. The Burned-Out Plastic Spheres composites involves a technique that proves to be a far simpler means of fabricating 3–3 composites. PZT powder is mixed with volatile polymethyl methacrylate (PMMA) spheres in a 30/70 volume ratio. The mixture is die-pressed using poly(vinyl alcohol) as a binder and the PMMA spheres are volatilised out by very slowly heating the pressed pellet to 400°C. After sintering, the cold highly porous pellets are vacuum impregnated with a suitable polymer: either a flexible silicone rubber elastomer or a stiff epoxy resin. The finished product contains PZT regions ranging from 20–120 μm that are randomly interconnected within the composite.

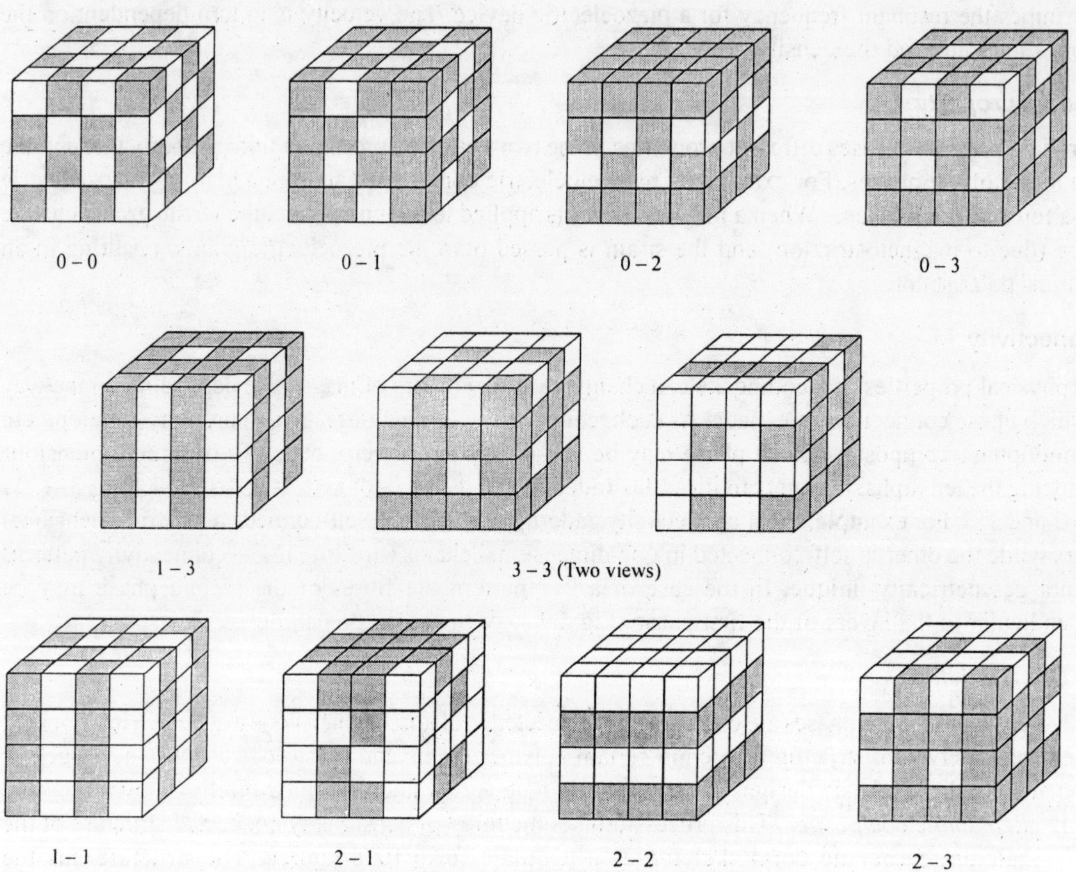

Fig. 29.4. Connectivity patterns in diphasic composites.

Both these composite types and the ones discussed below are shown schematically in Fig. 29.5.

3–2 and 3–1 Connectivity

Both 3–2 and 3–1 composites have a three-dimensionally connected ceramic phase. The second (polymer) phase is one- or two-dimensionally connected within the ceramic network. This tends to produce extremely rugged composites.

Perforated PZT-polymer composites have been fabricated by drilling holes in sintered PZT blocks and filling with epoxy. In some cases the perforation was left empty but capped with alumina plates and the whole structure encapsulated in epoxy — referred to as perforated 3–2–0 or 3–1–0 composites. The maximum hydrostatic piezoelectric coefficients attained for these composites are generally significantly greater than those associated with solid PZT and values of $d_h \cdot g_h$ have been reported that are greater than 200 times the figure of merit for solid PZT.

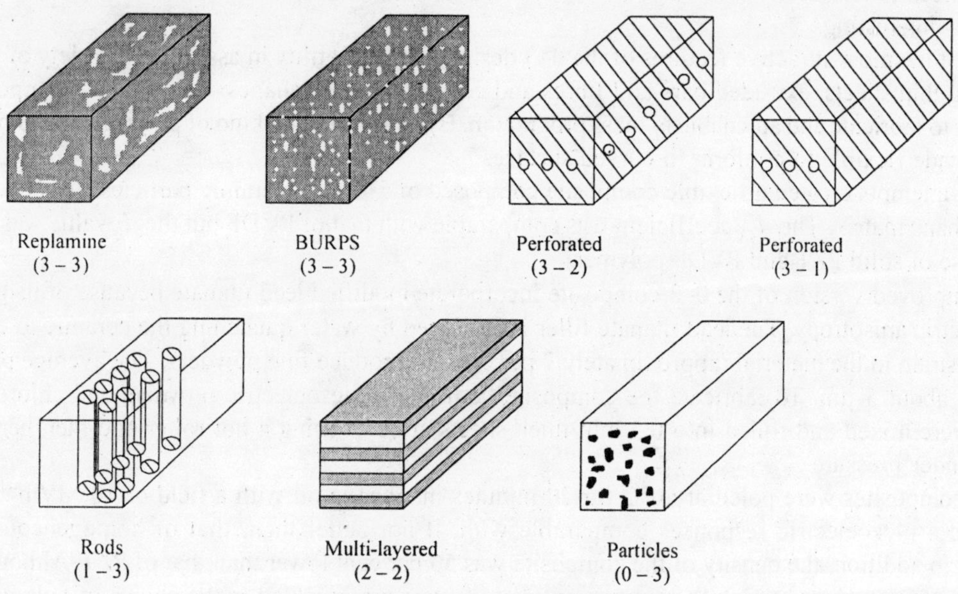

Fig. 29.5. Composite formations.

1–3 Connectivity

The 1–3 composite consists of a one-dimensionally connected ceramic phase contained within a three-dimensional polymer phase. The shape of the ceramic may be any that is continuous in one dimension between two opposing composite surfaces. The following designs have been investigated extensively:

1. PZT rod design: The rod shape was chosen to minimise the composite density and lower the dielectric coefficient ε_r by reducing the volume per cent of PZT. The composite is made by arranging the prepared rods in disc-shaped racks and covering them with epoxy.
2. Diced design: An alternative method of producing 1–3 composites is by dicing a block of PZT and subsequently back filling the spaces between the posts with the polymer phase.

One of the advantages of this connectivity design is that most of an applied stress is taken across the ceramic phase thereby amplifying the piezoelectric effect.

2–2 Connectivity

Piezoelectric composites with multilayer structures (2–2 connectivity) have been fabricated to produce high stability electrical filters.

Tape-casting provides a convenient way of making thin composite layers. Here, a ceramic powder is mixed with an organic binder to form a slurry which is spread in a continuous layer onto a substrate. After drying, the binder hardens to give a flexible plastic tape impregnated with the ceramic. This tape can be cut in the normal way. To fabricate multilayer composites, the component layers are stacked alternatively and pressed together at about 50°C to obtain a firm bond before firing.

0–3 Connectivity

Composites with 0–3 connectivity consist of a three-dimensional polymer matrix loaded with discrete ceramic particles. These are the simplest composites to make and although they do not have the desirable stress concentration factors found in other connectivity patterns, certain distinct advantages make them extremely interesting.

One of the most attractive features of the 0–3 design is its versatility in assuming a variety of forms, including thin sheets, extruded bars and fibres and certain moulded shapes. This type of composite is also easy to fabricate and amenable to mass production. By using the right kind of polymer, the composite can be made flexible to conform to curved surfaces.

Early attempts created a flexible composite composed of 5–10 μm ceramic particles suspended in a polyurethane matrix. The d_{33} coefficient was comparable with that of PVDF but the d_h value was lower than those of solid PZT and PVDF polymer.

An improved version of the 0–3 composite incorporate modified lead titanate because of its greater piezoelectric anisotropy. The lead titanate filler is produced by water quenching the ceramic to exploit the high strain in the material (approximately 7 per cent) to produce fine powders. The average particle size was about 5 μm. To fabricate the composite bodies, the piezoelectric powders and chloroprene rubber were mixed and rolled into 0.5 mm thick sheets at 40°C using a hot roller and then heated at 190°C under pressure.

The composites were poled at 60°C for 20 minutes in silicone oil with a field of 10 MVm^{-1}. This produced a piezoelectric responses comparable with, if not better than, that of homogeneous PZT samples. In addition, the density of the composite was 50 per cent lower than that of PZT. Although 0–3 composites appear to be simple in structure, many factors are involved in the choice of polymer and filler materials, fabrication and poling procedures. A large research effort has been mounted to optimise these factors and make better 0–3 composites.

PROPERTIES OF FERROELECTRIC MATERIALS

Ferroelectric materials are characterised by a range of electrical, thermal and mechanical properties. This section gives a brief description of the characteristics unique to ferroelectric materials before detailing the properties that are important in assessing an active polymer's suitability for transducer applications.

Ferroelectric Behaviour

Most electroactive polymers and polymer composites are insulating materials and so a film with electrodes on its top and bottom surfaces forms a parallel plate capacitor with a capacitance C where

$$C = \frac{\varepsilon_o \varepsilon_r A}{t} \quad \text{... (29.4)}$$

and ε_0 is the permittivity of free space, ε_r the permittivity of the materials, A the overlap area of the electrodes and t the film thickness. An applied voltage (V) generates a charge Q given by

$$Q = CV \quad \text{... (29.5)}$$

The surface charge density is called the electric displacement D therefore,

$$D = \frac{Q}{A} \quad \text{... (29.6)}$$

Combining equations 29.4 and 29.5 gives

$$D = \frac{\varepsilon_o \varepsilon_r V}{t} = \varepsilon_o \varepsilon_r E \qquad \text{... (29.7)}$$

In linear dielectric, D is directly proportional to E, increasing and decreasing linearly with E. In ferroelectric polymers, such as PVDF and the VDF:TrFE co-polymers, the surface charge density increases with field in a non-linear manner and exhibits a hysteresis when the field is decreased, with a significant amount of surface charge remaining when the field is completely removed. Subsequent applications and reversals of the electric field produce the characteristic D–E hysteresis loops shown in Fig. 29.6.

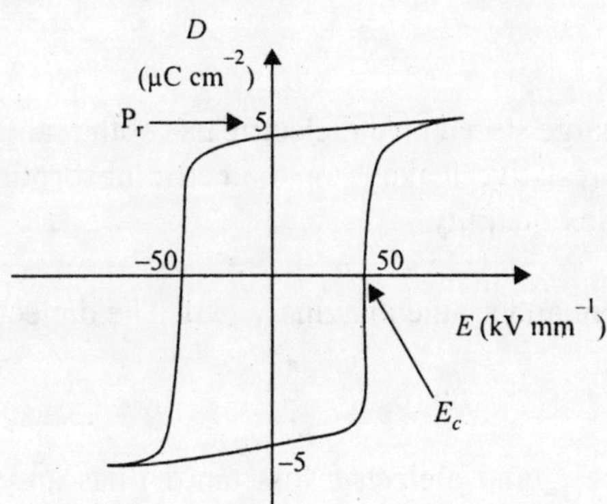

Fig. 29.6. Typical *D-E* hysteresis loop for an electroactive polymer.

The amount of surface charge density that remains is termed the remnant polarisation P_r and is caused by a net alignment of dipoles within the crystalline phase of the material. When the direction of the electric field is reversed the charge rapidly decreases and changes sign. The field at which the polarisation passes through zero is termed the coercive field E_c and the time taken to switch from positive to negative polarisation is called the switching time τ_s.

Measurement techniques are based on a circuit developed by *Sawyer* and *Tower*. For PVDF and the VDF:TrFE copolymers, typical values for P_r and E_c are in the region of 5 μCm^{-2} and 50 kVmm^{-1} respectively. *Furukawa* have shown that for PVDF, switching times of a few microseconds can be obtained at room temperature with an applied field of a few hundred kV mm^{-1}. Further investigations into VDF:TrFE have found that for a 65:35 molar composition at 20°C, τ_s decreases from 1 ms to 100 ns as field strength increases from 80 to 400 kV mm^{-1}.

Dielectric Properties

Dielectric constant

The dielectric properties of a material are charactetised by the permittivity ε, or more commonly by its relative permittivity (or dielectric constant) ε_r where

$$\varepsilon_r = \frac{\varepsilon}{\varepsilon_0} \qquad \text{... (29.8)}$$

and ε_0 is the permittivity of free space, 8.85 × 10^{-12} Fm^{-1}.

The value of ε_r can be calculated from equation 29.8 after determining the capacitance C of a sample with known area A and thickness t.

The values of ε_r quoted for PVDF range from 10–12 at 1 kHz, depending on the manufacturer. For VDF:TrFE copolymers, ε_r varies with molar composition, reaching a maximum at 55:45 VDF:TrFE. This maximum value at room temperature has been reported to be as high as 15, although values nearer that of PVDF are normally observed. These values are over two orders of magnitude lower than those of piezoelectric ceramics. This can be seen as a disadvantage in some applications, such as long-range hydrophones, which require materials with a high capacitance to allow the signals produced to travel large lengths of cable with the minimum of sensitivity loss due to cable impedance. Therefore, for piezoelectric polymer devices, it is necessary to have an external pre-amplifier close to the transducer to drive the cable.

Dielectric loss

With alternating voltages, the charge stored on a dielectric has both real (in phase) and imaginary (out of phase) components caused by resistive leakage or dielectric absorption respectively. The dielectric permittivity is therefore a complex quantity.

$$\varepsilon = \varepsilon' - j\varepsilon'' \qquad \text{... (29.9)}$$

where ε' is the real component and ε'' the imaginary part. The dielectric loss or $\tan(\delta_e)$ is defined as

$$\tan(\delta_e) = \frac{\varepsilon''}{\varepsilon'} \qquad \text{... (29.10)}$$

A dielectric with capacitance C and dielectric loss $\tan(\delta_e)$ has an equivalent conductance G at a frequency f where

$$G = \omega C \tan(\delta_e) \qquad \text{... (29.11)}$$

with ω, the angular frequency equal to $2\pi f$.

The noise voltage V_D produced by this conductance is given by the expression for *Johnson* noise in a resistor

$$V_D = \sqrt{\frac{4k_B T}{\omega C \tan(\delta_e)}} \qquad \text{... (29.12)}$$

and the equivalent noise current i

$$i_D = \sqrt{4k_B T \omega C \tan(\delta_e)} \qquad \text{... (29.13)}$$

kB is *Boltzmann's* constant (1.38×10^{-23} JK^{-1}) and T the absolute temperature of the material. Thus the value of $\tan(\delta_e)$ quantifies the amount of electrical energy dissipated into thermal energy. This is an important consideration when designing and evaluating piezoelectric transducers and normally needs to be as low as possible.

Values of $\tan(\delta_e)$ can be calculated from conduction measurements or obtained directly using modern capacitance bridges. Polymers are comparatively lossy dielectrics. Typical room temperature values of $\tan(\delta_e)$ for PVDF range from 0.015 to 0.020 at 1 kHz and for VDF:TrFE from 0.015 to 0.025, compared with values of 10^{-4} for some ferroelectric ceramics such as lithium tantalate and lead germanate. Composite materials tend to have $\tan(\delta_e)$ values somewhere between the values for the two pure phases.

Temperature dependence

Unlike PVDF, the VDF:TrFE co-polymers exhibit a distinct ferroelectric to paraelectric phase transition when heated between room temperature and the melting temperature. This transition can be seen in plots of ε_r as a function of temperature, where ε_r shows a peak (dielectric anomaly) around the *Curie* temperature, T_c. Fig. 29.7 shows plots of ε_r and tan(δ_e) vs temperature for co-polymers with molar ratios 56:44, 70:30 and 80:20 VDF:TrFE.

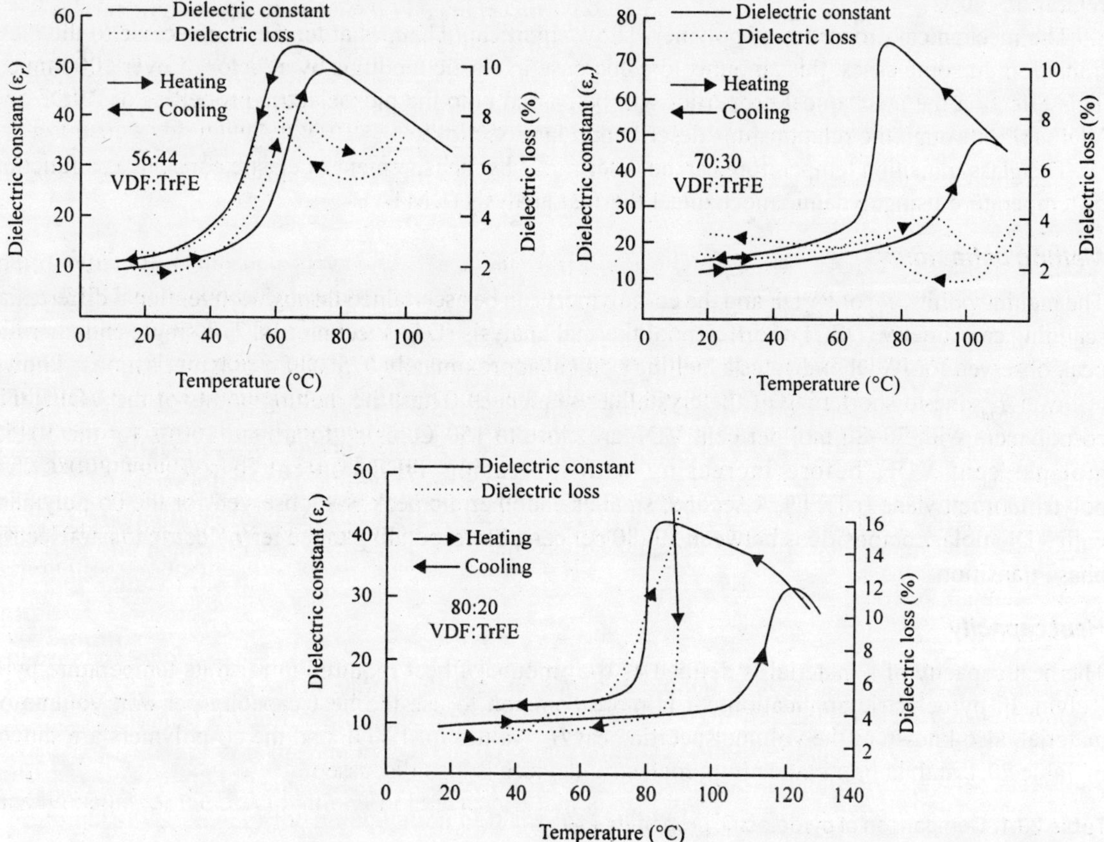

Fig. 29.7. Temperature dependences of ε_r and tan(δ) for 56:44, 70:30 and 80:20 VDF:TrFE co-polymers.

It can be seen that T_c increases with VDF content, ranging from 70°C for the 56:44 composition to 121°C for the 80:20 VDF:TrFE. Phase transitions for co-polymers with VDF contents higher than 80 mol per cent are difficult to detect as T_c is near the melting point. This offers an explanation to the fact that a phase transition has not been observed in PVDF. The T_c of co-polymers rich in VDF show a large thermal hysteresis, as seen with the 70:30 and 80:20 compositions, with T_c much higher on heating than on cooling. This hysteresis is also exhibited by tan(δ_e), however the peaks occur at temperatures 5–10°C lower than the peaks in ε_r. It is also interesting to note that co-polymers with VDF contents below 40 mol per cent show only a modest dielectric anomaly, indicating it is the VDF content that is responsible for the co-polymer's ferroelectric behaviour.

Thermal Properties

Glass transition

In common with most polymers that are completely or partly amorphous, ferroelectic polymers experience a phase change at the glass transition temperature T_g. Below T_g, the polymers are hard, rigid glasses. At

T_g and above, the polymers become flexible and elastomeric. Brittle, rigid polymers have a T_g above room temperature, for example, polystyrene has a T_g of 100°C. Rubbery or elastomeric materials have a T_g below room temperature. This is the case for PVDF and the co-polymers which have T_g in the region of –40°C.

The mechanical properties of polymers show significant changes at temperatures close to the glass transition. In some cases, this amounts to a decrease in elastic modulus by a factor of over 1000 times. This effect on the mechanical properties will be passed onto the piezoelectric properties of PVDF and VDF:TrFE through the relationships described in later sections.

The glass transition temperature can be determined by measurement of elastic constant as a function of temperature using dynamic mechanical thermal analysis (DMTA).

Melting behaviour

The melting points, T_m of PVDF and the co-polymers can be ascertained through conventional differential scanning calorimetry (DSC) or differential thermal analysis (DTA) techniques. The single endothermic peak observed for PVDF indicates a melting point at approximately 175°C. Co-polymerisation is known to lower T_m due to shortening of the crystalline sequences. Thus, the melting points of the VDF:TrFE co-polymers with 70–80 mol per cent VDF are close to 150°C, rising to around 160°C for the 50–55 mol per cent VDF, before increasing with increasing TrFE content to a T_m of 200°C for polytrifluoroethylene (pTrFE). A second, smaller endothermic peak was observed for the co-polymers with VDF molar compositions between 50–80 per cent, corresponding to the ferroelectric to paraelectric phase transition.

Heat capacity

The heat capacity of a material is defined as the amount of heat required to raise its temperature by 1 Kelvin. In pyroelectric applications, it is more common to use the heat capacity per unit volume of material, also known as the volume specific heat H_v. Values for PVDF and the co-polymers are shown in Table 29.1 and its importance in pyroelectric devices is also discussed.

Table 29.1. Comparison of pyroelectric properties.

Material	*p* ($\mu C\ m^{-2} K^{-1}$)	S_v ($MJ\ m^{-3}\ K^{-1}$)	F_v ($m^2\ C^{-1}$)	*FD* ($fm^2\ N^{-1}$)
$PVD_{uniaxial}$	27	2.43	0.1	0.88
VDF:TrFE 56:44	20	2.07	0.1	0.69
PZFNTU*	380	2.5	0.06	5.8

*PZFNTU: doped lead zirconate, a typical pyroelectric ceramic.

Thermal conductivity

As polymers have relatively loose molecular structures, their thermal conductivities are lower than those of ceramics. This should be considered when assessing materials for applications where heat conduction is important, such as pyroelectric detectors.

Values of thermal conductivity obtained for uniaxially stretched PVDF show there is a much higher effect parallel to the draw direction (5.5 $mWcm^{-1}\ K^{-1}$) compared to the value of 1.25 $mWcm^{-1}\ K^{-1}$ for the transverse '2' direction.

Piezoelectric and Pyroelectric Properties

In general, there are two characteristics to be considered when conducting an initial assessment of a material's suitability for a device; firstly the appropriate piezoelectric or pyroelectric coefficients to determine the degree of sensitivity and optimum mode of operation, e.g., thickness or hydrostatic; and secondly the effect of temperature on these properties, thus defining the maximum and minimum operating temperatures at which the material can be used efficiently.

Piezoelectric constants

Piezoelectric materials generate an electric charge in response to a mechanical stress. This is known as the direct piezoelectric effect, where the charge per unit area or electric displacement D is proportional to the applied stress T. There is also a converse effect, as an applied electric field E produces a proportional strain S in the material, resulting in either expansion or contraction, depending on the field's polarity. For both effects, the proportionality constant is termed the piezoelectric charge constant, d, thus

$$d = \frac{D}{T} = \frac{S}{E} \quad \text{... (29.14)}$$

There are three additional piezoelectric constants; the g constant gives the field produced by a stress and is useful in determining the voltage output of a stressed material; the other constants, e which relates T to E and h which relates S to E are only occasionally used. As T and S are related by the elastic constants c of the material and E and D are related by the dielectric constant, the piezoelectric constants are interrelated to one another. Therefore.

$$\frac{d}{g} = \frac{e}{h} = \varepsilon_o \varepsilon_r \quad \text{... (29.15)}$$

$$\frac{e}{d} = \frac{h}{g} = c \quad \text{... (29.16)}$$

The formal definition and derivation of these piezoelectric relationships may be found in *Berlincourt*.

The piezoelectric constants are tensor quantities, hence it is necessary to assign a co-ordinate system to produce a matrix of coefficients which relate to the symmetry of the crystal axis.

Anisotropy is introduced into ferroelectric polymer films by the poling process which aligns the dipoles in the direction of the poling field. By convention, this axis is known as the '3' direction, as illustrated in Fig. 29.8. For PVDF, an additional anisotropy is caused by the mechanical stretching procedure as already discussed. The 'I' axis is parallel to the stretch direction and the '2' axis is perpendicular to this in the plane of the film.

Thus, for polymer films with symmetry type 2 mm (i.e., uniaxially stretched PVDF) the components of the piezoelectric charge coefficient are:

d_{ij} = ... (29.17)

i \ j =	1	2	3	4	5	6
1	0	0	0	0	d_{15}	0
2	0	0	0	d_{24}	0	0
3	d_{31}	d_{32}	d_{33}	0	0	0

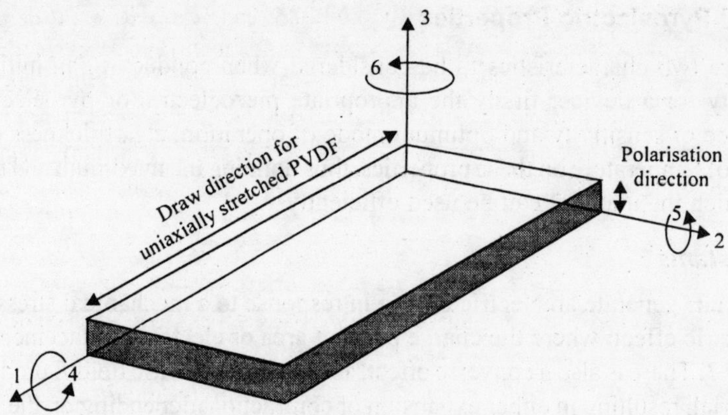

Fig. 29.8. Conventional axes for electrical and mechanical coefficients.

where the i subscript denotes the direction in which the charge is generated and the j subscript denotes the direction of applied stress (1 to 3 for uniaxial stresses and 4 to 6 for shear stresses). The matrix for ferroelectric polymers with symmetry type 2 mm, for example VDF:TrFE co-polymers and biaxially stretched PVDF, is similar except the l axis is equivalent to the '2' axis, thus $d_{15} = d_{24}$ and $d_{31} = d_{32}$.

The coefficients d_{15} and d_{24} associated with shear deformation are seldom used as there is a lack of practical applications using the shear mode. The most important piezoelectric coefficients for determining a material's suitability for a particular device are the d_{3j} and g_{3j} components and the hydrostatic coefficients d_h and g_h. The d_h coefficient is a measure of the charge generated due to application of a hydrostatic pressure. As the stresses are uniform in all three directions d_h is simply the sum of all d_{3j}.

$$d_h = d_{31} + d_{32} + d_{33} \qquad \text{... (29.18)}$$

The d_{33} coefficient describes the charge generated through the thickness of the film ('3' direction) due to a stress applied along the '3' axis. As it is difficult to stress a thin film in the thickness direction without contributions from d_{31} and d_{32}, the calculation of d_{33} is often done using equation 29.18 and measured values of d_{31}, d_{32} and d_h. However, it is possible to obtain d_{33} via the converse piezoelectric effect by using optical techniques, e.g., interferometry to measure the minute changes in thickness (less than a micron).

The transverse coefficients d_{31} and d_{32} can be obtained using comparatively simple strain removal techniques with generated charge passing through a capacitor to allow the subsequent voltage to be measured on a digital storage oscilloscope.

The d_h coefficient can be recorded by subjecting the sample to a hydrostatic pressure. Methods of achieving this include enclosing the sample in a uniform, low frequency (50 Hz or 50 Hz) sound field and application of a 1 atmosphere (0.1 MPa) pressure pulse by rapid air ingress of an evacuated chamber. In all cases, it is necessary to immerse the test sample in silicone fluid or oil to stabilise the temperature and eliminate compressional adiabatic heating effects. The g_h coefficient, calculated using equation 29.16, is important in comparing materials for hydrophones, particularly in the hydrophone figure of merit product FOM_h

$$FOM_h = \frac{d_h g_h}{\tan(\delta_e)} \qquad \text{... (29.19)}$$

A comparison of piezoelectric coefficients between uniaxially and biaxially stretched PVDF, the VDF:TrFE co-polymers and PZT 5 is shown in Table 29.2.

Table 29.2. Comparison of piezoelectric properties.

Material	d_{31}	d_{32}	d_{33}	d_h	g_{33}	g_h	$\frac{d_h g_h}{tan(\delta)}$
	(pC N⁻¹)				*(mV-m N⁻¹)*		*(pm² N⁻¹)*
$PVDF_{uniaxial}$(a)	23	3	–33	–7	–310	–66	23
$PVDF_{biaxial}$(a)	7	7	–20	–6	–188	–56	17
VDF:TrFE[74] 56:44	21	21	–51	–9	–524	–92	39
VDF:TrFE[74] 70:30	17	17	–40	–6	–422	–63	15
VDF:TrFE[74] 80:20	9	9	–24	–6	–263	–66	16
PZT-5(a)	160	160	–420	100	–28	–7	39

(a) Manufacturers' data

It is clear that the ceramic has very high piezoelectric charge coefficients (*d*) when compared to the polymers, but as ε_r is over 100 times greater for the ceramic (1700 compared to 12), *g* is an order of magnitude lower. Therefore, ferroelectric polymers will produce a higher voltage output per unit applied stress. However, this advantage is balanced out when figure of merits are employed for particular applications, such as FOM_h, as shown by the values of $d_h g_h/\tan(\delta_e)$. The *d* values for the co-polymers increase with increasing VDF content. This correlates with results obtained by Japanese researchers, who observed a maximum in piezoelectric constant at a VDF content of 55 mol per cent. It is also worth noting the d_{31}/d_{32} values of the co-polymers are higher than those of biaxially stretched PVDF, an additional advantage besides the elimination of the stretching stage required by PVDF. However, the highest d_{31} value, exhibited by uniaxially stretched PVDF, can be exploited in device configurations that are based upon the application of stresses along one transverse direction.

Electromechanical coupling factors

The electromechanical coupling factor *k*, quantifies the efficiency of a piezoelectric material as a transducer, converting mechanical energy into electrical energy and vice verse. Thus

$$k^2 = \frac{\text{converted energy}}{\text{input energy}} \quad \text{... (29.20)}$$

It can be defined from measurements of dielectric permittivity under different mechanical constraints, clamped (ε^s) or free (ε^T). It is therefore dependent on the shape of the specimen and its vibrational mode.

Analogous to the piezoelectric coefficients, k_{31} is the coupling factor related to the transverse vibration, perpendicular to the polarised direction and k_{33} is related to the vibration of thick specimens parallel to the direction of polarisation. Thus:

$$k_{31} = \frac{d_{31}}{\left[\varepsilon^T s_{11}^E\right]^{1/2}} \quad \text{... (29.21a)}$$

$$k_{33} = \frac{d_{33}}{\left[\varepsilon^T s_{33}^E\right]^{1/2}} \qquad \text{... (29.21b)}$$

where s is the compliance, i.e., the ratio of strain in the direction annotated by the first subscript to stress in the direction shown by using the convention in Fig. 29.8. The superscript E indicates this is the value under short circuit (constant electric field) conditions.

There are also another two coupling factors, k_p, for the radial vibration in the plane of the sample, and k_t for the vibration of thin specimens along the polarised direction; where

$$k_p = k_{31} \frac{2}{[1-\sigma^E]^{1/2}} \qquad \text{... (29.22a)}$$

$$k_t = h_{33} \frac{\varepsilon_{33}^5}{c_{33}^5} \qquad \text{... (29.22b)}$$

and σ is the *Poisson's* ratio under short circuit conditions, h_{33} is the piezoelectric constant in equation 29.16 and c_{33} is the ratio of stress to strain under open circuit (constant electric displacement) conditions.

Values of the coupling factors are determined by the piezoelectric resonance method. This analyses of the impedance Z of the material when it is excited by a voltage source and uses the resonant frequency at which Z = 0 and the anti-resonant frequency for which Z is infinite, to calculate the appropriate k. This method is satisfactory for ceramics, but large errors occur when applied directly to the polymers because of their large mechanical losses and subsequent high damping effects. It is necessary to use equivalent circuit models to select the parameters that produce resonance curves that are best fitted to observed ones.

The most practical coupling factors is k_t, which is important in evaluating a material for application such as ultrasonic imaging that exploit the converse piezoelectric effect to produce sound waves. The values of k_t for PVDF (0.2) and VDF:TrFE (0.3) are approximately half that of PZT (0.55). Thus for the same driving voltage, the ceramic will be a superior transmitting material. However, the peak driving voltage that can be applied without damaging the material is 25 V μm^{-1} for the polymers and 0.3V μm^{-1} for ceramics, almost 100 times greater. The smaller coupling factor can therefore be compensated for by using a larger driving voltage, although in some applications, high voltages are undesirable.

Pyroelectric constants

A pyroelectric material generates an electric charge when subjected to a change in temperature. It must be emphasised that at thermal equilibrium, the surface charges due to the material's internal polarisation are not observable as they are dissipated by internal space charges or electrical conduction. It is only when the temperature of the pyroelectric is altered externally by heat conduction or incident radiation that a change in polarisation occurs, resulting in a detectable surface charge.

This is quantified by the pyroelectric constant p with temperature T

$$p = \frac{\Delta P}{\Delta T} \qquad \text{... (29.23)}$$

Thus the charge Q produced by an electroded area A is given by

$$Q = pA\Delta T \qquad \text{... (29.24)}$$

A typical pyroelectric detector element with a capacitance of 10 pF produces a charge of 10^{-16} Coulombs. It is necessary to use a low noise, high impedance field effect transistor (FET) as a source follower amplifier to enable these small charges to be detected.

To assess pyroelectric detector performance it is necessary to ascertain its response to a changing source of radiation and compare this signal with the noise signals inherent in the detector and amplifier. The figures of merit used to compare pyroelectric materials are briefly explained below:

A simple detector consists of an electroded element of pyroelectric material with area A, thickness t, capacitance C_E and emissivity η. The element is exposed to a radiation of power W, which is modulated at a frequency, f. The temperature of the element will also be modulated at this frequency by an amount dependent on the material's heat capacity H and thermal conductance to its surrounding G, with a thermal time constant τ_T where

$$\tau_T = \frac{H}{G} \quad \text{... (29.25)}$$

The electrical time constant τ_E of the amplifier is given by

$$\tau_E = \frac{R_G}{(C_E + C_A)} \quad \text{... (29.26)}$$

where R_G is the gate resistance of the FET and C_A is the amplifier's capacitance. The frequency response of the detector is determined by these two time constants.

The pyroelectric current i generated per Watt of input power is defined as the current responsivity, R_i, thus

$$R_i = \frac{i}{W} \quad \text{... (29.27)}$$

Similarly for the voltage responsivity R_v

$$R_v = \frac{V}{W} \quad \text{... (29.28)}$$

At low modulation frequencies, i.e., $\omega < \tau_T^{-1}$, both R_i and R_v are proportional to frequency. For $\omega > \tau_T^{-1}$

$$R_i = \frac{\eta p}{H_v t} \quad \text{... (29.29)}$$

where H_v is the specific heat capacity per unit volume. The voltage responsivity R_v reaches a maximum at $\omega = (\tau_E \;\tau_T)^{-1/2}$ at a value

$$R_v(\max) = \frac{pAR_G}{G(\tau_e + \tau_T)} \quad \text{... (29.30)}$$

It is clear from equation (29.30) that R_v (max) is optimised by reducing the element's thermal capacity by using a thin sample and also by minimising G by thermally isolating it from its surroundings.

The high frequency ($\omega > \tau_E^{-1}$; $\omega > \tau^{-1}{}_T$) dependence of R_v is calculated from

$$R_v = \frac{\eta p}{H_v t(C_E + C_A)\omega} \quad \text{... (29.31)}$$

if the capacitance of the element is larger than the amplifier's capacitance ($C_E > C_A$) then

$$R_v = \frac{\eta p}{H_v \varepsilon_r \varepsilon_o A \omega} \qquad \text{... (29.32)}$$

Therefore, under these conditions, the response is proportional to a figure of merit F_v

$$F_v = \frac{p}{H_v \varepsilon_r \varepsilon_o} \qquad \text{... (29.33)}$$

However, if $C_A > C_E$ then a second figure of merit F_i is given by

$$F_i = \frac{p}{H_v} \qquad \text{... (29.34)}$$

Equations 29.33 and 29.34 clearly show that materials with a high pyroelectric constant and small specific heat capacity are required. However, the final choice of material is dependent on the capacitance of the amplifier to be used and the size of the detector element.

Materials with a small dielectric constant, such as the ferroelectric polymers, are best suited to large area elements and vice versa.

In evaluating a pyroelectric detector it is important to know the minimum power of incident radiation that can be detected. This is obviously related to the responsivity but the noise levels produced by the detector and amplifier must also be considered.

The major noise sources for a typical pyroelectric detector are the dielectric or Johnson noise, the amplifier current and voltage noise and the thermal noise, caused by fluctuations in the power flow from the element to its heat sink. Each of these has an equivalent voltage generated at the amplifier input, V_D (given by equation 29.12), V_i, V_a, V_r respectively, which combine to give the total equivalent input noise V_n by the equation

$$V_n^2 = V_D^2 + V_i^2 + V_a^2 + V_r^2 \qquad \text{... (29.35)}$$

The noise equivalent power (NEP) expresses the incident power required to produce a signal equal to the rms noise voltage, thus

$$\text{NEP} = \frac{V_n}{R_v} \qquad \text{(units W Hz}^{-1/2}\text{)} \qquad \text{... (29.36)}$$

The reciprocal of NEP is defined as the detectivity D, thus giving a figure of merit which increases, rather than decreases, with improved detector performance. The specific detectivity D^* defined as

$$D^* = \frac{\sqrt{A}}{\text{NEP}} \qquad \text{(units mHz}^{1/2}\ \text{W}^{-1}\text{)} \qquad \text{... (29.37)}$$

allows a comparison of detectors with different areas when the dielectric noise dominates, which is true for many devices. The D^* will then become

$$D^* = \frac{R_v \sqrt{A}}{V_D} \qquad \text{... (29.38)}$$

On substitution of R_v and V_D from equations 29.32 and 29.12 respectively, a further figure of merit F_D is derived, showing that D^* will be optimised by maximising F_D where

$$F_D = \frac{p}{H_v \sqrt{\varepsilon_r \varepsilon_o \tan(\delta_e)}} \quad \text{... (29.39)}$$

However, this does not apply for small detectors operating at very low or very high frequencies, as other noise sources dominate. In this case, the detectivity will be maximised by maximising F_v in equation 29.33.

All pyroelectric materials are piezoelectric. Thermal expansion of the ferroelectric polymer induces a small secondary pyroelectric effect due to the piezoelectric coefficients coupling to the strains produced on expansion. Another undesirable signal is generated when the pyroelectric detector operates in an environment that is acoustically noisy or highly vibrational. This microphonic signal can be significantly reduced by mounting the detector in a rigid, mechanically isolated package.

A typical technique for obtaining values of p involves the use of a chopped radiation source, such as a laser beam or an infra-red lamp. The measured charge is then compared with a known calibration sample.

Values of p, H_v and the figures of merit F_v and F_D for the ferroelectric polymers are shown in Table 29.1 against values for a typical pyroelectric ceramic based on doped lead zirconate. The pyroelectric constants of the polymers are at least an order of magnitude lower than the ceramic and the smaller F_D indicates an inferior detector performance. However, in terms of cost, the polymers offer an advantage as they are readily available in large area, thin sheets, eliminating slicing, lapping and polishing required by the ceramics. It is interesting to note that the increase in piezoelectric coefficients exhibited by the co-polymers does not manifest itself in the pyroelectric properties, although a room temperature value of 90 μC m^{-2} K^{-1} has been reported for Japanese 51:49 VDF:TrFE co-polymer.

Temperature dependence

For all device applications it is necessary to know the effects of temperatures on the piezoelectric and pyroelectric properties; firstly to ensure the material has sufficient thermal stability to withstand elevated temperatures without significant loss in activity and secondly to determine if the level of activity varies with changes in operating temperature.

Thermal ageing studies are carried out by measuring the room temperature activity after known period of time at an elevated temperature. Ageing curves for the VDF:TrFE co-polymers are illustrated in Fig. 29.9. The ageing temperature of 70°C is representative of the maximum temperature most devices could be subjected to under extreme (but conceivable) storage conditions, such as direct sunlight. The degree of thermal stability is directly related to the Curie temperature of the ferroelectric polymer, as this is the point where it experiences a permanent loss in ferroelectricity and hence piezoelectricity. Therefore, the 56:44 VDF:TrFE co-polymer with a T_c of 70°C depoles rapidly at this temperature, losing 65 per cent of its activity after 1 hour. With increasing VDF content and thus increasing T_c, losses in the coefficient reduce to 23 and 13 per cent after 24 hours for the 70:30 and 80:20 VDF:TrFE respectively. The 56:44 molar composition is clearly unsuitable for most device applications and it appears that the 70:30 VDF:TrFE offers the optimum combination of piezoelectric coefficient and thermal stability.

PVDF has a similar ageing curve, losing 10 per cent activity after 24 hours at 70°C. Tests with a 100 per cent relative humidity at 70°C result in a 20 per cent loss and a higher ageing temperature of 90°C produces a greater than 50 per cent reduction. However, it has been shown that after the initial loss in coefficient, reductions in activity are insignificant with further periods of ageing. Thus the use of pre-aged film is desirable in devices with stringent thermal stability specifications.

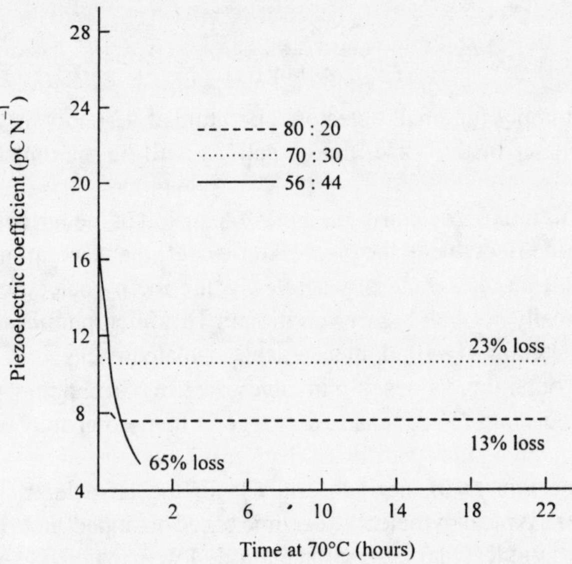

Fig. 29.9. Ageing curves for VDF:TrFE co-polymers.

Temperature dependence observations for PVDF and the co-polymers found an increase in piezoelectric charge coefficient beginning at the glass transition temperature. In the case of the co-polymers, this rise reaches a maximum at temperature T_{max}, just below T_c (50°C for 55:45 VDF:TrFE, 100°C for 72:28), before undergoing a sharp fall. This confirms that the temperature dependence is related to the ferroelectric to paraelectric phase transition. Similar behaviour has been observed for the pyroelectric constant and the electromechanical coupling factor k_{31}, although it must be noted that the vlaue of k_{33} is independent of temperature.

Providing the temperature does not go above T_{max}, this effect of temperature on piezoelectric and pyroelectric properties is reversible, particularly for pre-aged film. Rises in temperature from 20 to 50°C result in approximately a 60 per cent and 40 per cent increase in d_{31} and d_{33} respectively. Similarly for 56:44 VDF:TrFE p is raised from 20 to 40 μC m^{-2} K$^{-1.31}$. This could be a disadvantage in many applications and electronic compensation may be required.

Mechanical Properties

Elastic compliance and stiffness

It can be seen from the above section that the mechanical properties of ferroelectrics are closely related to the piezoelectric properties. The compliance or elastic flexibility s_{ij} is defined as

$$s_{ij} = \frac{\text{strain in direction } i}{\text{stress in direction } j} = \frac{S_i}{T_j} \quad \text{... (29.40)}$$

and the stiffness c_{ij} is defined as

$$c_{ij} = \frac{\text{stress in direction } i}{\text{strain in direction } j} = \frac{T_i}{S_j} \quad \text{... (29.41)}$$

The subscripts *i* and *j* acquire a value 1 to 6 following the convention in Fig. 29.8, where T_1, T_2, T_3 are tension stresses parallel to the 1, 2 and 3 axes respectively; T_4, T_5, T_6 are the shear stresses around the axes 1, 2 and 3. Similarly, S_1, S_2, S_3 are the relative tension strains and S_4, S_5, S_6 the shear strains.

Young's modulus

The *Young's* modulus (*Y*) describes the elastic compliance under tension.

$$Y_{11} = \frac{1}{s_{11}} \quad \text{... (29.42)}$$

This is the most common form of *Young's* modulus for polymers, as it is comparatively straightforward to meet the required conditions of a homogenous stress and strain. This condition is satisfied when the sample length is much larger than its width and thickness.

The simplest technique for measuring Y_{11} is to apply a tensile force *F* parallel to the '*1*' axis to a specimen of areas *A* and length *l*. The *Young's* modulus is proportional to the initial gradient of the resultant stress-strain curve. Thus

$$Y = \frac{F / A}{\Delta l / l} \quad \text{... (29.43)}$$

where Δl is the incremental length in the 'l' direction.

However, for most polymers under constant strain, *Y* will decrease with time due to relaxation processes, or crepe. Therefore, the value of Y is dependent on the rate of strain and the speed of testing. More realistic values are obtained under dynamic conditions, where the test sample is subjected to sinusoidal vibrations. The applied force and resultant deformations are detected by a combined load cell and strain gauge.

The *Young's* modulus of PVDF is typically 1.9–2.0 GPa although this varies with manufacturer. This is also true for the VDF:TrFE co-polymers where values of 1.84 and 1.15 GPa have been recorded for 70:30 VDF:TrFE from different suppliers. This difference becomes very important when evaluating material for certain piezoelectric polymer device configurations, such as those with rigidly supported membranes. It has been shown for these type of devices that the output or sensitivity V_o is proportional to the product of piezoelectric coefficient, *Y* and thickness of membrane *t*, i.e.,

$$V_o \propto d_{31} Y_{11} t \quad \text{... (29.44)}$$

Therefore, for the 70:30 VDF:TrFE with a *Y* of 1.15 GPa, which is almost half that of PVDF, the effect of the doubled d_{31} vlaue over biaxially stretched PVDF is cancelled, assuming the film thickness is constant. In the case of the co-polymer with a *Y* of 1.84 GPa, the sensitivity is predicted to double over an equivalent PVDF device.

Mechanical loss

Dynamic measurements of elastic modulus will yield a complex quantity defined as

$$Y = Y' + iY'' \quad \text{... (29.45)}$$

where *Y* is the complex *Young's* modulus, *Y'* the real part and *Y''* the imaginary part, also called the loss modulus. The angle reflecting the time lag between the applied stress and strain is δ. Analogous to the dielectric loss tangent, the mechanical loss tangent is

$$\tan(\delta_m) = \frac{Y''}{Y'} \quad \text{... (29.46)}$$

and the mechanical quality factor Q_m is equal to the reciprocal of $\tan(\delta_m)$. The $\tan(\delta_m)$ is a damping term and represents the ratio of energy dissipated as heat to the maximum energy stored in the material during one oscillation cycle.

The value of $\tan(\delta_m)$ for PVDF is typically 0.1. This is 10 times greater than the values of 0.01 for ceramics (PZT 5). As discussed earlier, it is this high mechanical loss or damping that prevents measurement of the electromechanical coupling factors using piezoelectric resonance methods.

Poisson's ratio

Poisson's ratio (σ) is defined as the ratio of negative lateral strain to longitudinal strain along the direction under tension. Thus:

$$\sigma_{21} = -\frac{s_{21}}{s_{11}} \quad \text{... (29.47a)}$$

$$\sigma_{31} = -\frac{s_{31}}{s_{11}} \quad \text{... (29.47b)}$$

$$\sigma_{32} = -\frac{s_{32}}{s_{22}} \quad \text{... (29.47c)}$$

(Note that the negative is conventionally present to ensure a positive *Poisson's* ratio).

APPLICATIONS

The discovery by *Kawai* of the strong piezoelectric effect in PVDF and later by *Wada* and *Bergman* of a corresponding pyroelectric response were important developments for the field of electromechanical and pyroelectric transducers. The enormous potential for a light, rugged, large area flexible materials with piezo- and pyroelectric properties was quickly realised and it inspired a world-wide effort on applications. The low mass of these materials is particularly important in aerospace and space applications, especially with the constraints placed on designers of satellites, probes and a possible space station.

The need for a combination of desirable properties that often cannot be obtained in single-phase materials has driven the development of composite materials. For example, in an electromechanical transducer, one may wish to maximise the piezoelectric voltage sensitivity, minimise the density (to obtain good acoustic matching), and also make the transducer mechanically flexible to conform to a curved surface. These properties are partially conflicting in nature, and a single-phase material that simultaneously satisfies these requirements is virtually non-existent. Thus in many applications, one might optimise conflicting requirements by combining the most useful properties of two or more phases that do not ordinarily appear together in nature.

It should be noted that often the major difficulty has been that of ousting entrenched, mature technologies rather than the shortcomings in piezoelectric performance.

Sound Transducers in Air

Sensitivity and impedance considerations dictate that, overall, microphones or converting sound to electrical energy are one of the most promising applications for PVDF in acoustic transducers. In most cases advantage is taken of the fact that PVDF can be used as a combined diaphragm and transducer that both collect the sound and converts it to an electrical signal without the need for armatures, coils, magnets, etc., which increase both complexity and cost. Typical structures take the form of self-supporting domes or bimorphs consisting of two sheets of thin PVDF film (10–30 μm) clamped between two flat plates. The domes or discs of PVDF are the active part of the device, with the response frequency and to some extent sensitivity, governed by the mechanical dimensions of the active elements. Their low mass contributes to the low level of inertially produced noise (a characteristic of PVDF based units), which is especially important in portable microphones. A unique advantage can be gained by using PVDF transducer elements in so-called noise cancelling microphones, designed for use in noisy environments. The great simplicity of the PVDF element permits the construction of microphones that have a very symmetrical response to the sound field, an essential requirement in this type of microphone.

The rather low sensitivity of PVDF devices when used in sound producing mode has limited the use of PVDF in the field to high audio frequency or close coupled (e.g., headphone) devices. PVDF based high frequency tweeters were also incorporated in Pioneer speaker systems and enjoyed a somewhat longer commercial life. Other sound producing devices have been proposed, but none have achieved commercial application.

Sound Transducers in Water

A particular feature required of most underwater transducers is the need to operate at significant depths, necessitating a transducer that will withstand a considerable static over-pressure and yet will be sensitive to the very low pressure fluctuations in sound waves.

Several approaches have been made to this difficult problem in devices using thin PVDF film backed by air cavities, with varying degrees of success. A more elegant use of PVDF underwater is in larger area underwater microphones. These are fabricated from large (150 mm × 150 mm) plates of PVDF and are used mounted flat against the hull of a boat below the water line. The plates operate in a compression mode and having no internal cavities will withstand high over-pressures. The large-area plates are a cheaper option than the conventional approach of using many small piezoceramic sensors.

A material of potentially even greater use underwater is the closely related co-polymer of vinylidene fluoride and trifluoroethylene. This material needs no orientation before poling and can be formed using conventional plastics processing methods into self-supporting shapes that only require activation by poling. Such techniques are of use in the production of shape PVDFTrFE underwater sensors.

Ferroelectric composites can also be used as an underwater transducer to detect (passive mode) and/or generate (active mode) sound. Transducers made from ceramics have the disadvantage that their densities make it difficult to obtain good impedance matching with water. One of the problems with PVDF is its low permittivity which produces a low element capacitance which loads the output. The equivalent circuit for a flank array transducer linked to an amplifier via a cable is shown in Fig. 29.10. The open circuit voltage sensitivity M_{oc} is loaded by stray capacitance C_s, and/or cable capacitance C_c. The end of cable sensitivity M_{ec} is given by

$$M_{ec} = \frac{C_e M_{oc}}{C_e + C_s + C_c} \qquad \text{... (29.48)}$$

where C_e is the element capacitance. If the element capacitance is much greater than the combination of $(C_s + C_c)$ then the sensor voltage output is important and g_h would be the appropriate device figure of merit. In the reverse case, d_h should be used and in general this would require amplifiers to be in close proximity to each transducer unit, which would be an added expense. Since composites have a better impedance matching (than ceramics) to water and their permittivity can vary from 20 to 1000, they would be an ideal choice for these transducers.

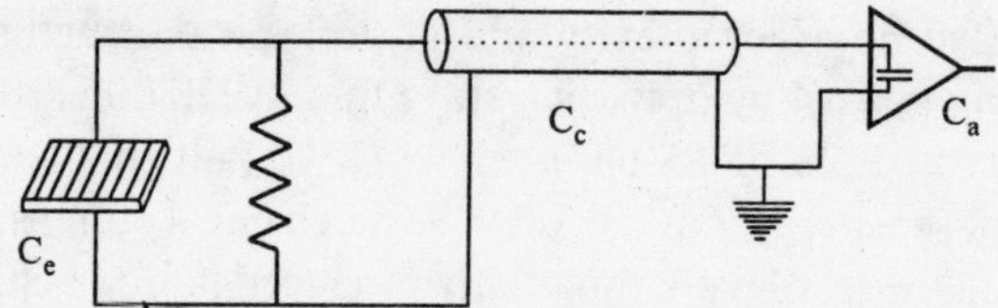

Fig. 29.10. Equivalent circuit for a flank array transducer.

Dispersions of powders in polymers with 0–3 connectivity offer the possibility of fabricating large area transducers at low cost. One of the major underwater uses for hydrophones would be in hull-mounted flank arrays. For good flow noise integration a large area must be covered. Ideally, the hydrophones pick up sound generated by other bodies and this information can be processed to detect and track the body. At present it is not certain whether the large area required will be achieved with large-area, or an array of small-area, tiles. *Ting* and *Geil* have investigated array structures based on PVDF and 0–3 composites. The major problems were due to coupling effects between PVDF elements, which were not present for the composite arrays.

The main sources of noise are:

1. Sea State Zero Noise (S_o), with this and its corresponding voltage (V_o) given by

$$S_o \approx 0.045 f^{-08} \qquad \text{... (29.49)}$$

$$V_o = g_h h S_o \xi \qquad \text{... (29.50)}$$

h = element thickness.

2. Amplifier Voltage Noise, which is component dependent.
3. Johnson Noise, with the voltage given by

$$V_J = \xi \left[\frac{4 k_B T h \tan(\delta)}{\omega \varepsilon A [1 + \tan^2(\delta)]} \right] \qquad \text{... (29.51)}$$

where

$$\xi = \frac{1}{1 + \frac{C_s + C_c}{C_e [1 - j \tan(\delta)]}} \qquad \text{... (29.52)}$$

The predicted noise spectra for a hydrophone is chosen in Fig. 29.11. It can be seen that the sea state zero noise dominates at low frequencies and the amplifier noise dominates at high frequencies. So for most practical applications g_h is the best figure of merit.

Mechanical Applications

A number of applications have been suggested where an applied force has been used to generate an electrical signal. One application is in telephone keypads. Other applications use the production of movement by an applied field to activate cooling fans or small information displays. None of these has progressed beyond a demonstration of feasibility, probably because satisfactory commercially available devices using other transducer techniques already exist. These devices do, however, show the great breadth of applications possible with PVDF. Recently, a range of sports practice equipment has been developed in the United States. In one application, PVDF is used as a sensor in a device to measure the power and accuracy of a baseball pitcher's throw, in another the reaction time and blocking power of football players is measured using PVDF sensors embedded in a football and blocking pad. This application gives a good illustration of the robustness of PVDF sensors.

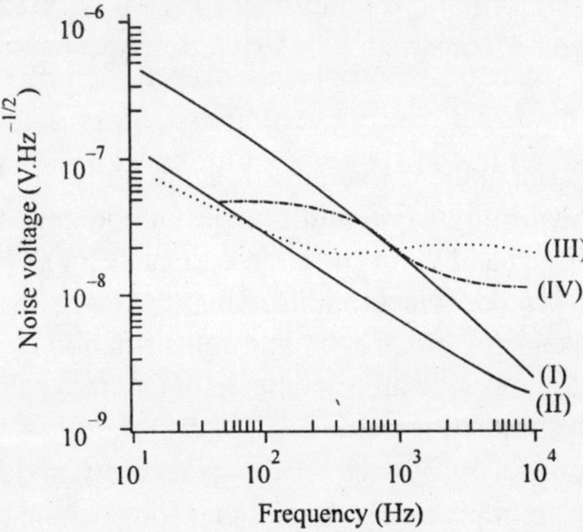

Fig. 29.11. Predicted hydrophone noise spectra: (I) sea-state zero; (II) Johnson noise of dielectric loss; (III), (IV) voltage noise of two existing pre-amplifiers. Materials constants were as follows: $\varepsilon_r = 44$, $\tan(\delta) = 0.032 + 0.47\sqrt{f} + 1.4f^{-1}$ (f = 20 Hz – 10 kHz), $d_h = 13.8$ pC N^{-1}, $g_h = 0.035$ V mM^{-1}.

Filter Bandwidth Modifications

A higher frequency applications of composites involves bandwidth modification in electrical filters. Electrical circuits operating at high frequency require some form of frequency control to limit the passband of frequencies. This control can take the form of a piezoelectric components shaped so that the frequency range of interest coincides with a resonant frequency of the piezoelectric element. At resonance the piezoelectric filter has minimum impedance, several orders of magnitude lower than its non-resonant impedance. Consequently, the element readily passes signals at frequencies close to its resonance, with the passband being defined by the mechanical Q of the device as

$$Q = \frac{f}{\Delta f_{3dB}} \quad \text{... (29.53)}$$

Here, f is equal to the central frequency and Δf_{3dB} is the 3 dB passband which can be 0.1 per cent of the central frequency.

It is possible to reduce or widen the bandwidth of a piezoelectric material by combining several active piezoelectric elements mechanically in series or parallel respectively. A reduction in bandwidth can be obtained by carefully grinding several piezoelectric discs to the same thickness and binding them together in a series configuration using an electrically conducting epoxy resin. For the fundamental thickness mode of a single disc, the resonant frequency is given by

$$f_L = \frac{1}{2L\sqrt{\rho s}} \quad \text{... (29.54)}$$

where ρ = sample density and s = elastic compliance.

The disc thickness L is equal to half the acoustic wavelength therefore, since the sign of the stress changes every half wavelength, in order to resonate in phase the polarity of adjacent discs must be reversed. Acoustic wave transmission occurs only very close to f_L and consequently the bandwidth is reduced.

One problem associated with series multilayer transducer designs is the extremely accurate dimensional tolerances which must be achieved if constructive interference is to be achieved.

Biomedical Applications

Work in this area has been carried out at the University of Padua (Italy) by *Rossi*. He has identified many applications where the tissue compatibility, easy conformability, good impedance matching to soft tissues, low water absorption, lightness, large bandwidth and large area make PVDF an ideal transducer. However, large scale applications are not envisaged. Typical applications are non-invasive cardio-pulmonary sensors, implantable transducers and sensors for prosthetics and rehabilitation devices. Forces available from piezo polymers in mechanical actuator mode are very small, but a practical micro-injector making use of thin PVDF tube working in radial contraction mode to dispense microlitre quantities has been described. An interesting medical use of piezopolymer in the charge generation mode is in the stimulation of bone healing. The charge generated by an implanted piezopolymer film around a long bone fracture has been shown to accelerate healing.

Although narrow band pass filters are readily available for most frequency ranges of interest, broadband filters, in which bandwidths up to 50 per cent of the central frequency are required, are more difficult to produce. Apart from electrical filters, broadband devices are being employed in acoustic imaging, especially for non-invasive flaw detection. Ultrasonic imaging has an advantage over its optical counterpart of improving contrast since it relies for contrast on elastic discontinuities which are generally of greater magnitude than changes in refractive index, particularly in biological tissue. In acoustic imaging the piezoelectric element acts as both source and detector in a 'pulse-echo' operational mode. At present the image is built up by mechanically scanning a single transducer over the area of interest and integrating its output to form a coherent picture.

The requirements for medical ultrasonic imaging transducers are met with 1–3 PZT-rod/polymer-matrix piezoelectric composites. The composites provide a higher electromechanical coupling and low impedance properties superior to those of the conventional piezoceramics and piezopolymers. Also, composite plates can be formed into complex shapes for beam focusing and steering.

Medical ultrasonic imaging is an important diagnostic tool for physicians as it provides images of the internal organs without subjecting the patent to hazardous ionising radiation. These images are formed by sending a short, directed pulse of high frequency (2–20 MHz) sound into the body and detecting the weak reflections from internal structures. Displaying the intensity of the echo as a brightness and scanning the beam direction in a plane yields remarkable tomographs of the soft tissue of the human body. Since the speed of sound in tissue ($\approx$ 1540 ms^{-1}) permits gathering of over one hundred images in the time it takes for a standard television scan, real time as well as static images can be formed. These views of the internal organs provide substantial diagnostic information in cases ranging from fetal examination through to cardiac assessment.

The key element in forming these pictures is the transducer which generates the acoustic pulse and detects its weak acoustic. At the heart of this transducer lies a thin piezoelectric plate, a few hundred microns thick, that accomplishes the energy conversions.

The requirements on this plate are:

1. It must be an efficient electromechanical energy converter, i.e., the thickness mode electromechanical coupling coefficient k_t should be as large as possible.

2. It must efficiently couple acoustic energy into and out of the tissue, i.e., its acoustic impedance must be close to that of tissue (1.5 Mrayl).
3. It must efficiently couple energy out of and into an electrical transmission line, i.e., the dielectric constant should be adjustable over a wide range from about 100 upwards.
4. Electrical and mechanical losses should be low, i.e., tan(δ) < 10 per cent.
5. It should be readily formed into curved shapes and sub-divided into acoustically and electrically isolated sub-sections for focusing and steering the accounting beam.

Table 29.3 compares the above criteria for piezoceramics, piezopolymer and 1–3 piezocomposites. It can be seen that the mix of properties provides the transducer designer with opportunities to enhance the performance of existing designs.

Table 29.3. Comparison of important ultrasonic parameters.

Parameter	*Ceramic*	*Polymer*	*Composite*
Thickness mode electromechanical coupling (%)	40–50	20–30	≤75
Acoustic impedance (Mrayl)	20–30	4	6
Dielectric constant	100–2400	10	20–1000
Dielectric loss (%)	≤3	15	5

Capacitance Thermocompensation

Some ceramic single phases are used for specific applications with apparently no need to resort to composite materials. For example, barium titanate ($BaTiO_3$) is the dominant material used in capacitor manufacture. However, electronic circuits are required to work under many conditions, not least of which is a changing environment. In many situations it may be possible to arrange a composite material so that the individual phase parameters vary with external conditions in such a way that the overall effect is to leave the composite parameter constant. A thermocompensated capacitor should have its capacitance almost independent of temperature. In general two methods are employed to attain this stabilisation:

1. A system of two or more capacitors in series or parallel connected so that the temperature coefficients of capacitance have opposite signs (one positive and the other negative).
2. One capacitor is made with a composite dielectric that is a mixture of two dielectrics either having different signs of temperature coefficient of permittivity or making it in such a way as to eliminate the Curie peak so that the permittivity is almost temperature independent.

The latter effect can be achieved with a composite made from $BaTiO_3$ and PVDF. The tetragonal distortion of $BaTiO_3$ is locked in below a particle size of around 5 μm because of a surface energy effect. This locked-in deformation persists to temperatures far in excess of the usual Curie temperature. This property may be used in making a composite of $BaTiO_3$ and PVDF. The $BaTiO_3$ particles should be around 1 μm in size and the above effect, in conjunction with fluoride ions occupying oxygen deficient sites in the $BaTiO_3$ lattice, will wash out the permittivity peak. This has been carried out by *Murali* where a $BaTiO_3$/PVDF (70 per cent/30 per cent weight) composite was made. For frequencies higher than 100 Hz, the temperature dependence of the permittivity was constant up to 190°C.

Pyroelectric Applications

Pyroelectric devices convert changing incident thermal radiation to an electrical output and are now much used in intruder detectors and thermal imaging systems. Conventionally, ceramics have been used in such applications; however, considering the desirable properties of large pyroelectric coefficient, high volume resistivity, low dielectric constant and loss and low specific heat, it can be seen that apart from the rather low value of pyroelectric coefficient, polymeric materials are superior to ceramics in several respects.

A number of experimental configurations have been described in this chapter and one or two simple devices based on the use of PVDF are commercially available (Siemens, Microwatt Applications). The relatively low pyroelectric coefficient of PVDF and, to some extent, the difficulty of handling small pieces of the thinnest films seem to have inhibited its more widespread adoption in practical devices. The availability of large active areas does not of itself seem to have provided sufficient advantage to device engineers.

SMART MATERIALS

This final section discusses what have been described for aerospace and space applications as the most important emergent materials technology area. SMART materials are more like a device than an intrinsic material. The term 'SMART' comes from the fact that these materials sense some effect and respond in some way as a result. In this way, they have both 'sensor' and 'actuator' properties. They are often modelled on biological systems where the sensor/actuator/process parts are likened to the nervous system/muscles/brain. This is a graphic analogic but not always very useful. There is a suggestion that the 'brain' of the SMART structure is some form of processing unit. Although this may be true, the most interesting area for materials is where all three functions are performed intrinsically by the material itself. Certainly, if a SMART structure consists of a sensor connected to an actuator by some electronic processing unit, then the structure is better described as a 'device' rather than a pure material.

Perhaps the simplest SMART materials can be used for vibration control. This is very important in space applications where it is important to keep satellite devices under tight control in order to work correctly. The SMART structure can be a simple sensor adhered to a structure and electrically connected to an actuator. The sensor could be a piece of 0–3 composite (with lead titanate as the active material) and the actuator could be a PZT disc. Any vibration produces an electrical signal in the composite and this is fed to the sensor through an inverting amplifier circuit so that the PZT is driven in antiphase. With the correct gain set, the vibration can be removed.

Such a system as the one described above would have problems operating over a wide range of frequencies because the composite, PZT and amplifier circuit all have a frequency response and so the optimum gain would depend on the type of frequencies expected for a particular operation.

The above principle is the basis of active vibration cancelling systems. Such systems have been researched by many automotive companies for the reduction of noise in the activity (i.e., where the driver sits) and for the production of a superior active suspension system.

One of the problems of such systems is that, the more complicated the structure, the more care has to be taken to ensure that it is secure and reliable over an extended period of time. For example, military specifications can often require 20–30 years in-service lifetimes. Reliability may increase if the sensor, actuator and interface are intrinsically part of the same piece of material. This could be achieved easily with 0–3 composite material in the form of a disc. The central area would be the sensor (using a lead titanate ceramic) and a ring of active material would surround this (using a PZT ceramic). If the polymer matrix were the same for both powders then the two regions of the material would be compatible and integrity should therefore be high. Such a structure has been used to actively cancel noise for underwater military applications. In such a system, the tuning of the sensor and actuator parts would be carried out by controlling the poling efficiency for the two different parts of the structure.

The above describes a basic form for SMART materials. A sensor and actuator are coupled together either via electronics or through the material itself.

Sensors can be based on acoustic waveguides, the piezoelectric effect or optical fibres. The mechanism depends on the input to be sensed. The actuator mechanisms fall within five areas:

1. Shape memory alloys: These are metals that revert to a pre-defined shape above a certain temperature. In doing so, they can generate large forces for actuation. The most prominent material is possibly nickel-titanium alloys. The main drawback is the slow response of these materials. The response depends on heating and cooling and so, the response depends on how fast the temperature can shift.
2. The piezoelectric effect: These materials produce a much smaller stress than shape memory alloys but they react much quicker. One of the problems is that piezoelectric actuators are non-linear and hysteretic.
3. Electrostrictive materials: The electrostrictive effect is a result of the quadratic dependence of the material permittivity with electric field. This is typified by lead magnesium niobate — it is non-hysteretic but also non-linear. No poling is required but a bias voltage may be necessary.
4. Magnetostriction: These are similar to piezoelectric materials except that the response is to a magnetic field rather than an electric one. The commonly known material is Terfenol-D — an alloy of terbium, dysprosium and iron ($\approx Tb_{.3}Dy_{.7}Fe_{1.9}$).
5. Electro rhelogical fluids: These substances contain micron-sized particles that form chains when placed in an electric field, resulting in increases an the apparent viscosity by several orders of magnitude. The main problems are abrasiveness and chemical instability.

Potential applications may include control of aerodynamic, hydrodynamic and optical surfaces, stealth, robotics, damping/tuning of structures and vibration suppression. In the field of civil engineering, applications are expected in the development of intelligent buildings and in bridge and road monitoring

The whole range of advanced composite materials is being considered as possible hosts for SMART materials technology.

But beyond the materials, however, this area may have a significant influence on the philosophy used for the design of structures. Engineers will not necessarily have to add mass (and the associated cost) in order to ensure safety. The incorporation of SMART materials will allow ongoing, continuous monitoring of a structure and may even allow the structure to heal itself when damaged.

Thus, this chapter has provided the reader with an introduction to a unique class of materials. From the discussion of properties and applications, it can be concluded that electroactive materials offer advantages over many alternative technologies, including flexibility, robustness and availability in large areas, thin sheets or unusual geometries.

However, it is widely believed that these materials will become established in applications that exploit their unique features, rather than through the substitution of existing transducer materials and requires an innovative approach.

Table 29.4 gives some of the points to consider when designing a system which will utilise an active material. It is observed that no one material is suitable for all applications. Further, many disadvantages can be engineered around with good system design.

The present position is such that technological and industrial advances demand more versatile and responsive transducer devices. In the development of such, a particular technology may well be used over others but each individual application is considered on its own merits and in some cases, a mixture of materials may well suit.

Table 29.4. Comparison of different electroactive materials.

Electroactive material		
Ceramics	*Polymers*	*Composites*
Very high active performance (d_{33})	Low d_{33} produces a low active performance	Reasonable acoustic coupling
Very poor hydrostatic activity	Large d_{31} produces only a reasonable hydrostatic performance	Adjustable buoyancy
Large ε_r reduces g_h	Low T_C leads to a strong temperature dependence and depoling problems	Low ε_r leads to a high g_h
Brittle	Low density leads to a good acoustic coupling	Good shock resistance
High density leads to poor acoustic coupling	Low ε_r lead to a high g_h	Good formability (e.g., domes)
Fabrication problems	Good shock resistance	Piezoelectric properties are easily adjusted
Shock resistance is very dependent on microstructure	Very good formability — can be formed into exotic shapes (e.g., domes)	Reasonably low dielectric loss (5%)
Only simple structures possible (e.g., discs)	Stretching can be problematic for PVDF	ε_r around 100–200 – good for many applications
Very low dielectric loss ≤ 3%	Relatively high dielectric loss (15%)	Interphase interactions and defects can cause problems
High ε_r leads to a high capacitance — good for hydrophone applications	Low ε_r (≈10) which is bad for some applications	

ACRONYMS AND SYMBOLISM

Chemical Acronyms used in this Chapter

PZT = lead zirconate titanate
VF = vinyl fluoride
TrFE = trifluoroethylene
PMMA = polymethylmethacrylate
PVDF = polyvinylidene fluoride
VDF = vinylidene fluoride
TeFE = tetrafluoroethylene
PTFE = polytetrafluoroethylene

Symbolism

C = capacitance
ε = permittivity
ε_o = permittivity of free space
Q = charge
E = electric field strength
$\tan(\delta_e)$ = dielectric loss
ω = angular frequency
k_B = *Boltzmann's* constant
i = current
S = strain

c	=	modulus
s	=	elastic compliance
p	=	pyroelectric constant
τ	=	time constant
R	=	resistance or responsivity
F	=	figure of merit or force
Y	=	*Young's* modulus
ρ	=	density
S	=	noise
A	=	surface area
ε_r	=	relative permittivity
t	=	thickness
V	=	voltage
D	=	electric displacement or detectivity
G	=	conductance
f	=	frequency
T	=	absolute temperature or applied stress
g	=	piezoelectric voltage coefficient
d	=	piezoelectric strain coefficient
k	=	electromechanical coupling factor
σ	=	*Poisson's* ratio
P	=	polarisation
H	=	heat capacity
W	=	power
η	=	emissivity
l	=	length
M	=	sensitivity
ω	=	volume flow rate

Note:

T is used for both absolute temperature and applied stress.
R is used for both resistance and responsivity.
F is used for both figure of merit and force.
D is used for both electric displacement and detectivity.

CHAPTER 30

Blowing Agents for Polymers

INTRODUCTION

The term blowing agents applies to substances that can produce pores or cells in polymeric compositions. If the cells are formed through a change in the physical state of the substance—i.e., through an expansion of a compressed gas, an evaporation of a liquid, or by the dissolving of a solid—the material is called a physical blowing agent. If the cells are formed by liberation of gases as products of the thermal decomposition of the material, the material is called a chemical blowing agent.

The introduction of bubbles of gas into a mass of rubber can modify the properties of the rubber in a number of ways, including the following:

1. Whereas solid rubber is virtually incompressible (however, soft), a cellular rubber has a low bulk compressibility. This can be very useful in some sealing applications.
2. The products have a low stiffness and high flexibility.
3. It is possible to obtain products of very low density.
4. The product can acquire very good shock absorption characteristics and show good cushioning behaviour.
5. Thermal insulation properties may be improved.
6. If there is an open cell structure, i.e., the cells are interconnected, the material can act as a sponge.

Bubbles may be introduced into a polymer by several methods, including:

1. The use of chemical blowing agent, i.e., chemicals that decompose on heating with the evolution of gas *in situ* in the polymer.
2. The use of volatile chemicals such as solid carbon dioxide or fluorocarbon gases which volatilise during the processing of the polymer. This technique is widely used with polyurethane foams and plasticised PVC.
3. Beating of air into polymer when in a liquid state. This process was the basis of latex foam technology, the foamed latex then being gelled in the cellular state prior to vulcanisation.
4. Impregnation of polymer with a gas, usually nitrogen, under very high pressures, and subsequent heating of the gas-saturated polymer at high temperatures, which softened the polymer and allowed the product to expand. One of the earliest applications for this approach was in the manufacture of expanded ebonite but today the process is largely restricted to plastics materials.

5. Incorporation of an additive such as a salt or starch which could be leached out by water was at one time (but not now) believed to be important with separators for battery boxes.
6. Use of a cross-linking system that evolved a gas during the cross-linking process. This was the original basis for the process of manufacture of polyurethane foams although today the foaming also involves volatiles.

Past practice has led to various terms being used somewhat haphazardly, but the following terminology is probably the most widely use and will be employed here:

1. Cellular refers to any product laden with bubbles.
2. Expanded refers to a closed cell product (i.e., a product where the bubbles or cells are not interconnected) made from a solid base.
3. Foam refers to an open cell product made from a liquid base such as rubber latex or a polyol-polyisocyanate mixture for making a polyurethane foam.
4. Sponge refers to an open cell product made from a solid base. An example would be where a chemical blowing agent was incorporated into a solid rubber compound which was then heated under pressure both to decompose the blowing agent and to vulcanise the compound, the resultant closed cell vulcanisate being cracked by passing between rollers.

This chapter is concerned with chemical blowing agents, which today are only used for a small part of the cellular rubber market. Such agents have a number of requirements, of which the following are of particular importance.

1. The need for gases to be evolved within a narrow but clearly defined temperature range and in a controlled and reproducible manner.
2. The decomposition range should be suitable for the polymer; for example, it should not be above the maximum possible processing temperature that can be used if significant polymer degradation is not to occur, neither should premature blowing occur, for example during mixing.
3. Gases evolved should not corrode processing equipment.
4. Any toxicity of blowing agents and their decomposition products must be catered for.

Many hundreds of materials have been investigated as blowing agents for polymers but the number used, particularly for rubbers, is very small.

Sodium carbonate, which gives off carbon dioxide, is still widely used for open cell products, being low in cost but rather erratic in decomposition. It decomposes in the range 100–130°C yielding 125–130 cm^3g^{-1}. Two other inorganic agents are ammonium carbonate and bicarbonate. Additional stearic acid may be incorporated into the compound to aid the decomposition.

Amongst the organic blowing agents, dinitrosopentamethylene tetramine, benzene sulphonyl hydrazide and azodicarbonamide are frequently quoted in the literature, with the first-named in particular being favoured for microcellular products such as shoe soling. Decomposing in the range 160–200°C, it is more powerful than sodium bicarbonate, yielding about 210 cm^3g^{-1}. Once again slightly acidic conditions are necessary for it to function at maximum efficiency and the rubber compounds should contain at least 4 pphr of stearic acid.

Early attempts to produce open-cell sponge by means of physical blowing agents (e.g., alcohol, benzene, and other volatile liquids) met with limited success. Eventually, blowing techniques which utilise volatile liquids, either aliphatic hydrocarbons or halogenated hydrocarbons, were perfected; these have attained considerable importance in the manufacture of cellular thermoplastic structures.

Parallel to the development of atmospherically blown sponge, experimental work was carried out with the aim of producing closed-cell rubber by gassing with elemental nitrogen under high pressure.

Diazoaminobenzene (DAB), or 1,3-diphenyltriazene, was the first commercially available organic blowing agent. Introduced as Unicel in 1940 by E. I. Du Pont de Nemours and Co., Inc., the compound was found extremely useful despite its toxicity and staining properties. For the first time it was possible to make cellular polymers with closed cells by conventional compounding and processing methods. As soon as the advantages derived from the use of diazoaminobenzene became apparent, an intensive search began for better organic blowing agents. In the early 1940s, a group of non-staining aliphatic compounds, the azonitriles, was placed on the market, and of these, 2,2′-azobisisobutyronitrile was used on a large scale in the production of flexible and rigid poly(vinyl chloride) compounds during World War II. Later, 2,2′-azobisisobutyronitrile, due to the toxicity of its residue, was abandoned in favour of non-toxic organic blowing agents. Neither the azonitriles nor the later suggested derivatives of azodicarboxylic acid could satisfy the need for a versatile, inexpensive blowing agent that would make the chemical blowing method competitive with the Rubatex process. The breakthrough came in 1946 with the introduction of dinitrosopentamethylenetetramine, one of the dinitrosoamines suggested earlier by *Briggs* and *Scharff*.

In the past forty years over a thousand chemicals have been suggested as blowing agents for high polymers. Some of these enjoyed a short period of commercial acceptance but were later replaced by better products; many others never reached full-scale industrial application, while yet others remained nothing but laboratory curiosities. Today no more than ten chemically different blowing agents are effectively used throughout the world. Presently, the main development effort in this field is directed toward extending the temperature spectrum of the chemical blowing agents by providing compounds with either higher or lower decomposition temperatures than already attainable.

PHYSICAL BLOWING AGENTS

Discussed here are those physical blowing agents which develop a cellular structure by passing from a liquid to a gaseous phase during the blowing operation. These compounds are liquids with boiling points ranging from room temperature to 110°C. Ideally, they should be odourless, non-toxic, non-corrosive, and non-flammable. At low temperatures, the liquids may have some swelling effect on the polymer to be expanded, or it may be desirable that they be soluble in the monomer, but they must possess no solvating power or otherwise affect the physical and chemical properties of the polymer. In the gaseous state, the physical blowing agents must be thermally stable and chemically inert at the temperature employed in the blowing process.

The efficiency of physical blowing agents depends upon a complete vapourisation of the compound and is directly related to the ratio of the volume occupied by equal weights of the vapour and the liquid. From the standpoint of the effective utilisation of the blowing agent, a high specific gravity combined with a low molecular weight is most desirable.

The change from the liquid to the gaseous state is a reversible and endothermic process. When the expanded polymer is cooled, a condensation of the physical blowing agent does not occur, inasmuch as the vapour is either under reduced pressure or diluted with air. The heat of vapourisation can be utilised to control the temperature of the expanding polymer as the latter undergoes a chain extension or cross-linking. This permits the preparation of large cellular bodies without the danger of thermal degradation of the polymer. The addition of nucleating agents (e.g., silica, silicates, sulphides) which serve as sites of bubble formation facilitates the development of small and uniform vapour-filled cells. Furthermore, chemical blowing agents, e.g., a bicarbonate-acid system, can be used in conjunction with volatile liquids to develop low-density cellular polymers.

Physical blowing agents are efficient and inexpensive gas-forming compounds but, in general, their use requires specialised equipment designed specifically for a predetermined blowing operation. The most widely used physical blowing agents are the aliphatic hydrocarbons, as well as the chloro-and fluoro-derivatives of aliphatic hydrocarbons. Of lesser importance are low-boiling alcohols, ethers, ketones and aromatic hydrocarbons. The physical properties of a number of selected physical blowing agents are given in Table 30.1.

Table 30.1. Properties of physical blowing agents.

Blowing agent	*Mol wt*[a]	*$d^{25,a}$ g/ml*	*Bp, °C*[a]	*Heat of vapourisation*[a,b] *cal (15°)/g*	*Blowing efficiency*[c]	
					At bp	*At 100°C*
Pentanes						
n-pentane	72.15	0.616^{30}	36.1	86.0	216	261
2-methylbutane	72.15	0.615	27.8	81.5^{25}	210	261
2,2-dimethylpropane	72.15	0.613	9.5	–	196	260
1-pentene	70.15	0.641^{20}	30.0	–	227	280
Cyclopentane Hexanes	70.15	0.740	49.2	99.6^{25}	279	323
n-hexane	86.17	0.655	68.7	80.4^{60}	212	232
2-methylpentane	86.17	0.653^{15}	60.2	76.5	207	232
3-methylpentane	86.17	0.660	63.3	–	211	234
2,3-dimethylbutane	86.17	0.657	58.0	75.0	207	233
2,2-dimethylbutane	86.17	0.645	49.7	73.8	204	229
1-Hexene	84.17	0.669	63.5	–	219	243
Cyclohexane Heptanes	84.17	0.774	80.8	94.8^{25}	266	281
n-heptane	100.20	0.679	98.4	77.0^{90}	206	207
2-methylhexane	100.20	0.674	90.0	–	200	206
2,2-dimethylpentane	100.20	0.670	79.2	77.3^{25}	193	204
2,3-dimethylpentane	100.20	0.691	89.7	81.7^{25}	205	211
2,4-dimethylpentane	100.20	0.668	80.6	78.4^{25}	193	204
3,3-dimethylpentane	100.20	0.689	86.0	–	202	210
3-ethylpentane	100.20	0.694	93.4	–	204	212
2,2,3-triethylbutane	100.20	0.686	80.8	69.0	198	209
1-Heptene	98.20	0.693	93.2	–	212	216
Benzene	78.11	0.874	80.1	94.1	324	342
Toluene	92.13	0.862	110.6	98.6^{25}	294	286
Dichloromethane	84.94	1.325	40.0	–	404	482
Trichloromethane	119.39	1.489	61.2	66.7^{20}	342	382
Trichloroethylene	131.40	1.466	87.2	–	330	342
Tetrachloromethane	153.84	1.584	76.7	46.6	296	316

(Contd...)

Blowing agent	*Mol wt*[a]	$d^{25,a}$ *g/ml*	*Bp, °C*[a]	*Heat of vapourisation* [a,b] *cal (15°)/g*	*Blowing efficiency*[c] *At bp*	*At 100°C*
1,2-dichloroethane	98.97	1.245	83.5	$77.3^{82.2}$	370	388
Trichlorofluoroethane	137.38	1.476	23.8	43.5	261	329
1,1,2-Trichlorotrifluoroethane	187.39	1.565	47.6	35.1	219	255
Methyl alcohol	32.04	0.787	64.6	263.0	679	752
Ethyl alcohol	46.07	0.785	78.3	204.0^{70}	491	521
Isopropyl alcohol	60.09	0.780	82.3	175.6^{20}	378	397
Ethyl ether	74.12	0.708	34.5	89.8^{30}	240	292
Isopropyl ether	102.16	0.725	67.5	68.2	198	217
Acetone	58.08	0.785	56.2	125.3	365	413
Methyl ethyl ketone	72.10	0.810	79.6	103.4	324	344

[a] Data taken from Timmermans, Physico-Chemical Constants of Pure Organic Compounds.
[b] At the boiling point, unless otherwise indicated.
[c] Data computed according to the formula.

$$\frac{22{,}440}{\text{mol wt}} \times (\text{density at } 25°\text{C}) \times \frac{(273+t)}{273}$$

where t = boiling point in °C for the next to the last column and 100°C for the last column. In the first approximation vapours were treated as ideal gases.

Aliphatic Hydrocarbons

The main source of liquid aliphatic hydrocarbons are low-boiling fractions of petroleum, variously called 'petroleum ether', 'ligroin', or 'light spirits'. The low-boiling solvents contain mainly isomers of pentane, hexane, and heptane and are available in distillation ranges closely corresponding to the boiling points of the various hydrocarbon isomers. These inexpensive liquids posses a high blowing efficiency and a comparatively low order of toxicity. Their value as blowing agents for polymers is limited by the high degree of flammability. The low-boiling pentanes are used extensively in production of expanded polystyrene.

Halogenated Aliphatic Hydrocarbons

The main advantage of chlorinated hydrocarbons is their non-flammability. Because of this property, these materials can be used as blowing agents in making fire-resistant cellular polymers. Chlorinated hydrocarbon blowing agents are more expensive than aliphatic hydrocarbons and present a serious toxicity problem in handling. Among the most commonly used compounds of this group are methylene chloride and tri- and perchlorethylene. Methylene chloride has been used for manufacturing flame-resistant cellular polystyrene and the chlorinated ethylenes have been suggested as auxiliary blowing agents in making cellular poly(vinyl chloride) and epoxy resins.

Fluorinated aliphatic hydrocarbons combine several desirable properties which make these organic liquids ideally suited for use as blowing agents. Among their most significant characteristics are non-flammability, an extremely low level of toxicity, excellent thermal and chemical stability, low thermal conductivity and good dielectric properties. The high density of fluorocarbon liquids contributes to a favourable blowing efficiency. Of the many commercially available aliphatic fluorocarbons, only two of the common ones are liquid at room temperature—trichlorofluoromethane and trichlorotrifluoroethane. Both products are presently used on a large scale as blowing agents for flexible and rigid polyurethan foam. Their use has been suggested in the preparation of cellular structures based on polyolefin resins, poly(vinyl chloride) compositions, epoxy resins and phenol-and urea-formaldehyde resins.

Chemical Blowing Agents

Chemical blowing agents decompose upon heating with liberation of gaseous products. This is usually an irreversible, exothermic reaction which occurs over a definite and short temperature range. Certain compounds such as bicarbonates, which decompose through thermal dissociation, evolve gas in a reversible and endothermic reaction.

The choice of a chemical blowing agent is dependent upon and limited by its rate of decomposition at elevated temperatures. A rapid rate of decomposition is desirable, but a violent breakdown of the compound must be avoided. The rate of decomposition should not be affected by pressure if the compound is to be used in the expansion process involving high pressures. To be useful commercially, the blowing agent must be stable over an extended period of time at temperatures encountered under ordinary storage conditions.

Chemical blowing agents preferably decompose with liberation of nitrogen, but other gaseous products such as carbon dioxide, carbon monoxide and hydrogen are frequently encountered. Corrosive gases (e.g., nitrogen dioxide, sulphur dioxide and hydrogen chloride) should be absent from the decomposition products. Also, the generated gas should be virtually insoluble in the polymer.

To obtain homogeneous distribution in the polymer, the blowing agent should be either easily dispersible by conventional methods of mixing, or should be soluble in the polymer. Furthermore, it is important that the blowing agent and its decomposition products exert no adverse effects on the physical and chemical properties of the polymer. For example, the rate of fusion (i.e., intimate blending of resin and plasticiser under the influence of heat) or cross-linking should not be impaired, the electrical properties should not deteriorate and the thermal and chemical resistance of the polymer should not be decreased by the presence of the residue.

The commercial acceptance of a chemical blowing agent quite often hinges on the properties of the residue. Ideally, the residue should be colourless, non-staining, non-discolouring, non-flammable, compatible with the polymer and should impart no objectionable odour. It is also essential that both the blowing agent and the residue be non-toxic and relatively harmless to the skin.

The efficiency of chemical blowing agents can be measured by the quantity of gas given off upon decomposition and is usually expressed as ml/g of blowing agent under given conditions. An inexpensive blowing agent is not necessarily the most economical if its gas yield is low. A gas yield based on unit cost better characterises the efficiency of a blowing agent than does a gas yield computed on the basis of unit weight. The major advantage derived from the use of chemical blowing agents is their ease of handling and their adaptability to processes requiring conventional equipment.

Testing methods

The thermal behaviour of chemical blowing agents has been a subject of many investigations, and numerous methods have been proposed for the determination of the decomposition rate, temperature, and gas yield.

To compare the efficiency of blowing agents, *Cooper* used an oil-heated, high-pressure gas bomb connected with a manometer which allowed the observation of pressure rise as the decomposition of the agent progressed. In another method, the relative performance of blowing agents and their behaviour in various liquids was studied by measuring the volume of the evolved gas as a function of the heating time at several temperature levels. Typical results obtained by this method are shown in Fig. 30.1

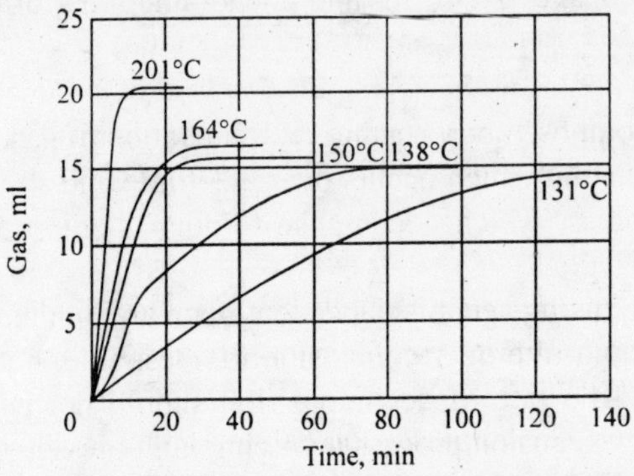

Fig. 30.1. Gas development curves of dinitrosopentamethylenetetramine.

In recent years differential thermal analysis and thermogravimetric analysis have been employed to investigate blowing agents. Differential thermal analysis permits the determination of the exotherm which is generated at the decomposition point of the blowing agent. A typical differential thermal analysis curve of the decomposition of a blowing agent is illustrated in Fig. 30.2.

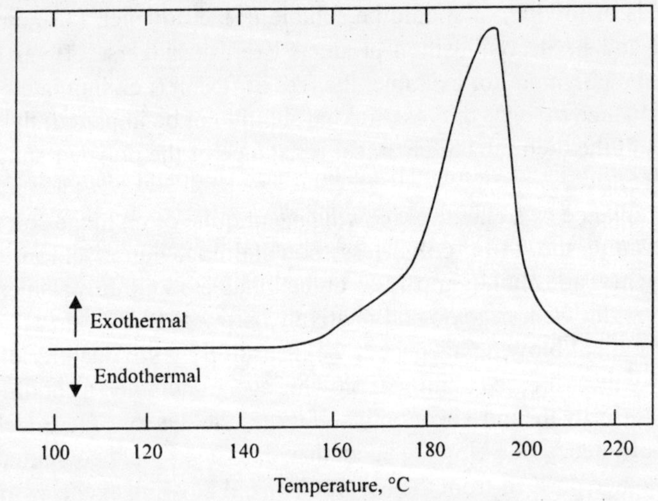

Fig. 30.2. Typical differential thermal analysis graph of 4-4′-oxybis-(benzenesulphonyl hydrazide). Rate of heating: approx. 10°C/min.

Using thermogravimetric analysis, the decomposition of a blowing agent can be followed by measuring the weight loss of a sample heated at a constant rate. Fig. 30.3 depicts results that can be obtained with thermogravimetric analysis.

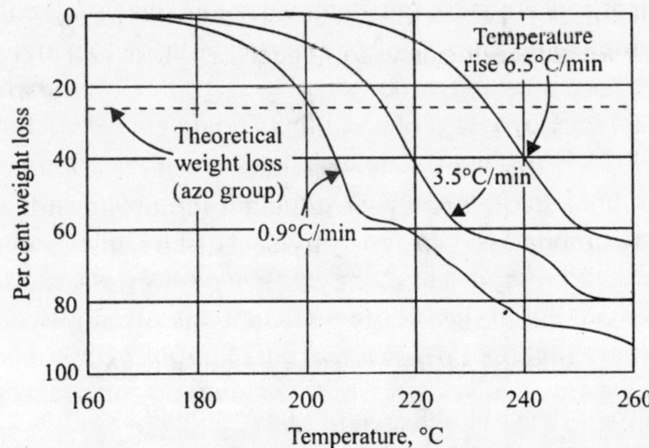

Fig. 30.3. Typical thermogravimetric analysis graph of azodicarbonamide.

The American Society of Testing and Materials in co-operation with the Society of the Plastic Industry has issued a tentative method, ASTM D 1715–60T, for determination of the total gas volume evolved from chemical blowing agents. This method is not applicable to measuring the rate of decomposition.

In all methods previously discussed, the rate of decomposition of the blowing agent is determined under environmental and thermal conditions significantly different from those encountered in the preparation of cellular polymers. Meaningful information on the thermal behaviour of blowing agents in polymers can be obtained when their performance is compared in a particular polymer system under controlled thermal conditions. For instance, in the absence of a cross-linking or stiffening effect of the decomposition products, the degree of expansion of the polymer is directly related to the quantity of gas given off by the blowing agent at a constant time-temperature cycle, provided no gas is lost by cell rupture or diffusion.

Carbonates

Ammonium carbonate

Ammonium carbonate is a white, crystalline powder with a strong ammoniacal odour. It may be prepared by reacting carbon dioxide with ammonia in the presence of water. The commercially available product is actually a mixture (or a double salt) of ammonium bicarbonate and ammonium carbamate (NH_2O–$CONH_4$). When exposed to air, the crystals of ammonium carbonate liberate ammonia and water and are transformed into a moist ammonium bicarbonate powder. The thermal dissociation of ammonium carbonate starts at 30°C and the reaction proceeds quite vigorously at 55–60°C.

The compound is a 'total' blowing agent in that it decomposes without leaving a residue. At 100°C the gas yield is 980 ml (STP)/g, one of the highest of all chemical blowing agents. The usefulness of ammonium carbonate as a blowing agent is severely limited by its poor storage stability, the indefinite composition of the commercial product, the objectionable ammonia odour, the difficulty with which the material can be dispersed in high polymers. The storage stability of ammonium carbonate can be improved, according to patent literature, by blending it with magnesium carbonate, zinc oxide, or alkylamines. Pastes consisting of the blowing agent and processing oils have better stability as well as improved. Ammonium carbonate has been used in the production of sponge rubber; however, it has been displaced by the more stable sodium bicarbonate. It has also been proposed as an auxiliary blowing agent in the preparation of polyurethan foam, as an additive to dinitrosopentamethylenetetramine for making poly(vinyl chloride) foam and as a blowing agent for urea–formaldehyde glues and chlorosulphonated polyolefin compounds.

Ammonium bicarbonate

Ammonium bicarbonate is more stable and easier to obtain in pure form than ammonium carbonate. Commercially, ammonium bicarbonate is prepared by passing carbon dioxide through an aqueous solution of ammonia. Under atmospheric pressure and in the absence of moisture, the salt decomposes slowly at 60°C yielding ammonia, carbon dioxide and water.

$$NH_4HCO_3 \rightleftharpoons NH_3 + CO_2 + H_2O$$

The gas yield is 850 ml (STP)/g assuming water to be available as vapour. It is more economical as a blowing agent than ammonium carbonate. Moisture accelerates the dissociation of the salt and high pressure suppresses the formation of the gaseous decomposition products. Ammonium bicarbonate suffers from the same disadvantages as the carbonate and can be used in the same applications.

Sodium bicarbonate

Among the inorganic blowing agents only sodium bicarbonate has remained in use over the years, although the introduction of organic blowing agents greatly diminished its importance in the manufacture of cellular polymers.

Sodium bicarbonate is an odourless non-toxic white powder (specific gravity 2.2), which is obtained as an intermediate in the manufacture of soda ash by the Solvay process, or by reacting sodium carbonate solution with carbon dioxide. The salt decomposes slowly at approximately 100°C with an attendant evolution of carbon dioxide; at 140°C the decomposition proceeds at a rapid but still controllable rate.

$$2NaHCO_3 \rightleftharpoons Na_2CO_3 + CO_2 + H_2O$$

Assuming water to be available as vapour, the total yield is 267 ml (STP)/g, which, combined with a low price, makes sodium bicarbonate the most economical chemical blowing agent. However, the residue of this decomposition is strongly alkaline and thus objectionable in many applications. Also, the degree of dissociation of sodium bicarbonate decreases rapidly with increased pressure and, therefore, the blowing agent cannot be used effectively in high-pressure expansion processes.

The storage stability of sodium bicarbonate is adequate but the material tends to pick up moisture and to cake badly. However, finely ground sodium bicarbonate remains free-flowing if conditioned with additives such as magnesium stearate. In blowing polymers, the efficiency of sodium bicarbonate is significantly lower than indicated by the gas yield. This is due to the rapid rate of diffusion of carbon dioxide from polymer, and to the poor dispersibility of the salt in elastomers and plastics. A paste of finely ground sodium bicarbonate (particle size less than 15 μ) in light process oil with lecithin as an emulsifier possesses improved dispersibility and can be mixed easily with elastomers. This product, has enjoyed commercial acceptance in the production of cellular rubber goods. Surface-active agents can also improve the dispersibility and efficiency of sodium bicarbonate paste. A co-precipitated master-batch of sodium bicarbonate containing rubber has been claimed to facilitate the dispersion of the blowing agent in rubber compositions. In the thermal decomposition of sodium bicarbonate, only half of the theoretically available carbon dioxide in the molecule is evolved. A total decomposition of the bicarbonate can be achieved by the use of acidic materials. In practice, weak acids are used in conjunction with sodium bicarbonate to obtain a higher gas yield from the blowing agent. For expansion of rubber, sodium bicarbonate is activated with stearic acid, oleic acid, or cottonseed oil acid. In thermosetting resins, the acidic curing catalyst also acts as an activator for the blowing agent.

Sodium bicarbonate has been extensively employed in the manufacture of opencell rubber sponge based on all types of natural and synthetic elastomers. Phenolic resin foams can be prepared by reacting the bicarbonate with an acid hardener and this process has become commercially important. Other polymers that can be expanded with sodium bicarbonate are natural and synthetic latexes, polyethylene, alkyd resins, poly(vinyl chloride), epoxy resins, polyamides and acrylic resins.

Organic carbon dioxide-generating chemicals

Ethylene carbonate is an example of organic blowing agents which decompose in polymers in the temperature range of 120–200°C with the liberation of carbon dioxide.

$$O{=}C\begin{matrix} \diagup OCH_2 \\ \diagdown OCH_2 \end{matrix}\Big|$$

The advantage claimed for this compound is its compatibility with a wide variety of polymers, which facilitates dispersion and thus leads to the formation of very fine cells.

Nitrites

Ammonium nitrite

Ammonium nitrite, through its thermal decomposition, offers a convenient source of elemental nitrogen. Prior to the introduction of organic blowing agents, this product has been suggested as an expanding agent, for pressure-blown cellular rubber.

In practice, ammonium nitrite is prepared *in situ* by incorporating into the rubber mix equimolar quantities of ammonium chloride and sodium nitrite. During the heating of the rubber, nitrogen is liberated.

$$NH_4Cl + NaNO_2 \longrightarrow N_2 + 2H_2O + NaCl$$

Unlike the thermal dissociation of carbonates, the decomposition of ammonium nitrite is an irreversible, exothermic reaction which proceeds independently of external pressure. The reaction produces nitrogen; but small quantities of nitric oxides which have a cure-accelerating effect and which cause corrosion of moulds and other processing equipment are also generated. Small quantities of water and polyhydric alcohols accelerate the evolution of gas. The hygroscopic sodium nitrite introduces enough moisture into the compounded polymer to initiate the reaction as soon as it comes into contact with ammonium chloride. End results are unpredictable. In common with other inorganic blowing agents, both sodium nitrite and ammonium chloride are difficult to disperse in polymers.

At the present time, ammonium chloride-sodium nitrite pills are used as an inflating agent during the curing of hollow rubber goods (e.g., children's balls).

Organic nitrites

Various organic nitrites have been used as blowing agents.

Alkylamine nitrites ($RNH_2{\cdot}HNO_2$) represent a group of organic nitrogen-releasing blowing agents which have been suggested for use in rubber and in poly(vinyl chloride). The compounds decompose with evolution of nitrogen and water.

$$RNH_2{\cdot}\,HNO_2 \longrightarrow ROH + N_2 + H_2O$$

Tert-butylamine nitrite, which was test marketed in the United States in 1955 as Blowing Agent X-950, decomposes at around 120°C and yields 374 ml (STP)/g (considering the evolved water as vapour).

Amidine nitrites (R—C(:NH)—$NH_2 \cdot HNO_2$), in particular guanidine nitrite (dec. 99–110°C), guanylurea nitrite (dec 139°C), and acetamidine nitrite (dec 155–160°C), have been claimed in the patent literature as blowing agents for rubber and plastics, but have not been used for this purpose commercially.

Hydrides and Peroxides

Alkali borohydrides

In recent years borohydrides, in particular potassium and sodium borohydrides, have been promoted as blowing agents. These compounds are convenient sources of hydrogen.

The rate of hydrolysis of alkali borohydrides is governed by the hydrogen ion concentration (pH) and increases rapidly with decreasing pH. In non-aqueous systems, acidic compounds (e.g., phthalic anhydride, stearic acid, or glycine) are added along with a small quantity of water. Under these conditions the decomposition of the borohydrides proceeds rapidly with the evolution of hydrogen.

This reaction is also catalysed by certain metallic salts, notably those of iron, cobalt, or nickel.

$$MBH_4 + 2H_2O \xrightarrow{(H^+)} MBO_2 + 4H_2$$

As a result of the decomposition of the blowing agent within the polymer matrix, hydrogen is entrapped in the cells; however, it diffuses rapidly and is replaced by air. This may create explosion hazards unless the concentration of hydrogen in and around processing equipment is kept below 4 per cent.

Potassium borohydride is a white crystalline solid which is non-hygroscopic under normal storage conditions. In air, it ignites from a free flame and burns quietly. The compound is readily soluble in water. Its rate of decomposition in an alkaline solution is very slow below 100°C. Under acidic conditions and/or at elevated temperatures, potassium borohydride decomposes rapidly, yielding 1660 ml (STP) hydrogen/g. Sodium borohydride is a hygroscopic solid which decomposes solely in moist air. In acidic media the compound generates 2370 ml hydrogen (STP)/g.

Alkali borohydrides have been proposed as blowing agents for several polymers in aqueous systems —rubber latex, poly(vinyl acetate), poly(vinyl alcohol), and melamine-formaldehyde resins. Poly(vinyl chloride) plastisols are expanded at room temperature with borohydrides in the presence of water and organic acids; the reaction can be delayed by the addition of isocyanates. A blend containing 5 per cent potassium borohydride extended with clay was offered commercially for blowing natural and synthetic elastomers in the presence of stearic acid, but the product failed in production trials and was withdrawn from the market.

Silicon oxyhydride

Silicon oxyhydride is a white, storage-stable, free-flowing powder; it liberates hydrogen either on heating (350°C) or under the influence of bases (Eq. 30.1).

$$\left[\begin{array}{c} \mathrm{H} \quad\quad \mathrm{H} \\ | \quad\quad\quad | \\ -\mathrm{O}-\mathrm{Si}-\mathrm{O}-\mathrm{Si}-\mathrm{O}- \\ | \quad\quad\quad | \\ \mathrm{O} \quad\quad \mathrm{O} \\ | \quad\quad\quad | \\ -\mathrm{O}-\mathrm{Si}-\mathrm{O}-\mathrm{Si}-\mathrm{O}- \\ | \quad\quad\quad | \end{array}\right]_x + H_2O \xrightarrow{(OH^-)} SiO_2 + H_2 \quad \text{... (30.1)}$$

The blowing agent is capable of generating 423 ml hydrogen (STP)/g and leaves as a residue silica filler which imparts some reinforcement to elastomers. Silicon oxyhydride has been suggested as an expanding agent for synthetic elastomer, mainly silicone rubbers.

Peroxides

Among the oxygen-releasing blowing agents, hydrogen peroxide is of particular commercial importance for foaming of rubber latex. In the Talalay process, use is of made of the ability of minute quantities of the enzyme catalase to decompose hydrogen peroxide rapidly at room temperature. Latex foam prepared by the Talalay process has a lower apparent density and a more uniform cell structure than foams obtained by the mechanical foaming method of *Dunlop*. Similar results are claimed for a foaming method utilising hydrogen peroxide in conjunction with polyamines. Latex foam can also be successfully produced with urea peroxide as blowing agent.

Oxalic Acid Derivatives

Oxalic acid derivatives comprise a group of carbon dioxide-generating blowing agents. Among the salts, ferrous oxalate (dec 160°C) has been specifically recommended for the blowing of rubber compounds. Of greater practical importance have been the oxalates of weak organic bases. Oxalates of urea, methylurea, biuret, dicyandiamide, and melamine, or physical mixtures of oxalic acid and the organic base in an equimolar ratio, were found to be efficient blowing agents for polymeric materials.

Urea oxalate has been offered commercially as Hucel (Hubron Rubber Chemicals, Ltd., Manchester, England) for use in the manufacture of rubber sponge. This salt can be prepared from aqueous solutions of the urea and oxalic acid in the presence of sodium alginate, casein, or oil emulsion, all of which reduce the crystal growth of the precipitated urea oxalate. Urea oxalate is relatively stable up to 130°C, but decomposes rapidly at 145°C. The gas yield is 140 ml (STP)/g. The gaseous phase consists mainly of carbon dioxide and carbon monoxide in the ratio 3.5:1, but some nitrogen, ammonia and water vapour are also present. Urea oxalate produces odourless non-staining rubber sponge with small and uniform cells. In conjunction with sodium bicarbonate, it has been used in the production of soft cellular-rubber shoe soling applied directly to canvas uppers.

In addition to urea oxalate, blends of urea and oxalic acid in molar ratios varying from 1:1 to 5:1 have been considered useful as blowing agents for rubber and plastics.

1,1′-Dithiodiformamidine (bis(methyliminoamino) disulphide oxalate), $NH_2C\text{–}(:NH)\text{—}S\text{—}S\text{—}C(:NH)NH_2(COOH)_2$, liberates carbon dioxide and carbon monoxide at 120–125°C. It can be easily prepared by the oxidation of thiourea with hydrogen peroxide in the presence of oxalic acid. But it has not been found attractive commercially. Rubber sponge expanded with this oxalate displays an unpleasant odour. Of more recent interest is the claim that oxamic acid, $H_2N\text{—}CO\text{—}COOH$, (dec 214°C) and its methyl and ethyl esters are useful in the expansion of polyamides having softening points between 170 and 220°C.

Urea and Related Compounds

Urea

Urea is a white, crystalline, highly hygroscopic material (sp. gr. 1.34) with a melting point of 132.7°C. When slowly heated beyond the melting point, it decomposes (Eq. 30.2) with the evolution of ammonia, leaving biuret as the residue. When heated rapidly to a temperature above 150°C, cyanic acid, which trimerises to cyanuric acid, is formed (Eq. 30.3).

$$2NH_2{-}\overset{\overset{\displaystyle O}{\|}}{C}{-}NH_2 \rightarrow NH_3 + NH_2{-}CO{-}NH{-}CO{-}NH_2 \qquad \text{....(30.2)}$$

$$3NH_2{-}\overset{\overset{\displaystyle O}{\|}}{C}{-}NH_2 \rightarrow 3NH_3 + (NCOH)_3 \qquad \text{... (30.3)}$$

The first decomposition route generates 187 ml (STP) ammonia/g; in the second decomposition reaction twice as much ammonia is liberated. Despite this high gas yield, urea is used only occasionally as a blowing agent for rubber. However, it plays an important role in the manufacture of cellular rubber as an activator for dinitrosoamine-type blowing agents and as a secondary cure accelerator in conjunction with thiazole- and thiuram-type accelerates. Because of the difficulty with which the crystalline material is dispersed in elastomers, finely ground, non-craking urea products have generally been employed. Also, a urea-metal salt complex (mp 100–115°C) is being offered for the same application.

Biuret

Biuret (dec 192–193°C) has been claimed as a blowing agent for elastomers, but only mixtures of urea and biuret are commercially exploited. Heating an equimolar blend of these two compounds liberates 275 ml (STP) ammonia/g.

$$NH_2{-}CO{-}NH{-}CO{-}NH_2 + NH_2{-}CO{-}NH_2 \longrightarrow 2NH_3 + (NCOH)_3$$

The commercial product is a non-caking powder (sp gr 1.45) which disperses fairly well in rubber. It produces odourless, non-staining, and non-discolouring cellular products and is recommended for use either alone or in combination with other blowing agents such as sodium bicarbonate or dinitrosoamines. The residue, cyanuric acid, has no deleterious effect on rubber and because it is virtually insoluble in water, cannot be leached when the cellular product is immersed in water. The efficiency of the urea-biuret blend at the usual blowing temperatures is higher than that of either component.

Aminoguanidine bicarbonate (NH_2—NH—C(:NH)—$NH_2.H_2CO_3$) and aminoguanyl urea (NH_2—NH—C(:NH)—NH—CO—NH_2) decompose with the evolution of ammonia and nitrogen (also carbon dioxide from the bicarbonate). Both chemicals have been considered as blowing agents for rubber, but their commercial application is economically impractical.

The *N*-nitro derivatives of urea and related organic bases have been described in the patent literature as blowing agents for plastics. The decomposition temperatures of these compounds range from as low as 120°C (*N*-nitro-*N'*-cyclohexylurea) to as high as 230°C (*N*-nitroguanidine). The thermal decomposition of the N-nitro compounds leads to the liberation of gases containing nitrous oxide and carbon dioxide as the main components. Most of the compounds of this group are odourless, colourless, easily dispersible solids which can be conveniently obtained by the nitration of the corresponding organic bases. It is claimed that the preparation and handling of these nitro compounds present no explosion hazards, but *N*-nitrourea, is highly sensitive to shock and dangerous when exposed to heat and flame.

N-Nitrourea (NH_2—CO—NH—NO_2) decomposes in air at 158–159°C giving off nitrous oxide, carbon dioxide, ammonia, and water vapour. When decomposed in paraffin at 129°C, the compound generates 380 ml gas/gram. It may be used as a blowing agent in the production of cellular polymers, both thermoplastic and thermosetting. The decomposition rate of *N*-nitrourea can be accelerated by the addition of organic bases (e.g., hydrazine, morpholine, or methylamine) with which it forms salts decomposing at temperatures below 100°C. The activation of *N*-nitrourea with alkyl-amines has been applied to the preparation of expanded potting compounds based on epoxy resins. *N*-Nitroguanidine (NH_2–C(:NH)– NH—NO_2) has been suggested as a blowing agent for high-melting polyolefins, in particular for liner polyethylene and polypropylene.

Azo Compounds

Aromatic azo compounds

Among the aromatic azo compounds, the triazenes have received the greatest attention and use as blowing agents. Other compounds that have been proposed as blowing agents are acrylazosulphones, arylazotriaryl-methanes (also the hydrazo compound and diazoethers).

Diazoaminobenzene

Diazoaminobenzene (1,3-diphenyltriazene, $C_6H_5NHN{=}NC_6H_5$), the first commercially available organic nitrogen-releasing blowing agent, can be obtained by diazotisation of aniline, followed by a coupling with the same base in a solution buffered with sodium acetate. The pure compound forms needlelike, golden-yellow crystals (sp. gr. 1.17), which melt with decomposition at 98°C. The commercial product, DAB is a brown-coloured, crystalline powder with a characteristic aromatic odour. It decomposes between 90 and 95°C, and liberates 113–115 ml nitrogen (STP)/g. It disperses easily since it is soluble in elastomers up to 8 per cent in highly plasticised rubber stocks. Larger quantities of the blowing agent tend to bloom from the uncured rubber compositions, but no blooming occurs from vulcanisates containing as much as 15 per cent diazoaminobenzene. In the presence of aldehyde-amine type accelerators, diazoaminobenzene has no effect on the rate of vulcanisation, a property which is of considerable importance in the manufacture of cellular hard rubber. The blowing agent imparts an intense orange discolouration, which precludes its use in light-coloured goods. Cellular products made with diazoaminobenzene stain many types of substrates, including paper, cloth, rubber and organic coatings. This blowing agent and cellular polymers made with it may cause skin irritation; the use of diazoaminobenzene is therefore restricted to the manufacture of industrial goods. Diazoaminobenzene is an efficient and economical blowing agent which produces expanded products with small, uniform cells. The material has been used in the production of cellular rubber. Its application in poly(vinyl chloride), phenolic resins and rubber-resin blends has also been claimed.

Attempts have been made to develop aromatic azo compounds free of the undesirable features of diazoaminobenzene. The elimination of staining properties has been claimed for a group of *sym-xeynyltriazenes*, of which the 1,3-*bis*(*o*-xenyl)-triazene is said to be non-discolouring as well. The latter decomposes at 130–135°C and yields 67 ml nitrogen (STP)/g. More efficient are the diarylpentazadienes of the general structure Ar—N=N—N(R)—N=N—Ar (where R=H, aryl, or alkyl), some of which liberate approximately the same quantity of gas as diazoaminobenzene. Because they are more costly to produce than the latter, pentazadienes have never been used commercially.

Among other triazenes, the substituted mono- and bissulfonyl derivatives have been found to be of some interest as blowing agents. In particular, *N*-cyanoalkylsulphonyltriazenes have been considered useful as non-staining blowing agents with low levels of toxicity. One of these compounds, 1-benzene-sulphonyl-3-cyanoethyl-3-phenyltriazene, was test marketed in the United States as Cyscocel a 40 per cent active blend.

It has been withdrawn from the market. The tan-coloured product decomposes at 110°C with a gas yield substantially lower than that of diazoaminobenzene. Cellular polymers prepared with this triazene are odourless and non-staining.

Aliphatic azo compounds

Blowing agents of this general chemical type include azonitriles and derivatives of azodicarboxylic acid.

Azonitriles

Azonitriles have an important advantage over diazoaminobenzene in that they permit the production of non-staining, white, and odourless cellular polymers. The most useful compound of this class is 2,2′-azobis (2-methylpropionitrile), better known as 2,2′-azobisisobutyronitrile. The compound may be synthesised by reacting acetone and alkali cyanide (or acetone cyanohydrin) with hydrazine sulphate, and oxidising the resulting hydrazo nitrile with chlorine in an aqueous medium. In an alternate method, acetone cyanohydrin and ammonia are converted to 2-aminoisobutyronitrile which is then oxidised with an alkali hypochlorite to yield 2,2′-azobisisobutyronitrile. The dry, pure product has a melting point of 105°C. The commercially offered blowing agent melts around 100°C and gives off 136 ml nitrogen (STP)/g. The thermal decomposition of 2,2′-azobisiobutyronitrile has been reviewed recently by *Hammond*. The following compounds have been identified as decomposition products: tetramethyl-succinonitrile (TMSN), isobutyronitrile, methacrylonitrile, and ketenimine. The main constituent of the residue, tetramethylsuccino nitrile, has a high level of oral toxicity (LD_{50} 60 mg/kg body weight). In solution, 2,2′-azobisisobutyronitrile decomposes by first-order kinetics in a wide range of solvents and yields a free radical, 2-cyanopropyl, which can be used as an initiator in vinyl-type polymerisations. A purified form of 2,2′-azobisisobutyronitrile is being offered for this purpose.

2,2′-Azobisisobutyronitiel is a white, fine crystalline powder (sp gr 1.10), readily soluble in elastomer, many conventional solvents (ketones, esters, alcohols, and aromatic hydrocarbons), and vinyl monomers (styrene, vinyl chloride, methyl methacrylate). It is virtually insoluble in water and aliphatic hydrocarbons. The storage stability of the compound is fairly good. At 90°C, 33 per cent of the blowing agent decomposes in an hour; at 40°C 2.5 per cent in a week. To preserve efficiency, it is preferably stored below 30°C. It is a flammable solid; it can be readily ignited with an open flame and continues to burn vigorously and completely when the source of flame is removed. Dry 2,2′-azobisisobutyronitrile can also be decomposed by a direct blow with a hammer, but no detonation occurs.

This blowing agent has no cure-activating effect on rubber and does not adversely affect the ageing properties of cellular vulcanisates. In poly(vinyl chloride) compositions, the residual tetramethylsuccino-nitrile acts as a hydrogen chloride acceptor, facilitating the heat stabilisation of cellular vinyl compositions expanded with 2,2′-azobisisobutyronitrile. The main disadvantage of 2,2′-azobisisobutyronitrile is the toxicity of the residue. The health hazards connected with its handling and processing can be minimised by the use of a non-dusting, oil-treated form of the blowing agent, and by providing good ventilation of the work area. Vulcanisation presses and expansion and annealing ovens must be equipped with adequate exhaust systems. Expanded products should be washed with warm water to remove from the surface the bloom of the toxic tetramethyl-sucicinonitrile, which is incompatible with the polymer.

During and shortly after World War II, large quantities of 2,2′-azobisisobutyronitrile found widespread applications in the manufacture of cellular elastomeric, thermoplastic, and thermosetting products. Since 1955, however, it has been forced aside in the blowing agent field as better compounds at lower prices became available. Currently, this product is mainly used as a polymerisation initiative and only minor quantities go into the expansion of poly(vinyl chloride) and polystyrene.

1-1′-Azobiscyclohexanecarbonitrile

1-1′-Azobiscyclohexanecarbonitrile, also known as azobis (hexahydrobenzonitrile) also know as azocyclohexylcyanide (CHDN) is less hazardous to process than 2,2′-azobisisobutyronitrile because the main component of its residue, dicyanodicyclohexyl, processes a lower level of toxicity than tetramethylsuccinonitrile. This homolog of 2,2′-azobisisobutyronitrile can be obtained by a process similar to that employed for making 2,2′-azobisisobutyronitrile, with the substitution of cyclohexanone for acetone. The pure compound decomposes at 114–115°C, and the commercially offered product at approximately 105°C. 1,1′-Azobiscyclohexanecarbonitrile decomposes more slowly than 2,2′-azobisisobutyronitrile and theoretically yields only 92 ml nitrogen (STP)/g. In spite of some advantages over 2,2′-azobisisobutyronitirle, 1,1′-azobiscyclohexanecarbonitrile has never become firmly entrenched in the field of blowing agents; it is more expensive and less efficient than the former. A high gas yield and a nontoxic residue characterise the oxime derivative of 2,2′-azobisisobutyronitrile.

Azobisisobutyramidoxime

Azobisisobutyramidoxime, obtained by reacting alcoholic solutions of hydroxylamine with 2,2′-azobisisobutyronitrile, is a white crystalline solid decomposing at 148°C. The compound, Genitron AO (Whiffen and Sons), was test marketed in Great Britain but was later withdrawn because of its high exotherm.

Among the derivatives of azodicarboxylic acid, the diamide is of major interest as a blowing agent for polymers.

Azobisformamide

Azobisformamide (ABFA), often called azodicarbonamide, is prepared by reacting hydrazine with an alkali cyanate or urea, followed by the oxidation of the resulting hydrazo compound with chlorine, chromic acid, a nitrate in the presence of a catalyst (salts of copper, cobalt, nickel, iron, and manganese), or sodium chlorate in the presence of ammonium metavanadate. The oxidation step, when carried out under closely controlled conditions, can yield a product of small particle size and narrow size distribution. When pure, azobisformamide forms orange-yellow crystals having a melting point of approximately 230°C. The commercial product is, depending upon the particle size, an orange-yellow to pale yellow powder (sp gr 1.65) decomposing in air at 195–205°C. Heated in a plasticiser above 210°C, the blowing evolves 220 ml gas (STP)/g, the highest gas volume of all organic blowing agents. Although azobisformamide is practically insoluble in common solvents and ester plasticisers, it is dispersed easily by conventional mixing methods in all elastomers and plastics. This blowing agent is slightly soluble in dimethylformamide, and more soluble in dimethyl sulphoxide (4–5 g/100 g solvent at 20°C). Its storage stability is unexcelled. Unlike other known organic blowing agents, azobisformamide does not support combustion and is self-extinguishing.

The thermal decomposition of dry azobisformamide in air yields the following products:

Gaseous products	32 per cent by wt
Nitrogen	65 per cent by vol
Carbon monoxide	32 per cent vol
Carbon dioxide	3 per cent by vol
Residue	68 per cent by wt

The composition of the residue varies depending upon the medium in which the blowing agent has been decomposed. In all instances, the following have been identified: hydroazobisformamide (biurea), cyanuric acid, urazole and under certain conditions, oxamide. The decomposition mechanism appears to be complex and no theory has been offered to account quantitatively for the presence of all decomposition products. *Reed* suggested that the primary decomposition of azobisformamide follows two main routes. In the first, shown in equations 30.4 and 30.5, nitrogen and carbon monoxide are liberated. In the second, shown in equation 30.6, nitrogen is evolved and biurea and cyanic acid are formed. Cyanuric acid and urazole, two constituents of the residue, are formed from cyanic acid and form biurea, respectively.

$$NH_2CO—N = N—CONH_2 \longrightarrow N_2 + CO + NH_2CONH_2 \quad ... (30.4)$$

$$H_2NCONH_2 \rightleftharpoons HCNO + NH_3 \quad ... (30.5)$$

$$2NH_2CO—N{=}N—CONH_2 \longrightarrow N_2 + NH_2CO—NH—NH—CONH_2 + 2HCNO \quad ... (30.6)$$

At elevated temperatures and in the presence of water, azobisformamide hydrolyses with the evolution of nitrogen, carbon dioxide, and ammonia (eq. 30.7).

$$2NH_2CO—N{=}N—CONH_2 + 2H_2O \longrightarrow$$

$$N_2 + 2CO_2 + 2NH_3 + NH_2CO—NH—NH—CONH_2 \quad ... (30.7)$$

In a neutral medium, the rate of hydrolysis is slow at temperatures below 100°C but in the presence of alkali carbonates a rapid decomposition occurs at 100°C. This reaction permits a quantitative determination of azobisformamide by the absorption of the generated ammonia in boric acid.

Cold aqueous alkalies react with azobisformamide to produce the alkali salts of azodicarboxylic acid which may be precipitated with an excess of the base (eq. 30.8). On heating, the solution of the alkali azodicarboxylate decomposes with the liberation of carbon dioxide and nitrogen (eq. 30.9).

$$NH_2CO—N{=}N—CONH_2 + 2NaOH \longrightarrow NaOCO—N{=}N—COONa + 2NH_3 \quad ... (30.8)$$

$$2NaOCO—N{=}N—COONa + 2H_2O \longrightarrow N_2H_4 + Na_2CO_3 + 2CO_2 + N_2 \quad ... (30.9)$$

Hot concentrated hydrochloric acid decomposes azoabisformamide with the evolution of carbon dioxide and nitrogen to yield biurea and ammonium chloride. *Diels* and *Paquin* reported the formation of hydrazine when azobisformamide is treated with warm, dilute sulphuric acid. As a weak oxidising agent, azobisformamide may react with many phenols and secondary amines used as stabilisers for polymers to produce colour-forming bodies of the quinoid and diazo types.

Although the decomposition point of azobisformamide in air is high, the blowing agent can be activated by many additives which are usually employed as compounding ingredients for polymers—e.g., protective agents or cure activators. These adjuncts lower the decomposition temperature and increase the decomposition rate of azobisformamide in polymers. Among the additives, certain vinyl stabilisers containing basic salts of lead, cadmium, or zinc are the most efficient activators (Fig. 30.4). Poly(vinyl chloride) stabilisers containing the same cation show increased activation with the increase in stabiliser solubility.

The activating effect of the basic metal salts has been explained by the instability of azodicarboxylates believed to be first formed during the reaction of azobisformamide with the cations of these vinyl stabilisers in alkaline medium.

The lead, cadmium, and zinc azodicarboxylates decompose at low temperatures and initiate the decomposition of the diamide. When barium stabilisers are present, no activation takes place, since barium azodicarboxylate is more stable thermally than the diamide.

Another group of additives that accelerate the decomposition of azobisformamide are polyols, e.g., glycols and glycerol. Also urea, alcoholamines, and certain organic acids, (e.g., oxalic and *p*-toluene-sulphonic acids) lower the decomposition temperature of azobisformamide due to their hydrolytic effect on the blowing agent.

In addition to dibasic acids, dinitrosopentamethylenetetramine has been claimed as an effective activator for azobisformamide. Biurea, proposed as a decomposition promoter does not decrease the decomposition temperature significantly.

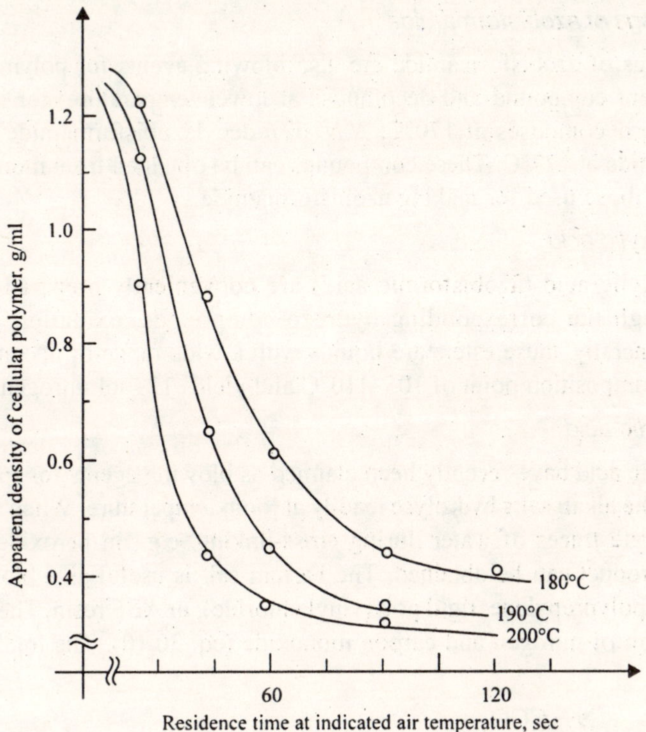

Fig. 30.4. Decomposition rate of 1,1´-azobisformamide in poly(vinyl chloride) stabilised with dibasic lead phthalate.

The degree of activation that can be attained with vinyl stabilisers depends upon the particle size of the azobisformamide. Since the activation occurs in a heterogeneous system, the surface area of the reagents has a direct bearing on the decomposition rate. Azobisformamide of small particle size is activated more readily than azobisformamide composed of large particles.

Azobisformamide of very fine particle size has been observed to possess a cure activating effect, probably due to the formation of a small quantity of free radicals during the decomposition. Aliphatic diazo compounds such as R—N=N—CO—NH$(CH_2)_n$ NH—CO—N=N—R, a diazodicarbonamide, and R—N=N—COO—$(CH_2)_n$OCO—N=N—R, a diazodicarbonate, are known to combine blowing efficiency with cross-linking properties. For example, diazodicarboxylates have been reported to cross-link natural rubber quantitatively.

Azobisformamide and its decomposition residue are considered non-toxic and non-irritating. The actual oral toxicity (LD_{50}) of azobisformamide, as measured in Wistar rats, is greater than 6400 mg/kg body weight. The compound has recently been suggested as a bleaching and maturing agent for flour and its has been accepted for this purpose by the Food and Drug Administration to a maximum level of 45 ppm.

Azobisformamide produces odourless, non-staining, non-discoluring cellular polymers. This blowing agent has been successfully used in the expansion of synthetic and natural rubber, poly(vinyl chloride) compounds, and polyolefin resins. It has been found useful in the manufacture of cellular acrylonitrile-butadiene-styrene plastics and high-impact polystyrene sheeting.

N-Substituted homologs of azobisformamide

N-Substituted homologs of azobisformamide are also blowing agents for polymers, but they are less efficient than the parent compound and decompose at lower temperature; for instance, *N,N'*-di-tert-butylazobisformamide decomposes at 170°C, *N,N'*-di-*n*-decylazobisformamide at 145°C, and *N,N'*-diphenylazobisformamide at 177°C. These compounds can be obtained from monosubstituted ureas by methods analogous to those used for making azobisformamide.

Esters of azodicarboxylic acid

Esters of azodicarboxylic acid (azobisformic acid) are conveniently prepared from hydrazine and chloroformates through the corresponding hydrazo compounds, oxidation of which yields the azodicarboxylates. Generally, these esters are liquids with a wide range of decomposition points. The diethyl ester has a decomposition point of 105–110°C and yields 129 ml nitrogen (STP)/g.

Salts of azodicarboxylic acid

Salts of azodicarboxylic acid have recently been claimed as blowing agents for special applications. In slightly acidic media, the alkali salts hydrolyse readily at room temperature. When these salts are used in polymers which generate traces of water during cross-linking, e.g., in peroxide cured polysulphide polymers, a cellular product can be obtained. The barium salt is useful as a blowing agent for high-melting polymers, e.g, polypropylene, rigid poly(vinyl chloride), or ABS resin. The dry salt decomposes at 245°C with evolution of nitrogen and carbon monoxide (eq. 30.10). The total gas yield is 177 ml (STP)/g.

$$\begin{matrix} N-COO \\ \| \\ N-COO \end{matrix} \!\!> Ba \longrightarrow BaCO_3 + CO + N_2 \qquad \text{... (30.10)}$$

Barium azodicarboxylate hydrolyses in presence of water and must be protected against moisture absorption to retain its efficiency. The alkali and alkaline earth azodicarboxylates are light yellow powders; they can easily be prepared from azobisformamide and the corresponding base.

Other aliphatic azo compounds suggested as blowing agents include sodium and potassium diazomethionates and azoacetic acid esters.

Hydrazine

Hydrazine salts

Hydrazine derivatives have long been recognised as potentially useful sources of nitrogen.

Phenylhydrazine

It decomposes upon heating with evolution of nitrogen and ammonia in equal volumes; the total gas yield is 207 ml (STP)/g. More recently, hydrazine and its salts, particularly oxalate and stearate, have been claimed as unusually effective blowing agents. Also, inorganic salts of hydrazine, e.g., the mononitrate activated with ferrous sulphate or molybdenum trioxide, have been recommended for making cellular polymeric compositions. None of these materials has been developed commercially, however.

Hydrazones

When reacted with hydrazine, many aromatic aldehydes and ketones form crystalline, high-melting compounds known as hydrazones, some of which have acquired practical significance as blowing agents.

Benzil monohydrazone

Benzil monohydrazone (C_6H_5—C(:N—NH_2)—CO—C_6H_5) a light yellow crystalline solid with a melting point of 150°C, yields nitrogen (100 ml (STP)/g) and leaves benzyl phenyl ketone as a nontoxic, yellow residue. Prior to the introduction of azobisformamide, benzil hydrazone was the only suitable blowing agent for calendered poly(vinyl chloride) sponge. Its relatively high decomposition temperature permitted the direct addition of the blowing agent to plasticised poly(vinyl chloride) compounds without the danger of decomposition of the blowing agent during the calendering operation.

Salicylaldehyde hydrazone

Salicylaldehyde hydrazone, which has a decomposition point of 210–220°C and a gas yield of 90 ml nitrogen (STP)/g, is another interesting blowing agent of this class.

Hydrazides

Many of the commercially used blowing agents fall within this category.

Hydrazides of organic sulphonic acid

Hydrazides of organic sulphonic acid constitute one of the most versatile and useful classes of organic blowing agents. Although derivatives of both aliphatic and aromatic sulphonic acids have been claimed in patents, only the latter are commercially important. The pure compounds of this class are mostly crystalline, nontoxic, odourless solids which can readily be obtained by reactive hydrazine with aliphatic or aromatic sulphonyl chlorides. Depending upon the sulphonic acid used, blowing agents, with decomposition points ranging from 80 (diethylaminosulphonyl hydrazide) to 245°C (bisbenzenesulphonyl hydrazide) can be prepared. One or both —NH_2 groups of the hydrazine molecule can be substituted with alkyl and/or aryl radicals and this further adds to the diversity of useful compounds. All sulphonyl hydrazides decompose on heating with a quantitative evolution of nitrogen. This decomposition may be accelerated, or even brought to completion at room temperature, by hydroxyl-bearing solvents—e.g., aliphatic alcohols or water—particularly if small quantities of alkaline substances or oxidising agents (ferric salts, iodine, hydrogen peroxide) are also present. Besides nitrogen, water vapour is also generated during the decomposition; the residue consists of alkyl or aryl disulphates and thiosulphoxides.

The decomposition reaction of sulphonyl hydrazides involves an internal reduction-oxidation of the sulphonyl hydrazide group.

$$4\ R—SO_2NH—NH_2 \longrightarrow 4N_2 + 6H_2O + R—S—S—R + R—S—SO_2—R$$

The decomposition extotherm is moderate since the reduction of the sulphonyl group proceeds endothermally and partially compensates for the exotherm generated during the oxidation of the hydrazine group. The decomposition residue is non-toxic and, in general, colourless; it does not influence the rate of cross-linking or fusion of polymers. In most instances they show adequate compatibility with polymers, and do not bloom to the surface of expanded products.

As amphoteric compounds, the sulphonyl hydrazides form crystalline salts with both acids, e.g., RSO_2NHNH_2.HCl, or alkalies, e.g., $RSO_2N(Na)NH_2$.

Sulphonyl hydrazides are flammable solids which ignite from a spark or flame. When ignited, they decompose rapidly in air, with or without a flame, leaving a voluminous residue. Under normal storage conditions, the compounds of this class are stable for a prolonged period of time.

Cellular polymers produced with sulphonyl hydrazides display a fine and uniform pore structure and are non-toxic and non-staining. In most cases, the cellular products are non-discolouring and odourless.

Sulphonyl hydrazides with an asymmetric molecular configuration have been found useful only in elastomeric applications, since in many other polymers, e.g., poly(vinyl chloride), they develop a mercaptan-like odour. The symmetrical hydrazides have been employed for all types of polymers.

Of the asymmetric sulphonyl hydrazides, benzensulphonyl hydrazide (BSH) is the most important product commercially. The pure material (sp. gr. 1.43) decomposes at 104°C and the commercially available product (Genitron BSH of Bayer) generates gas at about 100°C in air and at 90°C in rubber compositions. This blowing agent yields 115–130 ml nitrogen (STP)/g in addition to 195 ml steam. To improve the dispersibility of benzenesulphonyl hydrazide in elastomers, dispersions of the blowing agent in processing oils are being offered commercially. Benzenesulphonyl hydrazide is being extensively used in the production of atmospherically and pressure-blown soft rubber goods. The low decomposition temperature of the blowing agent requires careful control of processing (mixing, calendering, etc.) temperatures to prevent a loss of blowing efficiency. This limits the use of benzenesulphonyl hydrazide to soft, highly plasticised stocks which do not generate heat during mixing.

For the expansion of highly loaded rubber compounds such as are usually encountered in the manufacture of microcellular shoe soling, 1,3-benzenedisulphonyl dihydrazide has been recommended. The pure product melts with decomposition at 163°C; the commercially available blowing agent decomposes in air at 145°C and in rubber at 125°C. It liberates 170 ml nitrogen (STP/g, the highest volume of gas among the commercial hydrazide blowing agents. Due to its higher decomposition temperature, 1,3-benzenedisulphonyl dihydrazide can be processed more safely than benzenesulphonyl hydrazide. The blowing agent is marketed in the form of a 50 per cent paste in chloroparaffin.

Methyl carbanilate substituted with the sulphonyl hydrazide group in the para position also known as phenyl methylurethan-*p*-sulphonyl hydrazide, has been reported to be used in the manufacture of microcellular rubber in the former USSR. It decomposes in a temperature range of 140–160°C and liberates 95–100 ml (STP)/g.

$$H_3COCONH-C_6H_4-SO_2NHNH_2$$

Among the symmetric hydrazides, 4,4′-oxybis(benzenesulphonyl hydrazide) (OBSH) is the most widely used compound. It is a white, crystalline solid melting with decomposition at 164°C. The commercial product is sold as finely ground, oil-treated powder (sp. gr. 1.52), which decomposes slowly at 130°C. The most rapid and efficient gas evolution occurs at 150°–160°C. When heated in dioctyl phthalates, the material generates 125 ml (STP) of non-condensable gas per gram, over 98 per cent of which is nitrogen. In common with other difunctional sulphonyl hydrazides, 4,4′-oxybis(benzene-sulphonyl hydrazide) leaves a polymeric residue which is free of volatile, odouriferous mercaptides. The decomposition of the difunctional derivatives can be expressed by equation 30.11.

$$2nNH_2NH{-}SO_2{-}R{-}SO_2{-}NH{-}NH_2 \longrightarrow (-S-R-S-)_n + (-S-R-SO_2) + 4nN_2 + 6n\,H_2O \qquad \text{... (30.11)}$$

4,4′-Oxybis(benzenesulphonyl hydrazide) is useful as a blowing agent for all types of elastomers and plastics. Among the latter, branched polyethylene, poly(vinly chloride) plastisols, and epoxy and phenolic resins are currently expanded on a commercial scale with this blowing agent. It is most widely used, however, in the manufacture of thermal-insulation materials based on blends of synthetic rubbers and thermoplastics–e.g., nitrile-butadiene rubber-poly(vinyl chloride). In this application, 4,4′-oxybis-(benzenesulphonyl hydrazide) acts as both blowing and cross-linking agent in the absence of conventional curing agents. Other difunctional sulphonyl hydrazides posses similar cross-linking ability in varying degrees, depending upon the aryl or alkyl radical.

3,3′-Sulphonylbis(benzenesulphonyl hydrazide) is another difunctional hydrazide that has been available commercially. It decomposes at 148°C and gives off 110 ml nitrogen (STP)/g. The product has been primarily recommended for blowing poly(vinyl chloride) plastisol compositions.

Disulphonyl dihydrazide of undisclosed chemical composition with a gas yield of 120 ml nitrogen (STP)/g and a decomposition range of 175–180°C, has been offered for expanding polyethylene and calendered poly(vinyl chloride) compounds.

Trihydrazino-*sym*-triazene (THT), also called cyanuric trihydrazide, has been suggested as a blowing agent for making cellular products based on high-melting polymers, e.g., ABS resins, polypropylene, and rigid poly(vinyl chloride) compounds. It can be obtained by reacting cyanuric chloride or melamine with hydrazine. It decomposes at 230°C with the evolution of equal volumes of nitrogen and ammonia, and leaves melamine as residue. The theoretical gas yield is 262 ml (STP)/g; the actual gas evolution reaches 94 per cent of the theoretical.

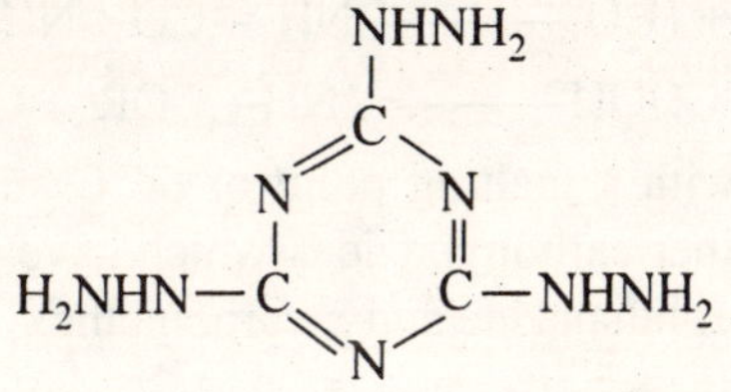

$$
\begin{array}{c}
NHNH_2 \\
| \\
C \\
N \quad\quad N \\
| \quad\quad \| \\
H_2NHN-C \quad\quad C-NHNH_2 \\
N
\end{array}
$$

Semi-carbazides

Sulphonyl semi-carbazides

Sulphonyl semi-carbazides of the general formula $R—SO_2—NH—NH—CO—NH_2$ constitute a distinctly new class of organic blowing agents. The individual members of this group can be synthesised by reacting the corresponding sulphonyl hydrazides with an alkali cyanate in an acidic medium. The semi-carbazides are soluble in alkalies, from which they can be precipitated in a pure form by acidification.

The semi-carbazides exhibit greater heat stability than the hydrazides from which they are derived. The conversion to semi-carbazides raises the melting and the decomposition point, and increases the gas yield. Liquid hydrazides, e.g., methane-sulphonyl hydrazide (dec 70°C), are transformed into crystallinc, stable semi-carbazides, e.g., methanesulphonyl semi-carbazide (dec 194°C). The decompostion of the blowing agents of this group leads to the evolution of nitrogen and carbon dioxide in quantities at least 30 per cent higher than one mole of gas for each sulphonyl semi-carbazide group in the molecule.

Typically, the semi-carbazides are high-temperature blowing agents, useful in the preparation of cellular polymers based on high-melting polymers. The compound decomposes in air at 213°C and liberates 145 ml gas (STP)/g. The evolved gas is composed of nitrogen (65 per cent), carbon dioxide (33 per cent, and carbon monoxide (2 per cent). This blowing agent has been recommended for the expansion of linear polyethylene, polypropylene, and ABS resins.

Azides

Blowing agents of the azide type are derivatives of either carboxylic or sulphonic acids. They can be prepared by treating the corresponding acid hydrazides with nitrous acid or by reacting the acid chlorides with alkali azides. The blowing agents of this group are white, crystalline products which give off nitrogen upon heating. The thermal decomposition of azides is accompanied by a high exotherm. The residue of the decomposition is non-staining and usually white. The azide blowing agents are flammable and shock-sensitive solids. They should be handled carefully to avoid spontaneous and rapid decomposition.

Carbonyl azides

Organic carbonyl azides, particularly those derived from aromatic acids, e.g., benzoyl azide, *p*-tert-butylbenzoyl azide, are soluble in ester plasticisers. This facilitates their incorporation into such thermoplastic resins as poly(vinyl chloride) homo- and co-polymers. They also disperse easily in all types of elastomers and millable plastics. Upon heating, carbonyl azides yield nitrogen and an isocyanate (eq. 30.12). The latter can be used as an additional source of gas when reacted with water (eq. 30.13) or acid (eq. 30.14). If the decomposition takes place in the presence of compounds containing hydroxyl or amine groups, residual diisocyanate may act as a cross-linking agent.

$$RCON_3 \longrightarrow RN=C=O + N_2 \qquad \text{... (30.12)}$$

$$2\,RN=C=O + H_2O \longrightarrow RNH—CO—NHR + CO_2 \qquad \text{... (30.13)}$$

$$RN=C=O + R'COOH \longrightarrow RNH—COR' + CO_2 \qquad \text{... (30.14)}$$

The *p*-tert-butylbenzoyl azide with a melting point of 64°C and a gas yield of 110 ml nitrogen (STP)/g has been test marketed. Other carbonyl azides which have been suggested as blowing agents for polymers include benzazide, terephthalazide and *p*-carbomethoxybenzazide.

Sulphonyl azides

Among the sulphonyl azides which have been claimed as blowing agents, diphenyl-4,4′-disulphonyl diazide has been of some value as a blowing and cross-linking agent for thermoplastic resins. The product is a white solid decomposing at 144–145°C. During the thermal decomposition, in addition to nitrogen (122 ml (STP)/g), free radicals are formed which are capable of producing solvent-resistance, cross-linked polymers. Cross-linked cellular polystyrene, polyethylene and alkyd resins have been successfully prepared by the use of dipheynyl-4,4′-disulphnoyl diazide. The blowing agent was test marketed in the United States but it failed to gain acceptance because the high decomposition exotherm led to partial degradation of the cellular polymers and has been withdrawn by the manufacturer.

N-*Substituted 5-amino-1,2,3,4-thiatriazoles*

N-Substituted 5-amino-1,2,3,4-thiatriazoles have been claimed as blowing agents for elastomers and plastics. The *N*-disubstituted derivatives of the general formula can be obtained by nitrosation of the corresponding thiosemi-carbazides.

```
      N—N
      ‖  ‖
RR'N—C   N
      \ /
       S
```

Depending upon the substituents, blowing agents with decomposition points from as low as 51°C (the dimethyl derivative) to as high as 162°C (the piperazine derivative) can be obtained. Unlike azides, the thiatriazoles give–off their nitrogen at a controllable rate. Simultaneously with the evolution of nitrogen, sulphur is also split off during the thermal decomposition. The liberation of sulphur is claimed to be advantageous in applications involving elastomers which are susceptible to sulphur vulcanisation. One compound of this class, Porofor TR (Farbcnfabriken Bayer), a morpholine derivative, has been used in Germany in the production of cellular poly(vinyl chloride) compositions based on plastisols. The blowing agent decomposes at 112°C and yields 130 ml nitrogen (STP)/g. Because of the presence of a minor quantity of sulphur among the decomposition products, the blowing agent should be compounded with stabilisers resistant to sulphur staining when light-coloured cellular vinyl plastics are to be produced. One of the disadvantages of this material is the fact that cellular products made with it may cause skin irritation. The unsubstituted 5-amino-thiatriazole, although claimed as a useful blowing agent for plastics, has not been offered commercially.

N-Nitroso Compounds

The *N*-nitroso derivatives of secondary amines and *N*-substituted amides form an important class of blowing agents which includes two commercial products—dinitro-sopentamethylenetetramine and *N,N'*-dimethyl-*N,N'*-dimetrosoterephthalamide.

N-dinitroso amines

Dinitrosopentamethylenetetramine (DNPT) can be obtained by the nitrosation of an ammonia-formaldehyde solution (molar ratio $NH_3:CH_2O = 2:3$) with sodium nitrite in the presence of acetic acid.

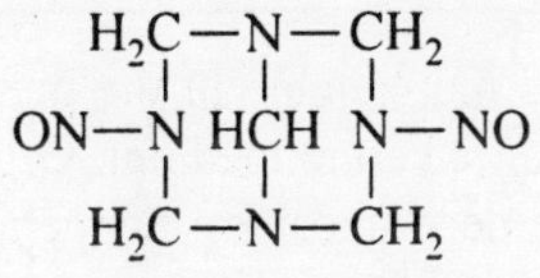

$$\begin{array}{ccccc} H_2C & - & N & - & CH_2 \\ | & & | & & | \\ ON-N & & HCH & & N-NO \\ | & & | & & | \\ H_2C & - & N & - & CH_2 \end{array}$$

The method of nitrosation involving the addition of an acid to a cold mixed solution of hexamethylene-tetramine and sodium nitrite is more convenient. In this instance, either organic acid (e.g., acetic acid or mineral acids (e.g., hydrochloric, nitric or sulphuric acids) may be used. Yields of between 75 and 79 per cent based on hexamethylenetetramine have been reported. An excess of nitrous acid leads to the formation of trimethylenetrinitrosoamine (mp. 105°C), an undesirable by-product. Because of thc instability of dinitrosopentamethylenetetramine in acidic medium, the nitrosation should be carried out with only the theoretical quantity of acid. Better yields and a more stable product have been claimed to be obtained by neutralising any excess of acid with alkaline materials.

After re-crystallisation from hot alcohol, dinitrosopentamethylenetetramine forms light yellow needles which melt with decomposition at 207°C. Commercial grades have lower decomposition point, usually between 190 and 200°C. The compound has a limited solubility in pyridine, morpholine, and ethyl acetoacetate, and is readily soluble in dimethylformamide and dimethyl sulphoxide. It is virtually insoluble in water, alcohols, aliphatic and aromatic hydrocarbons and most esters.

Dinitrosopentamethylenetetramine is particularly sensitive to mineral acids and certain salts of these acids (e.g., zinc chloride), which may cause a violent decomposition of the blowing agent at low temperatures and result in ignition of the mixture. In dilute hydrochloric acid, hydrolysis occurs at temperatures around 100°C.

$$C_5H_{10}(NO)_2N_4 + 3H_2O \longrightarrow 5CH_2O + 2N_2 + 2NH_3$$

The effect of acids on the decomposition temperature of dinitrosopentamethylenetetramine is proportional to their dissociation constants; the higher the dissociation constant, the lower the temperature, as shown in Table 30.2.

Table 30.2. Effect of acids on decomposition temperature of dinitrosopentamethylenetetramine.

Acid	*pKa at 25°C*	*Decomp. temp. °C*
Hydrochloric acid, conc.	<1.0	Room temp. with ignition
Oxalic acid		60–70
K_1	1.19	–
K_2	4.21	–
Salicylic acid	–	100
K_1	2.97	–
K_2	13.44	–
Phthalic acid		
K_1	2.89	–
Phthalic anhydride	5.41	115
Benzoic acid	4.20	130
Boric acid	9.24	130
Fatty acids	>10	180

Dinitrosopentamethylenetetramine is more stable in the presence of cold alkalies, but at elevated temperatures it decomposes readily in an alkaline medium. Organic bases such as urea, biuret-water, and polyhydric alcohols also lower the decomposition temperature. In an inert medium, a sustained thermal decomposition occurs above 180°C producing essentially nitrogen and solid white residue. Among the gaseous decomposition products, minor quantities of nitrous oxide, carbon dioxide, oxygen, water and formaldehyde have been identified. In acidic medium, trace amounts of amines are believed to be evolved. The solid residue formed upon decomposition has not been identified. The decomposition reaction is highly exothermic.

The compound is an oxidising agent and can be reduced quantitatively with titanous chloride in an acid medium. Reducing compounding ingredients (e.g., thiophenols or mercaptides) should not be brought in direct contact with it in order to avoid violent reaction at processing temperatures. It is a flammable solid, easily ignited by a free flame. In the dry state, it is sensitive to impact and fraction, but can be detonated only with considerable difficulty.

Dinitrosopentramethylenetetramine is an efficient, non-staining, non-discoluring blowing agent. It is commercially available in the form of blends consisting of the dinitrosoamine with inorganic diluents, processing aids, and/or stabilisers. In the United States, products containing 42 per cent dinitrosopentamethylenetetramine are most commonly used, but 80 per cent active compounds are also available. The latter products are more popular in Europe. The diluent facilitates handling and lessens the fire hazards associated with the use of the active ingredient. Ground non-reinforcing silica or washed natural clays are usually employed as inorganic diluents. Dinitrosopentamethylenetetramine is insoluble in elastomers and plastics, but disperses readily, particularly if treated with processing oils or plasticisers which are compatible with the polymer.

The dispersibility of dinitrosopentamethylenetetramine may be further enhanced by coating the blowing agent particles with selected surfactants. To stabilise it when it is in contact with strong acids, the addition of salts of weak organic acids and fatty acid esters of alkylhydroxylamines has been suggested.

Similar effects are obtained when an oxide, hydroxide, or carbonate of calcium, barium, zinc, or magnesium is used as a stabiliser.

The commercial 42 per cent blend of dinitrosopentamethylenetetramine is a cream coloured, odourless, non-hygroscopic solid in powder form (sp gr 1.91). Upon decomposition, the product liberates approximately 105 ml non-condensable gases (STP)/g. Under normal conditions, the storage stability of the commercial composition is excellent. The product has a very low level of actuate toxicity, and is only a minor skin irritant.

The processing stability of dinitrosopentamethylenetetramine blowing agents is good, particularly in neutral or basic media. The thermal decomposition is independent of pressure hence the blowing agent can be used for the manufacture of pressure expanded polymers. In the absence of activator, dinitrosopentamethylenetetramine blowing agents liberate gases at temperatures higher than those of conventional vulcanisation or fusing. In practice, to lower the decomposition temperature of the blowing agent, numerous additives have been used advantageously. Organic acids, e.g., salicylic or phthalic acids, substantially decrease the decomposition temperature and permit the development of vulcanisates with very fine cells. However, the end-products display a disagreeable 'fishy' odour. Another disadvantage of acidic promoters is their cure-retarding effect on rubber compounds. Organic sulphonic acids have been suggested as additives for the production of low-density sponge.

More important than the acidic additives are basic promoters for dinitrosopentamethylenetetramine. They decrease the decomposition temperature to a lesser degree than most acidic materials, but still bring it into a practical temperature range for a rapid gas generation. Of the basic materials, urea and urea-containing mixtures and complex have gained practical importance in controlling the decomposition of dinitrosopentamethylenetetramine. Cellular vulcanisates prepared with this compound in the presence of urea have a characteristic, faintly ammoniacal odour. This odour is fugitive and can be dissipated by a heat treatment (annealing) of cellular products. Urea in conjunction with *C*-cetyl betaine or melamine with diethylene glycol have also been recommended for odour control.

Neutral promoters also play an important role in the activation an odour control of dinitrosopentamethylenetetramine-blown sponge. The most commonly used additives of this class are glycol (diethylene, triethylene, polyethylene glycols). In addition to the promoters, deodorants are employed to counteract the readily discernible odour of vulcanisates expanded with dinitrosopentamethylenetetramine.

The basic and neutral promoters exert a pronounced activating effect on the vulcanisation of rubber. This effect, combined with a mild cure activation of dinitrosopentamethylenetetramine, particularly at higher concentrations, makes possible the development of fast-curing compounds with low levels of conventional accelerators.

Dinitrosopentamethylenetetramine is the most widely used blowing agent in the rubber industry, but has gained only limited acceptance in plastics. Cellular rubber goods based on natural and synthetic elastomers are being manufactured with dinitrosopentamethylenetetramine for a broad line of products such as microcellular shoe soling, automotive weatherstripping, thermal insulation, and playing balls for children.

Dinitrosopentamethylenetetramine has been suggested for use in a poly(vinyl chloride) but due to its high decomposition exotherm and its sensitivity to traces of hydrochloric acid, the resulting cellular products are discoloured, sometimes degraded and display an unpleasant odour. Better results have been claimed to be obtainable with poly(vinly chloride) compounds containing boric acid and urea as promoters. The low unit price, for the 42 per cent blend, and the high gas yield combine to make this compound the most economical nitrogen-releasing blowing agent.

N,N′-dinitrosoamides

In this group, *N,N′*-dimethyl-*N,N′*-dinitrosoterephthalamide has received the greatest commercial use.

The N*-methyl-*N*-nitrosoamides*

The *N*-methyl-*N*-nitrosoamides of aliphatic acids are low-temperature blowing agents; however, these compounds have not been commercially developed because of their poor storage stability. A more stable analog is that derived from terephthalic acid.

N,N′*-dimethyl-*N,N′*-diniltrosoterephthalamide*

N,N′-dimethyl-*N,N′*-diniltrosoterephthalamide (NTA) can be prepared by nitrosation of *N,N′*-dimethyl-terephthalamide in nitric acid solution. The product is an odourless, yellow crystalline solid melting with decomposition at 118°C. It liberates 180 ml nitrogen (STP)/g, with traces of water and carbon dioxide. The residue of the decomposition is dimethyl terephthalate, a white crystalline solid (mp 140°C). It has limited compatibility with most polymers, resulting in the formation of a white bloom on the surface of cellular polymers expanded with more than 5 per cent of the blowing agent. In the pure state, the compound evolves no gas during extended (140 hr) storage at 60°C. It can be kept indefinitely without significant decomposition at ambient temperatures (max 40°C). The thermal stability is reduced by mixing the blowing agent with numerous other materials. Alkalies decrease its stability in direct proportion to their alkalinity and solvating power. In the presence of strong alkalies, e.g., alcoholic solutions of sodium hydroxide, diazomethane may be formed. Strong amines, e.g., ethylenediamine, cause instantaneous decomposition. Dissolved in polar solvents (e.g., isopropanol or ethylene glycol), *N,N′*-dimethyl-*N,N′*-dinitrosoterephthalamide decomposes completely at room temperature within a few hours. The effect of mineral acids, which in general promote hydrolysis of the compound, is less pronounced. Many other materials normally considered inert (e.g., poly(vinyl chloride) resins or ester plasticisers), significantly decrease the decomposition temperature and bring about a slow liberation of nitrogen under ambient conditions. The blowing agent can be made less sensitive to these compounding ingredients by coating it with a modifier, e.g., glycolic esters of fatty acids. Exposure to sunlight and other sources of ultraviolet radiation results in a colour change from yellow to green. This colour change has no effect on the performance or thermal stability of the blowing agent.

N,N′-Dimethyl-*N,N′*-dinitrosoterephthalamide is a weak deflagrating explosive and ignites easily from spark or flame; its burning rate is extremely fast. It has a very low level of acute toxicity and is a minor skin irritant.

Commercially, this compound is available in the form of a blend (containing 70 per cent of active ingredient) with mineral oil as the diluent. The commercial product (Nitrosan, Du Pont) is an odourless, yellow-coloured, non-hygroscopic crystalline powder (sp. gr. 1.20) which melts with decomposition at 105°C. Prolonged heating at a lower temperature, e.g., 80–90°C, will result in a sustained and rapid decomposition. The gas yield of the blend is 126 ml (STP)/g.

N,N′-Dimethyl-*N,N′*-dinitrosoterephthalamide is an efficient, non-discolouring, non-hygroscopic blowing agent suitable for the of polymers that can be blended with it below 70°C. The material has found its widest applications exotherm in poly(vinyl chloride), compositions. Because of its low decomposition exotherm this blowing agent can be used in the preparation of thick sections of cellular vinyl compositions without the danger of polymer degradation. The most efficient utilisation of this blowing agent is achieved by the activation with mild alkaline materials, e.g., calcium sulphonates of paraffinic oils. It has been employed and has also been suggested for the expansion of liquid polyamide and epoxy resins.

Other dinitroso compounds

Closely related to the *N*-nitrosoamides already described are *N*-substituted nitrosourethans and ureas. The blowing agents of this class liberate a high volume of gas (2 moles for each —N(NO)CO)— group) consisting of equal parts of nitrogen and carbon dioxide. Among the most efficient blowing agents of this group are *N,N'*-dimethyl-*N,N'*-dinitrosoteramethylenediurathan (dec 133°C) *N,N'*-dimethyl-*N-N'*-dinitrosoteramethylenediurethan (dec 125°C, mp 123°C), both, where n=2 or 4, respectively, and *N,N'*-dinitrosoethyleneurea.

$$\begin{array}{c} CH_3 \diagdown \quad \overset{O}{\|} \qquad\quad \overset{O}{\|} \quad \diagup CH_3 \\ NCO(CH_2)_nCN \\ NO \diagup \qquad\qquad\qquad \diagdown NO \end{array}$$

The *N*-nitrosoamides of sulphonic acids (e.g., *p*-cresoldisulphonic acid, ethylenebis(benzenesulphonic acid), and *p,p'*-oxybis(benzenesulphonic acid) are known as gas-generating chemicals, but have not been used commercially.

An interesting group of liquid blowing agents are the *N*-nitroso-β-aminoketones, e.g., *N*-ethyl-*N*-nitrosodiacetoneamine (derived from mesityl oxide and ethylamine), and *N*-nitrosotriacetoneamine (derived from phorone and ammonia). The *N*-nitroso β-aminoketones decompose when heated with either alkali or acids with the evolution of nitrogen. In addition to the mononitroso compounds, *N*-dinitroso-β-aminoketones have been considered as blowing agents for polymers.

Since blowing agents are used exclusively in the preparation of expanded elastomers and plastics, their future is tied to new developments in cellular polymers. The wide acceptance of cellular polymers in industrial and consumer applications presages an increasing consumption of blowing agents in the future.

CHAPTER 31

Annealing of Plastics

INTRODUCTION

The annealing of plastics can be defined as a heat-treatment process aiming at the removal of strains and stresses set up in the material during its forming and fabricating operations. The plastic is brought up to a certain temperature, called the annealing temperature, kept at this temperature for a definite period of time, and then slowly cooled to room temperature.

In the plastics industry, annealing is restricted to thermoplastic polymers, either amorphous or crystalline. Generally, the main effect of annealing is to increase the density of the polymer thereby improving its heat resistance and dimensional stability when exposed to elevated temperature. Furthermore, annealing frequently improves the impact strength and prevents crazing and cracking of excessively stressed items. The magnitude of these changes depends on the nature of the polymer, the annealing conditions, and the past history of the specimen.

Amorphous, rigid polymers such as polystyrene and poly(methyl methacrylate) are more frequently annealed for stress relief than the other types of polymers. Annealing crystalline polymers, in addition to the usual stress relief, may also bring about profound changes in the nature of their crystalline state, which, in turn, is governed by the nature of the crystal structure, degree of crystallinity, size and number of spherulites, and orientation. In many cases, when proper temperature and pressures are maintained during the processing of polymers, the induced internal stresses may be insignificant, and additional heat treatment is not required. However, when the moulded parts are excessively stressed, or maximum dimensional stability and heat resistance are required, or when there is a need to enhance certain properties of the articles for specific applications, then annealing becomes necessary.

THEORETICAL BACKGROUND

Origin of internal stresses

In all forming operations, such as moulding and extrusion, the polymer flows into the mould and then solidifies. During this flow through the mould or die, shear stresses are developed, and a certain amount of orientation of the polymer chains takes place. If the polymer cools rapidly down to below its glass-transition temperature, molecular chains will essentially be frozen in a position of partial orientation. It is then said that the material contains internal or residual stresses and, associated with them, frozen-in strains.

Non-uniform cooling, which causes differential shrinkage of the material, also results in internal or residual stresses. The thick sections of the plastic cool slower, than the thin ones, so that pressure differentials are created which cause the plastic to flow from a thick to a thin section. If there is a wide variable in the thicknesses of different parts of a moulding, this internal flow may cause appreciable sink marks and frozen strains. Uneven cooling during moulding and the resulting non-uniform shrinkage also cause tensile stresses to develop in the surface of the moulding thus making the material susceptible to failure. Internal stresses can also be introduced during the machining operations of the moulded parts, i.e., during cutting, drilling, and sawing. These stresses act in a similar way to the frozen-in stresses set up during forming operations.

The origin of the internal stresses and frozen strains in the plastics lies in the viscoelastic nature of the polymers. According to the Maxwell model illustrating simple viscoelastic behaviour, the relaxation of stress at constant deformation is expressed by the exponential equation.

$$\tau = \tau_0 e^{-t/Trel}$$

in which τ is the shear stress at any time t, τ_0 is the initial shear stress within the material at time $t = 0$, and T_{rel} is the relaxation time, defined as the time at which the stress relaxes by the factor 1/e. The relaxation time is equal to η/G, η being the viscosity and G the modulus of elasticity. Since the viscosity decreases exponentially with increasing temperature, the relaxation time will rapidly decrease with increasing temperature, and consequently the stress relaxation will be very rapid.

The degree of stress relaxation is a controlling factor in resiliency, or the tendency of the polymer to return to its original shape. The relaxation or dissipation of the internal stress will cause the release of frozen strains within the plastic, which, in turn, will result in shrinkage and dimensional changes in the material. On the other hand, the presence of the internal stresses may cause the flow and deformation of the polymer at elevated temperatures. This also causes the internal stresses to dissipate.

At room temperature, which for amorphous polymers is usually much below the glass-transition temperature, the plastic is rigid and brittle and its viscosity is very high. Hence, stress relaxation cannot occur to a significant degree, and the presence of internal stresses may cause the part to fail at loads below those predicted from properties of the normal polymer. These residual internal stresses may also considerably lower the impact strength of the plastic or may cause severe crazing and cracking of the moulded part.

Crazing is the result of the development of a multitude of very small, almost infinitesimal platelike cracks in the plastic. This may occur in the presence of an organic liquid or its vapour, with or without the presence of applied mechanical stress. Crazing may also result from a tensile load acting on the plastic for a long time. These cracks appear on the surface of the plastic in a direction perpendicular to the stress, and apparently result from pulling apart small regions of the plastic under the action of stress. The tendency towards crazing and cracking is influenced by the value of the thermal coefficient of expansion and the thermal conductivity of the material.. The items that are subjected to thermal stresses due to alternate heating or cooling cycles, or to the chemical action of a specific environment, may crack at relatively low or even at no external stresses at all, if they contain high internal stresses.

Annealing Fundamentals

The fundamentals involved in the annealing of thermoplastic polymers can be most conveniently discussed by referring to amorphous and to crystalline polymers.

For typical amorphous polymers such as poly(methyl methacrylates), polystyrenes, and styrene–acrylonitrile co-polymers, the main effect of annealing is to change the density or specific volume. This in turn affects their mechanical behaviour, dimensional stability, and heat resistance. Fig. 31.1 illustrates the changes in the specific volume of an amorphous polymer with respect to temperature for different rates of cooling. The upper curve Q obtained by rapid cooling (quenching) results in material of a much greater specific volume (lower density) than that obtained by very slow cooling, curve S. The low density of the quenched polymer is the result of the frozen strains in the material. This represents an unstable state, and the polymer will tend to return to a more stable equilibrium state as represented by point C on curve S.

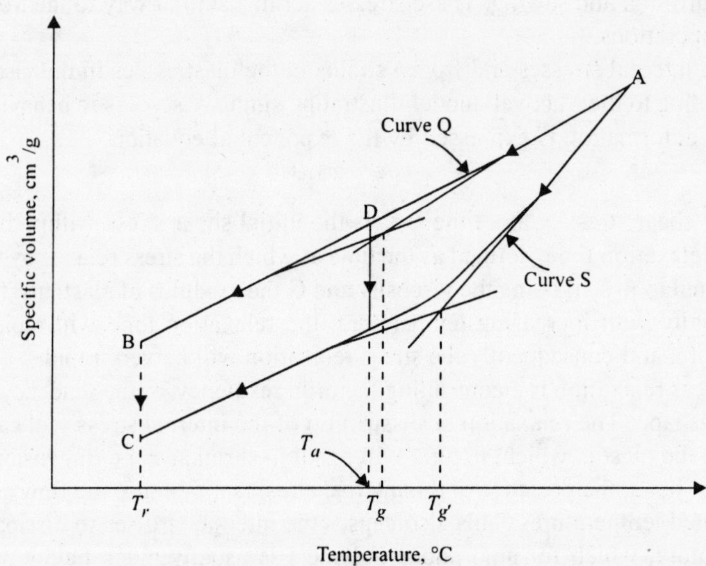

Fig. 31.1. Specific volume–temperature curves for an amorphous polymer. Curve Q results from quenching; curve S from slow (equilibrium) cooling. T_g and T_g' are the glass-transition temperatures of the polymer on quenching and slow cooling, respectively. DE represents the volume change on annealing the quenched polymer at the annealing temperature T_a, which is only slightly below the hypothetical glass-transition temperature T_g. Point B and C represent the specific volumes at room temperature T_r on quenching and slow cooling, respectively.

In the case of rigid amorphous polymers having a high glass-transition temperature, the return from point B to point C along the dashed line will take a very long time, and it is practically infeasible. At temperatures below the glass transition, the viscosity is very high and the movement of the polymer chains is restricted, so that large-scale molecular motion does not take place. However, some viscoelastic responses (which may cover a wide spectrum of relaxation times) exist due to the molar forces. This may account for the fact that upon very long storage at room temperature certain volume changes along the line BC may occur in the polymer.

When an amorphous polymer is heated, it changes at the glass-transition temperature from a hard and brittle material to a soft and rubbery one. This is usually accompanied by a change in the thermal coefficient of expansion, indicated by the changes in slope of the volume–temperature curves in the region of glass transition. The glass-transition temperature is affected by the cooling rate of the material. For slow cooling rates the glass-transition temperature increases, whereas for high cooling rates it decreases (Fig. 31.1). Consequently, the glass-transition temperature is intimately related to the free volume of the polymer, and the smaller the density of the polymer the smaller is its viscosity and the lower its glass-transition temperature. Since the glass-transition corresponds to the softening or heat-distortion temperature of the polymers, it is apparent that the quenched polymer specimen will have a lower softening point than the slowly cooled or annealed polymer. As a result the quenched polymer has a greater rate of stress relaxation, a smaller average relaxation time, and a higher creep rate than the slowly cooled or annealed polymer.

When a rapidly cooled polymer is exposed in service to elevated temperatures, the frozen-in strains will gradually be released. This is due to the greater thermal movement of the segments of the molecular chains, and causes a decrease in volume (increase in density) of the polymer. It is apparent that rapid recovery from strains, or rapid relaxation of stresses, can only occur at or a above the glass-transition temperature; at this temperature the mobility of the polymer chains becomes sufficiently great to accomplish quickly the necessary configurational changes. A considerable degree of stress relief can be secured at temperatures below the glass-transition temperature, but a much longer time is needed.

In most commercial, rigid, amorphous polymers, glass-transition temperatures occur between 40 and 150°C. At the glass transition, all amorphous materials show a certain characteristic value of viscosity equal to about 10^{13}P.

When the plastic part (which is usually rapidly cooled during moulding operations) is annealed, the volumetric changes occurring in the materials can be represented by the path DEC as shown in Fig. 31.1. The polymer is soaked at a temperature in the region of its glass transition for a sufficiently long time, during which its specific volume decreases along the line DE. At point E the polymer reaches the curve S, which represents the changes in the polymer upon very slow (equilibrium) cooling. The curve segment from point E to point C corresponds to the period of very slow cooling of the plastic after the annealing operation, has been performed.

If the specific volume–temperature curves for various rates of heating and cooling of a polymer were available from the experimental data, the time necessary for annealing to produce a desired density of the polymer at a certain temperature could be estimated from the relation

$$\frac{V_e - V}{V_e - V_i} = e^{-kt}$$

in which V_e is the specific volume at equilibrium cooling, V_i is the initial specific volume of the polymer, V is the desired specific volume of the polymer, k is the rate constant determined experimentally for an isothermal process of volume change with temperature, and t is the annealing time.

The annealing of crystalline polymers may bring changes in their degree of crystallinity, size of the crystallites, and their structural morphology. In polymers capable of a high degree of crystallinity, the internal structure changes markedly with decreasing temperature from the melting point down to a temperature where crystallinity approaches its maximum value.

When a crystalline polymer is cooled below its melting point, crystallisation does not occur immediately, but there is a period of time during which crystals begin to form. A molten polymer will crystallise at its maximum rate at a temperature equal to about nine tenths of its melting point on the absolute temperature scale. All crystalline polymers have branches or end groups, so that crystal imperfections must occur that result in certain amorphous segments between crystallites. The crystallites are homogenous single crystals of a polymer, are formed when the polymer chains are parallel and closely packed. Small crystallites are arranged in spherulites, but they also contain some amorphous material. The amount and degree of perfection of the crystallisation depends on the spatial relationship of the atoms in the molecules, the molecular weight, and the thermal treatment of the polymer.

Since most crystalline polymers contain ordered and disordered regions, they exhibit two characteristic temperatures: the melting point (usually defined as a temperature at which crystalline aggregates disappear) and the glass-transition temperature, but they maintain rigidity in the region between the glass temperature and the melting point due to the stiffening actions of crystallites. The glass-transition temperatures of most commercial crystalline polymers are usually between –100 and about +40°C. These temperatures are considerably lower than those of amorphous polymers. Because of the presence of some amorphous regions and different crystallite sizes, even highly crystalline polymers do not exhibit sharp melting points. For a carefully annealed polymer, most crystals melt approximately within a 5°C interval.

The crystalline structure of the polymer can be changed by the rate of cooling from the melt. Quenching a polymer by immersion in ice water may produce an amorphous polymer or one with small and imperfect crystallites of low crystallinity. On the other hand, slow cooling or annealing at a temperature slightly below the melting point tends to give a highly crystalline polymer with large spherulites. As the rate of cooling from the melt increases, the degree of crystallinity and the crystallite size decrease. Rapid quenching produces small spherulites, arranged at random, which give to the polymer a high transparency in thin sections. Slow cooling increases the spherulites's size and these scatter light much more than the very small crystallites, thereby causing the annealed polymer to appear cloudy or hazy. Crystallisation is accompanied by volume changes and consequently by changes in densities of the polymer. Since the crystallites have a higher density than the surrounding amorphous regions, the degree of crystallinity can be assessed from density values of the polymer. A direct consequence of the increasing density upon crystallisation is the shrinkage; for crystalline polymers this is usually much greater (1–4 percent than for amorphous (0.3–0.7 per cent polymers).

During the annealing of a crystalline polymer, deformed fibrillar crystals begin to melt, and recrystallisation of spherulites occurs. These spherulites rearrange themselves into a more ordered structure with a higher melting point. They grow thicker with time, but after a certain amount of time at a particular annealing temperature, no further change in density is observed. With an increased degree of crystallinity and crystallite size, the heat resistance and dimensional stability of the polymer increase.

Generally, annealing slightly improves the impact strength of crystalline polymers by lowering their brittleness temperature, as determined by ASTM method D 746-T. This is due more to the relaxation of internal stresses than to the morphological changes of the polymer structure. Experiments indicate that annealing of polypropylene having the melt index less then 1.5 improves its impact resistance at room temperature, but has little effect on its low-temperature properties. However, annealing of polypropylenes with the melt index greater than 1.5 has little or no effect on its impact resistance at room temperature (Table 31.1). The lower the melt index the higher the molecular weight of the polymer, the lower its brittleness temperature, and the greater its resistance to impact shock.

The brittleness temperature is usually lowered when tested within three hours after annealing. This is apparently due to the stress relief. Upon ageing at room temperature, the brittleness temperature as well as the density again increase, indicating that further crystallisation took place. This temporary improvement in the brittleness temperature, as well as in the impact resistance, has been observed after the annealing of slowly cooled specimens that are free from pronounced internal stresses.

Perhaps the most important practical change associated with annealing crystalline polymers is the reduction of the possible future shrinkage, which may occur on exposure to elevated temperatures encountered in service. On the other hand, annealing highly crystalline polymers such as high-density polyethylenes may increase the size of spherulites so that the polymer may become susceptible to environmental stress cracking. This latter phenomenon results in the failure of a highly crystalline specimen, when it is exposed to applied mechanical stress in the presence of an organic liquid, or an aqueous solution of a soap or other wetting agent. Thermal cracking may also occur when a high-density polyethylene is highly stressed and subjected to elevated temperatures.

Table. 31.1. Annealing conditions for various plastics.

Type of plastic and trade name	*Annealing medium*	*Annealing temperature, °C*	*Thickness, in.*	*Annealing time, hr*	*Remarks*
Polystyrene[a,b]	Air or water	60–70	Up to 1/4	1/2–1	General-purpose polystyrene
		70–77	Over 1/4	2–6	Heat-resistant polystyrene
Acrylic resins					
Lucite[c] 130	Air	60	Up to 1/8	2	Lucite 130 and 129 are general purpose type, Lucite 140 is heat-resistant type
Lucite[c] 129	Air	71	1/4	4	
Lucite[c] 140	Air	82	1/2	6	Each thickness and corresponding time for annealing applies to all types of Lucite listed
			1	8	
Plexiglas[d] G7 II	Air	80–110	Up-to 1/2	10	Formed parts may deform excessively at the annealing temperatures listed
Plexiglas[d] I-A	Air	70–90	1/2–3/4	11	
			1	11 1/2	Cast poly(methyl methacrylate) sheets are reheated to about 150°C and then cooled slowly
			4	22	Each thickness and corresponding time for annealing applies to both types of Plexiglas listed
Styrene-acrylonitrile co-polymer[b]	Air or water	77	Up to 1/8	1 min.	Heat-distortion temperature, ASTM D 648–45T, on 1/2 × 1/8 in. bar, in °C
			1/8–1/4	1–3 mn.	
			Ovcr 1/4	4–25 min	Unannealed 80–85
					Annealed 100–105
Acrylonitrile–butadiene–styrene (ABS) co-polymer					
Firestone X-2000[e]	Air	100	–	20 min	Heat-distortion temperature, ASTM D 648-56
					Unannealed 66 psi, 89°C
					12 psi, 97°C
					Annealed 66 psi, 95°C
					22 psi, 103°C
Cycloac brand[f]	Air	82	–	16	–
Polyamide					
Zytel[c] (nylon-6,6)	Oils or waxes	150	For each 1/8	1/4	For applications up to 70°C stress relief in boiling water; avoid the presence of air in the bath

(Contd ...)

Type of plastic and trade name	*Annealing medium*	*Annealing temperature, °C*	*Thickness, in.*	*Annealing time, hr*	*Remarks*
Fosta nylon[g] (nylon-6)	Glycowax S 932	130	Up to 1/2	1/4	Allow to cool in the bath to 55°C before removal
Nylatron GS[h] (nylon-6.6 containing dispersed MoS_2)	Gulf crest oil 196 or peanut oil	230	Up to 1/8	10 min, for maximum results 2 hr	Allow to cool slowly to 90°C in the bath before removal; bubble inert gas (N_2 or CO_2) through the bath
Polypropylene					
Profax[i]	Air or triethylene glycol	150	Up to 1/8 1/4	1/2–1 1	Recommended for the injection-moulded specimen; in the glycol bath only slight changes occur on annealing impact strength (120d), ft-lb per in. of notch, ASTM D 256–56
Firestone, X-3000[e]	Air	150		1	Temp, °C Unannealed Annealed 25 1.0 4.0 0 0.44 0.50 –10 0.34 0.47 –40 0.30 0.40
Polyethylene[j]					
Medium density 0.926–0.940 g/cm^3	Water	100	Up to 1/4 Over 1/4	15–30 min 1	If the melting point of the polymer is high enough, annealing is occasionally done in an autoclave at 15 psig for about 20 min[k]
Polychlorotrifluoroethylene					
Kel-F[l]	Air	125		24	Annealing at 175–195°C results in a highly crystalline polymer that is harder and less yielding
Polycarbonate resin					
Lexan[m]	Air or Gulf oil 205	125–130	per 0.1 in.	1	For thick sections a liquid bath is recommended; annealing is used mainly for machined parts or for extrusion mouldings of thick rods and plates.
Acetal resins					
Delrin[c]	Mineral oil, Primol D, or Nujol	160 150 ± 2	0.2–1 in. above 1in.	10–30 min. 1 hr/in.	Air-cool at room temperature in draft-free containers

[a] Koppers Co, Inc; [b] The Dow Chemical Co; [c] E. I. du Pont de Nemours and Co., Inc; [d] Rohm and Haas Co; [e] Firestone Plastic Co; [f] Marbon Chemical Div., Borg-Warner. Corp; [g] Foster Grant Co. Inc; [h] The Polymer Corp; [i] Hercules Powder Co; [j] Spencer Chemical Co; [k] Phillips Chemical Co; [l] Phillps Chemial Co; and [m] General Electric Co.

High-molecular-weight polymers show better resistance to stress cracking than the low-molecular-weight polymers, since their crystallisation rate is lower. For example, all highly crystalline polyethylene polymers (95 per cent crystalline) are susceptible to stress cracking, whereas propylene polymers (65–70 per cent crystalline) are not susceptible. Decreasing the density and increasing the molecular weight of polyethylene polymers considerably increases their resistance to stress cracking.

EQUIPMENT

The equipment used for annealing is essentially similar to that used for preheating thermoplastic sheets, rods, or tubes, prior to their shaping operations. Depending on the kind of annealing medium used, whether liquid or air, different designs of heating units are used.

When using liquid baths (water, oils, or waxes) for annealing, the equipment consists of tanks or troughs that should be of adequate size and shape to accommodate the parts to be annealed. To insure uniform heating and to maintain the required annealing temperature, a proper recirculating system and an adequate control of temperature should be provided. The temperature–control system should consist of a sensing element, a controller which interprets the information received from the sensing element, and a heating and/or cooling element regulated by the controller. Provision must be made for removing or recycling the liquid to prevent temperature build-up or localised hot spots in the system.

When using air as heating medium, air-flow ovens have proved to be most suitable. Basically, an air-flow oven consists of two compartments: one in which the air is heated to the required temperature either by gas or by electric heaters, and another compartment in which the material is heated by the hot air to the annealing temperature and soaked at this temperature for a required period of time. To avoid dead spaces or hot spots the hot air is rapidly circulated through the two compartments by means of adequate air compressors or blowers. The temperature must be very accurately controlled so that every part of the oven will have the required temperature and the material can be left in the oven for a considerable period of time without undue deterioration.

The other methods of heating involve the use of convention ovens, infrared ovens, and dielectric heating devices. However, the air-flow ovens are preferable, although they have relatively slow heating rates. For the annealing of large shapings, air-flow ovens are almost essential, as the temperature must be uniform over the entire shaping.

To secure uniform temperature throughout the oven space in circulating-air oven, careful design of racks and practical ways of handling the parts are very important. Furthermore, the clamping necessary for holding or suspending the parts in the oven should not exert too high a pressure, otherwise deformation of distortion or the parts will occur during annealing.

A proper design should permit free air circulation to avoid pockets of dead air where solvent vapours can accumulate. It should also provide for bleeding some part of the air and for making up fresh air so that the air in the oven is changed. This will prevent the accumulation of fumes and solvent vapours, if annealing of the cemented parts is being carried out.

After annealing, a proper method for very slow and gradual cooling has to be provided in order to avoid reintroducing stresses in the material which will defeat the purpose of the annealing. This may be accomplished in the following ways: (i) immediately packing the parts in a carton or in another material such as sawdust, to provide adequate insulation; and (ii) allowing the parts to cool in the air oven or in a liquid bath with the heating elements turned off.

Occasionally, as in the case of films, annealing is done on the hot rolls moving at very slow speed. This process is limited to thicknesses of 0.012 in. and over, since thinner materials will distort. In the blown-film extrusion operation, the film is cooled by air flowing through the 'annealing chamber' (a simple tube set between the die and the air ring). The cooling rate of the film is properly controlled so that the film surface can 'level-out' as opposed to cooling rapidly, which freezes in any surface roughness developed in the film during extrusion.

ANNEALING METHODS

The two most important variables affecting the annealing operation are the temperature and the time of annealing. These, however, do not only depend on the inherent properties of a particular polymer but are greatly influenced by moulding conditions, thermal history, size, thickness, weight, and shape of the parts to be annealed. Annealing operations will also vary greatly in technique, depending on the type of plastic and the purpose of its use.

Annealing Conditions

Theoretically, the most desirable annealing temperature for amorphous plastics would be above the glass-transition temperature. The stress relaxation and strain recovery would then be very rapid, and relatively short annealing times would be needed. However, such a temperature is not always practically feasible, since excessive distortion and warping of the plastic may occur. To anneal in the smallest amount of time possible, it is necessary to heat the plastic part to the highest temperature at which dimensional changes due to strain release are within permissible ranges. This temperatures is usually determined by exposing the assembly of the fabricated parts in an air oven or a liquid bath and increasing the oven temperature by steps of 3–5°C, until the greatest permissible change of shape or dimension occurs. This point is called the distortion temperature. It will vary with shape, thickness, weight, and size of the moulded part. The annealing temperature is then taken as about 5°C lower than the distortion temperature. This distortion temperature should not be confused with the heat-distortion temperature as determined by the ASTM standard method D 648 (264 or 64 psi).

The amount of time necessary for annealing at the annealing temperature will also vary with the thickness and geometry of the piece as well the amount of stress relief required. The annealing time frequently includes the time necessary for warming up the piece to a uniform temperature both at the surface and in the core. It is apparent that the thicker cross sections will require considerable longer annealing times than the thin ones. Usually, the optimum combination of the annealing temperature and the annealing time must be determined experimentally for a particular type of moulded part. Experience indicates that for the best results it is desirable to use somewhat lower annealing temperatures but longer annealing times.

The maximum permissible cooling rates after annealing vary with the thickness of the annealed parts. Thin sections can be cooled at a much higher rate, say, 60°C per hour, whereas thick sections are cooled at a very slow rate, 2–5°C per hour. The cooling rate should be very low in the neighbourhood of the glass-transition temperature, but below the glass transition cooling may proceed at a more rapid rate.

Annealing Media

The type of annealing medium also affects the time necessary for annealing. When liquid is used, relatively short annealing times are required, as the rate of heat transfer in a liquid bath is much greater than in the air. This, of course, is also affected by the thermal properties both of the liquid and of the annealed part. The other advantage of liquid as compared to air is that a more uniform distribution of temperature is achieved with no additional labour cost when the moulded part is immediately placed into the bath. There are several disadvantages to liquid media. Frequently, objectionable spotting occurs, which cannot be overcome without added cost. There is also the need for wiping when using oil, and difficulties in the handling of the parts. Although hot air is widely used, in many cases immersion in a suitable liquid gives better results. In fact, some plastics, such as nylon and acetal resins, cannot be annealed in hot air because oxidation may take place. Usually, they are annealed in an oil bath through which an inert gas is bubbled.

A suitable liquid for annealing should have an adequate heating range and stability, as well as complete inertness to the annealed plastic. The liquid should also be free of noxious fumes or vapours and should not present a fire hazard. Water is a very good annealing liquid, but it is limited to temperatures below its boiling point, which in many cases are too low. When annealing in water, a small amount of detergent is usually added to promote wetting of the part and rapid draining after removal. Actually, for maximum dimensional stability, the liquid bath should have a temperature 25–30°C above the temperature to which the parts will be exposed in service.

Effects of Annealing

The effectiveness and completeness of the annealing operation can be judged from the heat-distortion temperature test and from the results of the dip-solvent test. In fact, this latter test is also used to assess whether the moulded part requires annealing. The solvent test is accomplished by immersing the moulded pieces in a suitable solvent, or by keeping them wet with solvent by repeated strokes with a soft brush. Both the tested pieces and the solvent must be at room temperature.

Different solvents are used for different plastics. For example, moulded polystyrene articles are tested by immersion for one minute in either *n*-heptane or kerosene. The piece is removed without wiping, and observed at the end of 30 min. Highly stressed pieces develop extensive crazing or cracking, whereas unstressed pieces do not.

The solvent-crazing resistance of acrylic resins (e.g., Lucite) is determined in a solution of 5 per cent toluene in 2B denatured alcohol. If craze marks or cracks appear immediately after or within 30 sec. of immersion, the moulded part is considered to be in a highly stressed condition. If crazing occurs within 45–90 sec. after immersion, the part is relatively free from residual stresses. Polycarbonate parts are tested by immersion in carbon tetrachloride, and if the parts are highly stressed, cracking or crazing will occur immediately.

The solvent test is not only useful in determining the magnitude of residual stresses but it also helps to identify highly stressed regions of the moulded part. Another possible method is to determine the birefringence rating of a specimen between crossed polaroids. Such information, together with the heat-distortion data, can be used in modifying moulding conditions or making mould changes in order to reduce the magnitude and the distribution of the residual stresses. Properly formed parts should undergo no more than certain maximum dimensional changes when annealed at a selected temperature.

Annealing will eliminate any tendency of the plastic towards immediate crazing or cracking upon brief contact with solvents. Hence such plastic parts as polystyrene or poly(methyl methacrylates), which are to be joined by adhesive cements or are to be painted, are annealed prior to application. If the cemented or painted parts show some crazing due to certain solvents present in the adhesives or paints, they should be annealed after each application.

In many cases, fabricated parts can be obtained in a reasonably strain-free conditions provided attention is paid to details in the fabricating procedures. This can be achieved by selecting a suitable temperature and pressure for processing, extruding, moulding, or calendering the product, followed by cooling at a predetermined rate. This minimises the introduction of internal stresses, so there is no need for a special annealing operation.

Some practices in forming operations include a controlled heat treatment right on the line. This can be considered as a kind of annealing operation. For example, in extruding such cellulosic polymers as cellulose acetate–butyrate or propionate, even and uniform cooling of extrudates must be observed otherwise brittle products may result. Extruded cellulose acetate-butyrate pipes are blown with warm air at 80°C to prevent rapid cooling. Extruded rods are also gradually cooled in troughs containing water of different temperatures. The first section of a trough (about 20 ft long) contains water at 70–99°C, whereas the other two sections are maintained at 70 and 32°C, respectively. Vinyl and polyethylene parts are also often extruded into a water bath having closely controlled temperature zones. It is apparent that such processing methods already include a kind of annealing operation.

In a similar way, the cooling rate for calendered or extruded film and sheeting must be controlled. Cellulosic films or sheets extruded horizontally are passed through a series of chrome–plate steel rolls heated to different temperatures. The fist roll is held at about 65°C, whereas the second roll is held at about 80–100°C, the subsequent rolls are not heated. As the second 'cooling' roll is warmer that the first, any strains previously induced in the sheet are removed. In another example, poly(vinyl chloride) film is passed over a 138–145°C hot roll at a very slow speed of 3–4 yd/min to remove any strains which may later cause shrinkage.

In the above examples the film should not be stretched while it is hot. For some applications, however, such as in the growing field of shrink packaging, it is desirable to put strains in the film which are then relieved by annealing, thus causing the film to shrink.

In making filaments from polystyrene, the material is quenched in an ethylene glycol or glycerol bath at a temperature over 100°C, for the purpose of preserving a tough structure and cooling it enough to wrap the filament firmly without distortion around the rolls running at different speeds. Subsequently, the filaments are warmed while held firm to prevent shrinking. After orientation, the filaments pass through a liquid bath (water or glycerine) held at a proper temperature (e.g., about 132°C for the polystyrene filament). After leaving the annealing bath, the filament is wound on another pair of godet rolls running at the same speed as the pair just before it, so no further stretching is done. Another annealing method is to wind the filaments around a huge frame and place them in an air-circulating oven for about 10 min. at about 135°C.

When preparing moulded specimens for testing purposes, annealing is frequenlty used to obtain samples with uniform properties, or of desired density or crystalline structure. The standardised procedure involves cooling the specimens from the melt stage in an oven at a rate of 5°C per hour, as described in detail by ASTM method D 1928-62T.

Thermosetting plastics are usually not subjected to annealing operations because of their cross-linked, three-dimensional structure. The internal stresses in moulded thermosetting parts are frequently due to differential shrinkage in adjacent sections, rather than to cured-in flow stresses. These stresses can be minimised by keeping sections uniformly thick or by providing gradual changes in thicknesses. Occasionally, although not often, heat treatment is used for certain moulded thermosetting plastics to improve either their electrical properties, their modulus of elasticity, or their dimensional stability. At the Durez Plastic Company, for example, certain phenolic parts are annealed for stress relief in air-circulating ovens, at 121°C for 4 hrs for thin sections, and at 105°C for 24 hrs for thick sections. The Fiberite Corporation subjects moulded phenolics to an afterbaking operation with temperatures up to 150°C for up to 24 hrs.

The present technology of annealing various plastics is still in its development stage, particularly when new types of plastics are involved. Although the general principles involved in annealing are common for each family of polymeric materials, the specific annealing conditions for each particular plastic part must be determined in the laboratory or in the pilot plant, and must take into consideration the available fabrication techniques as well as the end use of the plastic part. This results in the great variety of annealing techniques used in the industry by different plastics manufacturers.

Table 31.1 summarises the annealing conditions for various plastics. It should be emphasised that these data should be regarded as suggestions for starting the proper experimental procedure to determine the specific working conditions for annealing a particular plastic part.

Economic Aspects

Annealing is an additional operation in the fabrication procedures of the plastic parts, and adds to the cost of the produced articles. For this reason, annealing is considered only if there is evidence of the part's failure due to the presence of res idual internal stresses, or due to objectionable dimensional changes which may occur under service conditions. Annealing is also used when there is need to enhance certain desired properties of the moulded parts for specific applications.

Every effort is being made by the fabricator to prevent the introduction of excessive stresses during moulding operations, through the use of a correct die design and adequate forming temperatures and pressures in fabrication operations. With the development of the newer and faster processing methods in moulding rigid plastics, and the continuous growth and expansion of plastics in numerous and varied applications, one can reasonably expect that annealing will play an increasingly more important role in the plastics technology than it does now.

Glossary

Abrasion loss. Volume of rubber abraded from a specified test piece under specified conditions.

Abrasion resistance. Resistance to wear resulting from frictional action upon a surface. It is usually expressed by the abrasion resistance index.

Accelerated ageing test. Test in which an attempt is made to produce the effects of natural ageing in a shorter time by increasing the rate of degradation.

Accelerator. Compounding ingredient used in small amounts with a vulcanising agent to increase the speed of vulcanisation and/or enhance certain properties of the vulcanisate.

Ageing. Irreversible change of material properties after exposure to an environment for a period of time.

Agglomeration. Reversible or irreversible joining together of latex particles.

Air oven ageing test. Test in which ageing proceeds in an oven in air at an elevated temperature and at atmospheric pressure in the absence of light.

Air trap. Void in a cellular polymer.

Anklet rubber. Rubber interlayer to reinforce the lower leg.

Antioxidant. Compounding ingredient used to retard deterioration caused by oxidation.

Bale cutter. Press in which a bale of raw rubber is forced against a knife, so cutting it into pieces for more convenient handling.

Banbury. An apparatus for compounding materials composed of a pair of contra-rotating rotors that masticate the materials to form a homogeneous blend. This is an internal-type mixer that produces excellent mixing.

Band pocket. Pre-assembly of some of the plies of the tyre carcass.

Bareness. Defect resulting from the failure of the rubber to fill out all the mould pattern detail.

Barrier plastics. A general term applied to a group of lightweight, transparent, and impact-resistant plastics, usually rigid co-polymers of high acrylonitrile content. The barrier plastics generally are characterised by gas, aroma, and flavour barrier characteristics approaching those of metal and glass.

Belt. Layer of material underneath the tread, laid substantially in the direction of the tread centre line, that restricts the carcass in a circumferential direction.

Bin curing. Unintentional vulcanisation of a mix during storage.

Blank slug. Piece of mix of suitable volume to fill the mould.

Blender. Vessel fitted with movable blades, to convert the contents, usually compounding ingredients, into a uniform, homogeneous mixture.

Blow pressure. The air pressure required to form the parison into the shape of the mould cavity in a blow moulding operation.

Blowing agent. A substance that alone or in conjunction with other substances is capable of producing a cellular structure in a plastic mass.

Blowing down. Removal of excess ammonia from latex by stirring the latex while passing a stream of air across the surface.

Blowing. Production of a cellular material by decomposition of an added ingredient.

Bolster. Space or filler in a mould.

Bound rubber. Portion of the rubber polymer in a mix that is so closely associated with compounding ingredients as to be insoluble in the usual rubber solvents.

Breakdown. Preliminary softening, by mechanical work, of raw rubber or a mix to render it more suitable for masticating, mixing or further processing.

Breaker. Intermediate ply between carcass and tread.

Breathing. The opening and closing of a mould to allow gases to escape early in the moulding cycle. Also called degassing. When referring to plastic sheeting, breathing indicates permeability to air.

Bulk density. Mass per unit volume of material including both open and closed cavities normally associated with the material.

Bulk moulding compound (BMC). Thermosetting resins mixed with stranded reinforcement, fillers, etc., into a viscous compound for injection or compression moulding.

Bump breathe, deprecated. Momentarily to open and close a mould at an early stage in the moulding process to expel gas.

Burst pressure test. Destructive pressure test of ultimate strength of a hose.

Calender. To prepare sheets of material by pressure between two or more counter-rotating rolls. Also, the machine performing this operation.

Carbon black. A black pigment produced by the incomplete burning of natural gas or oil. It is widely used as a filler, particularly in the rubber industry. Because it possesses useful ultraviolet protective properties, it is also used in moulding compounds intended for outside weathering applications.

Carcass casing. Rubber-bonded cord structure of a tyre integral with the bead, which contains the inflation pressure.

Cast. To form a 'plastic' object by pouring a fluid monomer-polymer solution into an open mould where it finishes polymerising.

Cavity. Portion of a mould that usually forms the outer surface of the moulded part. Depending on the number of such depressions, moulds are designated as a single-cavity or multi-cavity.

Cellular plastic. A plastic whose density is decreased substantially by the presence of numerous cells disposed throughout its mass.

Centrifugal casting. A method of forming thermoplastic resins in which the granular resin is placed in a rotatable container, heated to a molten condition by the transfer of heat through the walls of the container, and rotated so that the centrifugal force induced will force the molten resin to conform to the configuration of the interior surface of the container. It is used to fabricate large-diameter pipes and similar cylindrical items.

Centrifuged latex. Latex, the rubber concentration of which has been increased by the removal of serum by centrifugal force.

Coagulant. Substance used for causing coagulation.

Coagulation. Irreversible agglomeration of particles originally dispersed in a rubber latex to form a continuous phase of the rubber and a dispersed phase of the serum.

Coagulum. Immediate product of coagulation of latex consisting of agglomerates of rubber particles.

Coextrusion. Process of combining two or more layers of extrudate to produce a multiple-layer product in a single step.

Coinjection. Technique of injecting two materials into a single mould from two plasticating cylinders, either simultaneously or in sequence.

Cold cure. Vulcanisation at room temperature, commonly by means of sulphur monochloride solution or vapour.

Cold moulding. Procedure in which a composition is shaped at room temperature and cured by subsequent baking.

Cold parison blow moulding. Technique in which parisons are extruded or injection moulded separately and then stored for subsequent transportation to the blow moulding machine for blowing.

Cold runner moulding. Process in which sprue and runner system (the manifold section) is insulated from the rest of the mould and temperature-controlled to keep the plastic in the manifold fluid. This mould design eliminates scrap loss from sprues and runners.

Cold slug. The first material to enter an injection mould, so named because in passing through the sprue orifice it is cooled below the effective moulding temperature.

Compound. The plastic material to be moulded or blown into final form. This includes the resin itself, along with modifiers, pigments, antioxidants, lubricants, etc., needed to process the resin efficiently and to produce the desired properties in the finished article.

Compression moulding. Moulding process in which the blank is placed directly in the mould cavity and compressed to shape by closure of the mould.

Compression ratio. In an extruder screw, the ratio of volume available in the first flight at the hopper to that in the last flight at the end of the screw.

Compression set. Residual deformation of a test piece after removal of the compressive stress. Numerically, it is expressed as the percentage ratio of the residual deformation to the initial dimension of the test piece or to the compression strain.

Continuous vulcanisation/Continuous cure. Vulcanisation by the continuous passage of extruded, calendered or spread material over or through a unit which is capable of rapid and complete heat transfer (e.g., liquid or salt bath, fluid bed, microwave oven, rotary cure, hot air cure, steam cure).

Cord rubber. Soft rubber in which the tension element of a V-belt is fully embedded.

Cord. Consists of a number of strands of textile or metallic filaments twisted together and used as a reinforcement in rubber products such as tyres, belts and hoses.

Cover. External protective layer of polymeric material of a hose.

Crepe. Time-dependent change in dimensions resulting from an applied stress.

Cross-linking. The establishment of chemical bonds between the molecular chains in polymers. Cross-linking can be accomplished by chemical reaction, vulcanisation, degradation, or radiation.

Crown angle. Angle between tyre circumferential line and line of cords in the tyre at the crown of the tyre.

Crown. Road-contacting area lying between the shoulders of a tyre.

Cure. To change the physical properties of a material by chemical reaction, which may be condensation, polymerisation, or vulcanisation; usually accomplished by the action of heat and catalysts, alone or in combination, with or without pressure.

Curemeter. Instrument designed to determine the increase in stiffness of a vulcanisable rubber mix during heating. The instrument is used for determining the appropriate time for vulcanisation or for assessing the tendency to premature vulcanisation.

Curing temperature. Temperature at which a cast, moulded, or extruded product, a resin-impregnated reinforcing material, an adhesive, or other material is subjected to curing.

Curing time. In the moulding of plastics, the interval of time between the instant of cessation of relative movement between the moving parts of a mould and the instant that pressure is released. Also called moulding time.

Cycle. The complete, repeating sequence of operations in a process or part of a process. In moulding, the cycle time is the period, or elapsed time, between a certain point in one cycle and the same point in the next.

Deflashing. Any technique or method that removes excess, unwanted material from a moulded article. Specifically, the excess material is removed from those places on the article where parting lines of the mould that formed the article may have caused the excess material to be formed.

Deflection temperature. The temperature at which a specimen will deflect a given distance at a given load under prescribed conditions of test.

Degradation. A deleterious change in the chemical structure or physical properties of a plastic, caused by exposure to heat, light, or other agent.

Deliquescent. Capable of attracting moisture from the air.

Diaphragm gate. Gate used in moulding annular or tubular articles. The gate forms a solid web across the opening of the part.

Die. Detachable part of an extruder producing the profile of an extrudate.

Dielectric constant. (permittivity or specific inductive capacity) The ratio of the capacitance of an assembly of two electrodes separated solely by a plastics insulating material to its capacitance when the electrodes are separated by air.

Dielectric strength. The electric voltage gradient at which an insulating material is broken down or 'arced through,' in volts per mil of thickness.

Dimensional stability. Ability of a plastic part to retain the precise shape in which it was moulded, fabricated, or cast.

Dipping. Process for manufacturing seamless articles by dipping a former into a rubber solution or latex.

Drape forming. Method of forming thermoplastic sheet in which the sheet is clamped into a movable frame, heated, and draped over high points of a male mould. Vacuum then is pulled to complete the forming operation.

Drawing. The process of stretching a thermoplastic to reduce its cross-sectional area, thus creating a more orderly orientation of polymer chains with respect to each other.

Drive ability. Ability of a properly tensioned belt to function without undue slip on the pulleys when operating at its recommended maximum effective belt tension.

Dry blend. Term applied to a moulding compound, containing all necessary ingredients, mixed in a way that produces a dry-free flowing, particulate material. This term commonly is used in connection with polyvinyl chloride moulding compounds.

Durometer hardness. The hardness of a material as measured by the Shore Durometer.

Elastic deformation. The part of the deformation of an object under load that is recoverable when the load is removed.

Elasticity. That property of a material by virtue of which it tends to recover its original size and shape after deformation. If the strain is proportional to the applied stress, the material is said to exhibit Hookean or ideal elasticity.

Elastomer. Macro-molecular material that returns rapidly to approximately its initial dimensions and shape after substantial deformation by a weak stress and release of the stress.

Electroforming. Mould making method where by a thin layer of metal is deposited onto a pattern. Molten metal then may be sprayed on the back of the mould to increase its strength.

Electroplating. Deposition of metals on certain plastics and moulds for finish.

Elongation. Increase in length of a test piece produced by a tensile stress. This property is usually expressed as a percentage of the original test length.

Embossing. Techniques used to create depressions of a specific pattern in plastics film and sheeting. Such embossing in the form of surface patterns can be achieved on moulded parts by the treatment of the mould surface by photoengraving or another process.

Environmental conditioning. Storage of a test piece under specified conditions such as temperature, humidity, etc. for a specified time prior to testing.

Environmental stress cracking (ESC). The susceptibility of a thermoplastic article to crack or craze when stressed, in the presence of surface-active agents or in other environments.

Exotherm. (i) Temperature/time curve of a chemical reaction giving off heat, particularly the polymerisation of casting resins. (ii) The amount of heat given off. The term has not been standardised with respect to sample size, ambient temperature, degree of mixing, etc.

Extender. Organic material used as a replacement for a portion of the raw rubber required in a mix or for blending with the raw rubber (e.g., in oil-extended rubbers).

Extrusion coating. Process in which a resin is coated on a substrate by extruding a thin film of molten resin and pressing it onto or into the substrates, or both, without the use of an adhesive.

Extrusion. Process of compacting and melting a plastic material and forcing it through an orifice in a continuous fashion. Material is conveyed through the heated machine barrel by a helical screw or screws, where it is heated and mixed to a homogeneous state and then forced through a die of the shape required for the finished product.

Fatigue. Permanent deterioration in mechanical properties in a material subjected to fluctuating stresses and strains.

Filler. Inert substance added to a plastic to make it less costly. However, fillers may also improve physical properties, particularly hardness, stiffness, dimensional stability, and impact strength.

Flame spraying. Method of applying a plastic coating in which finely powdered fragments of the plastic, together with suitable fluxes, are projected through a cone of flame onto a surface.

Flash gate. Usually a long gate extending from a runner that runs parallel to an edge of a moulded part along the flash or parting line of the mould.

Flash line. A raised line appearing on the surface of a moulding and formed at the junction of mould faces.

Flash mould. A mould in which the mould faces are perpendicular to the clamping action of the press, so that the higher the clamping force, the tighter the mould seam.

Flash. Extra plastic attached to a moulding along the parting line; under most conditions it would be objectionable and must be removed before the parts are acceptable.

Flexible moulds. Moulds made of rubber or elastomeric plastics used for casting plastics. They can be stretched to remove cut pieces with undercuts.

Flocculant. Substance used for causing flocculation.

Flow Marks. Marks or lines on a moulded product, caused by imperfect fusion of flowing fronts.

Foamed plastics. Plastics with internal voids or cells. The foam may be flexible or rigid, the cells closed or connected, and the density anything from slightly below that of the solid parent resin down to, in some cases, 1 lb/ft cubed, or less.

Forming. A process in which the shape of plastic pieces such as sheet, rods, or tubes is changed to a desired configuration.

Friction welding. A method of welding thermoplastics materials whereby the heat necessary to soften the components is provided by friction.

Frost line. In the extrusion of polyethylene lay flat film, a ring shaped zone located at the point where the film reaches its final diameter. In this zone, the film has a "frosty" appearance caused by the film's temperature falling below the softening range of the resin.

Fusion bonding. Process for joining plastic parts where mating surfaces are heated to melting and held together under pressure. Also known as hot plate welding.

Gauge. Thickness of plastic film or sheet.

Gel rubber. Portion of raw rubber insoluble in a specified solvent.

Glass transition. Second order transition between a brittle glassy condition and a rubber like condition. The midpoint of the temperature range over which this transition takes place is commonly called the 'glass transition temperature'.

Granular structure. Non-uniform appearance of finished plastic material due to retention of, or incomplete fusion of, particles of composition, either within the mass or on the surface.

Granulator. Machine used for size reduction of plastic scrap for reuse. Also called grinder.

Grown tyre. Tyre subjected to expansion or growth due to service conditions.

Hardness. Resistance to indentation.

Haze. Cloudiness in plastic film.

Heat sensitiser. Gelling agent effective only at elevated temperature.

Homopolymer. The result of the polymerisation of a single monomer, a polymer that consists of a single type or repeating unit.

Hot air cure. Vulcanisation in heated air.

Hot runner mould. A thermoplastic injection mould in which the runners are insulated from the chilled cavities and remain hot so that the center of the runner never cools in normal cycle operation.

Hot water cure. Vulcanisation in hot water or aqueous solution.

Hysteresis. Ratio of energy lost to energy input in the rapid or instantaneous recovery of a deformed test piece.

Impact resistance/impact strength. Resistance to fracture under mechanical shock force.

Indentation hardness. Force required to produce a specified indentation in the cellular material under specified conditions.

Initiator. Peroxide used as source of free radicals. These substances are used in free radical polymerisations, in curing thermosetting resins, as crosslinking agents for elastomers and polyethylene, and for polymer modification.

Injection moulding. Moulding process in which a predetermined quantity of mix is forced into a closed headed mould from a separate heating chamber.

Insert. An integral part of a plastic moulding consisting of metal or other material that may be moulded into position or may be pressed into the part after the moulding is completed.

Insulating ply. Layer of rubber between braided or spiralled plies of reinforcement.

Iridescence. Loss of brilliance in metallised plastics and development of multicolour reflectance. It is caused by cold flow of plastic or coating and from extra heat during vacuum metallising.

Irradiation (atomic). As applied to plastics, term that refers to bombardment with a variety of subatomic particles, generally alpha-, beta-, or gamma-rays. Atomic irradiation has been used to initiate polymerisation and copolymerisation of plastics and in some cases to bring about changes in the physical properties of a plastic material.

Laminar flow. Flow of thermoplastic resins in a mould that is accompanied by solidification of the layer in contact with the mould surface, which acts as an insulating tube through which material flows to fill the remainder of the cavity. This type of flow is essential to duplication of the mould surface.

Latex. Colloidal aqueous dispersion of a polymeric substance.

Low-profile resins. Designation applied to special polyester resin systems for reinforced plastics. These systems are combinations of thermoset resins and thermoplastic resins used to minimise surface waviness in moulded parts.

Masterbatch. A plastics compound that includes a high concentration of an additive or additives. Masterbatches are designed for use in appropriate quantities with the basic resin or mix so that the correct end concentration is achieved. For example, colour masterbatches for a variety of plastics are used extensively, as they provide a clean and convenient method of obtaining accurate colour shades.

Mat. A fabric or felt of glass or other reinforcing fibre used in manufacturing plastic composite parts.

Mechanical conditioning. Prescribed programme of deformation of a test piece prior to testing.

Mechanically foamed plastic. A cellular plastic whose structure is produced by physically incorporated gases.

Memory. The tendency of a plastic article to return to a size and shape that existed during the manufacturing process.

Migration of plasticiser. Loss of plasticiser from an elastomeric plastic compound with subsequent absorption by an adjacent medium of lower plasticiser concentration.

Mix compound. Intimate mixture of a polymer or polymers with all the ingredients necessary for the finished article.

Mixer. Machine used for mixing together polymers and compounding ingredients.

Modulus of elasticity. The ratio of stress to strain in a material that is elastically deformed.

Molecular weight distribution. A measure of the relative amounts of polymers with different molecular weights within a batch of material. This measure may be indicated by the ratio of the weight-average molecular weight.

Monomer. A relatively simple compound that can react to form a polymer (i.e., polymerise)

Mould finish. Surface finish of a mould.

Mould lubricant. Material applied to a mould to prevent adhesion of the rubber to the mould.

Moulded product. Object produced in a closed mould (e.g. by compression moulding, injection moulding or transfer moulding).

Moulding material. Plastic material in varying stages of granulation, often compressing resin, filler, pigments, plasticisers, and other ingredients, ready for use in the moulding operation. Also called moulding compound or powder.

Moulding process. Process of shaping a material within a mould by applying pressure and usually heat.

Moulding shrinkage. Difference in dimensions between a moulded product and the mould cavity in which it was moulded, both the mould and the moulded product being at normal temperature when measured.

Narrow filling. Material under the insole to fill any cavity between insole and outsole.

Nucleating agent. A chemical substance that provides sites for crystal formation in polymer melts.

Open cell foam. A cellular plastic in which there is a predominance of interconnected cells.

Open steam cure. Vulcanisation, without a mould in steam.

Orientation. The alignment of the crystalline structure in polymeric materials so as to produce a highly uniform structure. It can be accomplished by cold drawing or stretching during fabrication.

Orifice. The opening in the extruder die formed by the orifice bushing (ring) and mandrel.

Peak cure. Vulcanisation during which the value of a major property proceeds to a maximum or minimum and then changes rapidly.

Peptiser. Compounding ingredient used in small amounts to accelerate by chemical action the permanent softening of raw rubber under the influence of mechanical action and/or heat.

Permanence. Resistance of a plastic to appreciable changes in characteristics with time and environment.

Permeability. The passage or diffusion of a gas, vapour, liquid, or solid through a barrier without being physically or chemically affected.

Plastic deformation. A change in dimensions of an object under load that is not recovered when the load is removed; opposed to elastic deformation.

Plastic, semirigid. A plastic with a stiffness or apparent modulus of elasticity between 10,000 and 100,000 psi at 23^{0}C.

Plastic. One of many high-polymeric substances, including both natural and synthetic products, but excluding the rubbers. At some stage in its manufacture, every plastic is capable of flowing, under heat and pressure if necessary, into the desired final shape.

Plasticate. To soften by heating or kneading. Synonyms are plastify, flux, and (imprecisely) plasticise.

Plasticise. To soften a material and make it plastic or mouldable, by means of either a plasticiser or the application of heat.

Plasticiser. A material incorporated in a plastic to increase its workability and its flexibility or distensibility; normally used in thermoplastics. The addition of the plasticiser may lower the melt viscosity, the temperature of the glassy transition, or the elastic modulus of the plastic.

Plasticity number. Measure of plasticity that varies inversely with the ease of deformation.

Plasticity retention index (PRI). Measure of the resistance of raw natural rubber to oxidation. It is expressed as the percentage ratio of the plasticity measured after air oven ageing for 30 min. at 140°C to the plasticity measured before ageing using a parallel plate plastimeter.

Plasticity. A property of a plastic that allows the material to be deformed continuously and permanently without rupture upon the application of a force that exceeds the yield value of the material.

Plastisols. Mixtures of vinyl resins and plasticisers that can be moulded, cast, or converted to continuous films by the application of heat. If the mixtures contain volatile thinners also, they are known as organosols.

Plastometer. Instrument for measuring plastic flow.

Platens. The mounting plates of an injection or compression moulding press, to which the entire mould assembly is bolted.

Plucking. Tearing out of isolated portions from a rubber surface during separation from a second surface.

Plunger. That part of a transfer or injection press that applies pressure on the unmelted plastic material to push it into the chamber, which in turn forces plastic melt at the front to the chamber out the nozzle.

Ply. Layer of rubber-coated parallel cords.

Polymer. A high molecular weight organic compound, natural or synthetic, whose structure can be represented by a repeated small unit, the monomer (e.g., polyethylene, rubber, cellulose). Synthetic polymers are formed by addition or condensation polymerisation of monomers. If two or more different monomers are involved, a copolymer is obtained. Some polymers are elastomers some, plastics.

Polymerisation. A chemical reaction in which the molecules of a monomer are linked together to form large molecules whose molecular weight is a multiple of that of the original substance. When two or more different monomers are involved, the process is called copolymerisation or heteropolymerisation.

Porosity. As applied to a defect, the accidental presence of numerous small cavities.

Postcure. Operation whereby thermoset parts are subjected to elevated temperatures for a period of time after being removed from the mould to attain maximum property levels.

Pot mould. Mould having a jacket through which a fluid may be circulated for controlling temperature.

Potting. A procedure similar to encapsulating except that here steps are taken to ensure complete penetration of all the voids in the object before the resin polymerises.

Preheating. The heating of a compound prior to moulding or casting in order to facilitate the operation, reduce the cycle, and improve the product.

Preplastication. Technique of premelting injection moulding powders in a separate chamber, then transferring the melt to the injection cylinder. The device used for preplastication commonly is known as a preplasticiser.

Prepolymer moulding. In the urethane foam field, term used for a system whereby a portion of the polyol is pre-reacted with the isocyanate to form a liquid prepolymer in a viscosity range suitable for pumping or metering. This component is supplied to end-users with a second premixed blend of additional polyol, catalyst, blowing agent, etc. When the two components are mixed together, foaming occurs.

Prepolymer. A chemical structure intermediate between that of the initial monomer or monomers and the final polymer or resin.

Prepreg. A term generally used in reinforced plastics to mean the reinforcing material containing or combined with the full complement of resin before moulding.

Profiling machine. Machine with two or more rolls of which at least one carries one or more circumferential grooves for converting rubber into strip having a predetermined contour.

Pultrusion. Automated method for producing continuous reinforced plastic shapes by pulling preimpregnated reinforcing fibres through a heated die where the resin is cured.

Recycle. Material from flash, trimmings, scrap, rejects, etc., that can be ground up or repelletised and fed back into the processing machine.

Reinforced plastics. Plastics with some strength properties greatly superior to those of the base resin, resulting from the presence of high strength fillers embedded in the composition. The reinforcing fillers usually are fibres fabrics, or mats made of fibres.

Reinforcing filler. Filler having a specific reinforcing effect, e.g., the ability to enhance abrasion resistance and mechanical strength properties.

Remoulding. Process for replacing the tread, shoulder and sidewalls of a worn tyre.

Resilience. Ratio of energy output to energy input in a rapid or instantaneous recovery of a deformed test piece.

Resin, liquid. An organic, polymeric liquid that, when converted to its final state for use, becomes a solid.

Retarder. Compounding ingredient used in small amounts to reduce the tendency of a mix to undergo vulcanisation prematurely.

Retreading. Generic term for reconditioning used tyres which covers replacement of the tread rubber only (shoulder to shoulder retreading, 'top capping' or 'recapping').

Rheology. The study of material flow under varying conditions of heat and pressure.

Rubber hose. Flexible pipe made of rubber with a reinforcement, generally textile or metallic.

Rubber polymer. Polymer or polymers constituting the basis of a rubber.

Rubber. A solid material with elastic properties made from latex of living plants, or synthetically, and used in the manufacture of rubber products.

Sample. One or more items taken at random from a lot and intended to provide information on the lot and possibly to serve as a basis for a decision on the lot or on the process that had produced it.

Scorch. Premature vulcanisation of a rubber mix.

Segregation. A separation of components in a moulded article, usually denoted by wavy lines and colour striations in thermoplastics. In thermosets, it usually means segregation of the resin and the filler on the surface.

Slip forming. Sheet-forming technique in which some of the plastic sheet material is allowed to slip through the mechanically operated clamping rings during a stretch-forming operation.

Slot extrusion. A method of extruding film sheet in which the molten thermoplastic compound is forced through a straight slot.

Slurry preforming. Method of preparing reinforced plastics preforms by wet processing techniques similar to those used in the pulp moulding industry.

Slush moulding. Method for casting thermoplastics, in which the resin in liquid form is poured into a hot mould where a viscous skin forms. The excess slush is drained off, the mould is cooled, and the moulding is stripped out.

Softener. Compounding ingredient used in small proportions to soften the vulcanisate or to facilitate processing or incorporation of fillers.

Softening range. The range of temperature in which a plastic changes from a rigid to a soft state. Actual values will depend on the method of test. Sometimes erroneously referred to as softening point.

Solvent moulding. Process for forming thermoplastic articles by dipping a male mould in a solution or dispersion of the resin and drawing off the solvent to leave a layer of plastic film adhering to the mould.

Spew flesh. Surplus material forced from a mould on closure under pressure.

Spew line. Line on the surface of a moulded product at the junction of the mould parts.

Spreader. Machine for distributing rubber dough or latex on the surface of sheet material by means of a blade.

Sprue gate. A passageway through which molten resin flows from the nozzle to the mould cavity.

Sprue. Feed opening provided in the injection or transfer mould; also, the slug formed at this hole. Spur is a shop term for the sprue slug.

Stabiliser. An ingredient used in the formulation of some plastics to assist in maintaining to physical and chemical properties of the compounded materials at their initial values throughout the processing and service life of the material.

Steam moulding. Process used to mould parts from pre-expanded beads of plastic, especially polystyrene using steam as a source of heat to expand the blowing agent in the material. The steam in most cases makes intimate contact with the beads directly, or it may be used indirectly to heat mould surfaces that are in contact with the beads.

Stock blender. Device comprising one or more driven rollers mounted above and used in conjunction with a mill to improve blending of a mix.

Strainer. Machine designed to force rubber through metal gauze screens to remove extraneous matter.

Stretch blow moulding. A blow moulding process in which the parison is stretched and blown at controlled temperature to cause orientation of the polymer molecules.

Sweating. Exudation of small drops of liquid, usually a plasticiser or softener, on the surface of a plastic part.

Take off. The mechanism for drawing extrude material away from the die.

Tear strength. Maximum force required to tear a specified test piece, the force usually acting substantially parallel to the major axis of the test piece.

Tensile strength. Maximum tensile stress applied during stretching a test piece to rupture.

Tensile stress at a given elongation. Stress required to stretch the test length of the test piece to the given elongation. In the rubber industry this definition is widely identified with the term 'modulus' and care should be taken to avoid confusion with the other use of modulus to denote the slope of the stress-strain curve at a given elongation.

T-head. Extruder head of T shape, to divert the flow of rubber to a direction at right angles to the axis of the screw, for direct extrusion round a core.

Thermoelasticity. Rubberlike elasticity exhibited by a rigid plastic, resulting from an increase of temperature.

Thermoforming. Any process of forming thermoplastic sheet that consists of heating the sheet and pulling it down onto a mould surface.

Thermoplastic elastomers (or thermoplastic rubbers). The family of polymers that display the rubberlike property of returning to their original shape after repeated stretching and the ability to be melted and processed like other thermoplastics.

Thermoset. A material that will undergo or has undergone a chemical reaction by the action of heat and pressure, catalysts, ultraviolet light, etc., leading to a relatively infusible state. Typical of the plastics in the thermosetting family are the aminos (melamine and urea), most polyesters, alkyds, epoxies, and phenolics.

Transfer moulding. A method of moulding thermosetting materials in which the plastic is first softened by heat and pressure in a transfer chamber and then is forced by high pressure through suitable sprues, runners, and gates into a closed mould for final curing.

Tread. Part of a tyre that normally come into contact with the ground.

Tumbling. Finishing operation for small plastic articles by which gates, flash, and fins are removed and/or surfaces are polished by rotating them in a barrel together with wooden pegs, sawdust, and polishing compounds.

Turn up. Free end of a ply, chafer or filler which is folded around the bead core.

Twin sheet thermoforming. Technique for thermoforming hollow objects by introducing high-pressure air in between two sheets and blowing the sheets in the mould halves (vacuum is also applied).

Ultraviolet (uv) stabiliser. A chemical additive that selectively absorbs or filters out light waves at the ultraviolet end of the spectrum, protecting plastics from their harmful effects (embrittlement, discolouration, crasing, and disintegration).

Undertread tread cushion. Thin layer of rubber, intermediate in hardness between the tread and the adjacent rubberised reinforcement ply, placed between them to provide a gradation in properties from tread to carcass.

Vacuum forming. Method of sheet forming in which the plastic sheet is clamped in a stationary frame, heated, and drawn down by a vacuum into a mould. In a loose sense, it sometimes is used to refer to all sheet-forming techniques, including drape forming involving the use of vacuum and stationary moulds.

Vacuum test. Test of the resistance of a hose to collapse under vacuum.

Venting. (a) Raised lines or dimples on the inner surface of the moulded tyre to prevent air being trapped during manufacture or service. (b) Air bleed holes or lines in the mould surface to prevent air being trapped between the tyre surface and the mould during vulcanisation.

Virgin material. Plastic resin or compound that has not been processed except to prepare it for its intended use.

Visco-elasticity. Stress response of a material as though it were a combination of an elastic solid and a viscous fluid with flow dependent on time, temperature, load and rate of loading.

Viscosity. Internal friction or resistance to flow of a liquid; the constant ratio of shearing stress to rate of a shear. In liquids for which this ratio is a function of stress, the 'apparent viscosity' is defined as this ratio.

Void. A space or bubble occurring in the center of a heavy thermoplastic part, usually caused by excessive shrinkage.

Vulcanisate. Main product of vulcanisation.

Vulcanisation. Normally irreversible process in which rubber, through a change in its chemical structure, usually brought about by cross-linking, is converted to a condition in which the elastic properties are conferred, re-established, improved or extended over a greater range of temperature.

Vulcanised rubber. Rubber that has undergone vulcanisation.

Water absorption. The ability of a thermoplastic material to absorb water from an environment.

Water extract. Material extracted from rubber by water under specified conditions.

Weatherometer. An instrument utilised to subject articles to accelerated weathering conditions (e.g., rich UV source and water spray).

References

Alfrey, T., *Testing of Polymers,* Interscience, New York, USA.

Baer, E., *Polymer Degradation,* Reinhold, New York. USA.

Beck, R. D., *Polymer Rheology*, Van Nostrand Reinhold, USA.

Bernhardt, E. C., *Co-polymerisation*, Pergamon, UK.

Bruins, P. F., *Polyblends and Composites*, Wiley Interscience, New York.

Brydson, J.A., *Principles of Polymer Chemistry*, Elsevier Applied Science, London.

Ceresa, R. J., *Block and Graft Co-polymerisation*, Wiley Interscience, New York.

Craig, A.S., *Polymer Applications and Technology,* Philosophical Library, New York.

Crochet, M. J., *Numerical Simulation of Non-Newtonian Flow*, Elsevier, USA.

Culbertson, B. and **Pittman, C.**, *New Monomers and Polymers*, Noyes, UK.

Dealy, J. M., *Kinetics and Mechanisms of Polymerisation Reactions*, Van Nostrand Reinhold, USA.

Donnet, J.B., *Molecular Weight Distributions in Polymers,* Marcel Dekker, New York.

Dorgham, M. A., and **Rosato, D. V.**, *Mechanism of Emulsion Polymerisation*, Interscience Enterprises, Geneva, Switzerland.

Du Bois, J. H. and **John, F. W.**, *Polymer Chemistry*, Van Nostrand Reinhold, USA.

Du Bois, J. H. and **Pribble, W. I.**, *Principles of Polymerisation*, Van Nostrand Reinhold, USA.

Dym, J. B., *Polymers Applications and Technology*, Van Nostrand Reinhold, USA.

Dym, J. B., *Thermodynamics of Polymerisation*, Industrial Press, UK.

Ehrenstein, G. W. and **Erhard, G.**, *Block and Graft Polymers*, Hanser, USA.

Ferry, J. D, *Viscoelastic Properties of Polymers*, Wiley Interscience, New York.

Fisher, E. C., *Contemporary Polymer Chemistry*, The Plastics and Rubber Inst., UK.

Fisher, E. G. and **Whitfield, J. L.**, *Emulsion Polymerisation*, Wiley Interscience, New York.

Frados, J., *Plastics Engineering Handbook*, Van Nostrand Reinhold, USA.

Gacher, R., *Plastics Additives*, Henser, USA.

Griff, A. A., *Plastics Extrusion Technology*, Van Nostrand Reinhold, USA.

Harper, C. A., *Handbook of Plastics and Elastomers*, McGraw-Hill, New York.

Hertzberg, R. W. and **Manson, J. A.**, *Suspension Polymerisation*, Academic Press, UK.

Kraus, G., *Fundamental Principles of Condensation Polymerisation,* Interscience, New York.

Kusy, P. F., *Trends in Polymer Characterisation*, Noyes, UK.

Levy, S. and **DuBois, J. H.,** *Electronic Properties of Polymers*, Van Nostrand Reinhold, USA.

MacDermott, C., *Selecting Thermoplastics for Engineering Applications*, Marcel Dekker, Reinhold, USA.

Mark-Bikales-Overberger-Menges, *Encyclopedia of Polymer Science and Engineering*, Vols. 1-15, Wiley Interscience, New York.

Martelli, F. G., *'Mass Spectrometry'*, Van Nostrand Reinhold, USA.

Matthews, G. A. R., *Polymer Mixing Technology*, Elsevier, USA.

McKelvey, J. M., *Polymer Processing*, Wiley Interscience, New York.

Middleman, S. S., *Fundamentals of Polymer Processing*, McGraw-Hill, New York.

Parker, S. P., *Free Radicals in Solution*, McGraw-Hill, USA.

Rauwendaal, C., *Polymer Extrusion*, Van Nostrand Reinhold, USA

Richardson, P. A., *Physical Chemistry of Polymers*, Noyes, UK.

Rosato, D. V., *Rheology–Theory and Applications*, Van Nostrand Reinhold, USA.

Rosato, D.V. and **Rosato, D.V.,** *Organic Polymers*, Van Nostrand Reinhold, USA.

Rosato, D.V. and **Rosato, D.V.,** *Polyamides*, Van Nostrand Reinhold, USA.

Rubin, I. I., *Crystallisation of Polymers*, Wiley Interscience, New York.

Schenkel, G., *Synthetic Polymers* , Elsevier, USA.

Schnabel, W., *Polymer Degradation; Principles and Applications*, Pergamon, UK.

Schwartz, S.S., and **Goodman, S.,** *Polymer-blends*, Van Nostrand Reinhold, USA.

Sors, L., *Physical Chemistry of High Polymers*, Van Nostrand Reinhold, USA.

Throne, J. L., *Plastics Processing Engineering*, Noyes, UK.

Titow, W. V., *Polymer Chemistry,* Elsevier, USA.

Tobin, W. J., *Principles of Polymer Morphology*, Noyes, UK.

Treloar, L.R., *Kinetics of Chain Polymerisation,* Clarendon Press, London.

Turi, E. A., *Thermal Characterisation of Polymer Materials*, Reinhold, USA.

Wienert, L.A., *Polymer Structure Properties and Applications,* Reinhold, New York. USA.

Winolow, R.M., *Environmental Effects on Polymeric Materials,* Interscience, New York, USA.

Williams, J. G., *Stress Analysis of Polymers*, Wiley, Interscience, New York.

Wiley, R.E., *Testing of Polymers,* Interscience, New York, USA.

Index